HOW THE UNIVERSE FORMED

SRINIVASAN KANNAN

Made with ♥ on the Notion Press Platform
www.notionpress.com

This book is dedicated with humility to:
my Acharyan,

Sri. U. Ve. K.K.V.A. Kumara Venkatachar Swamy,
Sri Mudaliyandan Thirumaligai,

and

my parents,
Srimati. S. Champakavalli

and

Sri. U. Ve. A. K. E. Srinivasan

Contents

Synopsis *ix*

1. Our Understanding 1
2. The True Source Of Knowledge 42
3. Who We Are 68
4. Our Origin (God) 96
5. Purpose Of Immortality (Purushartham) 122
6. Means To Attain The Goal (Upayam) 135
7. What Prevented Us From Attaining The Goal 160

Conclusion 181

Acharya
Sri U.Ve. KKVA Kumaravenkatachar Swamy
Mudaliyandan Thirumaligai

Sri U. Ve. A.K.E Srinivasan
Kooram

Synopsis

Life begins for each one of us when we start thinking about everything we see in the world - specifically the purpose of life and why we are born (a process in which we had no say). Till then, we are living in ignorance. We can know the truth only through inquiry, by questioning the things we don't understand. When we look around, we see many humans similar to us, along with other species and plant life that just seem to exist without the potential to understand who they are. As we grow up, we learn essential skills and subjects from the elders in the family and teachers at school, enabling us to understand and take charge of our lives.

We find that everything we see in the world already exists. We realize that our parents were instrumental in our birth. Similarly, it must be the same with the entire humanity and other living entities; which suggests that all life, including the first, have been caused. None of the 8.4 million species, known as living beings, could have created themselves.

We see our earth, planets, the sun, our galaxy and countless other dynamic bodies in the universe moving, active and functioning. So, the universe must be alive as we see it now and has obviously been caused as well.

In reality, we, the intelligent beings exist, the non sentient universe also exists. We didn't either create ourselves or the universe. However, nothing exists without being caused (Supreme Lord being the only exception). Obviously, there must be a greater power that must have caused this universe and all life forms.

Any creation - be it material energy or life forms, is possible only for an intelligent source. Non sentient matter, devoid of intelligence, cannot create itself or the life forms. Although humans are the 'naturally most intelligent' beings in our world, we have not created any of the material energy of this universe and creating something like our universe would be beyond even the collective potential of all humanity. In our experience, only humans can make products using the potential of the material energy available to us, but we cannot make something out of nothing.

We must realize that every embodied being has three entities enabling life possible in the body.

- The omniscient Supreme Lord, who is the origin for all existence, who dwells invisibly in every Self (Soul) and pervades everything, everywhere.
- The sentient invisible Souls, embodied to enjoy life experiences.
- And the non-sentient universe, that's fundamentally a form of material energy.

Our world and the universe have existed for billions of years, preceding the emergence of humanity and current life forms. Likewise, various life forms likely existed for millennia prior to this era. Just as birth and death are fundamental to life, they occur at a cosmic level; consequently, species re-emerge along with the birth and death of worlds. It is said that the formation and dissolution of the universe are cyclic, mirroring the life cycles within it.

A human carpenter makes furniture out of wood. A log of wood has no inherent potential to transform itself into a piece of furniture, nor does it have a purpose of its own; it is unaware of its own existence. Therefore, it is evident that all material products are designed by intelligent humans. Conversely, anything not man-made must be the creation of God.

We do not know everything about our universe, especially regarding dark matter and dark energy. Rather than knowing if they exist, we have given these titles to describe observed phenomena that we don't yet understand. Our current knowledge is limited to the observable universe, not the entire cosmos, which may be infinite. Furthermore, cosmologists speculate that other universes exist. Our understanding stems from observing matter and energy, then creating theories to match reality. While some of these theories are likely correct, others may not be.

Our understanding of the universe can never be final. The difficulty in achieving a correct understanding of the universe is evident from the fact that modern physicists frequently offer differing interpretations based on new, continuously evolving data. As we discover new insights daily, it is clear that learning about the universe is an endless pursuit.

It is not hard to realize why we cannot possibly know everything about our universe; it is potentially infinite, while human intelligence has limitations. Experience shows that the more we look, the more new realities we discover. Honestly, when we don't fully understand the observable universe, it's hard to comprehend how we can be certain of how it formed 13.8 billion years ago. I believe the true age of the universe is different, and that scientists will establish a new age in the future, just as we evolved from believing the earth was flat to knowing it is a sphere. It is likely not humanly possible to know exactly how the universe formed.

It is logical to surmise that only the Creator of the universe truly knows how it was formed. Evidence of this knowledge is revealed in our ancient sacred scriptures, such as the Vedas, Upanishads, and the Shrimad Bhagavad Gita. Scholars and sages widely accept the Vedas as a-paurushEya - knowledge revealed directly by the Creator rather than authored by any human. Similarly, no human has ever claimed authorship of the Vedas or the Bhagavad Gita.

How do we know that the knowledge contained in these texts is correct?

The validity of the knowledge in these texts can be established by a reality check, verifying the content against actual, observed reality. Because these texts accurately reflect objective reality, they have remained unchanged and logically consistent.

Sacred scriptures reveal that the only 'unborn' Supreme Power, without a second, is infinite in every aspect and the origin of countless worlds and conscious souls. Because reality perfectly matches the knowledge revealed by the Supreme Lord, we need only study these scriptures to understand the truth of existence.

We don't even need to say thanks to that Almighty God who is 'kindness personified' and everything He does for us, His children, is for our benefit. He dwells in everyone sentient and pervades everything non sentient; by empowering the sentient to experience life, by His will alone, without direct involvement.

Although all living entities and all universes originate from Brahman, the Supreme Lord, they are not Brahman themselves. Everyone and everything belong to God, yet God's belongings are not God. Just as a person's properties belong to him but are not the owner, the creation is distinct from the Creator.

All that we have to do is realize who we are and our relationship with the Creator. The moment we become enlightened about our immortal nature and why we are here experiencing mortality in this world, we could transcend repeated mortality experiences and attain divine existence, by the grace of God.

The God given knowledge has all the information about why and how the universe was formed, who we are, where we belong and how we can attain the ever-elusive bliss, in search of which we have come to this world.

tad vishnoh paramam padam sadA pasyannti sUrayah...

We all belong in the divine world Paramapadam. We can attain the divine world when we seek God's grace with unconditional devotion. We sought this universe and God gave us life experiences here. If we are happy here, we can continue going through the cycle of birth and death in all sorts of bodies, determined by our karma. If we are not happy with our experiences here, and like to be in the divine world experiencing divine happiness, we need to seek God's help, guided by an Acharya or Guru who is born in this world only for the purpose of elevating the Souls desirous of serving the Lord and attaining the immortal world. God would grant us immortal existence if our devotion is true and unconditional.

sarvam Sri krishnArpaNamastu...

CHAPTER ONE

Our understanding

How the Universe formed?

We still haven't figured out an answer to this question, on our own, but...

Scientists and physicists of this era have overwhelmed humanity with some of the most unthinkable technological innovations, covering every field useful to human existence. They have not left anything untouched, transforming imagination into reality. Be it healthcare, telecommunication, air travel, space exploration, defense technology, automation, AI, or quantum computing; they have exceeded our expectations in finding solutions to fulfill our needs, aspirations, and passions. All this was made possible by their imaginative thinking and understanding of physical laws in pursuit of mastering the material energy provided by nature. We couldn't ask for more. What they have achieved in such a short span of time, post-World War II, is nothing short of a miracle - exceptional and astonishing. All superlatives fall short. They have succeeded in achieving most of the goals they set, except for one thing.

One thing the modern physicists and cosmologists have been baffled by is how the universe formed? They have a fair understanding of some of the universal laws, yet there are some factors that remain beyond reach, such as the specifics of universe formation; how it is functioning the way it does and how it will end? We must acknowledge that our understanding remains inadequate despite all-out efforts, even with changing assumptions and mathematical equations; the results often mismatch reality. This remains an elusive challenge for all concerned.

Science, as we understand it, is supposed to reveal facts about reality - verified, tested, and confirmed knowledge. Once we classify something as scientifically proven, there should be no doubt about its accuracy or its reflection of reality, with repeated tests yielding the same results. However, our published scientific works concerning the universe have often been proven inaccurate. This may be because the universe is inherently mysterious and constantly changing. What we observe and record today may not yield the same results tomorrow, as new phenomena are constantly observed. Furthermore, the functional aspects of the universe are not something we can control or predict. We may discover new things about the universe every day, without end. There are many unknown factors about the universe, and change being the inherent character of material energy, adds to our woes. That's why the universe, born (manifested) of divine Prakruti, is known as 'maya', or indescribable or unknowable, though it is still material energy, some observable and some not.

But what is the purpose of knowing how the Universe formed? Is it to understand the specifics and harness nature's powers for the benefit of humanity? Or is it to explore the entire cosmos because we crave to experience every aspect of it? Perhaps it is our inability to know everything that bothers us. It is natural for us to want to master all knowledge, because we are made that way. We cannot give up, even knowing we did not create a single atom ourselves. We have certain natural tendencies that God Himself has. Being true children of God, we are within our rights to know all about our world. We are all indeed mini gods but with finite potential. We could claim to know all about the universe only when we successfully create a miniature universe out of nothing. But that's far-fetched even to imagine.

We cannot learn high school curriculum from someone with first grade knowledge, nor can we complete a masters or PhD course in elementary school. Knowledge has to come from a source that possesses greater understanding than what we seek to learn. However, even the world's best thinkers admit they do not know the origin of the universe. If we are seeking to learn anything unknowable to us, it has to be from the very source that knows; only that would be accurate knowledge.

We have no knowledge that pre-dates the formation of the universe. Just as only the inventor of a complex creation fully understands his product, others can only guess, analyze, and dissect it to partially understand the original design - provided it was made by someone of a similar caliber to us. In many cases, we cannot know the full design unless revealed by the inventor.

Products are only made when human intelligence is involved. It is possible for us to know, build, and innovate using material energy within our realm of understanding, but not beyond it. Sir Albert Einstein acknowledged that to understand a new concept, we might only achieve the correct outcome one out of a hundred times. In reality, there is no 'new' creation by humans; whatever we discover or invent is merely a product of available material energy. It is new to us because we did not know of it before.

Einstein said,

quote -

"I think and think for months, for years. Ninety-nine times the conclusion is false. The hundredth time I am right."

unquote.

This statement holds true for understanding new aspects of the material world that are observable and knowable. Einstein is not talking about things completely beyond our scope of imagination. This material world was created for the experience of the Souls, and it is within our power to make use of it. However, we have no power to control the universe or alter its laws, especially when we have trouble understanding everything here.

The speed of light has always been about 300,000 kilometers per second, and gravity has always existed. Space, time, and the fundamental elements - along with the laws of physics - have been present since the formation of the universe. Humans of earlier millennia (Yugas) were aware of these factors as well; while this knowledge may seem new to us born in this millennium, it would be a statement of ignorance to assume we are the first to discover the truths of our world.

For us, learning starts from the basics. We are not born with complete knowledge of everything, though we are naturally intelligent. We learn from elders in the family and teachers at school, to the extent they know. Even they learned from their preceding generations. However, no one truly knows everything because humans in all cycles of creation were born only after the universe was ready for our existence and procreation. Therefore, no one has complete knowledge about the formation of the universe - not even Lord Brahma, the first being of our universe and the architect of creation. Even He received knowledge about everything from the Supreme Lord.

By nature, we, as embodied souls, are naturally inquisitive and driven to acquire knowledge. We can attain higher knowledge beyond the grasp of our material senses either by studying authentic scriptures or by learning from those qualified to impart such knowledge. While self-study is useful, it can be misleading; without guidance from scholars who have mastered the original language of these texts, we may not be able to grasp their true meaning. There is no way to ensure our understanding of the universe is entirely accurate unless it is derived from an absolute source.

We can know only so much about our universe when we perceive it merely as material energy. In reality, the universe functions by the invisible (Avyakta) potency of the Creator, which we often fail to see. The universe makes better sense when viewed as part of the divine. It is obvious that we cannot know everything through observation alone due to our limited perception. Also, we have not yet built an instrument with the potential to reveal the functional aspects of our infinite universe in a single frame. Using the best gadgets to probe deeper into the far reaches of the cosmos will still not help us know everything.

This universe is primarily a form of material energy that has no intelligence or consciousness of its own; it cannot reveal knowledge of any sort, nor does it realize its own existence. Through observation and experiments, we can discover how some things work, but not what causes them to function in that manner. We still do not know how the forces of nature - such as space and gravity - and the great elements - air, fire, water, and matter/energy - were formed. We simply know that everything already exists.

Conditions are not the same across all planets, stars, or galaxies, which vary greatly in size and mass, though some characteristics may be similar. For instance, the atmosphere around Earth that protects us from harmful solar radiation and sun storms is not present on other planets in our solar system - perhaps because they do not need it. We can observe that gravity exists, yet we do not fully understand why it behaves the way it does, or why it is caused by the curvature of spacetime due to mass and energy. Then, the laws that work at the gross level do not apply to quantum levels. We still wonder how the expansion of spacetime can be faster than the speed of light if gravity exists; there must be something more powerful pushing massive galaxies hurtling away, exceeding the speed of light. Why is dark energy beyond our perception, yet causes space to expand? And why couldn't the things in our world have been any other way? The human race and life can exist and thrive here because the universe provides everything we need. While our bodies are composed of particles from this universe, they cannot survive on other planets if conditions are not similar to Earth. Life would not be possible here if even one crucial thing were missing or different.

Various life forms thrive in natural habitats suited to their existence - some in water, some on land, and some in the air. Hundreds of thousands of plant species are often overlooked when discussing life on Earth. We tend to wrongly assume that the importance of life is exclusive to humans, ignoring other species, despite the understanding that each of us has passed through countless life forms in previous existences. It is also said that there are countless worlds, different from our own, which are home to various other beings. Surely, all of this was not designed by mere nature, as nature is non-sentient, nor by life forms themselves, as they are caused. Neither matter nor life is self-originated.

We are learning through trial-and-error methods because we don't know how else, which means our answers are constantly changing. While many genius scientists have proposed brilliant theories about the universe - some of which may be partly correct - no single theory offers a true understanding of the whole. We lack a unified 'God equation' to describe the functional universe. We are trying to understand an infinite universe with finite intelligence. While our understanding holds true at a gross level, it fails at the quantum level. Our analyses and mathematical expressions often fail to align with reality because we observe only fragments rather than the whole. We miss the fundamental truth that though the universe is material energy, we are unable to see the divine power that's making it work by being all pervading; it surpasses our intelligence.

Every physicist wants to show that their theory is unique. Often, they win a Nobel Prize, only for a next-generation scientist to propose a new model that challenges the earlier discovery, earning another Nobel in the process. It seems we are having fun in the way we interpret the universe. Some theories even suggest our universe could be within a black hole! Yet, we know there are many black holes in our own galaxy, including the supermassive one, Sagittarius A*, at the center of the Milky Way. While some physicists propose these could be dense clusters of dark matter rather than black holes, we still don't know the exact nature of what's inside. Not all stars become black holes. There is no constant. Everything in the universe is differently dynamic.

We have not yet seen evidence of a gigantic black hole surrounding the universe, nor do we know what might lie outside such a structure. The idea that our universe is inside a black hole is often questioned because, by definition, light cannot escape a black hole's event horizon. Some cosmologists speculate we might be viewing another universe when they spot light from distant galaxies that do not appear to belong to our own. While they could be mistaken, light from a neighboring universe would not be visible if it were trapped within a black hole - unless it is escaping radiation, energy flares, or emission from a dying black hole. Ultimately, no one knows what is truly trapped inside a black hole, but it is uncertain if our universe could exist there, as the physical laws exhibited in our universe might not be permissible inside a black hole. Gravity and the speed of light within a black hole might work differently, or perhaps light cannot travel at all. We have no idea how atoms or subatomic particles behave inside a black hole, or if atomic structure remains the same.

The hypothesis that our universe exists within a black hole highlights the creativity of theoretical physics which is similar to many aspects explained about our observable universe, that's not in a black hole. Even what we can observe remains mysterious and deceptive. Some thinkers suggest the Big Bang was a 'white hole' - from which the universe emerged. Would a black hole with its inescapable gravity permit the existence of life forms? However, if we were in a conventional black hole, the universe could not expand as it currently does. Most importantly, if the universe were inside a black hole, there would be no 'outside' light or an external event horizon to define it, as nothing exists beyond it.

Electrons exhibit both wave-like and particle-like behavior depending on how they are measured. When unobserved, an electron's position is described by a wave function representing a probability distribution across a region of space. The moment a measurement is made, the electron interacts with the detector and is localized into a specific position, which is termed the 'collapse of the wave function.'

The interference patterns displayed when unobserved electrons pass through a double-slit, disappear when detectors are positioned at the slits. This fundamental principle of quantum mechanics states that it's impossible to simultaneously know both the exact position and momentum of an electron with accuracy. The more precisely one is measured, the more mysterious the other behaves.

However, recent observations of the double-slit experiment by some scientists are said to have revealed different results. This suggests we have yet to fully understand the potential of material energy at both the quantum and macro levels.

Quantum entanglement is again quite a fascinating phenomenon. It's about two particles getting so interconnected, they seem to share a single state! It's like having two magic coins always landing on the same side. If you flip one, the other does the same even if it's far away. The two particles have no individual properties. Mathematically they belong together though they are in completely different places.

Measuring one particle instantly affects the other, irrespective of how far apart they are, as though they communicate with each other instantly. It seems quite incomprehensible how it works in this way. There is a study going on as to how it begins, how two particles become quantum entangled. Scientists have gone beyond debating if the phenomenon is true. They are onto knowing how the entanglement develops. All our research is a continuous process.

How gravity works is quite complex. For instance, planets of the solar system orbit the Sun because of a coordinated balance between the Sun's gravity and the planet's forward speed or peripheral velocity, which must have started during the formation of the solar system. While the gravity of the Sun pulls planets toward its center, the planets move fast enough to 'dodge' crashing into the Sun, resulting in a continuous, elliptical orbital around the sun. Gravity also curves the path of a planet inward, while the speed of the planet keeps it moving around instead of falling straight into. And the Sun's movement around the center of the galaxy does not prevent planets from falling into it. Planets remain in orbit due to a balance between the Sun's gravity pulling them inward and their own forward velocity, creating a constant freefall, but never really falling, but orbiting around. Because the planets move forward fast enough, they miss falling into the Sun, creating an elliptical orbit.

While the sun also moves, the planets are tied to it, taken along, travelling with the sun through space orbiting around the center of our galaxy. Surprisingly, the movement of the solar system around the galactic center is different from the orbital dynamics of planets around the Sun.

The planets are positioned at ideal distance from the sun in relation to their orbital speeds, perfectly matching. If the planets stopped moving, they would plunge into the Sun.

Alternatively, if the Sun stopped moving and lost its gravity, planets would fly away into space as though the string to the sun's gravity was cut off.

Many physicists originally expected the formation of galaxies and the distribution of matter-energy across the cosmos to be relatively uniform. However, reality is quite different, as cosmic structure is remarkably diverse. Take, for instance, the planets of our solar system; each is distinct. Similarly, while our sun is just one of around 400 billion stars in our galaxy, it is likely that every star possesses unique characteristics. The Andromeda galaxy is far larger than our own, and the universe holds colossal galactic structures extending millions of light-years.

Yet we think the age of our universe is only 13.8 billion years and the expanse some 93 billion light years. It's probably the extent of the universe that is observable by the James Webb telescope but that's not the whole universe. We certainly do not observe everything clearly within that space, and due to accelerated expansion, what we can currently see at the edge of the observable universe may soon move out of sight.

Stars near our sun, in one of the outer spiral arms of our galaxy, are spaced about 4 to 5 light-years apart. In the central 'downtown' region of our galaxy, they are much closer together. However, we cannot easily observe this center because interstellar dust blocks our view.

In reality, the universe has no center, boundary, edge, or 'outside.' However, we often use these expressions as conceptual tools to form a mental picture of the cosmos. Space is expanding everywhere in all directions. While the standard cosmic inflation theory suggests a single initiation point, some alternative theories propose that expansion occurred at multiple places, possibly involving numerous 'big bangs.' Even this, however, remains largely speculative. The idea of a single, uniform inflation doesn't align perfectly with evidence suggesting the universe is expanding at varying speeds in different directions, making it unlikely that it will ever return to a singular state.

Some thinkers contend that our universe could be a simulation, which is an intuitive hypothesis. But recent studies by a team of physicists have concluded that mathematically the universe cannot be a simulation because computation has no potential to produce the reality in existence. It's based on the fact that some features of the universe cannot be generated by any algorithm, while simulations are algorithms. Further, the research revealed that it is impossible to describe all aspects of the universe as it is using computational potential. It is improbable to capture the universe in its entirety without some fundamentals left out.

Some believe the universe exists only because of an observer and is not real. However, there are countless sources of individual consciousness, plus that Supreme Consciousness who created everything, observing, even if one individual doesn't. The Universal Consciousness never shuts down. Individual consciousness does not control or overpower the Universal Consciousness. individual consciousness is limited in its perceptibility, hence cannot perceive the whole of reality in existence.

Normally a cat closes its eyes while eating food or drinking milk. It thinks that no one can see it eating because it has its eyes closed. It doesn't realize that others can still see.

In reality, the universe would exist even if there are no observers. The universe is material energy and the observers are intelligent, conscious beings. Likewise, consciousness (souls) can exist independently of the physical universe. However, it is true that the universe was created for the experience of these conscious souls.

No human can provide a truly comprehensive insight about the universe, for they did not create it. Only true knowledge from the very source this universe was caused by, can dispel our doubts. Everything about this universe and the Source it originated from is infinite. And knowledge is endless; we are in our infancy at learning. How can we ever get to understand the infinite through guesswork? While we can know certain things by reasoning and observing the present, we cannot know the past, as material energy is ever-changing. The saving grace is that the Creator of this universe also gave us the knowledge about it, as only He would truly know.

It is also true that we cannot fully predict how the universe will end. We know that everything born has a beginning and will inevitably have an end; in the case of humans, we call this end, 'death'. Similarly, since the universe was manifested, it surely will end. While we cannot predict the exact manner of its cessation - just as we cannot fully know how it was formed - we can understand that the manifested universe will dissolve and merge back into the Creator, the Supreme source from which it originated. According to Vedas, there are still roughly 155 trillion years before this happens, and many of us will not be around to experience the final dissolution - if we attain immortality. Ironically, even if we were present to experience the end of the universe, we would not exist in a human body to tell the story, as we could not survive the impact of the end. It is my understanding that the Creator would withdraw us (souls) first, lest we experience the pain caused by the dissolution of the universe.

Through experience, we know that non-sentient material energy cannot be the cause of intelligent life, nor can it be the cause of itself. In fact, scientists have recently ascertained why: material energy lacks the potential to create life. It is evident that the universe was formed by an infinitely intelligent, all-knowing, all-powerful, and unborn Original Living Source - the only one of its kind, without a second, equal, or superior. We know this because we did not create even a grain of sand, and we are incapable of creating a universe of such infinite magnitude.

While we are considered the most intelligent species on this planet, we often fail to realize that we are endowed with limited perception; we can observe only the material aspects of this universe using our physical senses. Even then, we do not fully understand certain material realities, such as dark matter and dark energy, which make up nearly 95% of the universe. The other 5%, even which we have not perceived fully, includes all living beings.

We do not understand how our DNA, unique to each one of us, was coded, which seems far more complex than any computer programming known to us. It is hard to believe that non-sentient matter coded countless individual DNA sequences, and every life form self-coded its own DNA at birth. It is a mystery how the DNA of a newly embodied soul differs from its previous life, suggesting a process beyond our understanding. While we know that we, as individual souls, are the power that uses the brain and body to their potential, we do not understand how our brains actually work. I wonder if each person's past karma is a factor reflected in our DNA, which determines our individual uniqueness. It is fair to say that our knowledge of the functional aspects of the human body is as limited as our understanding of the universe.

If we ponder over a little about our thought process, it might give us a better understanding of how little we know about the universe, or even human potential. A thought arises from the collective triggering of billions of neurons, resulting in trillions of interconnections (synapses) in the brain. Any given thought involves a dynamic network of neurons acting in complex patterns.

And who is the thinker here? Surely not the brain. To test this, if we remove the brain from the head and place it on a table, it cannot think. It is nothing more than hardware (made of soft tissue) without the thinker. The Soul is the thinker; it generates thoughts and uses the brain as a tool. Without the power of the Soul, our body would not function, and the body, brain, and mind have no purpose of their own. The body, with all its capabilities, is given to each Soul by the Supreme Lord to fulfill its purpose, but with certain limitations. The Soul needs a body to experience, express, and function. Just as electricity does not glow on its own but powers a lamp, a lamp cannot glow without electricity.

We are convinced from our assumptions that somehow an immense singularity of material energy, devoid of intelligence, knowledge or consciousness formed this infinite universe from Itself! (the big bang expansion). If we assume it was the Supreme God instead of infinite singularity, that would not be accepted. It had to be a non-sentient material energy that caused this universe, because some scientists think so.

Assuming there was a Big Bang, it is described as either a material explosion or an expansion of space. However, it seems improbable that a mere explosion, or even the expansion of space, could create a well-functioning universe with complex, perfectly working laws that are still beyond our comprehension. No scientist has conducted a Big Bang experiment to verify if such a creation is actually possible.

Yet, they think they are right and they don't need to prove a thing to anyone. But God has to prove He exists and that He created the universe. It would be even believable if they claimed the universe was formed by a human scientist. *What is astounding is that they give the honor of universe creation to a non-sentient, material energy called the big bang, which is devoid of intelligence.*

Then, there are some unanswered questions; as to how the spec of infinite energy came to exist? Who caused it? What was it before it became the singularity? etc. as from evidence we know that material energy needs someone else to cause its existence. Above all, what was the purpose? Matter has no purpose of its own existence. It exists only for the experience of sentient life forms.

Fundamentally, the general scientific understanding of the universe's origin (around 13.8 billion years ago) is accepted as the leading theory, despite dissenting opinions and the lack of a universally accepted alternative.

All the understanding stems from assumed theories based on observation, perceptions and perspectives, what they call scientific findings. Some of it is true though. However, because this event occurred so long ago, it is reasonable to conclude that definitive, absolute knowledge of the universe's inception remains beyond human reach, let alone knowledge of the multiverse.

The original explanation of cosmic inflation goes like this - A tiny, dense, and hot primordial singularity underwent an exponential expansion known as cosmic inflation, expanding the fabric of space itself - not merely an explosion in existing space. This rapid stretching allowed space to expand faster than the speed of light, carrying light-emitting energy to distant parts of the universe, instantly. It is technically the expansion of space itself, rather than objects traveling through space. Therefore, light did not need to travel through that distance; the expansion of space carried the 'photon-filled' space with it. Yet, we time the age of the universe based on the subsequent travel time of light, which has traveled at its standard speed.

What it means in real terms - There existed an infinite singularity. Then by a big bang explosion or space expansion, the infinite singularity formed into our universe, expanding exponentially in the initial moment. The initial expansion of space itself was said to be 100,000 light years in an instant. (Considering it as one second to understand the impact, it is - 100000 x 365.25 x 24 x 60 x 60 = 3,155.76 billion times faster than the speed of light) or light would take 100,000 years to cover the distance in its normal speed. There is a major anomaly here. The moment after cosmic inflation, there was light all around in all directions of the space expansion till the edge of space. Light should have taken at least 50000 years to reach the edge of the expanded space from the spot of the big bang event, considering it as the center. But there was light everywhere. It means, cosmic inflation carried light to all parts of the space. So, space expansion carries light at speeds greater than light can travel, without light having to travel all that distance? In that case, did the universe expand back in time? Because anything that can travel or be carried - faster than light - would go back in time according to the theory of special relativity. That is a core implication of Einstein's theory, often referred to as the 'causality violation problem'.

Then it wasn't all light photons, it contained infinite material energy that formed into our universe which is still expanding and growing.

Light is an interchangeable electromagnetic wave without mass; oscillating electric and magnetic fields. Changing electric field generates a magnetic field, which generates a new electric field. This self-sustaining cycle allows the wave to travel nonstop through a vacuum, which self-propagates its movement at the fastest speed anything can travel, of around 300,000 k.m. per second. Light doesn't need another external force to make it move or accelerate it. Everything else with mass needs a force to accelerate according to Newton's laws of motion.

Light photons have no mass, meaning they are not subject to friction or inertia.

Matter requires force to accelerate, and requires increasing energy to maintain or increase speed, with an impossible thrust (infinite force) to travel at the speed of light.

If light reached all around the space without travelling from the big bang spot, how are we assuming the big bang spot based on light emitted 13.8 billion years ago and how sure are we that it is the light from the big bang spot and not the light spread all across by the expanding space?

If light is now arriving from all directions, how can we claim a 'Big Bang location'? And how can we be sure the cosmic microwave background (CMB) is truly the light from 13.8 billion years ago, rather than just radiation scattered by expanding space? If this light is universal, how can we differentiate it from a single point of origin?

There are problems with the initial cosmic inflation model as it defies certain fundamentals of laws of physics.

Following initial cosmic inflation, expansion slowed down, allowing energy to cool and matter to form stars and galaxies. (So, it wasn't all light photons, but all the material energy that expanded with cosmic inflation) Despite this slowdown, the space between distant galaxies continues to expand, causing them to move away from us faster than light. This ongoing expansion is why the diameter of the observable universe is about 93 billion light-years (a radius of 46.5 billion light-years), while the age of the universe is only 13.8 billion years.

The 13.8-billion-year age refers to the time elapsed since the Big Bang, not the size of the entire cosmos. The 93-billion-light-year diameter only describes the observable portion - the region from which light has had time to reach us. The entire universe is likely much larger, or possibly infinite. While expansion stretches the fabric of spacetime, carrying matter with it, galaxies do not move through space at this superluminal speed; rather, the space between them is growing.

It is often claimed that matter does not 'travel' through space because it is not moving through the fabric of space itself, but is instead carried by the expanding space. However, star systems and galaxies continue to rotate and orbit at their own inherent speeds - for example, Earth orbits the Sun at approximately 107,000 km/h, while the solar system orbits the center of the Milky Way at 875,000 km/h. It takes our solar system about 230 million years for one orbit, completing approximately 20 to 21 'galactic years' since its formation. The Sun is pulled by our galaxy towards the Great Attractor at about 2.1 million km/h. This implies that, in addition to their local motion, celestial bodies are being transported by the expanding universe at effective speeds faster than light...

Let's look at relative speeds possible by various entities. As humans, we can walk or run up to a certain speed. Many other species can outrun us, and birds can fly much faster in the air. Except for humans, no other species uses vehicles or machines to travel faster. We use cars or trains to travel on the ground, or airplanes and spacecraft to travel faster in the air and in space. When we move, we call it speed; when we use other means, we call it transportation. If we were to use a method that travels at or faster than the speed of light, it would be considered 'teleportation'. Under this definition, space expansion is 'teleporting' massive galaxies across the cosmos! As a result, the light from these distant galaxies must be teleported too, as the pace of space expansion in some zones exceeds the speed of light.

Light travels at approximately 300,000 k.m. per second in a vacuum. And this is the fastest anything can travel on its own potential, without being transported or teleported.

But we now know that the speed of space expansion varies, ranging from slower than light or even faster than light. The speed of gravity and light are supposed to be the same. In space expansion, galaxies or light are not traveling, but being in space, everything is transported by the expanding space, when it is less than the speed of light or teleported when it exceeds the speed of light. That means, space is the supreme power in our universe when it comes to speeds.

We can visit or appear in any place within our minds, provided we have experienced it before, recalling from memories to be there instantly. In reality, our minds do not physically travel; rather, we visualize that place, similar to a virtual reality playback, where distance and speed are irrelevant. Similarly, we can mentally return to any place and time we have previously witnessed. All this occurs entirely within our mental perception.

To understand being part of space expansion, imagine sitting on an infinite train (instead of space) along with countless other travelers. The train is not moving along tracks, but rather, the floor of the train is constantly expanding. Because of this, I am moved to a new position farther away from my companion, even though I haven't taken a step. In this scenario, my companion is also being moved in the other direction by the expanding floor, causing the distance between us to increase, potentially at a rate faster than light. Because the space itself is expanding, all of us are constantly moving away from one another. The farther away another traveler is, the faster they are carried away. In this scenario, we start nowhere and arrive nowhere, unable to define our absolute position because we are not moving through the train - we are being moved by it. Within this, we can still move around locally, like the Earth orbiting the Sun, but our local movement is much slower than the expansion of the train. Consequently, our total velocity is a combination of our local movement and the rapid expansion of the space around us. Essentially, we are being physically 'teleported' away from each other by the expansion of space itself.

In the expanding train, if light emitted from my companion's position travelled towards me, it couldn't reach my position ever because the space (floor of the train) within the train was expanding faster than light. Overall, the infinite train itself wasn't traveling, though. It was stretching the space within.

On a funny note, if we want to travel faster than light, let's board a space platform and specify our destination (which we don't know though). Space would expand our position while we didn't move, and we would arrive at the spot anywhere in the universe, faster than light; even at the speed of cosmic inflation (3155.76 billion times faster than light). At this rate, we could explore much more of our universe, but still not the infinite universe. Future AI tools may enable some such thing when we get to master how space expansion can be harnessed to teleport humans.

If they accept matter being carried to a distance faster than light, by the pace of expansion, the theory of special relativity would be challenged as nothing can move faster than light and if it does, it will travel back in time as time is linked to the speed of light. Or this space expansion should also carry light with it, plus the speed at which light can go, so light would be faster by that extent.

If we travel at closer to the speed of light, time slows. Only light has no mass and hence any matter with mass cannot travel faster than light, is indeed logical and understandable. But the flaw here is that light can never reach farther away spaces from one position by traveling at its speed because it is slower than expanding space from the same position.

We have one more restriction. The speed of light is constant at about 300,000 k.m. per second. It does not slow down or go faster, unless interrupted by gravity or black holes.

While the expansion of space affects the entire universe, its impact varies locally; nearby galaxies are often gravitationally bound and moving toward one another, whereas distant galaxies are receding at accelerating speeds; some exceeding the speed of light, due to the expansion of space itself.

Where mass and gravity are more effective than dark energy, within the galactic clusters, galaxies orbit around the bigger mass pulled by gravity. Eventually, they may collide or merge or destroy each other. On the other hand, where the power of dark energy prevails more than the mass, galaxies are pushed away at speeds relative to the field of mass or gravity by the space expansion. Farther galaxies speed away even faster than light where the dark energy has the upper hand, which accelerates space expansion to that extent.

The universe is infinite in all directions. We keep announcing a new finding every time without real knowledge of the universe. Suddenly we might find stars in the universe older than 13.8 billion years. What does that mean? Something is surely wrong either with the age of the universe or about the star.

On a bigger scale, the Milky way galaxy is also traveling at around 2.10 million kilometers per hour towards the Great Attractor, which is part of the massive Laniakea Supercluster. We don't have recorded data of the Great Attractor or Supercluster. We have not been able to fully observe the Great Attractor which is only about 250 million light years from the Milky way galaxy. Probably most galaxies within the gravitational zone are moving closer to each other. What unified theory could define the functional universe when we have not yet been able to observe even the Great Attractor or even the central region of our galaxy only 25000 light years from our location, for whatever reasons. So, the faint light we see from the farthest part of the observable universe (93 billion light years wide) which we call the observable bubble, is not something we can claim to know all about. We don't really observe everything clearly to the edge even within the observable bubble.

The universe seems to function differently than our current understanding suggests. Our current models fail to provide a truly comprehensive picture of reality. Therefore, did the universe originate from multiple positions? Are galaxies continuously forming and dissolving in various locations? Truthfully, nobody knows for sure.

It is understood that gravity is not a force, but rather the curvature of spacetime caused by mass and energy. Consequently, gravity affects the path of light itself. While mass and energy are interchangeable, this framework fails at the subatomic scale. In the quantum realm, fundamentally different rules apply - such as wave-particle duality and the uncertainty principle. Our understanding is localized of bits here and pieces there. Therefore, current physics is likely incomplete at both the universal and quantum scales, or at the very least, we do not have a comprehensive model that bridges the two.

In the cosmic inflation model, the rapid expansion of space during the early universe resulted in regions of space separating faster than light could travel between them. This implies that light from beyond our current particle horizon - the maximum distance light could have traveled to us since the Big Bang - can never reach us. Because we observe the universe by analyzing redshift of light from distant galaxies, we are limited to observing up to a range. Consequently, we will never be able to detect the farthest extent of the universe that lies beyond this range.

While Newton described gravity as an attractive force between masses, Einstein's General Relativity explains it as a geometric effect: mass and energy warp spacetime, and objects move along these curves as if pulled. Just as visible matter and invisible dark matter generate this curvature, about 70 percent of the universe consisting of dark energy drives cosmic expansion, often at speeds effectively exceeding that of light.

Mass and energy curve the spacetime fabric inward to attract, whereas dark energy does not necessarily curve space outward to repel. Because the force of repulsion is greater than the force of gravity.

We have a similar phenomenon with electromagnetic fields.

Electromagnetic fields attract and repel due to the nature of electric charges and magnetic poles. Opposite charges (positive/negative) attract, while similar charges (positive/positive or negative/negative) repel; likewise, opposite magnetic poles (North/South) attract, and similar magnetic poles (North/North or South/South) repel. This happens due to the interactions of fundamental particles and their fields, triggered by virtual particles, where aligning fields (opposite poles) attract (causing lower energy), and opposing fields (like poles) cause repulsion (causing higher energy).

A positive charge is pulled towards a negative charge because their fields interact that minimizes energy. Two positive charges or two negative charges push each other away as their fields oppose, increasing the energy. Similarly north or south poles interact; opposite poles attract and like poles repel.

Light is trapped once it crosses the event horizon of a black hole. Does light travel only through newly created space within expanding bounds, or is space expansion stretching an already existing space that contains matter and energy?

There are many anomalies that stand out glaringly, telling us that the big bang couldn't be the cause of a beautifully functioning universe with unseeable dimensions and intricate laws.

- Any explosion or expansion cannot form into a well-functioning universe with so many unfathomable, perfectly working laws that are beyond our comprehension. Instead, if it were an explosion, it would end in colossal material debris. If it was space expansion, space - being non-sentient - is incapable of formulating effective functional laws. We have witnessed countless material energy explosions that did not form into anything meaningful. Even if a Big Bang expansion is accepted, was it an intelligent singularity that planned the universe's formation before the event? Did the material energy possess infinite intelligence to formulate laws governing mass, gravity, space-time, dimensions, and habitable conditions for life? How could it continue to expand for 13.8 billion years? Even now, when we witness a star forming into a black hole or neutron star, it is only possible because of the pre-existing laws at work. Without these laws, it would be a chaotic lump of matter.

- Was it a combination of material energy cum space which expanded? Because there was matter, energy, light, spacetime, gases and all that existed in our universe. Did space create matter and energy or matter and energy caused space and space expansion? There are more unanswered questions than known factors about the universe.

- Though we claim the age of our universe since the big bang event is about 13.8 billion years, based on the fact that the light from the point of big bang took 13.8 billion years to reach us, on the assumption that the big bang is the center of the universe. In that case, are we sure that we are in the farthest location of the universe and nothing of the universe exists beyond us? If not, the age of the universe should account for galaxies situated beyond us as well. There should be more galaxies on the farther side of the Milky way galaxy. Then the age of the universe should be more than 13.8 billion years? Is the age of the universe determined only based on our location from the big bang event?

- Now, does the space-time fabric shrink faster than the speed of light, when galaxies are pulled towards each other, just as space expands faster than light? Does not seem so. But gravity and the speed of light are supposed to be about the same speed. Or When galaxies are gravitationally bound and moving together, does spacetime contract faster than c (speed of light), analogous to superluminal cosmological expansion? Given that gravity propagates at c, this seems paradoxical that it does not.

- If the universe began with the Big Bang, how could that model potentially end? Could it be a 'Big Collapse'? It's likely that a collapse must have created the singularity that preceded the Big Bang. However, with distant galaxies moving away at an accelerating rate, a collapse seems impossible. In our currently expanding universe, both 'Big Expansion' and 'Big Collapse' seem highly unlikely.

- Does the expanding fabric of spacetime carry light with it at the same speed it expands? If so, the speed of light should vary depending on whether it is in an expanding or contracting zone. If light cannot escape when it enters a black hole, trapped due to gravity; similarly, light should also be able to go along with the faster space expansion when it enters the expansion zone, because it is the repelling force of dark energy that expands the space, just the opposite of gravity's pull.

- According to Sir Isaac Newton, one of the greatest scientific minds in history, matter at rest does not change its state unless an external force acts upon it. Newton's First Law of Motion, also known as the Law of Inertia, states that an object will remain at rest or continue moving in a straight line at a constant speed unless an external force acts on it. This fundamental principle explains an object's natural tendency to resist changes in its state of motion. But of course, forces (powers) of nature can act upon resting objects. If there was a Big Bang, was the initial singularity acted upon by another force?

- The constant formation and destruction of stars and galaxies throughout the cosmos raise the question: could there be multiple big bangs occurring in different places at different times? Furthermore, recent James Webb Space Telescope observations have detected well-formed galaxies, existing shortly after the theoretical Big Bang, that are unlikely to exist under current models. This has led some physicists to suggest our understanding of the initial Big Bang event may be flawed.

- Initially, scientists guessed there must have been around a hundred billion stars in the milky way galaxy and around a hundred billion galaxies in the universe. And that figure is continuously changing to trillions of galaxies based on observable extent of the universe and even that there might be countless more galaxies the light of which might never reach us, because they are so far away and expanding and the light from them might become diffused before reaching us.

- Earlier, the western scientists thought that the earth was flat, while Vedic scriptures and Bhagavad Gita declared that the earth was a sphere (bhUgol). Again, scientists felt earth was the center and the sun, planets of the solar system and the stars were going around the earth. Vedic knowledge revealed all along that earth was a planet along with other planets of the solar system, going around the Sun and that the sun was a star, like many other stars in our galaxy and even our galaxy was one of many in an 'anda tatAka' egg shaped pool of mini universe and countless such clusters of galaxies forming our 'brahmanda' (bruhat-anda, meaning infinite, egg shaped Cosmos/universe) governed by the architect of creation, Brahma. And there are countless Brahmas in the multiverse of varying magnitude.

- The modern scientific community had no idea about the galaxies or the multiverse, until recently. In the Srimad Bhagavatam, Lord Krishna explains in a conversation with Brahma about the existence of multiverses and different Brahmas governing them. This is evident enough that all the knowledge from the Vedic scriptures is indeed true and verifiable and that such knowledge can only be revealed by the Creator of the universe and not by humans. The Vedic knowledge has accurately explained all about the universe. Vedas would be revealed in all cycles of creation. Thanks to the Hubble and James Webb telescopes, we are able to see a few new realities about our universe. What we see now is still an incomplete picture, and the futuristic instruments too, cannot capture the whole secrets of our universe to tell the full story. But then we have the essential knowledge about the formation of the universe in the Vedas. Our own thinking about the universe thus far has not been accurate. It is understandable because we have neither created nor witnessed it.

- Nothing in the universe is static. Every planet, star, black hole, galaxy is spinning and moving around pulled by other greater bodies, as though propelled by the power of the universal laws (Prakruti). This is true even at the sub atomic levels where the electrons orbit around the nucleus comprising protons and neutrons. We have not understood what that power is, which drives the whole universe to go round in circles, spinning. Could it be the singularity before the big bang event which doesn't exist after it banged?

It would seem that what we are able to observe is a mini universe, like a bubble, merely a fraction of the whole universe. And what is beyond our observable limits, we would never be able to perceive forever, as all those far away galaxies are speeding away due to space itself stretching continuously. In millions of years from now, we might not be able to see even what is visible now as those galaxies in the edge of our visible bubble would also disappear from our view. It's not that we are missing out on a possibility. What we see now may not remain seeable in future.

And time itself is different to different worlds.

Far away galaxies closer to the edge of the observable universe, which is about 43.5 billion light years away from the middle, would mean that we are looking at it as it was 43.5 billion years in the past. When any change happens in the outer edge now, it would take much more than that for us to know, as that part of the space is stretching away faster than before. But time itself is supposed to have begun 13.8 billion years ago with the big bang event, but light from the big bang point would take longer than the age of the universe to reach farthest positions in space. The age of the universe is linked to the expansion of space-time, while the universe's age is calculated from its expansion timeline. And light-years are used to measure distance - the time it has taken light to reach us from the past.

It's like saying I have been traveling for 46.5 years since my birth but my age is only 13.8 years when I arrived here, after 46.5 years since the time I started. (billion years)

Time and the speed of light are linked through Einstein's theory of special relativity, which dictates that the faster you move through space, the slower you experience time relative to a stationary observer - a phenomenon termed 'time dilation.' While time doesn't entirely stop at light speed, its passage becomes extremely slow as an object approaches it. This shows that time itself warps with motion, based on the fundamental principle that the speed of light is constant

The most amazing aspect of spacetime is that nothing in the universe will ever be in the same spacetime coordinate it once was, for everything is continuously on the move. As Earth goes along, pulled by the sun, so too the sun and all the stars and black holes of our galaxy go along, pulled by the Great Attractor. Furthermore, all the far-away galaxies are moving away from one another as well. This means that nothing ever revisits the same spacetime coordinate it once passed. Everyone is in a new spacetime coordinate continuously for eternity.

On a humorous note, if we are cosmic travelers planning to meet aliens from another distant galaxy, we'll have to specify a meeting spot on a commonly known planet by name, rather than its coordinates, as there is no real '*space time address*' for anything. And naturally we'll need to travel faster than space expansion (obviously faster than light) to reach a faraway galaxy if it is in the fast-expanding space-zone. If we want to meet a second time in the future, we would have to figure out the location of the same planet as it would be in a different space time location.

We never experience the same spacetime twice; every moment is a unique, one-time occurrence. Once we pass through a moment, we can never return to it, constantly navigating new, uncharted spacetime. However, our daily lives are often characterized by monotony and repetition. While the universe is designed for constant, dynamic change, we often settle into boring routines, making our lives feel static and unremarkable. Our universe has been ingeniously designed by the Creator to inherently function with continuous change, which is not usually highlighted by the cosmologists.

Every planet might have its own unique characteristics like the DNA we possess. Every star could be distinctly different from one another. Not all stars would become a neutron star or become a black hole. They all have their prescribed lifetime. They are born and die just like the humans and other living species. Galaxies and even the universes have a lifetime. And every galaxy we look at is as it was in the past, because of the time it takes for light from it to reach us. If the Andromeda galaxy goes missing today, we would find that out after 2.5 million years. We are dealing with infinity when it comes to understanding the extent of the universe as well as it is also infinitesimal at quantum levels. Humans can't possibly possess such intelligence as to create a model universe with such extreme functional properties.

I wonder if there is a similarity between the universe and humans as we possess a fraction of this universe in our bodies. We have trillions of cells and microorganisms in each of our bodies. Billions die and new ones are regenerated. As we age, we end our life in this body and get into a new body. This could apply to other species as well. There is continuous activity and change in the universe which includes all of us. It is said that there is a striking structural and mathematical similarity between the pattern of neurons and electrical signals in the brain and the network of galaxies in the universe. This comparison reveals that both the brain and the universe function as web-like structures, despite working at different scales and under different physical laws.

The resemblance is more than visual and extends to specific characteristics as explained by astrophysicist Franco Vazza and neurosurgeon Alberto Feletti in a 2020 study published in Frontiers in Physics.

Both structures feature intersections (neurons in the brain and galaxies in the cosmos) interconnected by intricate net-like filaments. These filaments direct information or matter between the intersections, with empty spaces (fluids in the brain and dark energy in the universe) in between.

In both the brain or the universe, matter (neurons or galaxies) accounts for only about 30% of the total mass/energy, while the other 70% is made up of fluid in the brain, or dark energy in the universe.

This comparison is not to explain a biological or physical connection between a person's thoughts and galactic movements.

Instead, it indicates that universal connectivity, and information flow may be complex, regardless of their scale. The brain, spanning a mere few inches, and the universe, stretching across billions of light-years, both reveal complex, self-governing network patterns.

This similarity triggers for a deeper introspection that perhaps the same underlying consciousness that enables our minds also governs the vast cosmos, suggesting that a Supreme, universal power operates across all scales of existence. And that is none other than God.

Even after so many evidently apparent realities presented to humanity, we fail to understand the obvious truth that it is the Supreme Consciousness that pervades and empowers the universe including all the living entities. The neurons and electrical signals in the human brain resembling the galactic network connectivity is to show how the Supreme Lord empowers every person and the universe merely by His divine will, without directly getting involved in the functional aspects. He controls and empowers every universal law and law of karma by His divine powers. We can argue that there is no evidence for God's existence in any of this. We have no problem if we can't see material energy, but when it comes to God, our ego deters us from accepting God's existence. We are happy with material energy being the cause of the universe because material energy obeys when we use it in ways we like, but are not happy about God's mysterious, invisible, unperceivable powers, none of which we can control or comprehend even to a fraction.

Suppose we deny all scientific breakthroughs by so many brilliant scientists, and instead say they just happened by themselves, or claim that they were all formulated by their material bodies without life - rather than the living, intelligent humans who truly authored them; what would we say? Or claim that a single celled microbe derived the theories of gravitational waves, quantum mechanics, theory of relativity etc.; or all the scientific breakthroughs happened by accident; or the spacecraft, Hubble and James Webb telescopes launched themselves into space without human involvement; what would be our response? We would say it is unrealistic and unscientific. A person who thinks so must be out of his mind.

If it was not possible for material energy to think scientifically like humans, which is still finite in scope and intelligence, how can we imagine that the infinite universe with super intelligent functional universal laws can be caused by material energy such as a big bang event?

Human scientists would be upset if anyone denies them the credit for their work, but please be assured, God would not mind if we deny His existence or His creation. He knows only kindness and compassion for His created beings. He accepts them the way they are. And He doesn't have to prove anything to anyone.

When we are still unsure about our knowledge of the visible 5% of the bubble in our universe, how are we possibly going to find out more about the invisible, unknowable 95% of the infinite universe?

We are willing to accept that dark matter and dark energy cannot be seen or perceived but are sure they exist and accept this as a confirmed reality. But are not willing to accept that there is this divine Supreme Power existing everywhere but is beyond our power to know or see. For us, God doesn't exist because we are unable to perceive Him, but dark matter and dark energy exists even though we cannot perceive them. That's so glaringly prejudiced.

Scientists conducted tests at CERN, an underground facility in Geneva. They fired protons from opposite directions in a particle accelerator at close to the speed of light. The collision left behind some traces of decay, though there was no residual of any particle seen directly; and they detected the unseeable God particle, which disappeared after the collision of protons. They repeated this process many times to be sure they did not miss any data in the outcome or in the hope that something different might emerge. I doubt if the original intent was an effort to understand the big bang event or to know the origin of the universe and how the universe formed. The actual, assumed big bang that occurred 13.8 billion years ago, transformed into our universe. But the proton collision experiment resulted in the disappearance of material energy after the collision. This is proof enough that a material energy collision or explosion could not transform itself into an infinite universe. It would disperse or collapse into debris or disappear by the effect of impact.

Even at CERN, scientists (conscious living beings) performed these experiments. Protons did not conduct any experiment by themselves. So how could the big bang take place without any scientists conducting that event? Indeed, that scientist is God, but I doubt if there was a big bang experiment by God though.

Hundreds of scientists spent decades of their time and invested more than 20 billion dollars on this ongoing experiment. The happy outcome is that they accept there was a 'god particle' which could not be seen, but still won't accept that the real God cannot be seen. However, their intent for the experiment was also to study elementary particles, the Higgs boson, and they seem happy about the results obtained.

How does the material energy function? Does it understand its scope in terms of speed and time etc. What determines the speed of light to be around 300,000 k.m. per second? Time travels forward and not backward. So, we cannot revisit events of the past, except in our memories.

Our earth is a living entity, with the Supreme Lord being the inner self of the Soul of the earth. Similarly, our sun has a soul keeping the star functioning, known as the Sun God. The Supreme Lord is the indwelling power of the sun's Soul, empowering it. We worship Mother Earth, the Sun God, and many other powers of nature, as they have a presiding God power, with the Supreme Lord within them. This includes the great elements such as ether (Akasa), air, fire, water, planets, and stars. We are familiar with the story of Dhruva, who was bestowed the position of the presiding Soul of the polestar, known as Dhruva Nakshatra (Star of Dhruva).

We trust the scientists who proposed the Big Bang theory, partly because we do not yet have a widely accepted alternative. However, these scientists do not have direct confirmed knowledge of the event itself; none of them have conducted a scaled-down experiment to verify if a miniature universe could be formed in a lab via a similar event. Yet, they know their concept is accepted by most, largely because there is no known alternative to disprove it.

It is similar to saying, "I have seen God." Some people, unsure if God can be perceived through material senses, might believe me. Others, who already doubt the existence of God, would not accept the claim even if it were true.

We believe only when we do not know; belief is, in a way, blind. When we have realized knowledge, confirmed understanding, and awareness, we no longer need to believe, because we know. One great lesson from the Upanishads is that they encourage seekers of truth to question everything and think deeply until they are fully aware of reality. There should be no room for blind belief; rather we are encouraged to be enlightened seekers.

The same is true regarding God; we do not have to accept God's existence based on blind belief. Because we are naturally intelligent and consciousness is our nature, we must make the effort to know the truth rather than live in ignorance. It is within our power to realize the truth instead of merely believing in something we do not understand. Scientifically, we can understand that nothing exists without a cause. Everything we see is either man-made or God-made. To believe that non sentient material energy is a self-caused entity - and the creator of our universe and all life forms - without any basis for that realization, is, in fact, blind belief.

We often think we are smart when asking, 'If everything that exists has a cause, who caused the cause of everything?' The idea that the Supreme God must also have a cause is a naive argument born of ignorance. We must understand that God is the exception - the original, unborn, and self-existing cause that requires no cause Himself.

If matter is accepted as the original cause for the universe, we don't ask this question as to who caused the first matter that caused the universe?

Some scientists, considered to be possessing higher intelligence, believe they understand and know everything; they emphatically confirm that there is no such thing as God. God is a mystery for them as it is not logical from their understanding, experience and perception. They could perceive only visible matter and nothing higher or beyond. But they would not accept the fact that the idea of God is incomprehensible to them. If you tell them that they are also similar to God because they know everything for sure, probably they might accept the existence of God. Indeed, there is God in all of them, whether they know it or not, whether they accept God or not. Just that God is mysterious because He is beyond our perceptibility.

Some modern minds cannot understand God because they lack access to the Vedas, nor would they acknowledge the superior knowledge contained in the Vedas, even if it was accessible. Any source of knowledge, not their own, would be rejected. Probably because they might prefer to find out more about the universe in their own methods, without the help from God.

When someone is active and functioning, we say there is life in him, he is alive. But we are unable to perceive life directly as life is not visible to our material sight. But the existence of life can't be denied for being a mystery. Some scientists also deny the existence of Souls being the life power; they would rather accept the mind as life, electrical signals in the brain or anything else, but not a Soul.

Just because we cannot see the Conscious Souls in living beings, it does not mean people and other species are merely functioning dead bodies. We know dead bodies do not function. The proof of life is evident when we see people walking and performing actions; if there were no life in them, they would be dead. Similarly, we see planets, stars, galaxies, and the universe functioning, suggesting God is the invisible Super soul keeping the universe alive and active. Just as our bodies cannot function without souls, the universe cannot function without the Super soul (God) being present within it.

A straightforward way to understand God and souls as inseparable entities is by comparing them to the sun and its rays. The sun is the origin of the rays, and the rays cannot exist without the sun. While the sun's potency is present in the rays, the rays themselves are not the sun. Even if we assume the sun is infinite in all respects with an infinite number of rays, the individual rays remain finite; however, the power of the sun within those rays is infinite.

With the Hubble, the James Webb Space Telescope, and future advanced instruments, we may yet discover truly unbelievable, colossal structures in the universe - some spanning billions of light-years that defy current imagination. The more we look, the more we will find, for the universe may have no limit. Whatever we currently assume regarding the age and size of the universe (now estimated at 93 billion light-years) may prove to be far too small. As an immeasurable expanse, the universe likely holds endless new realities for us to uncover.

We have read from our scriptures such as Sri Ramayana, that Lord Brahma appeared before Ravana who performed penance wishing for boons to be granted by Brahma. Here Lord Brahma did not travel from one place to another. He just appeared.

In Dhruva's childhood incident, Lord Vishnu appeared before Dhruva, who was devotedly seeking to see the Lord. Lord Vishnu did not have to travel; no physical barriers such as speed or distance were involved. Similarly, Lord Shiva appeared before Arjuna and bestowed him with divine arrows that would help him in a war. Lord Shiva did not travel from somewhere to come to that place, because all dimensions were transcended. When there is travel involved, we are bound by limitations such as speed, mass, distance, time. For Gods and divine entities, they could appear in any place instantaneously.

It is suspected that the universe is multi-dimensional, as suggested by 'string theory'. Theoretical wormholes could shorten cosmic distances, potentially enabling travel faster than light could traverse that same distance through normal space. When considering cosmic distances, even the speed of light is slow and time-consuming; light travels only around six trillion miles in a year, and crossing the Milky Way galaxy takes roughly 100,000 years.

Souls can function only according to the potential of the bodies they inhabit. For instance, other species - such as animals and birds - cannot understand, analyze, or decode scientific accomplishments by humans because it is beyond their intelligence. While any human can generally learn to do what another human does, animals cannot emulate human capabilities.

This is best understood by comparing a high-performance F1 car to a regular street car: the engine, body, and dynamics of the F1 car enable it to far outperform the latter.

While other divine godly powers are derived from the Supreme Lord and function in various roles, they do not exhibit His infinite nature. To understand this, consider that a prime minister may exercise supreme powers while in office, yet displays normal human powers performing different duties at home as a father, husband, or friend. The key difference is that God is omnipresent, capable of managing different divine roles in all places at all times, simultaneously.

Or if we compare the Chairman of a company to God, all the management executives perform by the power of the chairman but they are not the chairman.

While we have to work to create things from available materials, the Supreme Creator does everything by a mere 'sankalpa' (Will or thought). What He wills becomes reality because He is Himself the source. His 'will' has the power to create. Whatever He does, it is to fulfill the wishes of the Souls, whom He loves so much, for the Souls are sentient beings, like Himself and are very dear to the Lord.

We build houses for our own use, and we cannot say a house built itself from earthly resources. Even if it did, what would be the purpose? A house has no self-use. Similarly, if the Big Bang formed this universe, it has no use for doing so, which would make it a senseless cause. The entire creation has a meaningful purpose, and it is for the enjoyment of sentient beings like us. The universe itself has no potential to enjoy.

God has given us all the knowledge about how He formed the universe. What is the problem in knowing from the Vedas? We don't have to give any credit to God for the IPR of the Vedas, nor does He expect us to acknowledge what He has done. We can take credit for learning all the knowledge in the Vedas; it is meant for us.

We had everything readily made available to us by an infinitely powerful Creator, beyond our comprehension. We did not create the plant life, which is a major source of food for humans and many other species, nor did we create the animals that serve as food for meat-eaters. Furthermore, most carnivores are either predators or prey. This is another reason why evolution could not have occurred in the way the popular theory was hypothesized or envisaged. The Creator had already designed some 8.4 million bodies for as many species, including humans; where, then, is the need for life forms to evolve from a single cell or for one species to turn into another?

Most importantly, all essential ingredients for life - food, water, air, clothing, and shelter - were provided by that Creator. All we have done is utilize these available resources to suit our needs for a comfortable existence. Moreover, we have been created with natural intelligence, consciousness, and awareness to observe, understand, and work to use every available resource for our potential benefit. The only thing lacking is our failure to recognize that nothing was created by us.

We are the only beneficiary in our part of the world. All other species are incapable of understanding about the universe or God.

Here is a quote from Albert Einstein.

In so far as theories of mathematics speak about reality, they are not certain, and in so far as they are certain, they do not speak about reality.

Unquote

Sir Einstein also said,

quote "I am not an atheist. The problem involved is too vast for our limited minds. We are in the position of a little child entering a huge library filled with books in many languages. The child knows someone must have written these books. It does not know how. It does not understand the languages in which they are written. The child dimly suspects a mysterious order in the arrangement of the books but doesn't know what it is. That, it seems to me, is the attitude of even the most intelligent human being toward God. We see the universe marvelously arranged and obeying certain laws but only dimly understand these laws" unquote

This shows the honesty and integrity of one of the most intelligent humans of recent times. There is nobility in his thoughts. There was no pretension in his expression though he knew he was the best of his times.

It is said that even Einstein was puzzled by the existence of the Soul and consciousness, often wondering who was truly seeing from behind the eyes. Only a questioning mind can find answers, and he was not a genius for nothing. He expressed that as humans, we possess a part of the universe in our bodies, sharing space and time. It takes special intelligence to imagine the existence of consciousness and a Soul in everyone.

It is not surprising that science has no clue about God; when they have problems understanding the existence of Souls in every living entity, how could they understand God? Science often only accepts what it can perceive, observe, or verify through instruments. Yet, science still does not realize it has not even been able to perceive certain material aspects of this universe. How, then, can we expect material science to understand divinity, which is beyond its realm of perception?

God owns everything in existence - the universes, the divine world, and all living entities. He manifested everything out of Himself, by a thought, when He alone existed. Naturally, only

He knows everything about His creation. He has not kept any secrets leaving us ignorant; rather, He has revealed knowledge in a way that is within our grasp, so we do not live life blindfolded. All the knowledge in the Vedas was simplified by Sage Veda Vyasa, sparing us from incomprehensible information or complex theories.

CHAPTER TWO

The True Source of Knowledge

Now, let's look at the original source of knowledge.

Vedas, meaning 'knowledge', represents a divine science that is verifiable if approached with an open mind and honesty. It is crucial to gain this knowledge under the guidance of a learned, authentic Guru. Because grasping the esoteric meanings of the Vedic scriptures on our own is difficult, texts like the Upanishads (Vedanta), the Srimad Bhagavad Gita, and other sacred scriptures help us understand the essence of this knowledge.

The Vedas are complete, eternal knowledge. The Supreme Source, who manifested our universe along with all beings and several gods, revealed the Vedas to Lord Brahma, the architect of creation. This knowledge was passed on orally to subsequent generations in every millennium, allowing us to know ourselves, the universe, and the Supreme Lord (chit, achit, and Iswara).

There are only three main entities, with the other two (chit and achit) originating from Iswara (Brahman). Chit, being sentient, exists in an unmanifest, subtle form invisible to material senses, while achit, being non-sentient matter, manifests with form

The Vedas are meant for us, because we are sentient beings. The universe is non sentient material energy that cannot comprehend the knowledge of the Vedas or know about us or the Supreme Lord. The universe has no use for the Vedas.

A natural question arises: can we know everything about the universe if we know the Vedas? The answer is both yes and no.

Yes, to the extent that we can grasp the knowledge within the Vedas based on our intellectual disposition. In this sense, we come to know that everything is caused by the original Supreme Power, rather than needing to search elsewhere or create our own assumed theories.

No, we cannot know everything about the universe, because the universe and its creator are infinite, and our finite intelligence cannot fully grasp the infinite. However, we do have the intelligence to understand that the infinite cannot be fully known.

While Albert Einstein is credited with the theory of general relativity (1915), was his physical brain truly the author? Surely not. The true power - the intelligent Soul that inhabited his body while he was alive - is no more. We still have a record of his theories, as it has only been about 110 years since they were conceptualized.

In contrast, the Vedas were revealed to humanity at the start of every cycle of creation. When the time calls for it to be scripted, a divine soul or God Himself incarnates to reveal this knowledge, either personally, as Lord Krishna did, or through written scriptures, like Sage Vyasa did over 5,000 years ago, which has stood the test of time into this millennium.

I guess things are different in every cycle of creation because that's what Lord Krishna hints when He says in the Srimad Bhagavatam, that there are countless universes, each one different from the other, different Brahmas in charge there. It confirms that God has infinite possibilities when it comes to His powers.

The Vedas, often considered *the **'fundamental knowledge manual'*** that accompanies the formation of the universe, reveal the process of creation, our origin, and the Source of all existence. They posit that God is the singular origin of everything, including this primordial knowledge. Complementing this, the Bhagavad Gita serves as a **'manual for life'** simplifying the profound essence of the Upanishads for humanity.

Let's see if it makes sense.

The one and the only self-existing, ever existing, unborn Supreme Lord is the cause of everything. You can name Him, call Him anything you like. For, He pervades everything, everywhere. He inhabits every sound and form but remains untouched and invisible. If we, the embodied Souls, reside in the body, we the Souls are the body for the Supreme Lord we originated from. He dwells within every soul, just as He pervades everywhere; every subatomic particle. He is the observer of our actions, witness to our karma, such as whatever we think, whatever we say and whatever we do. He keeps note of our karma by dwelling within every Soul. He doesn't influence us or intervene in anything we do or in anything we experience as a consequence of our actions. Similarly, all the matter and energy of this universe, behaves inherent to its nature due to the presence of the all-pervading Supreme Consciousness in them, though invisible.

The Taitriyopanishad, 'Brahmananda Valli' describes the metaphysical process of creation, not as a product from an external source, but as the self-expansion of Brahman (God). This sounds similar to the cosmic inflation of the universe at the instant after the big bang, but not really. The assumed big bang event sounds violent, while the real manifestation of the universe by the Supreme Lord is a blissful expression for the experience of the countless Souls it was created for. This means the universe, with all its dualities and forms, is fundamentally an expression of the same blissful consciousness.

Here's the expression from the Vedas:

"so'akAmayata, bahusyAm prajAyEyEti, sa tapo'tapyata, sa tapastaptvA, idam sarvamasrujata, yadidam kincha, tat srushtvA tadEvAnuprAviSat, tadanupraviSya, saccha tyacchAbhavat, niruktam chAniruktam cha, nilayanam chAnilayanam cha, vijnAnam chAnavijnAnam cha, satyam chAnrutam cha, satyamabhavat"

He, the Anandamaya, of the nature of infinite bliss, (Important to realize that the nature of the Supreme Brahman is infinite bliss) willed, 'May I cause many beings, may I manifest '. He deliberated. He, having thus deliberated, created / manifested all this, all that exists. Having created all that, Brahman entered into that Itself (anupravisya). Having entered into it, that became conscient and the inconscient; the defined and the undefined; the host of sentients (that forms the ground of non-sentients) and the non-sentients (that are dependent); the non-inert, and the inert; the true (Atma / Souls) and the changing (matter). The true Brahman entered all this that exists. (This concludes, Brahman is neither the Atma/Soul, nor the universe but is the indweller of both). Therefore, it is said that this universe has that true Brahman in it. (And hence the universe cannot be false)

This is pretty much the unified theory of everything in existence, or the God equation, that encompasses everything.

All this was in the beginning unmanifested without expression of names and forms. From that Brahman the manifested came into existence. That Brahman created all this from Himself. Hence, Brahman is called the one of good deeds.

It explains clearly that the universe formed from Brahman, but the universe is not Brahman.

The Supreme God of the nature of Bliss (Anandamaya) willed the creation of both collective and individual - chEtana (sentient Souls) and achEtana (non-sentient) matter/energy, the universe, forming ether and other elements.

The Supreme Lord's nature is defined - "sat, chit, Ananda swarupa" (existence, consciousness, bliss)

And He formed many of the powers of nature (gods), humans and others (some 8.4 million species).

'sat' similarity shared with God, signifies the sentient Souls / Atma, which are of the nature of consciousness, unborn, eternal, unchanging, in subtle, unmanifest form; 'tyat' signifies non-sentient matter / energy which is subject to change on account of giving up of the preceding state. The achEtana (non sentient having jAti (birth), guNa (quality) and kriya (action) and the chEtana (sentient Souls) who is devoid of jAti, guNa are the ground (AdhAra / support) for the non-sentient. Satyam signifies the Brahman / Supreme God who is the inner self of all entities, sentient and non-sentient, eternally. Brahman is both the material cause and instrumental cause for creation which is termed as 'sukrutam', the effect of itself.

As per the Upanishads, it is clear that the non-sentient matter inherently has jAti or birth and GuNa or quality born of this nature (Prakruti). The sentient Souls are devoid of jAti (birth) and GuNa (quality). The Souls are unborn, eternal. So where is the question of caste here? The jAti (birth), and Guna (quality) and kriya (action) are enabled in the material body when an embodied Soul enlivens taking birth in a body.

The Soul itself has no jAti or guNa. However, the Soul needs a body to enact Karma and the Soul is the beneficiary of Karma and is accountable for the effects of karma. The three guNAs such as sattva, rajas, tamas are the guNAs inherent to this Prakruti and affect all embodied beings. It is in the power of the Soul to transcend the effects of the three guNas born of Prakruti, by inculcating righteousness in its karma, remaining detached to cause and effect, through devotion to the Lord.

Taitriya Upanishad also describes the sequence of cosmic manifestation -

'tasmAdvA EtasmAd Atmana AkASah sambhUtah,

AkASAdvAyuh vAyoragnih agnErApah adbhyAm pruthivi,

pruthivyA Oshadhayah OshadhIbhyO'nnam

annatpurushah.'

From that Brahman, who is the Supreme Self, space (Akasa or ether) was formed. From space, air (Vayu), from air, fire (Agni), from fire, water (Apa), from water, earth (pruthivI or gross material energy) was manifested. From earth, plants/vegetation (Oshadhayah), from plants, food (annam) was produced. From food came the embodied living entities (purushas).

In essence, this verse explains the sequence of manifestation: from the subtlest element (space) to the gross (earth or material energy), and then the origin of food and the humans and all the species (embodied Souls). This also works in the reverse order of dissolution during cosmic extinction.

Mandukya Upanishad declares that creation begins with the vibration of the primal energy, Prakruti, which has 'Om' as its sound symbol. We also realize that no object can be thought of without the sound related to it. It is clear that 'sabda' (sound) and 'artha' (object) are inseparable. Thus 'Om' which is the most universal and all-inclusive sound, represents the whole creation; visible and invisible. The universe is merely a manifest form of the Divine. Om is the sound symbol of the supreme Brahman. Om represents all that exists in all times. Mandukya Upanishad-1 "om-iti-Etad-aksharam-idam sarvam", "All this world is the syllable Om".

It is important to note that the Vedic scriptures do not mention religion, or caste, when defining the nature of the soul. Humans are not born as Hindus, Christians, Muslims, or any other religion practiced today; rather, all are children of God with the freedom to choose a name and form for the Divine with whom they connect best. We are born as humans - or other species, based on karma.

There is no mention of caste or division based on birth. Sanatana Dharma is not a religion, but a way of life rooted in righteousness (dharma). Every conscious living entity is considered equal - none better or less than another - irrespective of where they are born or what they choose as their path. There is no 'superior or inferior' discrimination, even regarding other species or plant life, because the same God resides in everyone. We are who we are on account of our karma; our deeds define us, not our birth.

If we look at the formation of the universe from the Upanishad's point of view, the Supreme Lord (Brahman) is the source of all that exists. God did not need anything outside of Himself for the creation of the universe, the multiverse, or even the divine eternal world.

If we look at God as the origin for the universe, then there are no limitations to what could have happened. There was no need for the big bang explosion which would negate its very existence after the event, for the big bang transformed itself into this universe. It's like an egg hatching into a bird. Either the egg exists or the bird. The big bang causes the universe but doesn't exist to support the universe in its functioning or survival. Big bang is not the mother of the universe in our perspective as it doesn't survive to protect the newly formed universe. Like a child being abandoned by a parent who expires after the child was born.

We say our Sun is the life supporting energy for our existence on earth. Our earth and planets of the solar system are formed as part of the Sun, sustaining everything in the solar system. The Sun did not disappear after the planets were formed.

With God being the origin of the universe, He exists before the formation of the universe and exists after the formation of the universe and sustains it throughout its lifetime and withdraws it into Himself at the end of its tenure. He exists in the universe and outside of the universe as well. He is the one absolute reality at all times, everywhere and in everything. The universe manifests from the Lord and merges into the Lord. This cycle of formation and dissolution of the universe is cyclic as well. And the Supreme Lord creates, sustains and withdraws the universe at dissolution, without abandoning it or the countless Souls, the inhabitants of the universe.

Sri Parasara Bhattar, a divine Soul who incarnated to bestow knowledge on humanity, explains the entire story in one statement.

'akArArtho vishNuh jagadudaya rakshA praLaya krut.'

meaning, 'the Supreme Lord is the cause of creation, sustenance and annihilation of the universe'

The Supreme Lord is Himself the power in Lord Brahma, Lord Vishnu, Lord Shiva, lovingly known as the Trinity, taking care of creation, sustenance and annihilation of every universe in the multiverse. It is the one God doing the duties as different gods, without dilution of His original Power, as He remains independent of all His multiple godly identities.

Lord Krishna knew we would question the authenticity of the Vedic knowledge being the original from the Supreme Lord

Himself, and hence has given us verifiable proof in the Bhagavad Gita

Gita ch.4-1

'imam vivasvatE yogam proktvAnahamavyayam:

vivasvAn manavE prAha manurikshwAkavEbravIt.'

meaning "I taught (Lord Krishna being the incarnation of the Supreme Lord)

this imperishable knowledge (yoga) to VivasvAn; VivasvAn taught it to Manu; Manu declared it to IkshvAku"

Gita Ch. 4-2

'Evam paramparAprAptamimam rAjarshayo viduhu:

sa kAlEnEha mahatA yogo nashtah parantapa.'

meaning "Thus handed down in succession, the royal sages knew this knowledge (yoga). But with long lapse of time, O Arjuna, that knowledge (yoga) was lost to the world"

Gita Ch 4-3

'sa EvAyam mayA tEdyA yogA proktah purAtanah:

bhakto'si mE sakhA chEti rahasyam hyEtaduttamam.'

meaning "it is the same ancient knowledge (yoga) which is now taught to you by Me,

as you are My devotee and My friend. For, this knowledge is a supreme mystery"

But Arjuna questions the Lord as to how Krishna could have taught this knowledge to the first-born beings, when Arjuna knows that Krishna has taken birth now in his time and generation.

Gita Ch 4-4

'aparam bhavato janma param janma vivasvatah

kathamEtadvijAnIyAm tvamAdau proktavAniti.'

Arjuna asks - Later (in a different millennium) was your birth, and earlier the birth of VivasvAn. How then am I to understand that you taught it in the beginning (of creation)?

The Lord explains to clarify

Gita Ch 4-5

'bahUni mE vyatItAni janmAni tava chArjuna. tAnyahaṁ veda sarvAṇi na tvaṁ vEttha parantapa'.

Many births (incarnations) of Mine have passed, O Arjuna, and so is it with you also.

I know them all (as the Supreme Lord), but you do not know them.

The Lord explains about the purpose of His incarnations

'yada yada hi dharmasya glAnirbhavati bhArata....'

'paritrANAyaya sAdhUnAm vinASAya cha dushkrutAm...'

Whenever there is decline of Dharma (righteousness), and uprising of Adharma (unrighteousness or evil), then I incarnate Myself.

Lord Krishna explains about all the entities in the world

Gita 13.5

'mahAbhUtAnyahankAro budhir avyaktamEva cha...'

The great elements, the ahankara / ego, the budhi / intellect, the avyakta / Prakruti, maya, the ten senses and the one, as well as the five objects of the senses.

Five great elements: AkAsa / ether, Vayu / air, Agni / fire, Apa / water, Bhumi / earth, all material energy

Ahankara, ego

Buddhi, intellect

Avyakta or Prakruti or maya

Ten. five sense organs - ear, skin, eye, tongue, nose, and five motor organs; speech, hands, feet, organs of excretion and reproduction and the one, mind.

The five objects of senses. Sabda / sound, sparSa / touch, rUpa / sight, rasa / taste and gandha / smell.

25. Jivatma, Soul

26. Paramatma, the Supreme Lord

The five senses are connected with the five elements; sound to ether or space, touch to air, sight to fire, taste to water and smell to earth, representing all gross matter.

The Kathopanishad verses 1.3, 3-1. 3.9 explains vividly the metaphor of a chariot to describe embodied humans. If the body is the chariot, the Atman (Self) is the rider, the intellect (buddhi) is the charioteer, and the mind (manas) is the reins controlling the senses, which are the horses, guiding them towards sensory objects (roads). This analogy emphasizes that a strong, resolute intellect is important to control the senses (horses) by the mind (reins) to reach the ultimate spiritual goal (Supreme Abode) rather than getting lost in worldly attractions.

Sri Nammazhwar addresses the Lord thus, in his Thiruvaimozhi verse,

'muniyE, nAnmuganE, mukkaNnappA' (Muni, a human Sage, Lord Brahma, and Lord Shiva) to indicate that it is the Supreme Lord who is the power in all the divine manifestations.

Each Soul is responsible for its state of existence which it obtained in accordance with its desire or preference to experience such life. God being the caring father empowers every Soul and every world eternally. He has no purpose of His own to fulfill in any of this. All of creation is a fragment of His blissful expression. Many are in the material worlds because that's what they wanted to experience. And countless many chose to be with God and got to experience divine life in the company of God, sharing transcendental joy eternally.

We can relate this to our own experiences in every family. Many of us who are kind parents, unconditionally dedicate our lives in facilitating the fulfillment of our children's wishes without question. Parents have no other purpose. Some children prefer to stay with their parents and some would like to explore the outside world. But parents cannot be involved in the children's life outside because they chose to be on their own. Yet parents have to empower them in everything they do. We continue to support their endeavors in every way possible without getting involved directly. If we intervened or interfered, they would not like it. It's their life and they should take responsibility all the way. Only some children realize that it was the parents who facilitated everything, throughout.

In the cycle of creation, one day of Brahma takes 8.64 billion earth years, 4.32 billion years his day and 4.32 billion years his might. and 311.04 trillion earth years for Brahma's lifetime of 100 years. And the next Brahma will manifest to take care of a new universe. It is said that our Brahma has completed more than 50 years of his tenure. It would mean that our universe was formed more than some 155 trillion earth years ago! It is said that Lord Hanuman would be chosen to be the next Brahma.

Starting with the duration of millennia, the duration of the universe is stated.

Kali Yuga (present millennium). 432000 years

Dwapar Yuga 864000 years

Treta Yuga 1196000 years

Kruta Yuga 17280000 years

4 yugas (chatur yuga) make a 'tama' of 4.32 million years, 1000 'tama's or 4.32 billion years for a day and 4.32 billion years a night for Brahma.

We are now in the year 5126 (English calendar year - October 2025 AD) since this (28th) Kali Yuga began. The name of the year is 'visvAvasu' in a cycle of 60 years. We are in the 28th 'tama' or mahayuga out of 1000 tama's; and 972 x 4.32 million years, about 4.199 billion years to go before the solar system starts to end its existence. The end begins at night and the new sun and planets would be formed and ready to support new life before dawn. It means that we are in the early dawn of our Brahma's Day, known as SwEta Varaha Kalpa, Vaivasvata Manvantara, ashta-vimSati tama (28th part of 1000).

It is said that this millennium would end in some 427,000 years. Thousands of years before the end, it would rain non-stop for a year and there would be no rain for a similar duration (ati-vrushti and anA-vrushti), and duration would increase with incessant rain and no rain subsequently, ending most life forms including humans, as it would be impossible for survival in such conditions. And new generations of most life forms would begin at the dawn of a new millennium (kruta Yuga). The nature of kali Yuga is well explained by Sage Vyasa in the Srimad Bhagavatam. All the atrocities that happened from the beginning of the Kali Yuga for 5000 years so far, would progressively worsen towards the end of this millennium. Towards the end of the millennium at an appropriate time, the Lord would incarnate as 'Kalki' to punish the evil people and protect the righteous. Till then righteous people would manage to survive in accordance with their karma. If I deserve suffering based on my past karma, I have to endure the effects of my karma. If I don't deserve to suffer, that's when God manages situations within the karmic framework. When evil powers reign supreme, God incarnates personally.

It is important to understand that there are different time zones in the universe. A year on earth is a day in another world.

True knowledge of the universe has been passed down by the Supreme Lord. We don't believe it because we cannot see the Supreme Lord or Brahma, now. We are not willing to accept the scriptures as authentic, thinking it is written by someone like us. To check if it could be authored by humans, we should examine the content of the Vedas or Gita to see if it matches the reality of all existence. Only then would we realize that such knowledge is not humanly possible. It exceeds human potential to author such content. In fact, it is hard for us to grasp the correct perspectives of these scriptures on our own, let alone authoring something so transcendental.

The Vedas have been passed down in sound form, by the rishis and learned teachers. Only in our generations do we need printed texts and commentaries explaining the perspectives, as we are incapable of memorizing flawlessly what is taught verbally. But there is sufficient proof of how the original knowledge was written down in script.

Historically, Veda Vyasa, who is an incarnation of the Supreme Lord, retold the Vedas in writing consisting of around 100,000 verses, in four parts, for the benefit of humanity of this era. Sage Vyasa was born in Dwapara Yuga, the preceding millennium. Vedas are not authored by Sri Vyasa either; Vedas are 'a-paurushEya' (not authored by any human).

Sri Krishna has explained the process of creation of the universe in the Gita which is in recorded history.

We can definitively conclude that the Vedas are God given knowledge and not authored by a human by the simple fact that no one from any millennium has claimed credit for authoring the Vedas. We know, when a person authors any work, even if insignificant, would not allow denial of his work. I certainly would want to get the recognition for whatever little I have done. Likewise, if someone had authored the Vedas, surely, they would not miss having their name established in history for such a major authorship. Veda Vyasa retold the Mahabharata, as it happened. In that he says that Srimad Bhagavad Gita was preached by Lord Krishna to Arjuna, and Gita was not authored by Vyasa; he mentions that Sri Vishnu Sahasranama was composed by Bhishma, resting on the bed of arrows, again confirming that Vyasa was not the composer of Sri Vishnu Sahasranama. When that is the fact, how anyone would not take credit for the Vedas, if they had truly authored the Vedas?

When Arjuna desires to see Krishna's divine universal Form, the Lord explains,

Gita 11.8

'na to mAm SakyasE drashtum anEnaiva sva-chakshushA, divyaṁ dadAmi tE chakshuh paSya mE yogamaiSvaram.'

Meaning, 'But you will not be able to see Me with your own eye, I grace you with a divine sight. Behold My Lordly form'.

Lord Krishna revealed His Divine Universal Form to Arjuna on his request. Everyone saw Lord Krishna, the incarnation. But not the divine universal form. Arjuna was granted divine sight to be able to see His Universal Form.

Gita 11.19

Arjuna trembles in fear seeing the universal form (vishwarUpa), who witnesses countless worlds and beings formed from and withdrawn into by the Lord. Arjuna describes what he saw thus –

'The divine form of the Lord - 'having no beginning, middle, or end' (anAdi-madhyAntam), of infinite power (ananta-vIryam), and burning the universe with His own radiance'. Arjuna expresses in overwhelmed amazement, describing God's eternal, all-pervading nature.

In another incident during His childhood, Lord Krishna displays the universe in His mouth to mother Yashoda when she asks Him to open His mouth to check if He had eaten mud. Mother Yashoda was mystified by what she saw, as she accepted Krishna as her child and not as the Supreme Lord.

People who have no access to Vedic knowledge cannot understand this truth; they falsely presume it is modern thinking and intelligence to question the historical reality of the incarnations of Lord Krishna and Lord Rama. Consequently, they discredit the

Sri Ramayana and Mahabharata as mere mythology. However, both the epic scriptures are known as ITIHASA, meaning 'as it happened ', it is much more of an authentic reality of history in the earlier millenniums than the distorted modern history of our times taught in schools.

All our bodily functions operate automatically, independent of our conscious effort or thought. We are alive without actively striving to stay alive, yet we are powerless to prevent death when it inevitably happens. Our very existence serves as proof of the Soul and of God. Yet we have never seen a Soul out of trillions existing in every human body, including the Self, likewise, God remains invisible. Yet all planets, stars, galaxies, and universes serve as proof of a divine, super power present everywhere. While it can be argued that we are alive due to the presence of the Soul within us, the Soul does not keep the universe alive. Rather, the universe is kept alive and sustained by God's all-pervading power.

We have enough evidence to know that the Paramatma empowers our existence within our bodies, yet we are unable to sense His presence in us. Many of the vital functions of our body work on autopilot, without our knowledge, choice, effort, or involvement. Our heart keeps beating and pumping blood nonstop until death; we have no role in it. Food is digested without us having to even think about it. We are not even conscious of our being alive in deep sleep, yet we breathe, and all vital organs work throughout. A person stays alive even in a state of coma or under anesthesia because all the vital parts are kept working. If the self is not doing anything to stay alive in the body, it is obvious that someone else is taking care of us. And that is God, eternally present within us.

When awake, we make conscious choices and make an effort to do things using our physical body. But we (the Souls) have no involvement in the functioning of the body internally. Our immune system is active all the time. There is a self-healing mechanism in the body that works round the clock. Who is taking care of things in our body in which we are not involved? It is evident that either the Self or some other consciousness or power is needed to keep the body alive and functioning. And we know that we are not involved in everything. Is there a need to say that whatever things we are not actively involved in doing, it is God who empowers everything by His will alone?

Think about it: when we stop a car and turn off the engine, but it keeps moving, we assume someone else is driving, it's on autopilot, or it's a self-driving car. That 'somebody' or autopilot program is God. Clearly, material energy cannot function on its own. Whether it is our universe or our bodies, nothing works automatically without a super-intelligent programmer keeping it going. Since humans are not capable of such programming, obviously, it is God who empowers everything by His will, without direct involvement.

There have been many God-realized Souls irrespective of their religion. Even a Kabirdas, though a Muslim devotee, is revered as a God-realized Soul. Even a God-realized Soul cannot understand God fully, for God cannot be fully known. Those who think God can be fully known, do not know God. One who realizes this truth, truly knows God. Our goal is to know the reality of who we are, our origin, our nature, potential and relationship with our Creator and what is good for us. The created entity can never know the Creator unless revealed.

We blindly accept that single-celled microbes evolved into all 8.4 million species, including complex, intelligent humans, essentially giving microscopic organisms the honor of creation - yet we have no direct evidence of this. We have never witnessed material energy or life creating itself. Yet, we are willing to blindly believe that matter is the first cause and claim it is 'scientific' to think so.

It is naive to think that all reality can be perceived by material science, not realizing that anything divine is beyond the realm of material science or scientific perception. But do we have evidence to disprove the existence of God and His creation? Or evidence to prove material energy was the first cause of everything we see today?

Come to think of it, science starts learning things from being ignorant. There is an apparent flaw here. They don't know if what they learn is correct or not, unless it is from an authentic source. This method is fine in understanding a material reality we can observe. But not something exceeding our realm of understanding. Even science would agree that when you start from scratch, there will be flaws and uncertainty regarding accuracy. However, when you learn from the highest source, such as the Vedas, there is no doubting the accuracy. So, science might have a serious problem when it presents findings as knowledge without fully understanding reality; and it can lead to wrong conclusions. Assumption by science that the universe was formed by matter and all the species evolved from microbes, proves that science is ignorant and will accept anything presented to it as knowledge.

We face a similar problem with AI-generated solutions; We cannot know for sure if what the AI suggests is correct, as it lacks original knowledge, true intelligence, consciousness or understanding. It simply processes the data provided to it. Yet, because of our cognitive limitations, we are forced to rely on computer-driven systems. It is crucial for us to remain 'in the know' to avoid feeling lost and helpless.

Gita 10.3

'yo mAm ajamanAdim cha vEtthi lokamahESwaram.'

He who knows Me as unborn and without a beginning and the Supreme Lord of the worlds, he among the mortals is undeluded, is released from all karma.

Time –

Gita 10.33.

'ahamEvAkshayah kalo; dhAtAham viSwatomukhah.'

I am also inexhaustible, everlasting time; and of creators, I am Brahma, whose manifold faces turn everywhere.

This verse explains that the Lord is the origin of time as well, rather Time is eternal without beginning. We perceive time differently. For the universe - Time begins - with its birth and ends with dissolution. For each one of us Time begins with our birth in a body and ends with its death, every time.

Gita 10.39

'yacchApi sarva bhUtAnAm bIjam tadahamarjuna...'

I am that which is the seed of all beings, O Arjuna. Nothing that moves or does not move, exists without Me.

Gita 10.40

'nAnto'sti mama divyAnAmvibhUtInAm parantapa....'

There is no end to My divine manifestations....

Gita 10.41

'yadyadvibhUti matsattvam SrImadUrjitamEvavA....

Whatever being is possessed of power, or of splendor, or of energy, know that as coming from a fragment of My power.

Gita 10.42

athavA bahunaitEna kim jnAtEna tavArjuna...'

But of what use to you is all this extensive knowledge, O Arjuna, I stand sustaining this whole universe with a fragment of My power.

This explanation is very relevant to all those who question the veracity of God's role in creation. Lord Krishna clarifies that it doesn't matter how much He revealed, it is of no use as we cannot comprehend the whole truth about God. He affirms that the whole creation of our universe and multiverse is though infinite in our perspective; it is no more than a fragment of God's potential.

Probably even God hasn't used more than a fragment of His powers, ever.

The statements below give us ample understanding about how the Universe was formed

Vedanta sutra 1-1-2 defines Brahman thus

'janmAdy-asya-yatah'; that from which everything originates.

Isavasyopanishad 5

'tadantarasya sarvasya tad u sarvasyAsya bAhyatah'

that Brahman is within everything and also outside of everything.

Maha narayanopanishad 15

'vEnastatpasyan visvA bhuvanAni vidvAn.'

The omniscient Paramatman directly perceives all the worlds like the Sun; all this universe becomes one in Him. All this universe becomes united without the differentiation of names and forms in Him (when the universe is withdrawn or prior to formation), and also with the differentiation of names and forms, when the universe is formed. That all pervasive Lord is in all entities, sentient and non-sentient.

Prior to formation everything is merged into the Lord. Then He alone exists.

Ch. 9 / 8

'prakrutim swAmashtabhya visrujAmi punahpunaha

bhUthagrAmamimam kritsnam avaSam prakrutEr vaSAt'

Controlling the Prakruti (nature) which is My own, I send forth again and again all this multitude of beings, helpless under the sway of Prakruti.

This verse establishes the fact that all the life forms, some 8.4 million, are born in every cycle of creation. New generations of a multitude of species including humans are born in every millennium.

Sri Nammazhwar, an incarnation of a NityasUri from Paramapadam, the divine world of God, reveals the origin of our universe thus.

'Onrum, thEvum, ulagum, uyirum mattrum yAdumillAvanru, nAnmugan thannodu dEvarugoduyir padaithAn...'

When there were no gods, worlds, living entities, and nothing visibly existed, the Supreme Lord formed the worlds along with Brahma, other gods and living entities...

Mahopanishad affirms,

Eko ha vai nArAyaNa asIt,

na brahmA na ISAno, nApo, nAgnihi,

na vAyuh, nEmE dyAvA pruthivI,

na nakShatrANi na sUryah

sa EkAkI nara EvA...

Meaning, prior to manifestation of this universe, the Supreme Being Narayana alone existed. Neither Lord Brahma (God of creation) nor Lord Shiva (the God of destruction) manifested. That apart, earth, water, fire, air, ether, stars including sun etc., none existed. The definition of the name 'nArAyaNa' describes the power of the Supreme Lord (narANAm ayanah yasya saha) as the One, who is in everything and has everything in Him, and is the sustainer and upholder of everything when manifested, who is the refuge for everything (the universe / multiverse and beings), when He withdraws the whole creation. The word 'nArAyaNa' defines the functional powers of the Supreme Lord, rather than a name. All the names of God mentioned in the sacred scriptures define His infinite powers, and his countless glorious attributes.

Vedas have declared that God cannot be known fully, for everything about Him is infinite.

Brihadaranyaka Upanishad.

'PoorNamadah, poorNamidam, poorNAt poorNamudachyatE; poorNasya poorNamAdAya poorNamEvAvasishyatE'.

meaning "the Brahman, (the Supreme Lord), is infinite and complete, and this manifest world is also infinite and complete; when the complete is projected from the complete, only completeness remains, the all-pervading, and inexhaustible power of the Divine.

Every bit of information is available in the scriptures. There is no need to imagine or construct our own assumptions.

There is a saying in Tamizh. "katradu kaimaNNaLaVu, kalladadu ulagaLavu"

When we have learnt everything, we possibly could, we still only know so much as a fistful of sand in measure but what we don't know is still so much as this infinite universe in comparison.

If we find anything said in the Vedas or Bhagawad Gita that appears not to match reality, we are likely mistaken in our understanding. The Vedas and Srimad Bhagawad Gita cannot be wrong, as they are divine knowledge revealed by God.

To understand the greater significance of knowledge found in the Vedas, India gave the world the power of 'zero' in mathematics. It was Aryabhatta and Brahmagupta who formalized the use of zero in calculations and introduced the value of zero in the decimal system. In fact, even the numerals that are said to be Arabic are originally Indian. Historically 'Arabic numerals' are of Indian origin, developed in India between the 1st and 4th centuries CE. They are also known as the Hindu-Arabic numeral system because they were transmitted to Europe through the Arab world after being adopted by Arab mathematicians. The zero is also extremely significant in quantum computing. The normal computers use bits (0 or 1), but the quantum computers use qubits, (a 0, a 1 or both simultaneously), which gives the quantum computers the power to perform certain complex calculations faster by exploring multiple possibilities at once.

One of the greatest mathematical brains, Srinivasa Ramanujan, has honestly expressed that he was able to derive direct solutions to futuristic math equations by the divine grace of his family Deity Goddess Namagiri Thayar or Mahalakshmi. He attributed his mathematical insights to his family goddess, who is said to have appeared in his dreams and presented him with complex formulas.

He said, quote -

An equation for me has no meaning, unless it expresses a thought of God.

unquote.

Accepting that the universe manifested from God, we can ask how the universe would merge into God at dissolution, if it is spreading away nonstop? God is all pervading. Formation is possible simultaneously or intermittently, at all places in all directions and so can the dissolution, when withdrawn or absorbed by God at all places. Vedas explain that the Supreme Lord pervades the universes at manifestation. Formation and end of the universe or multiverse is unlikely from just one big bang event but it is feasible any which way we can imagine when God is the origin, as God has infinite possibilities.

And 'God's-will' is powerful enough to transform everything into a working reality.

Come to think of it, we had no control over our birth or even in selecting our parents or for that matter, being born as a human, animal, bird, insect, fish, plant life (trees) etc. as one of some 8.4 million species. Funnily, many children question their parents as to why they had the children and if the children had an option, they would not like to be born. *The reality though is that it is the karma of the Soul to be born that determines who the parents would be and not decided by parents. Just as a Soul has no power to choose parents, the parents have no power to choose who should be their children. It's their respective karma that unites the children and parents.*

And surely, we have no control over the time and manner of death of our bodies and what our next birth would be. Not to mention so many things that we plan and do, but have no control over the outcome, either. We are helpless in *taking birth and dying unwillingly*, in countless life cycles as different species determined by the law of karma. Life seems almost like a bubble, vulnerable, unpredictable and mysterious.

Formation and dissolution of the universe is explained in the Vedas, which is the knowledge from the Supreme Lord who is the self-existing first parent, as none of the embodied beings existed before creation. Everyone came to exist after the universe was formed. And so, there is no way we can know exactly how the universe formed based on our observations and equations now.

CHAPTER THREE

Who we are

Self-realization (swa swarupam)

Sri Pillai Lokachariyar, a divine Soul (NityasUri) from the eternal world who incarnated on earth to reveal the highest knowledge, has explained the secrets to knowing all about the purpose of our existence and what we have to do to attain the highest state. 'Artha panchakam' or five truths, that define five states of realization that would enable us to seek God's benevolent grace and attain the ultimate state of existence we have always wanted.

1. (swa swarUpam). Nature of the Self, who we are Or Self-realization.

1. (para swarUpam) Nature of Paramatma. our origin. Or God realization.

3. purushArtha swarUpam - purpose of immortality, what will I do go there? This is important because if the purpose is not worthy enough, why should we go there?

4. (upAya swarUpam) Means to attain the goal. How can we get there? What are the ways to attain our desired goal?

5. (virodhi swarUpam). Obstacles. What prevented us from attaining our destiny thus far?

These five steps define the essential aspects in knowing our present state of existence, and the ways to attain our goal. These five are applicable to everything we do, even in this world.

For instance, if I am in Bangalore, and would like to travel to Kanchipuram because that's where I belong, and my home there is the best place where my parents are waiting for me, and that I will be happy reaching home. Knowing the five things would help.

1. My current position (lost, helpless, suffering)
2. My destination where I belong (my true home, zero problems, happiness guaranteed)
3. What will I do after going there (be happy doing everything I want, devoid of suffering)
4. How can I reach the destination, (means to reach - seek help from others who know, learn from a Guru about the paths to reach our goal)
5. What prevented me thus far. (I chose to go away from home, got attached to outside life, till I encountered problems)

What is self-realization? Who is the Self that we are talking about? What is the proof of existence of the individual Self, whom we have never seen or perceived through senses? Do we simply believe in the existence of a Self? (Soul), though we have never seen the Self or Souls in other living entities either? How do we realize the Self that is beyond our perception?

The reason we have not been able to perceive the Atma or Soul is because Atma is a divine, subtle and invisible entity that's beyond the perceptibility of material senses. Just because we cannot see the Souls, does that not mean all the humans are lifeless? The fact that humans and other species are functional, is proof of their being alive and existence of the Souls in them. The bodies are considered dead without the presence of the Souls in them. Dead bodies don't function.

The first step to self-realization is to know, feel, understand and be aware who we truly are. We are able to know about things we see around us in this world because consciousness is our nature, awareness is inherent to us, knowledge is intrinsic in us. However, we could not know about the self if not for the Vedas and Bhagawad Gita. There is no scientific instrument that can reveal a Soul to us.

When we say 'I' or 'myself', am I talking about my body? Or are we different from our bodies? It's similar to a driver thinking he is the car. If the car is the body, the driver is not to be mistaken as the car. He is a different entity of the nature of consciousness who drives the car. The car does not know or understand anything about the driver. Similarly, we the Souls are the drivers in our bodies and clearly, we are not our bodies. Our bodies have no knowledge about us. As a matter of fact, our bodies don't know when we are in the body or when we leave the body. Similarly, our senses, mind and intellect are also non sentient and have no awareness of the Soul being their master.

There is a difference in the way we refer to ourselves or things that belong to us. When we say my body, it confirms that I am not the body. My body is not me. Every time we say 'my', it refers to everything else other than the self. If I say my hand, it means I am not my hand. The difference between 'I' and 'my' is distinguished here. The same is true about my mind, senses, and brain, which refer to my possessions which I am not. We also say I am conscious, which explains the awareness, nature of 'I' the Soul.

I think, therefore 'I am', is a line often stated to confirm that there can be no thoughts without the thinker, the Soul. The truth though is that 'I am', even without having to think. 'I am', even when I go unconscious. Our consciousness is not active in deep sleep. But the Self exists. But there is no thought without the thinker, and the Soul is the thinker.

We also use the expression 'my soul', or 'his soul' which is not the accurate phrase because I am the Soul, there is no my Soul, or he is the Soul. But we understand the essence of the expression though incorrectly stated. Similarly, we say I am sick, or are you unwell? In reality there is no sickness to the self or Soul; sickness is body related or health related. And medication is again for the body. We can see this difference in expression in all the Indian languages. Indians don't say I am not well or I am unwell. They say, 'sharir svaasthya nahin hai' in Hindi, 'maige usharilla' in Kannada, 'udambu sariyillai' in Tamizh or 'ontlo bagaledu' in Telugu which is supposed to mean my body is unwell or there is a problem with the health of the body. Ailments and sickness are body related. My head is aching or legs are hurting, is a correct expression. When a person dies, we say he or she has passed away. There is no death for the Souls.

My name is Krishna, a man, an engineer, a son, etc. refer to identities such as name, gender, profession, relationship. But all these are not the identity of the self. My true identity is that I am an Atma or Soul. We could even rightfully claim, we are a potency of God.

It is said that the pleasure or pain is experienced by us (the souls) and not the body, nor by the Supreme Lord who is present in the Souls but remains as the witness or observer but without involvement. The Supreme Lord remains untouched by the material energy though present in everything. God is not involved in the activities of the Souls. But it is the Souls that experience all activities in a body and a body itself does not experience anything.

The Upanishads declare that Jivatma (soul) is the experiencer and doer of all activities, possessing consciousness, knowledge, senses, and mind.

'esha hi drashta, sprashta, Srota, ghrAta, rasayita, manta, boddha, karta, vijnAnamaya, purusha sa parE akshara AtmanisampratishtatE'

The Jivatma is verily the seer, toucher, smeller, taster, thinker, knower, doer, the individual who is of the nature of pure knowledge, he becomes established in the transcendental divine nature of the Supreme Self, Paramatman.

It is also important to know that the real Self, the individual Souls, are able to perceive everything while being alive in a body using our senses, mind and intellect. We are not the senses, mind or intellect, either, which form our subtle body, not visible to us. Senses, mind, intellect are also non sentient. They do not function by themselves. The subtle body (sUkshma SarIra) is inseparable from the Self, even after death of the gross body. The individual Atma with the subtle body, finds another gross body on the merits of its karma, which we know as birth.

The Soul with the subtle body, goes on to get into the next body, carrying the impressions of karma from past lives. The subtle body remains attached to the Soul, till it attains the divine world where it acquires a divine subtle, gross body.

Prasnopanishad 6,

'hrudi hyEsha Atma'. The Jivatman is in the heart.

Lord Krishna is ever available right in our hearts and we don't have to go in search of God anywhere else.

Gita 15.15

'sarvasya chAham hrudi samnivishto mattah smrutir jnAnam apohanam cha, vEdais cha sarvair aham Eva vEdyo vEdAntakrud vEda-vidEva chAham.'

I am seated in the hearts of all beings. From Me are memory, knowledge and their removal as well. Indeed, I alone am to be known from all the Vedas. I grant the fruition of the rituals of Vedas; I alone am the knower of the Vedas.

The Souls are seated in the subtle heart or 'hruday'. People think, if there is a heart transplant procedure, the Soul from the donor would get into the body of the recipient? It's a sensible question to ask. But the reality is that the Soul, which is said to be in the space of every heart, is not the organ or heart muscle that beats to pump blood throughout one's lifetime. It is a reasonable doubt because we declare a person 'dead' when his heart stops beating.

The heart where the Soul resides with God within, is known as 'antahkaran' or inner heart and is subtle, invisible like the mind and intellect. It is similar to the 'mind' we use for thinking, being different from the brain. We generate emotions and feelings such as love, hate, like, dislike, affection, aversion, fulfillment etc. in our heart. The Soul, which is the subtle most and invisible to material senses resides in the emotional heart. But consciousness of the Soul pervades throughout the body from head to toe, like a lamp that fills its proximity with light.

Even within a lifetime, our body changes continuously. There are birth and death taking place continuously in our body. We can see this reality when we compare our photos taken at different ages from infancy to old age; we look different at every stage of life, with some semblance reminding us it's the same person. We change our body after death, when the Soul completes its term in the body or when the body becomes dysfunctional. The Soul would acquire a new body on the merits of accrued karma.

Lord Krishna explains this in the Gita 2.13

'dEhino'smin yathA dEhE kaumAram yavvanam jarA,

tathA dEhAntara prAptir dhIrAs tatra na muhyati.'

Just as the embodied soul passes through childhood, youth, and old age (pertaining the body), so too it acquires another body at death; a wise person (self-realized soul) is not deluded by that.

Lord Krishna explains to Arjuna that the physical body is changing and temporary. Just like we change clothes, so too at death of the body, Atma acquires a new body, and that the Soul is eternal. The Soul simply transmigrates into a new body, a truth understood by the wise. When we attain adulthood, giving up childhood we don't grieve at the death of childhood; we don't grieve when we change clothes. Death is inevitable for the born. Birth and death are for the bodies. The Soul is neither born nor would ever die.

The Souls are defined based on the life experience they seek.

- Badh'Atma - The embodied Souls in the material universe are considered 'badh'Atma' or Souls bound by karma, which is due to their own preference and choice. Some who are happy with the experience of life in the material world will continue here going through birth and death repeatedly in all sorts of bodies till they decide they have had enough and want no more of karmic life experiences.

- Mumukshu - 'badh'atmas' who are desirous of seeking liberation from existence in the material world, to attain God in the divine world, are known as 'mumukshu'.

- MuktAtma - 'mumukshu's', who succeed in attaining the divine world are 'MuktAtma'. They were 'badh'Atma' going through life cycles in the material world due to karma, before attaining divine immortal existence.

- NityasUris - Souls that chose to remain with the Lord in the divine world are 'NityasUris' ever in selfless service to God, who are beyond karma, experiencing the divine bliss of the Lord eternally. God shares some of His powers with every Nityasuri and the muktAtmas, alike. They will never be subjected to karma again. The scriptures define their state thus, "na cha punarAvartatE', meaning, attaining the divine world once, they don't have to return to the karmic material existence. They can, however, incarnate along with God to experience His incarnations. Many Nityasuris took birth on earth to experience the blissful pastimes of Lord Rama and Lord Krishna during incarnations. They also incarnate in this world to bestow divine knowledge to badh'Atmas, guiding them to attain immortality.

- KaivalyArthi - And there are some Souls who are fulfilled by experiencing peace within the Self, without any other experience, they are known as 'kaivalyarthi'. This is the most wasted state of existence because there is no return once a Soul decides to be unto himself. They remain ever deprived of the divine bliss of the Lord.

There is one thing that every living entity seems to want and strives to attain. Is it knowledge, good health, money, prosperity, success, fame, power, and position? Each one would have a preference based on what would satisfy them. When they are not content even after achieving what they wanted, they pursue other goals. There is no end to our wants and we are never satisfied with what we have. We go in search of more, hoping we may find true happiness in the next thing because we didn't find that fulfilling happiness thus far. In other words, we are in pursuit of happiness in everything we do. We all want to be happy every moment of life, doing everything we are capable of doing. We want to experience every possibility that could potentially make us happy. Then, who is seeking happiness. Is it ourselves (Atma, Soul) or is it our body, senses, mind that wants to be happy? It is not hard to realize that it is the Self (Soul) that's yearning for true happiness. It would seem we are not truly happy which is why we continue to look for more ways to be happy.

There are levels of experiences such as pleasure, joy, happiness and bliss. The body cannot feel pleasure or pain by itself. We (the Souls) experience all the emotions. The body is nothing more than a machine. At the physical level we (Souls) experience pleasure: more subtle is the joy perceived using the senses; even subtler and satisfying would be the happiness that we experience, which is a state of mind. Finally, it is none of these things that fulfills what we truly seek. What we want is bliss or 'Ananda', which we have never experienced in this world, as this world is devoid of the bliss we seek. We will keep pursuing till we attain bliss, because experience of bliss is not possible in this world. We all seek to experience bliss, naturally, because our origin (God) is bliss or Infinite Happiness.

In reality, there are no atheists or agnostics, though they don't believe in the existence of God. For them, seeing is believing. They can't see God being instrumental in whatever happens in their life. So, where is the need for God? It is also true that God isn't directly involved in our lives. Indeed, that was the purpose of creating the universe and empowering the Souls to acquire bodies of their merit and choice. That's what we wanted and that's what God gave us. However, even the atheists truly only love God and none else, nothing of this world, not even their close family, nor themselves. They love to be happy always, which is in the nature of every Soul. knowingly or unknowingly, it is indeed God that they love, who is the infinite ocean of bliss. When we seek happiness in everything that we do, it is God that we are seeking and trying to attain. If we realize this truth and reality, we would try to connect directly with God, instead of making futile efforts to seek happiness from material sources that are devoid of any happiness.

If we seek divine happiness from people, we fail because they are also similar to us and are in search of the same. When we are attached to our relationships, it lasts till the person is alive, bringing us grief at the death of the person, instead of happiness. When we look for happiness in material things, we are fooling ourselves, for matter being non sentient, cannot understand happiness, and so we get disappointed. The Supreme Lord is the only source where we can find the quality of joy we seek. Nowhere else or no one else can provide that bliss we have been in eternal search of.

Now it's important to realize the limits of human perceptibility. We cannot perceive something we cannot see or identify or recognize or understand or connect with. How possibly could we connect with God if we can't see Him in a form or visualize or imagine Him?

Sri Nammazhwar in his Thiruvaimozhi verse recounts the conversation Prahlada had with his father, Hiranyakashipu, when he questions the child where does his God exist and if he could reveal him? Prahlada responds by saying 'there is no place where God doesn't exist, He is everywhere, in everything; in me and in you as well'. And indeed, Lord Narasimha appears from a pillar when Hiranyakashipu breaks the pillar with his mace, on the assurance of Prahlada that God was also in that pillar. In fact, the Lord decided to incarnate as Narasimha (lion's head on a human body), only after Hiranyakashipu threatened Prahlada that he would maul him like a lion would maul its prey, if his Lord was not to be found in that pillar, which he had personally overseen as it was being built and was sure there was no God in that pillar. But of course, the Lord had to fulfill all the conditions of the boon Hiranyakashipu had obtained from Brahma, to remain an immortal.

We are made with five senses with which we perceive everything in our lives. In that 92 percent of all our perception is by sight. If we cannot see God in a form, how can we perceive His existence? Even in dreams, sight is the main perception tool. If someone shows us empty space to see a beautiful piece of art that is invisible, we would simply stare at the void and say we see nothing. Even though God is present in that empty space, we would not be able to see Him because our understanding is limited to our sensory perception. Therefore, to connect with God, we must perceive Him. We can perceive God in idols, deities and pictures once we realize His all-pervasiveness. Furthermore, we can even imagine His beautiful form in our minds, for God truly exists in all things and everywhere.

If the universe and the embodied beings have a form manifesting from God, surely God also has forms, for He is present in everyone and everything. Just that He is invisible to our material senses. If God has the power to manifest universes with form, He surely has the power to take any form He chooses. Also, He incarnates with all glorious attributes and powers as seen in Lord Rama and Lord Krishna. Originally, prior to creation, the Supreme Lord alone existed in His most magnificent divine form. We would be able to see God and interact with Him when we attain the divine world, possessing divine senses. We could also see God here and now if God reveals Himself.

Then, there is this question that keeps cropping up in our minds; if God truly exists, and He truly loves all the living beings, wants us to be ever happy, why doesn't God stop all the suffering here? Also prevent atrocities committed by cruel people; stop destruction caused by natural calamities; help people from suffering ailments. Also grant us perfect health, youth and strength, all the prosperity and comforts we are constantly in search of. And a life we can relate to, as the fairy tale saying goes 'and they lived happily ever after' granting us immortality as well. He could well have given us total protection from all the pain, misery, anguish that's ever prevailing here. Life seems to be one of grief and suffering here.

The answer to that is, we are seeking divine existence in the mortal world. There is a divine world where there is no suffering whatsoever. We would not have to suffer if we attain the divine world. In the divine world, there is no karma or consequences as all the divine Souls are noble and virtuous by the nature of their divine bodies, mind and senses, devoid of defilement. While the material bodies we are in, inherently possess the three guNAs (qualities) by which we are influenced in doing our karma. If we consciously develop noble qualities of the nature of sattva guNa, giving up rajasa guna and tamasa guna, we become qualified for God's grace and He would accept us to the divine world. Sattva guNa influences us to be dharmic and righteous.

Now let's ask ourselves a question: Why do we find so much suffering, grief, misery in this world? A lot of it is caused by humans or by natural calamities. Even if God gave us protection from all evil effects, it would not work unless God took away the freedom He has bestowed upon us. Endowed with the power of free will, evil minded people would still hurt the righteous because it is in their power to do so.

That's exactly what the colonizers and invaders did to prosperous, peaceful societies across our world. India is a great example. India was a prosperous, wealthy, advanced and righteous society, at one point contributing nearly a quarter of the world's economy. Yet, colonizers and invaders plundered our resources, killing millions of civilians and causing untold misery and poverty. Consequently, India became so poor that its economy was reduced to only about two or three percent of the world's total. Furthermore, the Moghuls killed tens of millions of people for not converting to their religion. However, India has begun recovering in recent decades, managed by some great political leaders enabled by the law of karma.

While some terrorists kill in the name of God, no God would support such acts of terror. Obviously, the perpetrators of these heinous atrocities will suffer the consequences of their evil karma. While we feel sorry for them when they go through that suffering, it is not caused by God.

To understand why many people, suffer in life, let us visit a prison where convicts are serving sentences for past crimes. The land's automated law system has found them guilty, and they are suffering the punishment for their actions (karma). Conversely, people outside the prison live happily, enjoying peaceful, righteous lives. They are not imprisoned or suffering because of their good karma. However, not all mistakes are crimes, and therefore, not everyone is imprisoned. Yet, no one can be sure they have accrued only good karma.

Even if we commit bad karma out of ignorance, we cannot escape the consequences. While those who suffer extreme misfortunes have likely done severe negative karma; those experiencing minor ailments are paying for lesser misdeeds. Ultimately, everyone bears the results of their actions - good karma brings happiness, while bad karma brings pain.

This world, as the popular saying goes - and as is well understood by everyone - has no such thing as a 'free lunch.' Nothing comes free, and nothing is given for free; every relationship often seems to work only on the principle of 'what's in it for me.' However, nothing has a cost attached to it when it comes to God. He created this world because we wished to experience existence on our own, and He did it unconditionally. He will grant us existence in the eternal world, again by His grace, at no cost. All that we need to do is seek His grace with unconditional love. He will know if our devotion is true, for He is right there within us. There is nothing we need to do, for He does everything.

God does not want anything in return from us; He does not do business deals with His children. There is no give-and-take. God knows only to give and bestow grace. He never hurts or punishes us, nor does He force us to do anything against our will. However, we must fend for ourselves, as He is not involved in our karma or its consequences. He has set the laws of karma to work without His direct intervention.

Then why did Lord Krishna and Lord Rama kill some bad people during incarnations?

God does not kill anyone in the literal sense, and certainly not an embodied soul born of karma.

All those killed by God were not born of karma like us; rather, they were people born due to a reason (kAraNa janma) to curses or boons - mainly obtained from Lord Brahma - destined to meet their end at the hands of the Lord. Some even obtained boons that only the Lord could defeat them. Even to Ravana, God was kind, offering his hand in friendship after the first day of battle. It was Ravana's ego that prevented him from accepting Lord Rama's grace. Ultimately, all those killed by God attained immortality.

When Lord Krishna declares, He would incarnate to protect the righteous, punish the evil doers and to establish dharmic order in the society, all the evil people he would kill would also be born in His time for a reason and not due to karma.

God does not incarnate to punish bad people born of karma. The law of karma prevails for all of us.

Everyone, including the bad people, belongs to God. God's love is equal to all. However, He has to remove the evildoers from the society when necessary to prevent them from spreading and destroying others. What is the purpose of the existence of such a society?

Just as we eliminate cancer cells to protect healthy cells and keep the body alive, the removal of evil is meant to safeguard the good. When our immunity successfully fights infection, we need no medication; but when immunity fails or is overwhelmed, we seek help from a doctor.

That is exactly what God does when He incarnates: He protects the righteous when they cannot defend themselves and is also meant to share His love with the devotees. While some of us may have seen Lord Rama or Lord Krishna during their incarnations, many of us have been reborn because we failed to recognize the Lord's supreme nature, assuming Him as an ordinary human. Those who realized Him as the Lord would have experienced His bliss and grace and would have attained the divine world after that life.

All of us possess certain negative qualities - such as Kama (insatiable desire), krodha (untamed anger), lobha (greed), moha (delusion), madha (arrogance, pride), mAtsarya (jealousy) - which influence us to do bad karma, binding us to suffering as a result of our actions. If we succeed in getting what we desire, it turns into greed; if we fail, we get angry and lose control, often harming anyone or anything we perceive as hindering our objective. When greed is successful, delusion sets in, resulting in the arrogance of power. Furthermore, when we cannot tolerate the success of others, jealousy arises, leading us to try and destroy them, which eventually causes our own destruction. Desires are like a burning fire; fulfilling them is akin to adding fuel, causing the fire to burn brighter and demand more fuel, endlessly. Like black holes, the more you feed desires, the more they consume. Consequently, our hearts and minds become totally polluted. Therefore, devotion to God and the ability to realize His presence in everyone is the only way to keep our hearts and minds pure.

Karma, good or bad, is not about what we do to ourselves; it is what we do to others that is accounted for by God, who remains an eternal witness to all our actions. Furthermore, we are duty-bound to fulfill our responsibilities without attachment to the fruits of our efforts. Failing to fulfill our duties also accrues to the karma account.

The law of karma is both complex and simple. It seems complex when we cannot figure out why certain incidents happen to us. Yet, it is simple if we understand this: good actions bring pleasant consequences, and if we experience unpleasant things, obviously it is a consequence of our own doing.

How can we recognize a new karma versus the effects of past karma? Whatever we do out of conscious choice is a new karma. Any effect we experience without active engagement in the present - whether pleasant or unpleasant - is the consequence of past karma, originating from this life or past lives.

For instance, if something falls on my head without me doing anything, it is surely an effect of past karma. However, if I choose to bang my head on a wall to test its strength and get injured, that is a new karma I have chosen to do. The same holds true for fortunate events occurring without our active effort.

That's why it is recommended to go about doing our duties righteously without seeking gains we imagine as ours. We might have sown a bitter guard plant seed, but expect a sweet mango out of it and become disappointed at the outcome when we don't get to taste the sweet mangoes. *God is kind even in formulating the law of karma. In that He rewards us more for our good karma and we face lesser consequences for the bad karma committed by us.*

Normally, we don't go in search of problems or grief, but do experience trouble even without seeking. However, we go all out looking for pleasurable, happy experiences, but we might not be successful every time. The reason for this is that the karma rewards would reach us anyways, and we don't have to pursue them. Just as we don't plan for problems, yet we face them;

likewise, we don't have to plan for rewards which will come our way anyways.

Sri Nammazhwar, clarifies in his verse thus about karma.

'karumamum karumapalanum Agiya kAraNan tannai.'

He says it is the Supreme Lord who empowers us to do all the karma, but without involvement. And He is the power by which the law of karma works. Even here He is not directly involved in granting us the fruits of our actions. The law of karma works as intended by God.

Lord Krishna clarifies about how He is beyond karma, in Gita 9.9

'na cha mAm tAni karmANi nibadhnanti dhananjaya.'

Lord Krishna reveals His divine nature as the unattached source and witness of all our activities, unlike us who are bound by karma.

God oversees the creation and sustenance of the material world (Prakruti), He is not personally involved as everything works by the power of His divine Will. His actions do not bind Him to consequences for the Lord has no karma nor does He desire for the rewards of karma. He does everything for the sake of this creation.

Lord Krishna also clarifies here that while He has no karma, He is not involved in our karma either and hence we are responsible for our actions and consequences we face.

The true nature of the Atma, is explained in the Gita 2.18

'antavanta imE dEhAnityasyoktAh SarIriNah.'

These bodies with the embodied self are said to have an end, while the Atma Itself is eternal indestructible and incomprehensible.

Gita 2.20

'na jAyatE mriyatE vA kadAchinnayam bhUtvA...'

This Atma is never born, it never dies, having come into being, it never ceases to be. Unborn, eternal abiding and primeval, the Atma is not slain when the body is slain.

Gita 2.22

'vAsAmsi jIrNAni yathA vihAya navAni gruhNAti....'

Just as a man takes off worn out garments and puts on the new ones, so does the embodied Atma cast-off worn-out bodies and enter into new ones.

Gita 2.28

'avyaktAdIni bhUtAni vyakta madhyAni bhArata.'

Beings have an unknown beginning, a known middle and an unknown end. What is there to grieve over in all this?

The aging process for all the species including plant life, is not something we can control. At the cellular level, which forms the basic structure of the human body, billions of cells are being replenished every day. Though new cells are formed by cell division, there is aging in progress continuously even with the formation of new cells. While not all cells die in a few days; cell lifespan varies significantly by cell type, like some specific brain cells, lasting an entire lifetime. But the overall age of the person is not impacted by different lifespan for cells. The whole body tells the same age. A phenomenon like this is a fundamental, seemingly predetermined, biological process. This is not as intended by human intelligence. It is working this way due to divine arrangement.

We haven't yet been able to control the aging process or death of the body. This involves rapidly dividing cells that wear out faster. The different rate of cell replacement is an important aspect of how the human body maintains functionality, and fights disease. The gradual degradation of stem cell function eventually causes the overall aging process. This balance between regeneration and cumulative degeneration is a subject of ongoing scientific research into aging and longevity.

Life is all about being happy and successful without causing harm, pain, or suffering to the Self or others. At the same time, we cannot claim to be happy by doing bad things to people. While many relationships are based on the potential benefits we can derive from one another, true relationships are unconditional, and not for the sake of benefit

Just as we learn science, mathematics and languages at school which nurtures our skill development, necessary for a prosperous livelihood, we must also learn the essential knowledge contained in the Vedic scriptures to understand the true purpose of human existence. *We did not get this human life just to complete our lifetime in a body, doing pretty much the same things every day of life.* The larger purpose is to realize why we have obtained this human birth, though we did not have any say in all this.

Vedic studies impart great virtues, as well as moral and ethical values that are currently lacking in our societies. A person rooted in these values is less likely to commit crimes or act immorally. The high crime rates in many countries stem from a lack of awareness regarding the benefits of leading a dharmic (righteous) life; people commit crimes because they have lost the ability to distinguish between right and wrong. Often, people are so obsessed with gratifying their senses and ego that they assume this is the purpose of life, hurting others without concern for the consequences. In contrast, those who lead a dharmic life are materially successful, emotionally fulfilled, and generally happy about life. Also, they are respected and liked by their family, friends, and society.

Sir Einstein said,

quote

Only a life lived for others is a life worthwhile,

unquote

His scientific contributions are timeless and benefitted humanity all over the world and humanity would remember him for a long time.

There is a similar statement from the scriptures.

'paropakArah puNyAya pApAya para pIdanam'

This verse by Veda Vyasa states that "helping others is noble (punya), and harming others is ignoble or sin (papa)".

This ancient wisdom is a fundamental teaching for living a virtuous, righteous life, emphasizing compassion and service to others, without causing harm to them.

Righteous karma is dharma. Any action or karma done against the principles of dharma would be a sin. *But no one is forced to be righteous. Dharmic scriptures explain the benefits of following dharmic life and also ill effects of being unrighteous.*

Some people justify unrighteous means to achieve the desired result in the name of doing one's profession. For instance, a practitioner of law shall not lie or deceive the legal system in order to save his client from being declared guilty of a crime, even after being aware of his guilt. The lawyer would have sworn to uphold the law and not undermine it. He should not be defending a known criminal to make a bigger income in the pretext of doing his professional duty.

People of our village remember and love my father even after 63 years after his passing, because he lived his life supporting the well-being of everyone in our small village, Kooram, near Kanchipuram. Being a landlord, he would share the foodgrain stock with the people during drought situations, earning their gratitude, affection and respect. My father was greatly influenced by Sri Azhwan, a prime disciple of Sri Ramanujacharya, as though he inherited certain virtues from Sri Azhwan, who was 'kindness personified'.

Sri Azhwan was truly special, in that he had transcended all pride, though he was endowed with a rare combination of royal birth, profound scholarship, divine personality, and prosperity; it was uncommon for one person to possess all such virtues. He had incarnated to impart noble qualities to humanity. It is said that Lord Rama appeared as Sri Azhwan to serve Sri Ramanujacharya in this millennium, to reciprocate the unparalleled services of Sri Lakshmana who incarnated as Sri Ramanuja.

My late maternal uncle resigned from a government position because he was forced to approve irregular commercial documents, though he refused to be corrupt. He suffered financially after resigning a well-paying job, and struggled to sustain his family accepting less paying private jobs. He navigated through the hardships successfully and all his children are well educated and prosperous. He never complained, and was happy and fulfilled throughout his life. Righteousness was not compromised.

I have a young friend, who did his masters in corporate law, in the UK. He gave up his job while working in a corporate firm because he had to defend unrighteousness. My brothers-in-law, who did masters in law, didn't accept even well-paying clients' cases, if they were known to be wrong. Making more profits is not wrong, as long as dharma is followed in doing business.

Originally, the caste system we follow today has never been in practice of the true followers of Sanatana dharma. The caste system being practiced now is contradictory to the value systems of sacred scriptures. The varnashrama dharma suggested in the scriptures have been intentionally misinterpreted to defame the dharma. Varnas and ashramas are a guideline to understand each one's life based on natural quality (skill) and karma or duties. We are defined by our actions and character. It has nothing to do with one's birth or caste or religion. There are both good and bad people amongst the followers of every religion. In fact, the caste system prevalent and practiced in every religion is wrong, though Hindus are singled out, without a basis.

God has not created the religions or castes or divisions amongst humans. It was designed by us. Even in the earlier millenniums there were Asuras who did not accept the Vedas and the principles of dharma, and went against the Vedic teachings. Those who abided by the Vedic scriptures were known as Devas.

Lord Krishna emphatically states,

chAturvarnyam mayA srushta ***guNa, karma*** vibhgashah.

The four stations of (varna) my creation (Prakruti) are based on Guna (quality, character) and karma (deeds). There is no mention of birth or caste or even religion. The Prakruti (universe) inherently possesses the three gunas such as sattva, rajasa and tamasa. And we possess a fragment of this universe in our bodies, which influences our thinking and behavior depending upon which Guna (quality) is in effect in our deeds. When we are overcome by uncontrolled anger, in a state of delusion, unable to discriminate between right and wrong -

tamasa Guna is at work; when we are attached, thinking only about the benefit of the self, greedy, rajasa guna is ruling us and when we exhibit kindness, pious nature, clarity and conscious decision making, we are influenced by the sattva guna.

The 14th chapter of Bhagawad Gita explains the nature of these gunas further - Nature of sattva being without impurity, is luminous and free from morbidity. Even sattva binds by attachment to pleasure and to knowledge, though.

The nature of rajas is of passion arising from greed and attachment. It binds the Atma with attachment to action.

Tamas is of false knowledge and delusion, it binds the Atma with negligence, indolence and sleep.

Sattva prevails on knowledge that illumines the senses; greed, longing enable rajas to prevail; and inactivity, negligence and delusion make tamas to prevail. The fruit of good deeds is pure, of the nature of sattva; pain is the fruit of rajas; and ignorance is the fruit of tamas.

One who realizes that the gunas are the agent of action and one who transcends beyond all three gunas attains the Lord. The traits of such a person are: he hates not illumination, nor activity nor delusion when these prevail; nor longs for them when they cease. He would be unconcerned, undisturbed by the gunas. Who remains the same in pleasure and pain, dear and hateful; blame and praise.

Gita chapter 16, enlists qualities we should nurture in order to excel beyond the three gunas for the Lord to recognize our eligibility to grant us divine existence. Qualities of a person seeking divine destiny; based on which he attains an existence beyond this mortal world - fearlessness, purity of mind, devotion to knowledge of the self, philanthropy, self-control, worship, study of scriptures, austerity, uprightness, non-violence, truth, freedom from anger, renunciation, tranquility, non-insulting others, compassion to all beings, freedom from desire, gentleness, sense of shame for wrong doing, freedom from fickleness, grandeur, patience, fortitude, purity, freedom from hatred and false pride. One who possesses these qualities irrespective of his birth would be considered a person of divinity which is what each one should aspire for.

Our bodies are nothing but the food we eat. Foods inherently possess the three guNAs (qualities) intrinsic to the Prakruti (nature) which influences our thinking. We should consume sattvic food, offered to God first, before eating it as God given grace (prasadam). After all, food is God given. Sattvic food enables righteous thoughts. Lord Krishna has suggested in the Gita, the types of food we could eat that would enable nurturing sattvic qualities (guNAs).

That apart, Sri Krishna assures His devotees that He accepts gracefully the food offered to Him with pure affection. He ate banana peel instead of the fruit, mistakenly offered by Vidura, out of pure devotion, who did not realize he was giving the Lord the peel instead of the fruit. Lord Rama ate fruits bitten and tasted by a devoted hunter lady, Sabari. She was keen that the Lord ate only the sweet fruits

Gita 9.26

'Patram pushpam phalam toyam yo mE bhaktyA prayacchati....'

Lord Krishna confirms that He partakes or tastes or accepts the food offered to Him with pure devotion.

Even a thousand years ago, Sri Ramanujacharya openly declared that all humans are equal. He invited devotees from all classes of the society to receive the highest knowledge about God, knowing which they would attain divine existence.

This historic incident in Sri Ramanujacharya's life epitomizes love for God, setting a profound example of non-discrimination and respect for true devotion

Ghajni Mohammed had snatched away the Deity of Melkote just as he did with thousands of other temples. Sri Ramanujacharya visits Melkote to find the Idol missing. Lord appears in his dreams and asks him to bring Him back from Delhi. Sri Ramanujacharya goes to Delhi and finds the son of Ghazni Mohammed in power, as Ghazni was dead. He requests the king to return the Idol. The king explains it was forbidden to keep or worship Idols of Hindu gods and that Sri Ramanujacharya was welcome to take his God back if he could find the Idol in the dump yard. Sri Ramanuja takes a look to find thousands of Idols of Gods lying in a heap, but not the Melkote God. He returns to the king and tells him that he knows his Idol was in the palace. He was allowed to go into the palace and Sri Ramanuja went to the chambers of the king's young daughter who was caringly holding the Idol. She had dressed up and adorned the Idol and seemed to be very attached to the Lord. Sri Ramanuja explains to her that the Lord belongs to thousands of devotees who miss Him and requests her to return the Idol. She explains how she has been taking good care of the Lord and that she could not live without Him. Sri Ramanuja explains that the Idol would be discarded after her lifetime and his rightful place was in the temple. He convinces her and takes the Idol from her and walks towards the palace gates. As he approaches the gates, she comes running to him, holds on to the feet of the Lord and gives up her life. The king let Sri Ramanuja take the Idol away saying that his daughter was suitably punished for blasphemy.

However, Sri Ramanuja realized that she was a true devotee of the Lord and established an idol for her alongside the main deity of the Lord Himself, naming her as 'Bibi nachiyar' or Tulukka nachiyar. This is the greatness of Sri Ramanuja and the Vedic culture which transcends beyond all religions, castes, or any other barriers dividing humans and Gods.

Our universe is known as 'brahmanda' or infinite cosmic egg. Not all life forms are mammals. There are plants, reptiles, birds, fish, insects etc. There are millions of species in plant life.

Chandogya Upanishad 6-3-1

'tEshAm khalvEshAm bhUtAnAm trINyEva bIjAni'

There are only three origins for all beings other than 'sat' Brahman who is the only unborn.

Beings are born of eggs, born of creatures (wombs) or born of seeds.

Gita 14-4

'sarva yonishu kauntEya mUrtayah sambhavanti yah.'

Whatever forms are produced in any womb, the Prakruti is their great womb and I am the origin.

Gita 10-39

'yacchApi sarva bhUtAnAm bhIjam tadahamarjuna.'

I am also that which is the origin of all beings, nothing that moves or does not move, exists without Me.

Sri Krishna confirms how all the beings are born.

a. Plant life - from seeds
b. Birds, fish and such species - from eggs
c. Humans and mammals are born from the wombs.

The beauty of life is in its diversity. It is colorful and enchanting, which is why, perhaps, God created different worlds and diverse life forms. There is no monotony or boredom. God with many names and forms, different species in various forms, humans adapting to different cultures, foods, dresses, art forms, each culture more beautiful in its diversity and difference. So many plant varieties, colorful flowers, diverse seasons with different climates. Everything is beautiful when we see reality in diverse existence. The greatest beauty is that all diversity shares one origin, God.

Life is all about happy experiences. Life is not meant to be a struggle. Life is also meant to be shared being together. There is no joy in experiencing happiness alone. Joy multiplies when shared. We feel happy seeing others happy. That's why God does everything out of divine bliss. Every experience would be a happy moment if humans don't ruin other's happiness by their senseless actions.

We might wonder as to why people of different religions worship different Gods if there is One Supreme God for all of creation? God does not differentiate between people of earth and countless other worlds. God worshipped in India in different names and forms is relevant to people here, where He is most celebrated. Similarly, different religions worship God in their own way. Gods worshiped in India are rooted in its culture. The word Bharat is not merely a name for the country. It derives significance from the roots of the word meaning 'bhAyA ratAm iti bhAratam'. 'bha' signifies light. Light denotes knowledge, wisdom and enlightenment. One who pursues light of knowledge is a Bharatiya. It is a society that is built on a value system - moral, ethical and righteousness. It has been a rich society financially and a wealthy society culturally, spiritually for thousands of years.

The other societies also belong to the same God, but they formed their own religions not based on the Vedic knowledge. They perceived God in their own way. In fact, each individual can perceive God in his own way. Even in India we have so many new religions and philosophies identifying God in their own ways.

It is similar to the languages and monetary systems prevalent in different countries. It is not about where God incarnated. It is about how people are able to relate to God. God is universal, not limited to earth alone. Some of the people who lived in the same time and society as Lord Krishna, refused to accept Lord Krishna as the Supreme Lord. But many others who realized the truth about Lord Krishna, benefitted by their devotion. God does not force anyone to do anything against their will, nor does He punish them for it.

New religions and Gods may become popular in future. That is not a problem. What is important is to understand knowledge-based religions to realize our nature and relationship with God that would lead us to attaining the highest state of existence we truly deserve.

What's the purpose of all this? What are we doing here? Are we born merely to live for a hundred years in a human body, then die and be born again, not knowing what we will be born as, if not as humans? At least we are aware of our existence as humans. But the other living species including plants do not have the awareness of who they are and why they acquired the bodies they are in, either. That's for God to worry about.

CHAPTER FOUR

Our Origin (God)

God realization (Para swarupam)

When it comes to God, both believers and non-believers often lack confirmed knowledge. Some believers accept the existence of the all-powerful God, because they were told so without actually knowing who He is. They lack any realized knowledge of God. While it is not wrong to believe, true devotion cannot be developed out of blind belief without knowing who God is and how we are related to Him. *Devotion must be an informed virtue.* God is not God because He is all powerful and can do us favors, but because He is our origin, our Father. No one else can take that position. Without the knowledge of who God is, we could easily fall prey to misguided concepts obeying anyone who offers benefits to lure us to carry out bad acts. It is better to know than merely accept what we were told, as we could end up believing in wrong concepts that can lead us to doing harmful things to fellow humans. We believe only because we don't know.

Believing in God is good as long as we remain righteous in our actions. However, people can be misled into engaging in terrorism in in the name of God when they lack true knowledge of God. If we knew even a little about God we would never hurt or harm anyone because there is God in everyone. The onus is on us to choose between right and wrong; knowledge and ignorance. No religion or philosophy preaches violence, cruelty, killing or unrighteous means for any purpose.

The same is true about the atheists who believe there is no God. They are ignorant because they have no basis for their thinking about why God cannot exist. Atheists may not be dangerous to others as long as they don't hate believers; they don't engage in hate crimes.

Knowing God is in our own benefit. When we realize that the power that created everything is our origin and eternal relation who dwells in us, it should naturally trigger that curiosity and inquiry in our minds to find out the truth about our origin. It's natural to want to know more about the same and go there, especially when we come to know that it is our home.

We are like the homeless travelers who do not know their roots or the place they truly belong. When they gain access to a message from home, and realize they belong to the most beautiful place in the universe, it is natural for them to look for ways and means to reach home. If their home is worse than their current existence, there is no point in going home. However, when we realize it is the best place, we shouldn't miss the opportunity. The important thing is that we don't have a home here we can call our own, and all the homes we get to stay in, while we keep looking for a good place to settle down are all rental and temporary (all sorts bodies). We can never settle down in a transient home to find lasting peace. However, the home we belong to, the home that's our own and we don't have to ever leave, is also full of prosperity and happiness. After knowing this truth, if we still miss the opportunity to go home, it is our loss.

Upanishads and Srimad Bhagawad Gita bring that message from home, directly from God, giving us directions to come home. In response, we just have to send a mental message to God that we want to come home and He would do the rest. We cannot get there on our own.

Kathopanishad 6.12

'naive vAchA na manasA prAptum Sakyo na chakshushA.'

Brahman cannot be perceived through speech, the mind, or the eyes.

Except from he who says 'He is,' how can that be known? How else can Brahman be realized except from the words of the Vedas that expresses that He is.

God is beyond the reach of sensory perception and intellect because God is the very basis of these faculties, not an object among them. Reality is divine and invisible to material perception, so it cannot be "seen" with physical eyes or articulated with "speech". The "mind" also cannot fully grasp it because the mind itself is a product of this ultimate reality.

Now, there have been several schools of thought expressing their perspectives on some of the statements of the Vedas, Upanishads and Srimad Bhagawad Gita. We can only refer to the source of knowledge given to us by God. All others are either commentaries by divine Souls or expressions of direct experiences of Azhwars and Acharyas with God, to help us in understanding the correct perspectives of the Vedas.

It is true that the Lord has graced the Souls with certain similarities with Himself such as

A. Eternality (nityatvam), or the nature of unborn, unchanging, immutable, existence.
B. consciousness or awareness (chetanatvam) or intelligence or knowledgeability or sentience.
C. That apart, the Lord Himself is the inner self of all the Souls empowering them, remaining inseparable in all states of existence.

There are certain dissimilarities between the Lord and the Souls as well, which is also explained. The problem arises when we assume the Souls as God because of certain similarities. If our father is infinitely rich and shares his wealth with all his children, it does not mean that the children are also the father. Only the father is infinitely rich and all the children can claim to have a share in the richness which is bestowed by the father, but they are not rich by themselves. This also explains how by being the potency of the Lord, Souls do not possess the infinite nature and capabilities of the Lord.

When it comes to consciousness that we all possess, it is not the same as the consciousness of the Lord. Only God is omniscient and none of the Souls are omniscient.

We are witnessing an AI revolution in India, a conference was organized by the honorable Prime Minister of India Sri Narendra Modiji, on Feb 19, 2026. Several thousand brilliant minds had assembled in one space. Some great leaders of countries, some brilliant CEOs of AI companies of the world, several intelligent participants, all sharing the best of their ideas and knowledge. As was observed, each one of them were differently brilliant in their expressions. Thousands of individual consciousness, witnessing the goings on, with individual understanding and perspective of the subject. There was an agreement of minds, with varied concepts. This shows, we all possess consciousness as our nature, but we do not perceive the consciousness of others unless shared. But God is the only one who is aware of every thought by everyone, everywhere and eternally. This reality is evident enough that though we possess consciousness as our nature, it is not the same as that infinite consciousness or omniscience, which is natural to God.

While on this subject, some statements by Sri Shankaracharya in his commentaries on the Vedas explaining Advaita Vedanta concepts can be confusing, if not understood correctly.

Sri Shankaracharya is the savior of the Vedic culture. If not for him, the Vedic school of thought would have become extinct momentarily till God would have revived it again.

Sri Shankaracharya incarnated at a time when new philosophies evolved creating different religious identities, different from the Vedic way of life (Sanatana dharma). Our ancient culture is not a religion; it is a way of life. We could say he protected Sanatana Dharma, which is by God's design.

Sri Shankaracharya's Vivekachudamani (v. 20) or Brahmajnanavalimala (v. 20)

'Brahma satyam, jagan mitya, jivo brahmaiva na parah'.

The literal translation is, Brahman is true, the universe is false or lie; Jiva, the Souls are not different from Brahman, or the Souls are indeed Brahman and none else.

Three entities are mentioned here. Brahman, Jagat, Jiva.

However, Sri Shankaracharya never declared himself to be Brahman or God (but some of us believe that we are also Brahman because we didn't understand the message of Advaita properly). This should give us an insight as to why he didn't claim he was God. However, Lord Krishna declared that He was the Supreme Lord, who incarnated for a purpose and demonstrated infinite powers that only God has. Jesus Christ did not claim he was God, but that he was the son of God; nor the great sages, prophets and spiritual leaders of various other philosophies, Azhwars, Acharyas and even Sri Ramanujacharya ever say they were God, though they were all God-realized-Souls; though we revere them all as divine men of God who revealed to humanity the truth about God. In that they are greater than God, because God couldn't do what they did.

Advaita Vedanta infers that countless individual Jivatma have always been Brahman. Just that they have forgotten their identity somehow due to ignorance and hence are born in the material world (Please do not question as to when they acquired ignorance, though their nature is knowledge). Karma is the cause of their experiencing life forms continuously in all sorts of bodies in the material universe (maya) which truly does not exist. This universe and all the life form they have been experiencing eternally is an appearance in their consciousness and not real. Everything is a dream, including whatever life forms they will be born into till the dissolution of the universe in some 155 trillion years. All the events and incidents we experience are not real. True knowledge of the Self (that I am Brahman) or enlightenment would bring them the realization that they have always been Brahman and the countless life and death they experienced is all a dream. None of it really happened. None of us truly ever existed in our bodies, neither in the past nor in the future.

That means there are different categories of Brahman.

1. The only Brahman (true Advaita prevails in this Brahman) without a second, unborn, who is infinite in every respect, omniscient, who manifested the universes and all the life forms.

2. Then, countless Souls who are also the same Brahman but somehow forget their reality and as a result are experiencing life as individual Souls due to ignorance. They will continue to experience all sorts of life forms due to their karma. Their eternal nature is consciousness and knowledge though; they somehow acquired ignorance even before they were born in the material universe. (There is no Advaita here because there are countless Souls, experiencing life differently from one another)

3. Those of us who realize they are Brahman and not a Jivatma (Soul), are considered enlightened. Now there is no ignorance in them when they attain enlightenment (cannot ask

how they acquired ignorance before being born and how they have overcome ignorance after countless births). But nothing changes for them even after becoming free from ignorance and realizing they are Brahman, as they continue to experience life in the same way, with the conviction that they are God, but still helplessly suffering effects of karma in this life. They just need to think that all the pain and suffering is not true because this universe and life experiences are a dream. There is no true liberation for them from mortality, as there never was a bondage. Knowing the self as Brahman is the only liberation, though continuing to experience life in bodies not knowing if they will be going through the life-death cycle like the others who are still ignorant, in not realizing

Truth is that there is only one Brahman without a second. If there are two, both of them cannot be Brahman. In that case, countless individual Souls (Jivatman) that exist, simply cannot be Brahman. Or if the countless Souls don't exist and Brahman alone exists, there is no purpose explaining about the universe and the living entities. What is the point of discussing things that are non-existent? Who are we explaining to if no one truly exists. That leaves Brahman alone, and there is no need for Him to say 'I am Brahman'. If the universe doesn't really exist, and it is merely an appearance in our consciousness, then even we and our consciousness should be false. Each one of us perceives the world differently based on our understanding because it is not one consciousness at work. If only Brahman exists, 'aham brahmAsmi' is senseless to say because the statement is not meant for Brahman to Himself, then who is it for, as nothing else exists.

Nowhere in the Vedas, Upanishads or Bhagawad Gita, is there a statement to suggest that individual Jivatma is Brahman, the Supreme Lord. In fact, there are many statements declaring that the Brahman or Paramatman is the inner self of the individual Jivatma, in all the living entities. There are statements to suggest that a Jivatman can attain Brahman only by the grace of Brahman. In the entire Bhagawad Gita, Lord Krishna has reiterated several times that He alone is the Supreme Lord, unambiguously. Not once has the Lord suggested that the Jivatman is Brahman. Why would Krishna impart knowledge recommending devotion as the only path to seeking God's grace if the individual Atma is already Brahman? And where is the need to attain liberation, if God alone exists and everything else is an illusion.

If the universe is false, then even the incarnations of the Supreme Lord and the knowledge contained in the Vedas, Upanishads and Bhagawad Gita would also have to be considered as non-existent. In that case, how can we accept statements such as 'aham brahmAsmi' and commentaries of Advaita Vedanta including Sri Shankaracharya's incarnation as true? If nothing else other than Brahman exists, where is the need to say that the individual Jivatma is Brahman and the existence of the visible, perceivable universe is false.

The Advaita principle is true and relevant to Brahman alone and is not applicable to Brahman's manifestations, such as His potency (individual Jivatma), the non-sentient material universes and the divine, eternal, non-material world, Paramapadam. We should not compare or equate the created with the creator. That's where the Advaita concept fails to explain the reality. In order to justify their school of thought, they have to deny the very existence of the created. If the creator is true, the creation is also true. If creation is false, where is the need to say that the creation exists as an appearance in consciousness but it is false.

The question here should be, if everything other than Brahman doesn't exist, why are the scriptures talking about different entities and identifying them by their different nature, characteristics and names and forms. Obviously because they are not Brahman but originate from Brahman. Here, mainly three entities are named even by Sri Shankaracharya in this commentary. Brahman (God), Jiva (Souls), and Jagat, (universe or material energy). If the universe is false why mention it as an entity and then say it doesn't exist. Even the Advaita concept is mentioning the three entities in denying the existence of the universe and life forms, but from our experience all three exist and are true. It has accepted the existence of the universe or maya, but narrated the universe or maya as inexplicable or indescribable (anirvachaniya). So, if you fail to understand a thing, should we say it doesn't exist? If the existence of living entities such as humans and universes is an illusion, why are we experiencing existence in reality and are able to perceive the universe being true in existence. If all this reality itself is a dream, why is there a dream that souls experience during sleep, which is exclusive to the dreamer alone and not a real experience or event in the real world?

We are not dreaming our existence nor do we exist in a dream because we also experience a dreaming state in sleep which is totally different from the experiences in the awakened state.

The statements of the Vedas are not easy to comprehend without guided help. That's why it's called 'marai' or hidden meanings, in Tamizh.The correct interpretation of this statement is explained by Sri Ramanujacharya which is - Brahman is the eternal truth in everything, but the material world formed by his power is ever changing, appears when formed and disappears when withdrawn by the Brahman and hence transient. Jiva, or Souls again one of His powers, is similar to God in nature, in some aspects, for the Soul is also unborn, eternal, unchanging, imperishable, immutable, for it is the inseparable potency of Brahman and not Brahman. And there are several dissimilarities between the Souls and Brahman as well.

Sri Ramanujacharya termed the essence of Vedic knowledge more aptly, realistically, as 'Visishta - advaita'. Or qualified non-dual Brahman. The One Supreme Lord, without a second but with infinite divine powers and auspicious qualities, virtues, and attributes, devoid of defilement, who has (everything as His body) these infinite universes and countless conscious entities (Souls) manifesting from Him. He is the indweller of both. Everything about Him is incomprehensible for us. He has everything in Him, while being in everything. There is nothing greater than Him or equal to Him.

Sri Ramanujacharya, being the incarnation of Lord Adisesha (Adisesha means, 'the one - ever in eternal service of the Lord'), also confirms the same Advaita concept of one Supreme Brahman without a second. The Supreme Lord has countless, unlimited powers. And His maya shakthi (the power of maya, Prakruti) that forms this universe/multiverse is beyond our scope to overcome. We remain bound by maya, until God releases us from the clutches of maya, as he is the origin and controller of maya which functions by His will. God manifests the material universe by bahiranga maya (external Prakruti), and He incarnates Himself as Sri Rama and Sri Krishna by the power of antaranga (internal) maya or yoga maya. According to Sri Ramanuja, Brahman is the ultimate reality, who is also present in the material energy, by being all pervading. Hence the universe is also a reality in existence. The Lord exists within the Souls as Antaryami. The Souls are the body for God and He is the inner self of Souls.

Brihadaranyaka Upanishad 1.4.10

"aham BrahmAsmi" literal meaning "I am Brahman"

Advaita Vedanta, considers this statement as being made by a Jivatma upon attaining enlightenment or self-realization. Should this be taken as a statement by just one Jivatma or does it include countless Souls? In that case, I know I have never thought so. I have the realization that I am a Soul with God dwelling in me just as He is in everyone else. I don't think I will ever change my perception of who I truly am, at least in this life.

Who could possibly be making such a statement? Brahman Himself. Only God can declare He is God.

Lord Krishna declares in the Gita that He alone is the Supreme Lord. No one else other than God can claim to be God and no one else has claimed to be God, as well.

One simple fact check would clear the illusion. We know that Brahman manifested the universes out of Himself and released His potency (souls) to take life forms in the universe, and we are all witnesses to the existence of the created worlds in which we also experience existence. If an enlightened Soul thinks he is Brahman, can that Soul also manifest or create universes? (But of course, they might argue that even the universes manifested by Brahman are also false or non-existent). Is there even a hint of evidence to show a Soul can do what the Brahman has done. The answer is an emphatic no; Souls do not possess the capabilities of the Brahman from whom they originated. The created beings cannot know the Creator unless the Creator creator reveals Himself. Even in Advaita concept, an enlightened Soul seeks liberation from repeated birth death cycles and does not claim to possess the potential of the Brahman. For that matter, the Vedas also cannot exist based on which the Advaita Vedanta was conceptualized. Even the incarnations of Sri Shankaracharya, Sri Ramanujacharya and great sages would be an illusion. Where is the need for all the knowledge revealed in the sacred scriptures if nothing exists. What are we trying to know when the existence of life forms and worlds are false?

'Sarvam Khalvidam Brahma' - a statement from Chandogya Upanishad 3.14.1.

It really means 'All this is indeed Brahman'. On one hand we say the universe is false and is an illusion and does not really exist and only Brahman is true. On the other hand, we also say everything in existence is Brahman, including the universe.

So, there is a problem in understanding this statement from the Upanishad. If the argument means, all the Souls in existence are Brahman, and it doesn't include the visible universe and visible life forms, even then there is a problem. If all the Souls are Brahman, why make a mention specifically that there are countless Souls and they are all Brahman. When you mention 'all this', it means 'many' and doesn't signify Advaita or the only One. The correct understanding of - sarvam khalvidam Brahma - is that the entire universe is pervaded by Brahman.

While the Advaita perspective highlights only a handful of statements, thousands of other Vedic declarations defining the nature of existence, individual souls, and their origin in Brahman remain largely unmentioned.

Even these few statements mean that the universe does not exist independently of Brahman and there is Brahman pervading everything. But Brahman also exists independently of the universes.

The Lord has not created the world to cheat us by causing an illusion; instead, it is a real universe. All our life experiences are real and most events and emotions stay in our memory as well. Only the experience in our dreams is an illusion as they are not real events in the world. If the universe is an illusion, if our life experiences as humans are not real, then we shouldn't be accusing anyone of committing crimes, because nothing ever really happened.

Vedas declare repeatedly that Brahman cannot be known by anyone, including enlightened Souls. Then how can they become Brahman without knowing what Brahman is?

Purusha suktam, Rig Veda verse explains.

'vedAhamEtam purusham mahAntam...

tamEvam vidvAn amruta iha bhavati, nA'nyah panthA vidyatE'ayanAya.'

'Only by knowing the Supreme Lord who is of divine magnificent effulgence, resplendent form, can one transcend mortality and attain immortal existence. For this there is no other way.

Kathopanishad (1.2.23) and Mundakopanishad (3.2.3)

'nAyamAtmA pravachanEna labhyo na mEdhayA na bahunA SrutEna,

yamyEvaisa vruNutE tEna labhyastssyaisa AtmavivruNutE tanum swAm.'

Brahman is not attained by knowledge (of speech) nor by intellect nor by learning scriptures. Brahman is attained only by him whom He chooses to reveal His divine form.

This verse explains that Brahman cannot be known by a Jivatma by his own effort. Knowing Brahman is possible only by His divine grace when sought with unconditional devotion, and Brahman would reveal Himself. Knowing Brahman is beyond our intellect.

No enlightenment or self-realization can reveal Brahman to us.

Countless Souls came to experience existence in the universe only after the universe was manifested. 'aham brahmAsmi' could not have been stated by a Soul, as knowledge of the Vedas are from God to the first being, Brahma, prior to the manifestation of the life forms. Countless Souls obtained life forms after the formation of the universe, every time. Nowhere has even Lord Brahma claimed that he is the Supreme Lord.

We know enlightenment is possible only for humans. However, there are infinite numbers of Souls in existence as microorganisms, cells in human bodies, millions of animals, insects, plant species which do not possess the potential for self-realization, how will they ever declare 'aham brahmAsmi' or realize themselves as Brahman? Because eventually every Soul has to realize it is Brahman, if Souls are Brahman.

Now it is important to note that everything said in the Vedas, Upanishads and Srimad Bhagawad Gita are Knowledge from God, as the Vedas are indisputably accepted by all as 'apaurushEya' or not authored by any human or a Jivatman. Everything said in the Bhagawad Gita is explicitly divine knowledge imparted by Lord Krishna. Krishna has confirmed in many verses of the Gita that He is the Supreme Lord and distinct from an individual Jiva.

Gita 4.6

'ajo'pi sannavyayAtmA bhUtAnAm ISwaro'pi san...'

Though I am birthless and of immutable nature, though I am the Lord of all beings, yet by employing my own nature (Prakruti) I incarnate out of My own free will.

(God incarnates, but a Jivatma like us goes through the birth death cycle helplessly due to karma).

Sri Shankaracharya recommends worship of Lord Krishna in his compositions 'bhaja Govindam' and Sri Krishnashtakam, and devotion to Lord Krishna alone would lead the Souls to liberation from this mortal world and to attain the divine world'. Even in the divine world, all the Souls would remain as Souls, but with divine senses and divine bodies, enlightened with God realization (knowledge about God) and would not be Brahman or God. The Lord would grace the Souls with certain similarities with Himself for being His children.

Now let's understand the truth expressly stated by Lord Krishna that He is the Supreme Lord.

Gita 7.7 'mattah parataram nAnyat kinchidasti dhananjaya'

There is nothing higher than Myself, O Arjuna. Everything rests in Me, as beads strung on a thread.

Gita 9.6

'YathAkASasthitho nityam vAyuh sarvatrago

thathA sarvANi bhUtAni matsthAnIdyupadhAraya..'

Just as the wind, though moving everywhere, is contained within the sky, similarly, all created beings and elements exist within Me.

It illustrates that God is the ordainer of all things, yet remains distinct from them.

Gita 9.7

'sarva bhUtAni kauntEya prakrutim yAnti mAmikAm

kalp akshaye punasthAni kalpAdau visrujAmyaham'

At the end of the millennium all material manifestations enter into My nature, and at the beginning of another millennium, I create them again.

Everything in the universe emerges from, exists within, and merges back into the divine material energy of God, in every cycle of creation.

Gita 9.10

'mayAdhyaksheNa prakruthi sUyatEsa characharam'

Under My direction, material nature (Prakruti) produces all moving and non-moving beings.

This emphasizes that matter is not independent or self-acting, but is a "product" that requires the will of the Divine to function.

Gita 10.8

'aham sarvasya prabhavO mattah sarvam pravartatE'

"I am the origin of all, from Me proceed everything. The wise who know this perfectly engage in My devotional service and worship Me whole heartedly.

This statement from Kathopanishad explains about God

"aNor aNIiyAn mahato mahIyAn"

describes God as "smaller than the smallest and greater than the greatest". This profound reality describes the nature of God being both the most infinitesimal and the most infinite, permeating everywhere and present within everything, from the tiniest particle to the vastness of the cosmos and multiverses. 'He is in everything and has everything else in Him '. A Jivatma surely does not fit this description.

To illustrate how each one would have to find the truth and reality of life by himself, Taitriyopanishad (Bhruguvalli) expounds how one could realize the truth about God. Bhrugu, the son of a God realized rishi (sage) Varuna, embarks on acquiring knowledge in the hermitage of his father, along with several disciples.

Upon completion of learning everything, all the disciples of Varuna return to their respective places, satisfied that they have learnt everything there is to know, except Bhrugu, the son of the teacher. He goes up to his father and confesses that he does not feel fulfilled with everything he has learnt, and requests the teacher to teach him about Brahman. Sage Varuna who is a God realized Soul did not tell his son that Jivatma is Brahman, because only God is God and countless individual Jivatma are not.

The teacher truthfully tells him that Brahman cannot be taught by explanations or shown or revealed. If you want to know Brahman, please do tapas or deep penance, insightful thinking, intense inner contemplation or meditation.

The teacher explains,

'annam, prANam, chakshuh, kshotram, mano vAchamiti, tam hovAcha'.

'Food, vital airs, the eye, the ear, the mind, the speech are Brahman, suggesting that everything is pervaded and empowered by Brahman. Tapas (intense, insightful thinking) is the path to knowing about Brahman.

Varuna the teacher, further explains -

'yato vA imAni bhUtAni jAyantE yEna jAtAni jIvanti

yatprayantyabhisamvisanti, tad vijijnAsasva, tad brahmEti'

'Desire to know that from which all these beings are born, that by which they live, and that towards which they move and enter into. That is Brahman'

In this statement it is clear that the Jivatma does not fit into the description of Brahman in any manner. Countless Jivatma are not the origin of everything but have their origin in Brahman. If Souls were Brahman there was no need to strain one's mind to know about Brahman.

The son comes back several times to check if his understanding of Brahman was correct, mentioning annam, prANam, manas, vijnAnam as Brahman, and every time the teacher asks him to do more tapas or introspection, saying, 'tapasA Brahma vijijnAsasva, tapo brahmEti'

desire to know Brahman through penance, penance is Brahman.

Here physical austerity is not suggested. Only deep thinking is advocated.

Finally, the disciple returns to his teacher, his face resplendent, reflecting the insight of knowledge, and his realization about Brahman, declaring 'Brahman is Bliss'.

'Anando brahmEti vyajAnAt

AnandAdhyEva khalvimAni bhUtani jAyantE

AnandEna jAtAni jIvanti

Anandam pratyabhisamvisantItii'

Ananda or bliss is Brahman and all existence is on account of bliss and eventually they move towards bliss and enter into bliss.

Here again if Jivatma is Brahman the bliss, why are the Souls constantly in pursuit of bliss?

Varuna the teacher, is overwhelmed with joy that he was able to guide his son in realizing the truth about God.

If bliss or happiness is the purpose of life experiences, do we experience bliss all our lifetime? Why is there so much suffering, crime, hate in our world instead? All those who truly learn and understand the essence of Vedic scriptures, about the benevolence of God's grace, purpose of existence, would not engage in doing evil things.

Some people contend that being the potency of the infinite (God), the Souls are also infinite in every aspect like God, because infinite divided is still infinite or of indeterminate value in mathematics and hence Souls are also infinitely powerful like God in every aspect. But this statement is true only for God. If you divide God's infinite powers, he still remains infinite and complete in every divided aspect. But Souls are not directly God. Here it is clarified that the Jivatma is the potency of God's Jiva Shakthi, and the universe is formed out of God's maya Shakthi or Prakruti. God who resides in every Soul as Antaryami is infinite but the Souls are finite.

If Souls are God, why are they helplessly caught up in mortality, experiencing pain and grief? If Souls are infinitely powerful, why are we ignorant of all-knowledge about even the Self which has to be known from the God given Vedas. Infiniteness is true only for God. God is infinite even when He incarnates. God is infinite even as the all-pervading self of the universe.

For instance, if the ocean is considered an infinite body of water, a drop - though a fragment of the ocean - is finite; that is why it is called a "drop" and not 'the ocean'. If the ocean itself is divided, it remains infinite. In this analogy, drops do not make the ocean; rather, drops originate from the ocean without changing the fundamental nature of the ocean. Even countless drops do not, in sum, form the ocean, but remain individual, finite drops.

Similarly, if the human body is considered a complete unit, a single cell within it is merely a finite part. If the universe is considered infinite, a subatomic particle is finite. Therefore, the infinitesimal part of the infinite is not, in itself, infinite.

By the same logic, the individual Jivatma is an infinitesimal part of the infinite Supreme Lord's Jiva Shakti and is, therefore, not infinite like God Himself. God's consciousness is infinite and all-pervading, but the consciousness and awareness of individual souls are limited to the extent of their perception. This is why only God is omniscient, while souls have limited knowledge.

Anything more than one of the same cannot be infinite, neither individually nor collectively.

As a matter of fact, all philosophies, without exception affirm the existence of only one Supreme God.

The Supreme Lord is also known as Brahman, Paramatman and Bhagawan. All three are one and the same God.

As Brahman, He is the source and origin for everything.

As Paramatman, He is the inner self of every Soul and is all pervading. He is the observer and witness to all karma by every Soul, in all the universes, eternally. The Souls have consciousness as their nature derived from the Lord (chEtanas chEtanAnAm)

As Bhagawan, He engages with his devotees, incarnates as Lord Rama, Lord Krishna as and when needed. The Lord in His Bhagawat swarupam (as Bhagawan) is omniscient, omnipotent and omnipresent, with infinite powers, infinite auspicious attributes having a most beautiful magnificent opulent form, beyond expression or description.

In the divine eternal world, we get to experience unending bliss shared by the Lord as Bhagawan.

Lord Krishna explains how He is distinct from His creation.

Gita 15-7

'mamaivAmSo jIvalokah jIva bhUtah sanAtanah.'

An everlasting part of Myself (Jivatma / Soul being an eternal 'amsh' / potency of the Supreme Lord) having come to bind the self in the world of life, attracts the senses of which the mind is the sixth, and which abide in Prakruti.

This reveals the Supreme Lord's omnipresence through His powers, the Jiva Shakthi and maya Shakthi. And God is both the basis of all existence and distinctly separate from it.

The Supreme Lord is not limited or contained by the created world. God is simultaneously one with His manifestations (His energy) and distinct from them, existing beyond the material realm. Which is why we can't perceive Him visibly. While God sustains everything, He is not limited by or dependent on material creation.

Gita 7-14

'daivI hyEshA gunamayI mama mAyA duratyayA, mAmEva yE prabadyantE mAyAm EtAm taranti tE'

means,

My transcendental energy composed of three gunas (modes of material nature) sattva / goodness, rajas / passion and tamas / ignorance bind the Soul to this world, my divine energy is insurmountable and difficult to overcome by oneself, but those unconditionally devoted to Me will crossover the effect of this maya, Prakruti, by My grace.

This verse emphasizes that it is impossible to overcome the power of this prakruti on our own. Only the Lord can get us out of this prakruti and accept us in the divine world. (God has to grant us a visa to go to the eternal world) Enlightenment or self-realization will still keep us confined to this world going in birth-death cycles.

Svetasvataropanishad 1-9

'jnAnau dvAvajAvIsanIsau ajA

hyEkA bhoktubhogArthayuktA'

Of the two, One (Paramatman) is omniscient and so the master of all, the other (Jivatman) is sentient and never the master. Both of them are unborn. The other one, the unborn prakruti, is associated with the enjoyer (Jivatman) for the purpose of his experience. (This is evident why the Supreme Lord created this universe). The Paramatman who has this universe as His body is yet infinite and a non-doer (he is not involved in the activities of the Jivatma who is the enjoyer). When one realizes the distinct nature of these three clearly, he becomes freed (attains immortality, by God's grace)

This statement in the Upanishads confirms unambiguously the following.

Paramatman is neither the prakruti which is the unborn power of the Lord by which the non-sentient universe manifested.

God is distinct from the Jivatman, who is the unborn sentient, indivisible potency of the Lord's Jiva Shakthi

Lord created the universe for the experience of the Souls. God is neither the material universe nor the sentient Souls.

God is the non-doer because He remains detached from the activities of the Souls. Creation of the universes and empowering the Souls to experience life in the universe is done by His will alone. (It's like a play for God, it's done out of bliss, joy for the enjoyment of the Souls)

Svetasvatara Upanishad 6.11

'Eko dEvah sarva bhUtRshu gUdhah sarva vyApi sarva bhUtAntarAtma'

Meaning "One God, who is hidden in all beings. all-pervading and the inner Self of all beings. He observes over all actions and is the home of all living things"

There is only one God (Eko dEvah). This One divine reality is not separate from the world but resides within all beings as the inner Self (sarva bhUtAntarAtma).

Despite being present in all beings, this divine reality is also completely independent of and beyond the material world, not bound by its qualities or actions.

It affirms God is all-pervading (sarva vyApi) and witnesses all actions of the Souls.

Svetasvatara Upanishad -

'dwA suparNa sayuja sakhAyA samAnam vruksham parishasvajAte.

tayoranyah pippalam svAdvattyanashnannanyo abhichAkashIti'.

Two birds, inseparable companions, are perched on the same tree; one eats the sweet fruit of the tree, but the other simply looks on without eating".

Of the two birds, one bird (the Jiva or individual soul) eats the fruits of the tree, while the other (the Paramatma or Supreme Being) simply watches without eating, thus the distinction between the soul and the divine witness. The Tree is the human body.

One bird, the individual self (Jivatma), which experiences the joys and sorrows of life, as an outcome of its actions (karma) in the world, and enjoys or suffers the fruits of its actions (karma). The Second Bird (Paramatma), the Supreme Lord, who is the divine witness, remains unattached and unaffected by the worldly experiences of the individual Soul.

Even matter would return to its origin, naturally. And the souls would like to be with God, who is their origin, to experience divine bliss. Our bodies fulfill this criterion by returning to the five elements of the universe after death. And the Souls can do the same by returning to our origin, the Supreme Lord, when we attain the divine world.

It is important to realize that merging into God at the end of the universe is not the desired state of existence for the Souls because we will be in a state of suspended animation till we take birth as life forms in the new universe, upon the next creation. Souls seek to experience the divine bliss of the Lord being in a divine body, in the divine world, like we experience life here in material bodies. We would like to exchange blissful experiences with the Lord and other Nityasuris, it is not in our nature to remain in a state of suspended animation. It is extremely painful for the Souls to be in a state not to be able to express and experience life, perceiving pleasure, joy, happiness which is not possible without being embodied. In the divine world all the Souls would be divinely embodied.

Non sentient matter is inherently dis-similar in nature to the Souls and the Supreme Lord.

Souls are different from the bodies they acquire, whether in the material world or in the divine world. However, God and His body are one and the same. There is no difference between the Paramatman, His mind, body and senses even when He incarnates in the material world.

Svetasvatara Upanishad 3.19

'apANipAdo javano grahItA paSyayya chakshuh

sa SruNotya karNAh.'

Without hands or feet, He is swift in movement and He grasps,

He sees without eyes, He hears without ears. He knows all that is to be known. There is none who knows Him. They say He is the first cause and the great Self.

This verse explains that the Supreme Lord is omniscient and infinite in His capabilities and functions without the limitations of the senses and mind which we possess with limited perceptibility. There is no soul and body differentiation in God. He, His body, His senses and mind are all one and the same, unlike the souls which are different from the bodies they acquire. That's why scriptures declare that God's body is divine and same as Himself, termed as "divya mangala vigrahaya". There is no such a thing as embodied God or Paramatma, for He is His body and His body is He.

God is omniscient and is all pervading, but consciousness of the Soul is limited to what it can perceive.

God is the origin for the Souls, and He is the only unborn. And everything else originated from God.

God is both with visible form and invisible form, but Souls are subtle, invisible, unmanifest, and are perceived when embodied.

God is the controller of prakruti or Maya, but Souls are bound by maya when in this material world.

Souls acquire bodies to function and experience life and are bound by karma but God incarnates by His will. God has no karma.

God decides the form, time and place of His incarnation but Souls have no control over the body they acquire or the duration of life in that body.

God's body is divine even when He incarnates but the material bodies the Souls acquire here are from this universe.

God designed the universe and formulated the law of karma but Souls have no control over the outcome of karma. Souls have the power of choice alone and are bound by the law of karma.

God is the only sarvajna, of the nature of infinite knowledge, eternal knowledge, not touched by ignorance, but the embodied Souls born in the material world are influenced by ignorance inherent to this prakruti, which can be dispelled by the knowledge of the Self and God.

Gita 7.26

'vEdAham samatItAni vartamAnAni chArjuna.'

I know all beings; past, present and those to come. But no one knows Me.

Eventually the prime goal is to understand God to an extent possible with our finite intelligence. If we show true intent, God would reveal Himself, there is no double in that.

CHAPTER FIVE

Purpose of immortality (Purushartham)

The immortal world is truly the place where we belong. We (the Souls) being unborn, divine and eternal, wished to experience independent existence in the mortal world, thinking this is where we would find true happiness. Now that we know we can't find eternal happiness here, it is time to take corrective measures and do the right thing. The Souls are immortal by nature and belong in the immortal world and not in mortal bodies that we have to change in countless lives experiencing suffering more often. Immortal Souls in immortal bodies would be a perfect match.

What is the true definition of bliss?

Chandogya Upanishad defines thus:

“yo vai bhUma tat sukham, nAlpE sukham asti, bhUmaiva sukham"

Meaning "That which is infinite (Bhuma) is happiness/bliss; there is no happiness in the finite; the infinite alone is bliss/ happiness”.

Bhuma is ‘plentitude’, not quantified but infinite.

he who experiences plentitude, that is infinite bliss. Immeasurable, not quantified.

Only the Supreme Lord is infinite in all respects. Naturally the infinite bliss or joy is available with God alone and none else or nowhere else.

Bliss that we can relate better to as 'boundless happiness' is defined in the Vedas as one that is,

- Unlimited
- Unending or eternal, and
- Ever growing.

Only such happiness is considered bliss, which is what we are ever in search of, but never experienced in this world. What we experience in deep sleep is a hint of bliss briefly, by the presence of the Lord dwelling in us, which we don't get to experience in a conscious state, and is not the infinite bliss, unending bliss, we seek. Infinite bliss is experienced only in direct involvement with God, in the divine world.

Sri Nammazhwar describes the bliss we can experience with the Lord:

'appozhudaikku appozhudu en ArA amudamE'

The unending bliss of the Lord we experience would increase every moment in the eternal world. It is new every moment for eternity.

Like the flowing water in the river that is ever new, every moment, the bliss of the Lord in the divine world would be ever new and increasing endlessly without limit.

'nalam andamillador nAdu puguvIr'

We would attain that world where there is no end to bliss.

However, the happiness that we have experienced in this world has been:

- Limited
- Momentary or temporary and
- Diminishing (if we persist for more)

Every living entity, (except the Supreme Lord) be it humans or other species of this universe; one thing is common - all the beings constantly seek divine happiness. They cannot find divine happiness in the material universe; it's just not available here.

Taitriyopanishad -

'raso vai saha' - Brahman is bliss.

'Esha hy Eva Anandayati' - this verily bestows bliss.

Lord Krishna explains in the Gita 9-22

'ananyaschintanto mAm yE janAh paryupAsatE

tEshAm nitya Hi yuktAnam yogakshEmam vahAmyaham'

There are those who, excluding all else, think of Me and worship Me, aspiring after eternal union with Me. Their prosperity and welfare are looked after by Me.

The Lord is saying that He literally takes care of a devotee, now in this life as well as on attaining the eternal world, after this life.

The Lord doesn't force anyone to be devoted to Him, but those who love Him, He is available for them. He loves everyone equally though, those who do not want to be a beneficiary of His love, it is not forced upon them.

The Vedas declare -

"tad vishnoh paramam padam sadA paSyanti sUrayah"

In the eternal world, all the immortal Souls are ever engrossed in experiencing the divine bliss the Lord shares with them.

In the divine world, countless Souls experience only happiness; devoid of pain, suffering, fear, karma, consequences etc. - totally under God's personal protection.

What we experience in this world is quite the opposite.

Lord Krishna says, Gita 9.33

'. anityam asukham lokamimam prApya bhajaswa mAm'

This world is transient, changing, filled with grief and suffering. Worship Me, be my devotee, to escape from repeated birth death cycles in this grief-stricken world.

Sri Nammazhwar declares 'vaigundam puguvadu maNnAvar vidhiyE'

Having been born as humans, we are destined to attain God's abode, Sri Vaigundam.

Some statements by Sri Ramakrishna Parama Hamsa reveal they are words of a realized Guru.

Quote

'He is born in vain who, having attained the human birth, so difficult to get, does not attempt to realize God in this very life'.

'You see many stars in the sky at night, but not when the sun rises. Can you therefore say that there are no stars in the heavens during the day? O man, because you cannot find God in the days of your ignorance, say not that there is no God'.

Unquote

Human life is the best opportunity for us to attain the eternal world. If we miss this chance, we may have to go through several life cycles before we get human birth again. Imagine a poor man, unhappy to be begging for a livelihood because he couldn't find work to earn a living; suddenly finds a treasure, but hides it and continues to beg. He was given the opportunity to come out of poverty, but did not use the treasure to his benefit. Similarly, when we get a human life, we shouldn't miss the valuable opportunity to attain the divine world, never to suffer the painful mortal existence again. Only humans are endowed with this awareness and the other species do not have the awareness of existence, even.

Svetasvataropanishad 3.15

'purusha EvEdam sarvam yadbhUtam yaccha bhavyayam

utAmrutatvasyESAnah yadannEnAti rohati'

The Supreme Lord has His divine abode, Paramapadam, which does not flourish on account of food; Amrutatva, 'eternality' signifies the non-material divine abode.

It is established that the Lord has both the worlds. Nitya vibhUti, eternal world and Leela vibhUti, mortal worlds.

Svetasvatara Upanishad (3.8)

'vEdAhamEtam purusham mhAntam Aditya varNam tamasah parastAt

tamEvam viditvAti mrutyumEti nA'nyah pantA vidyatE'yanAya.'

"I know this great Purusha (Supreme personality) who is resplendent like the sun and who is beyond all darkness. Knowing Him alone is the way to transcend death and attain immortality. There is no other way for liberation than this".

The Supreme Lord is described as the self-resplendent Person in the most magnificent form, who dispels all darkness and ignorance, transcending the limitations of the material world. Knowledge of this Supreme Purusha leads to freedom from the cycle of birth, death, rebirth, attaining immortality. There is no other way to achieve this ultimate state of existence.

Mundakopanishad. 2.2.7

'yah sarvajnah.... divyE brahmapurE hyEsha vyomanyAtma pratishtitah'...

The Supreme Being who is omniscient is established in the supreme world, in the luminous city of Brahman, signifies the non-material divine abode of the Lord, the city of Brahman is the divine world 'Paramapadam'.

Mundakopanishad 2.2.10

'hiranmayE parE koSE virajam Brahma nishkalam...'

In the supreme abode Paramapadam is Brahman. It is without rajas, transcending the three guNas (modes), it is without parts, pure and is the light of lights. It is which the knowers of the Paramatman realize.

PurushasUktam verse

'pado'sya viSwA bhUtani, tripAdasyAmrutam diwi

tripAdUrdhwa udait purushah, PadosyEhA bhavAt punah.'

All of creation (material universe) is but a fourth of Him. Three parts, divine and eternal, remain with Him alone.

'tripAd Urdhva udait purushah pAdo'syEhAbhavAtpunah

tato vishva'n vyakrAmat sASanAnaSanE abhi'

Three parts of His Are beyond all this. All of this is but a part. Again and again, all that eats, and that eats not, appeared from this one part of His.

This clarifies that the material universe has been created for the enjoyment of those who sought to experience life in this world. All living entities here survive on account of food.

Lord Krishna confirms in Gita 15-6 about the divine world.

"na tad bhAsayatE sUryo na SaSAnko na pAvakah, yad gatvA na nivartantE

tad dhAma paramam mama"

The supreme abode of Mine (Paramapadam) is not lit by the sun, moon, or fire. (This signifies that the spiritual divine world is self-luminous and transcends all material limitations). Those who reach this abode of Mine, never have to return to the material world, escaping the cycle of birth and death.

The scriptures say that those who attain the divine world would never have to return to the mortal worlds (na cha punarAvartatE). Sri Ramanujacharya is one of the few, whom the Supreme Lord has empowered to grant moksham (liberation) from the material world.

Sri Ramanujacharya in his 'Srivaikunta gadyam' describes the transcendental beauty of the divine world and the magnificence of the Lord's personality as ineffable, emphasizing His countless auspicious virtues (kalyana gunas) bestowing kind, benevolent grace upon the Souls in His presence.

Taitriya Upanishad 2.9.1,

'yato vAcho nivartantE aprApya manasA saha

Anandam brahmaNo vidvAn na bibhEti kutaschanEti.'

Signifies that the Supreme Lord cannot be described by speech or grasped by the mind. It emphasizes that the infinite magnificence of the Lord transcends verbal expression and intellectual comprehension. Human expressions have limitations and cannot fully encapsulate God's absolute beauty.

Sri Parasara Bhattar, son of Sri Koorathazhwan expresses his divine thoughts in a conversation with Sri Namberumal of Srirangam - in his composition 'Sri Rangaraja stavam, purva shatakam' thus -

'yadi mE sahasra vadanAdi vaibhavam....'

Lord Namberumal, after accepting the service of the temple devotees such as divine tirumanjanam (celestial bath), adornments with colorful decorative attire, most precious ornaments, fragrant floral garlands, graced everyone in His presence, with a divine experience of beholding His overwhelming beauty and they were all wonderstruck seeing such divine beauty in this world, never experienced before.

Lord Namberumal is titled 'AbharaNangaLukku azhagu kodukkum perumAL', the God who beautifies the ornaments He adorns or the ornaments gain transcendental beauty when the Lord adorns them.

The Lord reads Sri Bhattar's mind who stood stunned looking at the Lord, speechless. The Lord teases Bhattar with a question 'how do I look?' Can you describe what you see?

Bhattar was taken aback by that response from the Lord; he recovers realizing the fact that the Lord knows all his thoughts. He comes to senses and tells the Lord - surely, He could describe His beauty if the Lord bestows him with a thousand divine tongues (divine sense of expression). The Lord seemed amused with that suggestion by Bhattar and says He is granting him a thousand divine tongues and does so at once. Upon getting the divine sense expression, Bhattar tells the Lord, it was not possible to describe even a miniscule of His divinely glorious personality. That no expression or words can possibly describe His glorious auspicious form.

Now the Lord questions Bhattar as to why he sought a thousand divine tongues to say it was not possible. He could have said so even without that. Or did he not realize it was not possible before obtaining a thousand divine expressions?

Bhattar responded saying that even to say His beauty was beyond words or expression, he needed the divine capability and he could not have said so with material senses he possesses. The Lord was truly overwhelmed by the extent of Bhattar's affection for Him and wondered how this devotee defeated Him by his unparalleled love.

Lord Namberumal then tells Bhattar that He was overjoyed by his incomparable love and grants his return to the divine world whenever he is ready to go. To that, Bhattar surprises the Lord again with his intelligence. He says that he would accept going to the divine world on one condition. If the Lord did not grace him with this level of experience in the divine world, he would jump out of there and return to the Lord here in Srirangam.

There can be only one Parasara Bhattar!

There is an illustration of how the gods (devas) were overcome by ego and were brought to realize the reality of their powers.

Kenopanishad. 3rd chapter.

The gods of heaven (Indra loka) were victorious in a war against the asuras. The victory induced them to believe that they were the most powerful gods, invincible and that there was no greater power than them. They were convinced that their might was unequaled; the victory was totally their own doing.

They did not realize the truth that all their powers were bestowed by the Supreme Lord and not their own. They were deluded into believing that each one of them was independently superior and powerful.

3.1 - When the gods were installed with powers in the world, it was the Supreme Lord that verily achieved everything for the gods. But gods felt glorified thinking that it was their doing and their glory.

3.2 - Brahman (the Supreme Lord) knew of their egotism; He verily appeared before them as a Yaksha, bright as a thousand suns.

They did not know what He might be. (God appears in the form of a yaksha to remove their false pride. He was visible in the form of a bright light, but beyond their comprehension)

3.3 - The gods said to Agni (God of fire) "find out what that Yaksha is" who ventures out to know the Yaksha.

3.4 - He goes to the Yaksha who asks him "who are you? He replied "I am Agni, I am jAtavEdA" (one who grants the benefits)

3.5 - "What power do you possess?" asked the Yaksha, Agni said "I can burn up all this that is in the world"

3.6 - The Yaksha placed before him a 'dry blade of grass' and said "burn this". Agni tried with all his power; he could not burn it.

Agni felt drained of his power and was cold instead of his normal effulgence, standing in front of the resplendent Yaksha. Brahman appeared to remind them that all their powers originated from Him.

Agni returned to the gods and said "I could not know what this Yaksha is"

3.7 - Then the gods said to Vayu (God of air) 'find what this Yaksha is' he said 'so be it'

3.8 - He went to the Yaksha who said to him "who are you"? Vayu said 'I am known as Vayu or MAtarisvA (air)

3.9 - 'What power is there in you? The Yaksha questioned. Vayu said "I can carry away all this that is in the world"

3.10 - The Yaksha placed before him the same dry blade of grass and said 'take it up'. He tried with all his might but could not carry it away. (Vayu could not even move it and felt drained of power he normally possessed)

Vayu then goes back to the gods and said "I could not know what this Yaksha is"

3.11 - The gods said to Indra 'know what this Yaksha is, Indra said, so be it and approaches the Yaksha who disappeared. (the Yaksha disappeared lest Indra's pride be broken in the presence of his subordinate gods)

3.12 - Indra came across a resplendent woman adorned with many ornaments; it was the Divine Mother Uma (Parvati) who is the embodiment of Shakthi. He asked her 'what is this Yaksha?' (Divine Mother appears before Indra to do him a favor and by her supportive response Indra thought she probably knew all this and asks her about the Yaksha)

4.1 - She said 'it was Brahman. Attain glory in the victory achieved by Brahman'. On account of her revelation, he came to know it was Brahman. She explains that Brahman appeared before them in the form of a Yaksha to dispel their false pride of possessing all the powers. If they shed their ego that is the true victory they would have gained.

4.2 - That's how these gods - Agni, Vayu and Indra were ahead of other gods who were first to know of the Brahman, who appeared in the form of Yaksha.

When we attain the divine world, each one of us would be endowed with powers and pure-sattvic nature which is much beyond even the gods of the heavens (of material universe).

We should not seek existence in the divine world merely to become powerful, blissful, and free from misery, but rather to serve our Father for His joy. God expects nothing from us but true love, as nothing gives Him greater happiness than sharing experiences with His children, ultimately bringing infinite joy to everyone.

It is said in 'archirAdi mArgam' (path to the divine world) by Sri Pillai Lokachariyar, the Supreme Lord is filled with boundless joy to receive a soul that has finally arrived in the immortal realm. He is ecstatic to see a child come home after a long separation. While the Lord is, in reality, never separate from souls - residing within them eternally - souls often fail to realize His presence due to their association with material senses and body. Once in the divine world, they acquire divine senses and body; they would be able to share and express their feelings and love for God. He is ever inquisitive to hear from the newcomer and to know his expressions.

The Lord warmly initiates the conversation by holding the new visitor on His lap, asking, "Who are you, and what can I do for you?" He is overwhelmed with joy when the newcomer responds, "I am - here for your enjoyment; at your service; for your purpose" (aham annam). This is the ultimate joy the Supreme Lord feels, upon hearing these words from His child.

The Supreme Lord bestows some of His glorious attributes (gunas) and powers to every Soul on reaching the divine world, which is termed 'sAmyApatti' or similarity with Himself. Acquiring godly nature, the Souls experience 'divya premananda' or divine bliss, which is greater than anything else. The experience of engaging in service that pleases God is the greatest power the Souls inherently possess which even the Supreme Lord is overwhelmed by.

The divine world is where we can truly be serving God, supporting Him in every possible way. Imagine how it would make our parents happy if we get involved in assisting them in managing the family affairs and helping the other members needing support to overcome difficulties they might be encountering. It is not different with God, except there are no difficulties in the divine world.

In the divine world, there would be just love, affection, strength, beauty, boundless joy, without a dull moment, a threefold blissful existence, devoid of defilement or negative qualities, as nicely defined by the Vedas. 'Om Shanthi Shanthi Shanthihi...'

CHAPTER SIX

Means to attain the goal (Upayam)

All of us set goals in life and strive to achieve them. These objectives are often varied and time-bound, spanning different stages of our existence. Sometimes, goals shift according to our circumstances; we may even pursue new ones after successfully accomplishing our initial aims. Ultimately, these milestones are meant to be reached within our lifetime. We often feel disappointed as we approach the end of our lives if certain dreams remain unfulfilled or only partially realized. Our fundamental purpose is to reach goals that provide a sense of fulfillment, joy, and contentment. In the truest sense, there should be one ultimate goal - a supreme purpose - attaining which leaves nothing else to be desired. Achieving this greatest goal is the ultimate purpose of human life.

And that is to understand what we have to do in order to attain immortal existence and not be born again in mortal bodies. And why is this achievement so important? Because it is the best experience the Souls aspire for, and we have never tasted anything remotely close in all our lives in this world. How do we know it is true? Because Lord Krishna Himself, who is our greatest well-wisher, has assured us that the divine life is what we all deserve; and all the Azhwars and Acharyas who incarnated amongst us to explain the quality of life in the divine world, have confirmed that to be true.

Vedic knowledge has two important parts. Karma kanda, explaining the concept of karma, (righteous deeds/conduct), cause and effect; educating the seeker about the law of karma. Karma kanda is useful to those who love their life experiences in this world, or even more, if they desire to experience the joys of heaven, which is also part of the material universe. There is no role for violence, crime, cruelty and unethical conduct in the karma path. Only practice of dharma is advocated. If we wish to continue to experience righteous life here, because we find life is good in this world or in heaven, we would not be stopped. But the quality of joy experienced here is temporary, limited and diminishing. And we would not be fulfilled with such happiness. What we seek is unlimited, unending, divine happiness which is not possible in the material universe including the heavens.

The other part, the Upasana Kanda (devotional path), is for those who recognize their origin and relationship with the Lord and practice a devotional path to attain the eternal abode where they truly belong.

Broadly there are three paths suggested in scriptures to end the repeated birth death cycle.

1. Karma (doing our righteous duties)
2. Jnana (knowledge about the Self) and
3. Bhakthi (devotion) seeking God to reveal Himself to us.

Karma is often misunderstood. It does not mean we can do whatever we want to fulfill our desires. Instead, the suggested path is to perform our duties righteously, without attachment to our actions or their outcomes. While we remain the recipients of the fruits of our Karma, enjoying them without obsession keeps us detached from the cycle of cause and effect. This enables a sense of fulfillment in everything we do and every outcome we experience. Emphasis is on doing our prescribed duties that are natural to us, for the benefit of our dependents, friends, society, country and humanity at large.

We cannot claim to fulfill our duties by causing distress or harm to other humans or the other living entities of the world. We must perform our duties even if there is no self-benefit. A doctor cannot say he doesn't want to work because he doesn't need the income. A teacher shall not stop working because he is wealthy. Once we consciously undertake to perform our duties, we must not abandon because there is no self-benefit in continuing to do our karma.

Sri Krishna guides us in Gita verse 2.47

'karmaNyEvAdhikArastE mA phalEshu kadAchana...'

To work alone you have the right, and not to the fruits. Do not be impelled by the fruits of action. Nor have attachment to inaction.

You have the right to do your karma (duties), never to the fruits of karma. One should focus on performing their duties sincerely without attachment to the outcomes, as results are anyways not in our control. But we will not be deprived of the benefits which would come to us even without seeking.

Do not let the outcome motivate your efforts. Avoid Inaction. This verse warns not to become attached to inaction, just because you have no interest in the fruits.

Lord Krishna emphasizes the importance of doing one's duties, without like or dislike (raga or dwesha)

When we are focused on doing what's in our potential, the outcome or results would obviously be optimal. We may even achieve better than expected results when the focus is on doing our best without diverting our mind on the results. Results would always be an effect of our best performance. However, when we fixate on the outcome before executing the task correctly, we fail to perform to our potential, leading to sub-par results. We get what we deserve. Therefore, there is no point in wasting our energy on what's beyond our realm. To do our duties alone is in our control.

Lord Krishna further explains -

Gita 3.4

No man experiences freedom from activity (naishkarmya) by abstaining from works;

and no man ever attains success by mere renunciation of works.

Gita 3.5

No man can, even for a moment, rest without doing work; for everyone is caused to act, in spite of himself, by the Gunas born of nature.

Gita 3.8

You must perform your obligatory action; for action is superior to non-action or jnana yoga. For a person following non-action, not even the sustentation of the body is possible.

Gita 3.9

This world is held in the bondage of work only when work is not performed as a sacrifice, O Arjuna, you must perform work to this end, free from attachment.

It is suggested to do'Nishkama Karma' (desireless action), the results would anyway come our way and we don't have to go chasing the outcome of our actions. Problems arise when we expect good results from bad actions, a trap many of us fall into. However, when work is performed as a sacrifice or offering, such actions do not bind us to consequences, and we do not accrue negative karma. Consequently, we are not deprived of the good results of our labor, which we will get even without conscious pursuit.

Nishkama karma can be practiced if we understand how karma takes effect. We can understand karma if we renounce the thought that we have all the power to do karma, and follow three simple methods.

1. Kartrutva budhi tyaga - renounce ownership to karma. In a simple language don't claim 'I did it'. Give up the thought that it was possible because of the Self.

1. Karma phala tyaga - renounce the claim to the fruits of action, because the results are not singly our doing.

3. Mamata tyaga - renounce the sense of belonging for the fruits - it belongs to me, it's mine. There is a popular saying. To check if something belongs to us, set it free. If it comes to us, it is ours; if not, it never was.

For the simple reason that five entities are involved, including the Self, in everything we do. we would fail to carry out our actions without the cooperation of the other four.

Lord Krishna explains in the Gita 18, v 13 - 16.

That five distinct entities are involved in the completion of every action, whether performed by body, mind, or speech.

He who sees only the Self as the doer on account of uncultivated understanding, he of ignorant mind, does not see at all.

He who is free from the notion 'I am the doer' and whose understanding is not tainted, is not bound by his actions.

Let's look at a simple thing most of us are used to doing in the morning, say we wish to drink coffee... I can relate to my own experience here.

1. Self. Suppose I wish to drink a cup of coffee. I am the choice maker (choice is all that's in our power). I don't know how to make coffee and have to depend on my wife.

2. My wife. (who is not bound by my choice, makes her own choices) She would gladly make coffee for me, so her co-operation is a given. She normally offers me coffee even without asking. If she is not at home, I go without coffee.

3. My body, senses - which are different from me. If I get a stroke or my health fails, I cannot drink the coffee.

4. The prakruti, nature. If there is a fire alarm or earthquake at that moment, I would not drink the coffee.

5. Divine grace, the godly powers, law of Karma. I could even die before drinking the coffee.

All of us would have experienced this many times, even while travelling. We leave home on a long trip and at the airport our flight could be cancelled due to a strike, or weather not conducive for takeoff or due to some technical snag, or if we develop illness. It happens in many things that we undertake to do but would not be able to complete because one of the other four entities did not support our cause.

Lord Krishna also suggests how we can lead a healthy and fulfilling life, practicing karma yoga, when everything is done in moderation.

Gita 6.16

Yoga is not for him who over-eats, nor for him who fasts excessively; not for him O Arjuna who sleeps too much, nor for him who stays awake too long.

Sri Pillai Lokachariyar has a one-line solution to all our problems arising out of our actions or karma.

'karmam kainkaryattilE pugum'

Actions or karma done as service to God, in His devotion, would be considered as sacrifice, the effect of which would free us from karmic bondage.

If we perform all our duties and actions as a service to the Lord, we accrue no karma effects. Even if we fail to do as intended, it still doesn't matter, as we were not attached to the cause or outcome. Therefore, our spiritual progress continues from that point without having to start from the beginning.

This Acharya impresses upon us that being able to serve God in any way it pleases God, is the greatest joy or bliss a Soul can experience. And it's not easily obtained. Imagine the kindness of the almighty God who is the most powerful, accepting a little nothing we do in love, as service.

Sri Krishna suggests a way to relinquish karma.

Gita 9.27

'yat karoshi, yad asnAsi, yajjuhoshi, dadAsi yat,

yat tapasyasi kauntEya tad kurushva madarpaNam'

Whatever you do, whatever you eat, whatever you offer as sacrifice, whatever you give as charity, or perform as austerity, do it all as an offering to Me.

By dedicating our life in devotion to the Supreme Lord, all our actions are transformed into a spiritual practice and an act of worship. Even if our karma is imperfect, the Lord compensates for our deficiencies, as perfection in action is not human.

Good karma is like a bank balance; you can choose to spend it or relinquish it by renouncing the benefits with a sense of detachment. The merit is not wasted, as you can keep accruing and spending it without obsession. Conversely, bad karma is like a debt that must be repaid by exhausting its effects, as it cannot be renounced - we cannot escape its consequences. While debts must be repaid without option, God can help us repay our karmic debt within the framework if we seek His grace.

Not all bad karma leads to hell; only heinous and cruel actions result in punishment there. After completing that punishment, we receive a new life to resolve any remaining karma.

Gita 9.21

'tE tam bhuktvA swarga lokam vishAlsm, ksheeNE puNyE martya lokam vishanti...'

Those who perform noble deeds qualify to enjoy the vast heavenly realm; they return to the mortal earth when their merits are exhausted, signifying that heavenly pleasures are temporary.

When we accrue substantial noble karma, we become eligible to experience heaven, where we exhaust good karma before being reborn into prosperous human families. Accumulating significant good karma can qualify us to become gods, governing different worlds for millions of years until our term ends.

We don't qualify to choose good life forms if we have bad karma accrual. The law of karma determines automatically what life forms we deserve.

If you have a lot of money in the bank (good karma), you can spend it for anything that kind of money can buy. As soon as you spend up all the money, you return to earning more money, or be born as humans to collect more good karma.

If you have debts, you don't get to buy anything without money, but have to work hard just to repay the debts. You might take lower births as animals etc. to repay your debts of accumulated bad karma.

Let's say I have done only good things helping others or have never committed a crime or anything unlawful, no one is going to give me an award or reward for being righteous and law abiding. At the same time, if I commit even one crime, I will be punished for it. Being righteous and lawful is expected of all humans.

In order to attain the divine world, we need to renounce and nullify all accrued karma completely. It is not possible for us to do this without the help of God. God would enable us to exhaust our karma if we seek Him with true devotion. He will help us to get rid of bad karma by experiencing the consequences which will pass by quickly. When we do everything as an offering to God, that is the best way to not accrue new karma.

Many of material possessions we earned, may outlast our existence in this lifetime. So, we don't hold on to any material belongings after death. We own nothing in this life, neither material wealth nor relationships. Surprisingly, we do not even own our bodies. Our bodies, meant for this lifetime, are also a rented place we have to vacate at death and move on. However, there is something that we carry on from one life to the next, which we earn throughout our lifetime, and that is karma. We carry along the baggage of karma alone and nothing else. And karma cannot be shared with anyone.

Sri Krishna explains how the Lord helps us get rid of karma.

'yasyAham anugruhNAmi tasya vittham vahAmyaham.'

When I bestow my grace upon someone, I carry or bear his earnings (wealth or vittham).

This is normally (incorrectly) interpreted that the Lord takes away a devotee's wealth when the Lord bestows His grace on the devotee. Vittham literally means money.

The correct understanding of this statement is that the Lord nullifies our karma to accept us to the divine world. Because we have to relinquish all karma before we are eligible to go there. What we earn in several lifetimes that keeps accumulating is our 'karma-wealth' and not the material wealth which we have to leave behind anyways.

Sri Krishna is saying that He would take the responsibility of our accumulated karma and is not referring to our earned material wealth or money. Moreover, God has no need nor has use for our paper money or material wealth. By taking the responsibility for our karma, God enables us to attain the divine world.

It is karma that is instrumental in bringing us to this world, and it is karma again relinquishing which we gain eligibility for God's grace to attain immortal existence.

Sri Krishna explains in the Gita that karma works in three ways, good, neutral and bad. To understand this - let's say we start a business to earn money (legally); there is no guarantee we will make only profits. We could make profits or break-even or even incur losses. Similarly, the effect of our deeds which are not entirely in our control, can be good, neutral and bad.

When we do everything in devotion to God, the effects would not bind us with adverse consequences. However, we cannot do immoral deeds as an offering to God.

With good karma accrual we have a choice what we want to be; with bad karma balance, we don't get to choose; we have to undergo consequences commensurate with our bad karma.

- The path of Jnana or knowledge.

JnAna marga. knowledge is our nature, and is crucial in everything we do. If we lack knowledge or understanding in carrying out our duties without the sense of right and wrong, the consequences of such actions would bring us greater harm. When we are able to do our actions correctly, it implies that we have a proper understanding of our duties and how to accomplish them righteously.

Jnana or knowledge is also termed 'akarma' or non-action. Knowledge of self or 'Atma jnana' is the purpose of all knowledge. Atma jnana, is the first step to acquiring 'Brahma jnana' or knowing about the Lord. Study of scriptures is essential to understand the truth about Self or God. However true knowledge of God is possible only through devotion, when God recognizes our unconditional love for Him and reveals Himself.

Sri Krishna suggests that the karma path is natural and best suited for humans. Mere knowledge of the Self cannot elevate us to a higher state of realization or devotion. We cannot sustain our bodies merely by practicing jnana yoga. Importantly, Lord Krishna is highlighting that it is of no use to anyone when we run away from performing our duties, hoping to attain a higher state of realization. We can attain enlightenment even being a karma yogi.

Gita 3.35

"srEyAn swa-dharmo viguNah para-dharmAt sv-anushthitAt, swa-dharmE nidhanam srEyah para-dharmo bhayAvahah"

means, it is better to perform your own duty though imperfectly than trying to do another's well. Fulfilling your own responsibility is safer and better, even if it leads to failure, while following another's path is dangerous.

If we are not qualified doctors, we should not attempt to treat patients or prescribe medication, as that could potentially harm them, or even lead to death.

As humans, we are better off practicing 'Karma yoga' which is natural to us, then following the path of a 'jnana yogi', which is not natural to an embodied being.

It is better to adhere to one's own duties (swadharma) determined by their nature, rather than imitating another's, even if one's own tasks seem insignificant or difficult. Attempting to follow another's duty (paradharma) is dangerous and often causes more harm than good, as it does not align with your true nature.

There was a discussion between Sri Ramanujacharya and his disciple, Sri Koorathazhwan, regarding the inherent, prime nature of the Jivatma (soul). While conventional scriptural interpretation suggests that jnana (knowledge) is the primary, intrinsic attribute of the soul - a view generally accepted without debate by other Acharyas - Sri Koorathazhwan held a distinct perspective. He argued that the foundational nature of the soul is, first and foremost, its existence as an inseparable potency (sEsha) of the Lord. Therefore, he maintained that knowledge is not an independent, primary nature in itself, but rather an intrinsic, secondary aspect built into the soul's fundamental identity as the Lord's potency.

Just as a person's power is meant to serve him, souls - being the potency of the Lord - are meant to serve the Lord. Merely knowing our nature is not enough; actually, living according to that nature is essential. Souls that perform their prime nature (service to God) already possess the knowledge inherent to the purpose of their existence.

When Souls realize their imperishable relationship with the Lord and live up to it, knowledge is built into that. For instance, behaving as a son to a father implies knowledge of who I am. However, if I know I am a son but fail to fulfill my duties, that knowledge is useless. The act of supporting and serving my father - expected of a son - proves I already understand the relationship.

Sri Ramanujacharya agrees with Sri Azhwan that 'seshatvam' (service) to the Lord is the foremost nature of the Souls, practicing which reveals that they have the knowledge of who they are, rather than 'jnAtrutvam' (Atma jnAnam), which alone doesn't complete the potential of the Soul and its nature.

Sri Pillai Lokachariyar concludes how the Souls can attain the ultimate state.

'kainkaryam thAn nityam, kainkaryamum prArtithE peravEndum.'

Eternal service to the Lord, is the highest state a soul can attain, which is not easily obtained. When God allows a soul to engage in His service, He literally reduces Himself to a state of dependency at the mercy of the devotee. Here, the devotee binds God through sheer love to serve Him. Only the greatest of devotees succeed in getting the opportunity to serve the Lord. The Gopis of Vrindavan and the friends of Sri Krishna were considered some of the greatest devotees, as they wished to serve the Lord for His well-being without any expectations in return - not even the bliss Lord Krishna would share with them - because they would not want to waste a single moment without serving Him. It would be hard for normal devotees to understand such levels of devotion.

And devotees can reach that state only by yearning to serve the Lord and when God graces them to engage in His service. One such divine Soul is Sri Ramanujacharya who is the Adisesha or 'the one who is in eternal service of the Lord'. The Lord would grant anyone who seeks such a state, when their intentions are pure. And God forgets who He is, giving in or even surrendering to the extreme love of His devotees.

Scriptures seem to suggest that it is important for the Souls to know of their relationship with God, but Sri Azhwan implied that it is important for the Souls to practice, live, realize the relationship, rather than merely know we are the children of God. True knowledge of this relationship is demonstrated by performing our duties as children of God.

Atma jnani is still bound by the laws of karma and cannot escape from the clutches of this prakruti. We can attain immortality only by the grace of God and that happens only with unconditional devotion.

- Bhakthi or Devotion.

Sri Alavandar explains in this verse, the core essence of the Bhagavad Gita:

'Sva-dharma-jnAna-vairAgya-sAdhya-bhaktyEka-gocharaha'

The Lord is attainable only through Bhakti (devotional surrender).

This verse asserts that the Supreme Lord, as described in the Gita, is realized through devotion (Bhakti) that is developed by doing one's prescribed duties without attachment, accompanied by true knowledge of the Self and detachment to worldly desires, which would result in unconditional love for God.

We have to fulfill two aspects for our devotion to succeed.

- Unconditional devotion
- Undivided or exclusive devotion

Unconditional devotion defines our pure love for God without any expectations in return. True love would mean we are selfless, without any benefit or joy for the Self and going a step further, even if there is pain for the Self.

Undivided devotion signifies our exclusive love for God alone. When we are devoted to more than one person, we fail.

Sri Krishna gives a comprehensive solution to all our worldly problems.

Gita 9.65

'manmanA bhava madbhakto madyAji mAm namaskuru.'

'Focus your mind on Me, be My devotee, be My worshipper, you shall come to Me alone. I promise you, truly, for you are dear to Me.

Gita 18.66

'sarva dharmAn parityajya mAmEkam SaraNam vraja.

aham tvAm sarva pApEbhyo mokshayishyAmi mA suchah.'

Giving up all (self-prescribed) duties (dharmas) completely, seek Me alone for refuge, I will release you from all sins, grieve not.

Lord Krishna is directing us with a solution to life problems we all face; from self-conditioned duties to the highest duty of devotion to God. The promise of liberation from effects of karma brings freedom from the fear of sin and the anxiety of struggle in life. It is the concluding, ultimate advice given by Lord Krishna to Arjuna, after teaching various paths.

The Lord has a hidden nuance in this message. He says give up all the dharmas and come to me for protection. There is no mention of adharma here. Does it mean a sinner cannot surrender or would God not accept a sinner even if he surrenders. The implied meaning in this is very intricate, very subtle.

On one hand Lord Krishna says - give up all dharmas (ownership to righteous duties) before seeking my refuge. In the next line the Lord says - I will release you from all sins. If I have done only dharmic or righteous karma, where is the question of release from sins?

Sri Pillai Lokachariyar comes to our aid to explain what Lord Krishna meant: 'dharma shabdattil adharmam pugAdu.' He clarifies that 'righteousness does not include even a miniscule amount of unrighteousness, or adharma, any which way we look at it.

Let's, for instance, take a drug addict or an alcoholic seeking rehabilitation. The doctor would ask them to stop taking drugs, as only then could their body effectively respond to the treatment. If they continue to consume alcohol or drugs, the treatment will not be effective. Similarly, if we continue to engage in bad karma even after surrendering to God, we fail in our surrender, and there is nothing God can do to help us, as He does not forcibly intervene in our karma. He cannot protect us if we fail to truly seek His protection.

Then how is God helping us, if He doesn't help bad people? It implies that God is not directly interfering in our karma, good or bad. We have to take responsibility for our actions (karma). Only righteous people would be able to successfully surrender to God. Others would fail in their surrender as their sins would prohibit them from effective surrender. God cannot help us get rid of our karmic backlog if we do not succeed in our surrender. He does not force anyone against their will. If we continue to commit sins even after surrendering to God, it means that we are not interested in God's grace, just as a drug addict who continues to consume drugs after seeking rehabilitation is not interested in the treatment.

Although I might have sinned in the past, following a dharmic life in the present allows for effective surrender. However, if I continue to act wrongly, fully aware of it, my surrender will fail.

After successful surrender, God would find a way for us to relinquish all our past karma, good and bad, within the purview of laws of karma. And we would not accrue further karma after our successful surrender as it would be accepted by God as an offering to Him.

Surrendering to God is tricky as God knows if our intentions are true or not. Hence, it is suggested to develop true devotion to the Acharya, who has taken birth amongst us only for our benefit, to guide us to attain the Lord. Acharyas have no karma or consequences and do everything out of kindness, benefitting the Lord as well as the disciples. They have no other purpose of their own.

When we are devoted to the Acharya our success is guaranteed. If a person fails in his devotion to his Acharya (guru) or teacher, he would still succeed in reaching the Lord as the Acharya stands beyond the Lord. If we succeed in our devotion to Acharya, God comes to the acharya seeking us. God has positioned the Acharyas above Himself.

Sri Pillai Lokachariyar, in his Sri vachana bhushanam, clarifies:

'bhaktiyil asaktanukku prapatti

prapattiyil asaktanukku, idu'

One, who is inept in devotion to God, can still succeed by surrender, and he who is incompetent even in surrender, has an Acharya coming to his rescue. Even if we fail in devotion or surrender to God, we can still succeed in attaining God by the benevolent grace of Acharya.

Lord Krishna declares that He is indebted to all Acharyas for their selfless service in making devotees worthy of God's acceptance and in leading them to God. He tells Udhava, His dear friend that He considers Acharyas as Himself.

All that we have to do, if we wish to attain the divine world, is seek refuge with an Acharya, and reaching God's abode is a given.

By frequent visits to our favourite temple, we would develop love for God. It is important to know that we don't go to temples to seek anything from God. Also, God doesn't reward us because we seek benefits through prayers. We worship God only to express our unconditional affection for Him and expect nothing in return. We are here to give God our love and experience His divine grace in return, which would enable us with opportunities to serve God and his people.

In the temples, we can visually witness the whole existence - of the Paramatma (in the middle), Mahalakshmi (on the right side) being the source Jiva Shakthi, and Bhooma Devi (on the left side) being the power of prakruti or creation - all the three on one pedestal. We have to understand that they are the three realities of existence, the other two being the powers of the same Lord. God has countless powers, though.

- Jiva Shakthi or 'para' (superior, transcendental, divine) or 'akshara' (imperishable, eternal) and
- Prakruti / maya Shakthi or 'apsara' (external, material energy) or 'kshara' (changing, perishable).

We wanted to be in this world; God formed this world for our sake. We didn't come here on our own. Similarly, to go to the eternal world, we need to ask God and only He can take us to the divine world. Countless divine Souls exist in the immortal world experiencing eternal bliss. It is true because it is stated in the Vedas and reconfirmed by Lord Krishna in the Gita and by the great Azhwars and Acharyas.

We accept that we must obtain a visa to enter the USA if we are not U.S. citizens; merely knowing where the U.S. is and how to get there is not enough. We cannot enter that country without a visa, and that visa cannot be self-issued. Likewise, only God has the power to transcend this prakruti (maya) and permit our migration to the eternal world. Only the Soul, with its subtle, invisible body, can reach the divine world, not our material bodies. Upon arrival in the divine world, we would obtain a divine subtle body and a divine gross body before experiencing uninterrupted joy in the immortal world.

We experience so many things when alive and can describe our experiences. But no one can describe the experience of death for we won't return to tell about the death experience. So, what is the proof of existence after death? Again, Vedas enlighten us all about the existence of the self after death. There is enough proof of everything said in the Vedas about life and existence which can be perceived scientifically. We should not miss the opportunity to realize the truth of life and existence.

Sri Krishna says that all worship eventually reaches Him, Gita 9.23

'yEpyanya dEvatA bhaktah yajantE SraddhayAnvitAh; tEpi mAmEva kauntEya yajantyavidhipUrvakam'

those who worship other gods and goddesses with devotion, they worship Me alone, though improperly, without awareness.

How can you explain the taste of sweetness to someone who has never tasted anything sweet? The only way he or she can understand is by tasting something sweet, not from explanations. If the person lacks taste perception, there is no way he can ever experience taste or understand it. Similarly, the experience of God's grace cannot be explained but only understood when we taste God's grace.

Sri Poigai Azhwar confirms in his mudal thiruvanthadi.

'Tamar ungandu evvuruvam avvuruvam tAnE, tamar ugandadu eppEr mattrappEr....'

Lord takes that form as desired by His devotees, He becomes the name they call Him as, He becomes all that they imagine Him to be, that's how kind and easy He is.

This verse implies that God assumes a form pleasing to the devotee. Surely it would be difficult for a devotee to worship a formless God.

Sri Nammazhwar who understands God, declares thus -

'patrudai adiyavarkkeLiyavan, pirargalukkariya vithagan.'

The Lord is easily approachable even for simple devotees, yet remains beyond the reach of even the greatest masters of knowledge, if they think they know God, because they mastered all scriptures.

Vedas define God thus -

'Satyam, jnAnam, anantam, Brahma.'

Satyam - describes that Brahman possesses non conditioned existence and is absolute. In this, non-sentient matter that goes through change and the sentient Souls associated with the material universe are not absolute as both Souls and material energy have conditioned existence.

JnAnam - signifies eternally consistent consciousness without contraction. In this the liberated Souls, whose knowledge was once contracted when bound by mortality are not included. Does jnAnam mean that Brahman is consciousness or having consciousness as an attribute? JnAnam signifies both, just as Brahman signifies greatness by nature, as also greatness by attributes.

Anantam - denotes the nature of being without limitation of place, time or substance. His infiniteness applies to both nature and attributes. Anantam signifies that it is not limited spatially, temporally or with reference to substance. Brahman is free from all the limitations such as 'desha pariccheda' (this is here and not elsewhere), kala pariccheda (this is now and not any other time) and vastu pariccheda (this is and not that). Brahman is without any limitation, related to substance. This non limited nature in respect of substances would mean infinite auspicious qualities by which it has no equal or a superior and Brahman are the absolute in all respects.

We attain what we seek. Lord Krishna suggests an easy way to attain Him.

Gita 9.25

'yAnti dEvavratAh dEvAn, pitrUn yAnti pitruvratAh, bhUtAni yAnti bhutEbhuo, yAnti madya jinopi mAm.'

'Devotees of gods attain gods, devotees of manes go to manes, those worshipping ghosts and spirits go to them, and those who worship Me with devotion will come to Me.

In essence, this verse explains that individuals attain the state corresponding to the object of their devotion or the entity they worship.

God is the goal we want to attain, and He is the only means to attain Him. Nowhere else, the means to attain the goal and the goal we want to attain are one and the same. It's the Lord we want to be with in immortality and He is the only one who can grant us immortal existence.

Sri Andal, incarnation of the divine mother (Prakruti) as a five-year-old girl, expresses in Thiruppavai that even in devotion all of us need to approach God together for greater experience.

'Koodiyirundu kulirndElo' When you share God's grace being together with others, the joy is immeasurable.

'undannoduravEl namakkingu ozhikkavozhiyadu...'. She tells Sri Krishna 'The relationship we have with you is imperishable and cannot be severed (by anyone)'

'ettaikkum EzhEzhu piravikkum undannoduttomE yAvom unakkE nAmAtcheyvom'

Even in countless births, we remain eternally devoted to you, remain inseparable with you, serving you alone, there is nothing else we seek, please empower us to stay away from all distractions.

'sengaN chirucchiridE emmEl vizhiyAvo' Please bestow Your grace slowly on us, bit by bit, for we cannot bear the enormity of your love when in full flow. She tells the Lord that she knows He is magnanimous and His grace would be overwhelming; to be subtle with the effect of His kindness.

Lord Krishna explains that His incarnations are divine.

Gita 4.9

'Janma karma cha mE divyam Evam yo vetti tattwatah; tyaktwA dEham punarjanma naiti maameti so'rjuna'

He who knows in truth My divine birth and actions, upon leaving the body, does not take birth again; he comes to Me, O Arjuna!".

This verse highlights that understanding the transcendental nature of God and His pastime mysterious acts when He incarnates, leads to liberation from the cycle of rebirth, allowing one to attain His eternal abode.

Gita 8.5

'anta kale cha mAmEva smaran muktvA kalEvaram...'

He who at the last moment, while leaving the body, departs thinking of Me alone, attains Me. Of this there is no doubt.

But this is not mandatory for those who are devoted to an Acharya; they would attain God even if they miss to think of God at the time of leaving the body. Because God thinks of them, always.

True devotees, such as Prahlada, the Gopika and Gopala friends of Lord Krishna had no personal agenda. Their every thought was for the sake of Sri Krishna. They would lose their own self for Krishna. They would eat for Krishna and Krishna would feel He had eaten.

Sri Thiruppanazhwar is special in his own way. At the first opportunity he got to glimpse Lord Ranganatha, he witnessed the divine beauty of the Lord from His lotus feet to the crown. His expression says it all.

'en amudinaikkanda kaNgaL mattrondrinaik kANAvE.'

Meaning - 'having beheld the Lord's glorious divine beauty, my eyes desire to see nothing else'. Legend has it that this Azhwar never returned, having merged into the Lord.

Sri Nammazhwar gives a glimpse of what true devotion for God is.

While most devotees are filled with undivided devotion for the Lord, Sri Nammazhwar is different; he is not merely a great devotee, but the very embodiment of devotion - Krishna-bhakti personified.

Azhwar portrays his true nature in Thiruvaimozhi 6-7-1.

"uNNumsoru parugunIr tinnumvetrilaiyumellAm kannan emberumAn ennennE kaNgaL nIrmalgi...."

To him food, water and after-meal delicacy vetrilai (dessert) were all Krishna. Azhwar didn't need food, water or air. He was 'Krishna devotion - personified'. He ate, drank and lived Krishna bhakthi. He had the Lord filling every atom of his body from head to toe. In that all his senses competed to become engrossed in experiencing Krishna; the eye, in addition to seeing the beauty of Krishna, did all the functions of the other senses as well, of hearing, smelling, tasting and feeling the presence of the Lord. Similarly, the sense of hearing and ears did the functions of all the five senses. His divine experiences of the Lord were at least five-fold more than that of the other great devotees. The other devotees could only use the eye to see and other senses for their respective perception. But each of Nammazhwar's senses functioned as the other senses as well. As an expression of his unequalled devotion, it is said that Azhwar wanted to send out his mind as a messenger seeking Krishna's audience. And was immediately worried that his mind might betray him and remain with the Lord without returning to him.

It is said that Azhwar's mind was ever filled by Krishna and he would not forget Krishna even for a fraction of a moment. Just to play with Azhwar, Krishna would disappear from his mind for a second. And that was enough for Azhwar to faint for months. Azhwar couldn't bear missing the Lord even for a fraction of a second.

If you draw a portrait of 'Krishna bhakthi', it would be none other than Sri Nammazhwar.

Even Sri Krishna was so much in love with Azhwar, that He asked the Azhwar to go with Him to the divine world with his mortal body. Azhwar reminds the Lord that the laws of creation (nature) shall not be violated even though He was all powerful to do anything He wished. Yet it is true that God would go to any extent for the sake of a devotee.

Devotion is much simpler. We never truly forget the Self; even one who loses their memory still experiences the Self, questioning, "Who am I?" or "Where am I?"

Devotion to God is also as simple as being conscious of the Self. Just think of God or include God in everything you do. Devotion is the simplest and easiest thing to practice.

CHAPTER SEVEN

What prevented us from attaining the goal

(Virodhi swarupam)

We are ourselves the cause for our having been denied existence in the divine world. Our desire (Karma) to experience independent existence away from God is the reason why we are born in this world.

When we were given the choice between staying with God or independent existence, we turned away from God choosing to experience life in the material universe. God had to do a lot of work (though He did everything merely by His will) to create the universe and habitable life supporting planets and worlds for us to enjoy life here.

The moment we acquired material existence, we were under the influence of the material intellect, mind, senses and bodies born of this prakruti (nature). And this prakruti comprises the gunas or qualities that's intrinsic to it. We do our karma influenced by the nature of this prakruti.

Gita 18.40

'na tadasti prithivyAm vA divi dEvEshu vA punah.'

There is no creature, either on earth or again among the gods in heaven, that is free from these three gunas born of prakruti. (It is to be noted that heaven or hell are also worlds of this material universe).

Gita 7.25

'nAham prakASah sarvasya yogamAyA samAvrutah'

Obscured by My maya, I am not manifest (visible) to all. This deluded world does not recognize Me as the unborn and immutable.

Deluded by the desire to enjoy, influenced by the nature of the gunas, we have trapped ourselves in the life experiences of this world, which commensurate with our causes but not as decided by us but as determined by the law of karma. Does it mean we are not responsible? Indeed, we are the reason for all causes and effects we experience. If we do bad things, we might face some of the bad consequences soon and some later, depending upon the magnitude of bad karma. If I kill hundreds of innocent people, I cannot suffer the consequences in one life."

Problems come to us uninvited, while happy moments are elusive, even when we seek them. If we consistently do good, we can be confident that the consequences will also be good; we do not have to chase rewards, as they will find us automatically. While living a righteous life means we may not face negative consequences, any problems that do arise despite our righteousness are likely due to past karma. Nevertheless, righteous people are better equipped to overcome the consequences of past karma than wrongdoers.

Lord Krishna assures Gita 9.30

'apichEtsudurAchAro bhajatE mAm ananya bhAk.'

Even if the most sinful person worships Me with undivided devotion, he must be regarded as righteous for he has rightly resolved.

This verse emphasizes that no matter how sinful a person's past has been, if he seeks God with undivided, unconditional devotion (ananya-bhak), he should be considered holy because his inner resolve has changed firmly. It highlights that a devotee should be judged by his current, unwavering, and sincere resolve to follow the righteous path, rather than past misdeeds. This does not mean one could commit sins again. Rather, it means that if a sincere devotee had sinned in the past, those do not negate his devotion now. And he would not willfully sin again.

In essence, this phrase highlights the power of surrender to God to purify even the most - evil minded, provided they are determined to change, never to sin again. And God knows when the devotion is true.

Gita 9.31

'kshipram bhavati dharmAtmA.

kauntEya prati jAnIhi na mE bhaktahraNaSyati.'

Quickly he becomes righteous and obtains everlasting peace. **I*affirm O Arjuna, My devotee never perishes.***

This is Lord Krishna's promise that a sincere devotee never perishes. It means that devotion to God helps to be righteous (dharmic) in life. It highlights that even a person with a history of bad actions becomes righteous and attainș lasting peace (shashvat-shantim) by turning his mind to the Divine. A devotee who experiences divine grace once, would never be able to indulge in unrighteous acts.

'mana Eva manushyAnAm kAraNam bandha mokshayoh'

'The mind alone is the cause of bondage or liberation for human beings.

A mind attached to sensory objects leads to bondage, while a mind detached from worldly pleasures and devoted to God leads to liberation from the cycle of birth and death. The world binds us to our karma and its consequences. God does not bind us, but rather grants liberation and freedom from mortality. We can choose what we desire, and will attain that. God has simply facilitated our existence here.

It's interesting to know that when we are attached, we are selfish expecting benefits for ourselves. When we are affectionate, there is care and concern for others, devoid of selfish interests. We should stay away from attachment and nurture affection.

Sri Azhagiya Manavala Perumal Nayanar gives us an insight of reality, in his 'acharya hrudayam 1.23' end part.

'pagal viLakkum minminiyimAkkum...'

pagal viLakku (daylight) signifies the brightness of the Sun. And 'minmini' (firefly) is an insect that gives out a faint glow in the night. Fireflies produce a chemical 'luciferin' in their bodies, a pigment, emitting light, termed 'bioluminescence'.

If we assume that an insect which emits a feeble light in the night due to the presence of certain chemicals in its body is the same as the brightness of the Sun, we are in an illusion. Comparing and finding a similarity between the light of a firefly and brightness of the sun is like finding a similarity between an atma with the infinite God.

a. Though individual atma and the paramatma have certain similarities, such as consciousness, eternality, oneness with the Lord, it should not be mistaken that both God and the Souls are one and the same. Just as the brightness of the Sun cannot be compared to the light emitted by an insect, Souls do not possess infinite powers and capabilities that only God possesses.

a. Seeking happiness in the material world is like seeking the brightness of the Sun in a firefly. We can experience true happiness or bliss only with God.

This Acharya also explains how we ended up taking life forms in this world. He says -

1.10. Etan nimittam munnamE mudalmunnamEyAna achidayanAnAdi sambandhangal.

The reason why we are trapped, taking birth in all sorts of bodies is because we chose life in this universe (achit) which is non sentient. Had we chosen life with 'ayana' or the Lord, we would have been in the divine eternal world. There are just two entities for us to choose from. The material world or the Lord.

Achit or matter is ignorant by nature, devoid of consciousness or intelligence. Association with achit, which is the source of our bodies, has influenced us to acquire ignorance, assuming the Self as our bodies, while the nature of the Soul is 'jnana' or knowledge. Ignorance has engulfed the knowledgeable Souls like smoke covering the fire.

On the other hand, the Lord is the ocean of infinite knowledge. Ignorance cannot prevail where there is eternal knowledge. Knowledge dispels ignorance like light over darkness. Darkness cannot prevail where there is light. Souls being His inseparable potency, would find bliss with God and not in the material world.

We are bound by ignorance when we forget our true nature. Ignorance affects the knowledgeable Souls because the body, senses, mind and intellect we have are obtained from the non-sentient world, inherently comprising the three guNAs which is though inferior in nature to the divine Souls, bind us due to our attachment to experiencing the sense objects. When we are body conscious, we are under the influence of the guNAs of this prakruti. When we are conscious of our true nature, the material energy of the world cannot affect us.

Our bodies are made of the food we eat. Food is born of this prakruti (nature) inherent with the three guNAs. So, the food we eat also has its natural qualities such as sattva, rajas and tamas. We are familiar with the saying 'a sound mind in a sound body'. The state of body quality influences our mind and thinking. If we consume sattvic natured food, our thoughts would also be sattvic in nature. Likewise, rajasic and tamasic foods would influence our thinking, behavior and conduct of similar nature.

In the divine world, the bodies, senses, mind and intellect are all divine, transcending the guNAs prevalent in the material world. Ignorance is non-existent in the divine world.

Vedic culture advocates to look at all the people of our world as one family, 'vasudhaiva kutumbakam' signifies the understanding that all are children of God and belong as one family. Vasudha means 'earth' and Kutumbakam means 'family', a statement from 'Maha upanishad' signifies the essence of co-existence of all the people, animals, and nature as one global family, transcending geographical boundaries of nation, religion, and ethnicity.

God loves everyone equally without discrimination. But upholding dharma is His supreme consideration. Those who uphold dharma are very dear to God. In fact, God incarnates to establish dharma and to share His love with His devotees. Lord Krishna expresses to His friend Uddhava that He is fond of His devotees to the extent that He yearns to feel the dust from their feet.

God bestows highest grace on a demoness (rakshasi), Putana, who fed poisoned breast milk to Lord Krishna when he was an infant. Sri Krishna graced her with existence in the divine world that is meant for the greatest devotees. And there is nothing greater that can be granted. Uddhava, Sri Krishna's friend questions Him as to how He could justify granting the divine world (Paramapadam) to Putana whose intent was to kill infant Krishna and the same divine world for mother Yashoda, who has no equal to her, devoting her entire life for the sake of Krishna.

Sri Krishna's reply puzzles Uddhava. The Lord says, Putana also came to Me with undivided devotion though her motive was to kill Me, she was compelled to do it out of fear, for Kamsa. But the milk she wanted to feed was pure with motherly love exclusively for me. She deserved the highest I have and I have settled my debt with her. I don't owe her anything now. But I grant the same Paramapadam to mother Yashoda too, though there is no one equal to her when it comes to the love she bestowed upon Me. But I remain eternally indebted to her, for I have not been able to reciprocate in kind. I accept that I failed in her case, though I gave her all I have.

I would like to share my experience with the readers. I do sense the unconditional grace of God, as though life can be on autopilot mode. I have also been through times when I had no thinking about the Self or God and when I recall those days, I have a distinct memory of how things were different. His grace was unmistakable, then or now. I try talking to God within my mind, but I don't hear from Him. People might call this insanity or hallucination, but not so for me. For me it is real. It has been one sided though, but with an awareness that He is listening, and would respond when deserving. Or He has responded without direct involvement and makes us feel the difference in a way that situations resolve without us doing anything. How do I know God is listening? Scriptures and Azhwars and Acharyas have said so - that the Lord as Antaryami, is ever aware of all our thoughts. It is true for everyone, whether a devotee of God or not. But a devotee would sense God's grace as he would experience the difference in the way things happen. Things happen as though it was by some divine intervention, but yet they would have happened normally. Many things happen even without our effort or thinking about doing those things. It could be karmic or God is empowering the law of karma to take effect. Even without His direct response, He fills us with that joy not experienced in anything else. You cannot miss feeling His grace, though you cannot see Him directly. God is not easily seen, even in dreams.

If we want to be successful in this world, we can work to achieve our material goals. We don't need to ask God for worldly gains and success. He has already done everything needed for life here.

Merely thinking about God, we start feeling happy; free from wants; we feel fulfilled. The story of Dhruva is a great example, who devotedly seeks for the Lord to appear before him because the child had so many questions to ask the Lord. The moment the Lord appeared and lovingly held him on His lap, Dhruva felt totally fulfilled. The Lord tells the child to ask Him for anything he wanted without hesitation and that would be granted. Dhruva had planned to ask the Lord many questions. But the moment he saw the Lord, all his questions and desires vanished. The child tells the Lord he does not want anything, he had no questions to ask and that he was totally fulfilled.

Even in this world, the idea of life is all about happy relationships. There is no happiness without relationships. There is true joy in togetherness. Joy multiplies when shared. All the beings of our world are related through God who is the common factor present in all. We are so deeply attached to this life and temporary relationships here, but we have forgotten about our eternal relationship with God.

God knows everything we wish for, as He is ever present in us as the Antaryami. He is the first to know of our thoughts even before our mind can process the thoughts. Does it mean God will keep fulfilling everything we wish for. No, He would fulfill whatever we deserve, mostly within the karmic framework. Now it's up to each one of us to take responsibility for our actions and not blame God for our misdeeds and failures.

Sri Thondaradippodi Azhwar has expressed this aspect beautifully in his composition 'ThirumAlai' in Tamizh language, thus.

"ulluvAr ullitthellAm udanirundu aridiyenru velgippoy ennullE nAn vilavara chiritthittEnE"

Realizing you are in my heart, knowing my thoughts even before I do, fills me with wonderment. I cannot hide my hilarity, having foolishly considered it possible to deceive you.

Lord is the observer and witness in every Soul, noting every act of karma, (thought, speech and deed) automatically determining the consequence, governed by the law of karma, without His direct involvement. Except that God expected humans and all species to behave inherent to their nature. All other species have been fulfilling this expectation except the humans. Humans are expected to live a righteous life, benefitting themselves as well as other species. But only some adhere to righteousness and most others have violated this condition and are accruing unlimited bad karma which will have to be exhausted in countless lives in all sorts of bodies. The other species escape from karma consequences because their life is meant to square off certain karma effects and they don't accrue new karma.

Merely doing our duties without concern for others' suffering is not true righteousness. We must help others in need according to our disposition, doing what we can through material, financial, or personal service. This is the best way to serve God in this world. And we must thank God for enabling us to be of help to others. That is considered nishkama karma, where we remain desireless in our actions, claiming no fame or credit for the help rendered by us, to a person or society. God did not take any credit for creating the world for our sake or for empowering us or for providing all that is needed for life to thrive here. Truly service to humanity is the best way to serve God.

Prasnopanishad explains the different states that we experience in life.

1. Waking state
2. Dreaming state
3. Deep sleep state and
4. blissful state in deep sleep (turIya)

That apart, we have created an 'intoxicated state' in our awakened state, when under the influence of alcohol or drugs.

(vysvAnara, ordainer of waking state) In this state the consciousness of the atma relates to things external, the atma uses all his body parts, senses, mind, intellect, ego and antahkaraNa (inner heart), enjoys and experiences the gross objects enabled by the Paramatman who is the inner self.

In the awakened state, we are conscious of our thoughts, speech and actions, everything done out of choice. All our actions can be witnessed by others present around us (but of course God remains an eternal witness). We share our emotions and expressions with others and vice versa. The events we experience in the world are also experienced by others present around us. Everything we experience would be limited to what our senses, mind and intellect can perceive. We could not fly like a bird in the awakened state, but we could in our dreams. Everything we experience in the real world is in the awakened state.

Dreaming state - (taijasa) consciousness is internal and the atma experiences all that he does in the waking state but he enjoys subtle objects which are exclusive to the atma who is in the dreaming state.

Prasnopanishad chapter 4-5 declares that in the dreams, the Jivatman experiences greater forms of the same things he experiences in waking state. Whatever he has seen, heard and perceived in the waking state, he experiences again and again in the dreams assuming all forms. He experiences even things not experienced before; such as death of his own body.

Then, whatever we experience in the dreaming state, are not real incidents of the world. It happens only in our dreams. What we experience in the dreams cannot be experienced by the others we dream of, though present in our dreams. Dreams are perceived by our mind using the subtle senses. We see in the dream while our eyes are shut. If we play cricket with a full team, the real team members are not involved in the experience. No cricket is played by anyone in reality. What we experience in the dream is exclusive to us. Some, we may remember fully or partly or even nothing at all. We can exceed our imaginations in doing anything in the dream that might be impossible in reality. We could visit other planets or even unknown cosmic places, which may be totally different to what exists in reality. The self doesn't travel to the places we dream of; it plays out in our mind like in 'virtual reality'. Anything is possible in a dream. We could possess super powers like a superman. We realize it was all a dream when we wake up.

Deep-sleep state - (prAjna) in this state the atma does not desire any enjoyment nor sees any dreams. One-ness is attained with the Paramatman who causes the atman to experience bliss of deep sleep. Atma is not involved in any activities except experiencing the 'bliss'. He would not remember anything from this state for he has not experienced anything in a conscious state.

Prasnopanishad 4-6 - In the dreamless, deep sleep state - the Jivatman is in the company of the Paramatman and experiences happiness. It is to be noted that Jivatman experiences this happiness only when in a body. The body alone is the cause of worldly happiness. Desirable and undesirable do not affect one without a body.

The Paramatman who is the inner self of the Jivatman is never tainted with any defect in the waking state, dreaming state and dreamless state. Though, in the dreamless state, the Jivatman is in the presence of the Lord, but it would be without realization. The only outcome is happiness.

Of all the activities we do, we find that a good sleep is the most enjoyable experience. We do nothing during our deep sleep and yet that brings us the greatest joy. This in itself is evident that we do not have to do anything to remain happy or we are more joyful even when we don't do anything. Humans are harmless when sleeping, as we cannot hurt anyone during our sleep; unless we are sleep walkers who would transform dreams into reality. In fact, deep sleep is the greatest joy that all living beings experience. It is evident from this that we experience true bliss only with the Lord; from the Lord.

We also realize that all our problems, worries, suffering, pain and grief are experienced only in the awakened state. Every person does everything to experience happiness in the awakened state. Be it a religious person, a devotee, a commoner, an atheist; everyone, whether working or relaxing, active or lazing, playing or idle, is done seeking happiness, peace, bliss. While our attempt is to find happiness in everything we do, we also end up suffering pain and grief. And the moment we go into deep sleep, nothing can bother us. We could have experienced the greatest calamity and grief moments ago, but the moment we go into deep sleep, we do not feel any grief; at least not until we wake up. And the moment we wake up, we come to face all the reality; all our problems again.

'Turiya' - The fourth state cannot be perceived by the atma for it describes the Supreme Lord, who cannot be grasped by the mind as 'this much' for infinite cannot be known in its totality; while He is instrumental in the atman experiencing the other three states.

Our experiences in a lifetime may not be much different on a daily basis. It is pretty much the same every day of life, morning to night. We sleep and wake up in the morning to go through the routine things all over again.

Sri Thondaradippodi Azhwar says in his Thirumalai, the truth about life we experience in this world.

'vedanul prayam nooru.

piNi, pasi, mooppu, tunbam.

AdalAl piravi vEndEn...'

We live for around 100 years as humans if all goes well (depending upon our karma). If we get to live for 100 years, half of it is spent sleeping. In the years left over, we lose a few years in infancy being in a state where others have to care for us; we are incapable of doing anything ourselves and totally dependent on others. And more years go by in childhood, being playful all the time. Then we spend the years of teenage infatuation, seeking sense gratification. Nothing else matters in this phase of life. The rest of the years are lost in dealing with relationships, ailments, hunger, old-age and misery. *We lived through a whole precious human life totally wasted, without avail. There was no time spared for realizing the self, or our relationship with the Lord, purpose of human birth, realization of human potential etc.* Hence, I don't seek to be born here again O my Lord of Srirangam.

All the achievements, success, fame, prosperity would come to an end eventually as human life is short. Sometimes so short, we are uncertain as to the longevity, like a bubble that could burst anytime.

I remember the words of Steve Jobs, a modern-day business genius, in this context, who deserves a special mention. He expressed his disappointment about life being so short that he couldn't complete implementing his business plans. He loved what he was doing and liked to be able to continue his pursuit of business excellence.

We cannot carry forward any of the wealth or achievements to the next life. The only thing that brought us to this world, karma, would be carried into the next life, with added baggage, consequences of which have to be met in the future lives.

There are certain powers even divine Souls would not be able to handle for it would be beyond their potential to possess such powers and only the Supreme Lord possesses infinite powers.

There was an incident between Vali (the vanara king) and Hanuman (one of the greatest devotees of Lord Rama). Hanuman is the incarnation with Lord Shiva's potency.

Ravana, the Asura king, was on a mission to prove his unmatched power after obtaining boons from Lord Brahma. Seeking a fight with great warriors, he challenged Vali to a fight. Vali, amused, asked Ravana to wait until he finished his morning prayers, but Ravana insisted on fighting immediately. Vali, possessing a boon that granted him half of his opponent's strength, easily subdued Ravana, trapping him under his armpit like a toy. While holding Ravana, Vali flew around to complete his daily prayers at the shores of the four oceans. Later, when Vali released him and asked if he still wanted to fight, a defeated Ravana realized Vali was invincible and agreed to be his lifelong friend. Following this incident, Vali presumed he was invincible, and even the gods were no match for him.

However, in a conversation with his father, Indra - the King of Heaven - Vali learns that his boon only works against mortals, not godly beings like Lord Brahma, Lord Shiva, or Lord Hanuman, as he cannot snatch away divine power.

Due to his arrogance of power and jealousy over Sugriva's popularity, Vali drove his younger brother out of Kishkinda on a false pretext and took his wife, which was an unforgivable offense. Hanuman, who stands for dharma, righteousness, supports Sugriva, though Vali was more powerful than Sugriva.

Indra warned Vali that he could not win against Hanuman, who stood in support of Sugriva. Despite this, Vali boasted to Indra that he could easily kill Hanuman in a fight by absorbing half of his power. Indra explained that Vali could not handle even a minuscule portion of Hanuman's strength, as it was beyond his capacity to absorb such immense divine energy. Indra added that a fight against Hanuman would result in Vali's death because Hanuman was not only immensely powerful but also immortal. Finally, Indra clarified that Vali's boon only allowed him to absorb the powers of mortals, not the divine, godly powers possessed by Hanuman.

We are so used to asking things from God, but have we ever asked what we could do for Him, or what He wants from us? Try asking just once. God might not respond directly, but you will still experience a joy that only such an expression can bring.

The moment you involve God; we would naturally be conscious of our actions because the thought of God being in everyone would deter us from doing anything that would displease God and we would refrain from doing anything wrong. Without God, we lose the sense of right and wrong. With God in our minds, we can live life fearlessly, and successfully as well.

Without God, we lose direction in life, purpose of life, with the sense of fear and uncertainty prevalent in our hearts as to what would happen next. With the sense prevailing that God is observing all our actions, it would remind us to be better people.

Sri Koorathazhwan expresses that the best way to insulate ourselves from committing bad Karma is to do things God would be pleased of. Then how can we know what God loves and disapproves of? What God loves is noble karma and what God dislikes being ignoble karma or sins. Sri Krishna has expounded with clarity what He likes from the devotees. He loves to preserve dharma. When we are righteous in our lives, that is the best way to make God happy.

Just as we dedicate our life for the well-being of our loved ones, the entirety of creation - encompassing countless souls and universes - constitutes God's family. God's singular purpose is to ensure the welfare of His children. However, a distinction exists: while He does not intervene directly in the material universe, He engages personally with each soul in the divine realm.

God loves true devotees and takes total care of them. I can assure all the readers - this is true from my own experience. It is not imagined. I also felt guilty on several occasions that God is so kind that He overdoes the - taking care part, though sometimes I would feel undeserving of such grace. You can touch and feel God's grace. God does it unconditionally. We get to realize a sense of joy in our hearts when we are able to feel His grace. What did I do for His grace? Nothing. He bestows His grace even without asking for it. We are ordinary devotees who live our life for our happiness and are committed to taking care of our interests in life. Our goal is not a100% service to God or the purpose of all our activities is not for God. We live our life for our sake and in the bargain by the grace of Acharya, we realize our relationship with God. Our devotion cannot be honestly said to be 100 percent. We also think of God amidst all the things we do for ourselves and even that love is enough to melt His kind heart in accepting our nothing as devotion.

Sri Thyagaraja, one of the devotees of Lord Rama, says in his ‘pancha ratna kirtana’ - ‘endaro mahanubhavulu, andarIki vandanamulu’. Means, ‘there are countless devotees of the Lord and I offer salutations to all of them’.

We cannot truly consider ourselves devotees until God accepts us as such. Even from the beginning, I could sense the difference between being a ‘non-devotee’ and having the mere intent to ‘want to be a devotee’. Even such a subtle change in the mindset brings a perceivable taste for devotion. I have no doubt that with every degree of improved devotion, life becomes more beautiful. More than us, it is God who feels the joy of getting involved with a true devotee - as though He is on the hunt for them. Though He expects nothing, He has so much to share, offering infinite possibilities that He has not yet explored.

Thinking about God is the supreme form of meditation. You are ever in love with God without any fear or anxiety; filled with a deep sense of fulfillment and desirelessness. You can perform all the daily tasks more effectively and joyfully.

Our relationship with God is unique to God alone. Not because He will take care of us, but because we want to take care of Him. We are eternally related only to Him and none else. Even if God doesn’t respond, our relationship doesn’t change. Relationship with God is eternally imperishable, doesn’t matter what He does or doesn’t do. He thinks the same about each one of us. There are no motives, no intentions, no expectations, no restrictions, no barriers in this relationship. No rules, no boundaries in the way we can express our affection for Him. Everyone is eligible to seek His grace. All are equally qualified and eligible when it comes to reaching out to God. When we feel His grace, we just enjoy perceiving the relationship with Him just as He has been waiting for us to turn towards Him, recognizing this relationship. He knows our feelings and thoughts even before us. But He doesn't influence our thoughts or interfere in our actions. We are totally free to do what we like and He has no involvement whatsoever in what we do, except when we seek His grace and desire nothing else.

It is logical to question how everything expressed in this book is correct knowledge. Is it founded on my imaginative assumptions, or my individual perspectives or hearsay? If so, it cannot be correct knowledge.

The truth is that none of the content compiled in this book represents my own opinion; every statement and interpretation is sourced from authentic Vedic scriptures, including the Srimad Bhagavad Gita, the Upanishads, and the Arulicheyal (Divya Prabandham, or Tamizh Vedam) of the Azhwars, as well as the commentaries of the great Acharyas. My knowledge is strictly limited to the - little bit of study - of these scriptures, and I do not claim expertise in other philosophies. I can only share what I have learned about the Supreme Lord - who is known by different names across religions. I have shared my own life experiences influenced by the teachings of the divine Azhwars and Acharyas.

How do we know the knowledge contained in these scriptures is correct and true? The only way is to test the knowledge to see if it is realistic and makes sense; see if we derive the higher benefits of life by adapting to the recommended life practices or not. We have nothing to lose. As a matter of fact, we can see real, perceivable benefits, enjoy everything we normally do without having to give up on anything we like. In fact, we would enjoy every moment of life much more, when we have a proper understanding of life, which the scriptures define beautifully. There is no restriction in what you could do, as long as it is righteous. It's not so difficult to think of God in doing everything. Such thinking only brings us more inner strength. The thought of

God every time, would protect you from doing anything immoral or unethical. We would gladly die when the time comes, without the fear of the unknown after death. For God would fill our thoughts which would drive away the fear. Surely, I have felt the difference at least, which I submit with humility. In fact, I have been a materialist most of my life, and still am. In my realization, even the subtle change is due to the benevolent grace of our Acharya, upbringing by parents and elders in the family and influence of wise scholars well versed in the authentic scriptures that things changed for the better. Surely everything happens for a cause, though we might not be able to perceive the truth directly.

I cannot claim that I know the Vedas, Upanishads, Bhagawad Gita and all the sacred knowledge revealed by the great Azhwars and Acharyas, for it is humanly impossible to grasp the enormity of such devotional nuances, as it is. However, even learning a little bit of these scriptures from an Acharya or Guru is good enough to realize our imperishable relationship with the Lord. Trying to be an honest devotee of God is the easiest thing to do for it requires no special effort or skill. It is in our nature to be devoted to God (as affirmed by Sri Azhwan) because He is our Father. When you recognize this truth, you don't need to know anything else.

God bestows His grace by the kind benevolence of the Divine Mother who resides eternally in the heart of the Lord and the Acharya who is God sent to save us from eternal gloom. It is indeed by the influence of the Divine Mother that the Lord becomes lenient in accepting a devotee, overlooking his deficiencies if any.

We are reminded of a beautiful expression by Sri Puandaradasa who says,

'malagiddu eddAga, my murithu nintAga KrishnA enabAradE...'

Means, as you wake up from sleep, as you stretch your limbs, please remember to say Krishna; chant His name.

Puandaradasa contends that human life is rare and precious. He inspires people to chant the name of Krishna at all times - whether in joy or in trouble, or going through daily routine such as bathing or eating or playing. We must use the gift of speech to utter the divine name when alive and are capable of thinking about the Lord.

Sri Kulasekara Azhwar, in his devotional composition 'Mukunda Mala' v.20, affectionately asks all humans to become worthy of the precious human life obtained by the grace of God, thus:

We possess senses and body only to serve Lord Krishna. The tongue is meant to sing His glories, the mind to worship, the hands to serve, the ears to hear His pastime stories, and the eyes to behold His divine beauty.

Because God yearns for the love of His devotees and looks for little opportunities to bestow His divine grace upon them.

God is the only one we never have to fear, for He is the only one who does not hurt us - ever. Guaranteed. What we have to fear is evil actions by other humans and our own misdeeds, for it is our own bad karma that punishes us.

Even doubters should give it a chance by accepting that they belong to the same God. They lose nothing because there is no harm when they accept God in their life. Nothing really changes for them except the acceptance in their mind. And what if God truly exists? They gain because they accepted God. If they were determined not to accept they would have missed out realizing the truth and a precious human life with it. God knows when we will get a human birth again.

I find that devotion to God is the easiest thing to do. It's not complicated or hard to practice. Just remind the Lord in you - that you know He is there; and - you can feel His empowerment in everything you do. And you want nothing from Him except His love. Try this for a few days and you can feel the truth for yourself and your exclusive affection for God. God resonates that joy in you. And you are the happiest in the world.

May the kind Lord bestow His grace upon everyone.

Sarvam Sri Krishnarpanamasthu.

Conclusion

All scientific theories and equations concerning the formation of the universe are likely incomplete, as the cosmos is immense - perhaps infinite - and our understanding is limited by localized, observation-based physics. History shows that scientific revelations are often refined or replaced, proving that earlier, provisional knowledge was an approximation.

Our knowledge, either about how and when the universe formed or the vastness and extent of the universe is questionable when we compare it with the description of the universe formation in the sacred scriptures, such as the Vedas and Bhagawad Gita, which provide better details about the origin of the universe which seems to match the reality of all existence. More importantly, the knowledge contained in the scriptures is said to be from the Creator of the universe Himself. Who better than the Creator could know His creation?

One flaw in our understanding is that the universe and intelligent life forms were said to have been formed by a self-existing singularity of material energy devoid of intelligence. The greater anomaly is that the non-sentient singularity (big bang) is credited for formulating the universal laws of physics, keeping the universe functional for billions of years. It is obvious that there is something fundamentally wrong in our understanding of the universe and its origin.

Did you know that all of us are immortals trapped in mortal bodies? When we were offered a choice to either remain in immortality with God or choose independent existence away from God, we opted for the latter. We are experiencing mortality in countless life cycles eternally due to a choice we made in desiring to experience life in pursuit of greater joys and happiness.

We are constantly in search of happiness because the nature of the Supreme Lord, who is our origin, is infinite bliss, and it's natural that we seek that joy.

Now that we couldn't find that pure joy of life without the suffering in this world, we can attain immortal existence in the divine world where we will find the true bliss we all seek, when we seek God's grace with unconditional devotion. And God would do the rest.

Being the children of God, we deserve to live like the children of God.

www.ingramcontent.com/pod-product-compliance
Lightning Source LLC
LaVergne TN
LVHW101615210926
848480LV00010B/1098

Zugangscode

Nutzungsdauer 24 Monate

UF35FXN2

Zugangscode einlösbar bis 30.09.2026

https://www.pearson.de/mylab/experimentalphysik1

Experimentalphysik 1

Mechanik und Wärme

Martin Erdmann
Günter Flügge

Pearson

Bibliografische Information der Deutschen Nationalbibliothek

Die Deutsche Nationalbibliothek verzeichnet diese Publikation in der Deutschen Nationalbibliografie; detaillierte bibliografische Daten sind im Internet über *http://dnb.dnb.de* abrufbar.

Der Umwelt zuliebe verzichten wir auf Einschweißfolie.

10 9 8 7 6 5 4 3 2 1

25 24 23 22

ISBN 978-3-86894-416-7 (Buch)
ISBN 978-3-86326-311-9 (E-Book)

Zugangscode einlösbar bis 30.09.2026

www.pearson.de
A part of Pearson plc worldwide

Programmleitung: Birger Peil, bpeil@pearson.de
Korrektorat: Micaela Krieger-Hauwede, micaela.krieger@online.de
Coverabbildung: © wacomka, Shutterstock
Satz: Micaela Krieger-Hauwede, Leipzig
Druck und Verarbeitung: Drukkerij Wilco, Amersfoort

Printed in the Netherlands

Inhaltsverzeichnis

Teil I Kraft – Energie – Bewegung **13**

1 Kinematik **15**
1.1 Ort, Zeit, Geschwindigkeit, Beschleunigung 17
1.2 Bewegung in zwei Dimensionen 31

2 Grundgrößen, Messwert & Messgenauigkeit **39**
2.1 Grundgrößen: SI-Einheiten 41
2.2 Genauigkeit von Messwerten 44
2.3 Verteilung von Messwerten 46
2.4 Interpretation von Messwerten 50
2.5 Auswertung eines Experiments 53

3 Dynamik: Newtons Bewegungsgesetze **65**
3.1 Newtons Axiome, Kraft 68
3.2 Masse 70
3.3 Impuls 78
3.4 Newtons Bewegungsgesetz 79

4 Drehbewegungen **91**
4.1 Polarkoordinaten 93
4.2 Gleichförmige Kreisbewegung 95
4.3 Harmonischer Oszillator 98
4.4 Vektor der Winkelgeschwindigkeit 103
4.5 Drehimpuls 104
4.6 Drehmoment 105
4.7 Bewegungsgesetz 107

5 Arbeit, Energie, Leistung, Intensität **119**
5.1 Energieformen 121
5.2 Arbeit 121
5.3 Potentielle Energie 124
5.4 Kinetische Energie der Translation 124
5.5 Kinetische Energie der Rotation 125
5.6 Energieerhaltung 126
5.7 Leistung 131
5.8 Intensität 131

6 Gemeinsam verstehen **135**

Teil II Kollision – Gravitation – Bezugssysteme 139

7 Erhaltungssätze 141
7.1 Energieerhaltung 143
7.2 Impulserhaltung 143
7.3 Drehimpulserhaltung 144

8 Stoßprozesse 147
8.1 Elastische Kollision mit gleichen Massen 149
8.2 Elastische Kollision mit verschiedenen Massen 151
8.3 Inelastische Stöße 154

9 Koordinatensysteme 159
9.1 Kartesische Koordinaten 161
9.2 Zylinderkoordinaten 163
9.3 Kugelkoordinaten 167

10 Gravitationswechselwirkung nach Newton 173
10.1 Gravitationskraft 175
10.2 Gravitationspotential 185
10.3 Energie eines Himmelskörpers 189
10.4 Bahnkurven von Himmelskörpern 192
10.5 Historischer Bezug 203

11 Transformation zwischen Bezugssystemen 209
11.1 Inertialsystem 211
11.2 Galileitransformation 211
11.3 Beschleunigte Bezugssysteme 213
11.4 Lorentztransformation 220

12 Spezielle Relativitätstheorie 227
12.1 Lichtgeschwindigkeit 229
12.2 Zeitdilatation 230
12.3 Längenkontraktion 233
12.4 Raumzeit-Diagramme 233
12.5 Energie-Impuls-Raum 237
12.6 Anwendung: Elementarteilchenphysik 240

13 Gemeinsam verstehen 251

Teil III Schwingungen – Wellen – Körperdrehung 255

14 Dynamik starrer Körper 257
14.1 Dichte 259
14.2 Schwerpunkt 260
14.3 Trägheitsmoment 261

14.4 Drehschwingungen . . . 264
14.5 Satz von Steiner . . . 268
14.6 Rotationsenergie . . . 269
14.7 Trägheitstensor . . . 273
14.8 Kreisel . . . 278

15 Schwingungen **291**
15.1 Harmonischer Oszillator . . . 294
15.2 Gedämpfte Schwingungen . . . 298
15.3 Erzwungene Schwingungen . . . 303
15.4 Gekoppelte Schwingungen . . . 307
15.5 Fouriersynthese, -analyse, -transformation . . . 316

16 Wellen **329**
16.1 Wellenausbreitung . . . 331
16.2 Stehende Wellen . . . 340
16.3 Dispersion . . . 345
16.4 Interferenz . . . 349
16.5 Dopplereffekt . . . 353
16.6 Machscher Kegel . . . 357
16.7 Relativistischer Dopplereffekt . . . 358
16.8 Wellengleichung . . . 360
16.9 Transversale Wellen . . . 361

17 Gemeinsam verstehen **375**

Teil IV Hydromechanik – Wärme **377**

18 Hydromechanik **379**
18.1 Druck . . . 381
18.2 Auftrieb . . . 382
18.3 Luftdruck . . . 384
18.4 Kontinuitätsgleichung . . . 390
18.5 Bernoulli-Gleichung . . . 393
18.6 Innere Reibung in Flüssigkeiten . . . 397
18.7 Druckwellen . . . 404

19 Thermische Ausdehnung **417**
19.1 Längenausdehnung fester Körper . . . 419
19.2 Volumenausdehnung fester und flüssiger Stoffe . . . 422
19.3 Thermische Eigenschaften von Gasen . . . 423
19.4 Ideales Gasgesetz . . . 427

20 Kinetische Gastheorie **435**
20.1 Energie der Gasatome . . . 437
20.2 Absolute Temperatur und Energie . . . 441

20.3 Celsius- und Kelvinskala 442
20.4 Geschwindigkeitsverteilung 443

21 Wärme **447**
21.1 Wärmeenergie 449
21.2 Spezifische Molwärme 451
21.3 Wärmetransport 454

22 Thermodynamik **463**
22.1 Thermodynamische Zustände 465
22.2 Erster Hauptsatz der Thermodynamik 466
22.3 Spezielle Zustandsänderungen 467
22.4 Kreisprozesse 471
22.5 Entropie 479
22.6 Zweiter Hauptsatz der Thermodynamik 486
22.7 Dritter Hauptsatz der Thermodynamik 487

23 Thermodynamik realer Gase und Flüssigkeiten **495**
23.1 Van-der-Waals-Gleichung 497
23.2 Verdampfungswärme 498
23.3 Schmelzwärme 503

24 Gemeinsam verstehen **509**

Index **513**

Literaturverzeichnis **523**

Physik Denken

Im Physikstudium eignen Sie sich Fähigkeiten und Methoden an, um Naturvorgänge beschreiben zu können. Professionelles Arbeiten in der Physik erfordert fundierte Kenntnisse über experimentelle Methoden und theoretische Konzepte, die auf mathematischer Basis entwickelt werden. Mit unserem Lehrkonzept **Physik Denken** realisieren wir ein integratives Konzept, das Sie an das moderne Physikverständnis heranführt.

Die dreibändige Lehrbuchreihe zur Experimentalphysik orientiert sich an den Anforderungen des Bachelorstudiums Physik. Unterteilt in **Mechanik und Wärme**, **Felder und Wellen**, **Atome und Kerne**, vermittelt Ihnen diese Reihe die Themenbereiche die in Lehrveranstaltungen präsentiert und in Prüfungen abgefragt werden.

Jedes dieser Lehrbücher unterstützt Sie im Studium mit einem breit angelegten Instrumentarium:

- Passende Lernstoffportionen helfen, Ihren Lernfortschritt zu strukturieren.
- Das Lehrbuch gibt Ihnen ausführliche Berechnungen an die Hand, wie sie zum Verständnis der physikalischen Zusammenhänge erforderlich sind und in Übungsaufgaben und Klausuren gefordert werden.
- Zahlreiche präzise Abbildungen unterstützen Ihren Verständnisprozess.
- Das Buch hält mehrere hilfreiche Kästen für Sie bereit: In den Kästen **Beispiele** werden Anwendungen physikalischer Konzepte vorgestellt und in den Kästen **Experimente** werden Demonstrationsversuche beschrieben und ausgewertet. Kurze **mathematische Einschübe** erklären mathematische Symbole und ihre Bedeutung.
- Mit der *phyphox*-App werden moderne Smartphones mit ihrer Sensorik zum Laborgerät. Mit zahlreichen kleinen Experimenten können Sie sich selbst von den physikalischen Konzepten der Mechanik und Akustik überzeugen.
- Die Abschnitte **Das Wichtigste zum Mitnehmen** geben in Kurzform die wesentlichen Konzepte jedes Kapitels für die Vorbereitung von Klausuren und mündlichen Prüfungen wieder.

Gespräche über themenübergreifende Fragestellungen sind erfahrungsgemäß für den Lernfortschritt besonders wichtig und sollten frühzeitig mit Studienpartnern begonnen und praktiziert werden. In mündlichen Prüfungen sind sie der Gradmesser für den Wissensstand über physikalische Methoden und Konzepte:

- Damit Sie Ihren Lernfortschritt überprüfen können, geben wir Ihnen in den Kapiteln **Gemeinsam verstehen** Aufgabenstellungen für Partnerarbeit. Im Anhang des Lehrbuchs bzw. im Online-Bereich des Pearson Verlags finden Sie Skizzen möglicher Gesprächsverläufe.
- Außerdem haben wir für Selbsttests passende **Aufgaben** zusammengestellt, deren Lösungen Sie sämtlich im Anhang des Lehrbuchs bzw. im Online-Bereich des Pearson Verlags finden.
- Darüber hinaus haben Sie Zugriff auf zahlreiche weitere Onlineaufgaben in **MyLab | Physik** des Pearson Verlags.

Der vorliegende Band **Mechanik und Wärme** vermittelt Ihnen die grundlegenden Phänomene in den Bereichen Bewegung, Kraft und Energie (Teil I), Kollision, Gravitation und Bezugssysteme (Teil II), Körperdrehung, Schwingungen und Wellen (Teil III) sowie Hydromechanik und Wärme (Teil IV).

Wir danken dem Pearson Verlag für die hervorragende Unterstützung und Gestaltung dieses Lehrbuchs, insbesondere dem Lektor Herrn Birger Peil und der LaTeX-Setzerin Frau Micaela Krieger-Hauwede. Für die Unterstützung bei den zahlreichen Abbildungen danken wir Frau Christine Falter. Unser Dank geht ebenso an das Team der *phyphox*-App, insbesondere an Prof. Dr. Christoph Stampfer und Dr. Sebastian Staacks, für viele wertvolle Hinweise zu passenden Experimenten und Anmerkungen zu den Beschreibungen.

Schließlich wenden wir uns an unsere Leserinnen und Leser mit der Bitte um Ihre Nachricht (erdmann@physik.rwth-aachen.de), sollten trotz der großen Sorgfalt bei der Erstellung des Lehrbuchs noch Unstimmigkeiten auffallen.

Aachen

Univ.-Prof. Dr. rer. nat. Martin Erdmann
Univ.-Prof. Dr. rer. nat. Günter Flügge

MyLab | Experimentalphysik 1 – Learn a little...do a little

Ebenfalls enthalten, wie im Klassiker Giancoli Physik, ist die Umsetzung des bei Studierenden bewährten didaktischen Konzepts der digitalen Arbeitsblätter *Learn a little...do a little*. Hier werden mit Ihnen im zum Buch gehörigen **MyLab | Experimentalphysik 1** interaktive Aufgaben in Schritt-für-Schritt-Lösungen mit Zwischenergebnissen und Herleitungen durchgearbeitet. Sie können diese Aufgaben solange üben und mit neuen Zahlenwerten wiederholen, bis es wirklich sitzt. Knapp **130 Aufgaben** unterschiedlicher Schwierigkeitsgrade helfen Ihnen dabei, sich am Verständnis der Physik zu erfreuen.

Zugangscode vorne im Buch

Ebenso in **MyLab | Experimentalphysik 1** enthalten ist ein Klausurtrainer, mit dem Sie sich Aufgaben als Übungsklausur generieren lassen können und der Ihnen als Auswertung Empfehlungen zur Wiederholung von Kapiteln des Lehrbuchs gibt.

Das *Gradebook* in **MyLab | Experimentalphysik 1** gibt Ihnen eine Übersicht der Aufgaben, die sie absolviert haben, sodass Sie stets das Gelernte im Überblick haben. Eine detaillierte Anleitung finden Sie auf der Startseite zum **MyLab | Experimentalphysik 1**.

MyLab | Experimentalphysik 1 – Learning to go

In **MyLab | Experimentalphysik 1** kann Mensch sich mit jedem Browser einloggen, der JavaScript aktiviert hat. Wir empfehlen ein Laptop oder Tablet. Zusätzlich können Sie das Lehrbuch als E-Book zum Lesen überall mitnehmen.

Nutzen Sie diese einzigartige Möglichkeit zur Wiederholung des Stoffs und zur (Klausur-) Vorbereitung. Lernen Sie damit im Team, an der Uni oder im Rahmen des Selbststudiums. Fordern Sie bei Fragen die Unterstützung der Tutoren ein, sie helfen Ihnen gern. Lassen Sie sich nicht durch kleine Rückschläge entmutigen, sondern bleiben Sie dran. Dafür haben wir das **MyLab | Experimentalphysik 1** entwickelt, das Ihnen in kleinen *Learn a little...do a little*-Schritten zur Seite steht.

Bei technischen Fragen zu **MyLab | Experimentalphysik 1** können Sie sich jederzeit an unseren Support wenden.

Wir wünschen Ihnen viel Erfolg.

Ihr Pearson-Team

Kraft – Energie – Bewegung

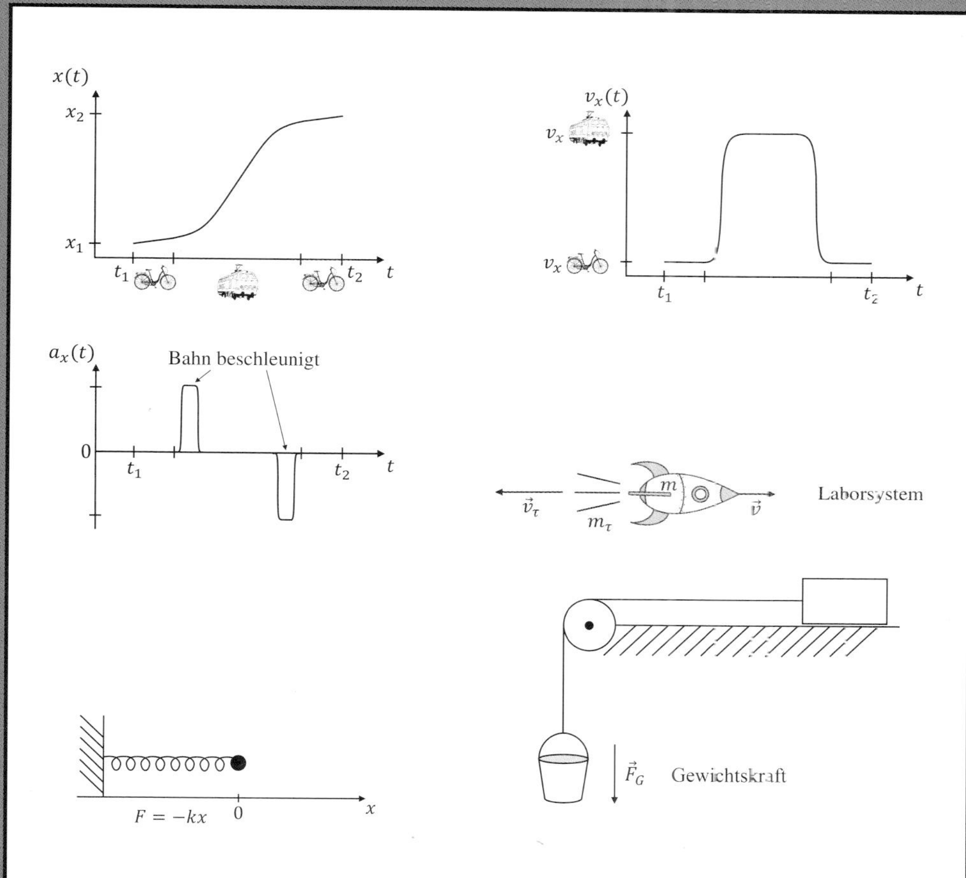

Teil I

Kinematik

ÜBERBLICK

1

1.1 Ort, Zeit, Geschwindigkeit, Beschleunigung **17**
1.1.1 Differentialrechnung in der Kinematik 18
1.1.2 Integralrechnung in der Kinematik 23

1.2 Bewegung in zwei Dimensionen **31**
1.2.1 Überlagerung von zwei Bewegungen 32
1.2.2 Bahnkurve beim Weitwurf 33

» Die Beschreibung von Bewegungen eines Körpers bezeichnen wir mit Kinematik. Die Begriffe **Ort**, **Zeit**, **Geschwindigkeit** und **Beschleunigung** haben dabei eine zentrale Bedeutung.

Den meisten Studierenden ist die Kinematik aus dem Physikunterricht bekannt. Wir wollen mit diesem Kapitel an die Schulkenntnisse anschließen und sicherstellen, dass alle die gleichen Startbedingungen im Physikstudium haben. «

1.1 Ort, Zeit, Geschwindigkeit, Beschleunigung

Um die physikalischen Konzepte zunächst einfach zu halten, betrachten wir Körper als punktförmige Objekte, die eine Masse tragen. In Kurzform bezeichnen wir sie als **Massenpunkte**.

Solange die Abstände oder die zurückgelegten Wege groß im Vergleich zu der Ausdehnung des Körpers sind, ist diese Vereinfachung eine gute Näherung. Zum Beispiel sind bei der Planetenbewegung die Abstände zwischen den Planeten groß im Vergleich zu der Größe der Planeten.

Bewegt sich ein Massenpunkt auf einer geraden Bahn im Raum, so genügt eine Dimension zur Beschreibung der Bewegung (▶ Abbildung 1.1).

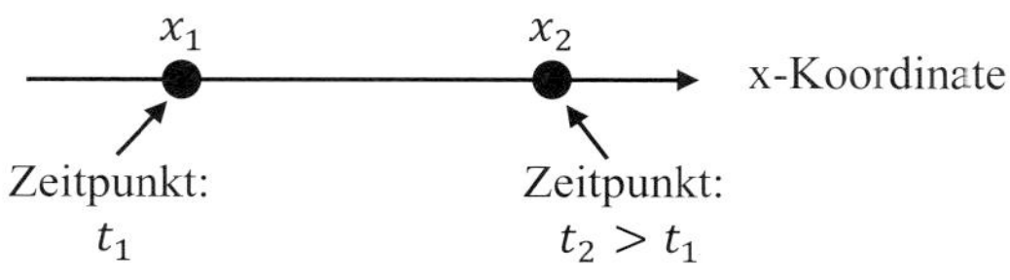

Abbildung 1.1 Eindimensionale Bewegung eines Massenpunkts entlang der x-Achse in Abhängigkeit von der Zeit t

Um die **Durchschnittsgeschwindigkeit** des Massenpunkts zu berechnen, benötigen wir die Wegdifferenz $\Delta x = x_2 - x_1$, die der Massenpunkt im Zeitintervall $\Delta t = t_2 - t_1$ zurückgelegt hat:

$$\overline{v}_x = \frac{\Delta x}{\Delta t} = \frac{x_2 - x_1}{t_2 - t_1} \tag{1.1}$$

Beispiel 1: Fahrt Köln Wohnung ↔ Aachen RWTH

Entfernung Köln–Aachen:	$\Delta x = 75\,\text{km}$
Gesamte Fahrzeit:	$\Delta t = 90\,\text{min}$

Die Durchschnittsgeschwindigkeit auf dieser Strecke beträgt also

$$\overline{v}_x = \frac{\Delta x}{\Delta t} = \frac{75\,\text{km}}{90\,\text{min}} = \frac{75\,\text{km}}{\frac{90}{60}\,\text{h}} = 50\,\frac{\text{km}}{\text{h}}\;.$$

Aus der Durchschnittsgeschwindigkeit kann man keine Informationen über den genauen Verlauf der Fahrt gewinnen. Eine vollständigere Beschreibung liefert die Momentangeschwindigkeit. Um die Momentangeschwindigkeit bestimmen zu können, müssen nicht nur Start- und Ankunftsort sowie Start- und Ankunftszeit bekannt sein, sondern der Ort zu jedem Zeitpunkt durch die sogenannte **Ortsfunktion** bzw. die **Bahnkurve**:

$$x(t) \tag{1.2}$$

↗ ↖

Ort in Abhängigkeit der Zeit

1.1.1 Differentialrechnung in der Kinematik

Die kinematischen Größen Ort, Geschwindigkeit und Beschleunigung sind über zeitliche Ableitungen miteinander verbunden. In diesem Abschnitt stellen wir ihre Zusammenhänge vor.

Mathematischer Einschub: Ableitung

Aus der Schule ist bereits bekannt, dass die Ableitung einer Funktion $f(x)$ nach x ihre Steigung $f'(x)$ an der Stelle x ist (linke ▶ Abbildung 1.2). Analog wird die Funktion $x(t)$ nach der Zeit t abgeleitet, diese Ableitung wird auch mit $\dot{x}$ bezeichnet (rechte ▶ Abbildung 1.2).

Abbildung 1.2 Links: Ableitung $f'(x)$ der Funktion $f(x)$ als Steigung an der Stelle x, rechts: Geschwindigkeit $v_x(t) = \dot{x}(t)$ als Steigung der Kurve $x(t)$ zum Zeitpunkt t

Die **Momentangeschwindigkeit** ergibt sich aus der Durchschnittsgeschwindigkeit für sehr kurze Zeitintervalle $\Delta t \to 0$

$$v_x(t) = \lim_{\Delta t \to 0} \overline{v}_x \tag{1.3}$$

und wird damit die Ableitung von $x(t)$ nach der Zeit (Gleichung (1.2), linke ▶ Abbildung 1.3). Verschiedene (zunehmend kurze) Schreibweisen dafür sind

$$\begin{aligned} v_x(t) &= \lim_{\Delta t \to 0} \frac{\Delta x}{\Delta t} \\ &= \frac{d}{dt} x(t) \\ &= \frac{dx}{dt} \\ &= \dot{x}\,. \end{aligned} \tag{1.4}$$

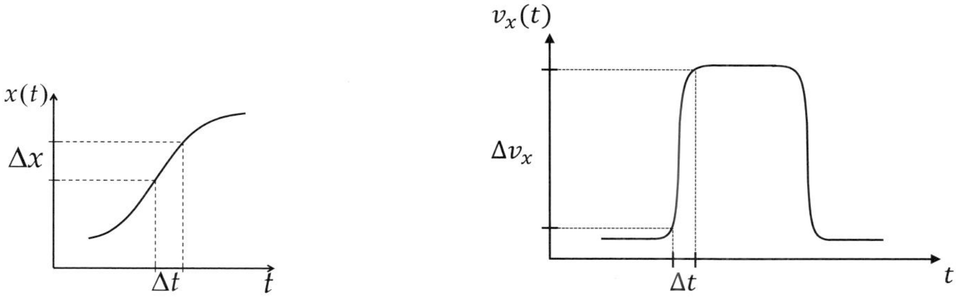

Abbildung 1.3 Links: Bestimmung der Geschwindigkeit $v_x(t)$ aus $x(t)$, rechts: Berechnung der Beschleunigung $a_x(t)$ aus $v_x(t)$

Mit der Geschwindigkeit $\dot{x}(t) = v_x(t)$ können wir nun auch die **Beschleunigung** einführen. Die Beschleunigung ist die Änderung der Geschwindigkeit in einem Zeitintervall. Die **Durchschnittsbeschleunigung** ist

$$\overline{a}_x = \frac{v_{x2} - v_{x1}}{t_2 - t_1} \,. \tag{1.5}$$

Analog zur Momentangeschwindigkeit ist die **Momentanbeschleunigung** definiert als (rechte ► Abbildung 1.3)

$$\begin{aligned} a_x(t) &= \lim_{\Delta t \to 0} \frac{\Delta v_x}{\Delta t} \\ &= \frac{d}{dt} v_x(t) \\ &= \frac{dv_x}{dt} \\ &= \dot{v}_x \,. \end{aligned} \tag{1.6}$$

Beispiel 2: Weg zur Vorlesung: Köln Wohnung $\leftrightarrow$ Aachen RWTH

Der nette Bahnbeamte hilft beim dynamischen Umsteigen zwischen Fahrrad und Zug, sodass beim Fahrzeugwechsel die Bewegungen nahtlos ineinander übergehen. In diesem Beispiel ergeben sich dann die folgenden in ► Abbildung 1.4 gezeigten Zeit-Diagramme: Oben für den Ort, unten links für die Geschwindigkeit und unten rechts für die Beschleunigung. In diesem Beispiel sehen wir, dass die Durchschnittsbeschleunigung $\overline{a}_x = 0$ ist, die Momentanbeschleunigung jedoch im Allgemeinen $a_x(t) \neq 0$ ist.

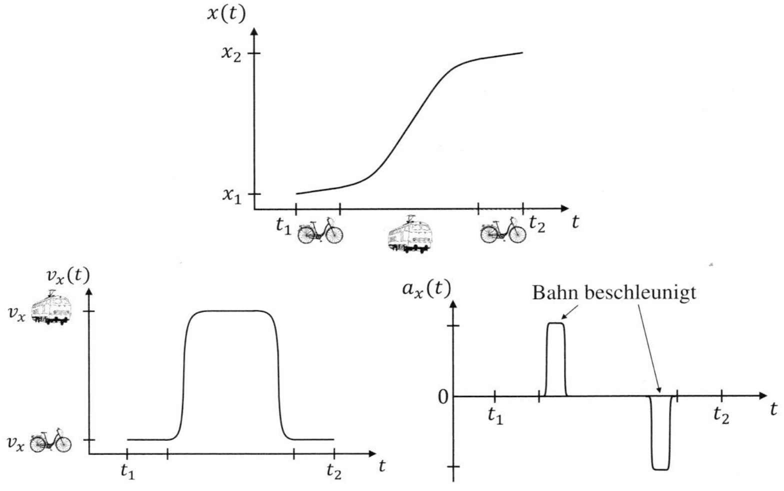

Abbildung 1.4 Oben: Beispiel für ein Ort-Zeit-Diagramm $x(t)$ mit langsamer-schneller-langsamer Bewegung, unten: zugehöriges Geschwindigkeits-Zeit-Diagramm $v_x(t)$ (links) und Beschleunigungs-Zeit-Diagramm (rechts)

Aufgabe 1: Ort, Geschwindigkeit, Beschleunigung

Gegeben: $x(t) = a \cdot t^2 + b$ mit $a = 1\,\mathrm{m/s^2}$ und $b = 1\,\mathrm{m}$

Gesucht: Skizzieren Sie Ort, Geschwindigkeit und Beschleunigung als Funkion der Zeit.

Um Schreibarbeit zu sparen, wird ab jetzt die folgende Konvention verwendet:

$$\text{»Geschwindigkeit«} = \text{Momentangeschwindigkeit}$$
$$\text{»Beschleunigung«} = \text{Momentanbeschleunigung}$$

Aus den Gleichungen (1.4) und (1.6) leiten wir folgenden Zusammenhang zwischen dem Ort $x(t)$ und der Beschleunigung $a_x(t)$ her:

$$a_x(t) = \frac{d}{dt} v_x(t) \tag{1.7}$$
$$= \frac{d}{dt}\left(\frac{d}{dt} x(t)\right) \tag{1.8}$$
$$= \frac{d^2}{dt^2} x(t) \tag{1.9}$$
$$= \frac{d^2 x}{dt^2} \tag{1.10}$$
$$= \ddot{x} \tag{1.11}$$

Die Beschleunigung ist also die 2. Ableitung der Ortsfunktion nach der Zeit.

Experiment 1: *phyphox*-App für Smartphones

Smartphones haben heutzutage eine Reihe von Sensoren, die sich hervorragend für die Durchführung von physikalischen Experimenten eignen:

- Beschleunigungssensor
- Magnetometer
- Gyroskop
- Lichtsensor
- Drucksensor
- Näherungssensor
- Mikrofon
- Lautsprecher
- GPS/Standort

Wie man die Messungen der Smartphone-Sensoren auslesen kann, haben die Entwickler der App »physical phone experiment« – kurz *phyphox* – gelöst (▶ Abbildung 1.5, https://phyphox.org). Wenn Sie ein Smartphone besitzen oder leihen können, können Sie die App aus den üblichen Stores herunterladen und installieren lassen. Beachten Sie, dass nicht jedes Smartphone alle Sensortypen besitzt. Im Menü »i« (oben) können die Sensoren unter »Geräte-Info« abgefragt werden. Falls der Sensor fehlt, kann das Experiment in der App nicht aufgerufen werden. Eine Liste der verfügbaren Sensoren für verschiedene Smartphones ist unter https://phyphox.org/news/sensor-database zu finden.

Abbildung 1.5 Ansicht der *phyphox*-App auf dem Smartphone (farbverändert)

Experiment 2: *phyphox* Aufzug

Zur Durchführung dieses Experiments ist ein Smartphone mit Drucksensor und Beschleunigungssensor erforderlich.

1. Suchen Sie in der *phyphox*-App das Experiment »Aufzug«. Berühren des Drei-Punkte-Symbols rechts oben zeigt ein Menü mit Informationen zum Experiment (»Experiment Info«).

2. Legen Sie das Smartphone auf den Boden des Aufzugs und starten Sie die Messung mit dem »Play button« (Dreieck rechts oben). Starten Sie die Fahrt des Aufzugs (z. B. nach oben).

3. Machen Sie sich die folgenden Zusammenhänge zwischen den drei Größen Höhe h, Geschwindigkeit v und Beschleunigung a des Aufzugs als Funktion der Zeit t noch einmal klar (► Abbildung 1.6):

   - Die Höhe h über der Starthöhe $h = 0$ ändert sich, solange die Geschwindigkeit v des Aufzugs von null verschieden ist. Vergleichen Sie auch dieses Ergebnis mit der mathematischen Beschreibung der Geschwindigkeit $v = dh/dt$ (1.4) und der linken ► Abbildung 1.3.

   - Eine gut erkennbare Beschleunigung a ergibt sich nur dann, wenn sich die Geschwindigkeit v des Aufzugs ändert. Die anfängliche Beschleunigung der Aufzugfahrt wird positiv notiert, das Abbremsen am Ende entspricht einer negativen Beschleunigung. Vergleichen Sie dieses Ergebnis auch mit der mathematischen Beschreibung der Beschleunigung $a = dv/dt$ (1.11) und der rechten ► Abbildung 1.3.

Vergleichen Sie Ihre Beobachtungen in ► Abbildung 1.6 mit dem Beispiel 2 »Weg zur Vorlesung«.

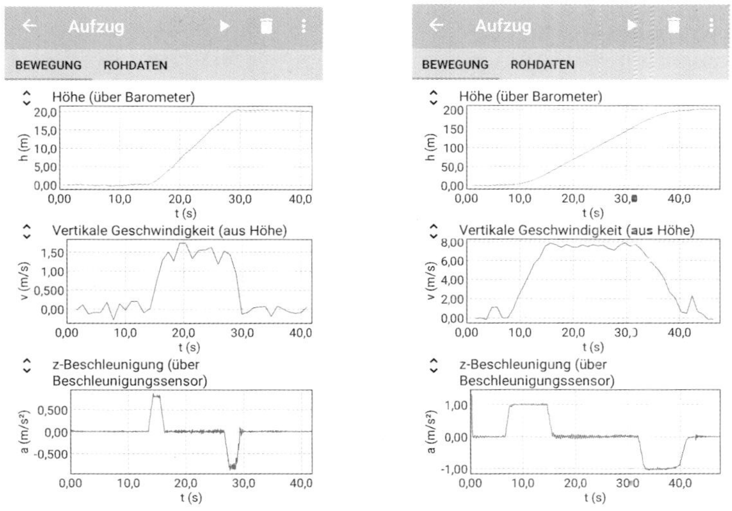

Abbildung 1.6 Experimente zum Aufzug mit Höhe h, Geschwindigkeit v und Beschleunigung a als Funktion der Zeit mit der *phyphox*-App: Die Aufzüge im Physikzentrum Aachen (links) und im Donau City (DC) Tower Wien (rechts) beschleunigen mit $a \sim 1\,\mathrm{m/s^2}$. Der DC-Aufzug erreicht höhere Geschwindigkeit durch länger andauernde Beschleunigung.

1.1.2 Integralrechnung in der Kinematik

Bisher haben wir Information über den Ort vorgegeben und daraus die Bewegung eines Massenpunkts beschrieben. Oft ist die Fragestellung aber umgekehrt. Beispiele dafür sind:

- Gesucht ist der Weg für eine gegebene Geschwindigkeitsverteilung $v_x(t)$.
- Gesucht ist der Weg für eine gegebene Beschleunigungsverteilung $a_x(t)$.
- Gesucht ist der Bremsweg eines Autos.
- Gesucht ist die Endgeschwindigkeit beim Sturz.

Ein allgemeiner Weg zur Beantwortung dieser Fragen ist die Integration. Dabei beachten wir den Zusammenhang zwischen Integral- und Differentialrechnung:

Integralrechnung $\leftrightarrow$ Umkehrung der Differentialrechnung

Gleichförmige Bewegung

Ein Spezialfall einer Geschwindigkeitsverteilung ist die **gleichförmige Bewegung**, d. h. der Massenpunkt bewegt sich mit konstanter Geschwindigkeit (► Abbildung 1.7).

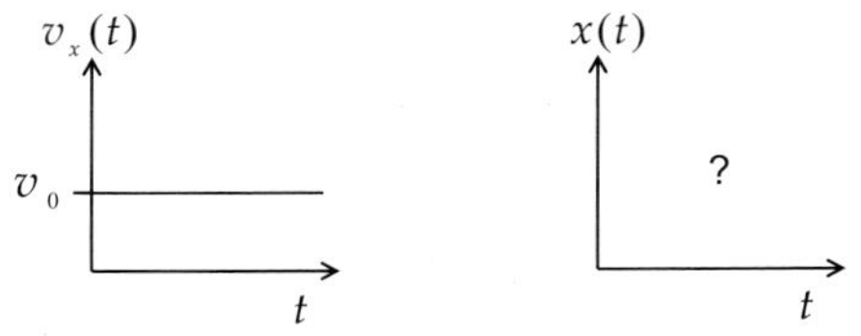

Abbildung 1.7 Welche Bahnkurve $x(t)$ gehört zu einer gleichförmigen Bewegung $v_x(t) = v_\circ$?

In diesem Fall ist die Momentangeschwindigkeit gleich der Durchschnittsgeschwindigkeit und es gilt

$$v_x(t) = \overline{v}_x = \frac{\Delta x}{\Delta t} = \frac{x_2 - x_1}{t_2 - t_1} = v_\circ \,. \tag{1.12}$$

Der im Zeitintervall $\Delta t = t_2 - t_1$ zurückgelegte Weg s ist damit

$$s = x_2 - x_1 = \overline{v}_x \cdot (t_2 - t_1) = v_\circ \cdot (t_2 - t_1)\,. \tag{1.13}$$

Allgemeine Bewegung

Im allgemeinen Fall einer nicht gleichförmigen Bewegung (► Abbildung 1.8) müssen wir ein Integral lösen, um von der Geschwindigkeit v auf die zurückgelegte Strecke $s \equiv x_2 - x_1$ schließen zu können.

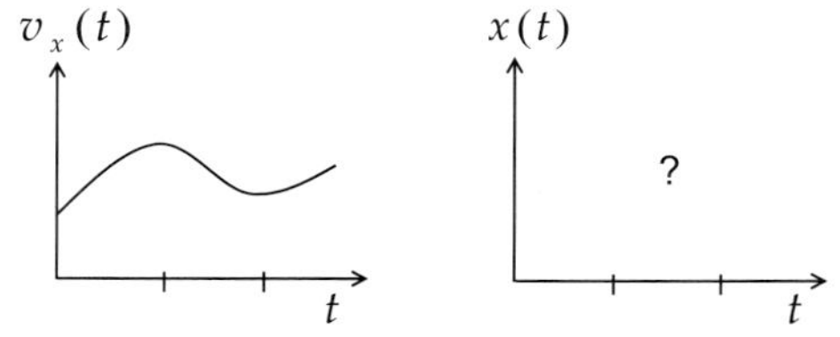

Abbildung 1.8 Gesucht ist die Bahnkurve $x(t)$ einer beliebigen Geschwindigkeitsverteilung $v_x(t)$.

Ist z. B. die Geschwindigkeitsverteilung $v(t)$ eines Massenpunkts mit

$$v_x(t) = \frac{dx}{dt} \tag{1.14}$$

gegeben, so kann man durch Integration den Weg s berechnen:

$$dx = v_x(t)\; dt \tag{1.15}$$

$$\int_{x_1(t_1)}^{x_2(t_2)} dx = \int_{t_1}^{t_2} v_x(t)\; dt \tag{1.16}$$

Die Integration der linken Seite ergibt

$$\int_{x_1(t_1)}^{x_2(t_2)} dx = x|_{x_1(t_1)}^{x_2(t_2)} \tag{1.17}$$
$$= x_2 - x_1 \equiv s\,.$$

Zur Bezeichnung der Integrationsgrenzen verwenden wir typischerweise die folgende Notation:

- Anfangszeit $t_1 = 0$
- Anfangsort $x_1(t_1) = x_\circ$
- obere Grenze variabel: $x_2 = x$, $t_2 = t$

Die Verwendung desselben Buchstabens t sowohl in der oberen Grenze des Integrals als auch in der Integrationsvariablen dt ist salopp, aber in der Physik üblich.

Mit dieser Notation ist die allgemeine Lösung des Integrals (1.16) die Bahnkurve

$$x(t) = \int_0^t v_x(t)\,dt + x_\circ\,. \tag{1.18}$$

Den Spezialfall einer gleichförmigen Bewegung $v_x(t) = v_\circ$ (1.12) können wir auch über diese allgemeine Form (1.18) lösen, weil wir die konstante Geschwindigkeit vor das Integral ziehen können:

$$x(t) = \int_0^t v_x(t)\,dt + x_\circ \tag{1.19}$$
$$= v_\circ \cdot \int_0^t dt + x_\circ \tag{1.20}$$
$$= v_\circ \cdot t + x_\circ \tag{1.21}$$

Beschleunigte Bewegung

Für eine gegebene Beschleunigung $a_x(t)$ eines Massenpunkts als Funktion der Zeit t bestimmen wir zunächst die Geschwindigkeit $v\,(t)$:

$$a_x(t) = \frac{dv_x}{dt} \tag{1.22}$$
$$dv_x = a_x(t)\,dt \tag{1.23}$$
$$\int_{v_\circ = v_x(0)}^{v_x(t)} dv_x = \int_0^t a_x(t)\,dt \tag{1.24}$$
$$v_x(t) - v_\circ = \int_0^t a_x(t)\,dt \tag{1.25}$$

Im Allgemeinen lautet die Lösung für die Geschwindigkeit $v_x(t)$ des Massenpunkts bei seiner bekannten Beschleunigung $a_x(t)$

$$v_x(t) = \int_0^t a_x(t)\, dt + v_\circ \,. \tag{1.26}$$

Um auch die Ortsfunktion des Massenpunkts zu erhalten, setzen wir $v_x(t)$ in Gleichung (1.18) ein. Beachten Sie, dass wir zweifach integrieren müssen und dementsprechend zwei Anfangsbedingungen – hier $x_\circ$ und $v_\circ$ – erhalten.

Gleichmäßig beschleunigte Bewegung

Ein Spezialfall ist die **gleichmäßig beschleunigte Bewegung** eines Massenpunkts, bei der die Beschleunigung $a(t) = a_\circ$ konstant ist (► Abbildung 1.9).

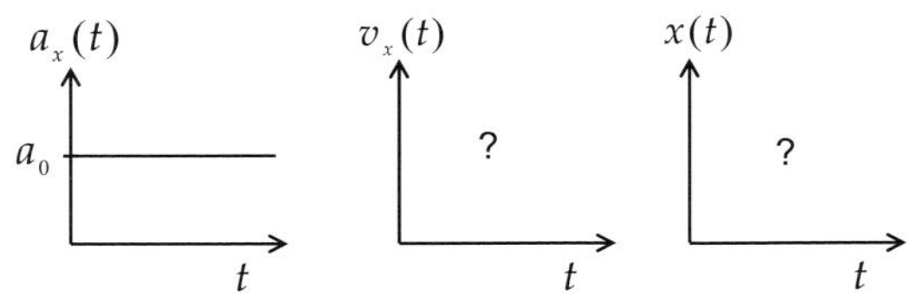

Abbildung 1.9 Wie verhalten sich $x(t)$ und $v_x(t)$ bei einer gleichmäßig beschleunigten Bewegung $a_x(t) = a_\circ$?

Wir starten von der allgemeinen Lösungsgleichung (1.26). In diesem Fall können wir $a_\circ$ vor das Integral ziehen:

$$v_x(t) = a_\circ \cdot \int_0^t dt + v_\circ \tag{1.27}$$

Damit erhalten wir bei gleichmäßiger Beschleunigung

$$v_x(t) = a_\circ \cdot t + v_\circ \,. \tag{1.28}$$

Analog zur gleichförmigen Bewegung (1.19) bis (1.21) bestimmen wir die Ortsfunktion:

$$x(t) = \int_0^t v_x(t)\, dt + x_\circ \tag{1.29}$$

$$= \int_0^t (a_\circ \cdot t + v_\circ)\, dt + x_\circ \tag{1.30}$$

$$= a_\circ \cdot \int_0^t t\, dt + v_\circ \cdot \int_0^t dt + x_\circ \tag{1.31}$$

$$= a_\circ \cdot \frac{1}{2}\, t^2 + v_\circ \cdot t + x_\circ \tag{1.32}$$

Somit ist der Ort $x(t)$ bei gleichmäßiger Beschleunigung $a(t) = a_\circ$ gegeben durch

$$x(t) = \frac{1}{2}\, a_\circ \cdot t^2 + v_\circ \cdot t + x_\circ\,. \tag{1.33}$$

In den Diagrammen der ► Abbildung 1.10 sieht man die Zusammenhänge.

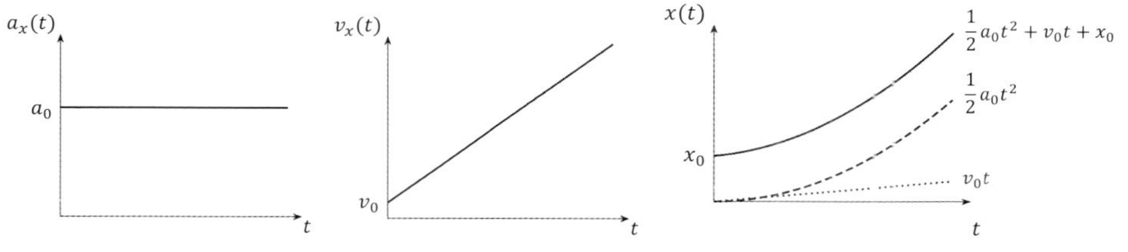

Abbildung 1.10 Links: $a(t) = \ddot{x}(t) = a_\circ = const.$, Mitte: $v(t) = \dot{x}(t) = a_\circ \cdot t + v_\circ$, rechts: $x(t) = \frac{1}{2}a_\circ \cdot t^2 + v_\circ \cdot t + x_\circ$

Im Zusammenhang mit der Beschleunigung wird uns häufig die **Erdbeschleunigung** g begegnen, also die Beschleunigung, die die Gravitationskraft der Erde an einem massebehafteten Körper bewirkt. Der Wert von g wird in Kapitel 10 genauer diskutiert. Er ist ortsabhängig und beträgt circa

$$g \approx 9{,}8\,\frac{\mathrm{m}}{\mathrm{s}^2}\,. \tag{1.34}$$

In Deutschland ist $g \approx 9{,}81\,\mathrm{m/s^2}$. Wir wollen den Wert für Abschätzungen meist einfach mit $g \approx 10\,\mathrm{m/s^2}$ annähern.

Aufgabe 2: Studienreise »Alte Brunnen«

Der Reiseführer hat ein 20 m langes Seil dabei (► Abbildung 1.11). Sie lassen einen Stein den Brunnen hinunterfallen. Den Aufschlag hören Sie nach 2 s. Reicht das Seil um hinunterzuklettern? Nutzen Sie als Erdbeschleunigung $g \approx 10\,\mathrm{m/s^2}$.

Abbildung 1.11 Studienreise »Alte Brunnen«

Gleichmäßig gebremste Bewegung

Ein Auto fährt mit $v_\circ = 50\,\mathrm{km/h}$ und soll zum Stillstand kommen. Die Verzögerung ist ebenfalls eine – in diesem Fall negative – Beschleunigung. Sie soll konstant sein $a = a_\circ = const.$. Für die aktuelle Geschwindigkeit v_x zur Zeit t gilt mit (1.28)

$$v_x\,(t) = a_\circ\, t + v_\circ\,. \tag{1.35}$$

Die Bedingung für den Stillstand ist $v_x\,(t_B) = 0$. Die rechte Seite der Gleichung ergibt damit die Bedingung für die Bremszeit t_B

$$a_\circ\, t_B = -v_\circ \tag{1.36}$$

$$t_B = -\frac{v_\circ}{a_\circ} > 0\,, \qquad \text{da } a_\circ < 0 \text{ ist.} \tag{1.37}$$

Durch Integration von (1.35) ergibt sich für die Strecke x wie zuvor (1.33)

$$x\,(t_B) = \frac{a_\circ}{2} t_B^2 + v_\circ\, t_B + x_\circ\,. \tag{1.38}$$

Mit der Anfangsbedingung $x_\circ \equiv 0$ und der Gleichung (1.37) folgt

$$x\,(t_B) = \frac{a_\circ}{2}\left(-\frac{v_\circ}{a_\circ}\right)^2 + v_\circ\left(-\frac{v_\circ}{a_\circ}\right) \tag{1.39}$$

$$= \frac{v_\circ^2}{2a_\circ} - \frac{v_\circ^2}{a_\circ} \tag{1.40}$$

$$= -\frac{v_\circ^2}{2a_\circ}\,. \tag{1.41}$$

Der Bremsweg, den man braucht, um bei einer Beschleunigung $a_\circ$ aus der Geschwindigkeit $v_\circ$ zum Stillstand zu kommen, ist

$$x_B = \frac{v_\circ^2}{2\,|a_\circ|}\,. \tag{1.42}$$

Zwischen Geschwindigkeit und Bremsweg besteht also eine quadratische Abhängigkeit (► Abbildung 1.12).

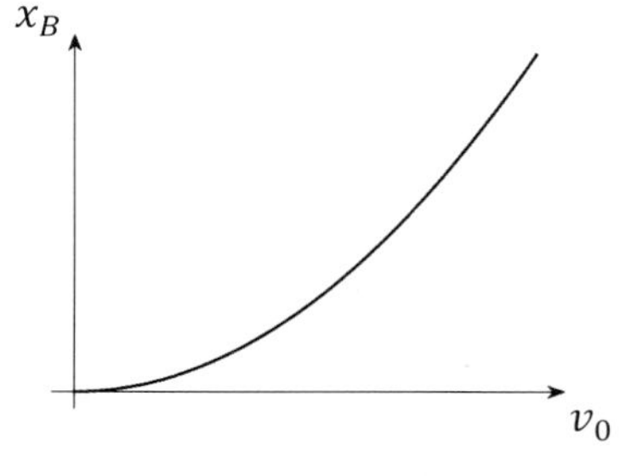

Abbildung 1.12 Der Bremsweg nimmt quadratisch mit der Geschwindigkeit zu.

Beispiel 3: Bremsweg eines Autos

Mit den typischen Werten $v_\circ = 50\,\mathrm{km/h} = 50\,(1000\,\mathrm{m}/3600\,\mathrm{s}) \approx 14\,\mathrm{m/s}$ und $a_\circ = -4\,\mathrm{m/s^2}$ ergibt sich für die Strecke x_B:

$$x_B = \frac{14^2\,\left(\frac{\mathrm{m}}{\mathrm{s}}\right)^2}{2 \cdot 4\,\mathrm{m/s^2}} = \frac{196}{8}\,\mathrm{m} \approx 25\,\mathrm{m}$$

Aufgabe 3: Endgeschwindigkeit beim Sprung

Ein Jugendlicher traut sich auf das 10 m-Brett und macht einen zaghaften Fußsprung (▶ Abbildung 1.13). Bestimmen Sie die Geschwindigkeit beim Eintauchen unter Verwendung der Erdbeschleunigung $g = 10\,\mathrm{m/s^2}$.

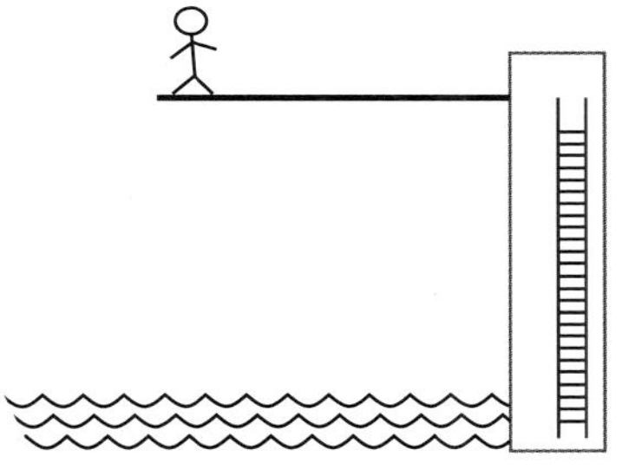

Abbildung 1.13 Turmspringer

Experiment 3: Fallschnüre

Mit den Fallschnüren kann ein zunächst qualitativer Nachweis der Ortsfunktion des freien Falls geführt werden. Wir verwenden gleichmäßige Abstände bzw. quadratische Abstände der Massestücke an einem Faden (▶ Abbildung 1.14). Die Zeiten bis zum Aufschlag sind mit $i = 1, 2, 3$

$$\Delta x_i = \frac{1}{2} g t_i^2 \qquad \Rightarrow \qquad t_i = \sqrt{\Delta x_i\; 2/g}\,. \tag{1.43}$$

Zwischen den drei Aufschlägen wird eine Zeitdifferenz von

$$t_2 - t_1 = \sqrt{2/g}\left(\sqrt{\Delta x_2} - \sqrt{\Delta x_1}\right) \tag{1.44}$$

$$t_3 - t_2 = \sqrt{2/g}\left(\sqrt{\Delta x_3} - \sqrt{\Delta x_2}\right) \tag{1.45}$$

wahrgenommen. Falls die Abstände gleich sind,

$$\Delta x_2 = 2\Delta x_1 \quad \Rightarrow \quad t_2 - t_1 = \sqrt{\frac{2}{g}\Delta x_1}\left(\sqrt{2}-\sqrt{1}\right) \approx \sqrt{\frac{2\Delta x_1}{g}}\,0{,}4 \tag{1.46}$$

$$\Delta x_3 = 3\Delta x_1 \quad \Rightarrow \quad t_3 - t_2 = \sqrt{\frac{2}{g}\Delta x_1}\left(\sqrt{3}-\sqrt{2}\right) \approx \sqrt{\frac{2\Delta x_1}{g}}\,0{,}3\,, \tag{1.47}$$

wird die Zeit zwischen den Aufschlägen immer kürzer. Wählt man die Faktoren für die Abstände aber quadratisch

$$\Delta x_2 = 4\Delta x_1 \quad \Rightarrow \quad t_2 - t_1 = \sqrt{\frac{2}{g}\Delta x_1}\left(\sqrt{4}-\sqrt{1}\right) = \sqrt{\frac{2\Delta x_1}{g}}\,1 \tag{1.48}$$

$$\Delta x_3 = 9\Delta x_1 \quad \Rightarrow \quad t_3 - t_2 = \sqrt{\frac{2}{g}\Delta x_1}\left(\sqrt{9}-\sqrt{4}\right) = \sqrt{\frac{2\Delta x_1}{g}}\,1\,, \tag{1.49}$$

so bleibt die Zeit zwischen den Aufschlägen konstant. Dies ist ein qualitativer Test für das Weggesetz des freien Falls $x(t) = \frac{1}{2}gt^2$.

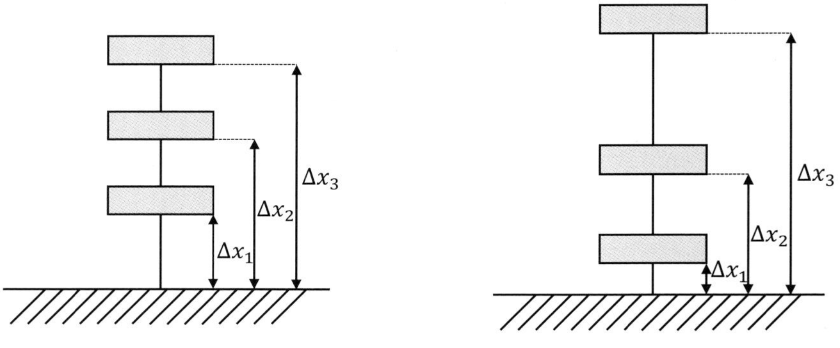

Abbildung 1.14 Fallschnur mit gleichmäßigen Abständen (links) und mit quadratisch zunehmenden Abständen (rechts)

Experiment 4: *phyphox* Fallschnüre

Befestigen Sie mindestens vier kleine Objekte (z. B. Schraubenmuttern oder kleine Wäscheklammern) mit gleichen Abständen an einem Faden. Wählen Sie die Abstände groß genug für die begrenzte Genauigkeit der Smartphone-Stoppuhr, z. B. 120 cm.

1 Suchen Sie in der *phyphox*-App die akustische Stoppuhr. Im Menü »Einfach« heben oder senken Sie die Empfindlichkeitsschwelle je nach Umgebungsakustik (»Schwelle«).

2 Legen Sie das Smartphone in sicherer Entfernung vom Auftreffpunkt der Objekte auf den Boden und starten Sie die Messung mit dem »Play button« (Dreieck oben rechts).

3 Halten Sie den Faden mit den Objekten hängend über eine Pappe oder Zeitung, sodass das unterste Objekt gerade die Pappe berührt. Lassen Sie den Faden fallen, sodass ein Objekt nach dem anderen aufschlägt. Sie hören die schneller werdende Abfolge der Klickgeräusche beim Auftreffen.

4 Im Menü »Parallel« sehen Sie die Zeiten der Klickgeräusche beim Auftreffen. Die erste Zeitdifferenz wird noch nicht gemessen, weil erst das Klickgeräusch des zweiten Objekts die Zeitmessung startet. Erst die Aufschläge des dritten und vierten Objekts werden als Zeitmessungen 1 und 2 registriert. Unter »Sequenz« sehen Sie die Differenzen der Auftreffzeiten, die immer schneller aufeinanderfolgen (linke ► Abbildung 1.15). Die Berechnung nach (1.46) und (1.47) stimmt mit 0,20 s und 0,16 s mit den gemessenen Zeitdifferenzen ungefähr überein und bestätigt das Weggesetz für den freien Fall $x(t) = \frac{1}{2}gt^2$.

Befestigen Sie nun die Objekte so an dem Faden, dass Sie den Abstand vom ersten Objekt – wie in ► Abbildung 1.14 und Gleichungen (1.48), (1.49) gezeigt – quadratisch vergrößern: z. B. $0, 40, 160, 360$ cm. Wiederholen Sie das Fallexperiment. Sie hören die gleichmäßigen Aufschläge und beobachten die gleichbleibenden Zeitabstände in der Messung (rechte ► Abbildung 1.15). Auch hier ergeben sich mit der Berechnung (0,29 s) über das Weggesetz des freien Falls ähnliche Zeitdifferenzen.

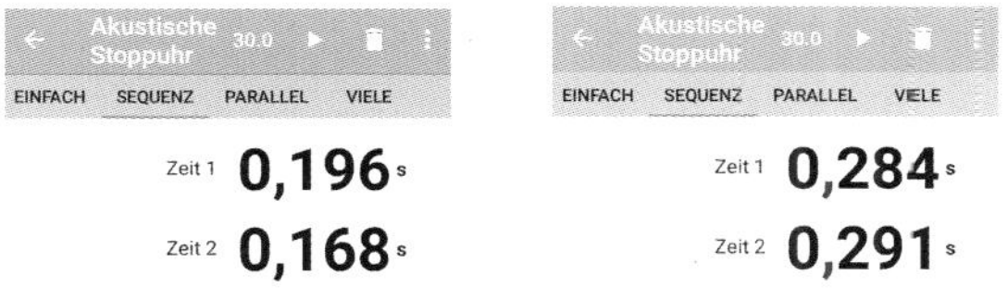

Abbildung 1.15 Links: Zeitdifferenzen zwischen den Aufschlägen bei Objektabständen von 120 cm, rechts: Zeitdifferenzen zwischen den Aufschlägen bei Objektabständen $0, 40, 160, 360$ cm vom unteren Objekt

1.2 Bewegung in zwei Dimensionen

Um eine Bewegung in zwei Dimensionen darzustellen, verwenden wir Vektoren. Es werden jeweils zwei räumliche Koordinaten zu einem Vektor zusammengefasst (► Abbildung 1.16). Die Zeit t bildet dabei einen kontinuierlich laufenden Parameter:

Ortsvektor $$\vec{r}(t) = \begin{pmatrix} x(t) \\ y(t) \end{pmatrix} \tag{1.50}$$

Geschwindigkeitsvektor $$\vec{v}(t) = \begin{pmatrix} v_x(t) \\ v_y(t) \end{pmatrix} = \begin{pmatrix} \dot{x}(t) \\ \dot{y}(t) \end{pmatrix} \tag{1.51}$$

Beschleunigungsvektor $$\vec{a}(t) = \begin{pmatrix} a_x(t) \\ a_y(t) \end{pmatrix} = \begin{pmatrix} \ddot{x}(t) \\ \ddot{y}(t) \end{pmatrix} \tag{1.52}$$

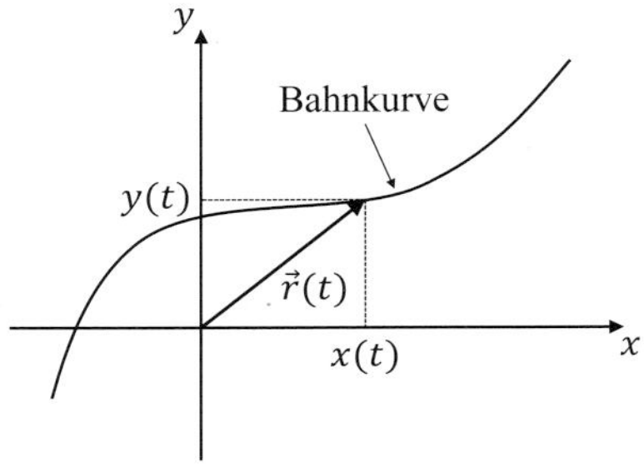

Abbildung 1.16 Bahnkurve in der x-y-Ebene als Funktion des Zeitparameters

1.2.1 Überlagerung von zwei Bewegungen

Wählt man in x-Richtung eine gleichförmig gradlinige Bewegung (1.21) mit $x_\circ = 0$

$$a_x(t) = 0 \tag{1.53}$$

$$v_x(t) = v_{\circ x} = const. \tag{1.54}$$

$$x(t) = v_{\circ x} \cdot t + x_\circ = v_{\circ x} \cdot t \tag{1.55}$$

und in y-Richtung eine gleichmäßig beschleunigte Bewegung (1.33) mit

$$a_y(t) = -g = const. \tag{1.56}$$

$$v_y(t) = -g \cdot t + v_{\circ y} \tag{1.57}$$

$$y(t) = -\frac{1}{2}\, g \cdot t^2 + v_{\circ y} \cdot t + y_\circ \, , \tag{1.58}$$

so erhält man als Überlagerung die Bahnkurve, die z. B. ein Ball bei einem schiefen Wurf durchläuft (► Abbildung 1.17).

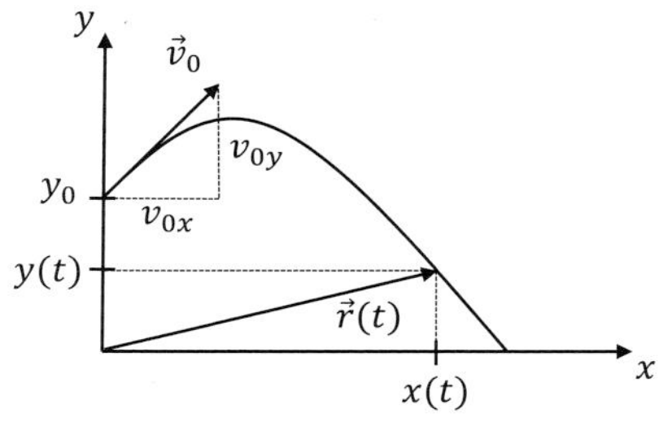

Abbildung 1.17 Bahnkurve als Überlagerung von Bewegungen entlang der x- und y-Achse

Mit den Vektoren (1.50), (1.51) und (1.52) können wir die Ballbewegung mit der Bahnkurve $\vec{r}(t)$, der Geschwindigkeit $\vec{v}(t)$ und der Beschleunigung $\vec{a}(t)$ beschreiben:

$$\vec{r}(t) = \begin{pmatrix} v_{\circ x} \cdot t \\ -\frac{1}{2}\, g \cdot t^2 + v_{\circ y} \cdot t + y_\circ \end{pmatrix} \tag{1.59}$$

$$\vec{v}(t) = \begin{pmatrix} v_{\circ x} \\ -g \cdot t + v_{\circ y} \end{pmatrix} \tag{1.60}$$

$$\vec{a}(t) = \begin{pmatrix} 0 \\ -g \end{pmatrix} \tag{1.61}$$

Aufgabe 4: Fall zweier Kugeln

Welche der beiden Kugeln schlägt zuerst auf dem Boden auf (► Abbildung 1.18)?

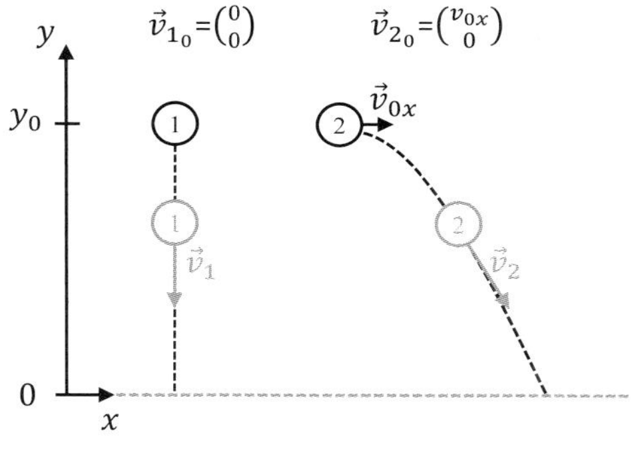

Abbildung 1.18 Fall zweier Kugeln mit verschiedenen horizontalen Anfangsgeschwindigkeiten $\vec{v}_{1\circ}$ und $\vec{v}_{2\circ}$

1.2.2 Bahnkurve beim Weitwurf

Um die Flugbahn $y(x)$ eines Balls in Abhängigkeit der x-Koordinate zu berechnen, haben wir zwei Gleichungen $x(t)$ und $y(t)$ mit insgesamt drei Variablen x, y und t. Das heißt, wir erhalten $y(x)$ durch Eliminieren der Zeit t.

Mit der x-Komponente von $\vec{r}(t)$ (1.59) ergibt sich

$$x = v_{\circ x} \cdot t \tag{1.62}$$

$$t = \frac{x}{v_{\circ x}} \,. \tag{1.63}$$

Einsetzen der Zeit t in die y-Komponente von $\vec{r}(t)$ ergibt

$$y = -\frac{1}{2}\, g \cdot t^2 + v_{\circ y} \cdot t + y_\circ \tag{1.64}$$

$$= -\frac{1}{2}\, g \cdot \left(\frac{x}{v_{\circ x}}\right)^2 + v_{\circ y} \cdot \left(\frac{x}{v_{\circ x}}\right) + y_\circ\,. \tag{1.65}$$

Die Bahnkurve $y(x)$ des Wurfs in Abhängigkeit des horizontal zurückgelegten Wegs x hat die Form einer Parabel ($y = k \cdot x^2 + m \cdot x + n$), der sogenannten **Wurfparabel**:

$$y(x) = -\frac{g}{2 \cdot v_{\circ x}^2} \cdot x^2 + \frac{v_{\circ y}}{v_{\circ x}} \cdot x + y_\circ \tag{1.66}$$

Beim Sonderfall des horizontalen Wurfs $v_{\circ y} = 0$ (linke ► Abbildung 1.19) vereinfacht sich die Gleichung (1.66) zu

$$y(x) = -\frac{g}{2 \cdot v_{\circ x}^2} \cdot x^2 + y_\circ\,. \tag{1.67}$$

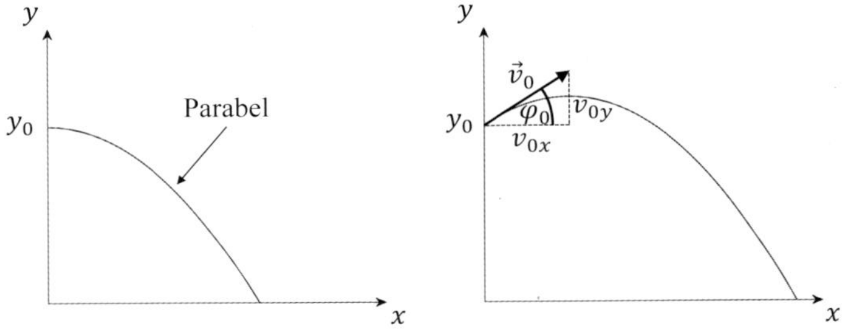

Abbildung 1.19 Links: horizontaler Wurf aus der Höhe $y_\circ$, rechts: schräger Wurf unter dem Winkel $\varphi_\circ$

Um die Abhängigkeit der Wurfweite x_w vom Abwurfwinkel $\varphi_\circ$ zu bestimmen, benötigen wir die horizontalen und vertikalen Komponenten der Anfangsgeschwindigkeit (rechte ► Abbildung 1.19):

$$|\vec{v}_\circ| = \sqrt{v_{\circ x}^2 + v_{\circ y}^2} \tag{1.68}$$

$$\frac{v_{\circ y}}{v_{\circ x}} = \tan \varphi_\circ \tag{1.69}$$

$$v_{\circ y} = |\vec{v}_\circ| \cdot \sin \varphi_\circ \tag{1.70}$$

$$v_{\circ x} = |\vec{v}_\circ| \cdot \cos \varphi_\circ \tag{1.71}$$

Um den **Scheitelpunkt** x_s der Wurfparabel zu berechnen, nutzen wir, dass die Ableitung im Maximum der Bahnkurve $dy/dx = 0$ ist (linke ► Abbildung 1.20):

$$\frac{dy}{dx} = -\frac{g}{v_{\circ x}^2} \cdot x_s + \frac{v_{\circ y}}{v_{\circ x}} = 0 \tag{1.72}$$

$$\Rightarrow \frac{g}{v_{\circ x}^2} \cdot x_s = \frac{v_{\circ y}}{v_{\circ x}} \tag{1.73}$$

$$x_s = \frac{v_{\circ y} \cdot v_{\circ x}}{g} \tag{1.74}$$

$$= \frac{|\vec{v}_\circ|^2 \cdot \sin \varphi_\circ \cdot \cos \varphi_\circ}{g} \tag{1.75}$$

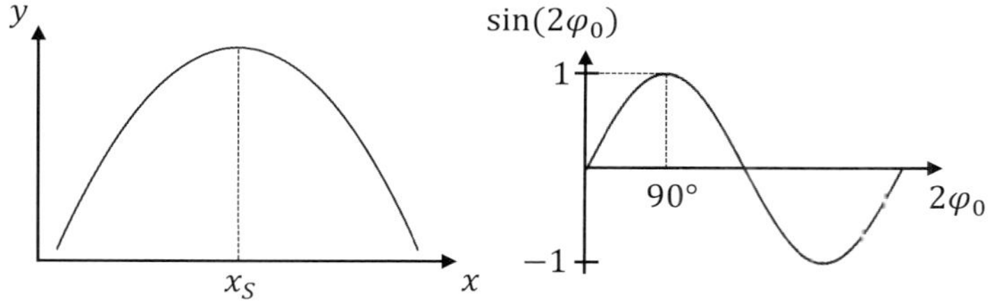

Abbildung 1.20 Links: Scheitelpunkt einer Wurfparabel, rechts: Sinusfunktion

Mit dem Additionstheorem $\sin(2\vartheta) = 2\sin\vartheta\cos\vartheta$ lautet der Ausdruck für den Scheitelpunkt x_s schließlich

$$x_s = \frac{|\vec{v}_\circ|^2}{2g}\sin(2\varphi_\circ)\,. \tag{1.76}$$

Der Scheitelpunkt x_s ist also maximal bei $2\varphi_\circ = 90°$ (rechte ► Abbildung 1.20) beziehungsweise

$$\varphi_\circ = 45°\,. \tag{1.77}$$

Die Wurfweite x_w ergibt sich aus der Rückkehrbedingung zur Erde mit $y(x_w) = 0$. Mit der Bahnkurve $y(x)$ aus Gleichung (1.66) ist für $x \neq 0$

$$y(x_w) = -\frac{g}{2v_{\circ x}^2}x_w^2 + \frac{v_{\circ y}}{v_{\circ x}}x_w + y_\circ = 0\,. \tag{1.78}$$

Mathematischer Einschub: Quadratische Ergänzung

Eine quadratische Gleichung der Form

$$kx^2 + mx + n = 0 \tag{1.79}$$

wird gelöst durch

$$x = -\frac{m}{2k} \pm \sqrt{\left(\frac{m}{2k}\right)^2 - \frac{n}{k}} \tag{1.80}$$

$$= \frac{m}{2k}\left[-1 \pm \sqrt{1 - \frac{4k^2}{m^2}\frac{n}{k}}\right] \tag{1.81}$$

$$= \frac{m}{2k}\left[-1 \pm \sqrt{1 - \frac{4kn}{m^2}}\right]. \tag{1.82}$$

Dies bedeutet für die Wurfweite x_w

$$x_w = \frac{v_{\circ y}}{v_{\circ x}} \cdot \frac{(-1)}{2\frac{g}{2v_{\circ x}^2}} \cdot \left[-1 \pm \sqrt{1 - \frac{4\frac{-g}{2v_{\circ x}^2}y_\circ}{\frac{v_{\circ y}^2}{v_{\circ x}^2}}}\right] \tag{1.83}$$

$$= \frac{v_{\circ y}v_{\circ x}}{g}\left[1 \mp \sqrt{1 + \frac{2gy_\circ}{v_{\circ y}^2}}\right] \tag{1.84}$$

$$= \underbrace{\frac{|\vec{v}_\circ|^2 \sin(2\varphi_\circ)}{2g}}_{=x_s}\left[1 + \sqrt{1 + \frac{2gy_\circ}{|\vec{v}_\circ|^2 \sin(\varphi_\circ)^2}}\right] . \tag{1.85}$$

Im letzten Schritt haben wir nur die physikalisch mögliche Lösung behalten.

Liegen Start und Landung auf gleicher Höhe $y_\circ = 0$, folgt dann mit (1.76) wie erwartet die Wurfweite $x_w = 2x_s$. Für die maximale Wurfweite x_w erhalten wir mit $\varphi_\circ = 45°$ (rechte ► Abbildung 1.20) einen einfachen Ausdruck mit einer quadratischen Abhängigkeit von der Abwurfgeschwindigkeit $v_\circ$:

$$x_w = 2x_s = \frac{|\vec{v}_\circ|^2}{g} \tag{1.86}$$

Beispiel 4: Der Wurf

Ein Ball wird von $y_\circ = 0$ mit $v_\circ = 30\,\mathrm{m/s}$ unter einem Winkel von $\varphi_\circ = 45°$ zur x-Achse abgeworfen. Durch die Variation [6] von $\varphi_\circ$ ergibt sich für die Wurfweite $x_w \propto \sin(2\varphi_\circ)$:

$$\begin{aligned} \varphi_\circ &= 25° & x_w &= 70\,\mathrm{m} \\ \varphi_\circ &= 45° & x_w &= 92\,\mathrm{m} \\ \varphi_\circ &= 65° & x_w &= 70\,\mathrm{m} \end{aligned}$$

Die Abhängigkeit der Wurfweite von der Startgeschwindigkeit $v_\circ$ folgt aus $x_w \propto |v_\circ|^2$:

$$\begin{aligned} v_\circ &= 30\,\frac{\mathrm{m}}{\mathrm{s}} & x_w &= 92\,\mathrm{m} \\ v_\circ &= 15\,\frac{\mathrm{m}}{\mathrm{s}} & x_w &= 23\,\mathrm{m} \end{aligned}$$

Die Abhängigkeit der Wurfweite von der Abwurfhöhe $y_\circ$ steigt langsamer als linear (Wurzelfunktion in (1.85)):

$$\begin{aligned} y_\circ &= 0\,\mathrm{m} & x_w &= 23\,\mathrm{m} \\ y_\circ &= 10\,\mathrm{m} & x_w &= 31\,\mathrm{m} \\ y_\circ &= 20\,\mathrm{m} & x_w &= 36\,\mathrm{m} \end{aligned}$$

Das Wichtigste zum Mitnehmen

Zum Mitnehmen: Kinematik 1

- Die **Kinematik** beschreibt Bewegungen in Raum und Zeit. Im einfachsten Fall wird die eindimensionale Bewegung eines **Massenpunkts** entlang einer **Ortsfunktion** oder **Bahnkurve** beschrieben:

$$x(t) \tag{1.2}$$

Mithilfe der Durchschnittsgeschwindigkeit des Massenpunkts

$$\overline{v}_x = \frac{x_2 - x_1}{t_2 - t_1} = \frac{\Delta x}{\Delta t} \tag{1.1}$$

definiert man die **(momentane) Geschwindigkeit** als die Ableitung von $x(t)$ nach der Zeit:

$$v_x(t) = \lim_{\Delta t \to 0} \frac{\Delta x}{\Delta t} = \frac{d}{dt} x(t) = \frac{dx}{dt} = \dot{x} \tag{1.4}$$

Entsprechend wird die **(momentane) Beschleunigung** als die Ableitung der Geschwindigkeit nach der Zeit definiert:

$$a_x(t) = \lim_{\Delta t \to 0} \frac{\Delta v_x}{\Delta t} = \frac{d}{dt} v_x(t) = \frac{dv_x}{dt} = \dot{v}_x = \ddot{x} \tag{1.6}$$

Die Beschleunigung ist also die 2. Ableitung der Ortsfunktion.

- Umgekehrt kann man durch Integration über die Geschwindigkeitsverteilung $v_x(t)$ die Bahnkurve $x(t)$ mit $x(0) = x_\circ$ berechnen, und zwar

$$x(t) = \int_0^t v_x(t)\, dt + x_\circ\,, \tag{1.18}$$

und aus der Beschleunigung $a_x(t)$ die Geschwindigkeit $v_x(t)$ bei einer Anfangsgeschwindigkeit $v_x(0) = v_\circ$:

$$v_x(t) = \int_0^t a_x\, dt + v_\circ \tag{1.26}$$

Speziell ist der Ort $x(t)$ bei gleichmäßiger Beschleunigung $a(t) = a_\circ$ gegeben durch

$$x(t) = \frac{1}{2}\, a_\circ \cdot t^2 + v_\circ \cdot t + x_\circ\,. \tag{1.33}$$

Zum Mitnehmen: Kinematik 2

- Eine **zweidimensionale Bewegung** lässt sich durch Vektoren beschreiben:

Ortsvektor $$\vec{r} = \begin{pmatrix} x(t) \\ y(t) \end{pmatrix} \quad (1.50)$$

Geschwindigkeitsvektor $$\vec{v} = \begin{pmatrix} v_x(t) \\ v_y(t) \end{pmatrix} = \begin{pmatrix} \dot{x}(t) \\ \dot{y}(t) \end{pmatrix} \quad (1.51)$$

Beschleunigungsvektor $$\vec{a} = \begin{pmatrix} a_x(t) \\ a_y(t) \end{pmatrix} = \begin{pmatrix} \ddot{x}(t) \\ \ddot{y}(t) \end{pmatrix} \quad (1.52)$$

Als Beispiel kann man einen Wurf als Überlagerung einer gleichmäßig beschleunigten Bewegung in y- und einer gleichförmigen Bewegung in x-Richtung schreiben:

$$\vec{r}(t) = \begin{pmatrix} v_{\circ x} \cdot t \\ -\frac{1}{2} g \cdot t^2 + v_{\circ y} \cdot t + y_\circ \end{pmatrix} \quad (1.59)$$

$$\vec{v}(t) = \begin{pmatrix} v_{\circ x} \\ -g \cdot t + v_{\circ y} \end{pmatrix} \quad (1.60)$$

$$\vec{a}(t) = \begin{pmatrix} 0 \\ -g \end{pmatrix} \quad (1.61)$$

Wenn man den Zeitparameter t eliminiert, erhält man eine Bahnkurve $y(x)$ in Form der sogenannten **Wurfparabel**:

$$y(x) = -\frac{g}{2 \cdot v_{\circ x}^2} \cdot x^2 + \frac{v_{\circ y}}{v_{\circ x}} \cdot x + y_\circ \quad (1.66)$$

Der Scheitelpunkt x_s liegt bei

$$x_s = \frac{|\vec{v}_\circ|^2}{2g} \sin(2\varphi_\circ) . \quad (1.76)$$

x_s ist also maximal bei $\varphi_\circ = 45°$. Liegen Start und Landung auf gleicher Höhe $y_\circ = 0$, ergibt sich die maximale Wurfweite x_w zu

$$x_w = 2x_s = \frac{|\vec{v}_\circ|^2}{g} . \quad (1.86)$$

Grundgrößen, Messwert & Messgenauigkeit

2

2.1 Grundgrößen: SI-Einheiten **41**

2.1.1 Zeiteinheit: Sekunde **s** 42

2.1.2 Längeneinheit: Meter **m** 42

2.1.3 Masseneinheit: Kilogramm **kg** 43

2.2 Genauigkeit von Messwerten **44**

2.3 Verteilung von Messwerten **45**

2.3.1 Diskrete Verteilung 45

2.3.2 Kontinuierliche Verteilung 47

2.3.3 Poissonverteilung 48

2.3.4 Gaußverteilung 49

2.4 Interpretation von Messwerten **50**

2.5 Auswertung eines Experiments **53**

2.5.1 Mittelwert 53

2.5.2 Statistische Fehler 55

2.5.3 Fehlerfortpflanzung 58

2.5.4 Systematische Fehler 60

ÜBERBLICK

» Für die Angabe eines experimentellen Resultats benötigen wir drei Größen: den Messwert, die Genauigkeit des Messwerts und die Einheiten, in denen der Messwert angegeben wird.

In diesem Kapitel stellen wir zunächst die Grundgrößen vor, aus denen sich die Einheiten für Messresultate ergeben. Anschließend führen wir in die Bedeutung der Messgenauigkeit ein, die unmittelbar mit der Reproduzierbarkeit von Experimenten zusammenhängt. Dafür erarbeiten wir einige grundlegende Verfahren, mit denen wir die Genauigkeit experimenteller Resultate quantitativ erfassen.

Wir erhalten so einerseits ein Maß für die Vergleichbarkeit der experimentellen Ergebnisse und können andererseits die Kompatibilität entsprechender theoretischer Rechnungen mit den Messungen überprüfen.

Wir geben hier eine kurze Einführung in statistische Verfahren, eine ausführliche Beschreibung zur Erarbeitung von experimentellen Resultaten und ihrer Genauigkeiten finden Sie im Buch *Statistische Methoden in der Experimentalphysik* [5]. Anwendungen dieser Methoden werden in den Praktika des Physikstudiums ausgiebig geübt.

In unserer Lehrbuchreihe dienen die Demonstrationsversuche und *phyphox*-Experimente vorwiegend dem Zweck, Zusammenhänge zwischen physikalischen Konzepten und experimentellen Beobachtungen zu zeigen. Sie enthalten daher meistens nur grobe Abschätzungen der experimentellen Genauigkeiten. «

Beispiel 5: Fallexperiment: Theorie

Für ein Fallexperiment (▶ Abbildung 2.1) haben wir die Ortsfunktion, Geschwindigkeit und Beschleunigung in Abhängigkeit von der Zeit in (1.2), (1.4) und (1.6) kennengelernt:

$$
\begin{aligned}
&x(t)\\
&v(t) = \dot{x}\\
&a(t) = \dot{v}(t) = \ddot{x}(t)
\end{aligned}
$$

Bei einer gleichmäßig beschleunigten Bewegung ist (1.33)

$$
\begin{aligned}
a &= const. = a_\circ\\
v(t) &= a_\circ t + v_\circ\\
x(t) &= \frac{1}{2}\, a_\circ t^2 + v_\circ t + x_\circ\,.
\end{aligned}
$$

In einem Kugel-Fallexperiment mit $v_\circ = 0$ können wir durch Messen von $\Delta x = x - x_\circ$ und $\Delta t = t - t_\circ$ mit

$$\Delta x = \frac{1}{2}\, g\, \Delta t^2 \tag{2.1}$$

die Erdbeschleunigung $a_\circ = g$ bestimmen und gleichzeitig einen quantitativen Test des Fallgesetzes durchführen.

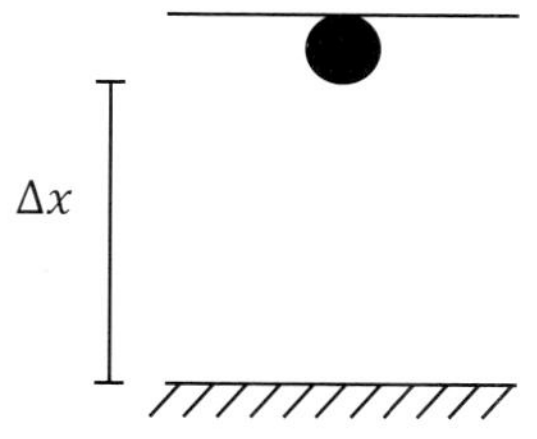

Abbildung 2.1 Fallexperiment

2.1 Grundgrößen: SI-Einheiten

Messen bedeutet einen Vergleich mit Standards. In diesem Abschnitt werden wir die in der Mechanik relevanten Grundgrößen nach dem internationalen Standard »SI« (Système International d'unités) einführen und die weiteren SI-Grundgrößen benennen [1].

Als Standards zur Definition der Grundgrößen wurden zunächst Artefakte (lat.: etwa »handgemacht«) wie das Urmeter oder Urkilogramm benutzt. Mit zunehmenden Anforderungen an die Meßgenauigkeit wurden diese nach und nach durch physikalische Konstanten ersetzt. Dies können Naturkonstanten wie die Lichtgeschwindigkeit c oder Festlegungen wie die Avogadro-Zahl N_A sein. Den Konstanten werden feste Werte zugeordnet und die Einheiten sind dadurch eindeutig definiert. Unsicherheiten resultieren nur aus der physikalisch-technischen Realisierung der Einheiten. Dieser Übergang ist seit 2019 abgeschlossen, sodass jetzt sieben Konstanten alle physikalischen Größen definieren. Aktuelle Entwicklungen zu den Grundgrößen kann man über die Webseite des »Bureau International des Poids et Mesures« (BIPM) einsehen [1].

2.1.1 Zeiteinheit: Sekunde s

- Ursprünglich war die Sekunde als der 86 400-te Teil eines Sonnentages definiert ($1\,\text{s} = 1\,\text{Tag}/(60 \cdot 60 \cdot 24)$). Bei dieser Definition unterliegt die Sekunde jedoch jahreszeitlichen Schwankungen und ist für Präzisionsmessungen ungeeignet.

- Heute ist die Sekunde über die Schwingungsfrequenz $\Delta\nu_{\text{Cs}}$ der elektromagnetischen Strahlung definiert, die beim Übergang zwischen zwei bestimmten Elektronenzuständen des natürlichen Caesium-Isotops ^{133}Cs entsteht (die sogenannte Hyperfeinaufspaltung $\Delta\nu_{\text{Cs}}$ des Grundzustands werden wir in der Quantenmechanik kennenlernen). Dieser Frequenz wird mit

$$\Delta\nu_{\text{Cs}} = 9\,192\,631\,770\,\text{s}^{-1} \tag{2.2}$$

als einer der **Naturkonstanten** des SI-Systems ein fester Wert zugeordnet. Danach ist 1 s die Zeitdauer von 9 192 631 770 Schwingungsperioden des Übergangs $\Delta\nu_{\text{Cs}}$.

Die experimentelle Reproduzierbarkeit der so definierten Sekunde gelang 2016 mit einer relativen Genauigkeit von $1/10^{16}$. Inzwischen (2021) ist die Sekunde mit anderen (»sekundären«) Frequenzstandards sogar auf $1/10^{18}$ genau und damit die am besten reproduzierbare Grundgröße [1].

2.1.2 Längeneinheit: Meter m

- Nach einer Messung des Erdmeridianquadranten (Entfernung vom Pol zum Äquator) durch die Astronomen Jean-Baptiste Delambre und Pierre Méchain wurde 1799 die Entfernung eines 10-Millionstels dieses Quadranten auf einem Platinstab markiert und als 1 m Länge deklariert. Auch mit verbesserten Varianten des Stabs liegt die Reproduzierbarkeit dieser Länge nur im µm-Bereich (10^{-6} m).

- 1983 wurde beschlossen, die Längenmessung auf eine Zeitmessung zurückzuführen. 1 m ist als die Strecke definiert, die Licht im Vakuum in der Zeit $1/299\,792\,458$ s zurücklegt. Der **Lichtgeschwindigkeit im Vakuum** c wurde damit als einer der Naturkonstanten des SI-Systems der feste Wert

$$c = 299\,792\,458\,\text{m/s} \tag{2.3}$$

zugewiesen. Der so definierte Meter ist mit einer relativen Genauigkeit von etwa $3/10^9$ reproduzierbar.

2.1.3 Masseneinheit: Kilogramm kg

- Das Kilogramm (1 kg) war als die Masse eines in Paris aufbewahrten Platin-Iridium-Zylinders definiert. Hauptsächlich durch ungeklärten Gewichtsverlust war die relative Genauigkeit der Definition auf etwa $1/10^5$ begrenzt.

- Seit 2019 wird die Einheit des kg wie alle Grundeinheiten an fest definierte Konstanten angeschlossen. Im Falle der Masse ist dies das **Plancksche Wirkungsquantum**

$$h = 6{,}62607015 \cdot 10^{-34}\,\frac{\mathrm{kg\,m^2}}{\mathrm{s}}\,. \tag{2.4}$$

An der Dimension von (2.4) kann man erkennen, dass in die Definition der Masseneinheit auch die Länge und die Zeit und damit auch der Anschluss an (2.2) und (2.3) einfließen.

Die physikalisch-technische Realisierung der Einheit des kg ist sehr anspruchsvoll und wird im Band »Felder und Wellen« unter dem Stichwort »Wattwaage« beschrieben [4]. Für das so definierte Kilogramm wird in der nächsten Zeit eine Verbesserung der Reproduzierbarkeit um 2 bis 3 Größenordnungen gegenüber der alten Definition erwartet [1].

Zwei weitere Grundgrößen werden wir in Kapitel 19 kennenlernen:

- **Die Stoffmenge mol** ist eine weitere **Basiseinheit des SI-Systems.** Bei der Neudefinition des SI-Einheitensystems über physikalische Konstanten wird das Mol über die **Avogadro-Konstante** N_A definiert:

$$N_A = 6{,}022\,140\,76 \cdot 10^{23}\,\frac{1}{\mathrm{mol}} \tag{19.29}$$

N_A ist als **Anzahl der Teilchen pro Mol** exakt festgelegt. Ein Mol eines Stoffs enthält dann genau $6{,}022\,140\,76 \cdot 10^{23}$ Teilchen, z. B. Atome oder Moleküle.

- Die Größe k ist als **Boltzmann-Konstante** eine weitere physikalische Konstante des SI-Systems. Sie hat den exakten Wert

$$k = 1{,}380\,649 \cdot 10^{-23}\,\frac{\mathrm{J}}{\mathrm{K}}\,. \tag{19.38}$$

Aus der Einheit $\mathrm{J} = \mathrm{kg\,m^2/s^2}$ kann man erkennen, dass sich mit der Boltzmann-Konstanten unter Verwendung der Einheiten Sekunde, Meter und Kilogramm die **Temperatureinheit K** definieren lässt (Abschnitt 20.2 und Abschnitt 20.3).

Als weitere Grundeinheit wird schließlich die **elektrische Ladung Coulomb C** im Band »Felder und Wellen« eingeführt [4]. Insgesamt gibt es sieben Grundgrößen. Die noch verbleibende ist die sogenannte Lichtstärke mit der Einheit Candela, die im Bachelorstudium der Physik eher selten vorkommt.

2.2 Genauigkeit von Messwerten

Experiment 5: *phyphox* Beschleunigungssensor

Zur Durchführung dieses Experiments ist ein Smartphone mit Beschleunigungssensor erforderlich.

1. Suchen Sie in der *phyphox*-App das Experiment »Beschleunigung mit g«. Wählen Sie den Menüpunkt »Betrag«, um die Messwerte darzustellen.
2. Legen Sie das Smartphone auf einen Tisch und starten Sie die Messung mit dem »Play button« (Dreieck rechts oben).
3. Beobachten Sie die wiederholten Messungen der Erdbeschleunigung: Die Schwankungen in der Nachkommastelle von g des Beschleunigungssensors zeigen Ihnen, dass die Genauigkeit dieses Experiments begrenzt ist.
4. Wir machen mehrere Messungen und stellen fest, dass der gemessene Wert von »Beschleunigung mit g« ungefähr $9{,}87\,\mathrm{m/s^2}$ beträgt. Der erwartete Wert von $9{,}81\,\mathrm{m/s^2}$ liegt deutlich darunter.

Die begrenzte Genauigkeit der Apparatur verhindert, dass bei jeder Messung der identische korrekte Messwert gefunden wird. Bei den hierdurch verursachten Messunsicherheiten spricht man in der Physik häufig von »Fehlern« (obwohl »Genauigkeit«, »Ungenauigkeit« oder »Unsicherheit« den Sachverhalt besser beschreiben) und unterscheidet zwei Arten:

Statistische Fehler sind zufällig auftretende Ungenauigkeiten. Sie streuen um einen Wert, wie wir bei der Messung der Erdbeschleunigung g gesehen haben.

Systematische Fehler sind Unsicherheiten, die auch bei Wiederholung des Experiments den Messwert immer in dieselbe Richtung verschieben (größere oder kleinere Messwerte). Im Falle der g-Bestimmung ist der Wert systematisch um die Fehleichung des Beschleunigungssensors verschoben. Wir können solche Verschiebungen korrigieren, allerdings nur mit begrenzter Genauigkeit.

Die Angabe der Messgröße inklusive ihrer Unsicherheiten wird üblicherweise in folgender Weise notiert:

Messwert $\pm$ statistische Fehler $\pm$ systematische Fehler

Im Folgenden stellen wir vor, wie die Fehler eines Experiments ermittelt werden.

Experiment 6: Fallexperiment

Am Beginn dieses Kapitels haben wir im Beispiel 5 die theoretischen Erwartungen für ein Fallexperiment angegeben. Dazu führen wir hier ein Fallexperiment durch, bei dem die Fallstrecke Δx einer Stahlkugel und ihre Fallzeit Δt gemessen werden. Die Kugel wird anfangs von einem Elektromagneten gehalten. Die Fallstrecke wird mit einem Maßband gemessen. Die Fallzeit wird mit einer elektronischen Stoppuhr bestimmt. Beim Start wird der Elektromagnet abgeschaltet und gleichzeitig die Uhr gestartet. Die Kugel fällt durch eine Lichtschranke, über die die Uhr gestoppt wird.

Wir wiederholen die Messung mehrfach und messen bei der Fallstrecke $\Delta x = 2\,\mathrm{m}$ folgende Werte für die Fallzeit Δt:

$$634\,\mathrm{ms},\ 630\,\mathrm{ms},\ 629\,\mathrm{ms},\ 631\,\mathrm{ms},\ 655\,\mathrm{ms},\ 633\,\mathrm{ms},\ 631\,\mathrm{ms} \tag{2.5}$$

Bevor wir diese Messung in Abschnitt 2.5 auswerten, stellen wir zunächst einige allgemeine Überlegungen zu Meßwerten und ihren Fehlern an.

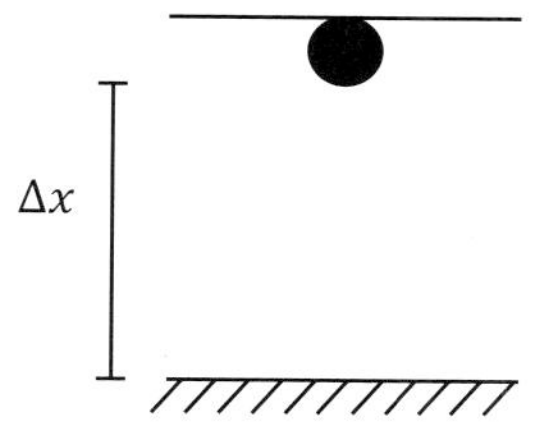

Abbildung 2.2 Fallexperiment

Beispiel 6: Fallexperiment: Ursachen für Fehler

In unserem Fallversuch sind Beispiele für mögliche Ursachen von Messfehlern:

Statistische Fehler entstehen durch zufällige Ungenauigkeiten beim Starten und Stoppen des Uhrwerks, Instabilitäten der Uhrelektronik bei der Zeitmessung und eine geschätzte Ablesegenauigkeit des Maßbands.

Zu *systematischen Fehlern* gehören Luftreibung, fehlerhafte Eichung der Stoppuhr bzw. des Maßbandes oder Verzögerungen im Auslösemechanismus der Kugel.

2.3 Verteilung von Messwerten

Die Frage, die wir im Folgenden beantworten wollen, ist: Mit welcher Wahrscheinlichkeit messe ich einen Wert nahe des wahren Werts?

Mathematischer Einschub: Wahrscheinlichkeit

Tritt ein Ereignis auf n verschiedene aber gleich wahrscheinliche Arten ein, wobei k dieser Arten die Eigenschaft A haben, dann ist die Wahrscheinlichkeit P für das Auftreten von A

$$P(A) = \frac{k}{n}\,. \tag{2.6}$$

Beispiel 7: Würfel

Das Ereignis A sei, eine 5 zu würfeln. Nur eine Seite des Würfels hat die 5, deswegen ist $k = 1$. Der Würfel hat $n = 6$ Seiten. Die Wahrscheinlichkeit, eine 5 zu würfeln, beträgt somit $P(A) = \frac{1}{6}$.

Eine **Zufallsvariable** kann aufgrund statistisch unkontrollierbarer Einflüsse verschiedene Werte annehmen. Man unterscheidet:

- **Diskrete Zufallsvariable**, z. B. Würfel
- **Kontinuierliche Zufallsvariable**, z. B. Zeitmessung, Temperaturmessung

Mathematischer Einschub: Bezeichnung von Zahlenmengen

Eine natürliche Zahl $n = 1, 2, \ldots$ notieren wir durch $n \in \mathbb{N}$, wobei $\mathbb{N}$ die Menge der natürlichen Zahlen beschreibt. Entsprechend notieren wir eine Zugehörigkeit zu ganzen Zahlen $r = \ldots, -2, -1, 0, 1, 2, \ldots$ durch $r \in \mathbb{Z}$. Reelle Zahlen x bezeichnen wir durch $x \in \mathbb{R}$ und komplexe Zahlen $c = a + ib$ durch $c \in \mathbb{C}$.

2.3.1 Diskrete Verteilung

Bei $n \in \mathbb{N}$ diskreten Zufallsvariablen r mit $r \in \mathbb{Z}$ ist die **Wahrscheinlichkeit** $P(r)$, dass bei einer Messung der Wert r auftritt, immer innerhalb des Intervalls

$$0 \leq P(r) \leq 1\,. \tag{2.7}$$

Ein Beispiel für eine diskrete Wahrscheinlichkeitsverteilung mit der Zufallsvariablen r im Bereich $a \leq r \leq b$ ist in ► Abbildung 2.3 gezeigt.

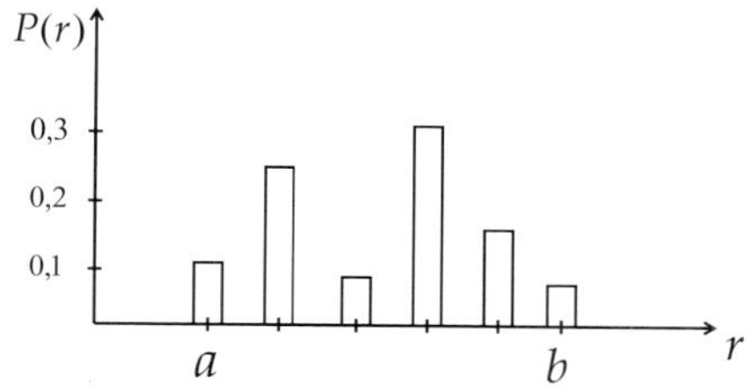

Abbildung 2.3 Diskrete Wahrscheinlichkeitsverteilung mit der Zufallsvariablen r im Bereich zwischen a und b

Die Wahrscheinlichkeit aller möglichen Fälle ist

$$\sum_{i=1}^{n} P(r_i) = 1 . \tag{2.8}$$

Der **Mittelwert** der diskreten Zufallsvariablen $\langle x \rangle$ ist gegeben durch

$$\langle r \rangle = \sum_{i=1}^{n} r_i \, P(r_i) . \tag{2.9}$$

2.3.2 Kontinuierliche Verteilung

Für eine kontinuierliche Zufallsvariable $x \in \mathbb{R}$ gibt es eine sogenannte **Wahrscheinlichkeitsdichteverteilung** $f(x)$. Ein Beispiel einer solchen Verteilung ist in ► Abbildung 2.4 gezeigt. Die Wahrscheinlichkeit, dass bei einer Messung der Wert x zwischen a und b liegt, erhalten wir durch das Integral über die Wahrscheinlichkeitsdichte:

$$P(a \leq x \leq b) = \int_a^b f(x) \, dx \tag{2.10}$$

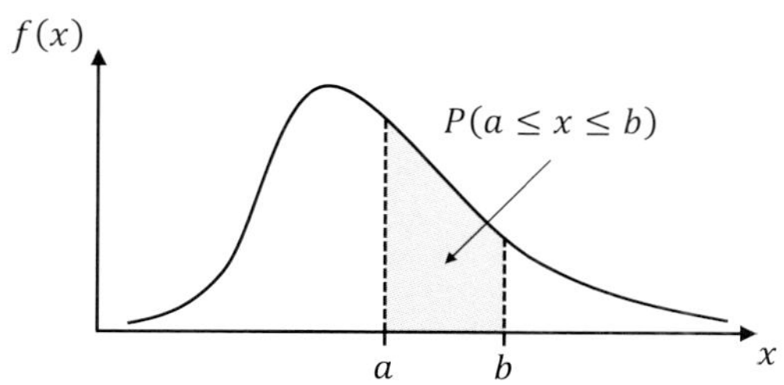

Abbildung 2.4 Wahrscheinlichkeitsdichteverteilung $f(x)$

Für die Wahrscheinlichkeitsdichte $f(x)$ gelten die Bedingungen, dass f stets positiv und das Integral über den gesamten Wertebereich von x auf 1 normiert ist:

$$f(x) \geq 0 \quad \text{und} \quad \int_{-\infty}^{\infty} f(x) \, dx = 1 \tag{2.11}$$

Das Integral über die Wahrscheinlichkeitsdichte bis zum Wert $x_\circ$ nennt man Wahrscheinlichkeitsdichtefunktion $F(x_\circ)$:

$$F(x_\circ) = \int_{-\infty}^{x_\circ} f(x)\,dx \quad \text{mit} \quad F(-\infty) = 0 \text{ und } F(\infty) = 1 \tag{2.12}$$

Sie gibt an, mit welcher Wahrscheinlichkeit ein Wert x mit $x \leq x_\circ$ auftritt.

Der Mittelwert für eine kontinuierliche Zufallsvariable ist

$$\langle x \rangle = \int_{-\infty}^{\infty} x\, f(x)\,dx\,. \tag{2.13}$$

Als Maß für die Breite einer Verteilung wird die **Varianz** $V = \sigma^2$ verwendet:

$$V = \sigma^2 \equiv \int_{-\infty}^{\infty} (x - \langle x \rangle)^2 f(x)\,dx \tag{2.14}$$

Hierbei ist σ die **Standardabweichung**. Sie ist ein Maß für die Größe der statistischen Schwankungen der Zufallsvariablen x um ihren Mittelwert $\langle x \rangle$. In der Physik nennen wir diese Größe Messunsicherheit oder einfach **Fehler**.

2.3.3 Poissonverteilung

Ein wichtiges Beispiel für eine diskrete Verteilung ist die Poissonverteilung (▶ Abbildung 2.5). Sie gibt die Wahrscheinlichkeit P für das r-fache Auftreten eines Ereignisses an, wobei die Ereignisse unabhängig voneinander zustandekommen:

$$P(r) = e^{-\mu} \cdot \frac{\mu^r}{r!}, \quad r = 0, 1, 2, \ldots \tag{2.15}$$

Die Poissonverteilung wird verwendet, wenn die Anzahl der Versuche extrem groß ist bei gleichzeitig sehr kleiner Wahrscheinlichkeit für ein Ereignis, oder auch wenn keine Anzahl von Versuchen feststellbar ist. Der Bereich kann z. B. ein Zeitintervall (radioaktive Zerfälle pro Sekunde) oder eine Fläche sein (Regentropfen pro Quadratmeter).

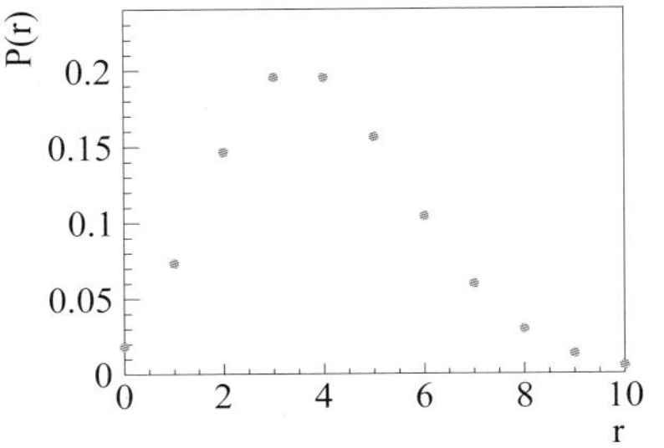

Abbildung 2.5 Poissonverteilung $P(r)$ einer diskreten Zufallsvariablen r mit Mittelwert und Varianz $\mu = 4$

Bei der Poissonverteilung entsprechen sich Mittelwert $\langle r \rangle$ und Varianz V. Beide werden üblicherweise mit dem Buchstaben μ notiert (2.15). Das Quadrat der Standardabweichung σ entspricht der Varianz:

$$\mu = \langle r \rangle = V = \sigma^2 \tag{2.16}$$

Die Standardabweichung der Poissonverteilung wird bei Zählungen häufig benötigt. Bei genügend vielen gezählten Ereignissen ($N \geq 10$) beträgt der statistische Fehler mit $\mu = N$ (2.16)

$$\sigma = \sqrt{\mu} = \sqrt{N}\,. \tag{2.17}$$

2.3.4 Gaußverteilung

Ein Beispiel für eine kontinuierliche Verteilung und gleichzeitig eine der wichtigsten Wahrscheinlichkeitsdichteverteilungen ist die Gaußverteilung (Normalverteilung) (► Abbildung 2.6). Sie ist folgendermaßen definiert:

$$f(x) = \frac{1}{\sqrt{2\pi}\,\sigma} \cdot e^{-\frac{(x-\mu)^2}{2\sigma^2}} \tag{2.18}$$

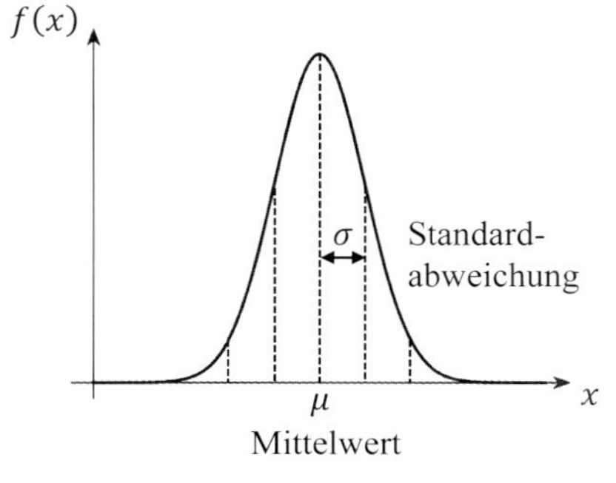

Abbildung 2.6 Gaußverteilung (Normalverteilung) $f(x)$ einer Zufallsvariablen x mit Mittelwert μ und Standardabweichung σ

Die Parameter μ und σ haben folgende Bedeutung:

Mittelwert (2.13) $$\mu \equiv \langle x \rangle = \int_{-\infty}^{\infty} x\, f(x)\, dx \tag{2.19}$$

Varianz (2.14) $$\sigma^2 = V = \int_{-\infty}^{\infty} (x-\mu)^2\, f(x)\, dx \tag{2.20}$$

Standardabweichung $$\sigma = \sqrt{V} \tag{2.21}$$

Die Gaußverteilung fällt in der Entfernung $|x-\mu| = \sigma$ auf den $1/\sqrt{e}$-ten Teil ihres Maximalwerts ab:

$$e^{-\frac{(x-\mu)^2}{2\sigma^2}} = e^{-\frac{1}{2}} = \frac{1}{\sqrt{e}} \tag{2.22}$$

Die Wahrscheinlichkeit, x im Intervall

$$\mu - \sigma \leq x \leq \mu + \sigma \tag{2.23}$$

zu finden, erhalten wir durch Integration über die Gaußverteilung in den entsprechenden Grenzen:

$$\frac{1}{\sqrt{2\pi}\,\sigma} \int_{\mu-\sigma}^{\mu+\sigma} e^{-\frac{(x-\mu)^2}{2\,\sigma^2}}\,dx = 68{,}26\% \tag{2.24}$$

Etwa zwei Drittel aller x-Werte liegen also innerhalb des $\pm 1\sigma$ Bereichs und etwa ein Drittel liegen außerhalb. Weitere Bereiche sind in ► Tabelle 2.1 aufgeführt.

Messwert innerhalb des Intervalls	Wahrscheinlichkeit
$\lvert x-\mu\rvert < 1\sigma$	68,26%
$\lvert x-\mu\rvert < 2\sigma$	95,45%
$\lvert x-\mu\rvert < 3\sigma$	99,73%

Tabelle 2.1 Fläche unter der Gaußkurve in Vielfachen der Standardabweichung

2.4 Interpretation von Messwerten

Jede einzelne Messung eines Experiments wird von vielen Fehlerquellen beeinflusst, die den wahren Wert zum Zeitpunkt der Messung verändern. Das gesamte Mess-Ergebnis des Experiments können wir als Wahrscheinlichkeitsdichteverteilung $h(x)$ auffassen, die in der Variation der Messwerte sichtbar wird.

In diesem Sinn können Messdaten $x_1, x_2, \ldots, x_n$ als Zufallsvariable aufgefasst werden, die der Wahrscheinlichkeitsverteilung $h(x)$ eines Experiments entnommen wurden. Gemeinsam bilden diese Messwerte die sogenannte **»Stichprobe«**. Eine Stichprobe besteht aus einer *diskreten* Anzahl n von Messungen x_i. Meistens ist $h(x)$ eine kontinuierliche Wahrscheinlichkeitsverteilung, sodass die Messwerte x_i selbst *kontinuierliche* Messgrößen sind.

Die Mathematik hilft uns mit dem sogenannten zentralen Grenzwertsatz, zumindest zufällige Fehler (= statistische Fehler) zu erfassen.

Mathematischer Einschub: Zentraler Grenzwertsatz

Die Summe vieler Zufallsvariablen x_i

$$y = \sum_{i=1}^{n} x_i \tag{2.25}$$

aus der gleichen beliebigen Wahrscheinlichkeitsverteilung $h(x)$ mit dem Mittelwert $\langle x \rangle$ und der Standardabweichung σ ergibt für eine große Anzahl $n \to \infty$ als resultierende Wahrscheinlichkeitsverteilung eine Gaußverteilung (2.18) mit dem Mittelwert

$$\langle y \rangle = n \langle x \rangle \tag{2.26}$$

und der Standardabweichung

$$\sigma_y = \sqrt{n}\,\sigma\,. \tag{2.27}$$

Ausgedrückt im Sinn eines Experiments mit seiner Wahrscheinlichkeitsverteilung $h(x)$: Viele Messwerte, die von unabhängigen Fehlerquellen zufällig in die eine oder andere Richtung gezogen werden, ergeben gemeinsam eine Gauß-Wahrscheinlichkeitsdichte $f(x)$.

Der zentrale Grenzwertsatz erlaubt damit eine gut verständliche Angabe von Fehlerbereichen für die Messresultate: Als statistische Fehlerangabe von Messresultaten geben wir eine Gaußsche Standardabweichung $1\,\sigma$ in positiver und negativer Richtung an. Damit liegen 68,26% (► Tabelle 2.1) der Messresultate im Bereich von ± 1 Standardabweichungen σ.

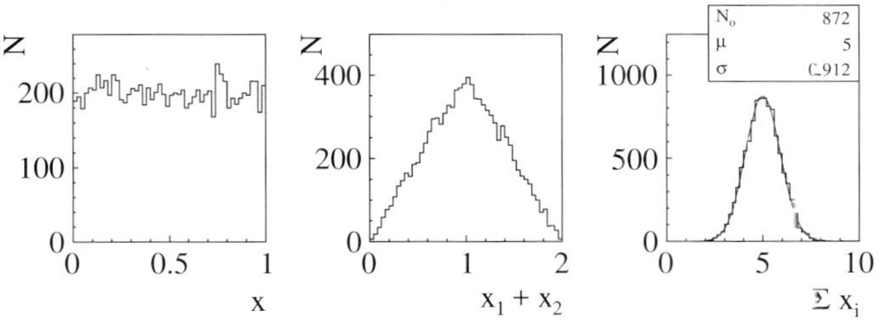

Abbildung 2.7 Zentraler Grenzwertsatz: Die Überlagerung von gleichverteilten Zufallsverteilungen (links) führt über eine pyramidenförmige (Mitte) zu einer Gaußverteilung (rechts).

Beispiel 8: Zentraler Grenzwertsatz

Die Überlagerung mehrerer Gleichverteilungen ergibt eine Gaußverteilung. In der linken ► Abbildung 2.7 ist eine gleichverteilte Zufallsverteilung mit Mittelwert 0,5 und Standardabweichung $1/\sqrt{12}$ gezeigt. Im mittleren Bild wurden 2 gleichverteilte Zufallszahlen addiert und in das Histogramm eingetragen. Im rechten Bild wurden

$n = 10$ Zufallszahlen addiert und dann eingetragen. Die Gaußverteilung ist hier schon gut zu erkennen. Die Anpassung an die Gaußform funktioniert sehr gut und liefert den erwarteten Mittelwert $\mu = 10 \cdot 0{,}5 = 5$ (2.26) und die Standardabweichung $\sigma = \sqrt{10}/\sqrt{12} = 0{,}913$ (2.27).

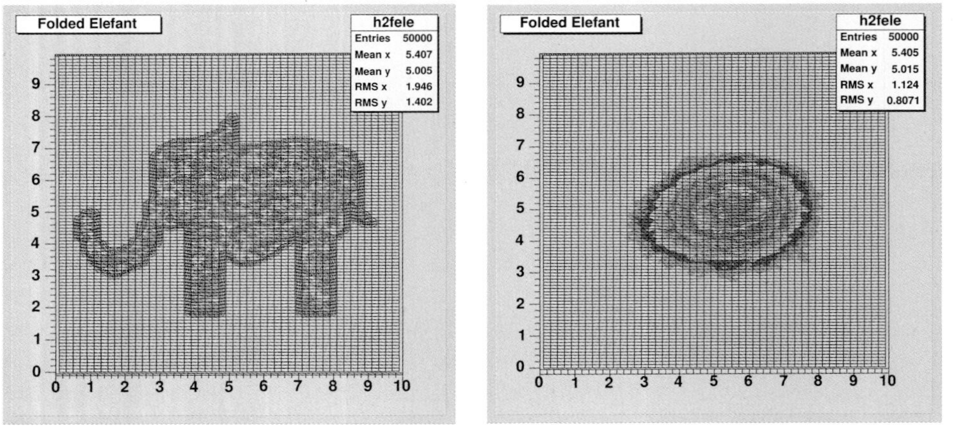

Abbildung 2.8 Zufällige Überlagerung von n zweidimensionalen Verteilungen: Der Elefant für $n = 1$ ist schon in der Verteilung für $n = 3$ nicht mehr zu erkennen.

Auch bei einer zweidimensionalen Verteilung lässt sich der Effekt beeindruckend zeigen (► Abbildung 2.8): Die Paare von Zufallszahlen sind im linken Bild aus dem Bereich des Elefanten gezogen. Rechts ist die dreifache Überlagerung von Paaren der Zufallsvariablen aus der Elefanten-Verteilung gezeigt. Vom Elefanten ist nichts mehr zu erkennen, die Entwicklung der Verteilung in Richtung einer Gaußverteilung kann man erahnen. Quelle: Prof. Dr. G. Quast, Universität Karlsruhe, Prof. Dr. C. Zeitnitz, Universität Wuppertal.

Aufgabe 5: Zentraler Grenzwertsatz

Erklären Sie die Addition zweier Zufallsvariablen am Beispiel aus ► Abbildung 2.9.

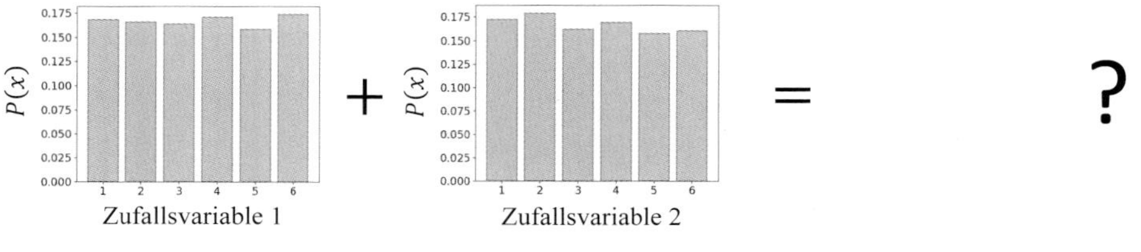

Abbildung 2.9 Wie ist die Wahrscheinlichkeitsverteilung der Summe der Augen auf zwei gleichzeitig geworfenen Würfeln?

2.5 Auswertung eines Experiments

2.5.1 Mittelwert

Am Beispiel des Fallversuchs (▶ Experiment 6) mit der Kugel führen wir jetzt eine Datenanalyse durch. Zunächst geht es darum, einen geeigneten Schätzwert $\overline{\Delta t}$ für die Fallzeit Δt zu finden.

Wir erhalten für den allgemeinen Fall von Messwerten x_i unter der Annahme, dass alle Werte gleichwahrscheinlich auftreten, den **Stichproben-Mittelwert**

$$\overline{x} = \frac{1}{n}\sum_{i=1}^{n} x_i\,. \tag{2.28}$$

Er ist eine Schätzung des wahren Mittelwerts $\langle x\rangle$ und kommt ihm für große n beliebig nahe:

$$\lim_{n\to\infty} \overline{x} = \langle x\rangle \tag{2.29}$$

Für unsere Fallzeitmessung bedeutet das

$$\overline{\Delta t} = \frac{1}{n}\sum_{i=1}^{n} \Delta t_i\,. \tag{2.30}$$

Beispiel 9: Fallexperiment: Mittelwert

Bei einer wiederholten Messung wurden bei einer Fallstrecke von 2 m für Δt im Experiment 6 die Werte (2.5) gemessen. Für $\overline{\Delta t}$ ergibt sich also

$$\begin{aligned}\overline{\Delta t} &= \frac{1}{n}\sum_{i=1}^{n} \Delta t_i\\ &= \frac{1}{7}(634\,\text{ms} + 630\,\text{ms} + 629\,\text{ms} + 631\,\text{ms} + 655\,\text{ms} + 633\,\text{ms} + 631\,\text{ms})\\ &\approx 635\,\text{ms}\,.\end{aligned}$$

Stellt man die Ergebnisse graphisch dar (linke ▶ Abbildung 2.10), so erkennt man eine Häufung der Messwerte um 631 ms. Der Mittelwert wird jedoch durch den Messwert bei 655 ms vom offensichtlichen Häufungspunkt auf 635 ms verschoben.

Als Test für die Stabilität des Mittelwerts wird der »getrimmte Mittelwert« berechnet. Dabei vernachlässigt man die m größten und kleinsten Werte bei der Bestimmung des Mittelwerts. Als getrimmter Mittelwert für $m = 1$ (nur den kleinsten und den größten Wert weglassen) ergibt sich $\overline{\Delta t} = 631{,}8$ ms (rechte ▶ Abbildung 2.10).

Nun stellt sich die Frage: Wie wahrscheinlich ist es, dass der Messwert bei 655 ms zufällig zustande kam? Aufgrund des zentralen Grenzwertsatzes sollten zufällig bedingte Schwankungen der Messwerte einer Gaußverteilung folgen.

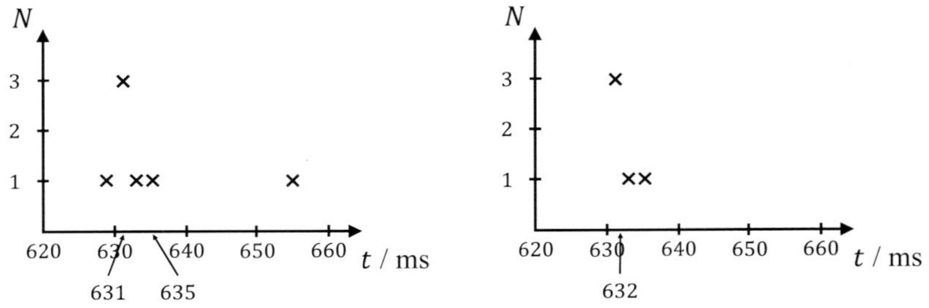

Abbildung 2.10 Links: Diagramm der Messwerte aus dem Fallexperiment der ► Abbildung 2.2, rechts: In der getrimmten Verteilung werden der niedrigste und höchste Wertcrop weggelassen.

Eine Anpassung der Gaußfunktion an das Histogramm (linke ► Abbildung 2.11) liefert den Mittelwert 631,7 ms und die Standardabweichung $\sigma \approx 2$ ms.

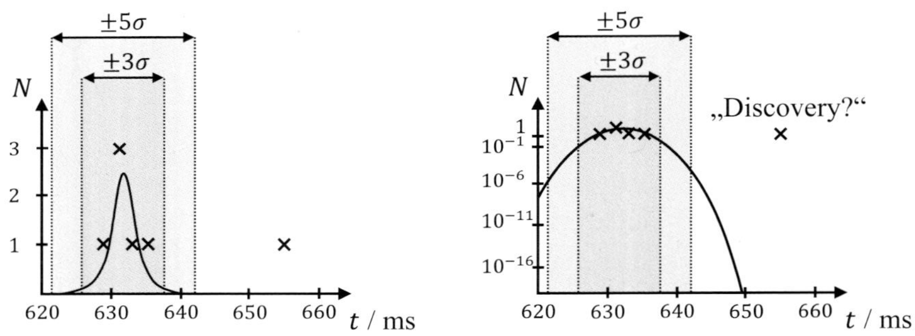

Abbildung 2.11 Links: Anpassung einer Gaußfunktion an die Daten der ► Abbildung 2.10, rechts: Die logarithmische Skala zeigt die 3σ- und 5σ-Grenzen. Der höchste Messwert liegt weit außerhalb der Gaußverteilung.

Die Distanz zwischen Mittelwert und Messwert beträgt also 23 ms $\approx 12 \cdot \sigma$. Die logarithmische Darstellung der Häufigkeiten zeigt in eindrucksvoller Weise die geringe Wahrscheinlichkeit, dass der Messwert 655 ms aus üblichen statistischen Schwankungen der Messung resultiert (rechte ► Abbildung 2.11). Solche Messungen außerhalb des erwarteten Wertebereichs sind für Suchen nach neuen physikalischen Phänomenen besonders interessant.

In unserem Experiment handelt es sich bei dem Messwert aber vermutlich um ein Apparaturproblem bei der Einzelmessung. Diese Vermutung könnten wir durch Wiederholungsmessungen absichern. In der folgenden Auswertung wird dieser Messwert nicht weiter berücksichtigt.

Beispiel 10: Fallexperiment: Korrigierter Mittelwert

Der Mittelwert im Häufungsbereich der Messungen beträgt

$$\overline{\Delta t} = \frac{1}{6}\,(634\,\mathrm{ms} + 630\,\mathrm{ms} + 629\,\mathrm{ms} + 631\,\mathrm{ms} + 633\,\mathrm{ms} + 631\,\mathrm{ms}) \approx 631\,\mathrm{ms}\,.$$

2.5.2 Statistische Fehler

Da alle Messgrößen fehlerbehaftet sind, müssen wir auch für unseren gemessenen Mittelwert einen Fehler angeben. Dazu berechnen wir als erstes die **»Stichproben-Varianz«**. Sie ergibt sich unter der Annahme gleichverteilter diskreter Werte x_i aus

$$V = \frac{1}{n-1}\sum_{i=1}^{n}(x_i - \overline{x})^2\,. \tag{2.31}$$

Die Varianz ist ein Maß für die Streuung der Einzelmessung. Da es sich bei dem Mittelwert $\overline{x}$ nicht um den wahren Wert, sondern um einen Schätzwert handelt, reduziert sich die Zahl der Freiheitsgrade (hier die Anzahl der Messungen bzw. der genommenen Stichproben) um 1, sodass nur durch $(n-1)$ dividiert wird.

Standardabweichung der Einzelmessung

Die Standardabweichung σ der Einzelmessung berechnen wir aus der Varianz $\sigma = \sqrt{V}$ (2.31):

$$\sigma = \sqrt{\frac{1}{n-1}\sum_{i=1}^{n}(x_i - \overline{x})^2} \tag{2.32}$$

Die Streuung gibt uns Auskunft über die Reproduzierbarkeit der Messung. Wegen des zentralen Grenzwertsatzes erwarten wir eine gaußförmige Verteilung der Messdaten, sodass der nächste neue Messwert mit einer Wahrscheinlichkeit von 68% in folgendem Intervall liegt:

$$\left[\,\overline{\Delta t} - \sigma;\ \overline{\Delta t} + \sigma\,\right] \tag{2.33}$$

Beispiel 11: Fallexperiment: Streuung der Messwerte

Mit den Zahlenwerten aus dem Beispiel ergibt sich

$$\sigma = \sqrt{\frac{1}{6-1}\left(3^2 + 1^2 + 2^2 + 0 + 2^2 + 0\right)\,\mathrm{ms}^2} \approx 2\,\mathrm{ms}\,.$$

Bei einer wiederholten Einzelmessung würden wir also mit 68%-iger Wahrscheinlichkeit einen Wert im Bereich $t = (631 \pm 2)$ ms erwarten.

Standardabweichung des Mittelwerts

So wie die Einzelmessungen x_i als Zufallsvariable aufgefasst werden können, sind auch die Schätzung des Mittelwerts $\overline{x}$ (2.28) und der Standardabweichung σ (2.32) selbst Zufallsvariable.

Aufgrund des zentralen Grenzwertsatzes sind sie daher gaußverteilt. Beachten Sie, dass beim zentralen Grenzwertsatz die Summe der Zufallsvariablen $y = \sum_{i=1}^{n} x_i$ (2.25) und die Standardabweichung der Summe $\sigma_y = \sqrt{n}\,\sigma$ (2.27) anders als beim Stichprobenmittelwert $\overline{x} = \sum_{i=1}^{n} x_i/n$ (2.28) nicht durch die Anzahl n der Zufallsvariablen geteilt werden. Um die Standardabweichung $\sigma_{\overline{x}}$ des Mittelwerts zu erhalten, muss die Standardabweichung der Grenzwertsatz-Summe noch durch n dividiert werden ($\sqrt{n}\,\sigma/n$):

$$\sigma_{\overline{x}} = \frac{\sigma}{\sqrt{n}} \tag{2.34}$$

Beispiel 12: Fallexperiment: Fehler des Mittelwerts

Angewandt auf unser Experiment ergibt sich mit (2.34)

$$\overline{\Delta t} = \frac{\sigma}{\sqrt{n}}\,. \tag{2.35}$$

Für die Zeitmessung in unserem Versuch können wir nun den Mittelwert, die Standardabweichung und den Fehler auf den Mittelwert angeben:

$$\overline{\Delta t} = 631\,\mathrm{ms}$$
$$\sigma = 2\,\mathrm{ms}$$
$$\sigma_{\overline{\Delta t}} = \frac{2\,\mathrm{ms}}{\sqrt{6}} \approx 1\,\mathrm{ms}$$

Die Fallzeit der Kugel ergibt sich zu

$$\overline{\Delta t} \pm \sigma_{\overline{\Delta t}} = (631 \pm 1)\,\mathrm{ms}\,.$$

Aufgabe 6: Tennisballmaschine

Bei einer Tennisballmaschine (► Abbildung 2.12) wurden 4 Messungen der Wurfweite durchgeführt: $x_w = 21\,\text{m}, 18\,\text{m}, 16\,\text{m}, 25\,\text{m}$.

Wie groß sind die mittlere Weite $\overline{x}_w$, die Standardabweichung σ und die Genauigkeit der mittleren Wurfweite $\sigma_{\overline{x}_w}$?

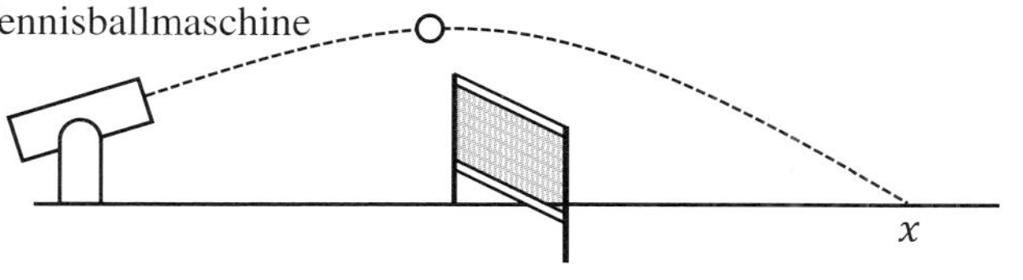

Abbildung 2.12 Messung der Wurfweite

Unser eigentliches Ziel ist, die Erdbeschleunigung aus der Zeitmessung und der Fallstrecke über

$$g = \frac{2\Delta x}{\overline{\Delta t}^2}$$

zu bestimmen (2.1). Um g und seinen Fehler berechnen zu können, muss auch die Längenmessung untersucht werden. Wie bei der Zeitmessung ist auch die Längenmessung fehlerbehaftet.

Liest man die Strecke Δx auf dem Bandmaß ab, so muss man zwei Punkte x_1 und x_2 bestimmen, aus denen sich aus der Differenzbildung die Fallstrecke $\Delta x = x_2 - x_1$ ergibt. Bei jeder der Messungen x_1 und x_2 kann man annehmen, dass der abgelesene Wert um $\sigma_{x_1} = \sigma_{x_2}$ schwankt.

Wie man diese beiden Fehler zum Fehler auf Δx kombiniert, erklärt das folgende Gesetz der Fehlerfortpflanzung.

2.5.3 Fehlerfortpflanzung

Mathematischer Einschub: Partielle Ableitung

Eine Funktion von zwei Variablen $f(x, y)$ wollen wir zunächst nur nach der Variablen x ableiten. Die Variable y wird dabei als Konstante betrachtet. Eine solche Ableitung wird als partielle Ableitung bezeichnet:

$$\left.\frac{\partial f(x,y)}{\partial x}\right|_{y=const.} \tag{2.36}$$

Genauso kann die partielle Ableitung nach der Variablen y durchgeführt werden:

$$\left.\frac{\partial f(x,y)}{\partial y}\right|_{x=const.} \tag{2.37}$$

Analog lassen sich auch höherdimensionale Funktionen $f(x, y, z)$ bezüglich ihrer einzelnen Variablen partiell ableiten.

Das Fehlerfortpflanzungsgesetz beschreibt, wie sich ein Fehler σ_{x_k} der Messgröße x_k auf eine aus x_k abgeleitete Messgröße y auswirkt. Ist $y = f(x_k)$ lediglich von x_k abhängig, z. B. $y = 2x_k$, wird der Fehler σ_y durch die Ableitung von $y(x_k)$ an der Stelle x_k und die Multiplikation mit σ_{x_k} bestimmt (► Abbildung 2.13):

$$\sigma_y = \frac{dy}{dx_k} \cdot \sigma_{x_k} \tag{2.38}$$

In unserem einfachen Beispiel $y = 2x_k$ ist dann $\sigma_y = 2\sigma_{x_k}$.

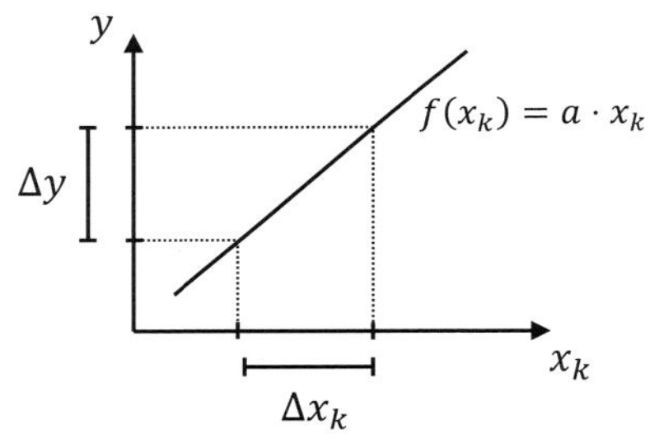

Abbildung 2.13 Verdeutlichung der Fehlerfortpflanzung am Beispiel der Funktion $y = a \cdot x_k$

Setzt sich eine Messgröße y aus n voneinander unabhängigen Messgrößen x_k zusammen, wird die quadratische Summe der Einzelbeiträge mit den partiellen Ableitungen gebildet:

$$\sigma_y^2 = \sum_{k=1}^{n} \left(\frac{\partial y}{\partial x_k}\right)^2_{\overline{x_k}} \sigma_{x_k}^2 \tag{2.39}$$

In dieser sogenannten Gaußschen Fehlerfortpflanzung sind $(\partial y/\partial x_k)_{\overline{x_k}}$ die partiellen Ableitungen an der Stelle der Mittelwerte $\langle x_k \rangle$. Dabei werden aus praktischen Gründen die Schätzungen der Mittelwerte $\overline{x_k}$ (2.28) als beste Näherung der wahren Mittelwerte $\langle x_k \rangle$ (2.29) benutzt.

Beispiel 13: Fallexperiment: Fehler der Längenmessung

Die Längenmessung $\Delta x = x_1 - x_2$ in unserem Beispiel hat mit $\sigma_{x_1} = \sigma_{x_2} = \pm 0{,}3\,\text{mm}$ den Fehler

$$\sigma_{\Delta x}^2 = \left(\frac{\partial \Delta x}{\partial x_1}\right)^2 \cdot \sigma_{x_1}^2 + \left(\frac{\partial \Delta x}{\partial x_2}\right)^2 \cdot \sigma_{x_2}^2 \tag{2.40}$$

$$= (1)^2 \cdot \sigma_{x_1}^2 + (-1)^2 \cdot \sigma_{x_2}^2 \tag{2.41}$$

$$= 2 \cdot \sigma_{x_1}^2 = 2 \cdot \sigma_{x_2}^2 \,. \tag{2.42}$$

Damit ist $\sigma_{\Delta x} = \sqrt{2 \cdot \sigma_{x_2}^2} = \sqrt{2}\sigma_{x_2} \approx 0{,}4\,\text{mm}$ und $\Delta x = (2{,}0 \pm 0{,}0004)\,\text{m}$.

Beispiel 14: Fallexperiment: Fehler der Erdbeschleunigung

Ebenso können wir mit dem Fehlerfortpflanzungsgesetz den Fehler auf die Erdbeschleunigung $g = 2\Delta x/\overline{\Delta t}^2$ berechnen:

$$\sigma_g^2 = \left(\frac{\partial g}{\partial \overline{\Delta t}}\right)^2 \cdot \sigma_{\overline{\Delta t}}^2 + \left(\frac{\partial g}{\partial \Delta x}\right)^2 \cdot \sigma_{\Delta x}^2 \tag{2.43}$$

$$= \left(2\Delta x \frac{-2}{\overline{\Delta t}^3}\, \sigma_{\overline{\Delta t}}\right)^2 + \left(\frac{2}{\overline{\Delta t}^2} \sigma_{\Delta x}\right)^2 \tag{2.44}$$

$$= \left(\frac{2\Delta x}{\overline{\Delta t}^2}\right)^2 \left[\left(-2\frac{\sigma_{\overline{\Delta t}}}{\overline{\Delta t}}\right)^2 + \left(\frac{\sigma_{\Delta x}}{\Delta x}\right)^2\right] \tag{2.45}$$

$$= g^2 \cdot \left[\left(2\frac{\sigma_{\overline{\Delta t}}}{\overline{\Delta t}}\right)^2 + \left(\frac{\sigma_{\Delta x}}{\Delta x}\right)^2\right] \tag{2.46}$$

Wir unterscheiden zwischen dem **absoluten Fehler**

$$\sigma_g = g \cdot \sqrt{\left(2\frac{\sigma_{\overline{\Delta t}}}{\overline{\Delta t}}\right)^2 + \left(\frac{\sigma_{\Delta x}}{\Delta x}\right)^2} \tag{2.47}$$

und dem **relativen Fehler**

$$\frac{\sigma_g}{g} = \sqrt{\left(2\frac{\sigma_{\overline{\Delta t}}}{\overline{\Delta t}}\right)^2 + \left(\frac{\sigma_{\Delta x}}{\Delta x}\right)^2} \,. \tag{2.48}$$

Wir erkennen aus dem Beispiel der Längenmessung, dass bei Differenzen (und Summen) die absoluten Fehler quadratisch addiert werden.

Im Beispiel der Erdbeschleunigung sehen wir, dass bei Quotienten (und Produkten) die relativen Fehler quadratisch addiert werden. Potenzen wie in diesem Fall $\overline{\Delta t}^2$ werden dabei als Faktoren berücksichtigt.

Allgemein gelten die wichtigen Regeln:

> *Bei Addition/Subtraktion $y = \sum \pm x_k$ werden die Quadrate der* **absoluten Fehler** *addiert $\sigma_y^2 = \sum \sigma_{x_k}^2$.*

Bei Produkten/Quotienten $y = \prod x_k^{\pm 1}$ *werden die Quadrate der* **relativen Fehler** *addiert* $(\sigma_y/\overline{y})^2 = \sum(\sigma_{x_k}/\overline{x}_k)^2$.

Beispiel 15: Fallexperiment Resultat mit statistischem Fehler

Im Versuch ergibt sich mit den berechneten Mittelwerten der Fehler auf die Erdbeschleunigung g zu

$$\Delta g = 10\,\frac{\mathrm{m}}{\mathrm{s}^2}\sqrt{\left(2\frac{1}{631}\right)^2 + \left(\frac{0{,}0004}{2}\right)^2} \approx \frac{3}{100}\,\frac{\mathrm{m}}{\mathrm{s}^2}\,.$$

Als Ergebnis erhalten wir mit unserem Versuch den Messwert für die Erdbeschleunigung

$$g = (10{,}05 \pm 0{,}03)\,\frac{\mathrm{m}}{\mathrm{s}^2}\,.$$

Damit haben wir bislang den statistischen Fehler unserer Messung von g berechnet. Der Wert stimmt innerhalb dieses Fehlers nicht mit dem Wert für Deutschland von $g = 9{,}81\,\mathrm{m/s}^2$ überein. Das wollen wir jetzt genauer untersuchen.

2.5.4 Systematische Fehler

Für eine vollständige Auswertung unseres Experiments wären auch die systematischen Fehler zu bestimmen. Anders als bei den statistischen Fehlern, bei denen grundlegende mathematische Konzepte wie der zentrale Grenzwertsatz (Abschnitt 2.4) unsere Datenauswertung leiten, ist die Bestimmung systematischer Fehler eine Detektivarbeit unter der Fragestellung: Welche Effekte könnten die Messung so beeinflusst haben, dass sie das Ergebnis merklich verändern? Für eine vertiefende Darstellung zur Bestimmung systematischer Fehler verweisen wir auf das Buch *Statistische Methoden in der Experimentalphysik* [5].

In unserem Fallexperiment würden wir z. B. die Eichgenauigkeit des Uhrwerks und des Längenmaßbands im Vergleich mit anderen, genaueren Apparaturen vermessen. Zusätzlich müssten wir die Effekte durch die Luftreibung an der Kugel bestimmen. Auch wenn wir solche Effekte korrigieren, verbleiben Unsicherheiten auf diese Korrekturen.

Bei Messgrößen, die aus Einzelvariablen x_k zusammengesetzt sind, können wir jeden Messwert um den systematischen Fehler jeweils einer Einzelvariablen verschieben und die Auswirkung auf das Endergebnis bestimmen. Die Änderung des Endergebnisses ist für jede Messgröße deren systematischer Fehler σ_k auf die gesamte Messgröße. Bei statistischer Unabhängigkeit der systematischen Einzelfehler können wir sie durch quadratisches Summieren zusammenfassen:

$$\sigma_{\mathrm{sys}} = \sqrt{\sum_{k=1}^{m} \sigma_k^2} \qquad (2.49)$$

Das Wichtigste zum Mitnehmen

Zum Mitnehmen: Messwert & Messgenauigkeit 1

- Die **Einheiten** für alle physikalischen **Grundgrößen** werden seit 2019 im internationalen Standard »SI« (Système International d'unités) über sieben **physikalische Konstanten** definiert. Den Konstanten werden feste Werte zugeordnet und die Einheiten sind dadurch eindeutig definiert. Unsicherheiten resultieren nur aus der physikalisch-technischen Realisierung der Einheiten.

  Für die drei in der Mechanik benötigten Grundeinheiten gilt:

  Die **Zeiteinheit Sekunde** s ist durch die Frequenz $\Delta\nu_{\mathrm{Cs}}$ des Übergangs zwischen zwei bestimmten Elektronenzuständen des natürlichen Caesium-Isotops $^{133}\mathrm{Cs}$ (Hyperfeinaufspaltung des Grundzustands) definiert:

  $$\Delta\nu_{\mathrm{Cs}} = 9\,192\,631\,770\,\mathrm{s}^{-1} \tag{2.2}$$

  Die **Längeneinheit Meter** m ist über die **Lichtgeschwindigkeit** c und die Zeiteinheit s definiert:

  $$c = 299\,792\,458\ \mathrm{m/s} \tag{2.3}$$

  Die **Masseneinheit Kilogramm** kg ist durch das **Plancksche Wirkungsquantum** h sowie über die Zeiteinheit s und die Längeneinheit m definiert:

  $$h = 6{,}62607015 \cdot 10^{-34}\,\frac{\mathrm{kg\,m^2}}{\mathrm{s}} \tag{2.4}$$

- Mit der **Wahrscheinlichkeit** $P(r)$ für die Messung der **diskreten Zufallsvariablen** r gilt für die Summe der Wahrscheinlichkeiten aller möglichen Fälle

  $$\sum_{i=1}^{n} P(r_i) = 1\,. \tag{2.8}$$

  Der **Mittelwert** ist gegeben durch

  $$\langle r \rangle = \sum_{i=1}^{n} r_i\, P(r_i)\,. \tag{2.9}$$

Zum Mitnehmen: Messwert & Messgenauigkeit 2

- Für eine **kontinuierlich** verteilte Zufallsvariable x mit der **Wahrscheinlichkeitsdichte** $f(x)$ gilt

$$f(x) \geq 0 \quad \text{und} \quad \int_{-\infty}^{\infty} f(x)\, dx = 1\,. \tag{2.11}$$

Der **Mittelwert** für eine kontinuierliche Zufallsvariable ist

$$\langle x \rangle = \int_{-\infty}^{\infty} x\, f(x)\, dx\,. \tag{2.13}$$

Ein Maß für die Schwankungen der Zufallsvariablen x um ihren Mittelwert $\langle x \rangle$ ist die **Varianz** $V = \sigma^2$:

$$V = \sigma^2 = \int_{-\infty}^{\infty} (x - \langle x \rangle)^2 f(x)\, dx \tag{2.14}$$

σ wird **Standardabweichung** oder einfach **Fehler** genannt.

- Ein wichtiges Beispiel für eine diskrete Wahrscheinlichkeitsverteilung ist die **Poissonverteilung**. Sie gibt die Wahrscheinlichkeit P für das r-fache Auftreten bestimmter Ereignisse an:

$$P(r) = e^{-\mu} \cdot \frac{\mu^r}{r!}\,, \quad r = 0, 1, 2, \ldots \tag{2.15}$$

Bei der Poissonverteilung entsprechen sich Mittelwert $\langle r \rangle$ und Varianz V:

$$\mu = \langle r \rangle = V = \sigma^2 \tag{2.16}$$

Die Standardabweichung der Poissonverteilung wird bei Zählungen häufig benötigt:

$$\sigma = \sqrt{\mu} = \sqrt{N} \tag{2.17}$$

Zum Mitnehmen: Messwert & Messgenauigkeit 3

- Eine der wichtigsten Wahrscheinlichkeitsdichten ist die **Gaußverteilung (Normalverteilung)**

$$f(x) = \frac{1}{\sqrt{2\pi\sigma^2}} e^{-\frac{1}{2}\left(\frac{x-\mu}{\sigma}\right)^2} . \tag{2.18}$$

Die Parameter μ und σ haben die Bedeutung

Mittelwert (2.13) $\quad \mu \equiv \langle x \rangle = \int_{-\infty}^{\infty} x\, f(x)\, dx\,, \quad$ (2.19)

Varianz (2.14) $\quad \sigma^2 = V = \int_{-\infty}^{\infty} (x-\mu)^2\, f(x)\, dx\,, \quad$ (2.20)

Standardabweichung $\quad \sigma = \sqrt{V}\,. \quad$ (2.21)

Die Wahrscheinlichkeit, x im Intervall $\mu - n\sigma \leq x \leq \mu + n\sigma$ zu finden, ergibt sich durch Integration über die entsprechenden Grenzen:

Messwert innerhalb des Intervalls	Wahrscheinlichkeit
$\lvert x-\mu \rvert < 1\sigma$	68,26%
$\lvert x-\mu \rvert < 2\sigma$	95,45%
$\lvert x-\mu \rvert < 3\sigma$	99,73%

- Der **zentrale Grenzwertsatz** führt zu einer Gaußverteilung:

  Die Überlagerung von zufällig verteilten Fehlerquellen führt zur Gaußfunktion als Wahrscheinlichkeitsdichteverteilung im Fall von vielen Messungen.

- Für den allgemeinen Fall von Messwerten x_i erhalten wir unter der Annahme, dass alle Werte gleichwahrscheinlich auftreten, den **Stichproben-Mittelwert**

$$\overline{x} = \frac{1}{n} \sum_{i=1}^{n} x_i\,. \tag{2.28}$$

Er ist eine Schätzung des wahren Mittelwerts $\langle x \rangle$ und kommt ihm für große n beliebig nahe:

$$\lim_{n\to\infty} \overline{x} = \langle x \rangle \tag{2.29}$$

Zum Mitnehmen: Messwert & Messgenauigkeit 4

- Die Standardabweichung der Einzelmessung ist

$$\sigma = \sqrt{\frac{1}{n-1}\sum_{i=1}^{n}(x_i - \overline{x})^2} \, . \tag{2.32}$$

Die Schätzung des Mittelwerts $\overline{x}$ und der Varianz σ^2 sind selbst Zufallsvariablen und damit gaußverteilt. Für die **Standardabweichung des Mittelwerts** ergibt sich

$$\sigma_{\overline{x}} = \frac{\sigma}{\sqrt{n}} \, . \tag{2.34}$$

- Hängt eine Größe $y = f(x_1....x_n)$ von n voneinander unabhängigen Messgrößen x_k mit den Fehlern σ_{x_k} ab, so ergibt sich der Gesamtfehler nach dem **Gaußschen Fehlerfortpflanzungsgesetz** zu

$$\sigma_y^2 = \sum_{k=1}^{n} \left(\frac{\partial y}{\partial x_k}\right)^2_{\overline{x_k}} \sigma_{x_k}^2 \, . \tag{2.39}$$

Darin sind $(\partial y/\partial x_k)_{\langle x_k \rangle}$ die partiellen Ableitungen an der Stelle der Mittelwerte $\langle x_k \rangle$. Aus praktischen Gründen werden die Mittelwertschätzungen $\overline{x_k}$ als beste Näherung der wahren Mittelwerte $\langle x_k \rangle$ verwendet.

Wichtige Spezialfälle sind:

Addition/Subtraktion $y = \sum \pm x_k$
Quadrate der **absoluten Fehler** werden addiert $\sigma_y^2 = \sum \sigma_{x_k}^2$

Produkt/Quotient $y = \prod x_k^{\pm 1}$
Quadrate der **relativen Fehler** werden addiert $\sigma_y^2 = \sum (\sigma_{x_k}/\overline{x_k})^2$

Unkorrelierte systematische Fehler können quadratisch zu einem Gesamtfehler addiert werden.

Dynamik: Newtons Bewegungsgesetze

3

3.1 Newtons Axiome, Kraft 68
3.1.1 1. Trägheitsgesetz 68
3.1.2 2. Aktionsprinzip 69
3.1.3 3. Actio = Reactio 70

3.2 Masse 70
3.2.1 Schwerkraft und schwere Masse 71
3.2.2 Beschleunigung und träge Masse 72
3.2.3 Träge und schwere Masse sind äquivalent 74
3.2.4 Reibung 76
3.2.5 Bewegungsgleichung 76

3.3 Impuls 78

3.4 Newtons Bewegungsgesetz 79
3.4.1 Bewegungsgleichung für konstante Massen 79
3.4.2 Bewegungsgleichung für veränderliche Masse 80

ÜBERBLICK

» Mit »Dynamik« bezeichnen wir die Lehre von den Ursachen der Bewegungen. Dabei stellen wir Fragen wie:

- Warum fallen Körper gradlinig beschleunigt zu Boden?
- Warum kreisen Planeten um die Sonne?

Heute erscheint es uns selbstverständlich, dass die Ursache für eine Änderung der Bewegung eines Körpers in den Wechselwirkungen des Körpers mit seiner Umgebung liegen. Historisch ist diese Überlegung eine richtungsweisende Abstraktionsleistung.

In diesem Kapitel stellen wir Newtons Axiome vor, die Wechselwirkungen von Körpern mit ihrer Umgebung auf Kräfte zurückführen. Als nützliche Größe führen wir den Impuls ein und formulieren darüber Newtons Bewegungsgesetz in einer allgemeinen Form. Als Anwendung dieses Bewegungsgesetzes berechnen wir unter anderem die Flugbahn einer Rakete unter der Einwirkung der Schwerkraft. «

Beispiel 16: Kräfte

Kräfte $\vec{F}$ können verschiedener Art sein und verschieden große Reichweiten aufweisen (▶ Abbildung 3.1):

Auf einen massiven Körper wirkt die Gravitation: **Gewichtskraft** $\vec{F_G}$. Auf eine gespannte Feder wirkt eine Rückstellkraft: **Federkraft** $\vec{F_D}$. Auf einer rauen Oberfläche wird die Bewegung durch die **Reibungskraft** $\vec{F_R}$ behindert.

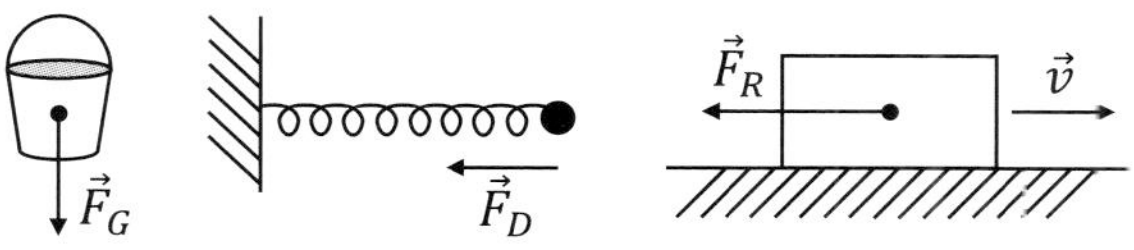

Abbildung 3.1 Links: Gewichtskraft $\vec{F_G}$, Mitte: Federkraft $\vec{F_D}$, rechts: Reibungskraft $\vec{F_R}$

Mathematischer Einschub: Vektoraddition

Kraft ist eine im Raum gerichtete Größe. Zu ihrer mathematischen Beschreibung nutzen wir Vektoren. Der Vektor zeigt in die Richtung der Kraft und hat als Betrag (Länge) die Größe der Kraft.

Vektoren lassen sich in Komponenten entlang der Raumrichtungen eines Koordinatensystems unserer Wahl zerlegen (z. B. kartesische Koordinaten, ▶ Abbildung 3.2):

$$\vec{F} = \begin{pmatrix} F_x \\ F_y \\ F_z \end{pmatrix} \tag{3.1}$$

$$\left|\vec{F}\right| = \sqrt{F_x^2 + F_y^2 + F_z^2} \tag{3.2}$$

Kräfte lassen sich nach den Regeln der Vektoraddition addieren und subtrahieren (▶ Abbildung 3.3, 3.4):

$$\vec{F_1} + \vec{F_2} = \begin{pmatrix} F_{x1} + F_{x2} \\ F_{y1} + F_{y2} \\ F_{z1} + F_{z2} \end{pmatrix}$$

$$\vec{F_2} = \begin{pmatrix} F_{x1} - F_{x2} \\ F_{y1} - F_{y2} \\ F_{z1} - F_{z2} \end{pmatrix}$$

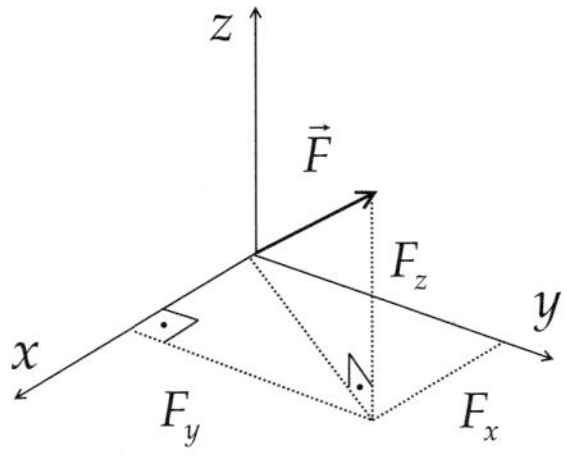

Abbildung 3.2 Kraftvektor und seine Komponenten im Raum

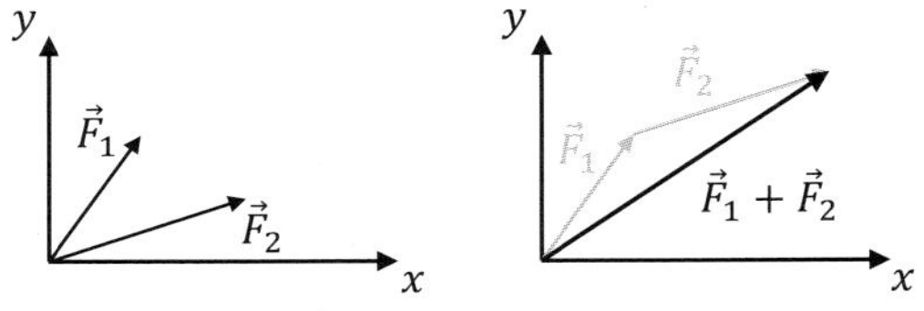

Abbildung 3.3 Addition zweier Kräfte (Vektoren)

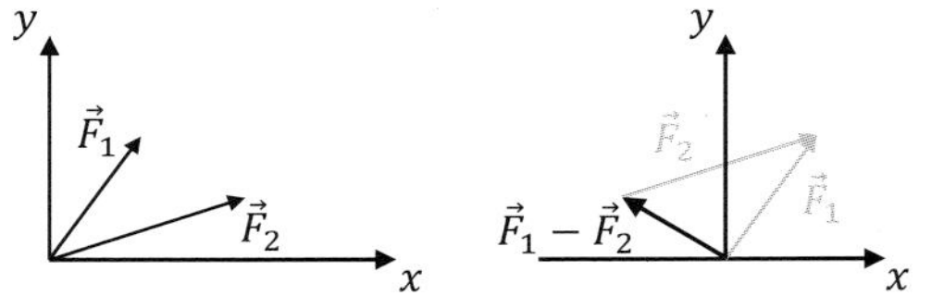

Abbildung 3.4 Subtraktion zweier Kräfte (Vektoren)

3.1 Newtons Axiome, Kraft

Isaac Newton veröffentlichte 1687 drei grundlegende Gesetze (Axiome) über Kräfte und Bewegung. Newtons Axiome bilden eine Grundlage zur mathematischen Beschreibung der Bewegung eines Körpers unter dem Einfluss von Kräften.

3.1.1 1. Trägheitsgesetz

Newtons Trägheitsgesetz lautet:

> **Ohne äußere Einwirkung verharrt jeder Körper im Zustand der Ruhe oder in gleichförmiger, gradliniger Bewegung.**

Auf Messgrößen bezogen bedeutet das Gesetz: Eine konstante Geschwindigkeit $\vec{v}$ eines Körpers ist gleichbedeutend damit, dass die Summe aller am Körper angreifenden Kräfte $\vec{F}_i$ verschwindet:

$$\vec{v} = const. \qquad \Leftrightarrow \qquad \sum_{i=1}^{n} \vec{F}_i = 0 \tag{3.3}$$

Um die Bedingungen zu erfüllen, gibt es z. B. folgende Möglichkeiten:

$$\vec{F} = 0 \tag{3.4}$$

$$\vec{F}_1 = -\vec{F}_2 \qquad \text{(linke ► Abbildung 3.5)} \tag{3.5}$$

$$\sum_{i=1}^{4} \vec{F}_i = 0 \qquad \text{(rechte ► Abbildung 3.5)} \tag{3.6}$$

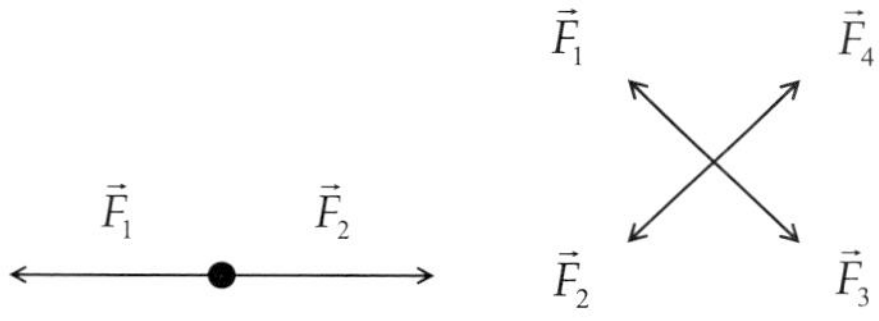

Abbildung 3.5 Entgegengesetzt gleiche Kräfte heben sich auf.

3.1.2 2. Aktionsprinzip

Newtons Aktionsprinzip heißt:

> **Eine Kraft, die auf einen massebehafteten Körper wirkt, ändert die Bewegung des Körpers.**

Ausgedrückt in Messgrößen erfährt ein Körper mit Masse m durch die Kraft $\vec{F}$ eine Beschleunigung $\vec{a} = \vec{F}/m$. Dieses Axiom kann auch als Definition der Kraft gesehen werden:

$$\vec{F} = m \cdot \vec{a} \tag{3.7}$$

Demnach wird eine Kraft durch die Beschleunigung $\vec{a}$ gemessen, die ein Probekörper der Masse m erfährt.

Die Kraft wird in der Einheit **Newton** angegeben:

$$\text{Newton} = \text{N} = \frac{\text{kg} \cdot \text{m}}{\text{s}^2} \tag{3.8}$$

3.1.3 3. Actio = Reactio

Newtons Gesetz von Wirkung und Gegenwirkung lautet:

Bei zwei Körpern, die nur untereinander wechselwirken, ist die Kraft $\vec{F}_1$ auf den einen Körper entgegengesetzt gleich der Kraft $\vec{F}_2$ auf den anderen Körper.

In der mathematischen Fomulierung heißt das Gesetz

$$\vec{F}_1 = -\vec{F}_2 \,. \tag{3.9}$$

Aufgabe 7: Trägheitsgesetz

Ein Körper hat die Geschwindigkeit $\vec{v} = \begin{pmatrix} 1{,}5 \\ 1 \end{pmatrix} \frac{\mathrm{m}}{\mathrm{s}}$.
Werfen ihn die fünf Kräfte (► Abbildung 3.6) aus seiner Bahn?

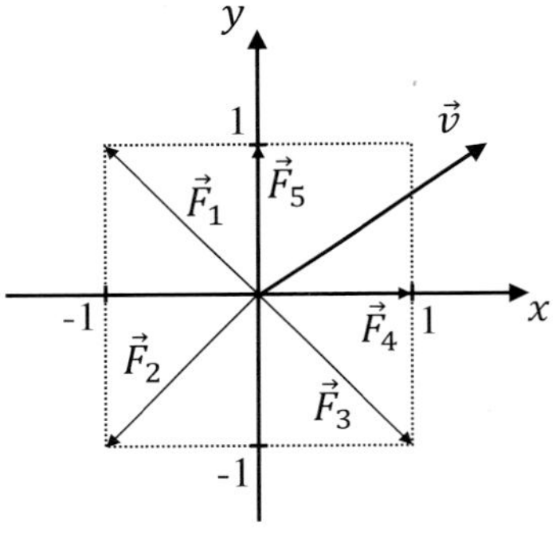

Abbildung 3.6 Fünf Kräfte wirken auf einen bewegten Körper.

3.2 Masse

Masse gehört zu den im Abschnitt 2.1.3 aufgelisteten Grundgrößen. Sie ist

- eine Körpereigenschaft[1],
- Ursache für die Schwere und die Trägheit eines Körpers,
- unverändert bei Deformation des Körpers,
- unabhängig vom Aggregatzustand (fest, flüssig, gasförmig).

1 Bei Elementarteilchen (d. h. Teilchen ohne messbare räumliche Ausdehnung) hat die Entdeckung des Higgs-Bosons am CERN-Beschleuniger ein physikalisches Verständnis für die Herkunft der Teilchenmasse ergeben: Ihre Masse beruht auf dem Kontakt des Teilchens mit dem sogenannten Higgs-Feld. Genaueres wird in Spezialvorlesungen zur Elementarteilchenphysik gelehrt.

Die Masseneinheit 1 kg wurde ursprünglich durch die Masse eines Platin-Iridium-Zylinders definiert, der in Paris aufbewahrt wird. Diese klassische Definition wurde 2019 durch eine atomare Definition ersetzt (Abschnitt 2.1.3).

3.2.1 Schwerkraft und schwere Masse

Die auf der Erde beobachtete Schwerkraft beruht auf der anziehenden Wechselwirkung zwischen Massen, die als Gravitation bezeichnet wird. Die Erde und ein darauf befindlicher Körper ziehen sich dabei gegenseitig mit einer Kraft an, die aufgrund des 3. Newtonschen Gesetzes (3.9) entgegengesetzt gleich groß ist.

Die Stärke der anziehenden Kraft F_G steigt mit der Masse m des Körpers. Die Wirkung der Erdanziehung können wir aufgrund der unveränderlichen Masse M_E der Erde in einer Konstanten g zusammenfassen und so die Masse m des Körpers über seine Gewichtskraft messen:

$$F_G = m\,g \tag{3.10}$$

Auf diesen Zusammenhang zwischen Schwerkraft und Gravitation werden wir in Kapitel 10 näher eingehen.

Als Messapparatur eignet sich eine Feder, die durch die Gewichtskraft des Körpers zusammengedrückt wird (► Abbildung 3.7).

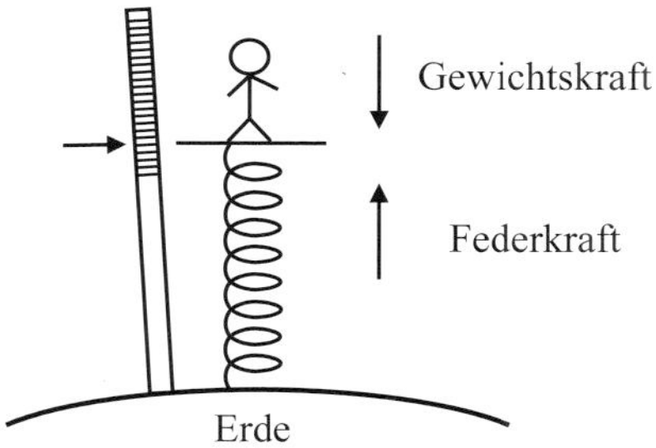

Abbildung 3.7 Massenbestimmung durch Messen der Gewichtskraft, z. B. durch Vergleich mit einer Federkraft

Nach dem aus der Schule bekannten Hookeschen Gesetz ist die rücktreibende Kraft $F_D = -k\,x$ einer Feder proportional zum Kompressionsweg x (die x-Koordinate zeigt nach unten). Damit können wir eine Feder zur Massenbestimmung eichen, indem wir z. B. die Auslenkung mit Vielfachen einer Testmasse bestimmen. Wenn sich Gewichts- und Federkraft ausgleichen, ist

$$F_G = -F_D \tag{3.11}$$

$$m\,g = k\,x \tag{3.12}$$

$$m = \frac{k}{g}\,x\,. \tag{3.13}$$

Beachten Sie: Masse ist nicht identisch mit dem Begriff der Gewichtskraft! Ein Astronaut behält seine Masse auch im Zustand der Schwerelosigkeit.

3.2.2 Beschleunigung und träge Masse

In dem folgenden Experiment untersuchen wir Kräfte, die eine beschleunigte Bewegung einer Masse verursachen. Bildet der von Newton als Kraft $\vec{F} = m \cdot \vec{a}$ (3.7) bezeichnete Begriff eine sinnvolle dynamische Größe?

Experiment 7: Luftkissenbahn

Eine Luftkissenbahn besteht aus einer Schiene mit vielen Löchern, durch die Luft geblasen wird. Kleine Wagen lassen sich damit wie auf einem Luftkissen nahezu reibungsfrei bewegen (► Abbildung 3.8).

Über eine Umlenkrolle am Ende der Bahn wird mit einem Faden die horizontale Bewegung des Wagens mit der vertikalen Bewegung eines Metallstücks mit der Masse m_G verbunden, das die Kraft F_G aufgrund der Erdanziehung erfährt:

$$F_G = m_G\, g \tag{3.14}$$

Diese Kraft ist die Ursache für die horizontale Bewegung des Wagens auf der Bahn. Wir beobachten, dass der Wagen durch die Gewichtskraft beschleunigt wird.

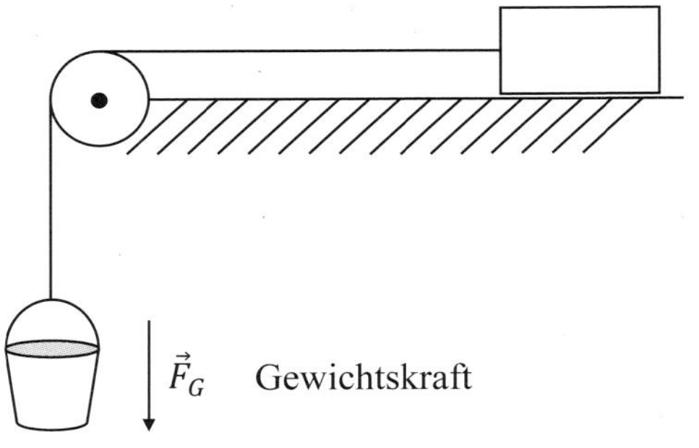

Abbildung 3.8 Überprüfung des Zusammenhangs von Masse und Beschleunigung an einer Luftkissenbahn

In unserem Experiment wird ein System bewegt, das aus dem Wagen, dem Massestück und dem Faden besteht. Wir bezeichnen die Masse des Wagens mit m_W und die des Fadens mit m_F. Zusammen mit der Masse m_G des herunterhängenden Metallstücks beträgt die gesamte Masse des beschleunigten Systems

$$M = m_W + m_G + m_F\,. \tag{3.15}$$

Die Beschleunigungskraft (3.7)

$$F_T = M\, a\,, \tag{3.16}$$

die das gesamte bewegliche System mit der Masse M erfährt, wird von der Kraft F_G des Metallstücks verursacht. Diese Gesamtmasse wird durch die Kraft F_G beschleunigt:

$$F_T = F_G \tag{3.17}$$

$$M\,a = m_G\,g \tag{3.18}$$

$$a = \frac{1}{M}\,m_G\,g \tag{3.19}$$

Der Bewegungsvorgang findet nahe des Erdbodens statt, wo die Beschleunigung g (3.14) und damit auch F_G in sehr guter Näherung konstant ist.

Lassen wir den Wagen am Punkt $x_\circ = 0$ aus der Ruhe heraus ($v_\circ = 0$) zur Zeit $t_\circ = 0$ starten, messen wir eine Bahnkurve, die sehr gut durch (1.33) beschrieben wird:

$$x = \frac{1}{2}\,a\,t^2 \tag{3.20}$$

Die Beschleunigung ist also über die zurückgelegte Strecke $\Delta x = x - x_\circ = x$ und die dabei vergangene Zeit $\Delta t = t - t_\circ = t$ konstant und messbar:

$$a = \frac{2x}{t^2} \tag{3.21}$$

Wir überprüfen jetzt das 2. Newtonsche Gesetz (3.7) bezüglich $F_T = M\,a$, indem wir die Masse m_W des Wagens variieren und die zugehörige Beschleunigung a messen. Mit dem Kehrwert der Beschleunigung erwarten wir eine leicht zu bestimmende Messgerade, deren Steigung dem Kehrwert von F_T entspricht:

$$\frac{1}{a} = \frac{1}{F_T}\,M \tag{3.22}$$

$$\frac{t^2}{2x} = \frac{1}{F_T}\,(m_W + m_G + m_F) \tag{3.23}$$

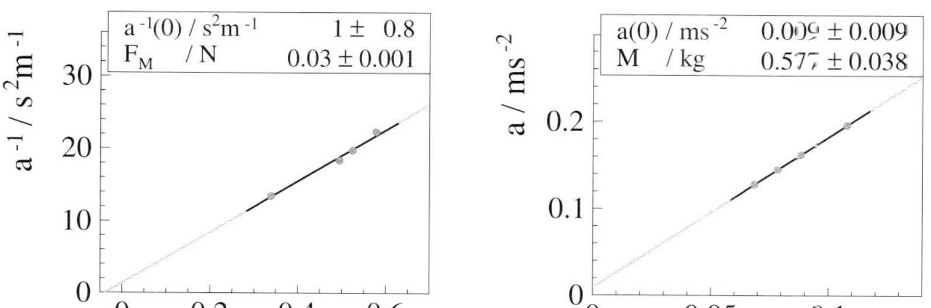

Abbildung 3.9 Lineare Zusammenhänge: links Kehrwert der Beschleunigung $1/a = t^2/(2x)$ in Abhängigkeit von der Wagenmasse m_W, rechts die Beschleunigung a aufgrund der angreifenden Kraft $F_G = m_G\,g$

Aus den Messwerten in der linken ► Abbildung 3.9 ergibt sich der erwartete lineare Zusammenhang zwischen der inversen Beschleunigung und der Wagenmasse. Damit erweist sich Newtons Kraftdefinition $\vec{F} = m \cdot \vec{a}$ (3.7) als sinnvolle dynamische Größe.

Im Experiment können wir alternativ die angreifende Kraft F_G mithilfe verschiedener Metallstücke m_G variieren und die Beschleunigung des Gesamtsystems messen (rechte ► Abbildung 3.9). Auch dieses Experiment zeigt Übereinstimmung mit Newtons Bewegungsgesetz.

In beiden Messungen geht die Anpassung nicht exakt durch den Ursprung. Die Achsenabschnitte mit ihren Fehlern spiegeln teilweise die Messungenauigkeiten wider, bringen aber auch zum Ausdruck, dass in beiden Fällen die beschleunigte Masse M zusätzlich zu m_W immer auch $m_G + m_F \ll m_W$ enthält.

3.2.3 Träge und schwere Masse sind äquivalent

In den beiden vorherigen Abschnitten haben wir Kräfte im Zusammenhang mit schwerer und träger Masse untersucht. Wir fassen hier die Unterschiede noch einmal zusammen:

- Gewichtskraft $F_G = m_G\, g$ (3.10):

  Ein Körper zeigt seine Masse als schwere Masse m_G aufgrund der anziehenden Schwerkraft zwischen dem Körper und der Erde. Je mehr Masse, desto größer ist die Gewichtskraft.

- Beschleunigungskraft auf einen Körper $F_T = m_T\, a$ (3.16):

  Ein Körper widersetzt sich aufgrund seiner Masse als träger Masse m_T einer Bewegungsänderung. Je mehr Masse ein Körper besitzt, desto größer muss die aufgewendete Kraft sein, um die gleiche Beschleunigung zu erreichen.

m_T und m_G sind zunächst zwei physikalische Größen, die durch verschiedene Messvorschriften definiert sind. m_G wird z. B. mit einer Federwaage gemessen und ist über g mit der Schwerkraft verbunden. m_T wird über den Zusammenhang zwischen Kraft und Beschleunigung bestimmt und ist damit mit Newtons 2. Axiom verbunden.

Bei dem Luftkissenbahnversuch in Abschnitt 3.2.2 haben wir zunächst stillschweigend für die kleine, ziehende Masse $m_T = m_G$ gesetzt. Seine Gewichtskraft m_G greift unter anderem seine eigene träge Masse m_T an, die ja Teil des gesamten bewegten Systems aus Wagen, Faden und Gewicht ist (3.15). Die Annahme $m_T = m_G$ bedarf aber streng genommen einer experimentellen Prüfung.

Galilei, Eötvös

Historisch wurde der Zusammenhang zwischen m_T und m_G in den erstmals von Galileo Galilei (1564–1642) durchgeführten Fallexperimenten untersucht, die er an einer schiefen Ebene verlangsamte. Galilei stellte im Widerspruch zur damals vorherrschenden Lehre von Aristoteles fest, dass alle Körper gleich schnell fallen. Das weist auf eine gleichförmig beschleunigte Bewegung fallender Körper hin (die Falldauer $t = \sqrt{2\,h/a}$ (1.33) aus der Höhe h ist nur von der konstanten Beschleunigung a abhängig).

In der Formulierung von Newtons 2. Axiom heißt das für die Kraft auf den fallenden Körper mit träger Masse m_T

$$F_T = m_T\, a \quad \text{mit} \quad a = const.\,. \tag{3.24}$$

Andererseits ist für denselben Körper der schweren Masse m_G die Schwerkraft mit der masse- und materialunabhängigen Konstante der Erdanziehung g

$$F_G = m_G\, g \quad \text{mit} \quad g = const.\,. \tag{3.25}$$

Sind die Kräfte F_T und F_G gleich groß, so gilt wegen der konstanten Faktoren a und g zumindest eine Proportionalität zwischen m_T und m_G für alle Körper unabhängig von Material und Masse:

$$m_T\, a = m_G\, g \tag{3.26}$$

Damit kann man m_T und m_G auch gleichsetzen.

Mit späteren Pendelexperimenten von Newton wurden immer ausgefeiltere Untersuchungen zur Äquivalenz von m_T und m_G angestellt, besonders vom ungarischen Physiker Eötvös und seinen Nachfolgern. Experimentell findet man mit einer Genauigkeit von 10^{-13} keinen Unterschied zwischen der schweren und der trägen Masse eines Körpers:

$$m_G = m_T = m \tag{3.27}$$

Einstein

Das folgende Gedankenexperiment wurde von Albert Einstein verwendet, um die Gleichheit von träger und schwerer Masse zu verdeutlichen.

Experiment 8: Gedankenexperiment Fahrstuhl

Nach der linken ► Abbildung 3.10 denken wir uns einen in Ruhe befindlichen Fahrstuhl ($\vec{v} = 0$) im Gravitationsfeld der Erde $\vec{a} = \vec{g}$. Wir vergleichen diese Situation in der rechten ► Abbildung 3.10 mit einem mit $\vec{a} = -\vec{g}$ gleichmässig beschleunigten Fahrstuhl ohne Gravitationsfeld, dessen Geschwindigkeit mit $\vec{v} = -\vec{g}\,t$ zunimmt.

In beiden Fällen wirkt auf den Eimer die durch die Federauslenkung gemessene Kraft

$$\vec{F} = m_G\, \vec{g} = -m_T\, \vec{a} = m_T\, \vec{g}\,. \tag{3.28}$$

Für den Beobachter im Fahrstuhl sind für $m_G = m_T$ beide Situationen nicht unterscheidbar!

Die Äquivalenz von träger und schwerer Masse ist eine der Grundlagen für die Allgemeine Relativitätstheorie, die im fortgeschrittenen Studium gelehrt wird

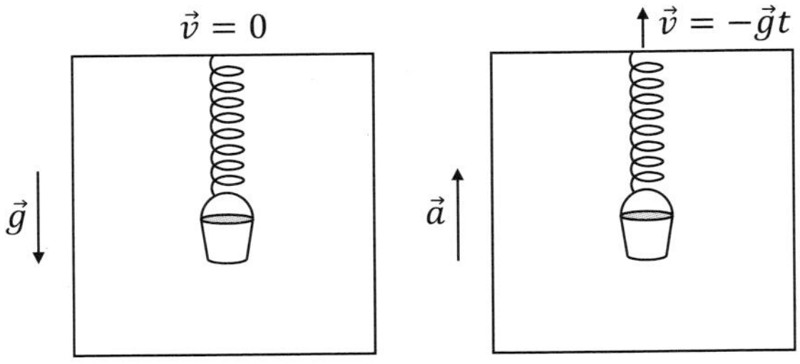

Abbildung 3.10 Links: Gewichtsmessung im ruhenden Fahrstuhl, rechts: Kraftmessung im beschleunigten Fahrstuhl

3.2.4 Reibung

Als Beispiele für Kräfte hatten wir in ► Abbildung 3.1 neben Schwerkraft und Federkraft auch Reibungskräfte genannt. Reibung ist im täglichen Leben allgegenwärtig und aus dem Schulunterricht bekannt. Dabei denkt man meist an die Reibungskräfte F_R zwischen festen Körpern, die durch die empirische Gleichung

$$\vec{F}_R = -\mu \cdot F_N \cdot \frac{\vec{v}}{v} \tag{3.29}$$

beschrieben werden. Darin ist F_R die Reibungs- oder Bremskraft, die als **Gleit- oder Rollreibung** einer Bewegung in Richtung $\vec{v}$ entgegenwirkt, F_N bezeichnet die Normalkraft, die senkrecht zur Berührungsfläche z. B. als Gewichtskraft wirkt. Der **Reibungskoeffizient** μ kann sehr unterschiedliche Werte annehmen, z. B. zwischen $\mu \approx 1$ für Autoreifen auf trockener Straße und $\mu \approx 0{,}01$ für Stahl auf Eis. Für **Haftreibung** bei $\vec{v} = 0$ ist μ im Allgemeinen etwas größer. F_R hängt nicht oder nur wenig von der Größe der Berührungsfläche ab.

Die beschriebenen Reibungskräfte hängen von der Richtung der Bewegung, nicht aber von der Größe der Geschwindigkeit ab. Geschwindigkeitsabhängige Reibung tritt vor allem in Flüssigkeiten und Gasen auf und wird in Abschnitt 18.6 eingehend beschrieben.

3.2.5 Bewegungsgleichung

Eine zentrale Bedeutung erhält das 2. Newtonsche Gesetz (3.7) bei folgender Verwendung: Bekannt seien die Kraft F und die Masse m des beschleunigten Körpers. Dann kann die Ortsfunktion des Körpers folgendermaßen berechnet werden:

$$m\,a = F \tag{3.30}$$

$$a = \frac{F}{m} \tag{3.31}$$

$$\frac{dv}{dt} = \frac{F}{m} \tag{3.32}$$

$$\frac{d^2x}{dt^2} = \frac{F}{m} \tag{3.33}$$

Die Kraft F kann sowohl vom Ort x (z. B. Federkraft $F = -kx$) als auch beispielsweise bei der Reibung von der Geschwindigkeit $v = dx/dt$ abhängen. Genauer können wir also

$$\frac{d^2x}{dt^2} = \frac{1}{m}F\left(x, \frac{dx}{dt}\right) \tag{3.34}$$

schreiben. Diese Gleichung bezeichnen wir als **Bewegungsgleichung**. Sie enthält physikalische Größen und deren Ableitungen und ist daher eine **Differentialgleichung**. Differentialgleichungen spielen in der Physik bei der Beschreibung von Vorgängen eine entscheidende Rolle.

Mathematischer Einschub: Differentialgleichung

Eine Differentialgleichung ist eine Gleichung für eine gesuchte Funktion, in der auch Ableitungen dieser Funktion vorkommen.

Nach der Integration von (3.32) erhalten wir

$$v(t) = \int_0^t \frac{F}{m}\,dt + v_\circ\,. \tag{3.35}$$

Die **Lösung der Bewegungsgleichung** ist die Ortsfunktion bzw. die **Bahnkurve**

$$x(t) = \int_0^t \left(\int_0^t \frac{F}{m}\,dt + v_\circ\right) dt + x_\circ\,. \tag{3.36}$$

Damit ist der Weg des Körpers als Resultat seiner Anfangskinematik ($x_\circ, v_\circ$) und der von außen auf ihn einwirkenden Kraft F bekannt.

Newtons 2. Gesetz (3.7) formuliert also einerseits die **Definition der Kraft**, *auf der anderen Seite ein fundamentales* **Bewegungsgesetz**, *mit dem physikalische Vorgänge vorhergesagt werden können.*

Aufgabe 8: Freier Fall

Stellen Sie mithilfe von Newtons 2. Gesetz (3.7)

$$m\,\vec{a} = \vec{F}$$

die Bewegungsgleichung für den freien Fall einer Kugel in der senkrechten Koordinate y auf (▶ Abbildung 3.11) und berechnen Sie die Bahnkurve $y(t)$.

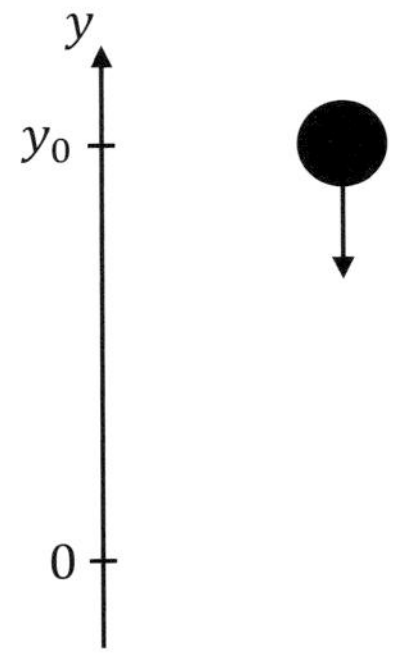

Abbildung 3.11 Freier Fall

3.3 Impuls

Mit dem Impuls führen wir eine weitere dynamische Größe ein, die Masse und Bewegung eines Körpers verbindet. Der Impuls ist als das Produkt aus Masse und Geschwindigkeit eines Objekts definiert:

$$\vec{p} = m\,\vec{v} \tag{3.37}$$

Da die zeitliche Ableitung der Geschwindigkeit einer Beschleunigung entspricht (1.6), gilt bei einer konstanten Masse $m = const.$ des Körpers

$$\frac{d\vec{p}}{dt} = m\,\vec{a}\,. \tag{3.38}$$

In diesem Fall ist die Impulsänderung direkt proportional zur Beschleunigung eines Körpers, wobei die Masse m die Proportionalitätskonstante ist. Wegen Newtons Kraftdefinition $\vec{F} = m\,\vec{a}$ (3.7) kann eine Impulsänderung durch eine von außen einwirkende Kraft F verursacht werden.

Ist die Körpermasse während des Bewegungsvorgangs verändert ($m \neq const.$), müssen wir die Produktregel bei der Ableitung beachten:

$$\frac{d\vec{p}}{dt} = \frac{dm}{dt}\,\vec{v} + m\,\frac{d\vec{v}}{dt} \tag{3.39}$$

$$= \frac{dm}{dt}\,\vec{v} + m\,\vec{a} \tag{3.40}$$

Der erste Term der rechten Seite muss z. B. bei einer Rakete berücksichtigt werden, die nach dem Start Masse zu ihrer eigenen Beschleunigung ausstößt.

Im folgenden Abschnitt 3.4 werden wir verschiedene Anwendungen untersuchen.

3.4 Newtons Bewegungsgesetz

Mit dem Impuls $\vec{p} = m\,\vec{v}$ (3.37) lässt sich Newtons 2. Gesetz (3.7) in allgemeiner Form als Bewegungsgleichung schreiben:

$$\frac{d\vec{p}}{dt} = \sum_{i=1}^{n} \vec{F}_i \tag{3.41}$$

Bei bekannten Kräften $\vec{F}_i$, die auf einen Körper mit Impuls $\vec{p}$ einwirken, ist die zeitliche Änderung seines Impulses $d\vec{p}/dt$ festgelegt. Durch Integration ist die Bahnkurve des Körpers vollständig berechenbar!

Die wichtigste Herausforderung besteht in der Formulierung von Newtons Bewegungsgesetz für den Anwendungsfall. Die Berechnung der Bahnkurve ist dann die zweite Herausforderung.

Bei unveränderlicher Körpermasse $m = const.$ können wir für die zeitliche Ableitung des Impulses $d\vec{p}/dt$ das Produkt $m\,\vec{a}$ einsetzen (3.38), die angreifenden Kräfte benennen und der Berechnung der Bahnkurve wie in Abschnitt 3.2.5 gezeigt folgen.

Bei veränderlichen Massen ($m \neq const.$) wird zunächst die linke Seite von Newtons Bewegungsgesetz – wie in (3.40) gezeigt – nach der Produktregel abgeleitet und auf der rechten Seite werden die angreifenden Kräfte eingesetzt. Anschließend erfolgt die Berechnung der Bahnkurve.

Das Gesetz gilt für makroskopische Systeme. Für mikroskopische Systeme (Atome, Elementarteilchen) benötigen wir andere Bewegungsgesetze, die wir im Band »Atome und Kerne« kennenlernen werden.

Das Bewegungsgesetz (3.41) eignet sich insbesondere für Translationsbewegungen, d. h. gerichtete Bewegungen, die nicht auf einen Raumpunkt bezogen sind wie z. B. Drehungen (Rotationsbewegungen). Dafür werden wir später ein weiteres geeignetes Bewegungsgesetz kennenlernen.

3.4.1 Bewegungsgleichung für konstante Massen

Wir zeigen hier die erforderlichen Schritte, um die Bahnkurve für das Experiment mit der Luftkissenbahn (Abschnitt 3.2.2) zu berechnen.

Beispiel 17: Anwendung Luftkissenbahn

Beim Experiment mit der Luftkissenbahn besteht das bewegte System aus Wagen, Gewicht und Faden. Seine Gesamtmasse M beträgt $M = m_W + m_G + m_F$ (3.15). Zusammen mit der Geschwindigkeit des Gesamtsystems v lautet der Impuls des Gesamtsystems

$$p = M\,v\,. \tag{3.42}$$

Die Kraft auf das Gesamtsystem wird vom Gewicht $F_G = m_G\,g$ (3.14) verursacht, sodass wir als Bewegungsgleichung (3.41) Folgendes formulieren:

$$\frac{dp}{dt} = F_G \tag{3.43}$$

Durch Einsetzen erhalten wir eine gleichmäßig beschleunigte Bewegung mit der effektiven Beschleunigung $a = dv/dt$ wie folgt:

$$\frac{d}{dt}(M\,v) = m_G\,g \tag{3.44}$$

$$M\,\frac{dv}{dt} = m_G\,g \tag{3.45}$$

$$a = \frac{m_G\,g}{M} = const. \tag{3.46}$$

Für die Anfangsbedingungen $x_\circ = 0$ und $v_\circ = 0$ des bewegten Systems ergibt sich die Bahnkurve als Lösung dieser Bewegungsgleichung nach zweifacher Integration genau wie bei (1.33) beziehungsweise (3.20):

$$x(t) = \frac{1}{2}\,a\,t^2 \tag{3.47}$$

3.4.2 Bewegungsgleichung für veränderliche Masse

Unser Ziel ist, die Bewegungsgleichung einer Rakete aufzustellen und ihre Bahnkurve zu berechnen.

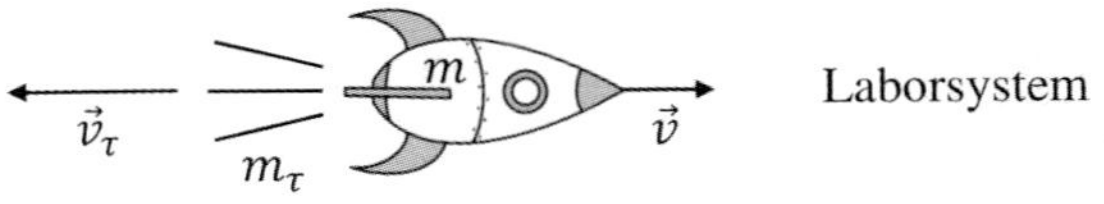

Abbildung 3.12 Eine Rakete der Masse m und Geschwindigkeit $\vec{v}$ stößt Treibgase der Masse m_T mit der Geschwindigkeit $\vec{v}_T$ aus.

Diese Aufgabe ist anspruchsvoll, aber mit unseren bisher gelernten Mitteln durchführbar.

Impulse der Rakete und des ausströmenden Gases

Zunächst untersuchen wir die Impulsänderung der Rakete. Ihr Impuls zum Zeitpunkt t beträgt

$$\vec{p}(t) = m(t) \cdot \vec{v}(t) \,. \tag{3.48}$$

Zum Zeitpunkt $t + dt$ ist der Impuls

$$\vec{p}(t + dt) = m(t + dt) \cdot \vec{v}(t + dt) \tag{3.49}$$

$$= (m - dm_T)(\vec{v} + d\vec{v}) \,. \tag{3.50}$$

Die Rakete hat die Treibstoffmasse dm_T verloren und dafür an Geschwindigkeit $d\vec{v}$ zugelegt. Die Impulsänderung der Rakete pro Zeitintervall dt ist

$$d\vec{p} = \vec{p}(t + dt) - \vec{p}(t) \tag{3.51}$$

und wird durch den Impuls des ausgestoßenen Gases

$$d\vec{p}_T = dm_T \, \vec{v}_T \tag{3.52}$$

ausgeglichen.

Der Impuls $\vec{P}$ des *Gesamtsystems* aus Rakete und ausströmendem Gas ändert sich dann folgendermaßen:

$$d\vec{P} = \vec{p}(t + dt) - \vec{p}(t) + d\vec{p}_T \tag{3.53}$$

$$= (m - dm_T)(\vec{v} + d\vec{v}) - m\vec{v} + dm_T \, \vec{v}_T \tag{3.54}$$

$$= m\vec{v} + m \, d\vec{v} - \vec{v} \, dm_T - dm_T \, d\vec{v} - m\vec{v} + dm_T \, \vec{v}_T \tag{3.55}$$

$$= m \, d\vec{v} - \vec{v} \, dm_T - dm_T \, d\vec{v} + dm_T \, \vec{v}_T \tag{3.56}$$

$$\approx m \, d\vec{v} + (\vec{v}_T - \vec{v}) \, dm_T \tag{3.57}$$

Im letzten Schritt haben wir den Term mit zwei differentiellen Änderungen $dm_T \, d\vec{v}$ vernachlässigt (eine Korrektur in zweiter Ordnung).

Die Geschwindigkeit $\vec{v}_T^*$ des Gases im Ruhesystem der Rakete ergibt sich aus der Differenz der Gas-Austrittsgeschwindigkeit $\vec{v}_T$ und der Raketengeschwindigkeit $\vec{v}$ (► Abbildung 3.12). Bei gleichmäßigem Abbrennen des Brennstoffs der Rakete ist

$$\vec{v}_T^* = \vec{v}_T - \vec{v} \approx const. \,, \tag{3.58}$$

wobei die Richtungen $\vec{v}_T^*$ und $\vec{v}$ entgegengesetzt sind.

Die aktuelle Masse der Rakete zum Zeitpunkt t resultiert aus der Differenz der Startmasse und der Masse des ausgestoßenen Gases:

$$m = \underbrace{m_0}_{\text{Startmasse}} - \underbrace{m_T}_{\text{verbrauchter Treibstoff}} \tag{3.59}$$

Die differentielle Abnahme der Raketenmasse entspricht der differentiellen Zunahme des ausgestoßenen Gases:

$$dm = -dm_T \tag{3.60}$$

Damit vereinfacht sich die Änderung des Gesamtimpulses (3.57) der Rakete und des ausströmenden Gases zu

$$d\vec{P} = m \, d\vec{v} - \vec{v}_T^* \, dm \,. \tag{3.61}$$

Raketengeschwindigkeit ohne Einwirkung äußerer Kräfte

Mit Newtons Bewegungsgesetz (3.41) ergibt sich ohne Einwirkung einer äußeren Kraft keine zeitliche Impulsänderung des Gesamtsystems:

$$\frac{d\vec{P}}{dt} = 0 \tag{3.62}$$

Der Impuls des Gesamtsystems aus Rakete und Gas bleibt somit unverändert bzw. erhalten: Wir sprechen von **Impulserhaltung**. Darauf werden wir in Abschnitt 7.2 näher eingehen.

Mit (3.62) müssen sich die Terme der rechten Seite von (3.61) ausgleichen und wir erhalten

$$m\, d\vec{v} = \vec{v}_{\tau}^{\,*}\, dm\,. \tag{3.63}$$

Wir fassen die Variablen jeweils auf einer der beiden Seiten der Gleichung passend zusammen (Trennung der Variablen) und führen eine Integration durch. Die Integrationsgrenzen ergeben sich links aus der Anfangsgeschwindigkeit $\vec{v}_\circ = 0$ und aus der Geschwindkeit $\vec{v}(t)$ zu einem bestimmten Zeitpunkt t. Auf der rechten Seite ist anfangs die Startmasse $m_\circ$ vorhanden und die verbleibende Masse $m(t) = m_\circ - m_\tau(t)$ (3.59) zum selben Zeitpunkt t:

$$\int_0^{\vec{v}(t)} d\vec{v} = \vec{v}_{\tau}^{\,*} \int_{m_\circ}^{m_\circ - m_\tau(t)} \frac{dm}{m} \tag{3.64}$$

$$\vec{v}(t) = -\vec{v}_{\tau}^{\,*} \cdot \underbrace{\ln\left(\frac{m_\circ}{m_\circ - m_\tau(t)}\right)}_{\geq 0} \tag{3.65}$$

Das Minuszeichen zeigt an, dass die Flugrichtung der Rakete und die Richtung des ausgestoßenen Gases entgegengesetzt sind. Neben der Gasgeschwindigkeit $\vec{v}_{\tau}^{\,*}$ hängt die Raketengeschwindigkeit auch von der ausgestoßenen Antriebsmasse $m_\tau(t)$ ab, wie das folgende Experiment eindrücklich zeigt:

> *Entscheidend für die Geschwindigkeit der Rakete ist der Rückstoß in Form des Impulses der ausgestoßenen Gasmasse!*

Experiment 9: Spielzeugrakete

Mit einer Spielzeugrakete kann man die Bedeutung des Treibstoffimpulses schön veranschaulichen.

Füllt man den Tank der Rakete halb mit Wasser der Masse $m_\tau = m_{\text{Wasser}}$ und setzt das verbleibende Volumen mit einer Luftpumpe unter Druck, so wird die Rakete vom Impuls des herausspritzenden Wassers angetrieben.

Pumpt man stattdessen ausschließlich Luft der Masse $m_\tau = m_{\text{Luft}}$ in den Tank der Rakete, reicht der Impuls der ausströmenden Luft nur für marginale Flugstrecken aus.

Aufgabe 9: Rakete

Begründen Sie anhand der Gleichung (3.65) das unterschiedliche Verhalten der Rakete beim Befüllen mit Wasser oder Luft.

Geschwindigkeit beim senkrechten Raketenstart

Beim senkrechten Start der Rakete wirkt die Gewichtskraft $\vec{F}_G$ gegen die Startrichtung (▶ Abbildung 3.13). Wir untersuchen hier eine eindimensionale Bewegung, bei der wir die positive z-Richtung als Koordinate in Richtung der Raketenspitze wählen. Dementsprechend zeigt die sich mit der Zeit ändernde Gewichtskraft der Rakete in die negative z-Richtung:

$$F_G = -m\,g \tag{3.66}$$

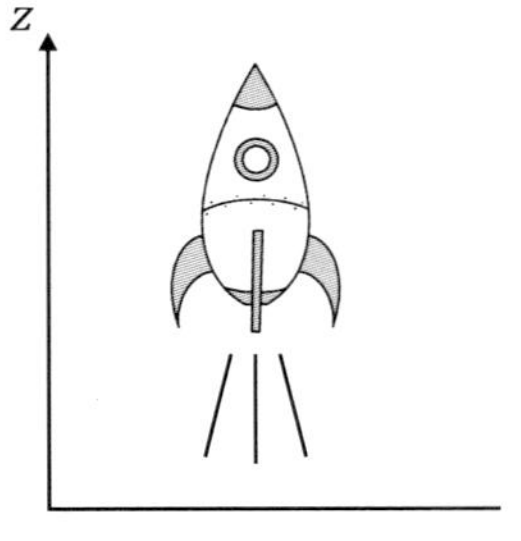

Abbildung 3.13 Senkrechter Raketenstart

In Newtons 2. Gesetz (3.41) ist die Gewichtskraft F_G die äußere angreifende Kraft. Mit der Impulsänderung des Gesamtsystems aus Rakete und ausströmendem Gas nach (3.61) ergibt sich folgende Zeitverteilung der Raketengeschwindigkeit $v(t)$:

$$\frac{dP}{dt} = F_G \tag{3.67}$$

$$m\,\frac{dv}{dt} - v_\tau^*\,\frac{dm}{dt} = -m\,g \tag{3.68}$$

$$dv = v_\tau^*\,\frac{dm}{m} - g\,dt \tag{3.69}$$

$$\int_0^{v(t)} dv = v_\tau^* \int_{m_\circ}^{m_\circ - m_\tau(t)} \frac{dm}{m} - g\int_0^t dt \tag{3.70}$$

$$v(t) = -v_\tau^*\,\ln\left(\frac{m_\circ}{m_\circ - m_\tau(t)}\right) - g\,t \tag{3.71}$$

Das erste Minuszeichen auf der rechten Seite von (3.71) ist wie in (3.65) der Indikator dafür, dass die Rakete entgegen der Richtung des ausströmenden Gases fliegt, also in positive z-Richtung. Der rechtsstehende Term zeigt, dass die Raketengeschwindigkeit durch die angreifende Gewichtskraft reduziert wird.

Bahnkurve der senkrecht startenden Rakete

Wir wollen nun auch die Bahnkurve der senkrecht startenden Rakete berechnen. Dafür beschreiben wir in der Raketengeschwindigkeit $v = dz/dt$ (3.71) das gleichmäßige Abbrennen des Treibstoffs mit dem Term $m_\tau(t) = q\,t$. Dabei ist q der Quotient des ausgestoßenen Gases pro Zeit:

$$q = \frac{\Delta m}{\Delta t} = const. \tag{3.72}$$

$$\frac{dz}{dt} = -v_\tau^* \ln\left(\frac{m_\circ}{m_\circ - q\,t}\right) - g\,t \tag{3.73}$$

$$= v_\tau^* \ln\left(\frac{m_\circ - q\,t}{m_\circ}\right) - gt \tag{3.74}$$

$$= v_\tau^* \ln\left(1 - \frac{q\,t}{m_\circ}\right) - gt \tag{3.75}$$

Um Schreibarbeit zu vermeiden, führen wir eine Kurzschreibweise für den Quotienten aus q (ausgestoßenes Gas pro Zeit) und der Startmasse $m_\circ$ der Rakete ein:

$$\mu \equiv \frac{q}{m_\circ} \tag{3.76}$$

Die Bahnkurve $z(t)$ erhalten wir durch Integration der Raketengeschwindigkeit:

$$\int_0^z dz = v_\tau^* \int_0^t \ln \underbrace{(1 - \mu\,t)}_{\equiv s}\, dt - g \int_0^t t\,dt \tag{3.77}$$

Das linke Integral und das Integral ganz rechts können wir sofort lösen. Das Integral über den logarithmischen Term führen wir auf die Integration der Logarithmusfunktion zurück:

$$\int \ln s\, ds = s\,(\ln s - 1) \tag{3.78}$$

Dafür nutzen wir folgende Transformation der Variablen t und dt:

$$s \equiv 1 - \mu\,t \tag{3.79}$$

$$\frac{ds}{dt} = -\mu \qquad \Rightarrow \qquad dt = -\frac{1}{\mu}\,ds \tag{3.80}$$

Beachten Sie bei der Berechnung des Integrals auch die Transformation der Integrationsgrenzen:

$$\int_0^t \ln(1 - \mu\,t)\,dt = -\frac{1}{\mu} \int_{s(0)}^{s(t)} \ln(s)\,ds \tag{3.81}$$

$$= -\frac{1}{\mu}\left[\, s(t)\,[\ln(s(t)) - 1] \;-\; s(0)\,[\ln(s(0)) - 1]\,\right] \tag{3.82}$$

$$= -\frac{1}{\mu}\,\left[\,(1 - \mu\,t)\;[\ln(1 - \mu\,t) - 1] + 1\right] \tag{3.83}$$

$$= -\frac{1}{\mu}\,(1 - \mu\,t)\,\ln(1 - \mu\,t) + \frac{1}{\mu}\,(1 - \mu\,t) - \frac{1}{\mu} \tag{3.84}$$

$$= -\frac{1}{\mu}\,(1 - \mu\,t)\,\ln(1 - \mu\,t) - t \tag{3.85}$$

Einsetzen in Gleichung (3.77) zusammen mit $\mu = q/m_\circ$ (3.76) liefert die Bahnkurve der Rakete:

$$z(t) = v_\tau^* \left[-\frac{1}{\mu} \left(1 - \mu\, t\right) \ln\left(1 - \mu\, t\right) - t \right] - \frac{g}{2}\, t^2 \tag{3.86}$$

$$= -v_\tau^* \left[\left(\frac{m_\circ}{q} - t \right) \ln\left(1 - \frac{q\, t}{m_\circ}\right) + t \right] - \frac{g}{2}\, t^2 \tag{3.87}$$

Der Term $-v_\tau^* > 0$ zeigt, dass die Rakete entgegengesetzt zur Richtung des ausströmenden Gases fliegt (siehe (3.65)).

Schubkraft der Rakete

Wird die Rakete beim Start wie gewünscht abheben? Die Bahnkurve (3.87) sieht kompliziert aus, wir untersuchen im Folgenden, wie die Startphase der senkrecht startenden Rakete verläuft. Direkt nach dem Start ist die Zeit t klein, d. h. der logarithmische Term lässt sich nähern.

Mathematischer Einschub: Taylor-Entwicklung

Das wesentliche Verhalten einer komplizierten Funktion $f(x)$ kann man in der Nähe eines vorgegebenen Funktionswerts $x_\circ$ durch eine Taylor-Entwicklung erfassen.

Die erste, zweite und dritte Ableitung der Funktion $f(x)$ nennen wir hier $f'(x)$, $f''(x)$ und $f'''(x)$. Den Wertebereich um $x_\circ$ beschreiben wir durch $\Delta x = x - x_\circ$.

Dann ist die Taylor-Entwicklung für die Funktion $f(x)$ um die Stelle $x_\circ$ einschließlich der 3. Ordnung:

$$f(x)|_{x_\circ} \approx f(x_\circ) + f'(x_\circ)\,\Delta x + \frac{1}{2!}\, f''(x_\circ)\,\Delta x^2 + \frac{1}{3!}\, f'''(x_\circ)\,\Delta x^3 \ldots \tag{3.88}$$

Der Einschluss von Termen noch höherer Ordnung (4. Ableitung etc.) ist möglich.

Uns interessiert hier das Verhalten des ersten Terms der Ortsfunktion $z(t)$ (3.86):

$$f(t) = -\frac{1}{\mu}\left(1 - \mu t\right) \ln\left(1 - \mu t\right) \tag{3.89}$$

Für die Taylor-Entwicklung von $f(t)$ benötigen wir den Funktionswert und die dazugehörigen Ableitungen mit ihrem Wert beim Start bei $t = 0$:

$$f(0) = 0 \tag{3.90}$$

$$f'(t) = -\frac{1}{\mu}\left(-\mu\right) \ln\left(1 - \mu t\right) - \frac{1}{\mu}\left(1 - \mu t\right) \frac{-\mu}{1 - \mu t} \tag{3.91}$$

$$= \ln\left(1 - \mu t\right) + 1 \;\Rightarrow\; f'(0) = 1 \tag{3.92}$$

$$f''(t) = \frac{-\mu}{1 - \mu t} \;\Rightarrow\; f''(0) = -\mu \tag{3.93}$$

$$f'''(t) = \frac{-\mu^2}{(1 - \mu t)^2} \;\Rightarrow\; f'''(0) = -\mu^2 \tag{3.94}$$

Die Taylor-Entwicklung lautet damit

$$f(t)|_0 \approx f(0) + f'(0)\,t + \frac{1}{2!}\,f''(0)\,t^2 + \frac{1}{3!}\,f'''(0)\,t^3 \tag{3.95}$$

$$= 0 + t - \frac{\mu}{2}\,t^2 - \frac{\mu^2}{6}\,t^3\,. \tag{3.96}$$

Mit dieser Näherung vereinfacht sich die Ortsfunktion der Rakete (3.86) zu einem zeitabhängigen Polynom dritten Grades. Wir setzen für μ (3.76) ein und erhalten die Bahnkurve am Beginn des senkrechten Raketenstarts:

$$z(t) \approx v_\tau^* \left[\left(t - \frac{\mu}{2}\,t^2 - \frac{\mu^2}{6}\,t^3\right) - t\right] - \frac{g}{2}\,t^2 \tag{3.97}$$

$$= v_\tau^* \left[\left(-\frac{q}{2m_\circ}\,t^2 - \frac{q^2}{6{m_\circ}^2}\,t^3\right)\right] - \frac{g}{2}\,t^2 \tag{3.98}$$

$$= -v_\tau^* \frac{q}{2m_\circ}\,t^2 - \frac{g}{2}\,t^2 - v_\tau^* \frac{q^2}{6{m_\circ}^2}\,t^3 \tag{3.99}$$

$$= \frac{1}{2}(\underbrace{-v_\tau^*}_{\geq 0} \frac{q}{m_\circ} - g\,)\,t^2 \underbrace{-v_\tau^*}_{\geq 0} \frac{q^2}{6{m_\circ}^2}\,t^3 \tag{3.100}$$

Die Brenndauer t_B der Rakete errechnen wir aus der gesamten Treibstoffmasse m_S und der Brennkonstante $q = \Delta m/\Delta t = m_S/t_B$ (3.72). Damit erhalten wir die Ortsfunktion

$$z(t) = \frac{1}{2}\left(\frac{m_S}{m_\circ}\frac{(-v_\tau^*)}{t_B} - g\right) t^2 + \frac{1}{6}\left(\frac{{m_S}^2}{{m_\circ}^2}\frac{(-v_\tau^*)}{t_B^2}\right) t^3\,. \tag{3.101}$$

Im ersten Term von (3.101) erkennen wir eine gleichmäßig beschleunigte Bewegung mit der Anfangsgeschwindigkeit $v(t=0)=0$ und dem Anfangsort $z(t=0)=0$ wieder (1.33). Zum Verständnis des zweiten Terms berechnen wir die Beschleunigung $\ddot{z}(t)$:

$$\ddot{z}(t) = \left(\frac{m_S}{m_\circ}\frac{(-v_\tau^*)}{t_B} - g\right) + \left(\frac{{m_S}^2}{{m_\circ}^2}\frac{(-v_\tau^*)}{t_B^2}\right) t \tag{3.102}$$

Wir erkennen zusätzlich zum konstanten einen linear zeitabhängigen Term, der die Rakete mit zunehmender Beschleunigung antreibt, bis beim Brennschluss $t = t_B$ der Treibstoff verbraucht ist und die Rakete unter dem Einfluss der Schwerkraft an Geschwindigkeit verliert und zurückfällt.

Die Bedingung für einen sofortigen Start lesen wir im ersten Term von (3.102) ab:

$$\frac{m_S}{m_\circ}\frac{(-v_\tau^*)}{t_B} - g > 0 \tag{3.103}$$

Die Rakete hebt also nur dann sofort ab, wenn diese Startbedingung erfüllt ist. Wir können die Startbedingung noch besser verstehen, wenn wir die Startmasse $m_\circ$ der Rakete zum Term g schreiben:

$$m_S\,\frac{(-v_\tau^*)}{t_B} > \underbrace{m_\circ\,g}_{=|F_G|} \tag{3.104}$$

Die linke Seite von (3.104) identifizieren wir als Schubkraft (Schubimpuls/Zeit):

$$F_S \equiv m_S \frac{(-v_\tau^*)}{t_B} > 0 \tag{3.105}$$

Damit wird sofort klar, dass die Schubkraft das Startgewicht der Rakete übertreffen muss:

$$F_S > |F_G| \tag{3.106}$$

Sonst brennt die Rakete solange im Stand und hebt erst ab, wenn (3.106) mit dem Restgewicht m der Rakete erfüllt ist.

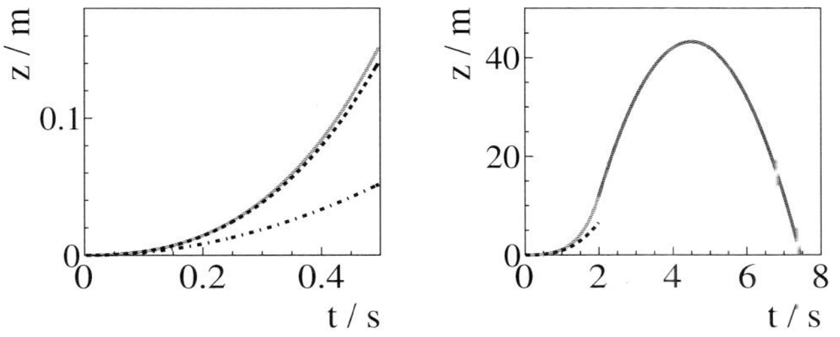

Abbildung 3.14 Steighöhe der Rakete z in Abhängigkeit von der Zeit t. Die durchgezogene Kurve zeigt die Bahnkurve (3.87) und den anschließenden Flug unter dem Einfluss der Erdanziehung (3.107), die weiteren Kurven sind Näherungen durch die Taylor-Entwicklung (3.101).

Beispiel 18: Raketenstart

Wir setzen ein für die

- Erdbeschleunigung $g = 10\,\mathrm{m/s}^2$,
- Brenndauer $t_B = 2\,\mathrm{s}$,
- Startmasse $m_\circ = 0{,}6\,\mathrm{kg}$,
- Gasmasse $m_S = 0{,}5\,\mathrm{kg}$ und für die
- Geschwindigkeit des ausströmenden Gases $v_\tau^* = -25\,\mathrm{m/s}$.

In der linken ► Abbildung 3.14 ist die Steighöhe $z(t)$ (3.87) in der Brennphase bis $t_B = 0{,}5\,\mathrm{s}$ mit der durchgezogenen Kurve gezeigt. Unsere Taylor-Näherung (3.101) in 2. Ordnung ist mit der Strichpunktkurve gezeigt und beschreibt nur den Anfang der Kurve zufriedenstellend. Die Näherung 3. Ordnung (gestrichelte Kurve) beschreibt den Verlauf ganz gut, aber weicht danach von der vollständigen Lösung ab (rechte ► Abbildung 3.14).

In der rechten ► Abbildung 3.14 ist auch der weitere Verlauf des Raketenflugs gezeigt. Die Endgeschwindigkeit nach Brennschluss ergibt sich aus (3.71) zu etwa $v_\circ = 25\,\mathrm{m/s}$. Die Steighöhe beträgt laut (3.87) nach der Brenndauer ca. $z_\circ = 12\,\mathrm{m}$. Danach handelt es sich um eine gleichmäßig beschleunigte Bewegung (1.33), bei der die Rakete zunächst noch bis zu einer maximalen Höhe steigt und dann herunterfällt:

$$z(t) = -\frac{1}{2}\,g\,t^2 + v_\circ\,t + z_\circ \tag{3.107}$$

Das Wichtigste zum Mitnehmen

Zum Mitnehmen: Dynamik: Newtons Bewegungsgesetze 1

1. Trägheitsgesetz

Ohne äußere Einwirkung verharrt jeder Körper im Zustand der Ruhe oder in gleichförmiger, gradliniger Bewegung:

$$\vec{v} = const. \quad \Leftrightarrow \quad \sum_{i=1}^{n} \vec{F}_i = 0 \tag{3.3}$$

2. Aktionsprinzip

Eine Kraft, die auf einen Körper wirkt, ändert die Bewegung des Körpers:

$$\vec{F} = m \cdot \vec{a} \tag{3.7}$$

Dieses Axiom kann als Definition der Kraft gesehen werden. Die Einheit der Kraft heißt **Newton**:

$$\text{Newton} = \text{N} = \frac{\text{kg} \cdot \text{m}}{\text{s}^2} \tag{3.8}$$

Eine Kraft kann durch die Beschleunigung $\vec{a}$, die ein Probekörper der Masse m erfährt, bestimmt werden.

3. Actio = Reactio

Bei zwei Körpern, die nur untereinander wechselwirken, ist die Kraft $\vec{F}_1$ auf den einen Körper entgegengesetzt gleich der Kraft $\vec{F}_2$ auf den anderen Körper:

$$\vec{F}_1 = -\vec{F}_2 \tag{3.9}$$

Träge und schwere Masse sind äquivalent

- Gewichtskraft $F_G = m_G g$:
  Ein Körper zeigt seine schwere Masse m_G bei der Anziehung auf der Erde.
- Beschleunigter Körper $F_T = m_T a$:
  Ein Körper zeigt seine träge Masse m_T, indem er sich einer Beschleunigung widersetzt:

$$m_G = m_T = m \tag{3.27}$$

Zum Mitnehmen: Dynamik: Newtons Bewegungsgesetze 2

- Der **Impuls** ist als das Produkt aus Masse und Geschwindigkeit eines Objekts definiert:

$$\vec{p} = m\,\vec{v} \tag{3.37}$$

Mit dem Impuls lässt sich **Newtons Bewegungsgesetz** in allgemeiner Form schreiben:

$$\frac{d\vec{p}}{dt} = \sum_{i=1}^{n} \vec{F}_i \tag{3.41}$$

Bei bekannten Kräften $\vec{F}_i$ ist die zeitliche Änderung des Impulses $d\vec{p}/dt$ und damit ist die Bahnkurve des Körpers vollständig festgelegt.

Ohne Einwirkung einer äußeren Kraft folgt aus (3.41) **Impulserhaltung**

$$\frac{d\vec{p}}{dt} = 0\,.$$

Die Änderung des Impulses kann sich sowohl auf die **Masse** dm als auch auf die **Geschwindigkeit** $d\vec{v}$ beziehen.

Das wird z. B. bei der Bewegungsgleichung der Rakete deutlich. Mit der um dm abnehmenden Masse m und (zunehmenden) Geschwindigkeit $\vec{v}$ der Rakete und der (konstanten) Geschwindigkeit $\vec{v}_T{}^*$ der ausgestoßenen Gase im mitbewegten System der Rakete gilt

$$m\,d\vec{v} = \vec{v}_T^*\,dm\,. \tag{3.63}$$

Daraus folgt durch Integration für die Geschwindigkeit der Rakete

$$\vec{v}(t) = -\vec{v}_T^* \cdot \underbrace{\ln\left(\frac{m_\circ}{m_\circ - m_T(t)}\right)}_{\geq 0} \tag{3.65}$$

mit der Anfangsmasse $m_\circ$ und der verbrauchten Antriebsmasse m_T. Unter Einschluss der Schwerkraft bei senkrechtem Start und mit $m(t) = m_\circ - m_T(t)$ ist dann

$$v(t) = -v_T^* \underbrace{\ln\frac{m_\circ}{m(t)}}_{\geq 0} - g\,t\,. \tag{3.71}$$

Das Minuszeichen zeigt an, dass die Flugrichtung der Rakete und die Richtung des ausgestoßenen Gases entgegengesetzt sind.

Drehbewegungen

4

4.1	**Polarkoordinaten**	**93**
4.2	**Gleichförmige Kreisbewegung**	**95**
4.3	**Harmonischer Oszillator**	**98**
4.4	**Vektor der Winkelgeschwindigkeit**	**103**
4.5	**Drehimpuls**	**104**
4.6	**Drehmoment**	**105**
	4.6.1 Rechte-Hand-Regel	106
4.7	**Bewegungsgesetz**	**107**
	4.7.1 Pendelschwingung	108

ÜBERBLICK

» Bisher haben wir Bewegungen von Körpern entlang von Raumrichtungen (Achsen) untersucht, die auch als *Translationsbewegungen* bezeichnet werden. Neu in diesem Kapitel sind Drehungen um Achsen, die auch *Rotationsbewegungen* genannt werden.

Zur Beschreibung der Rotationen stellen wir Polarkoordinaten sowie als kinematische Größen die Winkelgeschwindigkeit und die Zentripetalbeschleunigung vor. Als dynamische Größen führen wir den Drehimpuls und das Drehmoment ein, mit denen wir das Newtonsche Bewegungsgesetz für Rotationsbewegungen formulieren.

In Anwendungen erkennen wir beim Federpendel – ein sogenannter *harmonischer Oszillator* – den Zusammenhang zwischen Rotationen und Schwingungen. Außerdem untersuchen wir die Schwingungen des Fadenpendels. «

4.1 Polarkoordinaten

Zur Beschreibung eines Vektors benötigen wir ein Koordinatensystem wie z. B. kartesische Koordinaten mit den Komponenten x, y, z in ▶ Abbildung 3.2.

Wenn ein Vektor $\vec{r}$ um eine Achse gedreht wird, eignen sich Polarkoordinaten mit den zwei Variablen Radius r und Winkel φ (▶ Abbildung 4.1):

$$r, \varphi \tag{4.1}$$

Dabei werden nur die Komponenten des Vektors in der Rotationsebene erfasst (eine Übersicht zu den wichtigsten Koordinatensystemen folgt in Kapitel 9).

In Abbildung 4.1 wird der Vektor $\vec{r}$ durch Polarkoordinaten und alternativ mit kartesischen Koordinaten beschrieben. Mit der Sinus- und Kosinusfunktion lassen sich Polarkoordinaten in kartesische Koordinaten (x, y) umrechnen:

$$\vec{r} = \begin{pmatrix} r_x \\ r_y \end{pmatrix} = \begin{pmatrix} r \cdot \cos(\varphi) \\ r \cdot \sin(\varphi) \end{pmatrix} \tag{4.2}$$

Umgekehrt erfolgt die Umrechnung von kartesischen Koordinaten in Polarkoordinaten durch

$$r = |\vec{r}| = \sqrt{r_x^2 + r_y^2}, \qquad \varphi = \arctan \frac{r_y}{r_x}. \tag{4.3}$$

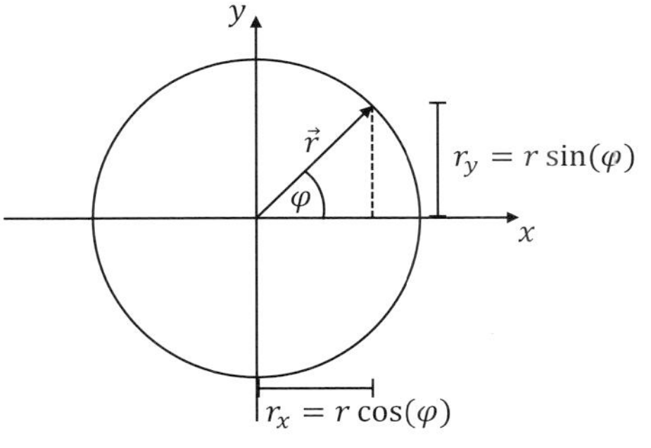

Abbildung 4.1 Zur Umrechnung zwischen kartesischen x, y und Polarkoordinaten r, φ in der x-y-Ebene

Aus dem Radius r und dem Winkel φ ergibt sich der **Bogen** bzw. **Bogenweg** (▶ Abbildung 4.2)

$$s = r \cdot \varphi\,. \tag{4.4}$$

Beachten Sie, dass der Winkel φ in der Einheit Radiant (rad) gemessen wird (nicht in Grad), sodass der Umfang $s_\circ$ eines Kreises mit Radius r insgesamt $s_\circ = 2\pi\, r$ ergibt.

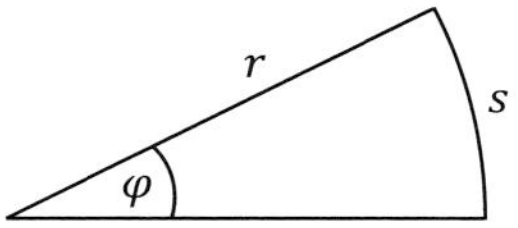

Abbildung 4.2 Zusammenhang von Winkel φ, Radius r und Bogen s

Den Winkel φ des Vektors $\vec{r}$ können wir als Funktion der Zeit verändern und damit $\vec{r}$ rotieren lassen. Die zugehörige **Winkelgeschwindigkeit** bzw. **Kreisfrequenz** lautet

$$\omega \equiv \frac{d\varphi}{dt} = \dot{\varphi} \tag{4.5}$$

und wird in der Einheit rad/s angegeben.

Damit lässt sich die Ortsfunktion $\varphi(t)$ eines Körpers analog zu (1.18) aus seiner Winkelgeschwindigkeit berechnen:

$$\varphi(t) = \int_0^t \omega(t)\,dt + \varphi_\circ \tag{4.6}$$

Für gleichförmige (gleichbleibende) Winkelgeschwindigkeit $\omega = const.$ gilt

$$\varphi(t) = \omega \int_0^t dt + \varphi_\circ \tag{4.7}$$

$$= \omega t + \varphi_\circ\,, \tag{4.8}$$

wobei $\varphi_\circ$ den anfänglichen Winkel bezeichnet. Durch Integration erhalten wir die Zeit T für einen Umlauf:

$$\varphi(T) - \varphi_\circ = \omega \int_0^T dt \tag{4.9}$$

$$2\pi = \omega T \tag{4.10}$$

$$T = \frac{2\pi}{\omega} \tag{4.11}$$

Die Umlaufzeit T wird auch als **Schwingungsperiode** bezeichnet. Wir werden diese Bezeichnung im Abschnitt 4.3 begründen.

Bisher haben wir die Geschwindigkeit ω im Bezug auf den Winkel φ erfasst (4.5), nun wollen wir auch die entsprechende Geschwindigkeit auf dem Bogen s beschreiben ((4.4) und ► Abbildung 4.2). Die Geschwindigkeit v eines Massenpunkts, der auf einer Kreisbahn mit dem Radius r rotiert, hängt mit der Winkelgeschwindigkeit ω folgendermaßen zusammen:

$$v = \frac{ds}{dt} = \frac{d(r\varphi)}{dt} = r\frac{d\varphi}{dt} = \omega\,r \tag{4.12}$$

4.2 Gleichförmige Kreisbewegung

Ein Massenpunkt soll eine gleichförmige Kreisbewegung mit der Winkelgeschwindigkeit ω ausführen (linke ► Abbildung 4.3). Der Abstand des Massenpunkts von der Drehachse bleibt dabei unverändert (Radius $r = \mathit{const.}$) und der Winkel folgt der zeitlichen Abhängigkeit $\varphi(t) = \omega t + \varphi_\circ$. Wir wählen im Folgenden als Anfangswinkel $\varphi_\circ = 0$.

Den Ortsvektor haben wir bereits in Gleichung (4.2) erwähnt. In kartesischen Koordinaten lautet er unter Verwendung von r und $\varphi = \omega\, t$

$$\vec{r}(t) = \begin{pmatrix} r_x(t) \\ r_y(t) \end{pmatrix} = \begin{pmatrix} r \cdot \cos(\omega t) \\ r \cdot \sin(\omega t) \end{pmatrix} . \tag{4.13}$$

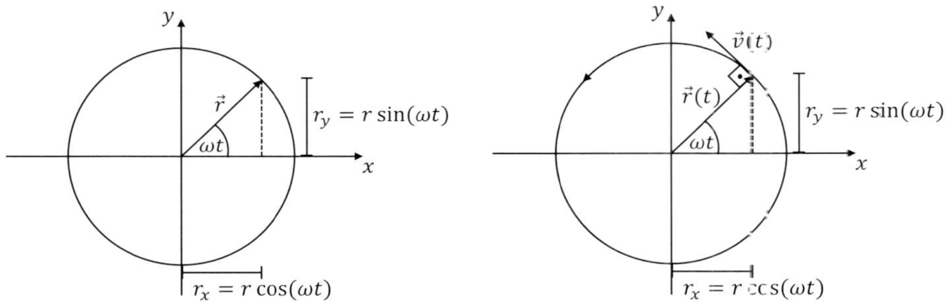

Abbildung 4.3 Links: Der Radius- oder Ortsvektor $\vec{r}$ rotiert mit der Kreisfrequenz ω Rechts: Die Geschwindigkeit $\vec{v}(t)$ ist die Tangente an den Kreis des Ortsvektors $\vec{r}(t)$.

Den zugehörigen Geschwindigkeitsvektor erhalten wir durch die zeitlichen Ableitungen der Komponenten (4.13):

$$v_x(t) = \frac{dr_x(t)}{dt} = \frac{d}{dt}(r \cdot \cos(\omega t)) = r\frac{d}{dt}\cos(\omega t) = -r\,\omega\,\sin(\omega t) \tag{4.14}$$

$$v_y(t) = \frac{dr_y(t)}{dt} = \frac{d}{dt}(r \cdot \sin(\omega t)) = r\frac{d}{dt}\sin(\omega t) = \quad r\,\omega\,\cos(\omega t) \tag{4.15}$$

Als Geschwindigkeitsvektor ergibt sich (weiterhin in kartesischen Koordinaten)

$$\vec{v}(t) = \begin{pmatrix} v_x(t) \\ v_y(t) \end{pmatrix} = r\,\omega \begin{pmatrix} -\sin(\omega t) \\ +\cos(\omega t) \end{pmatrix} . \tag{4.16}$$

Dabei zeigt $\vec{v}(t)$ in die Richtung der Tangente und hat den Betrag $|\vec{v}| = r\,\omega$ (rechte ► Abbildung 4.3). Auf die gleiche Weise können wir nun den Beschleunigungsvektor aus den zeitlichen Ableitungen der Komponenten des Geschwindigkeitsvektors berechnen

$$a_x(t) = \frac{dv_x(t)}{dt} = \frac{d}{dt}(-r\,\omega\,\sin(\omega t)) = -r\,\omega^2\,\cos(\omega t) \tag{4.17}$$

$$a_y(t) = \frac{dv_y(t)}{dt} = \frac{d}{dt}(\;\;r\,\omega\,\cos(\omega t)) = -r\,\omega^2\,\sin(\omega t) \tag{4.18}$$

und erhalten

$$\vec{a}(t) = \begin{pmatrix} a_x(t) \\ a_y(t) \end{pmatrix} = -r\,\omega^2 \begin{pmatrix} \cos(\omega t) \\ \sin(\omega t) \end{pmatrix} . \tag{4.19}$$

$\vec{a}(t)$ heißt **Zentripetalbeschleunigung** $\vec{a}_{ZP}$ und zeigt in Richtung des Kreismittelpunkts (► Abbildung 4.4). Der Betrag ist $|\vec{a}| = r\omega^2$. Die Zentripetalbeschleunigung wird im ruhenden Koordinatensystem gemessen. Da ihre Richtung entgegengesetzt zum Ortsvektor $\vec{r}$ verläuft, lässt sich $\vec{a}$ durch den Ortsvektor beschreiben:

$$\vec{a}_{ZP} = \vec{a} = -\omega^2\vec{r} \tag{4.20}$$

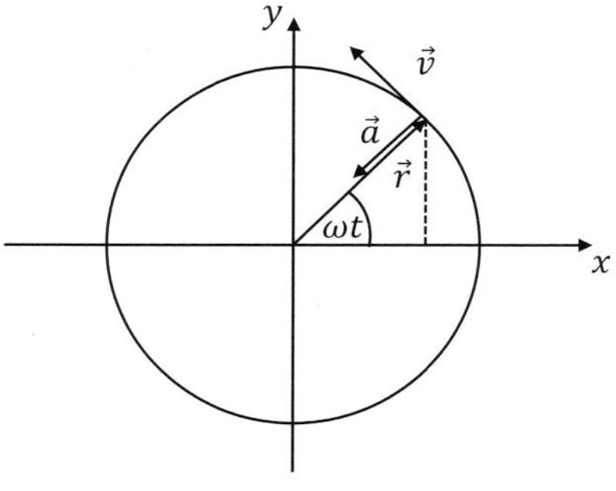

Abbildung 4.4 Die Geschwindigkeit $\vec{v}$ steht senkrecht auf $\vec{r}$, die Beschleunigung $\vec{a}$ zeigt entgegengesetzt zu $\vec{r}$ zum Kreismittelpunkt.

Die Zentripetalbeschleunigung hat denselben Betrag wie die **Zentrifugalbeschleunigung** $\vec{a}_{ZF}$, die im bewegten System des Massenpunkts gemessen wird und deren Richtung radial nach außen zeigt. Die Zentrifugalbeschleunigung wird im Zusammenhang mit Transformationen zwischen Bezugssystemen später ausführlich beschrieben (Abschnitt 11.3).

Die entsprechenden Kräfte heißen **Zentripetalkraft**

$$F_{ZP} = m\,\vec{a}_{ZP} = m\vec{a} = -m\,\omega^2\vec{r} \tag{4.21}$$

und **Zentrifugalkraft**

$$F_{ZF} = m\,\vec{a}_{ZF} = -m\vec{a} = m\,\omega^2\vec{r} . \tag{4.22}$$

Experiment 10: *phyphox* Zentripetalbeschleunigung

Zur Durchführung dieses Experiments werden das Gyroskop und der Beschleunigungssensor des Smartphones benutzt.

1 Suchen Sie in der *phyphox*-App das Experiment »Zentripetalbeschleunigung«. Berühren des Drei-Punkte-Symbols rechts oben zeigt ein Menü mit Informationen zum Experiment (»Experiment Info«). Wählen Sie das Menü »Verhältnis« zur Messung der Winkelgeschwindigkeit und der Beschleunigung.

2 Starten Sie die Aufzeichnung und lassen Sie das Smartphone (z. B. in einer Salatschleuder oder einem Spielplatzkarussell) mit verschiedenen Geschwindigkeiten rotieren. Stoppen Sie die Aufzeichnung.

3 Die Beschleunigung a wird gegen die Winkelgeschwindigkeit ω (► Abbildung 4.5, oben) und gegen das Quadrat ω^2 aufgetragen (► Abbildung 4.5, unten).

4 ► Abbildung 4.5 zeigt die Proportionalität zwischen der Beschleunigung a und dem Quadrat der Winkelgeschwindigkeit ω^2. Mit der Salatschleuder (rechts) werden viel höhere Beschleunigungen erreicht als mit dem Kinderkarussell (links).

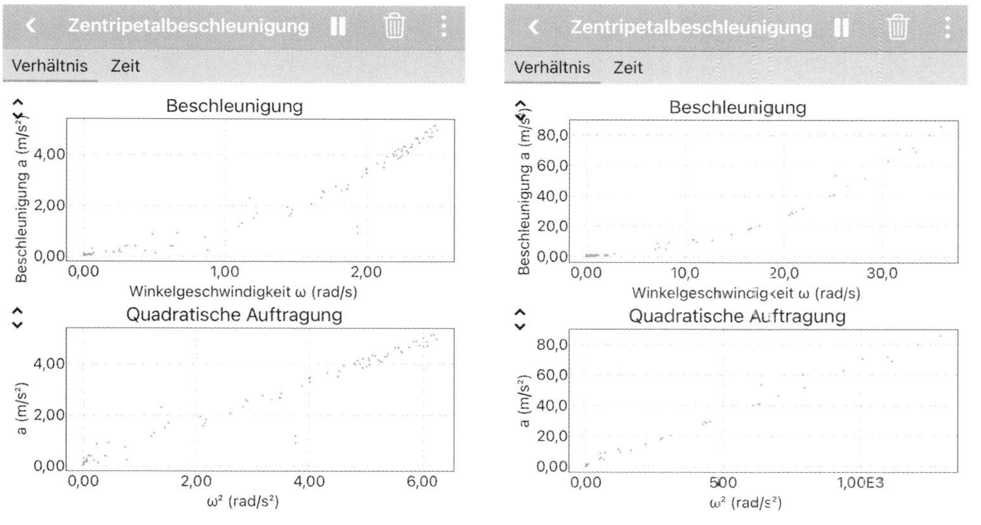

Abbildung 4.5 Die Beschleunigung a in Abhängigkeit der Winkelgeschwindigkeit ω (oben) und ihres Quadrats ω^2 (unten). Man erkennt die Proportionalität $a \propto \omega^2$.

4.3 Harmonischer Oszillator

In diesem Abschnitt stellen wir einen – zunächst überraschenden – Zusammenhang zwischen der gleichförmigen Kreisbewegung und Schwingungen her. Dafür untersuchen wir zuerst die Projektion der Kreisbewegung auf die x-Achse im Hinblick auf Ort, Geschwindigkeit und Beschleunigung. Anschließend stellen wir die Newtonsche Bewegungsgleichung für ein Federpendel auf.

In den ► Abbildungen 4.6 bis 4.8 betrachten wir Projektionen der gleichförmigen Kreisbewegung auf die x-Achse zu einigen ausgewählten Winkeln $\varphi = 0, 30, 90\,^\circ$ bzw. die entsprechenden Zeiten t in ► Abbildung 4.9.

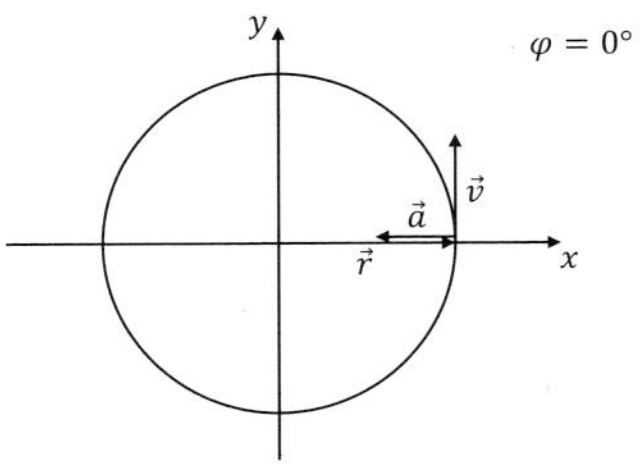

Abbildung 4.6 $\varphi = 0°$

$$r_x = \max = |\vec{r}|$$
$$v_x = 0$$
$$a_x = \min = -\,|\vec{a}|$$

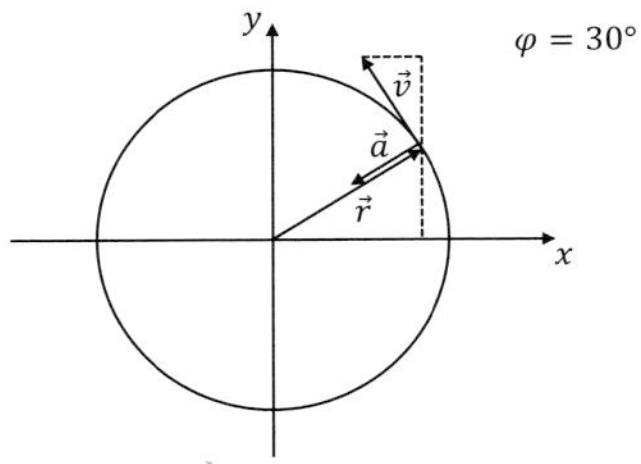

Abbildung 4.7 $\varphi = 30°$

$$0 < r_x < |\vec{r}|$$
$$-\,|\vec{v}| < v_x < 0$$
$$-\,|\vec{a}| < a_x < 0$$

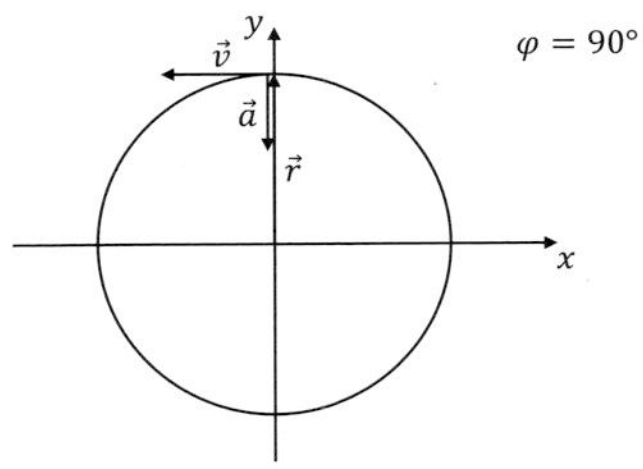

Abbildung 4.8 $\varphi = 90°$

$$r_x = 0$$
$$v_x = -\,|\vec{v}|$$
$$a_x = 0$$

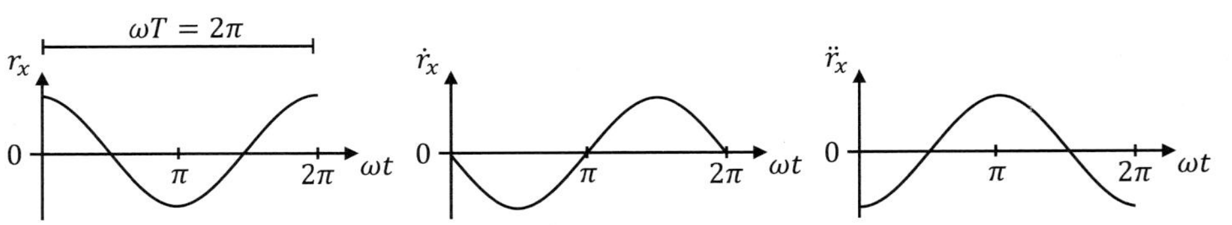

Abbildung 4.9 Zeitliche Verläufe von Projektionen der gleichförmigen Kreisbewegung auf die x-Achse: Ort $r_x(t)$ mit der Amplitude r, Geschwindigkeit $v_x = \dot{r}_x(t)$ mit der Amplitude $\omega\, r$, Beschleunigung $a_x = \ddot{r}_x(t)$ mit der Amplitude $\omega^2\, r$

Aus der x-Komponente der Ortsfunktion (4.13) bilden wir hier auch ihre zeitlichen Ableitungen analog zu (4.14) und (4.17) und erhalten so Ort, Geschwindigkeit und Beschleunigung als Funktion der Zeit:

$$r_x = x = r\,\cos \omega t \tag{4.23}$$

$$\dot{r}_x = \dot{x} = v_x = -\omega\, r\,\sin \omega t \tag{4.24}$$

$$\ddot{r}_x = \ddot{x} = a_x = -\omega^2\, r\,\cos \omega t = -\omega^2\, x \tag{4.25}$$

In ► Abbildung 4.9 ist eine Darstellung der zeitlichen Verläufe während einer Umdrehung gezeigt. Um die Zusammenhänge genau zu verstehen, verfolgen Sie die drei Winkel aus den ► Abbildungen 4.6 bis 4.8 in den drei ► Abbildungen 4.9.

Für die x-Projektion der gleichförmigen Kreisbewegung ergibt sich aus (4.23) und (4.25) folgender Zusammenhang zwischen der Beschleunigung und der Ortsfunktion:

$$\ddot{x} = -\omega^2 x \tag{4.26}$$

Dieser zunächst rein mathematische Zusammenhang zwischen einer Kreisbewegung mit der Kreisfrequenz ω und seiner Projektion auf die x-Achse wird sich im Folgenden als wichtige Bewegungsgleichung eines harmonischen Oszillators erweisen.

Beispiel 19: Harmonischer Oszillator: Federpendel

Ein Federpendel besteht aus einem Massenpunkt mit der Masse m, der sich wie in ► Abbildung 4.10 reibungsfrei horizontal bewegt, und einer Feder mit der Federkonstanten k, die eine rücktreibende Kraft bewirkt:

$$F = -k\,x \tag{4.27}$$

Anhand von Newtons 2. Gesetz (3.41) stellen wir die Bewegungsgleichung auf. Mit dem Impuls $p = m\,v = m\dot{x}$ des Massenpunkts ergibt sich die Bewegungsgleichung des sogenannten harmonischen Oszillators:

$$\frac{dp}{dt} = F \tag{4.28}$$

$$m\ddot{x} = -k\,x \tag{4.29}$$

$$\ddot{x} = -\left(\frac{k}{m}\right) x \tag{4.30}$$

Dabei bezeichnet $\ddot{x}$ die Beschleunigung und x die Auslenkung des Massenpunkts.

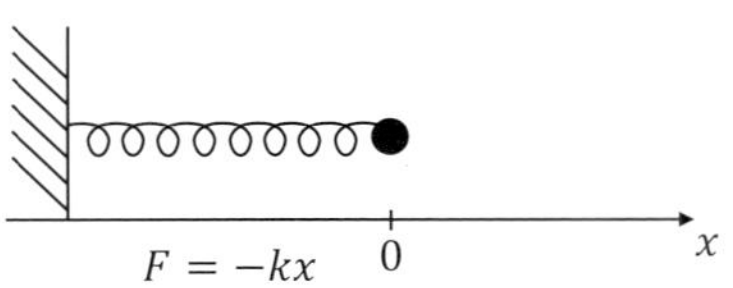

Abbildung 4.10 Auf eine Masse m wirkt die Federkraft $F = -kx$ mit der Federkonstanten k.

Den Zusammenhang (4.30) zwischen der Ortsfunktion und der Beschleunigung eines Massenpunkts kennen wir bereits aus der Kreisbewegung (4.26).

> *Das mathematische Konzept der Projektion einer Kreisbewegung beschreibt die beobachtete Federschwingung und löst die Bewegungsgleichung des harmonischen Oszillators.*

Die Kreisfrequenz ω ist durch die Federkonstante k und die Masse m des Massenpunkts festgelegt:

$$\omega^2 = \frac{k}{m} \tag{4.31}$$

$$\omega = \sqrt{\frac{k}{m}} \tag{4.32}$$

Die Bewegungsgleichung (4.30) ist eine Differentialgleichung, weil die Ortsvariable x und ihre Ableitung $\ddot{x}$ gemeinsam vorkommen. Durch die zweite Ableitung ist sie eine Differentialgleichung zweiter Ordnung.

Wir müssen also zweifach in der Zeit integrieren und haben dementsprechend zwei Anfangsbedingungen. Die Integration ist die Umkehrung der Differentialrechnung in den Gleichungen (4.23) bis (4.25). Ein allgemeiner Ansatz für die Ortsfunktion des Federpendels ist deswegen

$$x(t) = A \cdot \cos(\omega t + \varphi_\circ) \tag{4.33}$$

mit den Anfangsbedingungen maximale Auslenkung A und Anfangswinkel $\varphi_\circ$. Die maximale Auslenkung A wird auch als **Amplitude** bezeichnet. Der konstante Winkel $\varphi_\circ$ innerhalb der Kosinusfunktion wird häufig **Phase** genannt.

Die Bezeichnung des Federpendels als *harmonischer Oszillator* bezieht sich auf die trigonometrische Funktion (hier die Kosinusfunktion) für die Beschreibung seiner Bewegung. Ihr folgend schwingt das Federpendel gleichmäßig mit einer Kreisfrequenz ω. Die Zeitdauer einer Schwingung beträgt $T = 2\pi/\omega$ (4.11), was die Bezeichnung von T als Schwingungsperiode erklärt.

Experiment 11: Experiment zur Kreisbewegung und zum Federpendel

Lässt man im Licht einer Lampe eine Scheibe mit einem Stift so rotieren, dass der Schatten des Stifts die Projektion der Kreisbewegung zeigt, kann man mit einem dazu synchron schwingenden Federpendel die Korrektheit der Ortsfunktion $x(t)$ aus Gleichung (4.33) sehr schön veranschaulichen.

Produkte von Vektoren

Für die folgenden Abschnitte benötigen wir die wichtigen Produkte zwischen vektoriellen Größen.

Mathematischer Einschub: Skalarprodukt $\vec{a} \cdot \vec{b}$

Das Skalarprodukt zwischen zwei Vektoren

$$\vec{a} = \begin{pmatrix} a_x \\ a_y \\ a_z \end{pmatrix} \qquad \vec{b} = \begin{pmatrix} b_x \\ b_y \\ b_z \end{pmatrix} \tag{4.34}$$

ist definiert als

$$\vec{a} \cdot \vec{b} = \begin{pmatrix} a_x \\ a_y \\ a_z \end{pmatrix} \cdot \begin{pmatrix} b_x \\ b_y \\ b_z \end{pmatrix} \tag{4.35}$$

$$= a_x \cdot b_x + a_y \cdot b_y + a_z \cdot b_z \tag{4.36}$$

$$= |\vec{a}| \cdot |\vec{b}| \cdot \cos\vartheta\,, \tag{4.37}$$

wobei ϑ den Winkel zwischen den Vektoren $\vec{a}$ und $\vec{b}$ bezeichnet.

Aufgrund des Verlaufs der Kosinusfunktion $\cos(\vartheta)$ (linke ► Abbildung 4.11) ergeben sich Spezialfälle.

Abbildung 4.11 Verlauf von $\cos\vartheta$ (links) und $\sin\vartheta$ (rechts) für $0° < \vartheta < 360°$

Wenn zwei Vektoren senkrecht aufeinander stehen, verschwindet ihr Skalarprodukt ($\vartheta = 90°$ bedeutet $\cos\vartheta = 0$, linke ► Abbildung 4.12):

$$\vec{a} \perp \vec{b} \qquad \Rightarrow \qquad \vec{a} \cdot \vec{b} = 0 \tag{4.38}$$

Stehen die beiden Vektoren parallel, so ist das Skalarprodukt maximal ($\vartheta = 0°$, d. h. $\cos\vartheta = 1$). Das Skalarprodukt eines Vektors $\vec{a}$ mit sich selbst ergibt das Quadrat seines Betrags:

$$\vec{a} \parallel \vec{a} \qquad \Rightarrow \qquad \vec{a} \cdot \vec{a} = |\vec{a}|^2 \tag{4.39}$$

Das Skalarprodukt kann zur Bestimmung der Komponenten eines Vektors eingesetzt werden (Projektion in der rechten ► Abbildung 4.12). Wir definieren einen sogenannten

Einheitsvektor in der Richtung der x-Achse mit der Länge 1 durch

$$\vec{e}_x = \begin{pmatrix} 1 \\ 0 \\ 0 \end{pmatrix} . \tag{4.40}$$

Das Skalarprodukt des Vektors $\vec{a}$ mit dem Einheitsvektor $\vec{e}_x$ ergibt die x-Komponente des Vektors $\vec{a}$:

$$\vec{a} \cdot \vec{e_x} = |\vec{a}| \cdot \cos\vartheta = a_x \tag{4.41}$$

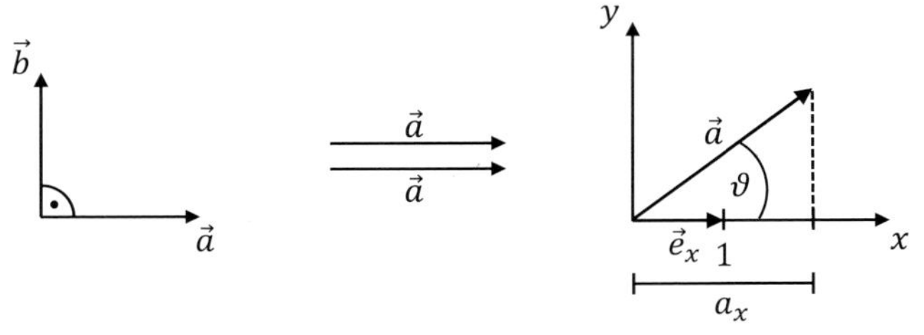

Abbildung 4.12 Anwendungen des Skalarprodukts: aufeinander senkrechte Vektoren $\vec{a} \perp \vec{b}$, parallele Vektoren $\vec{a} \parallel \vec{a}$, Projektion eines Vektors $\vec{a}$ mithilfe eines Einheitsvektors $\vec{e}_x$

Mathematischer Einschub: Vektorprodukt bzw. Kreuzprodukt $\vec{a} \times \vec{b}$

Das Kreuzprodukt von $\vec{a}$ mit $\vec{b}$ ist definiert als

$$\vec{a} \times \vec{b} = \begin{pmatrix} a_x \\ a_y \\ a_z \end{pmatrix} \times \begin{pmatrix} b_x \\ b_y \\ b_z \end{pmatrix} \tag{4.42}$$

$$= \begin{pmatrix} a_y b_z - a_z b_y \\ a_z b_x - a_x b_z \\ a_x b_y - a_y b_x \end{pmatrix} \tag{4.43}$$

$$= |\vec{a}| \cdot |\vec{b}| \cdot \sin\vartheta \cdot \vec{e}_{\perp \vec{a}, \vec{b}} \, . \tag{4.44}$$

Der Winkel zwischen den Vektoren $\vec{a}$ und $\vec{b}$ wird mit ϑ bezeichnet. Der Vektor $\vec{a} \times \vec{b}$ steht senkrecht auf $\vec{a}$ und $\vec{b}$ entlang eines Einheitsvektors $\vec{e}_{\perp \vec{a}, \vec{b}}$. Dabei bilden $\vec{a}$, $\vec{b}$ und $\vec{a} \times \vec{b}$ in dieser Reihenfolge ein Rechtssystem (► Abbildung 4.13).

Da hier die Sinusfunktion $\sin(\vartheta)$ maßgebend ist (rechte ► Abbildung 4.11), ergeben sich zwei Sonderfälle, die sich von den Fällen (4.38), (4.39) des Skalarprodukts unterscheiden. Stehen zwei Vektoren senkrecht aufeinander, ist der Betrag des Kreuzprodukts maximal ($\vartheta = 90°$ bedeutet $\sin\vartheta = 1$):

$$\vec{a} \perp \vec{b} \qquad \Rightarrow \qquad |\vec{c}| = |\vec{a}||\vec{b}| \tag{4.45}$$

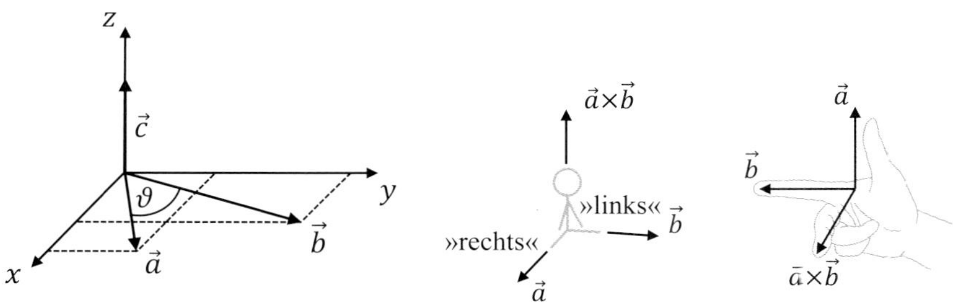

Abbildung 4.13 Kreuzprodukt $\vec{c} = \vec{a} \times \vec{b}$. Die Vektoren $\vec{a}$ und $\vec{b}$ stehen senkrecht auf $\vec{c} = \vec{a} \times \vec{b}$ und bilden zusammen in dieser Reihenfolge ein Rechtssystem, was in den beiden rechten Abbildungen veranschaulicht ist.

Für parallele Vektoren verschwindet das Kreuzprodukt ($\vartheta = 0°$, d. h. $\sin\vartheta = 0$):

$$\vec{a} \parallel \vec{b} \qquad \Rightarrow \qquad \vec{c} = 0 \tag{4.46}$$

Die Länge von $\vec{a} \times \vec{b}$ entspricht der Größe der von $\vec{a}$ und $\vec{b}$ aufgespannten Fläche (► Abbildung 4.14).

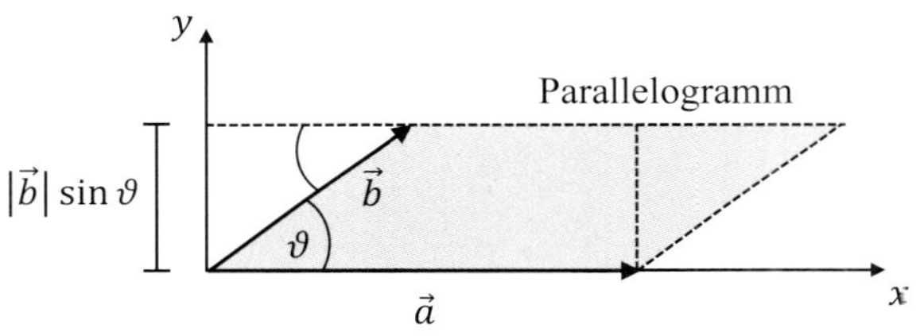

Abbildung 4.14 Die Vektoren $\vec{a}$ und $\vec{b}$ spannen das Parallelogramm $\vec{a} \times \vec{b}$ mit der Fläche $\left|\vec{a} \times \vec{b}\right| = |\vec{a}| \cdot |\vec{b}| \cdot \sin\vartheta$ auf.

Eine nützliche Rechenregel für ein Kreuzprodukt aus drei Vektoren ist die *bac* − *cab*-Regel:

$$\vec{a} \times \left(\vec{b} \times \vec{c}\right) = \vec{b}\,(\vec{a} \cdot \vec{c}) - \vec{c}\left(\vec{a} \cdot \vec{b}\right) \tag{4.47}$$

4.4 Vektor der Winkelgeschwindigkeit

In diesem und den folgenden Abschnitten definieren wir physikalische Größen, die sich zur Beschreibung von Drehbewegungen eignen.

Für die gleichförmige Kreisbewegung mit dem Radius r haben wir bereits den Zusammenhang $v = \omega r$ (4.12) zwischen den Beträgen der Geschwindigkeit v eines Massenpunkts und der Winkelgeschwindigkeit ω kennengelernt.

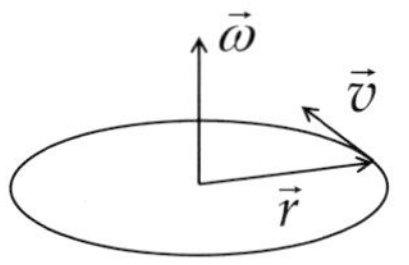

Abbildung 4.15 Zusammenhang zwischen Winkelgeschwindigkeit $\vec{\omega}$, Bahngeschwindigkeit $\vec{v}$ und Radius $\vec{r}$ der Kreisbewegung

Wir definieren nun einen Vektor der Winkelgeschwindigkeit $\vec{\omega}$. Er steht parallel zur Drehachse und senkrecht zu $\vec{v}$:

$$\vec{v} = \vec{\omega} \times \vec{r} \tag{4.48}$$

Für eine feste Drehachse können wir umgekehrt auch $\vec{\omega}$ bestimmen. Durch Vektormultiplikation mit $\vec{r}$ ergibt sich wegen Gleichung (4.47)

$$\vec{r} \times \vec{v} = \vec{r} \times (\vec{\omega} \times \vec{r}) \tag{4.49}$$

$$= \vec{\omega}\,(\vec{r} \cdot \vec{r}) - \vec{r}\,(\vec{r} \cdot \vec{\omega}) \ . \tag{4.50}$$

Bei einer *Kreisbewegung* stehen die Vektoren $\vec{r}$ und $\vec{\omega}$ senkrecht aufeinander (► Abbildung 4.15), sodass der hintere Term $\vec{r} \cdot \vec{\omega} = 0$ ist (4.38). In diesem Fall ist der Vektor der Winkelgeschwindigkeit zusammen mit (4.39)

$$\vec{\omega} = \frac{1}{r^2}\,(\vec{r} \times \vec{v}) \ . \tag{4.51}$$

4.5 Drehimpuls

Der Drehimpuls eines Körpers wird in Bezug auf einen bestimmten Raumpunkt über das Vektorprodukt aus Ortsvektor $\vec{r}$ und Impulsvektor $\vec{p}$ des Körpers definiert. Der Ortsvektor $\vec{r}$ hat seinen Ursprung in diesem Raumpunkt, z. B. dem Koordinatenursprung, und lokalisiert den Körper (linke ► Abbildung 4.16):

$$\vec{L} = \vec{r} \times \vec{p} \tag{4.52}$$

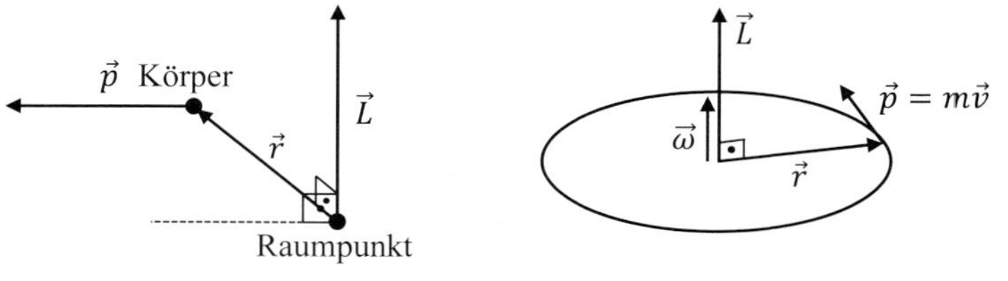

Abbildung 4.16 Links: Definition des Drehimpulses $\vec{L}$ bezüglich eines Raumpunkts, rechts: Drehimpuls $\vec{L}$ einer Kreisbewegung der Masse m mit der Winkelgeschwindigkeit $\vec{\omega}$

Für eine *Kreisbewegung* mit fester Drehachse ergibt sich der Drehimpuls bezüglich dieser Achse wegen (4.48) und (4.47) zu

$$\begin{aligned}\vec{L} &= \vec{r} \times \vec{p} && (4.53)\\ &= \vec{r} \times (m\vec{v}) && (4.54)\\ &= m\vec{r} \times (\vec{\omega} \times \vec{r}) && (4.55)\\ &= m[\vec{\omega}\,(\vec{r}\cdot\vec{r}) - \vec{r}\,\underbrace{(\vec{r}\cdot\vec{\omega})}_{=0}] && (4.56)\\ &= m\,r^2\,\vec{\omega}\,. && (4.57)\end{aligned}$$

Die Vektoren des Drehimpulses $\vec{L}$ und der Winkelgeschwindigkeit $\vec{\omega}$ stehen hier parallel zueinander (rechte ► Abbildung 4.16).

Der Vergleich der Drehbewegung mit der Translationsbewegung zeigt die Bedeutung des Terms $m\,r^2$:

$$\underbrace{\vec{p}}_{\text{Impuls}} = \underbrace{m}_{\text{Masse}} \cdot \underbrace{\vec{v}}_{\text{Geschwindigkeit}} \qquad (4.58)$$

$$\underbrace{\vec{L}}_{\text{Drehimpuls}} = \underbrace{m\,r^2}_{\text{Trägheitsmoment}} \cdot \underbrace{\vec{\omega}}_{\text{Winkelgeschwindigkeit}} \qquad (4.59)$$

Das **Trägheitsmoment**

$$I \equiv m\,r^2 \qquad (4.60)$$

hat bei Drehbewegungen die Bedeutung, die wir von der Masse bei Translationsbewegungen kennen. Mit ihm lässt sich der Drehimpuls von Massenpunkten folgendermaßen schreiben:

$$\vec{L} = I\,\vec{\omega} \qquad (4.61)$$

4.6 Drehmoment

Das Drehmoment bezüglich eines Raumpunkts ist durch das Vektorprodukt zwischen einer Kraft $\vec{F}$ und dem Ortsvektor $\vec{r}$ von diesem Raumpunkt (z. B. dem Koordinatenursprung) zum Angriffspunkt der Kraft $\vec{F}$ definiert (► Abbildung 4.17):

$$\vec{D} = \vec{r} \times \vec{F} \qquad (4.62)$$

Häufig wird zur Bezeichnung des Drehmoments auch $\vec{M}$ oder $\vec{\tau}$ (Torque) verwendet. Für den Betrag des Drehmoments gilt

$$|\vec{D}| = |\vec{r}| \cdot |\vec{F}| \cdot \sin\varphi\,. \qquad (4.63)$$

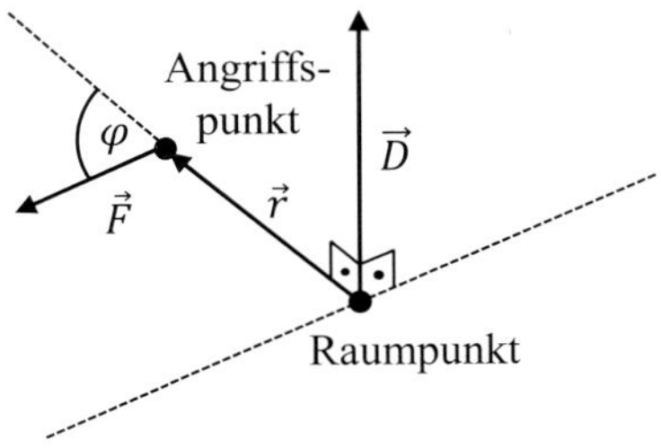

Abbildung 4.17 Zur Definition des Drehmoments $\vec{D}$ bezüglich eines Raumpunkts. Die Kraft $\vec{F}$ greift am Endpunkt vom Vektor $\vec{r}$ unter dem Winkel φ an.

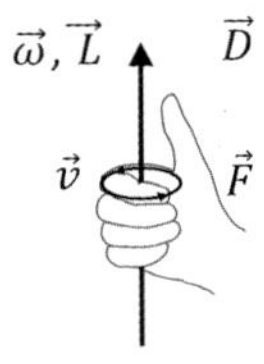

Abbildung 4.18 Zur schnellen Orientierung: Winkelgeschwindigkeit $\vec{\omega}$ und Drehimpuls $\vec{L}$ sind entlang des Daumens der rechten Hand gerichtet, wenn die Geschwindigkeit $\vec{v}$ längs der Finger zeigt. Dasselbe gilt für das Drehmoment $\vec{D}$ und die Kraft $\vec{F}$.

4.6.1 Rechte-Hand-Regel

Zur schnellen Orientierung für die Richtung der Vektorgrößen der Drehbewegung ist neben der Veranschaulichung des Kreuzprodukts als Rechtssystem in ► Abbildung 4.13 auch die »Rechte-Hand-Regel« von ► Abbildung 4.18 nützlich.

Beispiel 20: Balkenwaage

Die Funktion einer Balkenwaage beruht auf dem Konzept sich ausgleichender Drehmomente: Die Waage ist im Gleichgewicht, falls gilt

$$\vec{r}_1 \times \vec{F}_1 = -\vec{r}_2 \times \vec{F}_2 \tag{4.64}$$

$$r_1\, m_1\, g = r_2\, m_2\, g \tag{4.65}$$

$$\frac{m_1}{m_2} = \frac{r_2}{r_1}\,. \tag{4.66}$$

Aufgabe 10: Fortschrittliche Wippe

Zwei Kinder ($m_l = 30\,\text{kg}$, $m_r = 15\,\text{kg}$) sitzen im gleichen Abstand von der Drehachse (► Abbildung 4.19). Gesucht ist die Position der Masse mit $m = 20\,\text{kg}$, damit die Kinder wippen können.

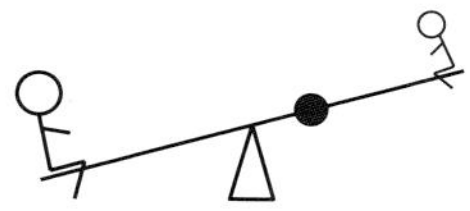

Abbildung 4.19 Fortschrittliche Wippe mit verstellbarem Zusatzgewicht

Experiment 12: Drehmoment der Garnrolle

Hält man eine Rolle am Faden, wie z. B. eine Garnrolle, so kann man je nach Winkel des Fadens beim Ziehen verschieden gerichtete Drehmomente erzielen und die Garnrolle in verschiedene Richtungen rollen lassen.

In der linken ▶ Abbildung 4.20 zeigt das Drehmoment $\vec{D_1} = \vec{r}_1 \times \vec{F}_1$ zum Betrachter hin, die Garnrolle vollführt eine Drehung nach links. In der mittleren Abbildung zeigt das Drehmoment $\vec{D_2} = \vec{r}_2 \times \vec{F}_2$ vom Betrachter weg, die Garnrolle rollt nach rechts. In der rechten Abbildung verschwindet das Drehmoment $\vec{D}_3 = \vec{r}_3 \times \vec{F}_3 = 0$, weil die Vektoren $\vec{r}_3$ und $\vec{F}_3$ parallel zueinander stehen (4.46).

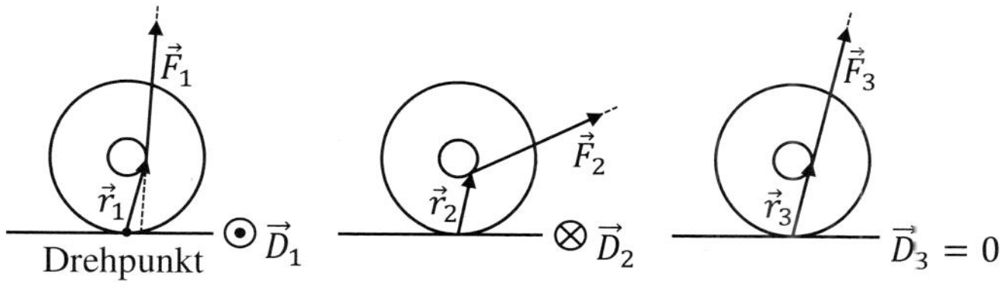

Abbildung 4.20 Abhängigkeit des Drehmoments von der Zugrichtung am Faden der Garnrolle. In den Bildern von links nach rechts rollt die Garnrolle nach links, nach rechts oder rollt nicht.

4.7 Bewegungsgesetz

Analog zum Bewegungsgesetz für Translationsbewegungen $d\vec{p}/dt = \sum_{i=1}^{n} \vec{F}_i$ (3.41) wollen wir nun ein Bewegungsgesetz für Rotationsbewegungen formulieren. Anstelle der zeitlichen Ableitung eines Impulses (3.38) führen wir dafür die zeitliche Ableitung des Drehimpulses $\vec{L}$ (4.52) durch, wobei die betrachtete Masse m unverändert bleiben soll. Dabei entspricht $\vec{r} \times d\vec{p}/dt$ einem Drehmoment:

$$\frac{d\vec{L}}{dt} = \frac{d}{dt}(\vec{r} \times \vec{p}) \tag{4.67}$$

$$= \frac{d\vec{r}}{dt} \times \vec{p} + \vec{r} \times \frac{d\vec{p}}{dt} \tag{4.68}$$

$$= \underbrace{\vec{v} \times m\vec{v}}_{=0} + \underbrace{\vec{r} \times \frac{d\vec{p}}{dt}}_{=\vec{D}} \tag{4.69}$$

Wegen der Drehmomentdefinition $\vec{D} = \vec{r} \times \vec{F}$ (4.62) wird eine Drehimpulsänderung durch von außen einwirkende Drehmomente $\vec{D}_i$ verursacht. Das Bewegungsgesetz für Rotationsbewegungen lautet damit

$$\frac{d\vec{L}}{dt} = \sum_{i=1}^{n} \vec{D}_i \,. \tag{4.70}$$

Bei bekannten Drehmomenten $\vec{D}_i$, die auf einen Körper mit Drehimpuls $\vec{L}$ einwirken, ist die zeitliche Änderung seines Drehimpulses $d\vec{L}/dt$ festgelegt. Durch Integration ist die Bahnkurve des Körpers vollständig berechenbar.

Für ein verschwindendes äußeres Drehmoment $\vec{D} = 0$ bleibt der Drehimpuls eines Körpers erhalten:

$$\frac{d\vec{L}}{dt} = 0 \qquad \Rightarrow \qquad \vec{L} = const. \tag{4.71}$$

Auf diese **Drehimpulserhaltung** werden wir in Abschnitt 7.3 und Gleichung (7.11) zurückkommen.

Experiment 13: Drehstuhl

Eine Person mit Trägheitsmoment $I_\circ$ sitzt mit zwei Hantelgewichten gleich großer Masse m auf einem Drehstuhl (▶ Abbildung 4.21) und verändert während der Drehbewegung die Position beider Hanteln. Dadurch verändert sie das Trägheitsmoment der beiden Hanteln $I = 2\,m\,r^2$ (4.60). Die Änderung vom Trägheitsmoment $I_\circ$ soll dabei vernachlässigt werden.

Die verschiedenen Abstände der Hanteln von der Drehachse bezeichnen wir mit r_1 und r_2, wobei $r_2 > r_1$ ist. Die Trägheitsmomente sind entsprechend $I_1 = 2\,m\,r_1^2$ und $I_2 = 2\,m\,r_2^2$ mit $I_2 > I_1$.

Da kein äußeres Drehmoment wirkt, ist der Drehimpuls $L = I\,\omega$ (4.61) erhalten:

$$\begin{aligned} L_1 &= L_2 \\ I_1\,\omega_1 &= I_2\,\omega_2 \\ \omega_1 &= \frac{I_2}{I_1}\,\omega_2 > \omega_2 \end{aligned}$$

Die Position mit dem kleineren Trägheitsmoment $I_1 < I_2$ (Hanteln in kleinerem Abstand von der Drehachse) ergibt somit die größere Winkelgeschwindigkeit $\omega_1 > \omega_2$.

4.7.1 Pendelschwingung

Ein mathematisches Pendel bezeichnet ein idealisiertes Pendel mit einer Punktmasse m und einem gewichtslosen Faden. Das Pendel vollführt Schwingungen, die entlang eines Kreises stattfinden, dessen Kreismittelpunkt in der Aufhängung liegt (▶ Abbildung 4.22).

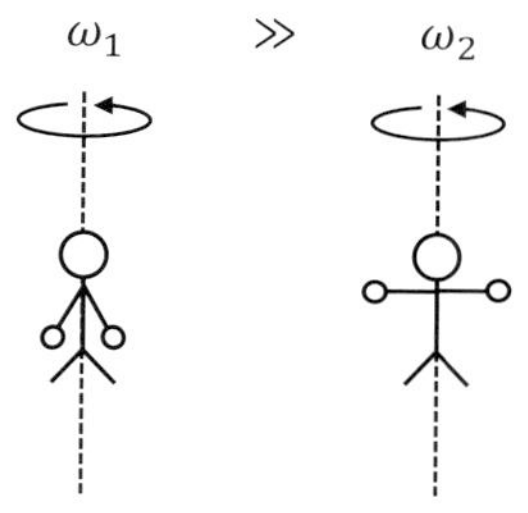

Abbildung 4.21 Vergrößerung des Trägheitsmoments $m\,r^2$ führt zur Verringerung der Winkelgeschwindigkeit $\vec{\omega}$ und umgekehrt.

Auf den Massenpunkt wirkt die Gewichtskraft $\vec{F} = m\,\vec{g}$ mit dem Vektor der Erdbeschleunigung (die z-Koordinate zeigt nach oben)

$$\vec{g} = \begin{pmatrix} 0 \\ 0 \\ -g \end{pmatrix} . \tag{4.72}$$

Durch die Aufhängung des Pendels an dem Faden entsteht mit der Gewichtskraft ein Drehmoment, das der Bewegung des Pendels nach Durchqueren der Ruhelage entgegenwirkt. Bezeichnen wir den Vektor von der Aufhängung zum Massenpunkt mit $\vec{r}$, so beträgt das von außen auf den Massenpunkt einwirkende Drehmoment (4.62)

$$\vec{D}_G = \vec{r} \times \vec{F} . \tag{4.73}$$

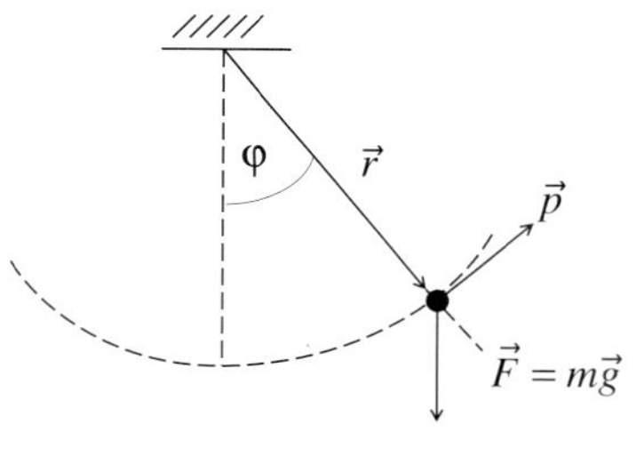

Abbildung 4.22 Mathematisches Pendel mit Punktmasse m

Mit dem Bewegungsgesetz (4.70) für Drehbewegungen können wir die Bewegungsgleichung für das Pendel aufstellen. Die zeitliche Änderung des Pendel-Drehimpulses wird durch das äußere Drehmoment seiner eigenen Gewichtskraft verursacht:

$$\frac{d\vec{L}}{dt} = \vec{D}_G \tag{4.74}$$

$$\frac{d}{dt}\left(\vec{r} \times \vec{p}\right) = \vec{r} \times \vec{F} \tag{4.75}$$

$$\underbrace{\frac{d\vec{r}}{dt} \times \vec{p}}_{=0} + \vec{r} \times \frac{d\vec{p}}{dt} = \vec{r} \times \vec{F} \tag{4.76}$$

Das Kreuzprodukt aus der Geschwindigkeit $d\vec{r}/dt = \vec{v}$ und dem Impuls $\vec{p} = m\,\vec{v}$ verschwindet, weil die Vektoren parallel stehen (4.46). Unter Berücksichtigung von $\vec{v} = \vec{\omega} \times \vec{r}$ (4.48) und (4.47) ergibt sich

$$m\,\vec{r} \times \frac{d\vec{v}}{dt} = m\,\vec{r} \times \vec{g} \tag{4.77}$$

$$\vec{r} \times \left(\frac{d}{dt}\left(\vec{\omega} \times \vec{r}\right)\right) = \vec{r} \times \vec{g} \tag{4.78}$$

$$\vec{r} \times \left(\dot{\vec{\omega}} \times \vec{r}\right) + \vec{r} \times \left(\vec{\omega} \times \dot{\vec{r}}\right) = \vec{r} \times \vec{g} \tag{4.79}$$

$$\dot{\vec{\omega}} \cdot \vec{r}^2 - \vec{r} \cdot \underbrace{\left(\vec{r} \cdot \dot{\vec{\omega}}\right)}_{=0} + \vec{\omega} \cdot \underbrace{\left(\vec{r} \cdot \dot{\vec{r}}\right)}_{=0} - \dot{\vec{r}} \cdot \underbrace{\left(\vec{r} \cdot \vec{\omega}\right)}_{=0} = \vec{r} \times \vec{g}\,. \tag{4.80}$$

Wir haben ausgenutzt, dass die Vektoren der Winkelgeschwindigkeit $\vec{\omega}$, bzw. deren zeitlicher Ableitung $\dot{\vec{\omega}}$, und des Pendelfadens $\vec{r}$ bei einer Kreisbewegung senkrecht aufeinanderstehen (► Abbildung 4.15) und deswegen ihr Skalarprodukt verschwindet (4.38). Die Vektoren beider Seiten von (4.80) stehen senkrecht zur Zeichenebene der ► Abbildung 4.22. Ihre Beträge ergeben sich mit (4.39) und (4.44) zu

$$\underbrace{|\dot{\vec{\omega}}|}_{\dot{\omega}=\ddot{\varphi}} \cdot |\vec{r}|^2 = -|\vec{r}| \cdot |\vec{g}| \cdot \sin\varphi\,. \tag{4.81}$$

Die Winkelgeschwindigkeit ω entspricht der zeitlichen Ableitung der Winkelkoordinate φ: $\dot{\varphi} = d\varphi/dt = \omega$ (4.5). Entsprechend ist $\dot{\omega}$ die zweite zeitliche Ableitung der Winkelkoordinate $\ddot{\varphi}$. Bezeichnen wir die Länge des Pendels weiterhin mit $l = |\vec{r}|$, so lautet seine Bewegungsgleichung

$$\ddot{\varphi} = -\frac{g}{l}\sin\varphi\,. \tag{4.82}$$

Für kleine Auslenkungen φ des Pendels können wir die Sinusfunktion nähern ($\sin\varphi \approx \varphi$) und finden als Bewegungsgleichung

$$\ddot{\varphi} \approx -\frac{g}{l} \cdot \varphi\,. \tag{4.83}$$

Diese Differentialgleichung kennen wir bereits, es ist die Bewegungsgleichung des harmonischen Oszillators (4.30): Analog zur Gleichung (4.33) ist die Lösung die Ortsfunktion

$$\varphi = A \cdot \cos\left(\omega t + \varphi_\circ\right). \tag{4.84}$$

Dabei ist ω die Winkelgeschwindigkeit der zugehörigen Kreisbewegung mit

$$\omega^2 = \frac{g}{l} \tag{4.85}$$

$$\omega = \sqrt{\frac{g}{l}}\,. \tag{4.86}$$

Die Periodendauer T der Pendelschwingung hängt mit ω über $\omega = 2\pi/T$ (4.11) zusammen. Die Schwingungsdauer des Pendels hängt also nur von der Pendellänge und der Erdbeschleunigung ab:

$$T = 2\pi\sqrt{\frac{l}{g}} \tag{4.87}$$

Experiment 14: *phyphox* Fadenpendel

Zur Durchführung dieses Experiments ist ein Smartphone mit Gyroskop erforderlich.

1 Suchen Sie in der *phyphox*-App das Experiment »Fadenpendel«. Berühren des Drei-Punkte-Symbols rechts oben zeigt ein Menü mit Informationen zum Experiment (»Experiment Info«). Nutzen Sie die Zeitautomatik (z. B. 10 s Startverzögerung, 20 s Messdauer), damit Sie das Pendel beim Schwingen nicht stören. Wählen Sie das Menü »Autokorrelation« zum Ablesen der Schwingungsperiode.

2 Befestigen Sie das Smartphone sicher an *zwei* Fäden (damit das Smartphone sich beim Schwingen nicht dreht), die möglichst lang sind, eventuell bis zur Zimmerdecke reichen. Starten Sie die Messung mit dem »Play button« (Dreieck rechts oben) und lassen Sie das Smartphone wie ein Pendel bei kleiner Auslenkung schwingen.

3 Die *phyphox*-App nutzt mehrere Schwingungen für die Berechnung der Schwingungsperiode T (► Abbildung 4.23). Notieren Sie die Länge l von der Aufhängung bis zum Schwerpunkt des Smartphones sowie die im Display angezeigte Schwingungsperiode T (»Periode«).

- Wiederholen Sie das Experiment für fünf verschiedene Fadenlängen l und messen Sie jeweils die Periode T.
- Prüfen Sie, ob die Schwingungsdauer quadratisch mit der Länge der Aufhängung steigt. Tragen Sie dazu die Schwingungsperiode T als Funktion der Wurzel aus der Fadenlänge l auf.
- Bestimmen Sie die Geradensteigung a und aus diesem Wert nach (4.87) die Erdbeschleunigung $g = (2\,\pi/a)^2$ (► Abbildung 4.23).

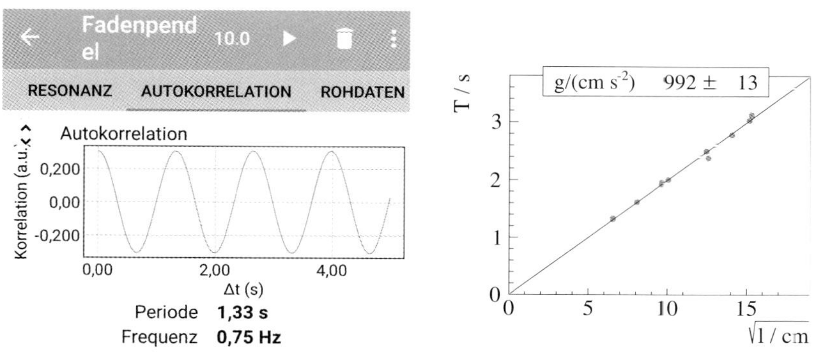

Abbildung 4.23 Experiment zum Fadenpendel mit der *phyphox*-App: Messung der Schwingungsperiode für ein 43 cm langes Pendel (links) und linearer Zusammenhang zwischen der Schwingungsdauer T und der Wurzel aus der Pendellänge $\sqrt{l}$ (rechts)

Aufgabe 11: Kugelbahn

Ein Massenpunkt rotiert anfangs mit einer Winkelgeschwindigkeit $\omega_\circ$. Sein Trägheitsmoment sei $I = m\,r^2 = const.$ Reibungskräfte (z. B. Luftwiderstand bei hoher Geschwindigkeit: Abschnitt 18.6.5 (18.123)) erzeugen ein bremsendes Drehmoment $D_b = -b\omega^2$ mit $b = const.$.

Gesucht ist die Bewegungsgleichung des Massenpunkts und die Winkelgeschwindigkeit $\omega(t)$ als Funktion der Zeit. Eine etwaige Eigendrehung soll vernachlässigt werden.

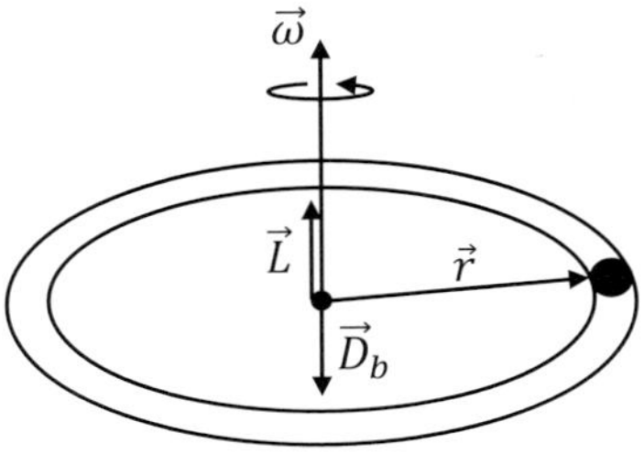

Abbildung 4.24 Ein Massenpunkt rotiert auf einer Kreisbahn mit Radius r und Winkelgeschwindigkeit $\vec{\omega}$.

Das Wichtigste zum Mitnehmen

Zum Mitnehmen: Drehbewegungen 1

- **Polarkoordinaten** (r, φ) mit

$$r = |\vec{r}| = \sqrt{r_x^2 + r_y^2}\,, \quad \varphi = \arctan \frac{r_y}{r_x} \tag{4.3}$$

eignen sich besonders zur Beschreibung von Kreisbewegungen. Für die Rückführung in kartesische Koordinaten (x, y) gilt

$$\vec{r} = \begin{pmatrix} r_x \\ r_y \end{pmatrix} = \begin{pmatrix} r \cdot \cos(\varphi) \\ r \cdot \sin(\varphi) \end{pmatrix}. \tag{4.2}$$

- Wichtige Größen zur Beschreibung von Bewegungen in Polarkoordinaten sind
  - der **Bogen** oder **Bogenweg** $s = r \cdot \varphi$
  - und die **Winkelgeschwindigkeit** bzw. **Kreisfrequenz**

$$\omega = \frac{d\varphi}{dt} = \dot{\varphi} \qquad \text{Einheit } \frac{\text{rad}}{\text{s}}. \tag{4.5}$$

Damit lässt sich die Ortsfunktion $\varphi(t)$ aus seiner gleichförmigen Winkelgeschwindigkeit $\omega = const.$ gemäß

$$\varphi(t) = \omega t + \varphi_\circ$$

berechnen, wobei $\varphi_\circ$ den anfänglichen Winkel bezeichnet.

Die **Schwingungsperiode** T entspricht der Zeit eines Umlaufs:

$$\varphi(T) - \varphi_\circ = \omega T = 2\pi$$

$$T = \frac{2\pi}{\omega} \tag{4.11}$$

Die Geschwindigkeit auf einer Kreisbahn mit dem Radius r hängt mit der Winkelgeschwindigkeit ω folgendermaßen zusammen:

$$v = \frac{ds}{dt} = \frac{d(r\varphi)}{dt} = r\frac{d\varphi}{dt} = \omega r \tag{4.12}$$

Zum Mitnehmen: Drehbewegungen 2

- Für eine **gleichförmige Kreisbewegung** mit $\omega = const.$ ergibt sich die Geschwindigkeit als zeitliche Ableitung aus (4.2) zu

$$\vec{v}(t) = r\omega \begin{pmatrix} -\sin(\omega t) \\ +\cos(\omega t) \end{pmatrix} . \tag{4.15}$$

  Dabei zeigt die **(Tangential-)Geschwindigkeit** $\vec{v}(t)$ in die Richtung der Tangente und hat den Betrag $|\vec{v}| = r\omega$.

- Die Beschleunigung ergibt sich als Ableitung der Geschwindigkeit (4.15):

$$\vec{a}(t) = -r\omega^2 \begin{pmatrix} \cos(\omega t) \\ \sin(\omega t) \end{pmatrix} \tag{4.19}$$

  $\vec{a}(t)$ heißt **Zentripetalbeschleunigung** und zeigt in Richtung des Kreismittelpunkts:

$$\vec{a}_{\mathrm{ZP}} = \vec{a} = -\omega^2 \vec{r} \tag{4.20}$$

- Die Zentripetalbeschleunigung wird durch eine **Zentripetalkraft** hervorgerufen:

$$F_{\mathrm{ZP}} = m\,\vec{a}_{\mathrm{ZP}} = m\vec{a} = -m\,\omega^2 \vec{r} \tag{4.21}$$

  Im mitbewegten System des Massenpunkts zeigt die der Zentrifugalbeschleunigung entgegengesetzt gleiche **Zentrifugalbeschleunigung** radial nach außen mit einer entsprechenden **Zentrifugalkraft**:

$$F_{\mathrm{ZF}} = m\,\vec{a}_{\mathrm{ZF}} = -m\vec{a} = m\,\omega^2 \vec{r} \tag{4.22}$$

  Die Zentrifugalbeschleunigung wird im Zusammenhang mit Transformationen zwischen Bezugssystemen später ausführlich beschrieben.

- Für die x-Projektion der gleichförmigen Kreisbewegung ergibt sich analog zu (4.20) folgender Zusammenhang zwischen der Beschleunigung und der Ortsfunktion:

$$\ddot{x} = -\omega^2 x \tag{4.88}$$

Zum Mitnehmen: Drehbewegungen 3

- Für eine Masse m an einer Feder mit der Federkonstanten k gilt die Bewegungsgleichung des harmonischen Oszillators:

$$\ddot{x} = -\left(\frac{k}{m}\right) x \tag{4.30}$$

Dabei bezeichnet $\ddot{x}$ die Beschleunigung und x die Auslenkung des Massenpunkts. Ein Vergleich mit (4.25) zeigt:

> *Die Projektion einer Kreisbewegung beschreibt die beobachtete Federschwingung und löst die Bewegungsgleichung des* **harmonischen Oszillators** *mit der Kreisfrequenz ω:*

$$\omega = \sqrt{\frac{k}{m}} \tag{4.32}$$

Die Differentialgleichung zweiter Ordnung (4.30) hat die allgemeine Lösung

$$x(t) = A \cdot \cos\left(\omega t + \varphi_\circ\right) \tag{4.33}$$

mit der Amplitude A und dem Anfangswinkel $\varphi_\circ$.

- Das **Skalarprodukt** zwischen zwei Vektoren ist definiert als

$$\begin{aligned} \vec{a} \cdot \vec{b} &= \begin{pmatrix} a_x \\ a_y \\ a_z \end{pmatrix} \cdot \begin{pmatrix} b_x \\ b_y \\ b_z \end{pmatrix} \\ &= a_x \cdot b_x + a_y \cdot b_y + a_z \cdot b_z \\ &= |\vec{a}| \cdot |\vec{b}| \cdot \cos\vartheta \, , \end{aligned} \tag{4.37}$$

wobei ϑ den Winkel zwischen den Vektoren $\vec{a}$ und $\vec{b}$ bezeichnet.

- Das **Kreuzprodukt** von $\vec{a}$ mit $\vec{b}$ ist definiert als

$$\begin{aligned} \vec{a} \times \vec{b} &= \begin{pmatrix} a_x \\ a_y \\ a_z \end{pmatrix} \times \begin{pmatrix} b_x \\ b_y \\ b_z \end{pmatrix} \\ &= \begin{pmatrix} a_y b_z - a_z b_y \\ a_z b_x - a_x b_z \\ a_x b_y - a_y b_x \end{pmatrix} \\ &= |\vec{a}| \cdot |\vec{b}| \cdot \sin\vartheta \cdot \vec{e}_{\perp \vec{a}, \vec{b}} \, . \end{aligned} \tag{4.44}$$

Der Vektor $\vec{a} \times \vec{b}$ steht senkrecht auf $\vec{a}$ und $\vec{b}$. Dabei bilden $\vec{a}$, $\vec{b}$ und $\vec{a} \times \vec{b}$ in dieser Reihenfolge ein Rechtssystem.

Zum Mitnehmen: Drehbewegungen 4

- Der **Vektor der Winkelgeschwindigkeit** $\vec{\omega}$ zeigt parallel zur Drehachse. Sein Vektorprodukt mit einem Ortsvektor $\vec{r}$ ergibt die Geschwindigkeit $\vec{v}$ am Endpunkt von $\vec{r}$:

$$\vec{v} = \vec{\omega} \times \vec{r} \tag{4.48}$$

Der **Drehimpuls** eines Massenpunkts bezüglich eines Raumpunkts ist als Vektorprodukt aus seinem Ortsvektor $\vec{r}$ von diesem Raumpunkt und seinem Impulsvektor $\vec{p}$ definiert:

$$\vec{L} = \vec{r} \times \vec{p} \tag{4.52}$$

Damit ergibt sich eine Analogie zwischen der Drehbewegung und der Translationsbewegung:

$$\underbrace{\vec{p}}_{\text{Impuls}} = \underbrace{m}_{\text{Masse}} \cdot \underbrace{\vec{v}}_{\text{Geschwindigkeit}} \tag{4.58}$$

$$\underbrace{\vec{L}}_{\text{Drehimpuls}} = \underbrace{mr^2}_{\text{Trägheitsmoment}} \cdot \underbrace{\vec{\omega}}_{\text{Winkelgeschwindigkeit}} \tag{4.59}$$

Mit dem **Trägheitsmoment**

$$I = mr^2 \tag{4.60}$$

gilt also bei Drehbewegungen in Analogie zu $\vec{p} = m\vec{v}$

$$\vec{L} = I\vec{\omega}\ . \tag{4.61}$$

- Das **Drehmoment** ist durch das Vektorprodukt zwischen einer Kraft $\vec{F}$ und dem Ortsvektor $\vec{r}$ von einem Raumpunkt zum Angriffspunkt der Kraft $\vec{F}$ definiert:

$$\vec{D} = \vec{r} \times \vec{F} \tag{4.62}$$

Die zeitliche Änderung des Drehimpulses eines Massenpunkts wird durch von außen angreifende Drehmomente verursacht, genauso wie eine Kraft eine Impulsänderung hervorruft ($d\vec{p}/dt = \sum_{i=1}^{n} \vec{F}_i$). Die **Bewegungsgleichung** für Drehbewegungen lautet

$$\frac{d\vec{L}}{dt} = \sum_{i=1}^{n} \vec{D}_i\ . \tag{4.70}$$

Für ein verschwindendes äußeres Drehmoment $\vec{D} = 0$ bleibt der Drehimpuls eines Körpers erhalten (**Drehimpulserhaltung**).

Zum Mitnehmen: Drehbewegungen 5

- Für kleine Auslenkungen φ des **mathematischen Pendels** können wir die Sinusfunktion nähern ($\sin\varphi \approx \varphi$) und finden

$$\ddot{\varphi} \approx -\frac{g}{l} \cdot \varphi \,. \tag{4.83}$$

Dies ist die Bewegungsgleichung des harmonischen Oszillators (4.30). Analog zur Gleichung (4.33) ist die Lösung

$$\varphi = A \cdot \cos\left(\omega t + \varphi_{\circ}\right). \tag{4.84}$$

Dabei ist ω die Winkelgeschwindigkeit der zugehörigen Schwingung mit

$$\omega = \sqrt{\frac{g}{l}} \,.$$

Die Periodendauer T der Pendelschwingung ω ergibt sich mit $\omega = 2\pi/T$ (4.11) zu

$$T = 2\pi\sqrt{\frac{l}{g}} \,. \tag{4.87}$$

Die Schwingungsdauer des Pendels hängt also nur von der Pendellänge und der Erdbeschleunigung ab.

Arbeit, Energie, Leistung, Intensität

5

5.1 Energieformen . . . 121
5.2 Arbeit . . . 121
5.3 Potentielle Energie . . . 124
5.4 Kinetische Energie der Translation . . . 124
5.5 Kinetische Energie der Rotation . . . 125
5.6 Energieerhaltung . . . 126
5.7 Leistung . . . 131
5.8 Intensität . . . 131

ÜBERBLICK

» Arbeit, Energie, Leistung und Intensität sind Begriffe des Alltags. In diesem Kapitel werden wir diese Begriffe in der Form einführen, wie sie in der Physik verwendet werden. Die physikalischen Definitionen unterscheiden sich von denen des Alltags. «

5.1 Energieformen

Wir unterscheiden verschiedene Formen, in denen Energie auftreten kann:

Potentielle Energie: Beispiele sind die gespeicherte Energie in einer zusammengedrückten Feder, in einem auf ein Hausdach gehobenen Dachziegel oder in oberhalb eines Wasserkraftwerks befördertem Wasser.

Kinetische Energie ist Bewegungsenergie. Beispiele sind die Bewegungsenergie eines Bobfahrers (Translationsenergie) oder die Bewegungsenergie der Drehbewegung eines Jo-Jos (Rotationsenergie).

Wärme-Energie (Thermische Energie) steckt z. B. in einem heißen Stein.

Der Energiebegriff ist für die Beschreibung von physikalischen Vorgängen oft sehr nützlich. In der Mechanik ist er eine alternative Möglichkeit zu Newtons Bewegungsgesetzen. In der Wärmelehre ist der Kraftbegriff nur bedingt hilfreich, dort führt der Energiebegriff weiter.

Verschiedene Energieformen können ineinander umgewandelt werden. Zum Beispiel kann potentielle in kinetische Energie übergehen.

5.2 Arbeit

Arbeit ist durch das Skalarprodukt zwischen der Kraft, die auf einen Körper wirkt, und dem Weg entlang seiner Bewegungsrichtung definiert. In differentieller Schreibweise ist Arbeit

$$dW = \vec{F}\, d\vec{r}\,. \tag{5.1}$$

Um die bei einem Vorgang insgesamt verrichtete Arbeit zu berechnen, integrieren wir über den Weg des Körpers:

$$W = \int \vec{F} \cdot d\vec{r} \tag{5.2}$$

$$= \int F_x\, dx + \int F_y\, dy + \int F_z\, dz \tag{5.3}$$

Die Arbeit wird in der nach dem Physiker James Joule (1816–1869) benannten Einheit Joule gemessen:

$$\text{Joule} \qquad \mathrm{J} = \mathrm{N} \cdot \mathrm{m} \tag{5.4}$$

Wenn in (5.1) die Kraft $\vec{F}$ und der Weg $d\vec{r}$ gleichgerichtet sind, ergibt sich $dW > 0$, bei entgegengesetzter Richtung ist $dW < 0$.

Allgemein gilt folgende **Vorzeichenkonvention**: Arbeit, die an einem System geleistet wird, z. B. Wasser der Masse m, das beim Herabfallen aus der Höhe h eine Turbine antreibt, wird im System »Turbine« positiv gerechnet:

$$W = \int_0^h \vec{F}\, d\vec{r} \tag{5.5}$$

$$= m \cdot g \cdot h > 0 \tag{5.6}$$

Das ist konsistent mit der Definition der Arbeit (5.1), denn die Kraft wirkt in Richtung des Weges und damit wird $W > 0$.

Oft wird Arbeit von außen an einem System gegen eine Kraft geleistet. Nach dem 3. Axiom von Newton (Actio = Reactio) treten dabei entgegengesetzt gleiche Kräfte auf, teilweise auch als Trägheitskräfte, und man muss sich genau überlegen, welche der Kräfte bei der Arbeit wirkt. Beispiele sind eine Feder, die wir zusammendrücken, oder ein Eimer Wasser, den wir gegen seine Gewichtskraft vom Boden hochheben. Hier muss dann wie im folgenden Beispiel »Möbelpacker« die der Feder- oder Gewichtskraft entgegenwirkende Kraft angesetzt werden. Damit ergibt sich konsistent das richtige Vorzeichen. Umgekehrt zählt die geleistete Arbeit im System »Arbeiter« negativ, Weg und Schwer- oder Federkraft sind dort entgegengesetzt gerichtet.

> Allgemein gilt für alle Formen von Energie, also auch für die Arbeit, die eingängig so genannte **»egoistische« Vorzeichenkonvention**: Energie, die einem System zugeführt wird, ist stets positiv. Energie, die einem System entzogen wird, ist negativ zu rechnen. Dabei muss man genau definieren, was das System ist.

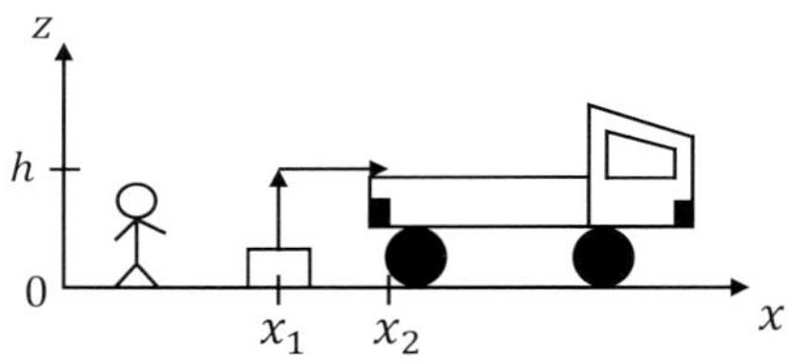

Abbildung 5.1 Arbeit beim Möbelpacken

Beispiel 21: Arbeit eines Möbelpackers

Ein Möbelpacker möchte eine Kiste auf einen LKW stellen (► Abbildung 5.1). Startpunkt und Endpunkt sind die Positionen

$$\vec{A} = \begin{pmatrix} x_1 \\ 0 \\ 0 \end{pmatrix} \quad \text{und} \quad \vec{B} = \begin{pmatrix} x_2 \\ 0 \\ h \end{pmatrix}. \tag{5.7}$$

Die Gewichtskraft der Kiste beträgt

$$\vec{F}_G = m \begin{pmatrix} 0 \\ 0 \\ -g \end{pmatrix} = const. \qquad (5.8)$$

Um die Kiste gegen ihre Gewichtskraft zu bewegen, benötigt er die Kraft

$$\vec{F} = -\vec{F}_G \,. \qquad (5.9)$$

Insgesamt wird Arbeit an der Kiste (dem System) geleistet, sodass die Arbeit konsistent mit der Vorzeichenkonvention positiv ist:

$$W = \int_{\vec{A}}^{\vec{B}} \vec{F}\, d\vec{r} \qquad (5.10)$$

$$= -m(-g) \cdot \int_0^h dz \qquad (5.11)$$

$$= m \cdot g \cdot h \qquad (5.12)$$

Die horizontale Verschiebung trägt nicht zur Arbeit im physikalischen Sinn bei. Der Weg verläuft senkrecht zur Kraftrichtung und damit ist das Skalarprodukt null. Setzt man $m = 50\,\text{kg}$, $h = 2\,\text{m}$ und $g = 10\,\text{m/s}^2$, dann ergibt sich als geleistete Arbeit

$$W = m \cdot g \cdot h = 50\,\text{kg} \cdot 10\,\frac{\text{m}}{\text{s}^2} \cdot 2\,\text{m} = 1000\,\text{J} = 1\,\text{kJ} \,. \qquad (5.13)$$

Aufgabe 12: Gustav oder Harry, wer arbeitet mehr?

Gustav und Harry bringen jeder drei Kisten mit einer Masse von jeweils 20 kg in das Lager (► Abbildung 5.2). Wer verrichtet mehr Arbeit?

Abbildung 5.2 Großer und kleiner Lagerarbeiter

Konservative Kraft

Kräfte, unter denen das Integral der Arbeit $\int \vec{F}\, d\vec{r}$ unabhängig vom Weg ist, nennt man **»konservative« Kräfte**. Die bekannteste konservative Kraft ist die Schwerkraft $F = m\,g$ auf der Erde.

Im Gegensatz hierzu sind z. B. Reibungskräfte nicht konservativ, da Reibungskräfte im Allgemeinen von dem zurückgelegten Weg und/oder der Geschwindigkeit abhängen (Abschnitte 3.2.4 und 18.6).

5.3 Potentielle Energie

Für eine konservative Kraft ist die geleistete Arbeit nur vom Anfangspunkt A und vom Endpunkt B abhängig.

Wir definieren die potentielle Energie über die Arbeit: Die Differenz der potentiellen Energien ΔE_{pot} zwischen den Orten B und A beschreibt die Arbeit, die man gewinnen kann, wenn der Körper von B nach A zurückkehrt:

$$\Delta E_{\text{pot}} = E_{\text{pot}}(B) - E_{\text{pot}}(A) = \int_B^A \vec{F}\, d\vec{r} \tag{5.14}$$

Im Beispiel von dem Beladen des Lastwagens gewinnt das System »Kiste« die Differenz der potentiellen Energien $\Delta E_{\text{pot}} = mgh$ (5.12). Beim Entladen wird sie dem System »Kiste« wieder entzogen, die potentielle Energie verringert sich.

5.4 Kinetische Energie der Translation

Wie die potentielle Energie ist auch die kinetische Energie eines Körpers über die Arbeit motivierbar. Die Arbeitsdefinition (5.1) lautet

$$dW = \vec{F}\, d\vec{r}\,. \tag{5.15}$$

Für Körper konstanter Masse gilt nach Gleichung (3.38)

$$\vec{F} = \frac{d\vec{p}}{dt} = m\,\frac{d\vec{v}}{dt}\,. \tag{5.16}$$

Damit lässt sich die differentielle Arbeit über die Geschwindigkeit eines Körpers ausdrücken, der von der Position B nach A zurückkehrt. Dafür erweitern wir $d\vec{r}$ durch $(d\vec{r}/dt)\,dt$ und kürzen dt anschließend wieder weg:

$$dW = m\,\frac{d\vec{v}}{dt}\cdot\frac{d\vec{r}}{dt}dt \tag{5.17}$$

$$= m\,\frac{d\vec{v}}{dt}\cdot\vec{v}\,dt = m\,\vec{v}\,d\vec{v} \tag{5.18}$$

Sind die Richtungen von $\vec{v}$ und $d\vec{v}$ parallel, so ist

$$dW = m\,v\,dv\,, \tag{5.19}$$

$$W = \int_{v(B)}^{v(A)} m\,v\,dv \tag{5.20}$$

$$= \frac{1}{2}\,m\,v^2\Big|_{v(B)}^{v(A)} \tag{5.21}$$

$$= \frac{1}{2}\,m\left(v^2(A) - v^2(B)\right). \tag{5.22}$$

Mit kinetischer Energie bezeichnen wir den Term

$$E_{\text{kin}} = \frac{1}{2}\,m\,v^2\,. \tag{5.23}$$

Die Differenz der kinetischen Energien des Körpers an den Stellen A und B entspricht der an ihm geleisteten Arbeit:

$$W = \int_B^A \vec{F}\,d\vec{r} = E_{\text{kin}}(A) - E_{\text{kin}}(B) \tag{5.24}$$

Dabei ist W je nach Richtung von $\vec{F}$ positive Beschleunigungs- oder negative Bremsarbeit. Dieser Zusammenhang gilt sowohl für konservative als auch für nicht konservative Kräfte.

5.5 Kinetische Energie der Rotation

Die kinetische Energie für die Kreisbewegung eines Massenpunkts können wir in eine zur Drehbewegung passende Form bringen. Setzen wir in die Definition der kinetischen Energie (5.23) den Zusammenhang zwischen der Winkelgeschwindigkeit und der Geschwindigkeit $v = \omega\,r$ (Gleichung (4.12)) ein, so erhalten wir

$$E_{\text{rot}} = \frac{1}{2}\,mr^2\,\omega^2\,. \tag{5.25}$$

Mit der Definition des Trägheitsmoments $I = mr^2$ des Massenpunkts nach Gleichung (4.60) lautet die **Rotationsenergie**

$$E_{\text{rot}} = \frac{1}{2}\,I\,\omega^2\,. \tag{5.26}$$

In dieser Schreibweise werden die Parallelen von Masse $\leftrightarrow$ Trägheitsmoment und Geschwindigkeit $\leftrightarrow$ Winkelgeschwindigkeit bei Translations- und Rotationsbewegungen wiederum offensichtlich.

5.6 Energieerhaltung

Für konservative Kräfte sind die beiden Werte für die Arbeit (5.14) und (5.24), also

$$W = \int_B^A \vec{F}\, d\vec{r} = E_{\text{pot}}(B) - E_{\text{pot}}(A) \quad \text{und} \tag{5.27}$$

$$W = \int_B^A \vec{F}\, d\vec{r} = E_{\text{kin}}(A) - E_{\text{kin}}(B)\,, \tag{5.28}$$

gleich groß und damit gilt

$$E_{\text{pot}}(B) - E_{\text{pot}}(A) = E_{\text{kin}}(A) - E_{\text{kin}}(B)\,, \tag{5.29}$$

beziehungsweise

$$\underbrace{E_{\text{kin}}(A) + E_{\text{pot}}(A)}_{\equiv E(A)} = \underbrace{E_{\text{kin}}(B) + E_{\text{pot}}(B)}_{\equiv E(B)}\,. \tag{5.30}$$

Die mechanische Gesamtenergie E eines Körpers setzt sich aus seiner kinetischen Energie der Translation und seiner potentiellen Energie zusammen:

$$E = E_{\text{kin}} + E_{\text{pot}} \tag{5.31}$$

Mit Gleichung (5.30) folgt

$$E(A) = E(B)\,. \tag{5.32}$$

Die Gesamtenergie bleibt also erhalten! Wir sprechen von **Energieerhaltung**. Die Summe aus kinetischer und potentieller Energie ist in einem abgeschlossenen System mit konservativen Kräften zu *jedem* Zeitpunkt konstant:

$$E = E_{\text{kin}} + E_{\text{pot}} = \mathit{const.} \tag{5.33}$$

Das Konzept der Erhaltung der Gesamtenergie gilt allgemein, also nicht nur bei Translationsbewegungen, sondern auch für Rotationsbewegungen.

Beispiel 22: Potentielle Energie bei der Dehnung einer Feder

Eine Feder habe die rücktreibende Kraft $F_F = -k\,x$ mit der Federkonstanten k (linke ► Abbildung 5.3). Die Arbeit W, die verrichtet werden muss, um die Feder um die Strecke $\overline{AB}$ zu dehnen, berechnet sich aus der Dehnungskraft $F = -F_F$:

$$W = \int \vec{F}\, d\vec{r} \tag{5.34}$$

$$= -\int_A^B F_F\, dx \tag{5.35}$$

$$= \frac{1}{2}\, k\, x^2 \Big|_A^B \tag{5.36}$$

$$= \frac{1}{2}\, k\, (B^2 - A^2) \tag{5.37}$$

Die verrichtete Arbeit können wir grafisch darstellen. Die Fläche unter der Kurve $F(x) = kx$ im Bereich $[A, B]$ entspricht dem Integral, also der verrichteten Arbeit (rechte ► Abbildung 5.3). Die potentielle Energie im Federsystem ist

$$\Delta E_{\text{pot}} = E_{\text{pot}}(B) - E_{\text{pot}}(A) \tag{5.38}$$
$$= \int_B^A \vec{F}_F \, d\vec{r} \tag{5.39}$$
$$= -\frac{1}{2} k \left(A^2 - B^2\right) \tag{5.40}$$
$$= \frac{1}{2} k \left(B^2 - A^2\right) . \tag{5.41}$$

Die geleistete Arbeit (5.37) wird also als potentielle Energie (5.41) in der Feder gespeichert.

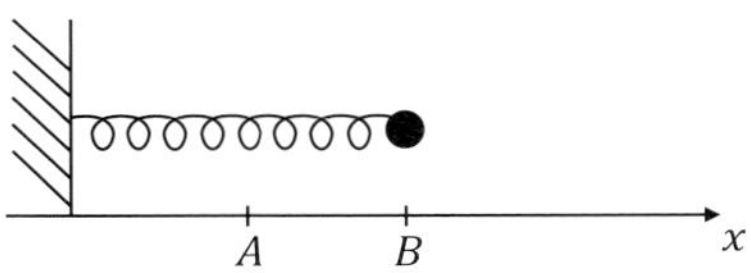

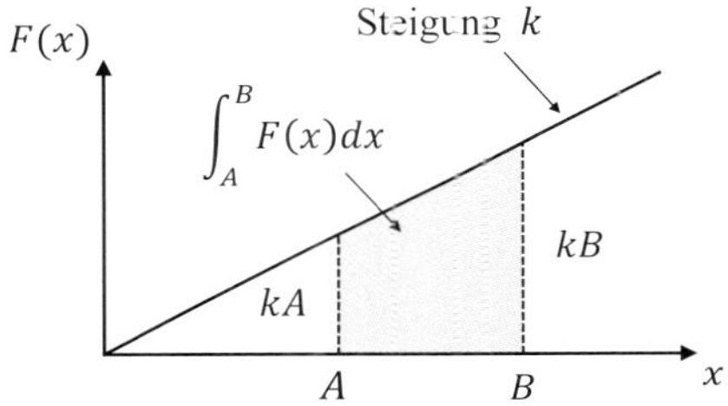

Abbildung 5.3 Links: Feder wird von A nach B gedehnt. Rechts: Arbeitsdiagramm, das Integral $\int_A^B F(x)dx$ (Schraffur) entspricht der geleisteten Arbeit.

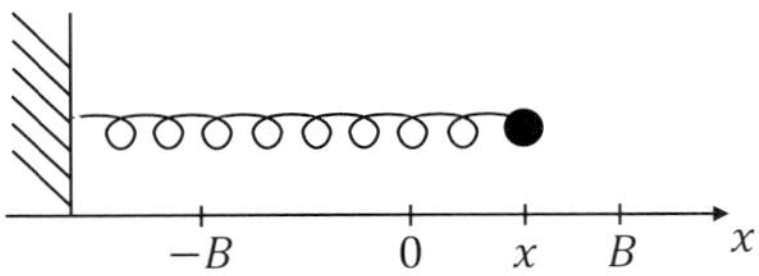

Abbildung 5.4 Feder wird aus der Ruhelage $x = 0$ nach $x = B$ ausgelenkt

Beispiel 23: Kinetische Energie bei der Entspannung einer Feder

Sei die Feder mit einem Massenpunkt zunächst bis zum Ort $x = B$ ausgelenkt. Wir lassen die Feder bis zu einem Ort x zwischen $A = 0$ und B entspannen (► Abbildung 5.4).

Durch die von der Federkraft $\vec{F}_F$ an der Masse m geleistete Arbeit

$$W = \int_B^x F_F \, dx \tag{5.42}$$
$$= \int_B^x (-kx) \, dx \tag{5.43}$$
$$= -\frac{k}{2} \left(x^2 - B^2\right) > 0 \tag{5.44}$$

verringert sich die potentielle Energie

$$W = E_{\text{pot}}(B) - E_{\text{pot}}(x) \tag{5.45}$$

und wird aufgrund der Energieerhaltung in kinetische Energie des Massenpunkts umgewandelt (5.29). Da der Massenpunkt am Ort B ruhte, ist

$$W = E_{\text{kin}}(x) - E_{\text{kin}}(B) \tag{5.46}$$

$$= \frac{1}{2}\, m\, v^2(x) - \frac{1}{2}\, m\, v^2(B) \tag{5.47}$$

$$= \frac{1}{2}\, m\, v^2(x)\,. \tag{5.48}$$

Aus den Gleichungen (5.44) und (5.48) folgt die Geschwindigkeit v des Massenpunkts an der Stelle x

$$-\frac{k}{2}\left(x^2 - B^2\right) = \frac{1}{2}\, m\, v^2\,(x) \tag{5.49}$$

$$v\,(x) = \pm\sqrt{\frac{k}{m}\,(B^2 - x^2)}\,, \tag{5.50}$$

wobei die negative Lösung der in ► Abbildung 5.4 beschriebenen Situation entspricht, während die positive Lösung die Gegenbewegung in positiver x-Richtung beschreibt.

Unser früheres Ergebnis zum Federpendel, das wir aus dem Kraftkonzept gewonnen hatten (4.33), lautete mit der Anfangsbedingung $\varphi_\circ = 0$

$$x\,(t) = B \cdot \cos\left(\sqrt{\frac{k}{m}}\, t\right)\,. \tag{5.51}$$

Die Frage stellt sich, ob das aus dem Energiekonzept erhaltene Ergebnis mit diesem Resultat vereinbar ist. Die Ableitung von (5.51) ergibt dieselbe Geschwindigkeit des Massenpunkts an der Feder:

$$v = -B\sqrt{\frac{k}{m}}\,\sin\left(\sqrt{\frac{k}{m}}\, t\right) \tag{5.52}$$

$$= \pm B\sqrt{\frac{k}{m}}\,\sqrt{1 - \cos^2\left(\sqrt{\frac{k}{m}}\, t\right)} \tag{5.53}$$

$$= \pm\sqrt{\frac{k}{m}\,(B^2 - x^2)} \tag{5.54}$$

Um einen physikalischen Vorgang zu beschreiben, können verschiedene konzeptionelle Ansätze verwendet werden, wie wir hier an dem Kraftansatz bzw. dem Energiekonzept sehen.

Orts-Geschwindigkeits-Diagramm

Die Zusammenhänge zwischen (5.50), (5.51), (5.52) und (5.54) lassen sich graphisch veranschaulichen, wenn man zur quadratischen Gleichung (5.49) zurückkehrt:

$$-\frac{k}{2}\left(x^2 - B^2\right) = \frac{1}{2}\,m\,v^2\,(x) \tag{5.49}$$

$$m\,v^2 + kx^2 = kB^2 \tag{5.55}$$

$$(\sqrt{m}\,v)^2 + (\sqrt{k}\,x)^2 = (\sqrt{k}\,B)^2 \tag{5.56}$$

(5.56) beschreibt einen Kreis in kartesischen Koordinaten $(\sqrt{k}\,x)$ und $(\sqrt{m}\,v)$ mit dem Radius $(\sqrt{k}\,B)$ (► Abbildung 5.5).

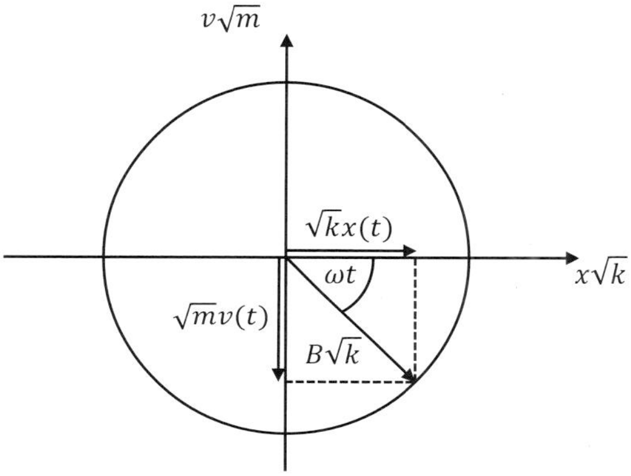

Abbildung 5.5 Orts-Geschwindigkeits-Diagramm einer Federschwingung: Der Radiusvektor der Länge $B\sqrt{k}$ rotiert im Uhrzeigersinn mit der Winkelgeschwindigkeit ω. Die Komponenten entlang der Achsen beschreiben dann Ort $x(t)$ und Geschwindigkeit $v(t)$.

Lässt man den Radiusvektor mit der Kreisfrequenz ω im Uhrzeigersinn umlaufen, kann man mit $\omega = \sqrt{k/m}$ die Gleichungen für Ort (5.51) und Geschwindigkeit (5.52) aus ► Abbildung 5.5 ablesen:

$$\sqrt{k}\,x(t) = \sqrt{k}\,B\cos\omega\,t \;\Rightarrow\; x\,(t) = B\cdot\cos\left(\omega\,t\right) \tag{5.51}$$

$$\sqrt{m}\,v(t) = -\sqrt{k}\,B\sin\omega t \;\Rightarrow\; v\,(t) = -B\cdot\omega\sin\left(\omega\,t\right) \tag{5.52}$$

Das Federpendel schwingt also mit der Amplitude $x_{\max} = B$ zwischen $\pm B$ hin und her. Die Geschwindigkeit oszilliert gegenphasig und erreicht den maximalen Wert von

$$v_{\max} = \pm B\cdot\omega = \pm B\cdot\sqrt{\frac{k}{m}} \tag{5.57}$$

beim Nulldurchgang $x = 0$. Als Zusammenhang zwischen $x(t)$ und $v(t)$ ergibt sich wieder (5.50).

Beispiel 24: Energieerhaltung beim Federpendel

Die Gesamtenergie E (5.33) in einem Federpendel ist aus seiner Anfangsausdehnung berechenbar:

$$E(B) = E_{\text{pot}}(B) \tag{5.58}$$

$$= \frac{1}{2}\,k\,B^2 \tag{5.59}$$

An jedem anderen Ort x zwischen $[-B, B]$ setzt sich die Gesamtenergie aus der kinetischen Energie (5.23) und der potentiellen Energie (5.41) zusammen:

$$E(x) = \frac{1}{2}\,m\,v^2(x) + \frac{1}{2}\,k\,x^2 \tag{5.60}$$

Am Ort $x = 0$ ist $E_{\text{pot}} = 0$. Daher ist dort die kinetische Energie $E_{\text{kin}} = \frac{1}{2}\,m\,v^2$ maximal und deswegen auch der Betrag der Geschwindigkeit maximal (5.50):

$$v_{\text{max}} = \pm\sqrt{\frac{k\,B^2}{m}} \tag{5.61}$$

In ► Abbildung 5.6 ist die potentielle Energie des Massenpunkts an jedem Ort x durch die Kurve $E_{\text{pot}} = \frac{1}{2}\,k\,x^2$ ablesbar. Die kinetische Energie ergibt sich in der Abbildung aus der Energiedifferenz zur Gesamtenergie E.

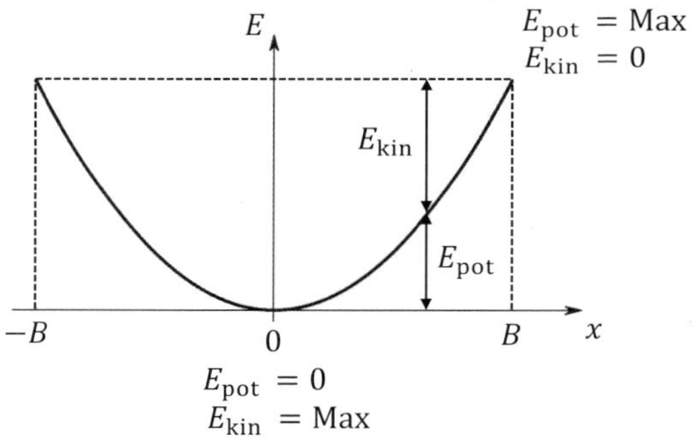

Abbildung 5.6 Die potentielle Energie E_{pot} und die kinetische Energie E_{kin} teilen sich die Gesamtenergie E und oszillieren dabei jeweils zwischen 0 und ihrem Maximalwert.

5.7 Leistung

Leistung ist definiert als die pro Zeiteinheit geleistete Arbeit:

$$P = \frac{dW}{dt} \tag{5.62}$$

$$= \frac{\vec{F}\,d\vec{r}}{dt} \tag{5.63}$$

$$= \vec{F}\,\vec{v} \tag{5.64}$$

Sie wird in der Einheit Watt $W = \mathrm{J/s}$ gemessen, die nach James Watt (1736–1819) benannt ist.

5.8 Intensität

Intensität beschreibt einen Strom von Energie E, der pro Zeiteinheit Δt durch eine zur Ausbreitungsrichtung der Energie senkrecht stehende Flächeneinheit ΔA transportiert wird:

$$I = \frac{E}{\Delta t\,\Delta A} \tag{5.65}$$

Die Einheit der Intensität ist $\mathrm{J/(s\,m^2)}$.

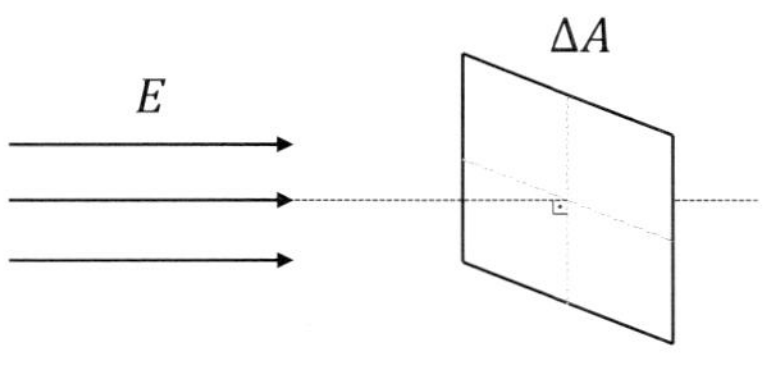

Abbildung 5.7 Die Intensität I ist die Energie E, die in der Zeit Δt senkrecht durch die Fläche ΔA fließt.

Das Wichtigste zum Mitnehmen

Zum Mitnehmen: Arbeit, Energie, Leistung, Intensität 1

- **Arbeit** W ist durch das Skalarprodukt aus der angreifenden Kraft und dem Weg in Bewegungsrichtung definiert, in differentieller Schreibweise

$$dW = \vec{F}\, d\vec{r}\,. \tag{5.1}$$

Die Arbeit hat die Einheit der Energie:

$$\text{Joule} \qquad \mathrm{J} = \mathrm{N} \cdot \mathrm{m} \tag{5.4}$$

Allgemein gilt für alle Formen von Energie, also auch für die Arbeit, die **Vorzeichenkonvention**:

> Energie, die einem System zugeführt wird, ist stets positiv, Energie, die einem System entzogen wird, ist negativ zu rechnen.
>
> Dabei muss man genau definieren, was das System ist.

- Kräfte, unter denen das Integral der Arbeit $\int \vec{F}\, d\vec{r}$ unabhängig vom Weg ist, nennt man **»konservative« Kräfte**. Die bekannteste konservative Kraft ist die Schwerkraft.

  Im Gegensatz hierzu sind z. B. von der Geschwindigkeit oder vom Weg abhängige Reibungskräfte nicht konservativ.

- Da für konservative Kräfte die geleistete oder aufgenommene Arbeit nur vom Anfangspunkt A und vom Endpunkt B abhängt, kann man für sie eine potentielle Energie über die Arbeit definieren:

$$\Delta E_{\text{pot}} = E_{\text{pot}}(B) - E_{\text{pot}}(A) = \int_B^A \vec{F}\, d\vec{r} \tag{5.14}$$

- Mit **kinetischer Translationsenergie** bezeichnen wir

$$E_{\text{kin}} = \frac{1}{2} m v^2\,. \tag{5.23}$$

Die Differenz der kinetischen Energien des Körpers an den Stellen A und B entspricht je nach Richtung von $\vec{F}$ der Beschleunigungs- oder Bremsarbeit:

$$\int_B^A \vec{F}\, d\vec{r} = E_{\text{kin}}(A) - E_{\text{kin}}(B) \tag{5.24}$$

Dieser Zusammenhang gilt sowohl für konservative als auch für nicht konservative Kräfte.

Zum Mitnehmen: Arbeit, Energie, Leistung, Intensität 2

- Mit der Definition des Trägheitsmoments $I = mr^2$ des Massenpunkts nach Gleichung (4.60) lautet die **Rotationsenergie**

$$E_{\text{rot}} = \frac{1}{2}\, I\, \omega^2 \,. \tag{5.25}$$

  Ein Vergleich von (5.23) und (5.25) verdeutlicht die Parallelen von Masse $\leftrightarrow$ Trägheitsmoment und Geschwindigkeit $\leftrightarrow$ Winkelgeschwindigkeit.

- Die **mechanische Gesamtenergie** E eines Körpers setzt sich aus seiner kinetischen Energie der Translation (eventuell auch Rotation) und seiner potentiellen Energie zusammen: $E = E_{\text{kin}} + E_{\text{pot}}$.

  Für konservative Kräfte gilt **Energieerhaltung**: *Die Summe aus kinetischer und potentieller Energie ist zu* jedem *Zeitpunkt konstant*:

$$E = E_{\text{kin}} + E_{\text{pot}} = const. \tag{5.33}$$

- Ein **Federpendel** mit der Federkonstanten k und der Masse m schwingt mit der Frequenz $\omega = \sqrt{k/m}$ (Anfangsbedingung $\varphi_\circ = 0$):

$$x\,(t) = B \cdot \cos \omega\, t \tag{5.51}$$

  Die Ableitung von (5.51) ergibt die Geschwindigkeit des Massenpunkts

$$v = -B\omega \sin\,(\omega\, t) \tag{5.52}$$

$$= \pm B\omega\sqrt{1 - \cos^2\,(\omega\, t)} = \pm\sqrt{\frac{k}{m}\,\left(B^2 - x^2\right)}\,. \tag{5.54}$$

  Um diesen physikalischen Vorgang zu beschreiben, können wir den Kraft- oder den Energieansatz verwenden.

- **Leistung** ist definiert als die pro Zeiteinheit geleistete Arbeit:

$$P = \frac{dW}{dt} \tag{5.62}$$

  Sie wird in der Einheit Watt W = J/s gemessen.

- **Intensität** beschreibt einen Strom von Energie E, der pro Zeiteinheit Δt senkrecht durch eine Flächeneinheit ΔA transportiert wird:

$$I = \frac{E}{\Delta t\, \Delta A} \tag{5.65}$$

  Die Einheit der Intensität ist J/(s m^2).

Gemeinsam verstehen

6

Wie gut können Sie bereits über die Themen sprechen? Führen Sie zu zweit ein wissenschaftliches Gespräch miteinander! Verteilen Sie dafür die Rollen: Anhand der Fragestellung erklärt Person 1 der Person 2 die wichtigsten physikalischen Zusammenhänge. Nehmen Sie Stift und Papier zur Hand, damit Sie bei Ihren Erklärungen Skizzen zeichnen und grundlegende physikalische Konzepte in Form von Gleichungen aufschreiben können.

Person 2 darf Verständnisfragen stellen, dabei z. B. Präzisierungen erfragen (was bedeutet der Buchstabe p in Deiner Gleichung genau?) oder auf größere Zusammenhänge hinweisen. Für die nächste Fragestellung tauschen Sie Ihre Rollen.

Bitte gehen Sie davon aus, dass Ihr Gespräch individuell ganz verschieden verlaufen wird. Die Skizzen über mögliche Gesprächsverläufe im letzten Teil des E-Books oder Online sind lediglich als Kurzzusammenfassung einiger Aspekte zur anschließenden (!) Ergänzung gedacht.

Bei wissenschaftlichen Gesprächen dieser Art werden Sie sehr schnell Lerneffekte verspüren. Außerdem dienen solche Gespräche auch als gute Vorbereitung für mündliche Prüfungen. Ergänzen Sie gerne die Themenpalette mit eigenen Fragestellungen.

Aufgabe 13: Partnerarbeit: Weitwurf

Welche Bahnkurve fliegt ein Ball beim Weitwurf und warum genau diese Kurve? Gibt es günstige Abwurfwinkel?

Aufgabe 14: Partnerarbeit: Massenmessung

Nennen Sie Möglichkeiten, die Masse eines Körpers zu messen.

Aufgabe 15: Partnerarbeit: Messung

Eine Messreihe zur Massenmessung lautet

m/g: $107, 105, 104, 105, 103, 110, 104$.

Was bedeuten diese Messungen in Bezug auf die wahre Masse des Körpers? In welchem Bereich erwarten Sie die Messwerte zukünftiger Messungen?

Aufgabe 16: Partnerarbeit: Newtons Bewegungsgesetz

Beschreiben Sie mit Ihren Worten Newtons Bewegungsgesetz $d\vec{p}/dt = \sum_{i=1}^{n} \vec{F}_i$ (3.41). Beantworten Sie dabei die folgenden Fragen:

- Was bedeutet $\vec{p}$?
- Auf wen bezieht sich $\vec{p}$?
- Welche Bedeutung hat die zeitliche Ableitung $d\vec{p}/dt$?
- Was ist $\vec{F}$?
- Auf wen bezieht sich $\vec{F}$?
- Welchen Sinn hat das Zeichen Σ?
- Fassen Sie zusammen: Wer reagiert auf was?
- Welchen Nutzen hat dieses physikalische Konzept?

Aufgabe 17: Partnerarbeit: Reibung

Welche Auswirkungen hat eine Reibungskraft $F_R = -\rho v$ auf die Geschwindigkeit eines Autos, das anfangs mit der Geschwindigkeit $v_\circ$ fuhr? (Details zu geschwindigkeitsabhängigen Reibungskräften stellen wir später in Abschnitt 18.6 vor.)

Aufgabe 18: Partnerarbeit: Garnrolle

Warum kann man eine Garnrolle durch fast vertikales Ziehen beziehungsweise fast horizontales Ziehen an ihrem Faden in verschiedene Richtungen bewegen?

Aufgabe 19: Partnerarbeit: Schwingung

Erläutern Sie den Zusammenhang zwischen der Schwingung eines Federpendels und der Kreisbewegung.

Aufgabe 20: Partnerarbeit: Schwingungsdauer

Welche Schwingungsdauer ist größer: Federpendel mit einer Feder oder Federpendel mit zwei Federn?

Aufgabe 21: Partnerarbeit: Energie im Federpendel

Erläutern Sie die Energieerhaltung beim Federpendel.

Aufgabe 22: Partnerarbeit: Aufstiegsarbeit

Wie ist Arbeit als physikalischer Begriff definiert? Erläutern Sie anhand der Besteigung des 8 848 m hohen Mount Everest durch Sir E. Hillary and T. Norgay. Wie groß ist die potentielle Energie eines Bergsteigers auf dem Gipfel?

Aufgabe 23: Partnerarbeit: Bergsteiger

Sir E. Hillary and T. Norgay beschließen, nach ihrer Mount Everest Besteigung (8 848 m) auf ihrem Schlitten hinunter zu fahren. Diskutieren Sie die Geschwindigkeit des Schlittens.

Kollision – Gravitation – Bezugssysteme

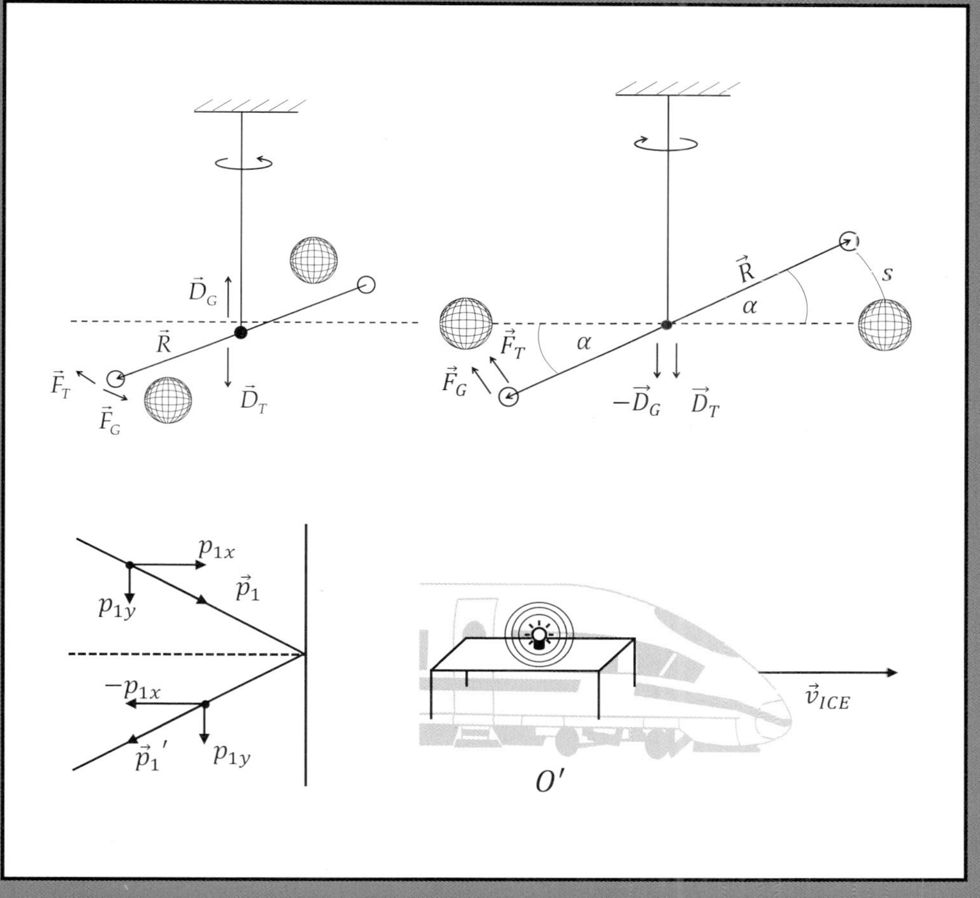

Teil II

Erhaltungssätze

7.1 Energieerhaltung 143
7.2 Impulserhaltung 143
7.3 Drehimpulserhaltung 144

ÜBERBLICK

7

» Erhaltungssätze haben fundamentale Bedeutung in der Physik. Die wichtigsten Erhaltungssätze in der Mechanik beziehen sich auf die Energie, den Impuls und den Drehimpuls eines Systems.

Erhaltungssätze gelten für geschlossene Systeme, d. h. Systeme, die nicht mit ihrer Umgebung in Wechselwirkung treten. Wenn allerdings der Energieaustausch mit der Umgebung explizit in der Energiegleichung berücksichtigt wird, ist auch hier eine konsistente Formulierung der Energieerhaltung möglich. «

7.1 Energieerhaltung

Die Gesamtenergie

$$E = \sum_{i=1}^{N} E_i \tag{7.1}$$

eines abgeschlossenen Systems bleibt zeitlich konstant:

$$E = const. \tag{7.2}$$

Dabei können verschiedene Energieformen innerhalb des Systems ineinander umgewandelt werden.

Für konservative Kräfte hatten wir bereits in Abschnitt 5.6 Gleichung (5.33) gesehen, dass die Gesamtenergie E mit der kinetischen Energie E_{kin} und der potentiellen Energie E_{pot} wie folgt zusammenhängt:

$$E = E_{\text{pot}} + E_{\text{kin}} = const. \tag{7.3}$$

Ist eine Kraft nicht konservativ, muss man andere Energieformen berücksichtigen, wie z. B. Wärmeenergie, die bei Reibungsphänomenen auftritt.

7.2 Impulserhaltung

Die Impulserhaltung für einzelne Körper oder Teilchen ist bereits in Newtons Gesetzen formuliert (Abschnitt 3.4, Gleichung (3.41)). Wenn keine äußeren Kräfte auf das Teilchen wirken, ist

$$\frac{d\vec{p}}{dt} = \sum_{j=1}^{k} \vec{F}_j = 0\,. \tag{7.4}$$

Dann bleibt der Impuls des Teilchens $\vec{p} = const.$.

Auch der Gesamtimpuls

$$\vec{\mathrm{P}} = \sum_{i=1}^{N} \vec{p}_i \tag{7.5}$$

eines abgeschlossenen Systems von N Teilchen bleibt zeitlich konstant, wenn keine äußeren Kräfte auf das System wirken:

$$\frac{d\vec{\mathrm{P}}}{dt} = 0 \tag{7.6}$$

$$\Rightarrow \quad \vec{\mathrm{P}} = const. \tag{7.7}$$

7.3 Drehimpulserhaltung

Auch die Drehimpulserhaltung für einzelne Körper oder Teilchen ist in der Formulierung der Bewegungsgesetze für Drehbewegungen enthalten (Abschnitt 4.7, Gleichung (4.70)). Für

$$\frac{d\vec{L}}{dt} = \sum_{j=1}^{k} \vec{D}_j = 0 \tag{7.8}$$

bleibt der Drehimpuls des Teilchens konstant: $\vec{L} = const.$

Ebenso bleibt auch der Gesamtdrehimpuls

$$\vec{\mathrm{L}} = \sum_{i=1}^{N} \vec{L}_i \tag{7.9}$$

eines abgeschlossenen Systems von N Teilchen zeitlich konstant. Wenn keine äußeren Drehmomente auf das System wirken, gilt

$$\frac{d\vec{\mathrm{L}}}{dt} = 0 \tag{7.10}$$

$$\Rightarrow \quad \vec{\mathrm{L}} = const.\,. \tag{7.11}$$

Das Wichtigste zum Mitnehmen

Zum Mitnehmen: Erhaltungssätze 1

- **Erhaltungssätze** sind von fundamentaler Bedeutung in der Physik. In der Mechanik beziehen sich die wichtigsten auf die Energie, den Impuls und den Drehimpuls eines **abgeschlossenen Systems**.
- Die **Gesamtenergie**

$$E = \sum_{i=1}^{N} E_i$$

eines abgeschlossenen Systems bleibt zeitlich konstant:

$$E = const. \tag{7.2}$$

Dabei können verschiedene Energieformen innerhalb des Systems ineinander umgewandelt werden. Für **konservative Kräfte** sind das nur die kinetische Energie E_{kin} und die potentielle Energie E_{pot}, also

$$E = E_{\text{pot}} + E_{\text{kin}} = const.\,, \tag{7.3}$$

sonst muss man andere Energieformen wie z. B. Wärmeenergie berücksichtigen.
- Die **Impulserhaltung** für einzelne Körper folgt bereits aus Newtons Bewegungsgesetz (3.41), wenn keine äußeren Kräfte auf den Körper wirken:

$$d\vec{p}/dt = \sum_{j=1}^{k} \vec{F}_j = 0 \quad \Rightarrow \quad \vec{p} = const.$$

Auch der **Gesamtimpuls**

$$\vec{\mathrm{P}} = \sum_{i=1}^{N} \vec{p}_i$$

eines abgeschlossenen Systems bleibt zeitlich konstant, wenn keine äußeren Kräfte auf das System wirken:

$$\vec{\mathrm{P}} = const. \tag{7.7}$$

Zum Mitnehmen: Erhaltungssätze 2

- Analog folgt die **Drehimpulserhaltung** für einzelne Körper aus

$$d\vec{L}/dt = \sum_{j=1}^{k} \vec{D}_j = 0 \quad \Rightarrow \quad \vec{L} = const.$$

Ebenso bleibt auch der **Gesamtdrehimpuls**

$$\vec{\mathrm{L}} = \sum_{i=1}^{N} \vec{L}_i$$

eines abgeschlossenen Systems von N Teilchen zeitlich konstant, wenn keine äußeren Drehmomente auf das System wirken:

$$\vec{\mathrm{L}} = const. \tag{7.11}$$

Stoßprozesse

8.1	**Elastische Kollision mit gleichen Massen**	**149**
8.2	**Elastische Kollision mit verschiedenen Massen**	**151**
	8.2.1 Elastischer Stoß an einer festen Wand	154
8.3	**Inelastische Stöße**	**154**

ÜBERBLICK

8

» Energie und Impuls eines Körpers können anderen Objekten weitergegeben werden. Die Übertragung erfordert eine Wechselwirkung, die im Fall von Stoßprozessen durch direkte Berührung der Kollisionspartner geschieht. In diesem Kapitel zeigen wir, wie sich Impulsübertrag und Impulserhaltung in Streuversuchen mit Kugeln sehr schön sichtbar machen lassen.

Wir werden weiterhin zwischen elastischen und inelastischen Prozessen unterscheiden. Bei beiden Prozessen ist die Energie erhalten, d. h. die Gesamtenergie der Kollisionspartner ist vor und nach dem Stoßprozess gleich groß. Bei elastischen Prozessen verbleibt die ursprüngliche Energie in der Bewegung der Kollisionspartner. Bei inelastischen Prozessen geht ein Teil der anfänglichen Energie in Formen innerer Energie über, wie z. B. Verformung oder Wärmeenergie der Stoßpartner. «

8.1 Elastische Kollision mit gleichen Massen

Elastische Stöße zwischen Kugeln, Teilchen etc. können wir mit den Konzepten der Energie- und Impulserhaltung verstehen.

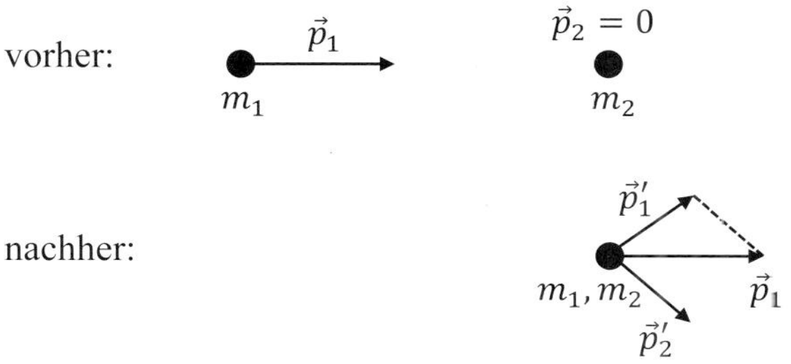

Abbildung 8.1 Eine Kugel der Masse m_1 mit Impuls $\vec{p_1}$ stößt elastisch auf eine ruhende Kugel der Masse m_2.

Bei einem elastischen Stoß zweier Kugeln muss der Gesamtimpuls des Systems aus den beiden Kugeln vor der Kollision

$$\vec{P} = \vec{p}_1 + \vec{p}_2 \tag{8.1}$$

genauso groß sein, wie der Gesamtimpuls nach der Kollision:

$$\vec{P}\,' = \vec{p}_1' + \vec{p}_2' \tag{8.2}$$

Da die Kugel mit der Masse m_2 anfangs ruht (▶ Abbildung 8.1, $\vec{p}_2 = 0$), vereinfacht sich die Impulserhaltung zu

$$\vec{P} = \vec{P}\,' \tag{8.3}$$

$$\vec{p}_1 + \underbrace{\vec{p}_2}_{=0} = \vec{p}_1' + \vec{p}_2' \,. \tag{8.4}$$

Ebenso muss die Energieerhaltung gelten. Die Gesamtenergien vor und nach der Kollision müssen gleich groß sein:

$$E = E' \tag{8.5}$$

Die Kugeln sollen sich auf einem Tisch reibungsfrei mit Energien E_{kin} bewegen und außer der direkten Kollision keine zusätzliche Wechselwirkung erfahren ($E_{\text{pot}} = 0$).

Die Gesamtenergie des Systems aus beiden Kugeln können wir aus dem Anfangszustand berechnen, bei dem nur die erste Kugel Bewegungsenergie besitzt:

$$E = E_{\text{kin},1} + \underbrace{E_{\text{pot}}}_{\text{hier } =0} \tag{8.6}$$

$$= \frac{\vec{p}_1^{\,2}}{2m_1} \tag{8.7}$$

Mit der anfangs ruhenden Kugel ($E_{\mathrm{kin},2} = 0$) lautet die Energieerhaltung beim Stoßvorgang (8.5)

$$E_1 = E_1' + E_2' \,. \tag{8.8}$$

Betrachten wir zunächst den Fall, dass beide Massen gleich groß sind ($m_1 = m_2 = m$). Quadrieren wir die Gleichung der Impulserhaltung (8.4), so ergibt sich

$$\vec{p}_1^{\,2} = \left(\vec{p}_1^{\,\prime} + \vec{p}_2^{\,\prime}\right)^2 \tag{8.9}$$

$$= \vec{p}_1^{\,\prime\,2} + \vec{p}_2^{\,\prime\,2} + 2 \cdot \vec{p}_1^{\,\prime} \cdot \vec{p}_2^{\,\prime} \,. \tag{8.10}$$

Aus der Energieerhaltung (8.8) folgt

$$\frac{\vec{p}_1^{\,2}}{2m} = \frac{\vec{p}_1^{\,\prime\,2}}{2m} + \frac{\vec{p}_2^{\,\prime\,2}}{2m} \tag{8.11}$$

$$\vec{p}_1^{\,2} = \vec{p}_1^{\,\prime\,2} + \vec{p}_2^{\,\prime\,2} \,. \tag{8.12}$$

Beim Vergleich von (8.10) und (8.12) sehen wir, dass das Skalarprodukt $\vec{p}_1^{\,\prime} \cdot \vec{p}_2^{\,\prime}$ verschwinden muss (den uninteressanten Fall $\vec{p}_1 = 0$ betrachten wir hier nicht):

$$\vec{p}_1^{\,\prime} \cdot \vec{p}_2^{\,\prime} = 0 \tag{8.13}$$

$$|\vec{p}_1^{\,\prime}| \cdot |\vec{p}_2^{\,\prime}| \cdot \cos\left(\vartheta_{12}'\right) = 0 \tag{8.14}$$

Diese Bedingung ist dann erfüllt, wenn $\cos(\vartheta_{12}') = 0$ ist, wenn also der Winkel der beiden Flugrichtungen der Kugeln im Endzustand $\vartheta_{12}' = 90°$ beträgt (▶ Abbildung 8.2).

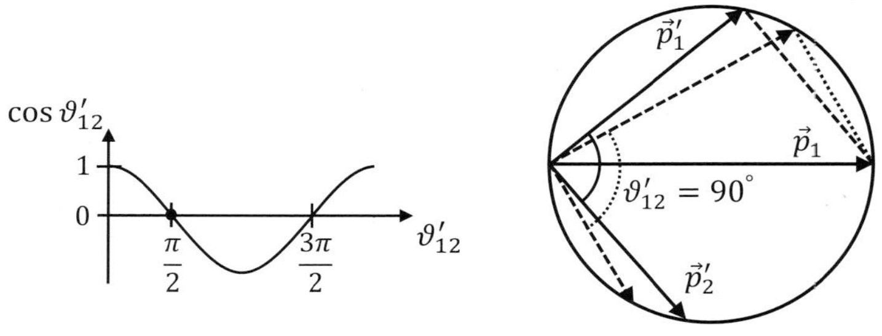

Abbildung 8.2 Links: Kosinus des Winkels ϑ_{12}' zwischen den Kugeln nach dem Stoßprozess. Rechts: Die Impulse $\vec{p}_1^{\,\prime}$ und $\vec{p}_2^{\,\prime}$ der beiden Kugeln gleicher Masse ($m_1 = m_2$) stehen nach dem Stoß senkrecht aufeinander ($\vartheta_{12}' = 90°$), außer beim zentralen Stoß, wo die Bewegungen auf einer Linie verlaufen. Die Endpunkte von $\vec{p}_1^{\,\prime}$ und $\vec{p}_2^{\,\prime}$ liegen auf einem Thaleskreis über $\vec{p}_1$.

Ein Spezialfall ist das zentrale Auftreffen der ersten Kugel auf die zweite Kugel (**zentraler Stoß**), sodass Bewegungen vor und nach dem Stoß entlang einer Linie erfolgen. Für $\vartheta_{12}' = 0°$ ist $\cos(\vartheta_{12}') = 1$. Die Bedingung (8.14) ist dann erfüllt, wenn $p_1' = 0$ ist, d. h. die erste Kugel nach dem Stoß liegen bleibt ($p_2' = 0$ ist unphysikalisch, m_1 müsste m_2 durchdringen).

Experiment 15: Stoß zweier Kugeln

Lässt man eine Kugel über eine schräg gestellte Schiene auf eine zweite, ruhende Kugel treffen, wobei beide anschließend in einen mit Sand gefüllten Kasten fallen, so kann man die Differenz der Flugwinkel von $90°$ sehr schön nachweisen. Für verschiedene Stoßparameter (Abweichung vom zentralen Auftreffen auf die zweite Kugel) lässt sich die Kreisform (Thaleskreis) demonstrieren.

8.2 Elastische Kollision mit verschiedenen Massen

Wählen wir Kugeln mit ungleichen Massen, so lassen sich die Massenterme in der Energieerhaltung (8.11) nicht einfach eliminieren:

$$\frac{\vec{p}_1^{\,2}}{2m_1} = \frac{\vec{p}_1^{\,\prime\,2}}{2m_1} + \frac{\vec{p}_2^{\,\prime\,2}}{2m_2} \tag{8.15}$$

Wir wählen das Koordinatensystem so, dass sich die erste, einlaufende Kugel in der x-Richtung auf die zweite, ruhende Kugel zubewegt. Im Allgemeinen haben beide Kugeln nach der Kollision einen von Null verschiedenen Impuls, den wir in seine p_x'- und p_y'-Komponenten zerlegen.

Da vor der Kollision keine Impulskomponente in der p_y-Richtung vorhanden war, impliziert die Impulserhaltung, dass die p_y'-Komponenten der Kugelimpulse nach dem Stoß entgegengesetzt gleich groß sind:

$$p_{1y}' = -p_{2y}' \tag{8.16}$$

In Abhängigkeit der Impulskomponenten sind dann die Quadrate der Impulse nach dem Stoß (► Abbildung 8.3)

$$\vec{p}_2^{\,\prime\,2} = p_{2x}^{\prime\,2} + p_{2y}^{\prime\,2} \tag{8.17}$$

$$\vec{p}_1^{\,\prime\,2} = \left(|\vec{p}_1| - p_{2x}'\right)^2 + p_{2y}^{\prime\,2}\,. \tag{8.18}$$

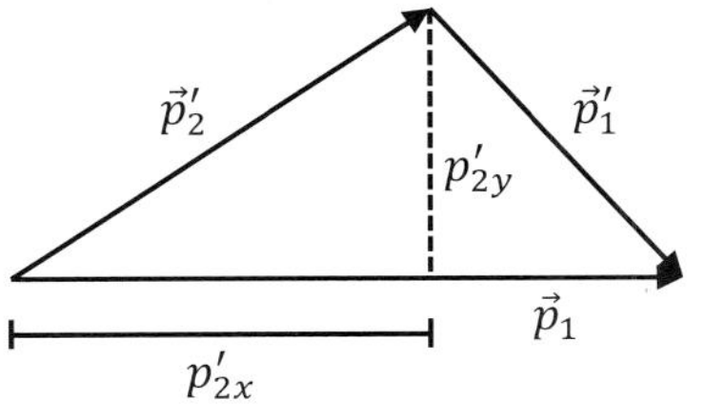

Abbildung 8.3 Impulserhaltung beim Stoß: Die Impulse $\vec{p}_1'$ und $\vec{p}_2'$ addieren sich zum Anfangsimpuls $\vec{p}_1$. Eingezeichnet sind auch die x- und y-Komponenten von $\vec{p}_2'$.

Durch Einsetzen in die Energieerhaltungsgleichung (8.15), Umsortieren der Terme nach $p_{2x}^{\prime 2}$ und $p_{2y}^{\prime 2}$ und Verwendung der Impulsdefinition $p = m\,v$ (3.37) erhalten wir

$$\frac{\vec{p}_1^{\,2}}{2m_1} = \frac{\left(|\vec{p}_1| - p_{2x}'\right)^2 + p_{2y}^{\prime 2}}{2m_1} + \frac{p_{2x}^{\prime 2} + p_{2y}^{\prime 2}}{2m_2} \tag{8.19}$$

$$0 = -2p_{2x}'\frac{|\vec{p}_1|}{m_1} + p_{2x}^{\prime 2}\frac{1}{m_1} + p_{2y}^{\prime 2}\frac{1}{m_1} + p_{2x}^{\prime 2}\frac{1}{m_2} + p_{2y}^{\prime 2}\frac{1}{m_2} \tag{8.20}$$

$$0 = -2p_{2x}'v_1 + p_{2x}^{\prime 2}\left[\frac{1}{m_1} + \frac{1}{m_2}\right] + p_{2y}^{\prime 2}\left[\frac{1}{m_1} + \frac{1}{m_2}\right] \tag{8.21}$$

$$0 = -2p_{2x}'v_1\left[\frac{m_1 + m_2}{m_1 m_2}\right]^{-1} + p_{2x}^{\prime 2} + p_{2y}^{\prime 2}\,. \tag{8.22}$$

Wir definieren die sogenannte **reduzierte Masse** μ durch

$$\mu \equiv \frac{m_1\,m_2}{m_1 + m_2}\,. \tag{8.23}$$

Damit verkürzt sich die Gleichung (8.22) zu

$$p_{2x}^{\prime 2} - 2p_{2x}'v_1\mu + p_{2y}^{\prime 2} = 0\,. \tag{8.24}$$

Die Interpretation der Gleichung können wir durch quadratische Ergänzung vereinfachen:

$$\left(p_{2x}' - \mu v_1\right)^2 + p_{2y}^{\prime 2} - (\mu v_1)^2 = 0 \tag{8.25}$$

$$\left(p_{2x}' - \mu v_1\right)^2 + p_{2y}^{\prime 2} = (\mu v_1)^2 \tag{8.26}$$

Diese Gleichung hat die Form einer Kreisgleichung ($x^2 + y^2 = R^2$).

Auch beim Stoß zweier Kugeln ungleicher Massen liegen die Impulskomponenten der zweiten Kugel auf einem Kreis mit dem Radius $R = \mu v_1$, dessen Mittelpunkt um $\mu\,v_1 = R$ verschoben ist (► Abbildung 8.4). Der Impuls der ersten Kugel nach dem Stoß lässt sich aus (8.26) und der Impulserhaltung

$$\vec{p}_1' = \vec{p}_1 - \vec{p}_2' \tag{8.27}$$

berechnen und in ► Abbildung 8.4 jeweils aus dem großen Dreieck ablesen.

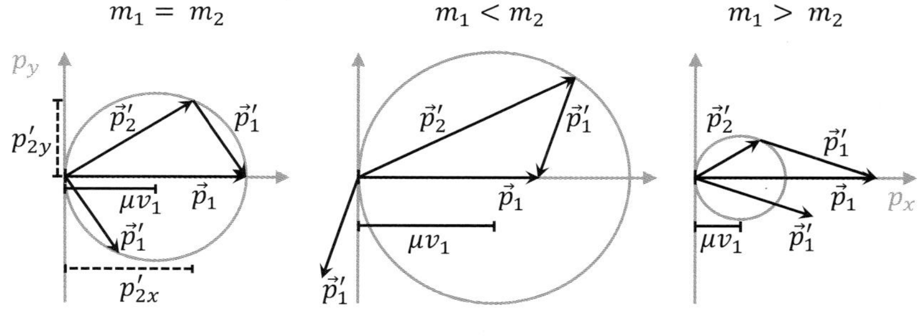

Abbildung 8.4 Geometrischer Zusammenhang der Impulse und der reduzierten Masse $\mu = m_1\,m_2/(m_1 + m_2)$ (8.23) bei Stoßprozessen mit verschieden großen Massen: links $m_1 = m_2$ (vergleiche ► Abbildung 8.2), Mitte $m_1 < m_2$, rechts $m_1 > m_2$.

In der linken ► Abbildung 8.4 ist die Situation für zwei gleich große Massen $m_1 = m_2$ gezeigt. In diesem Fall ist der Kreisdurchmesser $2R$ gleich der Länge des einlaufenden Impulsvektors $\vec{p_1}$. Der Kreis schneidet also die p_x-Achse bei p_1. Daher ist für die Fälle, bei denen die Masse m_1 der einlaufenden Kugel ungleich der Masse m_2 der ruhenden Kugel ist, das Verhältnis von $2R$ zum einlaufenden Impuls p_1 interessant. Dieses Verhältnis führen wir mit $R = \mu v_1$ und $p_1 = m_1\, v_1$ auf die Massen der Stoßpartner zurück:

$$\frac{2R}{p_1} = \frac{2\mu v_1}{m_1\, v_1} = \frac{2m_2}{m_1 + m_2} \tag{8.28}$$

Aus (8.28) können wir ablesen:

- Für $m_1 < m_2$ ist $2R > p_1$ und der Kreis schneidet rechts von $\vec{p_1}$.

  Die mittlere ► Abbildung 8.4 zeigt diesen Fall für das Massenverhältnis $m_2 = 3\,m_1$ mit der reduzierten Masse $\mu = m_1\, 3/4$. Der Kreisradius $R = \mu\, v_1 = m_1\, v_1\, 3/4 = p_1\, 3/4$ ist damit fast so lang wie der einlaufende Impulsvektor. Für Streuprozesse nahe eines zentralen Stoßes wird die einlaufende Kugel in den rückwärtigen Bereich gestreut.

- Für $m_1 > m_2$ ist $2R < p_1$ und der Kreis schneidet links von $\vec{p_1}$.

  In der rechten ► Abbildung 8.4 ist dieser Fall für $m_2 = m_1/3$ gezeigt. Hier ist die reduzierte Masse $\mu = m_1/4$, sodass sich ein kleiner Kreisradius $R = \mu\, v_1 = m_1\, v_1/4 = p_1/4$ im Vergleich zum einlaufenden Impuls p_1 ergibt. Bei diesem Massenverhältnis wird die einlaufende Kugel immer nach vorne gestreut.

Beispiel 25: Zentraler Stoß

Der zentrale Stoß zweier Kugeln mit $p_{1y} = p'_{1y} = p'_{2y} = 0$ lässt sich aus dem allgemeinen Ergebnis der ► Abbildung 8.4 ablesen:

$$p'_2 = p'_{2x} = 2\mu v_1 = \frac{2m_1\, m_2}{m_1 + m_2} v_1 \tag{8.29}$$

Mit (8.27) erhalten wir den Endimpuls der ersten Kugel:

$$p'_1 = p_1 - p'_2 = m_1 v_1 - 2\mu v_1 = m_1 \left(1 - \frac{2m_2}{m_1 + m_2}\right) v_1 \tag{8.30}$$

$$v'_1 = \left(1 - \frac{2m_2}{m_1 + m_2}\right) v_1 \tag{8.31}$$

Daraus ergibt sich eine Fallunterscheidung, die auch der Intuition entspricht:

- $m_2 < m_1 \rightarrow$ Kugel 1 bewegt sich verlangsamt vorwärts
- $m_2 = m_1 \rightarrow$ Kugel 1 bleibt liegen
- $m_2 > m_1 \rightarrow$ Kugel 1 wird rückwärts gestreut

8.2.1 Elastischer Stoß an einer festen Wand

Wir wollen jetzt Stöße an einer ebenen, glatten festen Wand untersuchen, wie beispielsweise Billardstöße an der Bande. Solche Vorgänge spielen auch in der Physik eine wichtige Rolle, z. B. in der kinetischen Gastheorie (Kapitel 20).

Wir stellen uns die feste Wand als unendlich große und schwere Kugel vor und können damit an elastische Stöße mit $m_2 \to \infty$ anknüpfen. Wir betrachten dazu den Kehrwert von (8.23) im Grenzübergang sehr großer Masse m_2:

$$\lim_{m_2\to\infty} \mu^{-1} = \lim_{m_2\to\infty} \frac{m_1 + m_2}{m_1\, m_2} = \lim_{m_2\to\infty} \frac{m_1/m_2 + 1}{m_1} = \frac{1}{m_1} \tag{8.32}$$

Wir können also für die zur Wand senkrechte Komponente p'_{1x} von $\vec{p_1}'$ (► Abbildung 8.5) die Formel (8.31) mit $\mu = m_1$ verwenden:

$$p'_{1x} = p_1 - p'_{2x} = m_1 v_1 - 2m_1 v_1 = -m_1 v_1 = -p_{1x} \tag{8.33}$$

Die zur Wand parallelen y- und z-Komponenten erfahren an der glatten, ebenen Wand keine Kräfte und bleiben unverändert:

$$p_{1y} = p'_{1y} \qquad p_{1z} = p'_{1z} \tag{8.34}$$

Es ergibt sich insgesamt die intuitiv einleuchtende Situation von ► Abbildung 8.5.

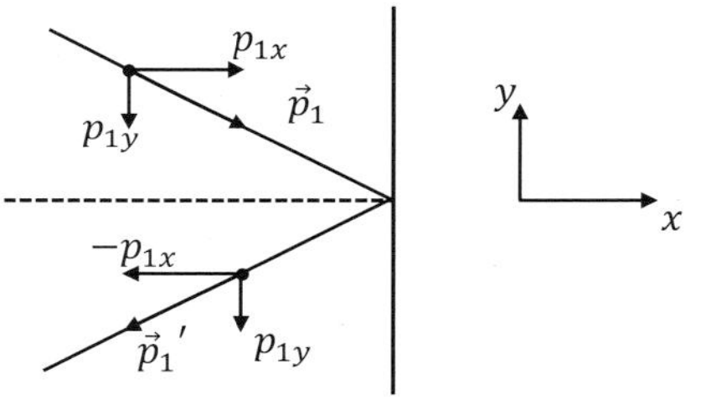

Abbildung 8.5 x-y-Projektion eines elastischen Stoßes an einer festen Wand in der y-z-Ebene

8.3 Inelastische Stöße

Bisher haben wir den genauen Vorgang, der sich bei der Berührung der Kugeln abspielt, unberücksichtigt gelassen und stattdessen über die Erhaltungssätze direkt das Endergebnis berechnet. Elastische Stöße mit mechanischer Energieerhaltung sind dabei ein Idealfall. In der Realität treten bei Stößen immer Verformungskräfte und damit auch mehr oder weniger große Wärmeverluste auf.

Experiment 16: *phyphox* (In)elastischer Stoß

Zur Durchführung dieses Experiments werden das Mikrofon und die Stoppuhr des Smartphones benutzt.

1 Suchen Sie in der *phyphox*-App das Experiment »(In)elastischer Stoß«. Berühren des Drei-Punkte-Symbols rechts oben zeigt ein Menü mit Informationen zum Experiment (»Experiment Info«). Wählen Sie das Menü »Höhen« zur Messung der Aufschlagzeiten.

2 Starten Sie die Aufzeichnung. Lassen Sie einen Ball oder eine Kugel aus verschiedenen z. B. mit einem Lineal oder Zollstock bestimmten Höhen auf eine feste Unterlage fallen. Das Smartphone registriert die Aufprallzeiten des Balls. Dabei sollte sich das Smartphone möglichst nahe der Aufprallstelle befinden. Je nach Elastizität des Materials springt der Ball mehrmals zurück (»Auftitschen«). Stoppen Sie die Aufzeichnung.

3 Das Smartphone registriert die Aufprallzeiten und berechnet daraus nach (1.33) die Höhen $h_i = (g/2) \cdot (t_i/2)^2, i = 1, 2...$ des jeweiligen Rückpralls und extrapoliert damit auf die Anfangshöhe (► Abbildung 8.6 links). Vergleichen Sie mit der anfangs gemessenen Höhe.

4 Je nach Beschaffenheit der Kugel und der Unterlage wird bei jedem Stoß ein Anteil der kinetischen Energie verloren gehen. Die prozentuale Restenergie lässt sich wegen $E_{\text{pot}} = mgh$ aus der verbleibenden Höhe berechnen (► Abbildung 8.6 rechts).

5 Die Anfangsenergie wird auf 100% gesetzt. Man erkennt, dass bei jedem Aufprall ein etwa gleicher Anteil der Energie verloren geht. Der mittlere Anteil der erhaltenen Energie wird in ► Abbildung 8.6 rechts unten berechnet.

6 Mit verschiedenen Materialien zeigt sich, dass der Versuch besser gelingt, wenn die Stöße nicht zu inelastisch sind.

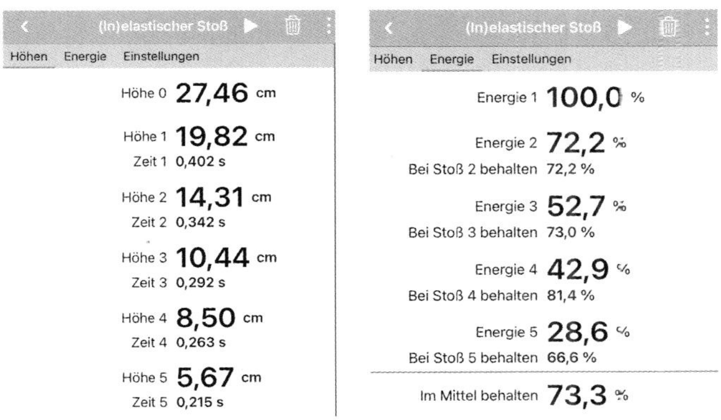

Abbildung 8.6 Aufprallzeiten und -höhen (links) und restliche Energie(anteile) nach dem Aufprall (rechts)

Im Extremfall kommt es zu vollkommen inelastischen Stößen, wie beispielsweise bei der Kollision von Knetgummikugeln. Wir betrachten dazu wieder die Situation der ► Abbildung 8.1, wo eine Masse m_1 auf eine ruhende Masse m_2 trifft. Nach dem Stoß sollen die Massen aneinander haften. Der Impulssatz liefert dann

$$m_1 v_1 = (m_1 + m_2) v_1' \tag{8.35}$$

$$v_1' = \frac{m_1}{m_1 + m_2} v_1 \,. \tag{8.36}$$

Mit Gleichung (8.36) kann beispielsweise in der Ballistik durch Absorption einer Gewehrkugel in einem dichten Material die Geschossgeschwindigkeit bestimmt werden.

Das Wichtigste zum Mitnehmen

Zum Mitnehmen: Stoßprozesse 1

- Bei einem **elastischen Stoß** einer Kugel mit dem Impuls $\vec{p_1}$ auf eine ruhende Kugel gelten **Impuls- und Energieerhaltung**:

$$\vec{p}_1 + \underbrace{\vec{p}_2}_{=0} = \vec{p}_1' + \vec{p}_2' \tag{8.4}$$

$$E_1 = E_1' + E_2' \tag{8.8}$$

Dabei sind die gestrichenen Größen die Energien und Impulse nach dem Stoß und $E_i^{(\prime)} = E_{\text{kin},i}^{(\prime)}$ die kinetischen Energien.

- Bei **gleicher Masse** der beiden Kugeln ($m_1 = m_2 = m$) stehen die Impulse $\vec{p}_1'$ und $\vec{p}_2'$ senkrecht aufeinander, außer beim zentralen Stoß (siehe (8.31)):

$$\vec{p}_1' \cdot \vec{p}_2' = 0 \tag{8.13}$$

Die Endpunkte von $\vec{p}_1'$ und $\vec{p}_2'$ liegen auf einem Thaleskreis über p_1.

- Beim Stoß zweier Kugeln **ungleicher Massen** zeigt der Impuls der zweiten Kugel auf einen Kreis mit dem Radius $R = \mu v_1$, dessen Mittelpunkt um $\mu v_1 = R$ verschoben ist (▶ Abbildung 8.4):

$$\left(p_{2x}' - \mu v_1\right)^2 + p_{2y}'^2 = (\mu v_1)^2 \tag{8.26}$$

Dabei ist μ die **reduzierte Masse**

$$\mu \equiv \frac{m_1 m_2}{m_1 + m_2}\,. \tag{8.23}$$

- Für den **zentralen** Stoß mit Massen m_1 und m_2 gilt

$$p_1' = p_1 - p_2' = m_1 v_1 - 2\mu v_1 = m_1 \left(1 - \frac{2m_2}{m_1 + m_2}\right) v_1\,. \tag{8.31}$$

Die Kugel 1 bewegt sich für $m_2 < m_1$ verlangsamt vorwärts, für $m_2 = m_1$ bleibt sie liegen und für $m_2 > m_1$ wird sie rückwärts gestreut.

Zum Mitnehmen: Stoßprozesse 2

- Für einen elastischen Stoß an einer ebenen, glatten **festen Wand** in der y-z-Ebene gilt

$$p'_{1x} = -p_{1x} \quad p'_{1y} = p_{1y} \quad p'_{1z} = p_{1z}\,. \tag{8.33}$$

- Beim **inelastischen** Stoß einer Masse m_1 auf eine ruhende Masse m_2 liefert der Impulssatz

$$v'_1 = \frac{m_1}{m_1 + m_2} v_1\,. \tag{8.36}$$

Koordinatensysteme

ÜBERBLICK

9

9.1 Kartesische Koordinaten 161
9.2 Zylinderkoordinaten 163
9.3 Kugelkoordinaten 167

» Wenn wir allgemeine Drehbewegungen, Planetenbahnen, Sternpositionen etc. beschreiben wollen, erweisen sich die gewohnten kartesischen Koordinaten oft als wenig vorteilhaft. Wir verwenden flexibel die Koordinatensysteme, die für die jeweilige Fragestellung am besten geeignet sind.

In ► Abbildung 9.1 sind drei Koordinatensysteme gezeigt, die besonders häufig verwendet werden: Kartesische Koordinaten x, y, z, Zylinderkoordinaten r, φ, z und Kugelkoordinaten r, φ, ϑ. Ein Spezialfall der Zylinderkoordinaten für $z = 0$ sind die Polarkoordinaten, die wir schon in Abschnitt 4.1 vorgestellt hatten. Kugelkoordinaten werden auch als sphärische Koordinaten bezeichnet.

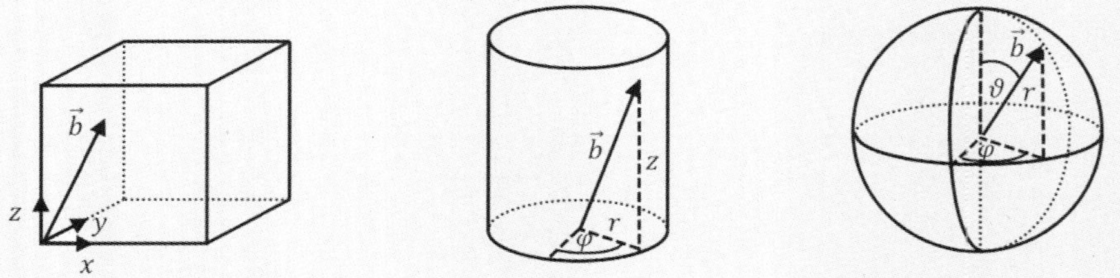

Abbildung 9.1 Die drei gebräuchlichsten Koordinatensysteme: Kartesische Koordinaten (x, y, z), Zylinderkoordinaten (r, φ, z) und Kugelkoordinaten (r, φ, ϑ)

Wir stellen im Folgenden ihre wichtigsten Eigenschaften und Relationen zusammen und beginnen mit oft verwendeten Aspekten der kartesischen Koordinaten. «

9.1 Kartesische Koordinaten

Kartesische Koordinaten sind aus der Schule bekannt und wurden in den vorigen Kapiteln schon viel verwendet. Wir stellen hier noch einmal kurz die wichtigsten Merkmale und Eigenschaften zusammen.

Ein Ortsvektor $\vec{b}$ (linke ► Abbildung 9.2) wird in kartesischen Koordinaten beschrieben durch

$$\vec{b} = \begin{pmatrix} b_x \\ b_y \\ b_z \end{pmatrix} . \tag{9.1}$$

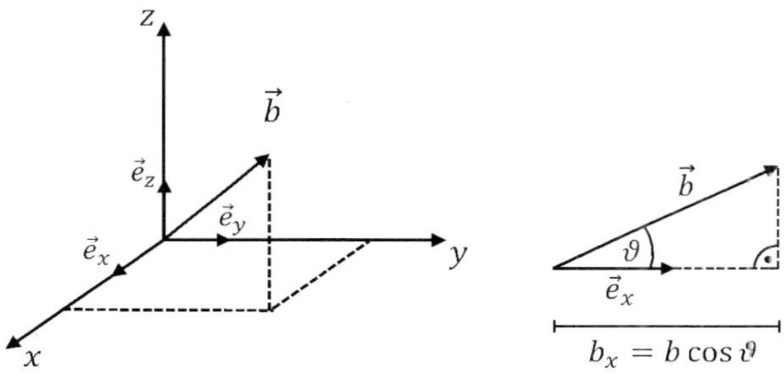

Abbildung 9.2 Links: Einheitsvektoren $\vec{e_i}$ und Komponenten b_i eines Vektors $\vec{b}$ in kartesischen Koordinaten $i = x, y, z$, rechts: Komponente b_x als Skalarprodukt $\vec{b} \cdot \vec{e_x}$

Der Betrag des Vektors ist

$$b = |\vec{b}| = \sqrt{b_x^2 + b_y^2 + b_z^2} . \tag{9.2}$$

Die **Basis-** oder **Einheitsvektoren** im kartesischen Koordinatensystem sind

$$\vec{e}_x = \begin{pmatrix} 1 \\ 0 \\ 0 \end{pmatrix}, \quad \vec{e}_y = \begin{pmatrix} 0 \\ 1 \\ 0 \end{pmatrix}, \quad \vec{e}_z = \begin{pmatrix} 0 \\ 0 \\ 1 \end{pmatrix} . \tag{9.3}$$

Sie bilden ein **rechtshändiges orthogonales** (rechtwinkliges) **Dreibein**.

Das Skalarprodukt eines Vektors $\vec{b}$ mit einem der Einheitsvektoren ergibt die Projektion von $\vec{b}$ auf die jeweilige Achse (rechte ► Abbildung 9.2). Damit erhält man die Komponente von $\vec{b}$ entlang dieser Richtung:

$$b_x = \vec{b} \cdot \vec{e}_x = |\vec{b}| \cdot \underbrace{|\vec{e}_x|}_{=1} \cdot \cos\vartheta_{\vec{b},\vec{e}_x} = b \cdot \cos\vartheta_{\vec{b},\vec{e}_x} \tag{9.4}$$

Analog bekommt man die beiden anderen Komponenten, sodass sich der Vektor durch die Einheitsvektoren folgendermaßen ausdrücken lässt:

$$\vec{b} = b_x\vec{e}_x + b_y\vec{e}_y + b_z\vec{e}_z \tag{9.5}$$

Im kartesischen Koordinatensystem ist ein räumliches Linienelement (▶ Abbildung 9.3) definiert durch

$$d\vec{s} = \begin{pmatrix} dx \\ dy \\ dz \end{pmatrix} = \vec{e}_x dx + \vec{e}_y dy + \vec{e}_z dz\,. \tag{9.6}$$

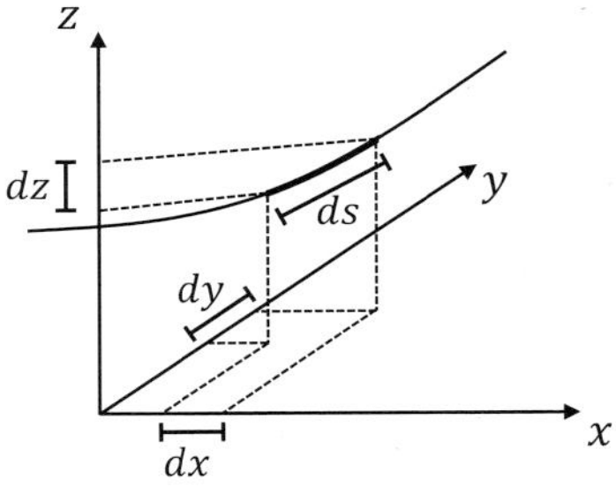

Abbildung 9.3 Komponenten dx, dy, dz eines Linienelements $d\vec{s}$

Die Geschwindigkeit beträgt

$$\vec{v} = \frac{d\vec{s}}{dt} = \begin{pmatrix} \frac{dx}{dt} \\ \frac{dy}{dt} \\ \frac{dz}{dt} \end{pmatrix} = \begin{pmatrix} \dot{x} \\ \dot{y} \\ \dot{z} \end{pmatrix}. \tag{9.7}$$

Die Beschleunigung lautet

$$\vec{a} = \frac{d\vec{v}}{dt} = \begin{pmatrix} \frac{d^2x}{dt^2} \\ \frac{d^2y}{dt^2} \\ \frac{d^2z}{dt^2} \end{pmatrix} = \begin{pmatrix} \ddot{x} \\ \ddot{y} \\ \ddot{z} \end{pmatrix}. \tag{9.8}$$

Ein Flächenelement ist in der linken ▶ Abbildung 9.4 gezeigt:

$$dA = dx \cdot dy \tag{9.9}$$

$$A = \int dA = \iint dx\, dy\,. \tag{9.10}$$

Das Volumenelement ist (rechte ▶ Abbildung 9.4)

$$dV = dx \cdot dy \cdot dz \tag{9.11}$$

$$V = \int dV = \iiint dx\, dy\, dz\,. \tag{9.12}$$

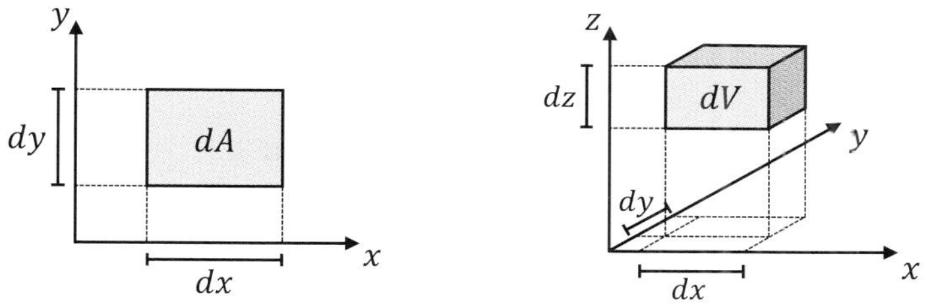

Abbildung 9.4 Links: Flächenelement $dA = dx \cdot dy$, rechts: Volumenelement $dV = dx \cdot dy \cdot dz$

Der sogenannte Gradient bildet einen Vektor aus den partiellen Ableitungen in die drei Koordinatenrichtungen. Wir werden ihn im Mathematischen Einschub: Gradient in Abschnitt 10.2 genauer einführen:

$$\text{grad} = \vec{e}_x \frac{\partial}{\partial x} + \vec{e}_y \frac{\partial}{\partial y} + \vec{e}_z \frac{\partial}{\partial z} \tag{9.13}$$

9.2 Zylinderkoordinaten

Hat man ein zu einer Achse rotationssymmetrisches System, so eignen sich die Zylinderkoordinaten. Betrachtet man nur die x-y-Ebene ($z = 0$), entsprechen die Zylinderkoordinaten den Polarkoordinaten.

In dem zylindrischen System wird ein Ortsvektor $\vec{b}$ durch den Radius r, den Azimutwinkel φ und die kartesische Koordinate z beschrieben (linke ► Abbildung 9.5).

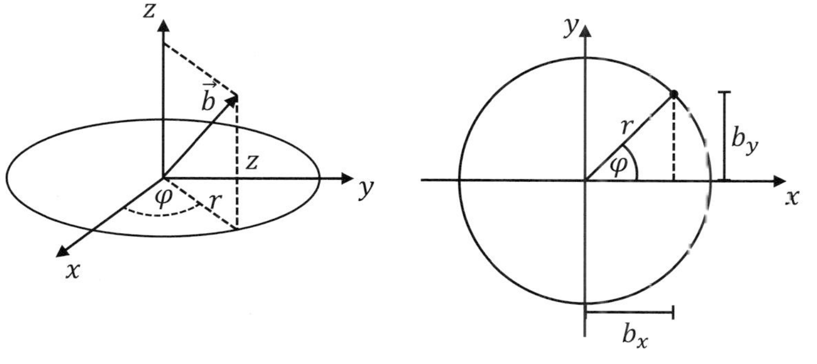

Abbildung 9.5 Links: Zylinderkoordinaten r, φ, z im kartesischen System, rechts: Umrechnung zwischen kartesischen und Zylinderkoordinaten

Um vom zylindrischen System in das kartesische System zu gelangen, kann man die folgenden Umrechnungen verwenden (► Abbildung 9.5):

$$\begin{aligned} b_x &= r\cos(\varphi) \\ b_y &= r\sin(\varphi) \\ b_z &= z \end{aligned} \tag{9.14}$$

Mit

$$b_x^2 + b_y^2 = r^2 \left(\cos^2(\varphi) + \sin^2(\varphi)\right) = r^2 \tag{9.15}$$

$$\frac{b_y}{b_x} = \frac{\sin(\varphi)}{\cos(\varphi)} = \tan(\varphi) \tag{9.16}$$

kann man für die Umkehrung schreiben

$$r = \sqrt{b_x^2 + b_y^2} \tag{9.17}$$

$$\varphi = \arctan\left(\frac{b_y}{b_x}\right) \tag{9.18}$$

$$z = b_z\,. \tag{9.19}$$

Die Einheitsvektoren der Zylinderkoordinaten sind

$$\vec{e}_r = \begin{pmatrix}\cos(\varphi)\\ \sin(\varphi)\\ 0\end{pmatrix}, \qquad \vec{e}_\varphi = \begin{pmatrix}-\sin(\varphi)\\ \cos(\varphi)\\ 0\end{pmatrix}, \qquad \vec{e}_z = \begin{pmatrix}0\\ 0\\ 1\end{pmatrix}. \tag{9.20}$$

Sie bilden in dieser Reihenfolge ein rechtshändiges orthogonales Dreibein (linke ► Abbildung 9.6).

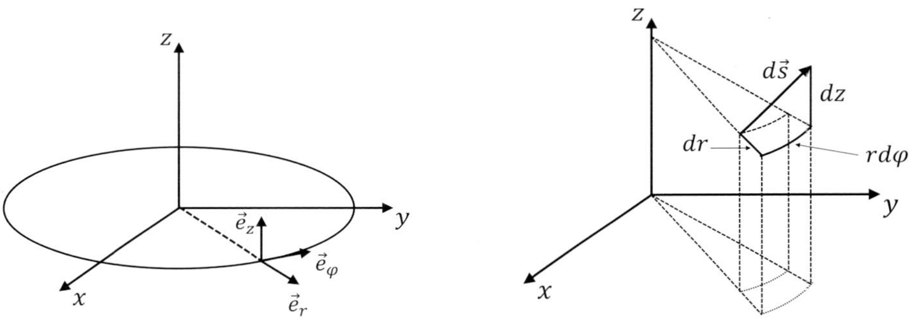

Abbildung 9.6 Links: Die Einheitsvektoren der Zylinderkoordinaten $\vec{e}_r, \vec{e}_\varphi, \vec{e}_z$ bilden im kartesischen System ein rechtshändiges orthogonales Dreibein. Rechts: Linienelement $d\vec{s}$ in Zylinderkoordinaten r, φ, z

Während die Einheitsvektoren im kartesischen System konstant sind, ändern sich die Einheitsvektoren $\vec{e}_r$ und $\vec{e}_\varphi$ mit dem Winkel φ. Durch Differentiation von (9.20) nach φ erhalten wir

$$d\vec{e}_r = d\varphi\,\vec{e}_\varphi\,, \quad d\vec{e}_\varphi = -d\varphi\,\vec{e}_r\,. \tag{9.21}$$

Das räumliche Linienelement ist (rechte ► Abbildung 9.6)

$$d\vec{s} = \begin{pmatrix}dr\\ r\,d\varphi\\ dz\end{pmatrix} = dr\,\vec{e}_r + r\,d\varphi\,\vec{e}_\varphi + dz\,\vec{e}_z\,. \tag{9.22}$$

Die Geschwindigkeit lautet damit

$$\vec{v} = \frac{d\vec{s}}{dt} = \dot{r}\,\vec{e}_r + r\,\dot{\varphi}\,\vec{e}_\varphi + \dot{z}\,\vec{e}_z\,. \tag{9.23}$$

Die Beschleunigung ist unter Berücksichtigung von (9.21)

$$\vec{a} = \frac{d\vec{v}}{dt} = \frac{d}{dt}\left(\dot{r}\,\vec{e}_r + r\,\dot{\varphi}\,\vec{e}_\varphi + \dot{z}\,\vec{e}_z\right) \tag{9.24}$$

$$\ldots = \left(\ddot{r} - r\dot{\varphi}^2\right)\vec{e}_r + \left(2\,\dot{r}\dot{\varphi} + r\ddot{\varphi}\right)\vec{e_\varphi} + \ddot{z}\vec{e_z}\,. \tag{9.25}$$

Das Flächenelement auf den Zylinderdeckeln (linke ► Abbildung 9.7, $z = const.$) ist

$$dA = dr \cdot r\,d\varphi = r\,dr\,d\varphi \tag{9.26}$$

$$A = \iint r\,dr\,d\varphi\,. \tag{9.27}$$

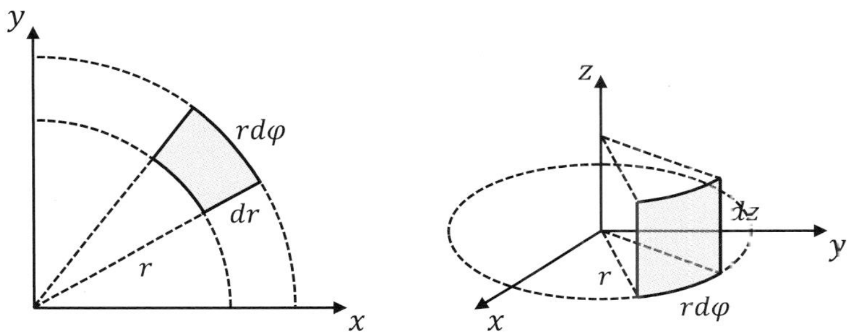

Abbildung 9.7 Links: Flächenelement dA auf den Zylinderdeckeln, rechts: Flächenelement dA auf dem Zylindermantel

Das Flächenelement auf dem Zylindermantel ist (rechte ► Abbildung 9.7)

$$dA = r \cdot d\varphi\,dz \tag{9.28}$$

$$A = r\iint d\varphi\,dz\,. \tag{9.29}$$

Das Volumenelement beträgt (► Abbildung 9.8)

$$dV = dr\,rd\varphi\,dz \tag{9.30}$$

$$dV = rdr\,d\varphi\,dz \tag{9.31}$$

$$V = \iiint r\,dr\,d\varphi\,dz\,. \tag{9.32}$$

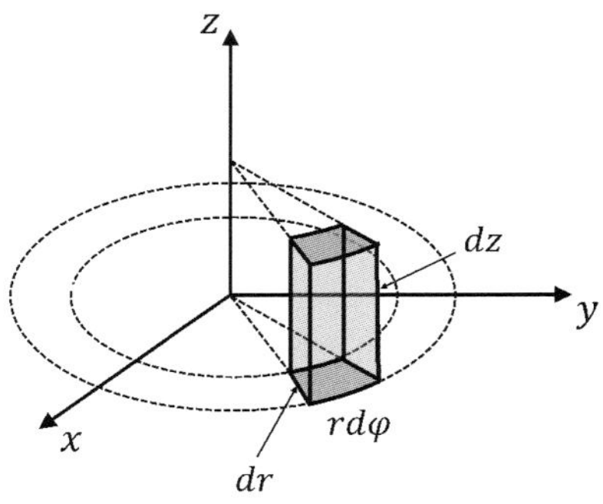

Abbildung 9.8 Volumenelement dV in Zylinderkoordinaten r, φ, z

Der Gradient bildet einen Vektor aus den partiellen Ableitungen in die drei Koordinatenrichtungen (Komponenten des Linienelements $d\vec{s}$ (9.22)):

$$\text{grad} = \begin{pmatrix} \frac{\partial}{\partial r} \\ \frac{1}{r}\frac{\partial}{\partial \varphi} \\ \frac{\partial}{\partial z} \end{pmatrix} = \vec{e}_r \frac{\partial}{\partial r} + \vec{e}_\varphi \frac{1}{r}\frac{\partial}{\partial \varphi} + \vec{e}_z \frac{\partial}{\partial z} \tag{9.33}$$

Beispiel 26: Gleichförmige Kreisbewegung

Für eine gleichförmige Kreisbewegung ($\varphi = \dot{\varphi} t, \dot{\varphi} = const., \ddot{\varphi} = 0$) in der Ebene $z = 0$ mit $r = const.(\rightarrow \dot{r} = 0)$ ist die Geschwindigkeit

$$\vec{v} = r\,\frac{d\varphi}{dt}\,\vec{e}_\varphi = r\dot{\varphi} \begin{pmatrix} -\sin(\dot{\varphi}t) \\ \cos(\dot{\varphi}t) \\ 0 \end{pmatrix}. \tag{9.34}$$

Die Beschleunigung ist hier die Zentripetalbeschleunigung

$$\vec{a} = \vec{a}_{ZP} = -r\,\dot{\varphi}^2\,\vec{e}_r = -r\,\dot{\varphi}^2 \begin{pmatrix} \cos(\dot{\varphi}t) \\ \sin(\dot{\varphi}t) \\ 0 \end{pmatrix}. \tag{9.35}$$

Aufgabe 24: Strohhalm

Ein Strohhalm habe die Maße (► Abbildung 9.9)

$$r_1 = 2\,\text{mm}\,, \qquad r_2 = 2{,}5\,\text{mm}\,, \quad h = 200\,\text{mm}\,.$$

Wie viel Material wird zur Herstellung benötigt?

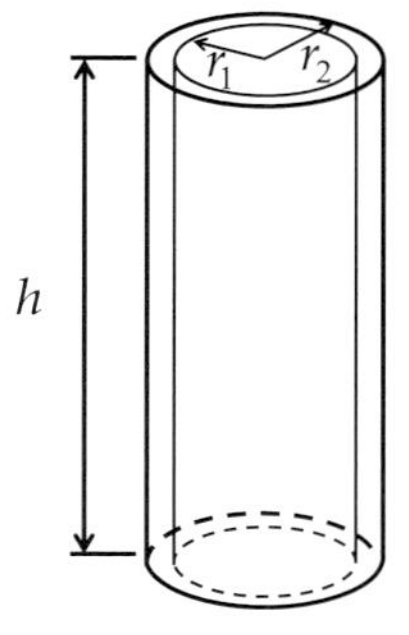

Abbildung 9.9 Strohhalm aus Aufgabe 24

9.3 Kugelkoordinaten

Im Koordinatensystem der Kugelkoordinaten (auch sphärische Koordinaten genannt, linke ► Abbildung 9.10) wird ein Ortsvektor $\vec{b}$ durch den Radius r, den Polarwinkel ϑ, und den Azimutwinkel φ beschrieben. Der Polarwinkel ist $\vartheta = 0$, wenn der Vektor $\vec{b}$ parallel zum Einheitsvektor in z-Richtung $\vec{e}_z$ steht.

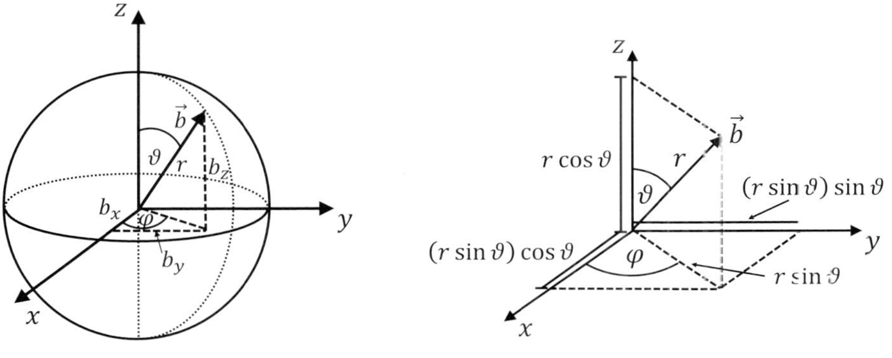

Abbildung 9.10 Links: Kugelkoordinaten r, ϑ, φ im kartesischen System, rechts: Umrechnung zwischen kartesischen und Kugelkoordinaten

In der x, y-Ebene entsprechen die Kugelkoordinaten den Polarkoordinaten. Die Umrechnung von Kugelkoordinaten in kartesische Koordinaten lautet (rechte ► Abbildung 9.10)

$$\begin{aligned} b_x &= r\cos(\varphi)\ \sin(\vartheta) \\ b_y &= r\sin(\varphi)\ \sin(\vartheta) \\ b_z &= r\cos(\vartheta)\ . \end{aligned} \tag{9.36}$$

Die Rücktransformation ist

$$r = \sqrt{b_x^2 + b_y^2 + b_z^2} \tag{9.37}$$

$$\vartheta = \arccos\left(\frac{b_z}{r}\right) = \arccos\left(\frac{b_z}{\sqrt{b_x^2 + b_y^2 + b_z^2}}\right) \tag{9.38}$$

$$\varphi = \arctan\left(\frac{b_y}{b_x}\right)\ . \tag{9.39}$$

Die Einheitsvektoren der Kugelkoordinaten

$$\vec{e}_r = \begin{pmatrix} \cos(\varphi)\sin(\vartheta) \\ \sin(\varphi)\sin(\vartheta) \\ \cos(\vartheta) \end{pmatrix}, \quad \vec{e}_\vartheta = \begin{pmatrix} \cos(\varphi)\cos(\vartheta) \\ \sin(\varphi)\cos(\vartheta) \\ -\sin(\vartheta) \end{pmatrix}, \quad \vec{e}_\varphi = \begin{pmatrix} -\sin(\varphi) \\ \cos(\varphi) \\ 0 \end{pmatrix} \tag{9.40}$$

bilden in dieser Reihenfolge ein rechtshändiges orthogonales Dreibein (linke ► Abbildung 9.11). Dabei ist $\vec{e}_r$ der radiale Vektor, $\vec{e}_\vartheta$ die Tangente an den Längenkreis und $\vec{e}_\varphi$ die Tangente an den Breitenkreis.

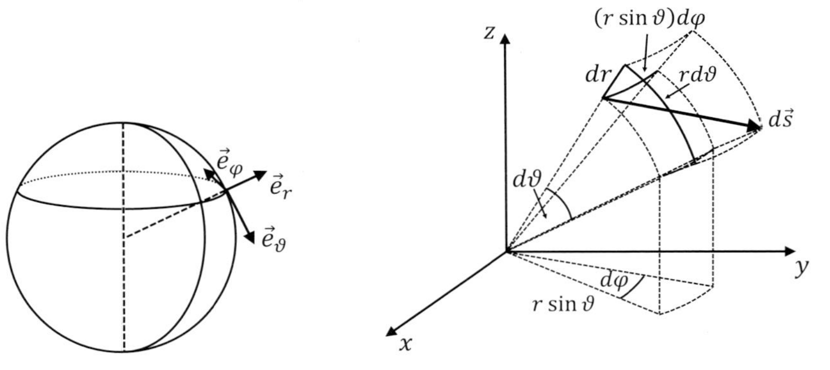

Abbildung 9.11 Links: Die Einheitsvektoren der Kugelkoordinaten $\vec{e}_r, \vec{e}_\vartheta, \vec{e}_\varphi$ bilden im kartesischen System ein rechtshändiges orthogonales Dreibein. Rechts: Linienelement in Kugelkoordinaten $\vec{s}$

Das räumliche Linienelement ist (rechte ► Abbildung 9.11)

$$d\vec{s} = \begin{pmatrix} dr \\ r\,d\vartheta \\ r\ \sin(\vartheta)\ d\varphi \end{pmatrix} = dr\,\vec{e}_r + r\,d\vartheta\,\vec{e}_\vartheta + r\ \sin(\vartheta)\ d\varphi\,\vec{e}_\varphi\,. \tag{9.41}$$

Die Geschwindigkeit lautet

$$\vec{v} = \frac{d\vec{s}}{dt} = \dot{r}\,\vec{e}_r + r\,\dot{\vartheta}\,\vec{e}_\vartheta + r\,\sin(\vartheta)\,\dot{\varphi}\,\vec{e}_\varphi\,. \tag{9.42}$$

Die Beschleunigung wird selten benötigt. Nach einer längeren Rechnung, die hier nicht vorgeführt wird, erhält man

$$\vec{a} = \left[\ddot{r} - r\dot{\vartheta}^2 - r\sin^2(\vartheta)\,\dot{\varphi}^2\right]\cdot\vec{e}_r \tag{9.43}$$

$$+\left[2\dot{r}\dot{\vartheta} + r\ddot{\vartheta} - r\sin(\vartheta)\cos(\vartheta)\,\dot{\varphi}^2\right]\cdot\vec{e}_\vartheta \tag{9.44}$$

$$+\left[(r\ddot{\varphi} + 2\dot{r}\dot{\varphi})\sin(\vartheta) + 2r\dot{\varphi}\dot{\vartheta}\cos(\vartheta)\right]\cdot\vec{e}_\varphi\,. \tag{9.45}$$

Das Flächenelement der Kugeloberfläche ist (linke ► Abbildung 9.12)

$$dA = r d\vartheta\,(r\sin(\vartheta))\;d\varphi \tag{9.46}$$

$$dA = r^2\sin(\vartheta)\;d\vartheta\,d\varphi \tag{9.47}$$

$$A = r^2\iint\sin(\vartheta)\;d\vartheta\,d\varphi\,. \tag{9.48}$$

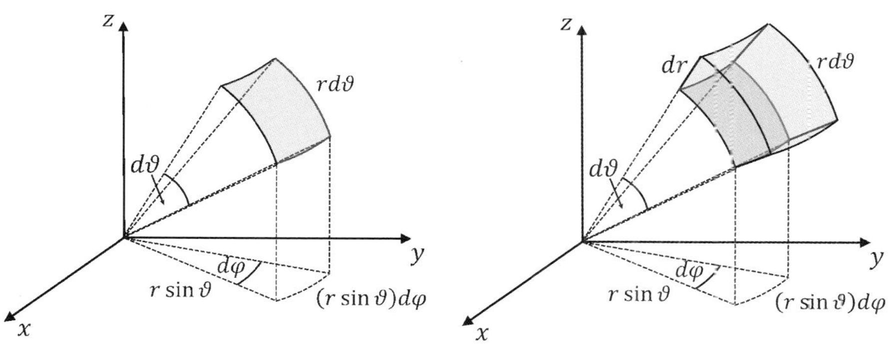

Abbildung 9.12 Links: Flächenelement dA der Kugeloberfläche in Kugelkoordinaten r, ϑ, φ, rechts: Volumenelement dV in Kugelkoordinaten r, ϑ, φ

Das Volumenelement ist (rechte ► Abbildung 9.12)

$$dV = dr\,r\,d\vartheta\,r\sin(\vartheta)\;d\varphi \tag{9.49}$$

$$dV = r^2 dr\sin(\vartheta)\;d\vartheta\,d\varphi \tag{9.50}$$

$$V = \iiint r^2 dr\sin(\vartheta)\;d\vartheta\,d\varphi\,. \tag{9.51}$$

Der Gradient bildet einen Vektor aus den partiellen Ableitungen in die drei Koordinatenrichtungen (Komponenten des Linienelements $d\vec{s}$ (9.41)):

$$\text{grad} = \begin{pmatrix} \frac{\partial}{\partial r} \\ \frac{1}{r}\frac{\partial}{\partial \vartheta} \\ \frac{1}{r\sin\vartheta}\frac{\partial}{\partial \varphi} \end{pmatrix} = \vec{e}_r \frac{\partial}{\partial r} + \vec{e}_\vartheta \frac{1}{r}\frac{\partial}{\partial \vartheta} + \vec{e}_\varphi \frac{1}{r\sin\vartheta}\frac{\partial}{\partial \varphi} \tag{9.52}$$

Beispiel 27: Gleichförmige Kreisbewegung

Die Geschwindigkeit für eine Kreisbewegung auf einem Breitenkreis ($\dot{r} = 0$ und $\dot{\vartheta} = 0$) ist mit (9.42) und $\varphi = \dot{\varphi}\, t$

$$\vec{v} = r \sin(\vartheta)\, \dot{\varphi} \begin{pmatrix} -\sin(\dot{\varphi}t) \\ \cos(\dot{\varphi}t) \\ 0 \end{pmatrix} . \tag{9.53}$$

Für den Längenkreis mit $\varphi = 0$ gilt mit $\vartheta = \dot{\vartheta}\, t$

$$\vec{v} = r \cdot \dot{\vartheta} \begin{pmatrix} \cos\left(\dot{\vartheta}t\right) \\ 0 \\ -\sin\left(\dot{\vartheta}t\right) \end{pmatrix} . \tag{9.54}$$

Aufgabe 25: Kugeloberfläche

Berechnen Sie die Oberfläche der Kugel in ► Abbildung 9.13.

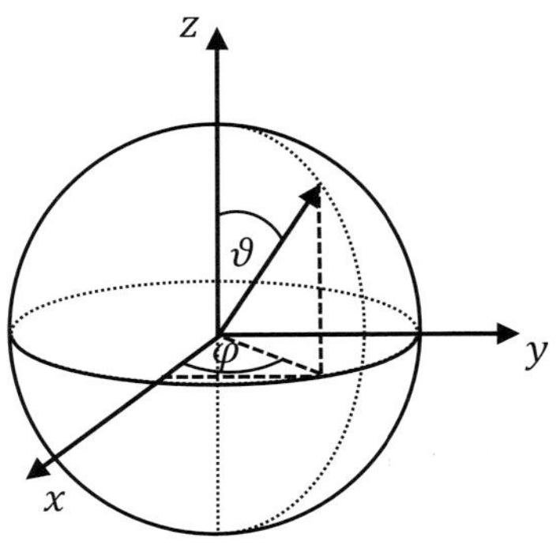

Abbildung 9.13 Berechnung der Kugeloberfläche mit Kugelkoordinaten

Das Wichtigste zum Mitnehmen

Zum Mitnehmen: Koordinatensysteme 1

- Ein Ortsvektor $\vec{b}$ wird in **kartesischen Koordinaten** beschrieben durch

$$\vec{b} = \begin{pmatrix} b_x \\ b_y \\ b_z \end{pmatrix} . \qquad (9.1)$$

Die **Basis-** oder **Einheitsvektoren** entlang der Achsen des kartesischen Koordinatensystems $\vec{e}_x, \vec{e}_y, \vec{e}_z$ bilden ein **rechtshändiges orthogonales Dreibein**.

Mit dem räumlichen Linienelement

$$d\vec{s} = \vec{e}_x dx + \vec{e}_y dy + \vec{e}_z dz \qquad (9.55)$$

lassen sich die Geschwindigkeit $\vec{v} = d\vec{s}/dt$ und die Beschleunigung $\vec{a} = d\vec{v}/dt$ definieren.

- Zwischen den **zylindrischen Koordinaten** r, φ, z (Spezialfall $z = 0$: **Polarkoordinaten**) und den Komponenten eines Vektors $\vec{b}$ im kartesischen System besteht der Zusammenhang

$$\begin{aligned} b_x &= r \cos(\varphi) \\ b_y &= r \sin(\varphi) \\ b_z &= z . \end{aligned} \qquad (9.14)$$

Auch die Einheitsvektoren der Zylinderkoordinaten $\vec{e}_r, \vec{e}_\varphi, \vec{e}_z$ bilden ein rechtshändiges orthogonales Dreibein.

Mit dem räumlichen Linienelement

$$d\vec{s} = dr\, \vec{e}_r + r\, d\varphi\, \vec{e}_\varphi + dz\, \vec{e}_z \qquad (9.22)$$

lassen sich wieder $\vec{v} = d\vec{s}/dt$ und $\vec{a} = d\vec{v}/dt$ beschreiben.

Das Flächenelement auf den Zylinderdeckeln ($z = const.$) ist

$$dA = r\, dr\, d\varphi . \qquad (9.26)$$

Das Flächenelement auf dem Zylindermantel ist

$$dA = r \cdot d\varphi\, dz . \qquad (9.28)$$

Das Volumenelement ist

$$dV = rdr\, d\varphi\, dz . \qquad (9.31)$$

Zum Mitnehmen: Koordinatensysteme 2

- Zwischen den **Kugelkoordinaten** (auch: sphärische Koordinaten) mit dem Radius r, dem Azimutwinkel φ, und dem Polarwinkel ϑ bezüglich $\vec{e}_z$ und den kartesischen Komponenten eines Vektors $\vec{b}$ besteht der Zusammenhang

$$\begin{aligned} b_x &= r\cos(\varphi)\ \sin(\vartheta) \\ b_y &= r\sin(\varphi)\ \sin(\vartheta) \\ b_z &= r\cos(\vartheta)\ . \end{aligned} \tag{9.36}$$

Die Einheitsvektoren $\vec{e}_r, \vec{e}_\vartheta, \vec{e}_\varphi$ bilden ebenfalls ein rechtshändiges orthogonales Dreibein.

Mit dem räumlichen Linienelement

$$d\vec{s} = dr\,\vec{e}_r + r\,d\vartheta\,\vec{e}_\vartheta + r\ \sin(\vartheta)\ d\varphi\,\vec{e}_\varphi \tag{9.41}$$

lassen sich die Geschwindigkeit $\vec{v} = d\vec{s}/dt$ und die (recht komplizierte) Beschleunigung $\vec{a} = d\vec{v}/dt$ beschreiben.

Das Flächenelement ist

$$dA = r^2\sin(\vartheta)\ d\vartheta\,d\varphi\,. \tag{9.47}$$

Das Volumenelement ist

$$dV = r^2 dr\sin(\vartheta)\ d\vartheta\,d\varphi\,. \tag{9.50}$$

- Der **Gradient grad** ist definiert als ein Vektor aus den partiellen Ableitungen in die drei Koordinatenrichtungen (Komponenten der Linienelemente $d\vec{s}$ (9.55) (9.22) (9.41)):

$$\text{grad} = \vec{e}_x\frac{\partial}{\partial x} + \vec{e}_y\frac{\partial}{\partial y} + \vec{e}_z\frac{\partial}{\partial z} \tag{9.13}$$

Der Gradient in Zylinderkoordinaten lautet

$$\text{grad} = \vec{e}_r\frac{\partial}{\partial r} + \vec{e}_\varphi\frac{1}{r}\frac{\partial}{\partial \varphi} + \vec{e}_z\frac{\partial}{\partial z}\,. \tag{9.33}$$

Der Gradient in Kugelkoordinaten lautet

$$\text{grad} = \vec{e}_r\frac{\partial}{\partial r} + \vec{e}_\vartheta\frac{1}{r}\frac{\partial}{\partial \vartheta} + \vec{e}_\varphi\frac{1}{r\sin\vartheta}\frac{\partial}{\partial \varphi}\,. \tag{9.52}$$

Gravitationswechselwirkung nach Newton

10

10.1 Gravitationskraft **175**
10.1.1 Messung der Gravitationskonstanten 178

10.2 Gravitationspotential **185**
10.2.1 Potentielle Energie 185
10.2.2 Potential 187
10.2.3 Zusammenhang zwischen Kraft und Potential 188
10.2.4 Äquipotentialflächen 189

10.3 Energie eines Himmelskörpers **189**
10.3.1 Potentielle Energie 190
10.3.2 Kinetische Energie 190
10.3.3 Gesamtenergie 192

10.4 Bahnkurven von Himmelskörpern **192**
10.4.1 Radiale Bewegung im Zentralpotential 193
10.4.2 Kegelschnitte 194
10.4.3 Berechnung der Bahnkurven 198
10.4.4 Bindungszustände und Streuprozesse 200

10.5 Historischer Bezug **203**
10.5.1 Meilensteine 203
10.5.2 Kepler-Gesetze 203

ÜBERBLICK

»Gravitation bezeichnet die Anziehung zwischen massebehafteten Körpern und gehört zu den fundamentalen Wechselwirkungen der Natur. In Form der Erdanziehung ist sie eine unmittelbare physikalische Lebenserfahrung für uns.

Im größeren Zusammenhang basieren die Planetenbewegungen unseres Sonnensystem auf Gravitationswechselwirkung. Auch die Sterne unserer Milchstraße sind durch Gravitation gebunden. Schließlich ist die Entstehungsgeschichte von Galaxien, Sternen und Planeten unseres Universums in wesentlichen Aspekten eine Folge der Gravitation.

In diesem Kapitel stellen wir Newtons Gravitationsgesetz vor. Wir berechnen einerseits die Bahnkurven der Planeten, die durch Gravitation im Sonnensystem gebunden sind. Andererseits zeigen wir für sich frei bewegende Himmelskörper, dass ihre Flugrichtungen durch Streuprozesse an anderen Himmelskörpern aufgrund der Gravitationswechselwirkung verändert werden.«

10.1 Gravitationskraft

Zwischen zwei massebehafteten Körpern M und m wirkt eine anziehende Kraft, die Gravitationskraft (► Abbildung 10.1).

Abbildung 10.1 Zwei Massen (M) und (m) ziehen sich aufgrund der entgegengesetzt gleichen Gravitationskräfte F_1 und F_2 gegenseitig an.

In folgendem Kraftgesetz von Newton ist die Anziehungskraft der Massen formuliert:

$$\vec{F} = -G\,\frac{m\,M}{r^2}\,\vec{e}_r \tag{10.1}$$

Dabei sind die Kräfte auf die jeweilige Masse entgegengesetzt gerichtet und gleich groß: $\vec{F} = \vec{F}_1 = -\vec{F}_2$. Die verschiedenen Terme werden wir im Folgenden motivieren.

Stellen wir uns eine punktförmige Quelle der Intensität $I_\circ$ vor (z. B. eine Lampe), die gleichmäßig in alle Richtungen strahlt (► Abbildung 10.2). Im Abstand r verteilt sich die Intensität ihres Lichts auf die Kugeloberfläche $4\pi r^2$. Auf einer Einheitsfläche im Abstand r ist die registrierte Intensität der Quelle deswegen um folgenden geometrischen Faktor geringer:

$$I \propto I_\circ \frac{1}{r^2} \tag{10.2}$$

Für Punktquellen entsteht demnach durch Geometrie eine Abnahme der Intensität mit dem Quadrat des Abstands zur Quelle.

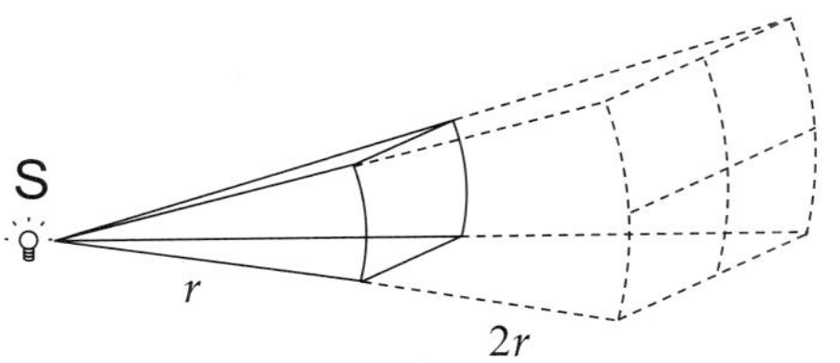

Abbildung 10.2 Verdeutlichung der $1/r^2$-Abhängigkeit der Gravitationskraft

Mit dieser Überlegung haben die verschiedenen Terme der Gravitationskraft (10.1) folgende Bedeutungen:

$F \propto \frac{1}{r^2}$	Die Kraft ist aus geometrischem Grund abnehmend,
$\vec{F} \propto \vec{e}_r$	sie wirkt entlang der Verbindungsachse,
$F \propto mM$	sie steigt mit der Größe der involvierten Massen,
$F < 0$	sie entspricht der Vorzeichenkonvention für anziehende Kräfte.

Der Parameter G ist eine wichtige Größe der Physik und heißt **Gravitationskonstante**. G hat den Wert [8]

$$G = (6{,}67430 \pm 0{,}00015) \cdot 10^{-11} \frac{\mathrm{Nm}^2}{\mathrm{kg}^2} . \tag{10.3}$$

Die Gravitationskraft ist eine **Zentralkraft**, da sie geometrisch nur vom Abstand r der beiden massebehafteten Körper abhängt. Denkt man sich eine Kugelschale um die Masse M und verschiebt die Masse m auf dieser Kugeloberfläche, bleibt die Gravitationskraft zwischen den beiden Massen gleich groß (► Abbildung 10.3).

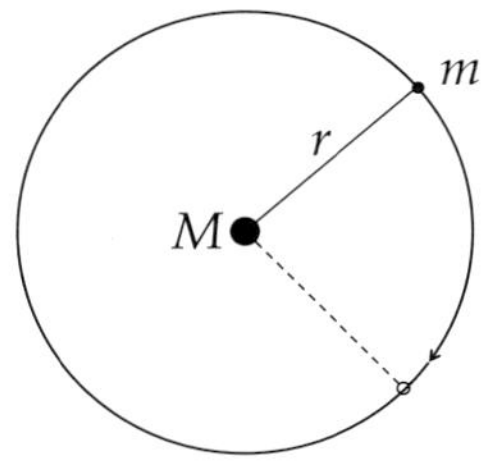

Abbildung 10.3 Die Gravitationskraft zwischen den Massen M und m ist auf einer Kugelschale um M für m konstant.

Wenn man die Masse m von der anderen Masse M entfernt, ist die dabei geleistete Arbeit nur vom Abstand r zwischen den beiden Massen abhängig und unabhängig vom exakten Weg. Die Gravitationskraft ist somit eine **konservative Kraft**.

Befindet man sich in Erdnähe ($h \ll R_{\text{Erde}}$), können wir näherungsweise alle Werte zusammenfassen, die sich nicht (G, M) oder kaum ($r = R_{\text{Erde}} + h \approx R_{\text{Erde}}$) verändern . Aus (10.1) ergibt sich dann

$$g \equiv \frac{G \cdot M_{\text{Erde}}}{R^2_{\text{Erde}}} \tag{10.4}$$

$$= \frac{6{,}67430 \cdot 10^{-11} \cdot 5{,}977 \cdot 10^{24}}{(6371)^2} \cdot \frac{\mathrm{Nm}^2}{\mathrm{kg}^2} \frac{\mathrm{kg}}{\mathrm{km}^2} \tag{10.5}$$

$$\approx 9{,}8 \, \frac{\mathrm{m}}{\mathrm{s}^2} . \tag{10.6}$$

Auf der Erde kann man also in guter Approximation mit

$$\vec{F} = -\frac{G \, M_{\text{Erde}} \, m}{R^2_{\text{Erde}}} \, \vec{e}_r = -m \, \vec{g} \tag{10.7}$$

rechnen. Aufgrund der Erdabplattung und unterschiedlicher Masseverteilungen, Höhen und Zentrifugalkräfte variiert der Wert der Erdbeschleunigung regional zwischen $\approx 9{,}83\,\mathrm{m/s^2}$ an den Polen und $\approx 9{,}78\,\mathrm{m/s^2}$ am Äquator. Der mittlere Wert in Deutschland beträgt $\approx 9{,}81\,\mathrm{m/s^2}$.

Aufgabe 26: Astronaut auf dem Mond

Wie viel stärker ist die Erdbeschleunigung g als die »Mondbeschleunigung« g'? Verwenden Sie die folgenden Größen:

$$M_{\text{Mond}} = 7 \cdot 10^{22}\,\text{kg}\,, \qquad R_{\text{Mond}} = 1\,800\,\text{km}\,, \qquad G = 7 \cdot 10^{-11}\,\frac{\text{Nm}^2}{\text{kg}^2}$$

Aufgabe 27: Schaukel auf der Sonne

Auf einer (abgekühlten) Sonne soll eine Schaukel mit der Pendellänge L installiert werden. Die Schwingungsperiode beträgt somit $T = 2\pi\sqrt{L/g}$ (4.87). Verwenden Sie $L = 1\,\text{m}$, $M_{\text{Sonne}} = 2 \cdot 10^{30}\,\text{kg}$, $R_{\text{Sonne}} = 700\,000\,\text{km}$, $g = GM/R^2$ und $G = 7 \cdot 10^{-11}\text{Nm}^2/\text{kg}^2$, um die Schwingungsperiode auf der Sonne zu berechnen. Vergleichen Sie die Schwingungsperiode auf der Sonne mit der Schwingungsperiode auf der Erde.

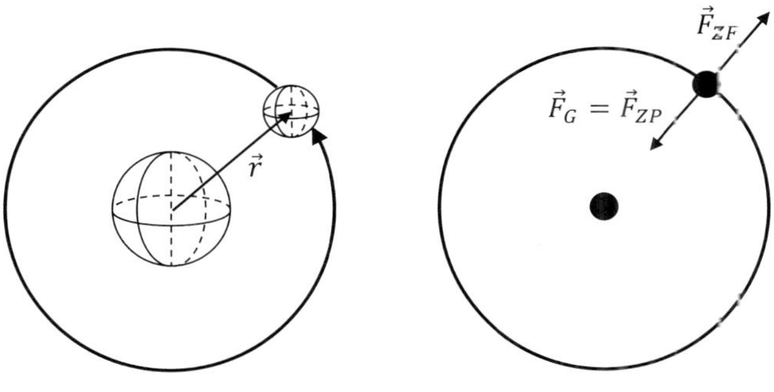

Abbildung 10.4 Der Mond umkreist die Erde im Abstand r. Die Gravitationskraft $\vec{F}_G$ entspricht dabei der zur Kreisbewegung gehörenden Zentripetalkraft $\vec{F}_{ZP}$ (und steht der Zentrifugalkraft $\vec{F}_{ZF}$ entgegen).

Beispiel 28: Mondbewegung, Erdmasse

Die Gravitationskraft $\vec{F}_G$ (10.1) hält das System aus Erde und Mond zusammen (► Abbildung 10.4). Eine Kollision der beiden Himmelskörper wird nur deswegen verhindert, weil der Mond um die Erde kreist. Die Kreisbewegung impliziert die Existenz einer Zentripetalkraft $\vec{F}_{\text{ZP}} = -m_M\omega^2\vec{r}$ (4.21), die radial und entgegengesetzt zur Zentrifugalkraft $\vec{F}_{ZP}$ steht.

Gravitationskraft $\vec{F}_G$ und Zentripetalkraft $\vec{F}_{ZP}$ entsprechen dabei einander:

$$\vec{F}_{ZP} = \vec{F}_G \tag{10.8}$$

$$-m_M\, r\, \omega^2\, \vec{e}_r = -\frac{G\, M_E\, m_M}{r^2}\, \vec{e}_r \tag{10.9}$$

Die Kreisfrequenz ω des Mondes können wir in die Umlaufzeit T des Mondes umschreiben $T = 2\pi/\omega$ (4.11) und erhalten

$$\left(\frac{2\pi}{T}\right)^2 = \frac{G\, M_E}{r^3} \tag{10.10}$$

$$T = 2\pi\sqrt{\frac{r^3}{G\, M_E}}\,. \tag{10.11}$$

Die Umlaufzeit steigt also etwas stärker als linear mit dem Abstand ($r^{3/2}$).

Mit Kenntnis der Mondumlaufzeit $T = 27$ Tage können wir mit (10.11) eine Bestimmung der Erdmasse M_E durchführen. Mit dem Abstand zwischen Mond und Erde von $|\vec{r}| = 400\,000$ km und der Gravitationskonstanten $G \approx 7 \cdot 10^{-11}\,\mathrm{Nm^2/kg^2}$ beträgt die Erdmasse

$$M_E = \frac{4\pi^2 r^3}{GT^2} \approx \frac{4\pi^2\,(4\,10^8)^3}{7\cdot 10^{-11}\,(27\cdot 9\,10^4)^2} \cdot \frac{\mathrm{m}^3}{\frac{\mathrm{Nm}^2}{\mathrm{kg}^2}\,\mathrm{s}^2} \approx 6\cdot 10^{24}\,\mathrm{kg}\,. \tag{10.12}$$

Der korrekte Wert ist $M = 5{,}97219 \cdot 10^{24}$ kg [9].

10.1.1 Messung der Gravitationskonstanten

Mit der sogenannten Gravitationswaage führen wir einen Versuch durch, bei dem wir Massen aufgrund der Gravitationskraft beschleunigen. Aus ihrer beschleunigten Bewegung können wir die Gravitationskonstante G (10.3) bestimmen.

Versuchsaufbau

Eine Hantel mit zwei kleinen Massen m an ihren Enden ist an einem Faden aufgehängt (► Abbildung 10.5). Kleine Drehungen der Hantel können über einen Spiegel mithilfe eines Laserstrahls sichtbar gemacht werden.

Auf gleicher Höhe wie die beiden kleinen Massen befinden sich zwei große Massen M. Wegen der $1/r^2$-Abhängigkeit der Gravitationskraft $\vec{F}_G$ (10.1) wirkt $\vec{F}_G$ in guter Näherung nur zwischen den jeweils benachbarten Massen M und m.

Entscheidend für die Versuchsdurchführung ist, dass die beiden großen Massen M entweder von der einen oder von der anderen Seite auf die kleinen Massen m wirken (► Abbildungen 10.6). Durch den Wechsel wird die am Faden hängende Hantel gedreht. Dabei wirken zwei Drehmomente auf die Hantel.

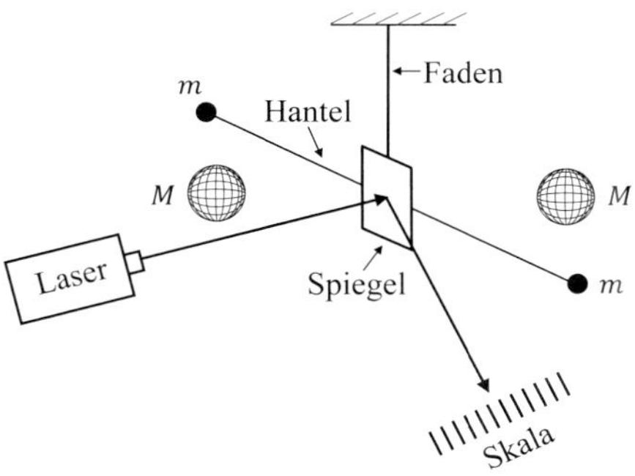

Abbildung 10.5 Gravitationswaage: 2 bewegliche Testmassen m werden von 2 großen festen Massen M angezogen. Die Bewegung wird über einen mitbewegten Spiegel gemessen, der einen Laserstrahl auf eine Skala reflektiert.

Drehmoment durch Gravitation

Die Gravitationskraft wirkt jeweils zwischen den beiden Massen m und M, die sich im Abstand r voneinander befinden und die beide den Abstand $\vec{R}$ von der Drehachse haben. Der Einheitsvektor $\vec{e}_r$ zeigt horizontal vom Zentrum der großen Masse zum Zentrum der kleinen Masse. Das gesamte Drehmoment auf die Hantel der kleinen Massen, das durch die Gravitationskraft zwischen den Kugeln verursacht wird, lautet (linke ► Abbildung 10.6)

$$\vec{D}_G = 2\vec{R} \times \vec{F}_G \tag{10.13}$$

$$= -2\vec{R} \times \frac{G\,m\,M}{r^2}\vec{e}_r\,. \tag{10.14}$$

Der Faktor 2 berücksichtigt die Drehmomente beider Kugelpaare. Der resultierende Drehimpulsvektor $\vec{D}_G$ steht immer parallel oder antiparallel zum Faden.

Drehmoment durch Torsion

Die Torsion (Verdrehung) des Fadens, an dem die Hantel mit den kleinen Massen aufgehängt ist, bewirkt ein rücktreibendes Drehmoment entlang des Fadens. Für kleine Auslenkungen ist es proportional zum Drehwinkel α:

$$\vec{D}_T = -D_F\,\alpha\,\vec{e}_F \tag{10.15}$$

In der Ausgangsstellung des Versuchs (linke ► Abbildung 10.6) zeigt der Einheitsvektor $\vec{e}_F$ entlang des Fadens nach oben. Das sogenannte Richtmoment D_F charakterisiert die Steifigkeit des Fadens und kann für kleine Winkel α als konstant angenommen werden.

Drehimpuls der Hantel

Zu Beginn des Versuchs (linke ► Abbildung 10.6) wirkt die Gravitationskraft so, dass sich die Hantel mit den kleinen Massen m gegen den Uhrzeigersinn den beiden großen Massen M entgegendreht. Da aber die Torsion des Fadens der Drehung entgegenwirkt, bleibt die Hantel irgendwann stehen.

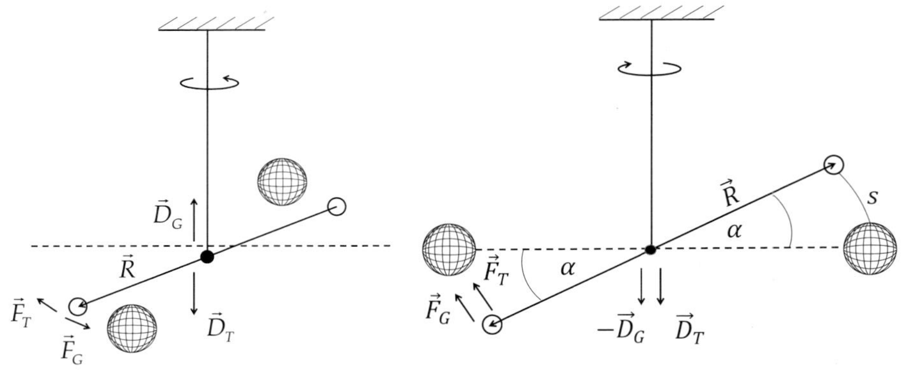

Abbildung 10.6 Links: In der Ausgangsposition heben sich die Kräfte der Gravitationsanziehung und der Torsion der Hantel mit den Testmassen sowie die zugehörigen Drehmomente $\vec{D}_G = -\vec{D}_T$ gerade auf. Rechts: In der Endposition werden die schweren Massen M auf die andere Seite der Testmassen geschwenkt. Die gleich großen Drehmomente $-\vec{D}_G = \vec{D}_T$ wirken jetzt beide im Uhrzeigersinn.

Solche Drehbewegungen der Hantel können wir mit ihrem Drehimpuls $\vec{L}_H$ (4.52) beschreiben. Nach dem Bewegungsgesetz für Drehbewegungen (4.70) verursachen die beiden äußeren Drehmomente $\vec{D}_G$ (10.14) und $\vec{D}_T$ (10.15) die zeitliche Änderung des Drehimpulses $\vec{L}_H$:

$$\frac{d\vec{L}_H}{dt} = \sum_{i=1}^{2} \vec{D}_i = \underbrace{\vec{D}_T}_{\text{Torsion}} + \underbrace{\vec{D}_G}_{\text{Gravitation}} \tag{10.16}$$

Der Drehimpuls der Hantel auf der linken Seite

$$\vec{L}_H = \vec{R} \times \vec{p} \tag{10.17}$$

zeigt ebenfalls je nach Bewegungsrichtung entlang des Fadens oder entgegengesetzt zum Faden. Den Drehimpulsbetrag und seine zeitliche Änderung drücken wir – wie beim Drehmoment der Torsion $\vec{D}_T$ (10.15) – über den Drehwinkel α aus. Dafür verwenden wir den Bogenweg $s = R\,\alpha$ (4.4) und seine zeitliche Ableitung $v = ds/dt = R\,d\alpha/dt$:

$$L_H = 2\,m\,R\,v = 2\,m\,R^2\,\frac{d\alpha}{dt} \tag{10.18}$$

$$\frac{d\vec{L}_H}{dt} = 2\,m\,R^2\,\frac{d^2\alpha}{dt^2} \tag{10.19}$$

Startposition

Für die Ausgangsstellung des Versuchs wird ein Ruhezustand erreicht, wenn sich die Drehmomente aus Gravitation und Torsion ausgleichen. Dann verschwindet der Hanteldrehimpuls $\vec{L}_H = 0$. Die Drehmomente $\vec{D}_G$ und $\vec{D}_T$ stehen dabei antiparallel (linke ► Abbildung 10.6) und haben denselben Betrag:

$$|\vec{D}_G| = |\vec{D}_T| \tag{10.20}$$

Messposition

Bringt man wie in der rechten ► Abbildung 10.6 die großen Massen M auf die andere Seite der kleinen Massen m in denselben Abstand r, entspannt sich der verdrehte Faden: Die Hantel dreht sich im Uhrzeigersinn. Das von der Gravitationskraft verursachte Drehmoment auf die Hantel ist dem Betrag nach gleich groß wie zuvor, wirkt aber nun wie das Torsionsdrehmoment ebenfalls im Uhrzeigersinn.

Im Versuch sind die Bewegungen der Hantel mit den kleinen Massen m so klein, dass wir näherungsweise beide Drehmomente als konstant annehmen. Der Kugelabstand r wird nur sehr gering verändert, d. h. $\vec{F}_G = const.$, und damit ist auch das Gravitationsdrehmoment konstant:

$$\vec{D}_G = const. \tag{10.21}$$

Ebenso ist die Rückdrehung so gering, dass das Drehmoment durch Torsion ungefähr konstant ist:

$$\vec{D}_T = const. \tag{10.22}$$

Den Betrag des Torsionsdrehmoments kennen wir aus dem Anfangszustand $\vec{D}_T = -\vec{D}_G$ (10.20). Unmittelbar nach dem Positionswechsel der großen Massen wirkt das Torsionsdrehmoment gemeinsam mit dem Gravitationsdrehmoment im Uhrzeigersinn. Damit ist das gesamte auf die Hantel wirkende Drehmoment näherungsweise das Zweifache des von der Gravitation verursachten Drehmoments.

> *Die entscheidende experimentelle Maßnahme ist also das Umklappen des Drehmomentvektors der Gravitation bei näherungsweise konstanten Drehmomentbeträgen.*

Im Bewegungsgesetz (10.16) wechselt das Vorzeichen des Gravitationsdrehmoments und wir erhalten

$$\frac{d\vec{L}_H}{dt} = \vec{D}_T - \vec{D}_G = -2\,\vec{D}_G\,. \tag{10.23}$$

Durch Einsetzen der beiden Gleichungen (10.19) und (10.14) in das Bewegungsgesetz (10.23) können wir die Bewegungsgleichung aufstellen:

$$2\,m\,R^2\,\frac{d^2\alpha}{dt^2} = -2\left(-2\,R\,\frac{G\,m\,M}{r^2}\right) \tag{10.24}$$

$$\frac{d^2\alpha}{dt^2} = 2\,\frac{G\,M}{R\,r^2} \tag{10.25}$$

In der Näherung konstanter Beschleunigung $d^2\alpha/dt^2 = const.$ handelt es sich um eine gleichförmig beschleunigte Bewegung, deren Bahnkurve wir analog zu (1.33) durch zweifache Integration lösen.

Der Zeitpunkt, an dem wir die großen Massen auf die andere Seite der kleinen Massen bringen, startet den Messvorgang zur Zeit $t = 0$. Als Anfangsbedingung stellen wir bei der Versuchsdurchführung durch eine entsprechende Eichung $\alpha(t = 0) = 0$ ein. Bei $t = 0$ ruht zudem die Hantel, sodass keine anfängliche Winkelgeschwindigkeit vorliegt

$(d\alpha/dt(t=0)=0)$. Die Bahnkurve der Hantel in Form des Drehwinkels $\alpha(t)$ folgt daher einer quadratischen Zeitabhänigkeit:

$$\alpha(t) = \frac{1}{2}\left(2\,\frac{G\,M}{R\,r^2}\right)t^2 + d\alpha/dt\,(t=0)\;t + \alpha(t=0) = \frac{G\,M}{R\,r^2}\cdot t^2 \qquad (10.26)$$

Durch Messung des Winkels, der Zeit, der Massen und der Abstände können wir damit die Gravitationskonstante G bestimmen.

Messgrößen

Um eine akzeptable Messgenauigkeit zu erreichen, messen wir den Winkel α über die Wegstrecke x, die der vom Spiegel der Hantel reflektierte Laserstrahl auf der Messskala eines weit entfernten Schirms zurücklegt.

Den Zusammenhang zwischen dem Winkel α und der Wegstrecke x können wir mithilfe der linken ► Abbildung 10.7 erarbeiten. Die Richtung des einfallenden Laserstrahls ist durch den von oben kommenden Pfeil zu sehen. Im Anfangszustand der Hantel wird er vom Spiegel unter dem Winkel 2β reflektiert (der Einfallswinkel β zur Spiegelflächen-Normalen entspricht dem Reflexionswinkel β).

Nach der Drehung der Hantel um den Winkel α wird der Laserstrahl in die Richtung des nach rechts zeigenden Pfeils reflektiert. Der Laserstrahl erfährt demnach eine Winkeländerung von 2α.

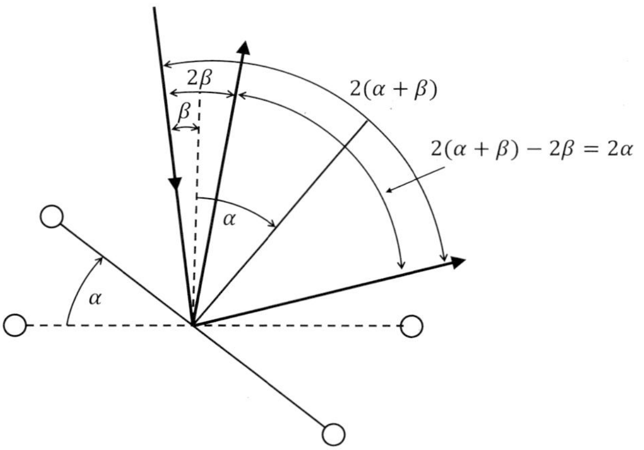

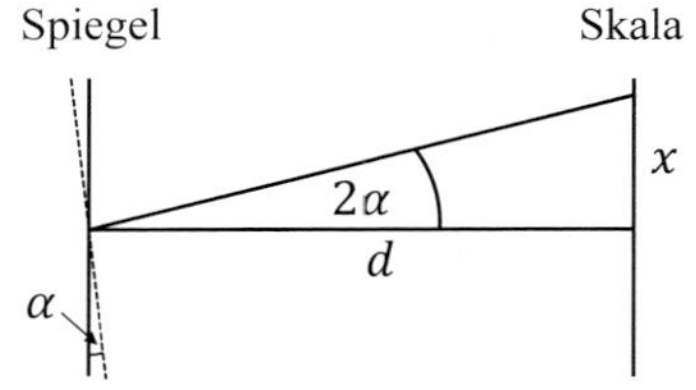

Abbildung 10.7 Links: Die Pfeile zeigen den Lichtstrahl. Im (gestrichelten) Anfangszustand wird er um den Winkel 2β reflektiert. Wenn sich der Spiegel um den Winkel α dreht, bewegt sich der reflektierte Strahl um den Winkel 2α. Rechts: Zusammenhang zwischen dem Drehwinkel α des Spiegels und der Wegdifferenz x des Lichtstrahls auf der Skala

Auf der Skala des weit entfernt stehenden Schirms (► Abbildung 10.5) können wir die Wegdifferenz x ablesen. Bei großer Entfernung d zwischen Spiegel und Schirm ist (rechte ► Abbildung 10.7)

$$\frac{x}{d} = \tan(2\alpha) \approx 2\,\alpha\,. \tag{10.27}$$

Setzen wir $\alpha = x/(2d)$ in (10.26) ein, folgt

$$\frac{x}{2\,d} = \frac{G\,M}{R\,r^2}\,t^2 \tag{10.28}$$

$$x = \underbrace{\frac{2\,G\,M\,d}{R\,r^2}}_{\equiv a}\,t^2\,. \tag{10.29}$$

Messergebnis

In ► Abbildung 10.8 sind die Messdaten auf der Meterskala x als Funktion der Zeit t in Sekunden gezeigt. Die durchgezogene Kurve zeigt eine Anpassung an die Messdaten mit einer quadratischen Funktion $x(t) = a\,t^2$ (10.29). Näherungsweise ergibt sich in den ersten zwei Minuten der quadratische Anstieg einer gleichförmig beschleunigten Bewegung aufgrund der angreifenden Drehmomente.

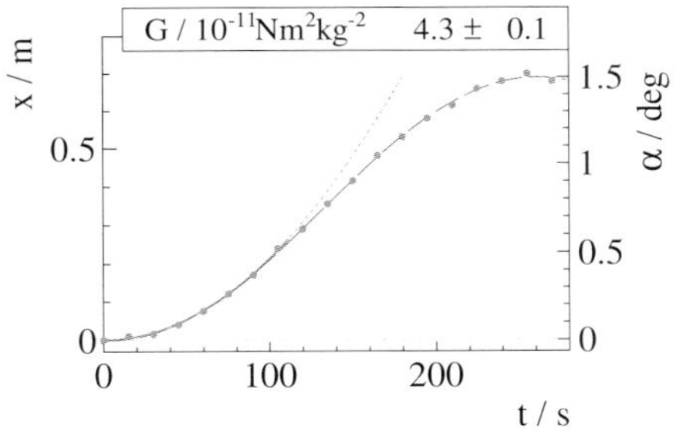

Abbildung 10.8 Messung der Wegdifferenz x in Abhängigkeit der Zeit t. Die durchgezogene Kurve zeigt eine quadratische Funktion im Anpassungsbereich (ihre Verlängerung ist als gestrichelte Kurve zu sehen), deren Parameter a nach Einsetzen der Massen M, m und der Abstände R, r, d die Gravitationskonstante G ergibt (10.29). Die gestrichelte Kurve ist eine Anpassung durch die Kosinusfunktion.

Unter Berücksichtigung der Werte

$R = 5\,\text{cm}$	$r = 4\,\text{cm}$	$d = 13{,}15\,\text{m}$
$M = 1{,}5\,\text{kg}$	$m = 0{,}02\,\text{kg}$	(m geht nicht in die Rechnung ein)

ergibt sich durch Einsetzen des angepassten Faktors $a = (2{,}14 \pm 0{,}04)\,10^{-5}\,\text{m/s}^2$ in (10.29) die Gravitationskonstante zu

$$G_{\text{Messung}} = (4{,}3 \pm 0{,}1) \cdot 10^{-11}\,\frac{\text{Nm}^2}{\text{kg}^2}\,. \tag{10.30}$$

Der hier angegebene Fehler zeigt zunächst nur die Unsicherheit in der Anpassung an die Messdaten. Zum Vergleich: Genaue Messungen ergeben für die Gravitationskonstante den Wert $G = (6{,}67430 \pm 0{,}00015) \cdot 10^{-11}\mathrm{Nm}^2/\mathrm{kg}^2$ (10.3).

Der Abweichung unseres Messwerts von (10.3) um mehr als 30% möchten wir nachgehen. Neben der Anpassung an die Messdaten ergeben sich statistische Fehler von G durch die Messgrößen in (10.29). M, d, R und t^2 lassen sich recht genau bestimmen und tragen nur im Promillebereich bei. Dagegen liegen die Unsicherheiten in der Ablesegenauigkeit x des Strahls auf der Skala und in dem Abstand r beim Umklappen der Apparatur mit seiner quadratischen Abhängigkeit r^2 jeweils im Prozentbereich. Über Fehlerfortpflanzung (Abschnitt 2.5.3) erhält man insgesamt einen Beitrag von ca. 10%, der die Abweichung allein nicht erklärt (Berechnungsdetails dazu sind hier nicht vorgeführt).

Viel größer sind systematische Fehler (Abschnitt 2.5.4). In ► Abbildung 10.8 ist eine zweite Analyse der Messdaten durch die Anpassung einer Kosinusfunktion

$$x(t) = -A \cos\left(B\,(t - C)\right) + A \tag{10.31}$$

gezeigt, die Messpunkte auch bei größeren Zeiten t berücksichtigt. Dabei interpretieren wir die Bewegung der Hantel als Schwingung eines harmonischen Oszillators mit einer rücktreibenden Kraft, die aus der Überlagerung der Gravitation und des Torsionsdrehmoments resultiert. Der angepasste Parameter $C = (4{,}2 \pm 1{,}4)\,\mathrm{s}$ zeigt, dass die Hantel bei $t = 0$ noch nicht ganz in Ruhe war, sondern eine Bewegung nach außen vollführte. Dadurch haben wir mit dem angepassten Faktor a in (10.29) eine etwas zu geringe Beschleunigung gemessen. Die alternative Auswertung mit der Kosinusfunktion (10.31) berücksichtigt diese anfängliche Bewegung. Über eine Taylor-Entwicklung von $x(t)$ um die Zeit $t = C$ bis zur zweiten Ordnung findet man die Beschleunigung $a = 0{,}5\,A\,B^2$ (10.29) wieder und erhält mit $A = (0{,}3451 \pm 0{,}0015)\,\mathrm{m}$ und $B = (0{,}01220 \pm 0{,}00014)\,\mathrm{s}^{-1}$ den Wert $G_{\mathrm{Messung}} = 5{,}2 \cdot 10^{-11}\mathrm{Nm}^2/\mathrm{kg}^2$.

Weitere systematische Fehler ergeben sich aus der mitbewegten Masse des Hantelstabs (siehe Abschnitt 14.3), den wir im Drehimpuls (10.18) vernachlässigt haben. Reibungskräfte in der Aufhängung und durch die Luft, die bei der langen Messzeit gegenüber der schwachen Gravitationskraft nicht zu vernachlässigen sind, führen ebenfalls zu geringeren Messwerten für G.

Insgesamt ist es trotz der erheblichen Unsicherheiten dennoch bemerkenswert, dass wir mit diesem vergleichsweise geringen experimentellen Aufwand und den durchgeführten Näherungen zumindest die Größenordnung dieses wichtigen physikalischen Parameters bestimmen können.

10.2 Gravitationspotential

10.2.1 Potentielle Energie

Die Änderung der potentiellen Energie eines Körpers der Masse m, der im Einflussbereich der Gravitation eines zweiten Körpers der Masse M ist, können wir mithilfe der Arbeit berechnen, die wir zum Verschieben der Masse m leisten müssen.

Nach unserer Vorzeichenkonvention (Abschnitt 5.2) wird die am Probekörper (System) gegen die Schwerkraft $\vec{F_G}$ also mit der Kraft $\vec{F} = -\vec{F_G}$ geleistete Arbeit positiv gezählt. Wir bringen den Probekörper der Masse m von r_1 nach $r_2 > r_1$:

$$W = \int_{r_1}^{r_2} \vec{F}\, d\vec{r} \tag{10.32}$$

$$= -\int_{r_1}^{r_2} \frac{-G\, M\, m}{r^2}\, \vec{e}_r\, d\vec{r} \tag{10.33}$$

$$= G\, M\, m \int_{r_1}^{r_2} \frac{dr}{r^2} \tag{10.34}$$

$$= G\, M\, m \left.\frac{-1}{r}\right|_{r_1}^{r_2} \tag{10.35}$$

$$= G\, M\, m \left(\frac{1}{r_1} - \frac{1}{r_2}\right) > 0 \tag{10.36}$$

Mit der von außen gegen die Schwerkraft an ihm geleisteten Arbeit gewinnt der Körper potentielle Energie (Abschnitt 5.3):

$$\Delta E_{\text{pot}} = E_{\text{pot}}(r_1) - E_{\text{pot}}(r_2) = W \tag{10.37}$$

Bei der Rückkehr von r_2 nach r_1 gibt der Körper die potentielle Energie zurück. Er könnte dabei Arbeit leisten, indem er z. B. eine Feder gegen die Kraft $\vec{F} = -\vec{F_G}$ langsam zusammendrückt. Nach unserer Vorzeichenkonvention zählen wir die vom Körper geleistete Arbeit negativ:

$$W' = \int_{r_2}^{r_1} \vec{F}\, d\vec{r} \tag{10.38}$$

$$= -\int_{r_2}^{r_1} \frac{-G\, M\, m}{r^2}\, dr \tag{10.39}$$

$$= G\, M\, m \left(\frac{1}{r_2} - \frac{1}{r_1}\right) < 0 \tag{10.40}$$

Konsistent ergibt sich der Verlust des Körpers an potentieller Energie durch

$$\Delta E_{\text{pot}} = E_{\text{pot}}(r_2) - E_{\text{pot}}(r_1) = W' \,. \tag{10.41}$$

Beispiel 29: Potentielle Energie der Raumstation ISS

Zu umfangreichen Forschungszwecken wurde die internationale Raumstation ISS mit der Masse $m \approx 450\,000\,\mathrm{kg}$ auf eine Umlaufbahn in der Höhe $h \approx 350\,\mathrm{km}$ gebracht. Ihre potentielle Energie relativ zur Erdoberfläche (Erdradius $R_{\text{Erde}} = 6\,380\,\mathrm{km}$) beträgt mit der Erdmasse M und $g = G\,M/R^2_{\text{Erde}}$ (10.4) näherungsweise

$$\begin{aligned}
\Delta E_{\text{pot}} &= E_{\text{pot}}\,(R_{\text{ISS}}) - E_{\text{pot}}\,(R_{\text{Erde}}) \\
&= m\,G\,M\left(\frac{1}{R_{\text{Erde}}} - \frac{1}{R_{\text{ISS}}}\right) \\
&= m\,G\,M\left(\frac{1}{R_{\text{Erde}}} - \frac{1}{R_{\text{Erde}} + h}\right) \\
&= m\,G\,M\left(\frac{(R_{\text{Erde}} + h) - R_{\text{Erde}}}{(R_{\text{Erde}} + h)\cdot R_{\text{Erde}}}\right) \\
&\approx m\,\underbrace{\frac{G\,M}{R^2_{\text{Erde}}}}_{=g}\,h = m\cdot g\cdot h\,.
\end{aligned}$$

Damit ergibt sich die potentielle Energie der ISS zu $\Delta E_{\text{pot}} \approx 1{,}5\;10^{12}\,\mathrm{J}$.

Potentielle Energie ist nur in Bezug auf einen Referenzwert oder, wie im Beispiel der Raumstation ISS, als Differenz zweier potentieller Energien definiert.

Eine übliche Konvention ist, als Referenzwert die potentielle Energie bei $r = \infty$ auf $E_{\text{pot}}\,(\infty) = 0$ zu setzen. Wenn wir den Probekörper der Masse m von $r_1 = r$ nach $r_2 = \infty$ bringen, leisten wir nach Gleichung (10.36) die Arbeit

$$W = G\,M\,m\left(\frac{1}{r} - \frac{1}{\infty}\right) \tag{10.42}$$

$$= \frac{G\,M\,m}{r}\,. \tag{10.43}$$

Gleichzeitig erhöhen wir die potentielle Energie des Körpers von

$$E_{\text{pot}}\,(r) = -\frac{G\,M}{r}\,m \tag{10.44}$$

auf den der Konvention entsprechenden Wert von

$$E_{\text{pot}}\,(\infty) = 0\,. \tag{10.45}$$

Als Differenz der potentiellen Energien ergibt sich konsistent der positive Wert

$$\Delta E_{\text{pot}} = E_{\text{pot}}\,(\infty) - E_{\text{pot}}\,(r) \tag{10.46}$$

$$= \frac{G\,M}{r}\,m\,. \tag{10.47}$$

10.2.2 Potential

Man nennt den ersten Term der rechten Seite der Gleichung (10.44) das Gravitationspotential $V(r)$ des Körpers mit der Masse M (► Abbildung 10.9):

$$V(r) = -\frac{G\,M}{r} \tag{10.48}$$

Da V geometrisch nur vom Abstand r abhängt, handelt es sich beim Gravitationspotential um ein sogenanntes Zentralpotential.

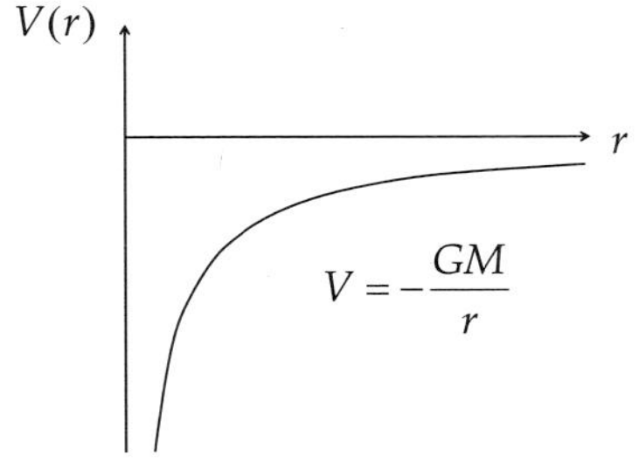

Abbildung 10.9 Das Potential $V(r)$ einer Masse M in Abhängigkeit vom Abstand r

Die Einheit für das Gravitationspotential ist $\mathrm{m}^2/\mathrm{s}^2$. Das Gravitationspotential hängt wegen $g = G\,M/R_{\mathrm{Erde}}^2$ (10.4) über die einfache Beziehung

$$V_{\mathrm{Erdoberfläche}} = -g\,R_{\mathrm{Erde}} \tag{10.49}$$

mit der Erdbeschleunigung an der Erdoberfläche zusammen. Mit $R_{\mathrm{Erde}} \approx 6\,380\,\mathrm{km}$ ergibt sich ein Potentialwert von $V_{\mathrm{Erdoberfläche}} \approx -62{,}6 \cdot 10^6\ \mathrm{m}^2/\mathrm{s}^2$, der in der Geodäsie (Erdvermessung) eine wichtige Rolle spielt.

Mathematischer Einschub: Gradient

Für die räumliche Erweiterung der (eindimensionalen) Ableitung d/ds wird der **Gradient grad** als ein Vektor aus den partiellen Ableitungen in die drei Koordinatenrichtungen definiert (9.13):

$$\mathrm{grad} = \vec{e}_x \frac{\partial}{\partial x} + \vec{e}_y \frac{\partial}{\partial y} + \vec{e}_z \frac{\partial}{\partial z} \tag{10.50}$$

Angewandt auf eine skalare Funktion $f(x, y, z)$, zeigt der Vektor grad f

$$\mathrm{grad}\, f(x, y, z) = \left(\vec{e}_x \frac{\partial}{\partial x} + \vec{e}_y \frac{\partial}{\partial y} + \vec{e}_z \frac{\partial}{\partial z}\right) f(x, y, z) \tag{10.51}$$

in Richtung des stärksten Anstiegs von f und seine Länge (Betrag) entspricht der Stärke dieses Anstiegs.

Der Gradient in Zylinderkoordinaten lautet (9.33)

$$\text{grad} = \vec{e}_r \frac{\partial}{\partial r} + \vec{e}_\varphi \frac{1}{r} \frac{\partial}{\partial \varphi} + \vec{e}_z \frac{\partial}{\partial z} \,. \tag{10.52}$$

Der Gradient in Kugelkoordinaten lautet (9.52)

$$\text{grad} = \vec{e}_r \frac{\partial}{\partial r} + \vec{e}_\vartheta \frac{1}{r} \frac{\partial}{\partial \vartheta} + \vec{e}_\varphi \frac{1}{r \sin \vartheta} \frac{\partial}{\partial \varphi} \,. \tag{10.53}$$

Als Merkregel kann man die Gradienten (10.50) (10.52) (10.53) mit den Linienelementen $d\vec{s}$ (9.55) (9.22) (9.41) durch »grad $= d/d\vec{s}$« verbinden.

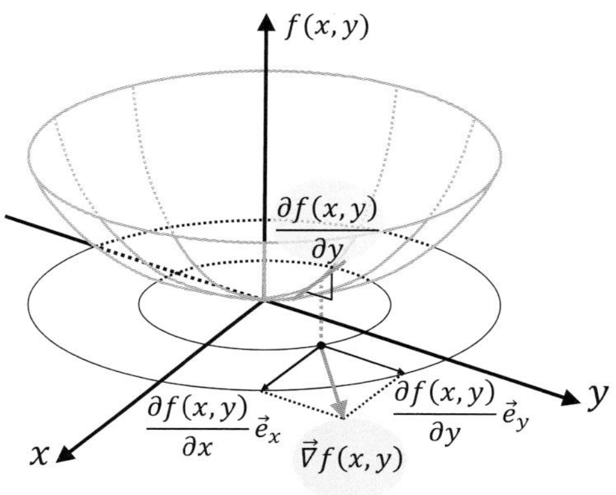

Abbildung 10.10 Veranschaulichung des Gradienten $\vec{e}_x \frac{\partial}{\partial x} + \vec{e}_y \frac{\partial}{\partial y}$, angewandt auf ein Höhenprofil $f(x, y)$

Eine anschauliche Bedeutung hat der Gradient für eine (zweidimensionale) Landkarte mit einem Tal im Zentrum (▶ Abbildung 10.10). Eine (skalare) Höhenfunktion $f(x, y)$ soll an jedem Punkt (x, y) die zugehörige Höhe darstellen. Der Gradient grad $= \vec{e}_x \frac{\partial}{\partial x} + \vec{e}_y \frac{\partial}{\partial y}$, angewandt auf f, weist als der Vektor grad $f(x, y)$ in Richtung des steilsten Anstiegs des Höhenprofils. Der Betrag des Gradienten entspricht der Stärke dieses Anstiegs.

10.2.3 Zusammenhang zwischen Kraft und Potential

Die **Gravitationskraft** und das **Gravitationspotential** beschreiben Aspekte desselben physikalischen Phänomens:

$$\vec{F}(r) = -\frac{G M m}{r^2} \vec{e}_r \tag{10.54}$$

$$V(r) = -\frac{G M}{r} \tag{10.55}$$

Kraft und Potential hängen über den Gradienten miteinander zusammen. In Kugelkoordinaten gilt nach Gleichung (9.52) und (10.53)

$$\mathrm{grad} = \vec{e}_r \frac{\partial}{\partial r} + \vec{e}_\vartheta \frac{1}{r} \frac{\partial}{\partial \vartheta} + \vec{e}_\varphi \frac{1}{r \sin\vartheta} \frac{\partial}{\partial \varphi} \,. \tag{10.56}$$

Die Ableitung der radialen Komponente des Potentials ist

$$\frac{\partial V}{\partial r} = \frac{\partial}{\partial r}\left(-\frac{G\,M}{r}\right) = \frac{G\,M}{r^2} \,. \tag{10.57}$$

Die Winkelableitungen sind beide null, weil das Potential (10.55) weder von φ noch von ϑ abhängt:

$$\frac{\partial V}{\partial \varphi} = \frac{\partial V}{\partial \vartheta} = 0 \tag{10.58}$$

Durch Vergleich von (10.57) mit (10.54) können wir die Gravitationskraft und das Gravitationspotential in Beziehung bringen. Der Gradient des Gravitationspotentials entspricht der Kraft auf eine Einheitsmasse:

$$\frac{\vec{F}}{m} = -\,\mathrm{grad}\, V \tag{10.59}$$

Anschaulich bedeutet Gleichung (10.59), dass die Gravitationskraft in Richtung des stärksten Potentialgefälles zeigt (▶ Abbildung 10.11).

10.2.4 Äquipotentialflächen

Da die Gravitationskraft konservativ ist, spielt für die geleistete Arbeit der exakte Weg, auf dem wir den Körper auf eine größere Distanz bringen, keine Rolle.

Auch für den Betrag des Potentials ist die genaue Position des Körpers mit der Masse m auf einer Kugelschale (Sphäre) mit dem Abstand r von der Masse M unerheblich. Solche Sphären nennt man deswegen **Äquipotentialflächen** (▶ Abbildung 10.11).

10.3 Energie eines Himmelskörpers

Um die Bewegung von Planeten und Kometen im Sonnensystem zu verstehen, untersuchen wir zunächst allgemein die Energie eines Körpers im Gravitationspotential der Sonne. Seine Gesamtenergie setzt sich aus potentieller und kinetischer Energie zusammen:

$$E = E_{\mathrm{pot}} + E_{\mathrm{kin}} \tag{10.60}$$

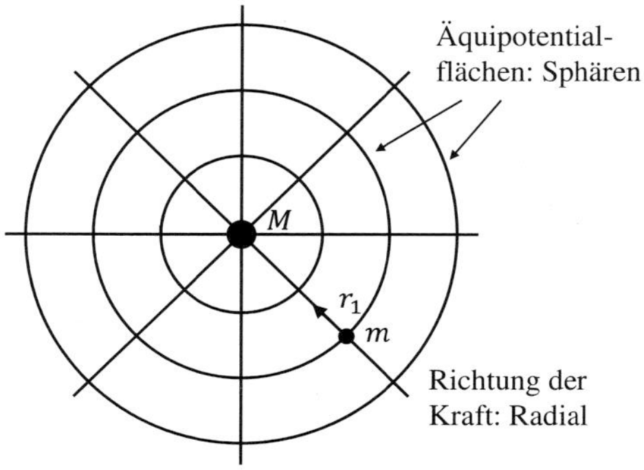

Abbildung 10.11 Sphärische Äquipotentialflächen und radiale Richtung der Gravitationkraft einer Masse M auf eine sehr kleine Testmasse m

10.3.1 Potentielle Energie

Die potentielle Energie des Körpers ergibt sich aus dem Produkt seiner Masse m und dem Gravitationspotential V (10.48), siehe auch (10.44):

$$E_{\text{pot}} = m\,V \tag{10.61}$$

$$= -\frac{G\,M\,m}{r} \tag{10.62}$$

Die Masse M der Sonne ist groß im Vergleich zur Masse m aller anderen Himmelskörper im Sonnensystem. Daher können wir uns mit ► Abbildung 10.11 die Sonne im Zentrum vorstellen und dazu Himmelskörper, die sich in ihrem Gravitationsfeld bewegen.

10.3.2 Kinetische Energie

Die kinetische Energie des Himmelskörpers mit Masse m, Geschwindigkeit v und Impuls $p = m\,v$ ist

$$E_{\text{kin}} = \frac{1}{2} m\,v^2 = \frac{p^2}{2\,m}\,. \tag{10.63}$$

Der Himmelskörper (z. B. die Erde) kann sich relativ zur Sonne bewegen, wie in ► Abbildung 10.12 beispielhaft für zwei Raumpunkte gezeigt ist. Dabei hat der Körper den Drehimpuls $\vec{L} = \vec{r} \times \vec{p}$, den wir hier durch Zylinderkoordinaten (r, φ, z) beschreiben.

Zunächst stellen wir fest, dass bei einem Zentralpotential wie der Gravitationskraft kein Drehmoment $\vec{D} = \vec{r} \times \vec{F}$ entsteht, weil die Gravitationskraft $\vec{F}$ immer in Richtung der Verbindungsachse $\vec{r}$ zwischen Himmelskörper und Sonne zeigt.

Der Körper soll keinem weiteren äußeren Einfluss ausgesetzt sein und damit auch keinen externen Drehmomenten $\vec{D}_i = 0$. Die Bewegungsgleichung für den Körper lautet somit

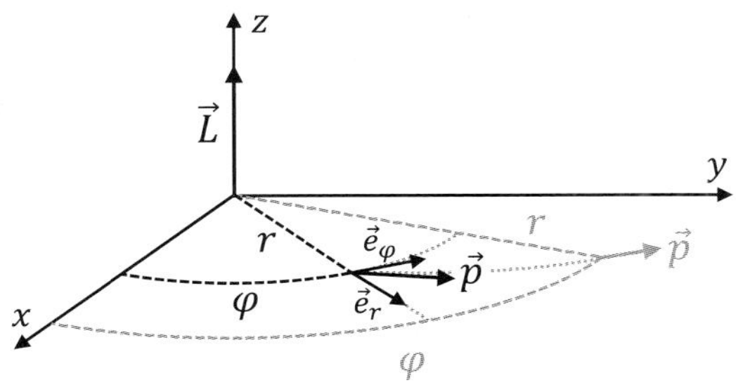

Abbildung 10.12 Impuls $\vec{p}$, beschrieben in Zylinderkoordinaten an zwei verschiedenen Raumpunkten, und zugehöriger Drehimpuls $\vec{L}$

$d\vec{L}/dt = \sum_{i=1}^{n} \vec{D}_i = 0$ (4.70). Sein Drehimpuls $\vec{L}$ innerhalb des Gesamtsystems von Sonne und Erde ist erhalten (Abschnitt 7.3):

$$\vec{L} = \vec{r} \times \vec{p} = const. \tag{10.64}$$

Die Bewegung des Körpers findet daher in der Ebene senkrecht zu $\vec{L}$ statt (▶ Abbildung 10.12). Wir legen die Drehimpulsrichtung entlang der z-Achse, sodass wir zur Beschreibung des Körpers nur die Koordinaten r und φ benötigen (die Polarkoordinaten) und seine z-Koordinate immer $z = 0$ ist.

Die Geschwindigkeit v des Körpers lesen wir aus Gleichung (9.23) ab:

$$\vec{v} = \dot{r}\,\vec{e}_r + r\,\dot{\varphi}\,\vec{e}_\varphi \tag{10.65}$$

Im Unterschied zur Kreisbewegung kann sich hier der Körper zu kleineren und größeren Radien r begeben (▶ Abbildung 10.12).

Die kinetische Energie wird dementsprechend in zwei Terme aufgeteilt:

$$E_{\text{kin}} = \underbrace{\frac{1}{2}\,m\,\dot{r}^2}_{\equiv E_{\text{rad}}} + \underbrace{\frac{1}{2}\,m\,r^2\,\dot{\varphi}^2}_{\equiv E_{\text{rot}}} \tag{10.66}$$

Der erste Term beschreibt die kinetische Energie in Richtung der radialen Komponente r, der zweite Term zeigt die Bewegungsenergie in der Rotation um den Winkel φ (Rotationsenergie (5.25)).

Der konstante Drehimpuls des Körpers entspricht in Zylinderkoordinaten unter Verwendung der Geschwindigkeit (10.65)

$$\vec{L} = \vec{r} \times m\,\vec{v} = m\,(\dot{r}\,\underbrace{\vec{r} \times \vec{e}_r}_{=0} + r\,\dot{\varphi}\,\underbrace{\vec{r} \times \vec{e}_\varphi}_{\vec{r}\times\vec{e}_\varphi = r\,\vec{e}_z}) \tag{10.67}$$

$$= m\,r^2\,\dot{\varphi}\,\vec{e}_z\,. \tag{10.68}$$

Die zeitliche Ableitung $\dot{\varphi}$ der Winkelkoordinate in der Rotationsenergie (10.66) können wir damit durch den (konstanten) Drehimpuls und die Radialkomponente r ausdrücken:

$$\dot{\varphi} = \frac{L}{m\,r^2} \tag{10.69}$$

Die Rotationsenergie hängt dann bis auf einige Konstanten nur von der radialen Koordinate r ab:

$$E_{\text{rot}} = \frac{1}{2}\, m\, r^2\, \dot{\varphi}^2 \tag{10.70}$$

$$= \frac{1}{2}\, m\, r^2\, \frac{L^2}{m^2\, r^4} \tag{10.71}$$

$$= \frac{L^2}{2\, m\, r^2} \tag{10.72}$$

Mit dem Argument der Drehimpulserhaltung $\vec{L} = const.$ ist es uns gelungen, die Bewegungsenergie (10.66) des Körpers auf eine Variable – die Radialkomponente r und ihre zeitliche Ableitung $\dot{r}$ – zu reduzieren:

$$E_{\text{kin}} = \frac{1}{2}\, m\, \dot{r}^2 + \frac{L^2}{2\, m\, r^2} \tag{10.73}$$

10.3.3 Gesamtenergie

Als Gesamtenergie des Körpers ergibt sich mit der kinetischen Energie (10.73) und der potentiellen Energie (10.62), sortiert nach Termen mit der Radialkomponente r und ihrer zeitlichen Ableitung $\dot{r}$,

$$E = E_{\text{kin}} + E_{\text{pot}} \tag{10.74}$$

$$= \underbrace{\frac{1}{2}\, m\, \dot{r}^2}_{\equiv E_{\text{rad}}(\dot{r})} + \underbrace{\left[\frac{L^2}{2m^2r^2} - \frac{G\, M}{r}\right] m}_{\equiv V_{\text{eff}}(r)} = const.\,. \tag{10.75}$$

Ohne äußeren Einfluss auf das System aus Sonne und Himmelskörper ist diese Gesamtenergie erhalten (7.2).

Die Gesamtenergie setzt sich aus zwei Termen zusammen: Der erste Term E_{rad} ist die Bewegungsenergie in radialer Richtung mit der Geschwindigkeit $\dot{r}$. Der zweite Term ist nur vom radialen Abstand r des Körpers abhängig und kann deswegen als »effektives Potential« bezeichnet werden:

$$V_{\text{eff}}(r) = \frac{L^2}{2m^2r^2} - \frac{G\, M}{r} \tag{10.76}$$

10.4 Bahnkurven von Himmelskörpern

Unser Ziel ist, die Bahnkurven von Himmelskörpern zu berechnen. Dafür werden wir zunächst die radiale Bewegung ohne die Rotationsbewegung untersuchen. Anschließend werden wir Kandidaten für Bahnkurven kennenlernen (Kegelschnitte) und schließlich die vollständigen Bahnkurven $r(\varphi)$ in Polarkoordinaten für Planeten und Kometen berechnen.

10.4.1 Radiale Bewegung im Zentralpotential

Um die prinzipiellen Bewegungsmöglichkeiten eines Körpers mit der Gesamtenergie E im effektiven Potential V_{eff} (10.76) zu verstehen, betrachten wir einen Probekörper mit der Masse $m = 1\,\text{kg}$.

Das effektive Potential V_{eff} setzt sich aus dem Rotationsterm $\sim +\frac{1}{r^2}$ und dem Gravitationsterm $\sim -\frac{1}{r}$ zusammen (► Abbildung 10.13). Das Potential zeigt daher ein Minimum bei $E < 0$.

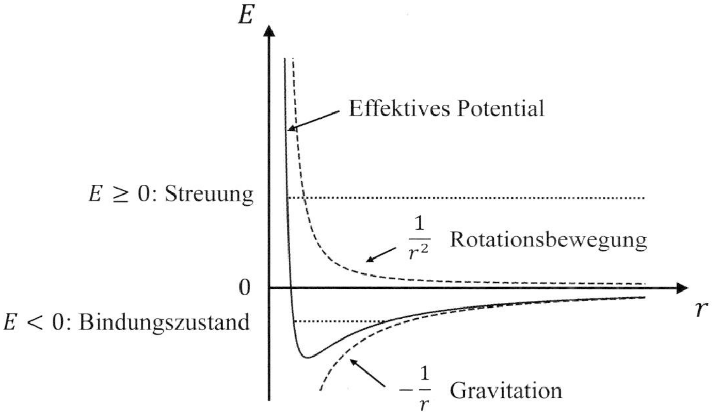

Abbildung 10.13 Das effektive Potential (durchgezogene Kurve) setzt sich aus dem Gravitationspotential ($\propto -1/r$) und dem Potential der Rotationsbewegung ($\propto 1/r^2$) zusammen. Bindungszustände (Ellipsen und Kreise) gibt es nur für $E < 0$, sonst tritt Streuung auf (Parabeln und Hyperbeln).

Da seine Gesamtenergie erhalten ist, kann sich der Körper bei gegebenem Wert $E = const.$ in ► Abbildung 10.13 nur auf einer horizontalen Linie bewegen.

Körper mit einer Gesamtenergie $E < 0$ oszillieren zwischen einer minimalen und einer maximalen Entfernung $[R_{\text{min}}, R_{\text{max}}]$ vom Zentrum des Gravitationspotentials. Sie befinden sich in einem **Bindungszustand**. Das ist die Situation für die Planeten unseres Sonnensystems.

Körper mit $E \geq 0$ kommen aus dem Unendlichen ($r = \infty$), nähern sich bis zu einer minimalen Entfernung R_{min} und verschwinden wieder im Unendlichen. Diese Prozesse heißen **Streuprozesse**.

Damit haben wir die Radialbewegung des Körpers im Einfluss der Gravitation qualitativ beschrieben. In den folgenden Abschnitten werden wir die vollständige Lösung für die Bahnkurven erarbeiten.

10.4.2 Kegelschnitte

Aufgrund der Drehimpulserhaltung bewegen sich die Körper im Gravitationspotential in einer Ebene. Aus der Gesamtenergie und der Radialbewegung der Körper haben wir auf die Existenz von Bindungszuständen und Streuprozessen geschlossen.

In einem ebenen kartesischen Koordinatensystem mit den Variablen x und y sind die Lösungsmöglichkeiten für Bahnkurven der Himmelskörper Kurven zweiter Ordnung. Diese Kurven kann man als Kegelschnitte verstehen. Wenn man in der linken ► Abbildung 10.14 die Ebene verkippt und die Schnittkurven mit dem Kegel beobachtet, erhält man die Kurven in ► Abbildung 10.14 rechts und ► Abbildung 10.15. Wir erleichtern uns das Verständnis beim Berechnen der Bahnkurven für Planeten, wenn wir diese Kurvenformen zunächst kennenlernen.

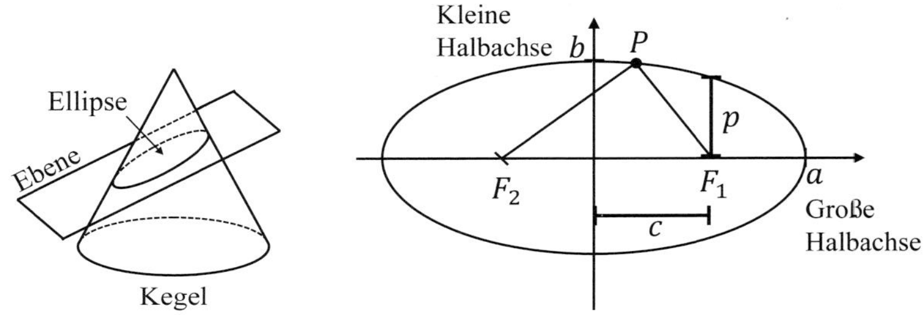

Abbildung 10.14 Links: Kegelschnitte als Schnittlinien eines Kegels mit einer Ebene, rechts: Ellipse mit großer Halbachse a und kleiner Halbachse b und den Brennpunkten F_1 und F_2

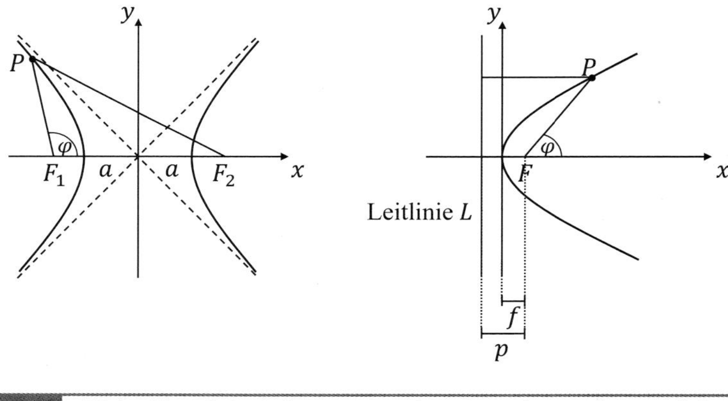

Abbildung 10.15 Links: Hyperbel mit den Brennpunkten F_1 und F_2, rechts: Parabel mit Leitlinie L und Brennpunkt F

Mathematischer Einschub: Ellipse

Eine Ellipse besitzt zwei Brennpunkte F_1 und F_2 (rechte ► Abbildung 10.14). Die Summe der Abstände $\overline{F_1P} + \overline{PF_2} = const. = 2a$ ist für alle Punkte P auf der Ellipse gleich groß. Die Darstellung einer Ellipse in Normalform lautet

$$\frac{x^2}{a^2} + \frac{y^2}{b^2} = 1\,, \qquad (10.77)$$

wobei a die große und b die kleine Halbachse bezeichnet.

In Polarkoordinaten folgt eine Ellipse folgender Parametrisierung, mit der x-Achse als Symmetrieachse und dem Pol für $+/-$ im Brennpunkt F_1/F_2 (► Abbildung 10.16):

$$r = \frac{p}{1 \pm \epsilon \cos\varphi} \qquad (10.78)$$

Die (numerische) Exzentrizität ϵ beschreibt den Abstand c des Brennpunkts vom Zentrum relativ zur Länge der großen Halbachse a:

$$\epsilon = \frac{c}{a} = \frac{\sqrt{a^2 - b^2}}{a} = \sqrt{1 - \frac{b^2}{a^2}} \qquad (10.79)$$

Für die Halbachsen gilt immer $a > b$, sodass die Exzentrizität den Wert 1 nie überschreitet:

$$\epsilon < 1 \qquad (10.80)$$

Der zweite Parameter p wird als Halbparameter bezeichnet, er lässt sich ebenfalls aus den Halbachsen berechnen:

$$p = \frac{b^2}{a} = a\left(1 - \epsilon^2\right) \qquad (10.81)$$

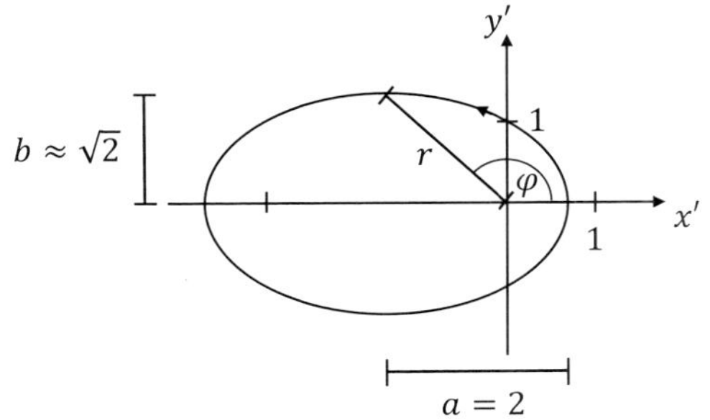

Abbildung 10.16 Konstruktion einer Ellipse in der Polarkoordinatendarstellung $r(\varphi)$. Die Halbachsen a und b sind eingezeichnet, ebenso das kartesische System (x', y') im rechten Brennpunkt F_1.

Beispiel 30: Ellipse

Für eine Ellipse mit $a = 2$, $\epsilon = 0{,}7$ ergibt sich für den Halbparameter $p = a\left(1 - \epsilon^2\right) \approx 1$ (10.81). Die radiale Koordinate r verläuft also in Abhängigkeit des Winkels φ mit

$$r(\varphi) = \frac{1}{1 + 0{,}7 \cos\varphi} \,. \tag{10.82}$$

Strategische Werte zum Zeichnen der Ellipse (▶ Abbildung 10.16) sind wie folgt:

φ	$0°$	$90°$	$135°$	$180°$
$\cos\varphi$	1	0	$-\frac{1}{2}\sqrt{2}$	-1
r	$\frac{1}{1+0{,}7} \approx 0{,}6$	1	$\frac{1}{1-0{,}7^2} \approx 2$	$\frac{1}{1-0{,}7} \approx 3{,}3$

Mathematischer Einschub: Kreis

Der Kreis ist ein Sonderfall der Ellipse. In der Normalform wird er dargestellt durch

$$x^2 + y^2 = a^2 \,. \tag{10.83}$$

Die beiden Brennpunkte der Ellipse liegen beim Kreis im Mittelpunkt aufeinander. Für die Parameter gilt Folgendes:

Die Exzentrizität ist

$$\epsilon = 0 \,. \tag{10.84}$$

Die Parameter p, a und b sind gleich groß:

$$p = \frac{b^2}{a} = \frac{a^2}{a} = a = b \tag{10.85}$$

Die Darstellung in Polarkoordinaten (10.78) lautet für alle Werte von φ

$$r = \frac{p}{1 \pm 0 \cdot \cos\varphi} = a \,. \tag{10.86}$$

Mathematischer Einschub: Hyperbel

In der linken ► Abbildung 10.15 haben alle Punkte P der Hyperbel gemeinsam, dass der Betrag der Differenz der Abstände von den zwei Brennpunkten konstant ist: $|\overline{F_1\,P} - \overline{P\,F_2}| = const. = 2a$

Ihre Darstellung lautet in der Normalform

$$\frac{x^2}{a^2} - \frac{y^2}{b^2} = 1\,. \tag{10.87}$$

In Polarkoordinaten mit dem Pol im Brennpunkt F_1 und der x-Achse als Symmetrieachse schreiben wir

$$r = \frac{p}{1 + \epsilon\,\cos\varphi}\,. \tag{10.88}$$

Die Gleichung liefert nur den linken Ast, für den rechten muss in Gleichung (10.88) das Vorzeichen von + auf − gesetzt werden. Die Zusammenhänge (10.80) und (10.81) gelten auch für die Hyperbel, jedoch mit

$$\epsilon > 1\,. \tag{10.89}$$

Mathematischer Einschub: Parabel

Alle Punkte P einer Parabel (rechte ► Abbildung 10.15) haben gemeinsam, dass für sie die Abstände zum Brennpunkt F und zur Leitlinie L gleich sind: $\overline{F\,P} = \overline{P\,L}$

Dargestellt wird die Parabel in der Normalform durch

$$y^2 = 2px \tag{10.90}$$

und in Polarkoordinaten mit dem Pol im Brennpunkt und der x-Achse als Polarachse durch

$$r = \frac{p}{1 + \epsilon\,\cos\varphi} = \frac{p}{1 \pm \cos\varphi}\,. \tag{10.91}$$

Beim Vorzeichen (+) ist die Parabel nach links, bei (−) wie in der rechten ► Abbildung 10.15 nach rechts geöffnet. Der Halbparameter p hat hier die Bedeutung der doppelten Brennweite f

$$p = 2f\,, \tag{10.92}$$

und die (numerische) Exzentrizität ist

$$\epsilon = \pm 1\,. \tag{10.93}$$

Alle Kegelschnitte lassen sich also in Polarkoordinaten in der gleichen Form $r(\varphi)$ mit nur zwei Parametern ϵ und p beschreiben:

$$r(\varphi) = \frac{p}{1 + \epsilon \cos \varphi} \tag{10.94}$$

Dabei haben wir die Vorzeichen ($\pm$) in den Parameter ϵ eingeschlossen. Durch ihn werden die Kurvenformen Ellipse $|\epsilon| < 1$ (10.80), Kreis $\epsilon = 0$ (10.84), Hyperbel $|\epsilon| > 1$ (10.89) und Parabel $\epsilon = \pm 1$ (10.93) voneinander unterschieden.

Im Folgenden wollen wir den Zusammenhang zwischen dem Gravitationspotential und diesen Bahnkurven genauer untersuchen.

10.4.3 Berechnung der Bahnkurven

Die Bahnkurven von Planeten und anderen Himmelskörpern im Gravitationseinfluss der Sonne können wir auf der Basis der Energieerhaltung $E = const.$ (10.75) und der Drehimpulserhaltung $\vec{L} = const.$ (10.64) berechnen.

Wir nutzen dafür Polarkoordinaten r und φ, die bei der Bahnbewegung Funktionen $r(t), \varphi(t)$ der Zeit t sind. Hier sind wir an der grundsätzlichen Form der Bahnkurven $r(\varphi)$ interessiert, sodass wir die Zeit eliminieren wollen. Die Situation ist ähnlich wie bei der Bahnkurve des schiefen Wurfs, dort hatten wir die zeitabhängigen Koordinaten $x(t), y(t)$ zu einer Wurfparabel $y(x)$ zusammengefasst (1.66).

Die Energiegleichung (10.75)

$$E = E_{\text{rad}}(\dot{r}) + E_{\text{rot}}(r) + E_{\text{pot}}(r) \tag{10.95}$$

$$= \frac{1}{2} m \dot{r}^2 + \frac{L^2}{2\, m\, r^2} - \frac{G\, M\, m}{r} \tag{10.96}$$

ist eine Differentialgleichung, die sowohl die radiale Koordinate r als auch ihre zeitliche Ableitung $\dot{r}$ enthält. Für die Berechnung der Bahnkurven wählen wir hier einen Ansatz über die Ableitung $d\varphi/dr$, die wir formal durch Erweitern mit dt auf die zeitlichen Ableitungen dr/dt und $d\varphi/dt$ zurückführen können:

$$\frac{d\varphi}{dr} = \frac{d\varphi}{dt}\frac{dt}{dr} = \frac{d\varphi}{dt}\left[\frac{dr}{dt}\right]^{-1} \tag{10.97}$$

Die zeitliche Ableitung $\dot{\varphi} = d\varphi/dt$ können wir der Drehimpulsgleichung (10.69) entnehmen:

$$\frac{d\varphi}{dt} = \frac{L}{m\, r^2} \tag{10.98}$$

Die zeitliche Ableitung $\dot{r} = dr/dt$ können wir aus der Energiebilanz (10.96) extrahieren:

$$\frac{dr}{dt} = \sqrt{\frac{2}{m}\left(E - E_{\text{rot}}(r) - E_{\text{pot}}(r)\right)} \equiv \sqrt{E'} \tag{10.99}$$

Wir setzen in Gleichung (10.97) ein und trennen die Variablen r und φ auf die linke und rechte Seite der Gleichung. Durch Einsetzen aller Energieterme erhalten wir für $\varphi(r)$

$$\frac{d\varphi}{dr} = \frac{L}{m\,r^2}\,\frac{1}{\sqrt{E'}} \tag{10.100}$$

$$\int_{\varphi_\circ}^{\varphi} d\varphi = \int_{r_\circ}^{r} \frac{L}{m\,r^2}\,\frac{1}{\sqrt{E'}}\,dr \tag{10.101}$$

$$\varphi\,(r) = \int_{r_\circ}^{r} \frac{L}{m\,r^2\sqrt{\frac{2}{m}\left(E - \frac{L^2}{2\,m\,r^2} + \frac{G\,M\,m}{r}\right)}}\,dr - \varphi_\circ \tag{10.102}$$

$$= \int_{r_\circ}^{r} \frac{L}{r\,\sqrt{2\,m\,E\,r^2 - L^2 + 2\,G\,M\,m^2\,r}}\,dr - \varphi_\circ\,. \tag{10.103}$$

Das hier auftretende Integral ist mithilfe sogenannter »Elliptischer Integrale« analytisch lösbar. Das Resultat geben wir hier an:

$$\varphi(r) = \arccos\left(\frac{\frac{L^2}{r} - G\,M\,m^2}{\sqrt{(G\,M\,m^2)^2 + 2\,m\,L^2\,E}}\right) - \varphi_\circ \tag{10.104}$$

Wir versichern uns, dass die Lösung korrekt ist, indem wir (10.104) nach r ableiten ($d\arccos(x)/dx = -1/\sqrt{1-x^2}$). Dadurch erhalten wir wieder Gleichung (10.103). Als Anfangswinkel wählen wir $\varphi_\circ = 0$.

Um die Bahnkurve $r(\varphi)$ zu erhalten, verwenden wir die Abkürzungen

$$p \equiv \frac{L^2}{G\,M\,m^2} \tag{10.105}$$

$$k \equiv \frac{2\,m\,L^2}{(G\,M\,m^2)^2} > 0 \tag{10.106}$$

$$\epsilon \equiv \sqrt{1 + k\,E} \tag{10.107}$$

und formen damit Gleichung (10.104) um:

$$\cos\varphi = \frac{\frac{L^2}{r} - G\,M\,m^2}{\sqrt{(G\,M\,m^2)^2 + 2\,m\,L^2\,E}} \tag{10.108}$$

$$= \frac{\frac{L^2}{G\,M\,m^2}\,\frac{1}{r} - 1}{\sqrt{1 + \frac{2\,m\,L^2}{(G\,M\,m^2)^2}\,E}} \tag{10.109}$$

$$= \frac{p\,\frac{1}{r} - 1}{\epsilon} \tag{10.110}$$

$$r(\varphi) = \frac{p}{1 + \epsilon\,\cos\varphi} \tag{10.111}$$

Offensichtlich erhalten wir als Lösung Bahnkurven vom Typ der Kegelschnitte (10.94), deren grundlegende Kurvenformen durch den Parameter ϵ (10.107) festgelegt werden.

Da der Parameter k (10.106) immer positiv ist ($k > 0$), ist allein das Vorzeichen der Gesamtenergie E des Himmelskörpers entscheidend:

Die Gesamtenergie E des Himmelskörpers bestimmt die Form der Bahnkurve.

Außerdem ist mit dem Vorzeichen der Energie E festgelegt, ob der Himmelskörper im Sonnensystem gebunden ist oder lediglich einen Streuprozess erfährt. Der Parameter p bestimmt die genaue Form der Bahnkurve. Im Folgenden wollen wir die verschiedenen Bahnformen und ihre Energien genauer untersuchen.

10.4.4 Bindungszustände und Streuprozesse

In ► Abbildung 10.17 ist das effektive Potential (10.76) für Himmelskörper als Funktion des Abstands r von der Sonne gezeigt. Zusätzlich sind verschiedene Projektionen möglicher Bahnkurven $r(\varphi)$ (10.111) der Himmelskörper eingezeichnet, die von der Energie E des Himmelskörpers abhängen.

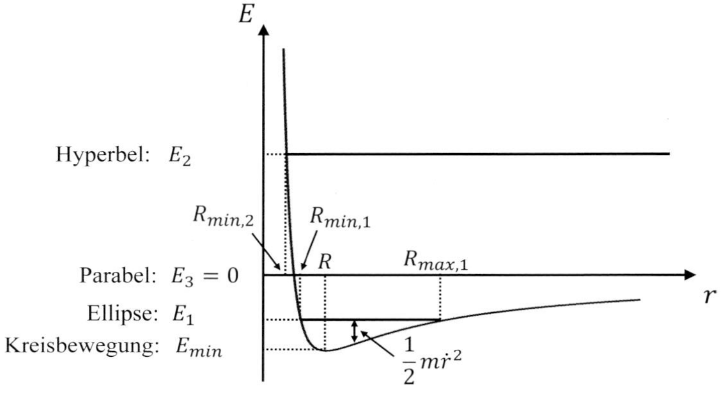

Abbildung 10.17 Effektives Potential (10.76) für Himmelskörper im Gravitationsfeld der Sonne. Eingezeichnet sind Energien E für die möglichen Bahnkurven mit den zugehörigen Radien R: Hyperbeln ($E_2 > 0, R_2 > R_{\text{min},2}$), Parabel ($E_3 = 0$, R_{min} ist der Schnittpunkt der Potentialkurve mit der r-Achse), Ellipsen ($E_{\text{min}} < E_1 < 0$, $R_{\text{min},1} < R_1 < R_{\text{max},1}$), Kreis ($E = E_{\text{min}}, R$).

► Tabelle 10.1 zeigt den zugehörigen Parameter ϵ, den wir weiter unten im Detail für jede Bahnkurvenform ausführen. Bei den Erläuterungen empfehlen wir, die Bahnkurve $r(\varphi)$ (10.111) sowie die Kurven der Kegelschnitte nach den ► Abbildungen 10.14 und 10.15 im Blick zu behalten.

	Hyperbel	Parabel	Ellipse	Kreis
E	$E > 0$	$E = 0$	$E < 0$	$E = -1/k$
$\epsilon^2 = 1 + kE$	$\epsilon^2 > 1$	$\epsilon^2 = 1$	$0 < \epsilon^2 < 1$	$\epsilon^2 = 0$

Tabelle 10.1 Energie von Himmelskörpern im Gravitationsfeld der Sonne und Parameter ihrer Bahnkurven $r(\varphi) = p/(1 + \epsilon \cos\varphi)$ (10.111)

Streuprozess Hyperbel

Für eine positive Gesamtenergie $E > 0$ ist der Parameter ϵ (10.107)

$$\epsilon^2 = 1 + k\,E > 1\,. \tag{10.112}$$

Die Bahnkurve $r(\varphi)$ ist nach der Bedingung (10.89) eine Hyperbel: Der Körper kommt aus dem Unendlichen, wird im Gravitationspotential gestreut und verschwindet wieder im Unendlichen (► Abbildung 10.17, Himmelskörper mit Energie E_2). Die minimale Entfernung zum Streuzentrum ergibt sich aus $r(\varphi = 0)$ (10.111). Unter Verwendung von p, k, und ϵ ((10.105) – (10.107)) finden wir

$$R_{\text{min}} = r(\varphi = 0) = \frac{p}{1 + \epsilon \cos(0)} = \frac{1}{1 + \sqrt{1 + k\,E}} \cdot \frac{L^2}{G\,M\,m^2}\,. \tag{10.113}$$

Streuprozess Parabel

Für den Fall, dass die Gesamtenergie $E = 0$ ist, ist der Parameter ϵ (10.107)

$$\epsilon^2 = 1 + k\,E = 1\,. \tag{10.114}$$

Die zugehörige Bahnkurve $r(\varphi)$ ist nach (10.91) eine Parabel. Auch hier handelt es sich um einen Streuprozess: Der Körper kommt aus dem Unendlichen, wird im Gravitationspotential gestreut und verschwindet wieder im Unendlichen (► Abbildung 10.17, Himmelskörper mit Energie E_3). Die minimale Entfernung zum Streuzentrum ergibt sich aus $r(\varphi = 0)$ (10.111) mit p (10.105):

$$R_{\text{min}} = r(\varphi = 0) = \frac{p}{1 + \epsilon \cos(0)} = \frac{1}{2} \cdot \frac{L^2}{G\,M\,m^2} \tag{10.115}$$

Bindungszustand Ellipse

Für den Fall, dass die Gesamtenergie $E < 0$ ist, ergibt sich der Parameter ϵ (10.107) zu

$$\epsilon^2 = 1 + k\,E < 1\,. \tag{10.116}$$

Aus der Bedingung (10.80) lesen wir ab, dass die Bahn eine Ellipse ist. Die minimale und die maximale Entfernung zum Zentrum ergeben sich aus $r(\varphi)$ (10.111) für $\varphi = 0$ und $\varphi = \pi$, d. h. $\cos 0 = 1$ und $\cos \pi = -1$. Mit p, k, und ϵ ((10.105) – (10.107)) erhalten wir

$$R_{\text{min}\,/\,\text{max}} = r(\varphi = 0/\pi) = \frac{p}{1 \pm \epsilon} = \frac{1}{1 \pm \sqrt{1 + k\,E}} \cdot \frac{L^2}{G\,M\,m^2}\,. \tag{10.117}$$

Die Entfernungen $R_{\text{min}\,/\,\text{max}}$ sind in ► Abbildung 10.17 für den Himmelskörper mit Energie E_1 eingezeichnet.

Bindungszustand Kreis

Ein Kreis ist laut (10.84) als Sonderfall der Ellipse durch $\epsilon = 0$ charakterisiert. Diese Bedingung erfordert, dass die Energie E des Himmelskörpers in ϵ (10.107) dem negativen Kehrwert des Parameters k (10.106) entspricht:

$$\epsilon = \sqrt{1 + k\,E} = 0 \tag{10.118}$$

$$E = -\frac{1}{k} = -\frac{(G\,M\,m^2)^2}{2\,m\,L^2} = -\frac{G^2 M^2 m^3}{2L^2} \tag{10.119}$$

Die Bahnkurve reduziert sich auf den Parameter p (10.105), woraus wir den Kreisradius berechnen:

$$r(\varphi) = \frac{p}{1 + \epsilon\,\cos\varphi} = p = \frac{L^2}{G\,M\,m^2} \tag{10.120}$$

In ► Abbildung 10.17 ist die Kreisbahn für den Himmelskörper mit der minimal möglichen Energie $E_{\min}$ eingezeichnet.

Beispiel 31: Planetenbahnen

Interessant ist, dass für die meisten Ellipsenbahnen der Planeten unseres Sonnensystems der Parameter $\epsilon \approx 0$ ist. Damit sind die Planetenbahnen annähernd kreisförmig. Große Werte für die Exzentrizität findet man bei manchen Kometenbahnen.

Planet	ϵ
Merkur	0,21
Venus	0,01
Erde	0,02
Mars	0,09
Jupiter	0,05
Saturn	0,06
Uranus	0,05
Neptun	0,01

Komet	ϵ
Halley	0,97
Encke	0,85

Beispiel 32: Bindungsenergie

Die Bindungsenergie der Erde im Sonnensystem können wir mit Gleichung (10.81) abschätzen. Dafür benötigen wir die große Halbachse a der Erdumlaufbahn sowie die Parameter p (10.105), k (10.106) und ϵ (10.107):

$$p = a\left(1-\epsilon^2\right) = a\left(1-(1+k\,E)\right) = -a\,k\,E \tag{10.121}$$

$$E = -\frac{1}{k}\frac{p}{a} = -\frac{\left(G\,M\,m^2\right)^2}{2\,m\,L^2}\frac{L^2}{G\,M\,m^2\,a} = -\frac{G\,M\,m}{2\,a} \tag{10.122}$$

$$= -\frac{6{,}7\,10^{-11}\cdot 2\,10^{30}\cdot 6\,10^{24}}{2\cdot 1{,}5\,10^{11}}\,\mathrm{N\,m} \approx -3\cdot 10^{33}\,\mathrm{J} \tag{10.123}$$

Damit sich die Erde aus dem Sonnensystem entfernen kann, müsste die entsprechende positive Energie aufgewendet werden, um als Gesamtenergie (10.75) mindestens $E = 0$ zu erreichen (► Abbildung 10.17).

10.5 Historischer Bezug

Die historische Entwicklung im Verständnis des Sonnensystems ist in Dokumentarfilmen eindrücklich nachvollziehbar, sie sind in Mediatheken zugänglich. Hier geht es um die drei in der Schule unterrichteten Kepler-Gesetze. Wir erklären sie mit den bislang vorgestellten physikalischen Konzepten.

10.5.1 Meilensteine

Wesentlich zum Verständnis des Sonnensystems beigetragen haben:

Nikolaus Kopernikus **(1473–1545)** setzte die Sonne ins Zentrum der Planetenbewegungen.

Tycho Brahe **(1546–1601)** führte eine genaue Vermessung der Planetenbewegungen durch.

Galileo Galilei **(1564–1642)** entdeckte Mondkrater und Jupitermonde mit seinem Teleskop.

Johannes Kepler **(1571–1630)** war Assistent Brahes und stellte die Kepler-Gesetze auf.

Isaac Newton **(1642–1727)** formulierte das Gravitationsgesetz.

10.5.2 Kepler-Gesetze

Die drei Kepler-Gesetze ordnen wir bereits kennengelernten physikalischen Konzepten zu.

> 1. Kepler-Gesetz: *Die Planeten bewegen sich auf Ellipsen, in deren Brennpunkt die Sonne steht (linke ► Abbildung 10.18).*

Diese Beobachtung haben wir in Abschnitt 10.4 mit den Konzepten der Energie- und Drehimpulserhaltung zusammen mit dem Gravitationspotential von Newton erklärt (10.116).

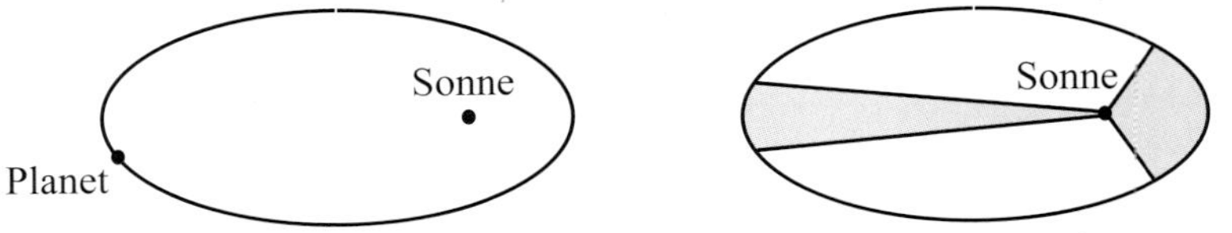

Abbildung 10.18 Links: 1. Keplersches Gesetz, Planetenbahnen sind Ellipsen mit der Sonne in einem der Brennpunkte. Die Exzentrizität ist in dieser und den folgenden Abbildungen zur Verdeutlichung übertrieben. Rechts: 2. Keplersches Gesetz, der Fahrstrahl von der Sonne zum Planeten überstreicht in gleichen Zeiten gleiche Flächen.

2. Kepler-Gesetz: *Der Radiusvektor (auch Fahrstrahl genannt) von der Sonne zu einem Planeten überstreicht in gleichen Zeiten gleiche Flächen (rechte ► Abbildung 10.18).*

Dieses Flächengesetz können wir in den Zusammenhang mit dem Drehimpuls $\vec{L} = \vec{r} \times \vec{p} = m\,\vec{r} \times \vec{v}$ bringen. Bewegt sich ein Planet auf der Bahnkurve s (► Abbildung 10.19), dann bleibt die Fläche dA wegen $\vec{L} = const.$ (10.64) konstant:

$$dA = \frac{1}{2}\,|\vec{r} \times d\vec{s}| = \frac{1}{2}\,|\vec{r} \times \vec{v}|\,dt = \frac{L}{2m}\,dt \tag{10.124}$$

$$\Rightarrow \frac{dA}{dt} = \frac{L}{2m} = const. \tag{10.125}$$

Das Gesetz entspricht demnach der Drehimpulserhaltung.

3. Kepler-Gesetz: *Die Quadrate der Umlaufzeiten T_i der Planeten verhalten sich wie die dritte Potenz ihrer großen Halbachsen a_i:*

$$\frac{T_i^2}{a_i^3} = const. \tag{10.126}$$

Dieses Gesetz lässt sich aus Newtons Gravitationskraft und der Zentripetalkraft der Planetenbewegung herleiten. Für den Spezialfall der Kreisbewegung haben wir das im Beispiel »Mondbewegung/Erdmasse« (10.11) gezeigt.

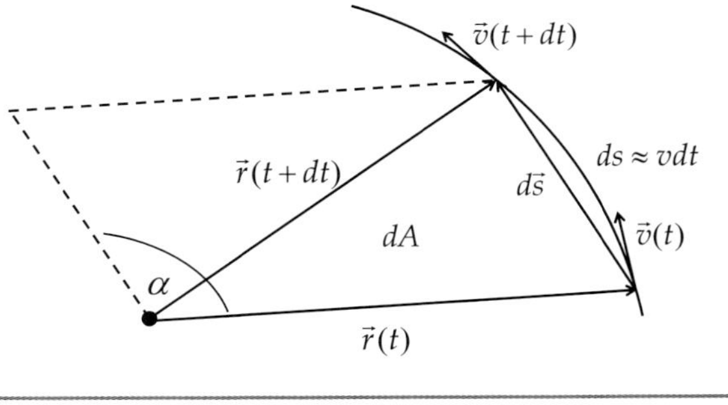

Abbildung 10.19 Zur Erklärung des 2. Keplerschen Gesetzes

Das Wichtigste zum Mitnehmen

Zum Mitnehmen: Gravitationswechselwirkung nach Newton 1

- Zwischen zwei Körpern mit den Massen M und m im Abstand r wirkt die anziehende Newtonsche **Gravitationskraft** $\vec{F}$:

$$\vec{F} = -G \frac{m\,M}{r^2}\,\vec{e}_r \tag{10.1}$$

Die **Gravitationskonstante** G ist eine wichtige konstante Größe der Physik mit dem Wert

$$G = (6{,}67430 \pm 0{,}00015) \cdot 10^{-11}\,\frac{\text{Nm}^2}{\text{kg}^2}\,. \tag{10.3}$$

Die Gravitationskraft ist eine **konservative Kraft**.

Der Zusammenhang mit der Erdbeschleunigung g ergibt sich aus der Erdmasse M_{Erde} und dem Erdradius R_{Erde}:

$$g \equiv \frac{G \cdot M_{\text{Erde}}}{R^2_{\text{Erde}}} \approx 9{,}8\,\frac{\text{m}}{\text{s}^2} \tag{10.4}$$

- Zur konservativen Gravitationskraft (10.1) gehört das **Gravitationspotential** $V(r)$ eines Körpers der Masse M:

$$V(r) = -\frac{G\,M}{r} \tag{10.48}$$

V ist ein **Zentralpotential**, das nur vom Abstand r abhängt, mit Sphären als Äquipotentialflächen.

Mit dem **Gradienten grad** im kartesischen System

$$\text{grad} = \vec{e}_x \frac{\partial}{\partial x} + \vec{e}_y \frac{\partial}{\partial y} + \vec{e}_z \frac{\partial}{\partial z} \tag{9.13}$$

und entsprechenden Ausdrücken (9.33) für Zylinder- und (9.52) für Kugelkoordinaten können wir die Gravitationskraft und das Gravitationspotential miteinander verknüpfen:

$$\frac{\vec{F}}{m} = -\,\text{grad}\,V \tag{10.59}$$

Zum Mitnehmen: Gravitationswechselwirkung nach Newton 2

- Die **potentielle Energie** eines Körpers im Gravitationsfeld einer Masse M ergibt sich aus seiner Masse m und dem Gravitationspotential V (10.48) mit $E_{\text{pot}} = 0$ für $r \to \infty$:

$$E_{\text{pot}} = m\,V = -\frac{G\,M\,m}{r} \tag{10.62}$$

Mit der Drehimpulserhaltung $\vec{L} = \textit{const.}$ kann man die Abhängigkeiten der **kinetischen Energie** des Körpers auf die Radialkomponente r und ihre zeitliche Ableitung $\dot{r}$ reduzieren:

$$E_{\text{kin}} = \frac{1}{2}\,m\,\dot{r}^2 + \frac{L^2}{2\,m\,r^2} \tag{10.73}$$

Damit erhalten wir als Gesamtenergie aus (10.73) und (10.62)

$$\begin{aligned} E &= E_{\text{kin}} + E_{\text{pot}} = \textit{const.} \\ &= \underbrace{\frac{1}{2}m\dot{r}^2}_{\text{radiale kinetische Energie}} + \underbrace{\left[\frac{L^2}{2m^2r^2} - \frac{G\,M}{r}\right]}_{\text{effektives Potential } V_{\text{eff}}} m\,. \end{aligned} \tag{10.75}$$

Ohne äußeren Einfluss ist diese Gesamtenergie erhalten.

- Die Lösungen der Differentialgleichung (10.75) sind die möglichen **Bahnkurven** von Körpern im Zentralpotential. Sie haben die Form von Kegelschnitten in Polarkoordinaten:

$$r\,(\varphi) = \frac{p}{1 + \epsilon\,\cos\varphi} \tag{10.111}$$

Der Parameter ϵ entscheidet über die Kurvenform. Mit

$$\epsilon \equiv \sqrt{1 + k\,E} \tag{10.107}$$

und wegen $k > 0$ wird diese wiederum vom Vorzeichen und von der Größe der Energie E bestimmt:

	Hyperbel	Parabel	Ellipse	Kreis
E	$E > 0$	$E = 0$	$E < 0$	$E = -1/k$
$\epsilon^2 = 1 + k\,E$	$\epsilon^2 > 1$	$\epsilon^2 = 1$	$0 < \epsilon^2 < 1$	$\epsilon^2 = 0$

Die Gesamtenergie E des Körpers bestimmt die Form der Bahnkurve.

Über die genaue Form der Kurven entscheiden die Parameter

$$p \equiv \frac{L^2}{G\,M\,m^2} \quad (10.105) \qquad \text{und} \qquad k \equiv \frac{2\,m\,L^2}{(G\,M\,m^2)^2} > 0\,. \quad (10.106)$$

Zum Mitnehmen: Gravitationswechselwirkung nach Newton 3

Streuprozess Hyperbel

Für eine positive Gesamtenergie $E > 0$ ist mit (10.107)

$$\epsilon^2 = 1 + k\,E > 1\,. \tag{10.112}$$

Die Bahnkurve ist nach (10.89) eine Hyperbel. Der Körper wird aus dem Unendlichen vom Gravitationspotential ins Unendliche gestreut. Die minimale Entfernung zum Streuzentrum ist mit (10.111)

$$R_{\min} = r(\varphi = 0) = \frac{p}{1+\epsilon} = \frac{1}{1+\sqrt{1+k\,E}} \cdot \frac{L^2}{G\,M\,m^2}\,. \tag{10.113}$$

Streuprozess Parabel

Für die Gesamtenergie $E = 0$ ist der Parameter ϵ (10.107)

$$\epsilon^2 = 1 + k\,E = 1\,. \tag{10.114}$$

Die Bahnkurve $r(\varphi)$ ist nach (10.93) eine Parabel. Auch hier wird der Körper aus dem Unendlichen ins Unendliche gestreut. Die minimale Entfernung zum Streuzentrum ist jetzt

$$R_{\min} = r(\varphi = 0) = \frac{p}{1+\epsilon} = \frac{1}{2} \cdot \frac{L^2}{G\,M\,m^2}\,. \tag{10.115}$$

Bindungszustand Ellipse

Für eine negative Gesamtenergie $E < 0$ ist

$$\epsilon = \sqrt{1 + k\,E} < 1\,. \tag{10.116}$$

Nach der Bedingung (10.80) ist die Bahn eine Ellipse mit dem Halbparameter p. Die minimale und die maximale Entfernung zum Zentrum ergeben sich aus $r(\varphi)$ (10.111) für $\varphi = 0$ und $\varphi = \pi$:

$$R_{\min\,/\,\max} = r(\varphi = 0/\pi) = \frac{p}{1 \pm \epsilon} = \frac{1}{1 \pm \sqrt{1+k\,E}} \cdot \frac{L^2}{G\,M\,m^2} \tag{10.117}$$

Zum Mitnehmen: Gravitationswechselwirkung nach Newton 4

- **Bindungszustand Kreis**

  Ein Kreis ist laut (10.84) als Sonderfall der Ellipse durch $\epsilon = 0$ charakterisiert:

$$\epsilon = \sqrt{1 + k\,E} = 0 \quad \Rightarrow \tag{10.118}$$

$$E = -\frac{1}{k} = -\frac{G^2 M^2 m^3}{2L^2} = E_{\min} \tag{10.119}$$

  $E_{\min}$ ist die Energie im Minimum des effektiven Potentials mit festem Radius $r = const.$. Die Bahnkurve reduziert sich auf den Parameter p (10.105), woraus wir den Kreisradius berechnen:

$$r(\varphi) = \frac{p}{1 + \epsilon\,\cos\varphi} = p = \frac{L^2}{G\,M\,m^2} \tag{10.120}$$

- Die drei Keplerschen Gesetze fanden in der Newtonschen Mechanik ihre Erklärung:

  – **1. Kepler-Gesetz:** *Die Planeten bewegen sich auf Ellipsen, in deren Brennpunkt die Sonne steht.*

    Diese Beobachtung haben wir mit den Konzepten der Energie- und Drehimpulserhaltung zusammen mit dem Gravitationspotential von Newton erklärt.

  – **2. Kepler-Gesetz:** *Der Radiusvektor (auch Fahrstrahl genannt) von der Sonne zu einem Planeten überstreicht in gleichen Zeiten gleiche Flächen.*

    Dieses Flächengesetz entspricht der Drehimpulserhaltung.

  – **3. Kepler-Gesetz:** *Die Quadrate der Umlaufzeiten T_i der Planeten verhalten sich wie die dritte Potenz ihrer großen Halbachsen a_i:*

$$\frac{T_i^2}{a_i^3} = const. \tag{10.126}$$

    Dieses Gesetz lässt sich für eine Kreisbewegung einfach aus der Gravitationskraft und der Zentripetalkraft herleiten (siehe Beispiel 28 in Abschnitt 10.1).

Transformation zwischen Bezugssystemen

11

11.1 **Inertialsystem** 211

11.2 **Galileitransformation** 211

11.3 **Beschleunigte Bezugssysteme** 213

11.3.1 Gradlinige, gleichförmig beschleunigte Bewegung 213

11.3.2 Rotierendes Bezugssystem 214

11.3.3 Erdrotation: Foucaultsches Pendel 218

11.4 **Lorentztransformation** 220

11.4.1 Bewegte Lichtquelle: Einsteins Postulat 221

11.4.2 Galileitransformation einer Kugelwelle 222

11.4.3 Lorentztransformation einer Kugelwelle 222

ÜBERBLICK

» Physikalische Gesetze sollen unabhängig davon gelten, wo wir die entsprechenden Experimente dazu durchführen. Wir werden zunächst die Bedingungen für geeignete Bezugssysteme festlegen (Inertialsysteme) und die Galileitransformationen zwischen solchen Koordinatensystemen kennenlernen.

Anschließend werden wir Verfahren kennenlernen, mit denen wir auch in anderen Bezugssystemen konsistent physikalische Gesetzmäßigkeiten beschreiben können. Dabei werden wir sogenannte Scheinkräfte wie z. B. die Zentrifugalkraft und die Corioliskraft kennenlernen.

In einem weiteren Schritt werden wir Transformationen zwischen den Koordinatensystemen sehr schnell bewegter Bezugssysteme durchführen, bei denen wir die Lorentztransformation benötigen. «

11.1 Inertialsystem

Ein Inertialsystem ist ein Bezugssystem, das keine Beschleunigung erfährt. In jedem Inertialsystem finden wir dieselben physikalischen Gesetze.

Durch ihre Rotation ist die Erde als unser natürliches Bezugssystem kein Inertialsystem. Für viele unserer Experimente ist die Beschleunigung durch die Erdrotation jedoch unerheblich und die Ergebnisse sind so gut wie in einem Inertialsystem. Mit dem Experiment des Foucault-Pendels werden wir allerdings den Beweis nachvollziehen, dass die Erde kein Inertialsystem ist.

Bessere Näherungen für Inertialsysteme sind unser Sonnensystem oder die Milchstraße.

11.2 Galileitransformation

Die Bewegung eines Massenpunkts können wir durch die Ortsfunktion $\vec{r}(t)$ in einem Inertialsystem O beschreiben (► Abbildung 11.1). In einem zweiten Inertialsystem O', das sich mit der Geschwindigkeit $\vec{u} = const.$ relativ zu O bewegt, wird die Bewegung des Massenpunkts durch $\vec{r}'(t)$ beschrieben.

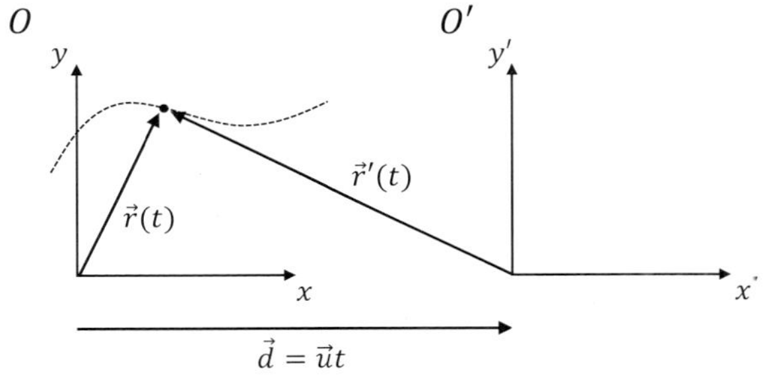

Abbildung 11.1 Zwei Inertialsysteme O und O' bewegen sich mit der Geschwindigkeit $\vec{u}$ relativ zueinander.

Um zwischen den Bezugssystemen hin und her zu transformieren, müssen wir die relative Geschwindigkeit $\vec{u}$ zwischen den beiden Systemen berücksichtigen:

$$\begin{aligned} \vec{r}' &= \vec{r} - \vec{u}\cdot t \\ \vec{v}' &= \vec{v} - \vec{u} \\ \vec{a}' &= \vec{a} \\ t' &= t \end{aligned} \qquad (11.1)$$

$$\begin{aligned} \vec{r} &= \vec{r}' + \vec{u}\cdot t \\ \vec{v} &= \vec{v}' + \vec{u} \\ \vec{a} &= \vec{a}' \\ t &= t' \end{aligned} \qquad (11.2)$$

Den Zusammenhang zwischen $\vec{r}$ und $\vec{r}'$ können wir aus ► Abbildung 11.1 ablesen, die Geschwindigkeiten und Beschleunigungen ergeben sich mit $t = t'$ durch Differentiation.

Die Beschleunigungen, die in Inertialsystemen gemessen werden, sind also gleich groß. Daher sind auch die zugehörigen Kräfte gleich groß:

$$\vec{F} = \vec{F}\,' \tag{11.3}$$

Objekte und ihre Bewegungen können also mithilfe der »Galileitransformationen« (11.1) und (11.2) gleichermaßen in verschiedenen Inertialsystemen beschrieben werden.

Beispiel 33: Galileitransformation: Wurf

Die Transformation zwischen zwei Bezugssystemen kann bei einem horizontalen Wurf mit der Startgeschwindigkeit $\vec{v}_\circ = v_\circ\,\vec{e}_x$ verdeutlicht werden. Wählt man ein Koordinatensystem O' so, dass es sich mit der Geschwindigkeit $\vec{u} = v_\circ\,\vec{e}_x$ relativ zu O bewegt (► Abbildung 11.2), so erscheint der Wurf in dem Koordinatensystem O' als freier Fall (► Abbildung 11.3).

In beiden Bezugssystemen des Beispiels gilt natürlich trotzdem dasselbe Fallgesetz (1.33), wie wir aus den y-Komponenten der Fallbewegung ablesen können. In unserem Beispiel sind die Geschwindigkeitskomponenten in (11.1) $u_x = v_\circ$ und $u_y = 0$:

System O :	$x = v_\circ t$	$y = h - \frac{1}{2}gt^2$
System O' :	$x' = x - u_x t$ $= v_\circ t - v_\circ t = 0$	$y' = h - \frac{1}{2}gt^2 - u_y t$ $= h - \frac{1}{2}gt^2$

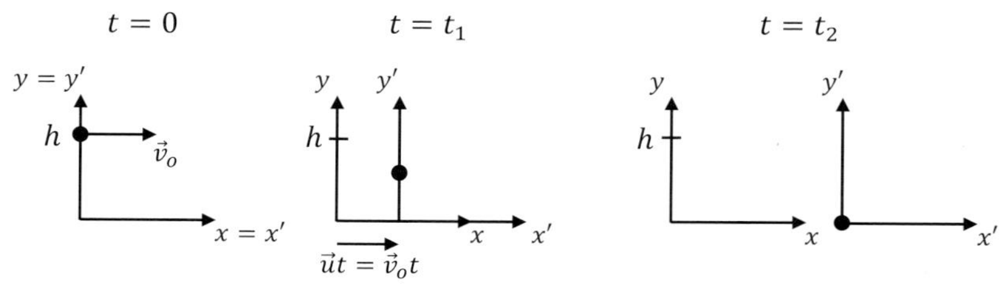

Abbildung 11.2 Das System (x', y') bewegt sich mit der Geschwindigkeit $\vec{v}_\circ$ in x-Richtung gegen das System (x, y).

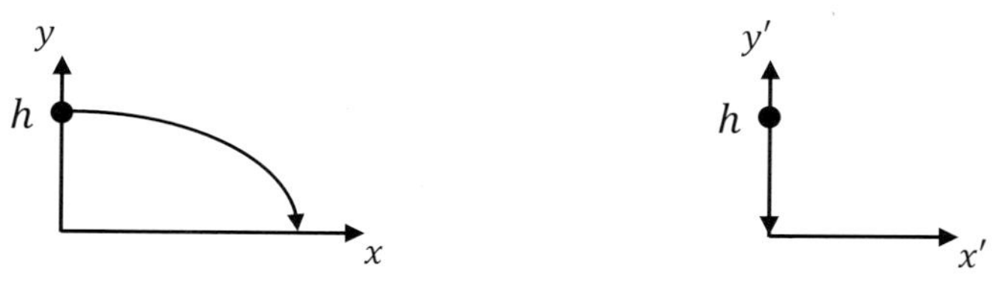

Abbildung 11.3 Ein horizontaler Wurf mit der Anfangsgeschwindigkeit $\vec{v}_\circ$ im System (x, y) (links) erscheint im System (x', y') als freier Fall (rechts).

Experiment 17: Galileitransformation

Ein Tischtennisball wird mit einer Spiralfeder senkrecht nach oben katapultiert und am selben Ort wieder eingefangen (senkrechter Wurf). Montiert man dieses Abschusssystem auf den Wagon einer Spielzeugeisenbahn und lässt diese mit konstanter Geschwindigkeit fahren, erweist sich im Laborsystem der senkrechte Abschuss als schiefer Wurf. Auch hier wird der Ball wieder eingefangen, da Ball und Eisenbahn dieselbe horizontale Geschwindigkeit haben. Im Bezugssystem der Eisenbahn wird natürlich weiterhin der senkrechte Wurf beobachtet.

11.3 Beschleunigte Bezugssysteme

In diesem Abschnitt wollen wir die Ursache von Scheinkräften untersuchen, die durch beschleunigte Bezugssysteme zustande kommen. Dazu gehören Trägheitskräfte wie die Zentrifugalkraft und die Corioliskraft.

11.3.1 Gradlinige, gleichförmig beschleunigte Bewegung

Ein Beobachter in einem beschleunigten Bezugssystem beschreibt Kräfte auf seinen und andere Körper, die allerdings nur aufgrund seiner eigenen Bewegung zustandekommen. Solche Kräfte bezeichnen wir als Scheinkräfte, da sie nicht auf einer fundamentalen Wechselwirkung beruhen, sondern durch ein beschleunigtes Bezugssystem verursacht werden.

Beispiel 34: Scheinkraft

Auf dem Billardtisch in einer luxuriösen Eisenbahn liegt reibungsfrei eine Kugel (► Abbildung 11.4). Beim Anfahren im Bahnhof beschleunige der Zug mit konstanter Beschleunigung $\vec{a} = const.$. Im Inertialsystem O des Bahnsteigs bleibt die Kugel aufgrund ihrer Trägheit liegen.

Für den Beobachter im Zug sieht es so aus, als wirke eine Kraft auf die Kugel, die sie beschleunigt. Tatsächlich ist aber die Ursache für die scheinbare Bewegung der Kugel die Beschleunigung des Systems O' des anfahrenden Zugs.

Wir sprechen hier von einer Scheinkraft, da sie nicht durch fundamentale Wechselwirkung, sondern durch das beschleunigte Bezugssystem und die Trägheit der Kugel entsteht.

Der Vergleich der beiden Bezugssysteme ergibt Folgendes für die jeweiligen Beschreibungen der Kugel durch Vektoren $\vec{r}$ und $\vec{r}'$ mit ihren Komponenten x, y und x' y':

System O :	$x(t = t_\circ) = x$	$y(t = t_\circ) = y$
	$x(t > t_\circ) = x$	$y(t > t_\circ) = y$
System O' :	$x'(t = t_\circ) = x$	$y(t = t_\circ) = y$
	$x'(t > t_\circ) = x - \frac{1}{2}\,a\,t^2$	$y(t > t_\circ) = y$

Im System $O(x, y)$ ruht die Kugel, im System O' dagegen wird sie scheinbar gleichmäßig beschleunigt (1.33).

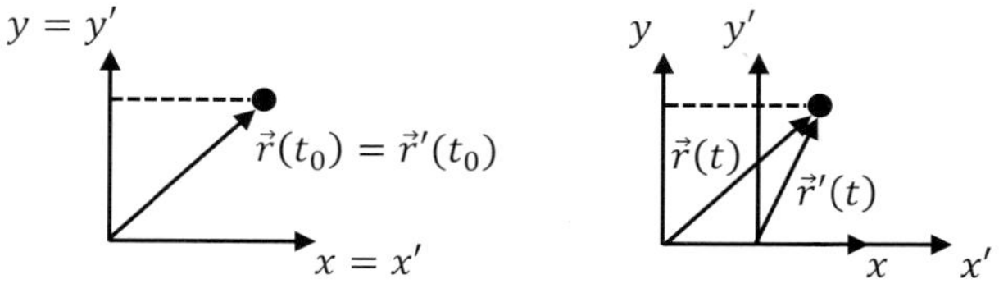

Abbildung 11.4 System $O(x, y)$: ruhender Bahnsteig, System $O'(x', y')$: anfahrender Zug. Links: Abfahrt bei $t = t' = t_\circ$, rechts: $t = t' > t_\circ$

11.3.2 Rotierendes Bezugssystem

Sei O ein Inertialsystem und O' ein mit konstanter Winkelgeschwindigkeit $\vec{\omega}$ um die z-Achse rotierendes Bezugssystem (▶ Abbildung 11.5). Die Winkelgeschwindigkeit $\vec{\omega}$ wird dabei im System O gemessen.

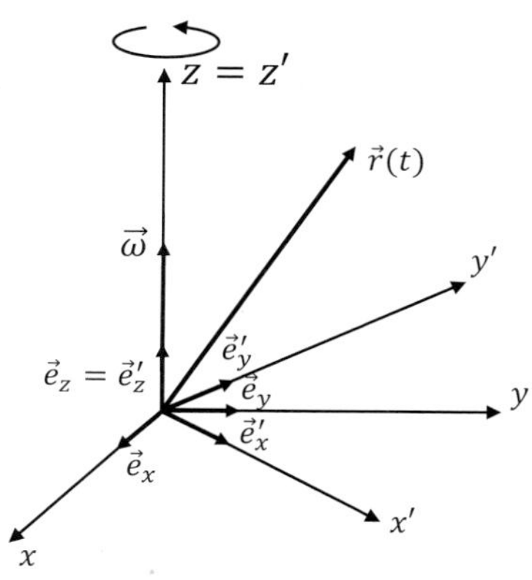

Abbildung 11.5 Das System $O'(x', y', z')$ rotiert mit der Winkelgeschwindigkeit $\vec{\omega}$ um die z-Achse des Inertialsystems $O(x, y, z)$.

Der Ortsvektor eines Massenpunkts zum Zeitpunkt t lautet im System O

$$\vec{r}\,(t) = x\,(t)\;\vec{e}_x + y\,(t)\;\vec{e}_y + z\,(t)\;\vec{e}_z\,. \tag{11.4}$$

In O' ist $\vec{r}\,'$ derselbe Vektor, er wird durch andere Koordinaten dargestellt:

$$\vec{r}\,'(t) = \vec{r}\,(t) \tag{11.5}$$

$$= x'(t)\,\vec{e}_{x'} + y'(t)\,\vec{e}_{y'} + z'(t)\,\vec{e}_{z'} \tag{11.6}$$

Für die Transformation der Geschwindigkeit und der Beschleunigung des Massenpunkts müssen wir zwischen den internen Bewegungen im rotierenden System O' und der Gesamtbewegung im Inertialsystem O unterscheiden. Dazu berücksichtigen wir die Rotation der Einheitsvektoren des Systems O'. Zur Eindeutigkeit benutzen wir dazu für die zeitlichen Ableitungen die Bezeichnungen

$$\frac{d_I}{dt} \quad \text{im Inertialsystem und} \quad \frac{d_R}{dt} \quad \text{im Rotationssystem.} \tag{11.7}$$

Die folgenden Berechnungen erscheinen zunächst aufwendig, sind aber konzeptionell nicht schwierig (Befolgen der Ableitungsregeln und Kenntnis des allgemeinen Zusammenhangs $\vec{v} = \vec{\omega} \times \vec{r}$ (4.48) bei der Kreisbewegung). Konzentration ist dafür erforderlich, welches Bezugssystem gerade betrachtet wird.

Transformation der Geschwindigkeit

Ein Beobachter in O nimmt die Geschwindigkeit eines Massenpunkts als

$$\frac{d_I}{dt}\vec{r} = \vec{v}(t) = \dot{x}\,\vec{e}_x + \dot{y}\,\vec{e}_y + \dot{z}\,\vec{e}_z \tag{11.8}$$

wahr. Die Geschwindigkeit im rotierenden System O' ist entsprechend

$$\frac{d_R}{dt}\vec{r} = \frac{d_R}{dt}\vec{r}\,' = \vec{v}\,'(t) = \dot{x}'\,\vec{e}_{x'} + \dot{y}'\,\vec{e}_{y'} + \dot{z}'\,\vec{e}_{z'}\,. \tag{11.9}$$

Um die Transformation der Geschwindigkeit zu berechnen, lassen wir den Beobachter im System O die Geschwindigkeit des Massenpunkts in den Koordinaten von O' beschreiben. Die Geschwindigkeit unter Beachtung der Produktregel bei der zeitlichen Ableitung von (11.6) ist

$$\frac{d_I}{dt}\vec{r}\,' = \underbrace{\dot{x}'\,\vec{e}_{x'} + \dot{y}'\,\vec{e}_{y'} + \dot{z}'\,\vec{e}_{z'}}_{\vec{v}\,'} + \underbrace{x'\,\frac{d\vec{e}_{x'}}{dt} + y'\,\frac{d\vec{e}_{y'}}{dt} + z'\,\frac{d\vec{e}_{z'}}{dt}}_{\vec{u}} \tag{11.10}$$

$$= \vec{v}\,' + \vec{u}\,. \tag{11.11}$$

Die Relativgeschwindigkeit $\vec{u}$ der beiden Systeme können wir für die gleichförmige Rotationsbewegung über die Winkelgeschwindigkeit $\vec{\omega}$ und den Ortsvektor des Massenpunkts $\vec{r}$ berechnen.

Bei der Rotation führt die Spitze des Einheitsvektors $\vec{e}_{x'}$ in ► Abbildung 11.5 eine Bewegung mit der Geschwindigkeit $d\vec{e}_{x'}/dt$ durch, die wir aus $\vec{\omega} \times \vec{e}_{x'}$ (4.48) berechnen können. Die zeitlichen Ableitungen der Einheitsvektoren können wir also folgendermaßen umschreiben:

$$\frac{d\vec{e}_{x'}}{dt} = \vec{\omega} \times \vec{e}_{x'} \qquad \frac{d\vec{e}_{y'}}{dt} = \vec{\omega} \times \vec{e}_{y'} \qquad \frac{d\vec{e}_{z'}}{dt} = \vec{\omega} \times \vec{e}_{z'} \tag{11.12}$$

Mit diesen Zusammenhängen und den Gleichungen (11.10) und (11.5) können wir für die Relativgeschwindigkeit $\vec{u}$ der beiden Bezugssysteme

$$\vec{u} = \vec{\omega} \times (x'\vec{e}_{x'}) + \vec{\omega} \times (y'\vec{e}_{y'}) + \vec{\omega} \times (z'\vec{e}_{z'}) \tag{11.13}$$

$$= \vec{\omega} \times \vec{r}\,' \tag{11.14}$$

$$= \vec{\omega} \times \vec{r} \tag{11.15}$$

schreiben. Eingesetzt in (11.11) ergibt sich für die Transformation der Geschwindigkeit des Massenpunkts:

$$\underbrace{\frac{d_I}{dt}\vec{r} = \vec{v}}_{\text{Geschwindigkeit im System } O} \quad = \quad \underbrace{\frac{d_R}{dt}\vec{r} = \vec{v}\,'}_{\text{Geschwindigkeit im System } O'} \quad + \quad \underbrace{\vec{\omega} \times \vec{r}}_{\text{Rotation des Systems } O'} \tag{11.16}$$

Der Zusammenhang (11.16) wurde für die Geschwindigkeiten gezeigt, die Herleitung gilt aber allgemein für die Ableitungen beliebiger Vektoren und wird im Folgenden sowohl für die Geschwindigkeit $\vec{v}$ als auch ihre Ableitungen verwendet.

Transformation der Beschleunigung

Die Transformation zwischen Zeitableitungen von Vektoren des Inertial- und rotierenden Systems (11.16) wenden wir jetzt auf die Geschwindigkeit an, d. h. wir ersetzen $\vec{r}$ durch $\vec{v}$:

$$\frac{d_I}{dt}\vec{v} = \frac{d_R}{dt}\vec{v} + (\vec{\omega} \times \vec{v}) \tag{11.17}$$

Einsetzen der Transformationsgleichung $\vec{v} = \vec{v}\,' + \vec{\omega} \times \vec{r}$ (11.16) ergibt

$$\frac{d_I}{dt}\vec{v} = \frac{d_R}{dt}(\vec{v}\,' + \vec{\omega} \times \vec{r}) + \omega \times (\vec{v}\,' + \vec{\omega} \times \vec{r})\,. \tag{11.18}$$

Wir identifizieren jetzt $d_I\vec{v}/dt = \vec{a}_I = \vec{a}$ als die Beschleunigung im Inertialsystem und $d_R\vec{v}\,'/dt = \vec{a}_R = \vec{a}\,'$ als die Beschleunigung im Rotationssystem und erhalten

$$\vec{a} = \vec{a}\,' + \frac{d_R}{dt}(\vec{\omega} \times \vec{r}) + (\vec{\omega} \times \vec{v}\,') + \omega \times (\vec{\omega} \times \vec{r}) \tag{11.19}$$

$$= \vec{a}\,' + \left(\frac{d_R}{dt}\vec{\omega} \times \vec{r}\right) + \left(\vec{\omega} \times \frac{d_R}{dt}\vec{r}\right) + (\vec{\omega} \times \vec{v}\,') + \omega \times (\vec{\omega} \times \vec{r})\,. \tag{11.20}$$

Bei der konstanten Rotationsgeschwindigkeit $\vec{\omega}$ und mit $d_R\vec{r}/dt = \vec{v}\,'$ ergibt sich dann

$$\vec{a} = \vec{a}\,' + 2(\vec{\omega} \times \vec{v}\,') + \vec{\omega} \times (\vec{\omega} \times \vec{r})\,. \tag{11.21}$$

Das Umstellen der Terme erleichtert uns die Interpretation dieser Transformationsvorschrift:

$$\vec{a}\,' = \vec{a} + \underbrace{2\,(\vec{v}\,' \times \vec{\omega})}_{\equiv \vec{a}_C} + \underbrace{\vec{\omega} \times (\vec{r} \times \vec{\omega})}_{\equiv \vec{a}_{ZF}} \tag{11.22}$$

Der zweite Term auf der rechten Seite ist die sogenannte Coriolisbeschleunigung $\vec{a}_C$, und der dritte Term ist die Zentrifugalbeschleunigung $\vec{a}_{ZF}$. Der rotierende Beobachter muss zusätzlich zur Beschleunigung, die in einem Inertialsystem gemessen wird, die Coriolisbeschleunigung $\vec{a}_C$ und die Zentrifugalbeschleunigung $\vec{a}_{ZF}$ berücksichtigen.

Selbst wenn im System O keine Beschleunigung des Massenpunkts $\vec{a} = 0$ gemessen wird, misst der Beobachter im System O' aufgrund seiner eigenen Rotationsbewegung eine von null verschiedene Beschleunigung $\vec{a}\,' \neq 0$ für den Massenpunkt in der Form der Coriolis- und Zentrifugalbeschleunigungen.

Durch Multiplikation mit der Masse m des Massenpunkts erhalten wir aus diesen Beschleunigungen die entsprechenden **Scheinkräfte**:

$$\begin{aligned} &\text{Corioliskraft} && \vec{F}_C = 2m(\vec{v}\,' \times \vec{\omega}) \\ &\text{Zentrifugalkraft} && \vec{F}_{ZF} = m\vec{\omega} \times (\vec{r} \times \vec{\omega}) \end{aligned} \tag{11.23}$$

Der Zusatz »Schein« soll hier auch darauf hindeuten, dass die Kräfte nicht auf fundamentalen Wechselwirkungen zwischen Körpern beruhen, sondern sich durch den Effekt der Messung in einem beschleunigten Bezugssystem ergeben.

Experiment 18: Der Fliehkraftregler

Ein Anwendungsbeispiel für rotierende Systeme ist der Fliehkraftregler. Dieses mechanische Gerät schaltet bei großen Drehzahlen ein Gerät aus und schützt so z. B. Motoren.

Die linke ► Abbildung 11.6 zeigt den Fliehkraftregler aus der Sicht eines Beobachters im Laborsystem. Die Kugel wird durch die Zentripetalkraft F_{ZP} auf eine Kreisbahn gezwungen. F_{ZP} setzt sich zusammen aus

$$\vec{F}_{ZP} = \vec{F}_G + \vec{F}_{Stab} \,. \tag{11.24}$$

Fliegt man im rotierenden System mit (rechte ► Abbildung 11.6), ruht die Kugel, d. h. die Summe aller an der Kugel angreifenden Kräfte verschwindet: $\sum_{i=1}^{n} \vec{F}_i = 0$. Demnach herrscht ein Kräftegleichgewicht. Aus der Skizze sehen wir, dass

$$\vec{F}_G + \vec{F}_{Stab} \neq 0$$

ist. Die resultierende Kraft aus diesen beiden Kräften wird durch die Zentrifugalkraft F_{ZF} kompensiert:

$$\vec{F}_G + \vec{F}_{Stab} = -\vec{F}_{ZF} \tag{11.25}$$

Dabei ist die Zentrifugalkraft $\vec{F}_{ZF}$ (11.23) entgegengesetzt gleich der Zentripetalkraft $\vec{F}_{ZP}$ (4.21):

$$\vec{F}_{ZF} = m\vec{\omega} \times (\vec{r} \times \vec{\omega}) = -\vec{F}_{ZP} \tag{11.26}$$

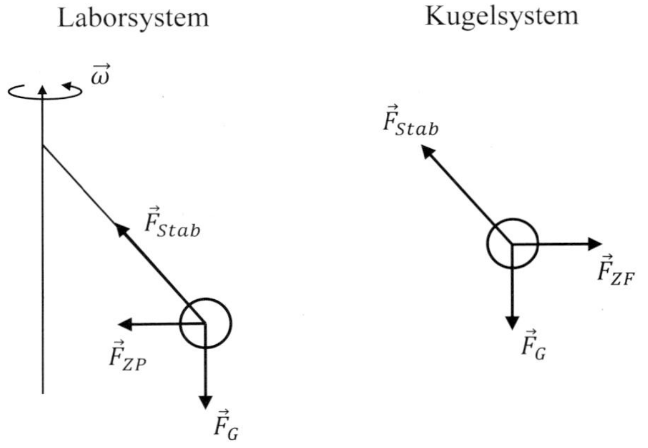

Abbildung 11.6 Fliehkraftregler: links im Laborsystem, rechts im mitbewegten Kugelsystem

11.3.3 Erdrotation: Foucaultsches Pendel

In einem eindrucksvollen Versuch gelang es dem französischen Physiker Jean Foucault 1851, die Erdrotation ohne Zusatzinformationen aus dem Universum mithilfe eines Pendels nachzuweisen. In einem Inertialsystem ist die Schwingungsebene eines Pendels durch die Anfangsauslenkung festgelegt und bleibt unverändert. Ein Pendel in einem rotierenden Bezugssystem erfährt zusätzlich die Corioliskraft mit der Konsequenz, dass sich die Pendelebene dreht.

Das Foucaultsche Pendel besteht aus einer Kugel an einem mehrere Meter langen Faden (große Schwingungsperiode) und einer kardanischen Aufhängung, sodass die Pendelebene sich problemlos drehen kann.

Bringen wir ein Pendel am Nordpol zum Schwingen, so bleibt von außerhalb der Erde gesehen die Schwingungsebene des Pendels fest, während sich die Erde darunter weiterdreht.

Ein Beobachter am Nordpol orientiert sich für die Zeit einer vollen Erdumdrehung an einem weit entfernten Fixstern. Sie dauert genau wie die Drehung der Pendelebene 23,93 h. Das ist etwas kürzer als unsere gewohnten 24 h des Sonnentags. Durch die elliptische Erdbahn um die Sonne benötigt die Erde etwas mehr als $360°$, damit ein fester Längengrad nach einer Erddrehung wieder der Sonne zugewandt ist.

Für den Beobachter am Nordpol wirkt der Vektor der Winkelgeschwindigkeit $\vec{\omega}$ der Erde maximal in der Corioliskraft $\vec{F}_C = 2m\vec{v}\,' \times \vec{\omega}$ (11.23). Die Geschwindigkeit $\vec{v}\,'$ beschreibt die Bewegung des Pendels im Erdsystem (► Abbildung 11.7 links). Schwingt das Pendel z. B. nach Osten ($\vec{v}_1'$), wird von oben betrachtet seine Bahn durch $\vec{F}_C$ im Uhrzeigersinn gedreht. Bei der Rückschwingung nach Westen ($\vec{v}_2'$) ist die Drehrichtung durch $\vec{F}_C$ ebenfalls im Uhrzeigersinn.

Am Südpol dreht sich die Pendelebene gegen den Uhrzeigersinn. Am Äquator findet keine Drehung statt.

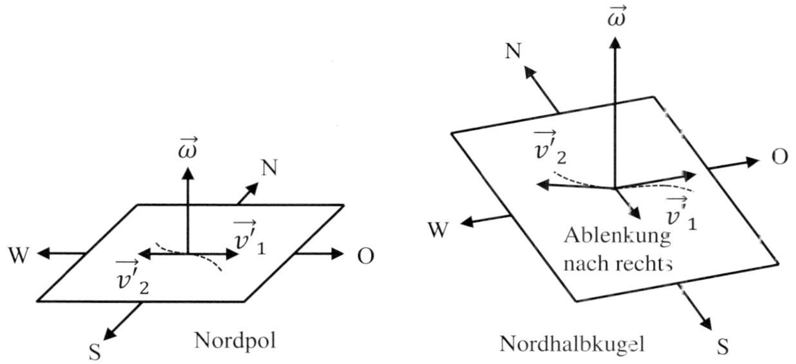

Abbildung 11.7 Foucaultsches Pendel am Nordpol und auf der Nordhalbkugel. $\vec{\omega}$ ist die Winkelgeschwindigkeit der Erddrehung, $\vec{v}_i'$ sind die Pendelgeschwindigkeiten im Erdsystem. Die Pendelbahnen sind gestrichelt.

In den Bereichen zwischen den Polen und dem Äquator dreht sich die Pendelebene langsamer als an den Polen. Entscheidend für die Drehgeschwindigkeit der Pendelebene ist die Winkelgeschwindigkeitskomponente $\vec{\tilde{\omega}}$ in radialer Richtung durch den Aufhängepunkt des Pendels (► Abbildung 11.8). Bezeichnen wir den Breitengrad mit λ ($\lambda = 0$ am Äquator, $\lambda = \pm\pi/2$ an den Polen), so ergibt sich die Projektion der Winkelgeschwindigkeit $\vec{\omega}$ der Erde zu

$$\vec{\tilde{\omega}} = \vec{\omega}\,\sin\lambda\,. \tag{11.27}$$

Die entsprechende Corioliskraft beträgt

$$\vec{F}_C = 2\,m\,\sin\lambda\,\left(\vec{v}\,' \times \vec{\omega}\right). \tag{11.28}$$

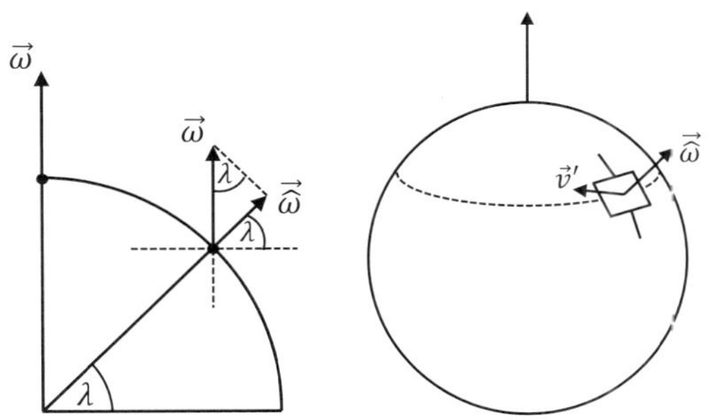

Abbildung 11.8 Links: Wirksame Komponente $\vec{\tilde{\omega}}$ der Winkelgeschwindigkeit $\vec{\omega}$ am Breitengrad λ, rechts: Foucaultsches Pendel in Aachen

Experiment 19: Foucaultsches Pendel in Aachen

Die Messung mit dem Foucaultschen Pendel in Aachen ergab eine Drehung der Pendelebene um $\Delta\varphi = 1{,}15°$ in $\Delta t = 345\,\text{s}$. Damit erhalten wir eine Winkelgeschwindigkeit von

$$\omega_{AC} = 2\,\pi\,\frac{1{,}15}{360}\,\frac{1}{345}\,\frac{\text{rad}}{\text{s}} = 5{,}8\,10^{-5}\,\frac{\text{rad}}{\text{s}}\,. \tag{11.29}$$

Am Nordpol beobachten wir eine Erdumdrehung um $360°$ relativ zu einem Fixstern, sie dauert $86\,164\,\text{s}$. Die Winkelgeschwindigkeit beträgt damit

$$\omega_{\text{Nordpol}} = \frac{2\,\pi}{86\,164}\,\frac{\text{rad}}{\text{s}} = 7{,}3\,10^{-5}\,\frac{\text{rad}}{\text{s}}\,. \tag{11.30}$$

Der Breitengrad λ_{AC} von Aachen ergibt sich damit aus (11.27) zu

$$\lambda_{AC} = \arcsin\left(\frac{\omega_{AC}}{\omega_{\text{Nordpol}}}\right) = 52°37'\text{N}\,. \tag{11.31}$$

Dieser Wert stimmt mit dem geodätischen Breitengrad $50°46'\text{N}$ ungefähr überein. Die Abweichung um 4% wird durch die Ablesegenauigkeit der Winkelskala bei vergleichsweise kurzer Messzeit erklärt.

11.4 Lorentztransformation

Für Transformationen zwischen zwei sehr schnell bewegten Bezugssystemen reichen die Konzepte, die uns zur Galileitransformation geführt haben, nicht aus. Für solche Transformationen benötigen wir die Lorentztransformation.

Zu ihrer Einführung betrachten wir zunächst ein Gedankenexperiment, das wir mit der Galileitransformation berechnen können und erweitern dann die Überlegungen.

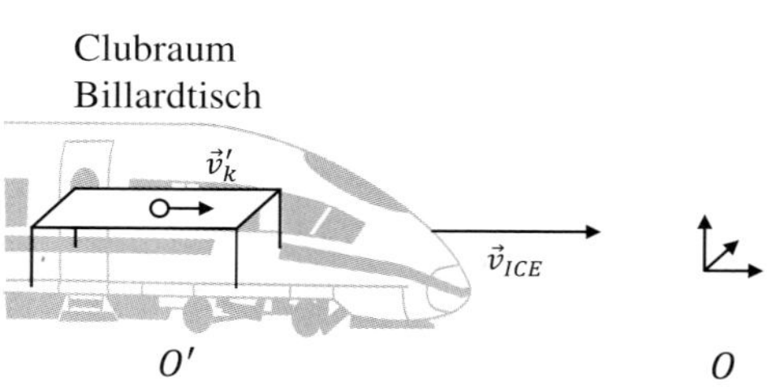

Abbildung 11.9 Billardkugel mit der Geschwindigkeit $\vec{v}_k'$ im fahrenden Zug (System O'). Die Geschwindigkeit $\vec{v}_{ICE}$ wird im ruhenden System O beobachtet.

In ► Abbildung 11.9 fährt ein Zug mit der Geschwindigkeit v_{ICE} durch einen Bahnhof, der das ruhende Beobachtungssystem O definiert. Beobachter im Zug befinden sich im

Bezugssystem O'. Im Clubraum des Luxuszugs steht ein Billardtisch, auf dem eine Kugel in Fahrtrichtung rollt.

Der Beobachter im Zug misst die Kugelgeschwindigkeit v'_k. Der Beobachter auf dem Bahnsteig misst die Kugelgeschwindigkeit v_k. Durch die Galileitransformation (11.1) lassen sich Ort der Kugel und Zeit zwischen den Systemen hin und zurück transformieren:

$$x'_k = x_k - v_{ICE}\, t \tag{11.32}$$

$$y'_k = y_k \tag{11.33}$$

$$z'_k = z_k \tag{11.34}$$

$$t' = t \tag{11.35}$$

11.4.1 Bewegte Lichtquelle: Einsteins Postulat

Wir ersetzen nun den Billardtisch im Zug durch eine Lampe und ein Gerät zur Messung der Lichtgeschwindigkeit. Auf dem Bahnhof befindet sich ebenfalls ein Messgerät für die Lichtgeschwindigkeit des Lampenlichts (► Abbildung 11.10).

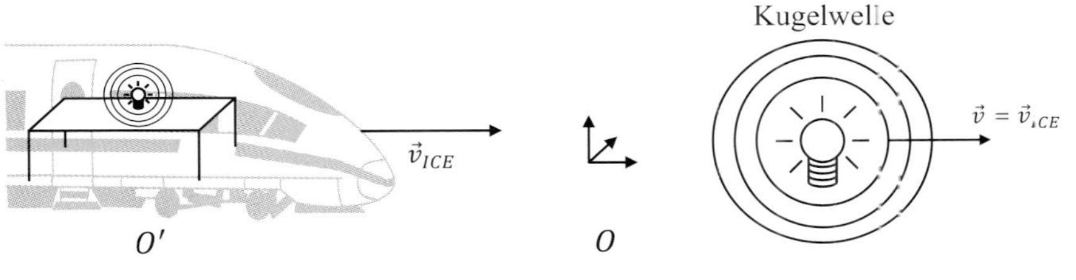

Abbildung 11.10 Links: Statt der Bewegung der Billardkugel wird jetzt die Ausbreitung von Licht in den beiden Systemen O und O' beobachtet. Rechts: Die Lichtwellen breiten sich in beiden Systemen O und O' kugelförmig aus.

Experimentelle Untersuchungen, z. B. im Michelson-Morley-Experiment, zeigen keine Abhängigkeit der Lichtgeschwindigkeit von der Bewegung eines Inertialsystems. Die Funktionsweise des Michelson-Interferometers und das Michelson-Morley-Experiment werden im Band »Experimentalphysik 2: Felder und Wellen« vorgestellt [4]. Aufgrund solcher experimenteller Befunde hat Albert Einstein folgendes Postulat aufgestellt, das entscheidende Konsequenzen für unser Gedankenexperiment hat:

Die Lichtgeschwindigkeit im Vakuum ist in jedem Inertialsystem gleich groß. In keinem Bezugssystem können wir die Bewegung der Lichtquelle aus der Kugelwelle selbst nachweisen:

$$c = c' \tag{11.36}$$

11.4.2 Galileitransformation einer Kugelwelle

Wir testen im Folgenden, ob die Galileitransformation konzeptionell mit dem Postulat von Einstein vereinbar ist.

Eine Kugeloberfäche beschreiben wir durch $x^2 + y^2 + z^2 = r^2$, wobei der Kugelradius mit r bezeichnet ist. Für Licht, das sich mit der Zeit t kugelförmig ausbreitet, ändert sich der Radius mit der Lichtgeschwindigkeit anhand von $r(t) = ct$.

In beiden Systemen, Zug und Bahnhof, kann die Kugelwelle folgendermaßen beschrieben werden:

$$O: \quad x^2 + y^2 + z^2 = c^2 t^2 \tag{11.37}$$

$$O': \quad x'^2 + y'^2 + z'^2 = c^2 t'^2 \tag{11.38}$$

Bei einer Transformation von (11.38) nach Vorschrift der Galileitransformation (11.1)

$$x' = x - vt\,, \qquad t' = t \tag{11.39}$$

müsste die Welle kugelförmig bleiben:

$$x^2 - \underbrace{2vtx + v^2t^2}_{\text{i. A.}\neq 0} + y^2 + z^2 = c^2t^2 \tag{11.40}$$

Im Allgemeinen sind aber der zweite und dritte Term auf der linken Seite nicht null, sodass die Welle nicht kugelförmig erscheint. Damit ist die Galileitransformation mit Einsteins Postulat nicht vereinbar. Gesucht ist eine Koordinatentransformation, die mit dem Postulat konstanter Lichtgeschwindigkeit kompatibel ist.

11.4.3 Lorentztransformation einer Kugelwelle

Bislang haben wir Zeit als universelle physikalische Größe aufgefasst, die kontinuierlich in positive Richtung verläuft. Um Einsteins Postulat zu erfüllen, müssen wir die These von einer universell geltenden Zeit $t' = t$ aufgeben.

Die Wellenfronten bewegen sich in beiden Systemen mit konstanter Geschwindigkeit fort, sodass die räumliche Koordinate x eine lineare Transformation in der Zeit erfordert ($x' = x - vt$). Genauso muss die Zeittransformation linear sein. Mit dem Ansatz $t' = t + \alpha x$ erhalten wir jetzt aus (11.38)

$$x^2 \underbrace{-2vtx} + v^2t^2 + y^2 + z^2 = c^2(t + \alpha x)^2 \tag{11.41}$$

$$= c^2(t^2 \underbrace{+2\alpha tx} + \alpha^2 x^2)\,. \tag{11.42}$$

Damit die beiden gemischten Terme auf beiden Seiten entfallen, ist $\alpha = -v/c^2$ erforderlich. Nach Umsortieren der Terme vereinfacht sich (11.42) zu

$$x^2 + v^2t^2 + y^2 + z^2 = c^2(t^2 + \frac{v^2}{c^4}x^2) \tag{11.43}$$

$$x^2 \underbrace{(1 - \frac{v^2}{c^2})}_{\equiv \gamma^{-2}} + y^2 + z^2 = c^2t^2 \underbrace{(1 - \frac{v^2}{c^2})}_{\equiv \gamma^{-2}}\,. \tag{11.44}$$

Den Faktor

$$\gamma = \frac{1}{\sqrt{1-\beta^2}} \quad \text{mit} \quad \beta \equiv \frac{v}{c} \tag{11.45}$$

müssen wir passend in die Transformationsbedingungen $x' \to x$ und $t' \to t$ einfügen, sodass die Kugelform im bewegten Bezugssystem (11.38) beim Übergang nach (11.37) bestehen bleibt. Die gesuchte **Lorentztransformation** zwischen Bezugssystemen lautet für eine eindimensionale Transformation entlang der x-Achse wie folgt:

$$\begin{aligned} x' &= \gamma(x - vt) \\ y' &= y \\ z' &= z \\ t' &= \gamma\left(t - \frac{vx}{c^2}\right) \end{aligned} \tag{11.46}$$

$$\begin{aligned} x &= \gamma(x' + vt') \\ y &= y' \\ z &= z' \\ t &= \gamma\left(t' + \frac{vx'}{c^2}\right) \end{aligned} \tag{11.47}$$

Wir vergewissern uns, dass Einsetzen der Lorentztransformation (11.46) in die Kugelwelle des Systems O' (11.38) nach Umsortieren der Terme wieder die Kugelwelle des Systems O (11.37) ergibt:

$$x'^2 + y'^2 + z'^2 = c^2 t'^2 \tag{11.48}$$

$$\gamma^2(x^2 - 2xvt + v^2t^2) + y^2 + z^2 = c^2\gamma^2\left(t^2 - 2t\frac{vx}{c^2} + \frac{v^2x^2}{c^4}\right) \tag{11.49}$$

$$\gamma^2 x^2\left(1 - \frac{v^2}{c^2}\right) + y^2 + z^2 = c^2 t^2 \gamma^2\left(1 - \frac{v^2}{c^2}\right) \tag{11.50}$$

$$x^2 + y^2 + z^2 = c^2 t^2 \tag{11.51}$$

Die Lorentztransformation ist demnach mit Einsteins Postulat vereinbar.

Der Faktor γ (11.45) wird auch **Lorentzfaktor** genannt. Die Geschwindigkeit wird in der Relativitätstheorie meist in Einheiten der Lichtgeschwindigkeit c ausgedrückt und als $\beta = v/c$ bezeichnet (11.45).

Die Abhängigkeit von γ von der Geschwindigkeit v der Bezugssysteme ist im Vergleich zur Lichtgeschwindigkeit c in ▶ Abbildung 11.11 dargestellt. Für kleine Werte von $\beta = v/c$ ist $\gamma = 1$, sodass sich für die Raumkoordinaten wieder die Galileitransformation

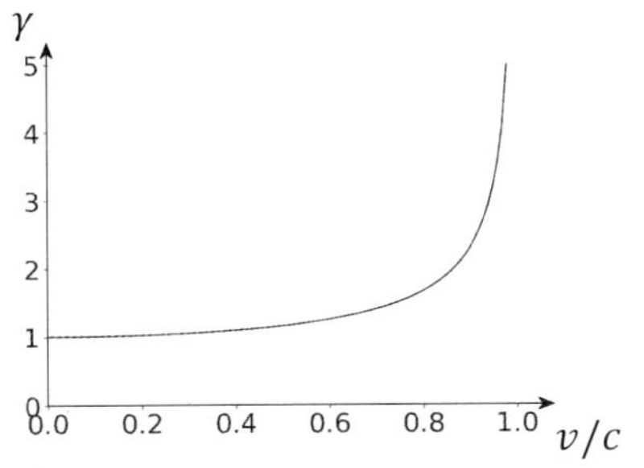

Abbildung 11.11 Der Lorentzfaktor γ geht für $\beta = v/c \to 1$ gegen ∞.

(11.1) ergibt. Dasselbe gilt für die Zeit, da der Korrekturterm v/c^2 klein wird. Für große Geschwindigkeiten $\beta = v/c \rightarrow 1$ divergiert der Faktor γ und beeinflusst die Transformation entsprechend stark.

Folgende Aspekte sind bemerkenswert:

- In der Lorentztransformation sind Raum und Zeit miteinander verknüpft.
- Die Lichtgeschwindigkeit c im Vakuum hat die Bedeutung einer Grenzgeschwindigkeit. c ist die maximale Geschwindigkeit, mit der Signale übertragen werden.
- Die Maxwellgleichungen (Elektrodynamik) sind invariant bzgl. der Lorentztransformation. In diesem Zusammenhang wurden sie 1895 von ihrem Namensgeber H. A. Lorentz zum ersten Mal aufgestellt. Die Maxwellgleichungen werden im Band »Experimentalphysik 2: Felder und Wellen« vorgestellt [4].
- Die Lorentztransformation hat große aktuelle Bedeutung in der Elementarteilchenphysik.

Das Wichtigste zum Mitnehmen

Zum Mitnehmen: Transformation zwischen Bezugssystemen 1

- Ein **Inertialsystem** ist ein Bezugssystem, das keine Beschleunigung erfährt. In jedem Inertialsystem finden wir dieselben physikalischen Gesetze.

  Die Bewegung eines Massenpunkts wird im System O durch $\vec{r}(t)$ beschrieben, in einem Inertialsystem O', das sich mit der Geschwindigkeit $\vec{u} = const.$ relativ zu O bewegt, durch $\vec{r}'(t)$. Die **Galileitransformationen** zwischen den Bezugssystemen lauten

$$\begin{aligned} \vec{r}' &= \vec{r} - \vec{u} \cdot t \\ \vec{v}' &= \vec{v} - \vec{u} \\ \vec{a}' &= \vec{a} \\ t' &= t\,, \end{aligned} \qquad (11.1) \qquad \begin{aligned} \vec{r} &= \vec{r}' + \vec{u} \cdot t \\ \vec{v} &= \vec{v}' + \vec{u} \\ \vec{a} &= \vec{a}' \\ t &= t'\,. \end{aligned} \qquad (11.2)$$

- Das System $O'(x', y', z')$ rotiert mit der Winkelgeschwindigkeit $\vec{\omega}$ um die z-Achse des Inertialsystems $O(x, y, z)$. Die Relativgeschwindigkeit $\vec{u}$ eines Punkts mit dem Ortsvektor $\vec{r}$ ist dann

$$\vec{u} = \vec{\omega} \times \vec{r}' = \vec{\omega} \times \vec{r}\,. \qquad (11.13)$$

  Damit ergibt sich für die Transformation der Geschwindigkeiten $\vec{v}$ und $\vec{v}'$ des Massenpunkts

$$\vec{v} = \vec{v}' + \underbrace{\vec{\omega} \times \vec{r}}_{\text{Rotation des Systems } O'} \qquad (11.16)$$

  und für die Beschleunigung

$$\vec{a}' = \vec{a} + \underbrace{2\left(\vec{v}' \times \vec{\omega}\right)}_{\equiv \vec{a}_C} + \underbrace{\vec{\omega} \times \left(\vec{r} \times \vec{\omega}\right)}_{\equiv \vec{a}_{ZF}}\,. \qquad (11.22)$$

  Ein mitrotierender Beobachter spürt also zusätzlich zur Beschleunigung $\vec{a}$, die im Inertialsystem O gemessen wird, die Coriolisbeschleunigung $\vec{a}_C$ und die Zentrifugalbeschleunigung $\vec{a}_{ZF}$.

Zum Mitnehmen: Transformation zwischen Bezugssystemen 2

- Den Beschleunigungen $\vec{a}_C$ und $\vec{a}_{ZF}$ entsprechen **Scheinkräfte**. Das sind Kräfte, die von beschleunigten Bezugssystemen herrühren:

$$\begin{array}{ll} \text{Corioliskraft} & \vec{F}_C = 2m\,(\vec{v}\,' \times \vec{\omega}) \\ \text{Zentrifugalkraft} & \vec{F}_{ZF} = m\vec{\omega} \times (\vec{r} \times \vec{\omega}) \end{array} \tag{11.23}$$

  Die Corioliskraft führt beim Foucaultschen Pendel zu einer Drehung der Schwingungsebene gegenüber der Erdoberfläche.

- Die Lichtgeschwindigkeit im Vakuum ist in jedem Inertialsystem konstant und gleich groß:

$$c = c' \tag{11.36}$$

  Das hat zur Folge, dass für große Geschwindigkeiten die Galileitransformation durch die **Lorentztransformation**

$$\begin{aligned} x' &= \gamma(x - vt) \\ y' &= y \\ z' &= z \\ t' &= \gamma\left(t - \frac{vx}{c^2}\right) \end{aligned} \tag{11.46}$$

$$\begin{aligned} x &= \gamma(x' + vt') \\ y &= y' \\ z &= z' \\ t &= \gamma\left(t' + \frac{vx'}{c^2}\right) \end{aligned} \tag{11.47}$$

  ersetzt werden muss mit dem **Lorentzfaktor** γ

$$\gamma = \frac{1}{\sqrt{1-\beta^2}} \quad \text{mit} \quad \beta \equiv \frac{v}{c}\,. \tag{11.45}$$

  Die Geschwindigkeit wird in der Relativitätstheorie meist als $\beta = v/c$ in Einheiten der Lichtgeschwindigkeit c ausgedrückt.

Spezielle Relativitätstheorie

12

12.1 **Lichtgeschwindigkeit** 229

12.2 **Zeitdilatation** 230

12.3 **Längenkontraktion** 233

12.4 **Raumzeit-Diagramme** 233
12.4.1 Abstand in der Raumzeit 234

12.5 **Energie-Impuls-Raum** 237

12.6 **Anwendung: Elementarteilchenphysik** 240
12.6.1 Produktion von Teilchen am Beschleuniger 240
12.6.2 Zerfall eines Lambda-Teilchens 242

ÜBERBLICK

» In der speziellen Relativitätstheorie werden wir die teilweise überraschenden Auswirkungen der Konstanz der Lichtgeschwindigkeit im Vakuum (Einsteins Postulat (11.36)) für Objekte untersuchen, die sich mit hohen Geschwindigkeiten relativ zu anderen Objekten bewegen.

Im Bereich der Elementarteilchenphysik ist die spezielle Relativitätstheorie eine wesentliche Voraussetzung für das Verständnis vieler physikalischer Phänomene. Wir werden deshalb unsere Anwendungsbeispiele bevorzugt aus diesem Forschungsbereich wählen. «

12.1 Lichtgeschwindigkeit

Die Lichtgeschwindigkeit c im Vakuum ist eine der fundamentalen Konstanten in der Physik (Abschnitt 2.1). Wir werden eine Messung von c in dem folgenden Experiment durchführen, das auf Foucault zurückgeht.

Experiment 20: Lichtgeschwindigkeit

In ► Abbildung 12.1 trifft ein Laserstrahl zunächst auf einen Strahlteiler, dann auf einen sich schnell drehenden Spiegel, der einen weiteren Spiegel in der Entfernung d beleuchtet. Die Zeit t, die der Strahl für die Distanz $2\,d$ benötigt, bestimmen wir mithilfe der Spiegeldrehung: Während der Strahl die Distanz $2\,d$ zurücklegt, dreht sich der Spiegel um den Winkel α, sodass der reflektierte Strahl versetzt zum Referenzstrahl auf den Strahlteiler und dann auf den Schirm trifft.

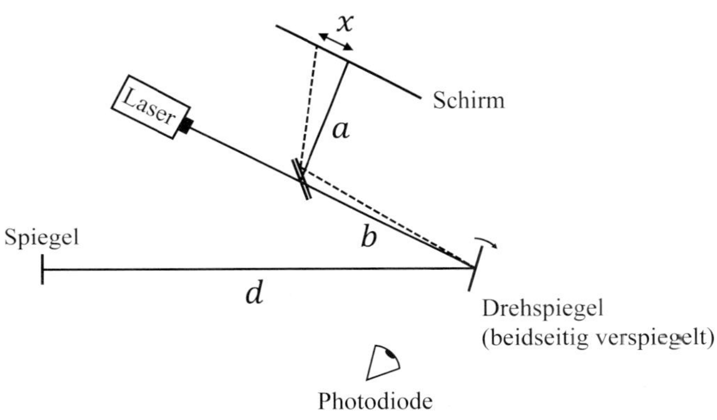

Abbildung 12.1 Messung der Lichtgeschwindigkeit c mit einem schnell rotierenden Drehspiegel. Der vom festen Spiegel reflektierte Strahl wird vom rotierenden Spiegel versetzt reflektiert (gestrichelte Linien). Die Rotationsgeschwindigkeit des Spiegels wird mit einer Photodiode bestimmt.

Die Winkelgeschwindigkeit des Drehspiegels wird mithilfe einer Fotodiode gemessen. Der Drehspiegel des Experiments ist beidseitig verspiegelt, sodass der Laserstrahl die Diode zweimal pro Umdrehung trifft. Gemessen wurde eine Trefferrate von 1 127/s, das entspricht der Umlaufzeit $T = 1/563{,}5\,\mathrm{s}$ für eine vollständige Spiegeldrehung. Die Winkelgeschwindigkeit beträgt mit $\omega = 2\,\pi/T$ (4.11)

$$\omega = 2\,\pi\,\frac{1}{T} = 2\,\pi\,563{,}5\,\frac{1}{\mathrm{s}} = 3\,541\,\frac{1}{\mathrm{s}}\,. \tag{12.1}$$

Aus ihr ergibt sich die Spiegeldrehung um den Winkel α in der gesuchten Laufzeit t des Lichts für die Strecke $2\,d$:

$$\alpha = \omega\,t \tag{12.2}$$

Den Winkel α messen wir über die Ablenkung x auf dem Schirm. Analog zum Gravitationswaagenversuch (Abschnitt 10.1.1) ergibt sich für den Ablenkwinkel des Laserstrahls $2\,\alpha$ (linke ► Abbildung 10.7) und für die Ablenkung x nach Gleichung (10.27)

$$\frac{x}{a+b} \approx 2\,\alpha\,. \tag{12.3}$$

Die Laufzeit t des Lichts auf der Strecke $2\,d$ erhalten wir durch Einsetzen von (12.3) in (12.2):

$$t = \alpha\,\frac{1}{\omega} = \frac{x}{2(a+b)}\,\frac{1}{\omega} \tag{12.4}$$

Die Lichtgeschwindigkeit c auf der Strecke $2\,d$ ergibt sich mit der Winkelgeschwindigkeit $\omega = 3\,541\,(1/\mathrm{s})$ (12.1), den Abständen $d = 15\,\mathrm{m}$ und $a + b = 3\,\mathrm{m}$ sowie der Ablenkung $x = 2\,\mathrm{mm}$ auf dem Schirm zu

$$c = \frac{2\,d}{t} \tag{12.5}$$

$$= 2\,d\,\frac{2(a+b)}{x}\,\omega = 2\cdot 15\cdot \frac{2\cdot 3}{2\,10^{-3}}\cdot 3\,541\,\frac{\mathrm{m}}{\mathrm{s}} = 3{,}19\,10^{8}\,\frac{\mathrm{m}}{\mathrm{s}}\,. \tag{12.6}$$

Der genaue Wert für die Lichtgeschwindigkeit im Vakuum ist als Naturkonstante festgelegt (Abschnitt 2.1, Gleichung (2.3)):

$$c = 299\,792\,458\,\frac{\mathrm{m}}{\mathrm{s}} \tag{12.7}$$

Unsere Messung weicht nur etwa 6% davon ab. Die Drehgeschwindigkeit des Spiegels ω und die Längenmessungen a, b und d sind mit sehr kleinen Fehlern im Promillebereich behaftet. Die Ablesegenauigkeit der Verschiebung x auf dem Schirm ist auf etwa 0,1 mm genau und bestimmt die Unsicherheit der Messung.

12.2 Zeitdilatation

Durch die Verknüpfung von Raum und Zeit werden Zeitmessungen in verschiedenen Bezugssystemen unterschiedlich ausfallen. In dem folgenden Gedankenexperiment werden wir die Umrechnung zwischen zwei Zeitmessungen bestimmen.

Experiment 21: Zeitdilatation

In der linken ► Abbildung 12.2 misst ein Astronaut in seinem Raumschiff die Zeit $\Delta t'$, die das Licht für die Strecke zum Spiegel und zurück benötigt.

Die rechte ► Abbildung 12.2 zeigt die Messung der Lichtlaufzeit durch das Kontrollzentrum, bei der sich das Raumschiff um die Strecke Δx weiterbewegt. Das Kontrollzentrum misst eine entsprechend längere Laufzeit Δt des Lichts.

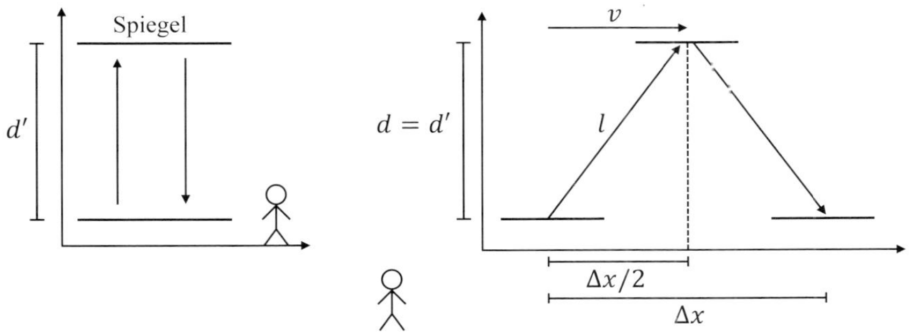

Abbildung 12.2 Laufzeitmessung zwischen zwei Spiegeln, links im mitbewegten System (Raumschiff) und rechts im Kontrollzentrum, gegen das sich das Raumschiff mit der Geschwindigkeit $\vec{v}$ bewegt

Aus der Sicht des Kontrollzentrums in der rechten ► Abbildung 12.2 ergibt die Geometrie des linken rechtwinkligen Dreiecks nach dem Gesetz von Pythagoras

$$l^2 = d^2 + \left(\frac{\Delta x}{2}\right)^2 . \tag{12.8}$$

Wir verwenden die folgenden Argumente, um die verschiedenen Terme in (12.8) in Zeiten umzurechnen:

- Die Lichtgeschwindigkeit c ist nach dem Einstein-Postulat (11.36) in beiden Systemen gleich groß.
- Aus der Sicht des Kontrollzentrums legt das Licht in der halben Laufzeit $\Delta t/2$ die Strecke $l = c\,\Delta t/2$ zurück.
- Das Raumschiff hat sich in der halben Laufzeit $\Delta t/2$ mit der Geschwindigkeit v um die Strecke $\Delta x/2 = v\,\Delta t/2$ weiterbewegt.
- Aus der Sicht des Astronauten hat das Licht in der halben Laufzeit $\Delta t'/2$ die Strecke $d' = d = c\,\Delta t'/2$ zurückgelegt.

Einsetzen aller Argumente in Gleichung (12.8) ergibt

$$\left(c\frac{\Delta t}{2}\right)^2 = \left(c\frac{\Delta t'}{2}\right)^2 + \left(v\frac{\Delta t}{2}\right)^2 \tag{12.9}$$

$$\Delta t^2(c^2 - v^2) = c^2 \Delta t'^{\,2} \tag{12.10}$$

$$\Delta t^2 \underbrace{\left(1 - \frac{v^2}{c^2}\right)}_{=\gamma^{-2}} = \Delta t'^{\,2} . \tag{12.11}$$

Mit $\gamma^2 = 1/\left(1 - v^2/c^2\right)$ aus Gleichung (11.45) ergibt sich als Zusammenhang zwischen den Zeitmessungen

$$\Delta t = \gamma\,\Delta t' . \tag{12.12}$$

Diese Gleichung formuliert die **Zeitdilatation**. Sie impliziert, dass Uhren in schnell bewegten Systemen aus der Sicht eines ruhenden Beobachters langsamer laufen.

Als **Eigenzeit** bezeichnet man die Zeit $\Delta t'$ zwischen zwei Ereignissen, die in einem Bezugssystem an demselben Ort stattfinden.

Beispiel 35: Myon-Lebensdauer

Myonen sind Elementarteilchen mit einer kurzen mittleren Lebensdauer von $\Delta t' \approx 5 \cdot 10^{-6}$ s. Bewegen sie sich mit Lichtgeschwindigkeit c, so würde sich damit eine maximale Reichweite der Myonen von $x' = c\,\Delta t' \approx 1{,}5$ km ergeben. Messungen der kosmischen Höhenstrahlung zeigen aber, das Myonen weitaus größere Entfernungen zurücklegen können.

Myonen, die von der Erde aus beobachtet z. B. $\gamma = 100$ aufweisen, leben im Mittel sogar $\Delta t = \gamma\,\Delta t' = 5 \cdot 10^{-4}$ s lang. Ihre Geschwindigkeit v beträgt wegen $\beta = v/c = \sqrt{1 - \gamma^{-2}} \approx 1$ (11.45) nahezu Lichtgeschwindigkeit c. Daraus berechnen wir eine Reichweite von $x = c\,\Delta t \approx 150$ km.

Aufgabe 28: Lorentztransformation

Entwickeln Sie die Lorentztransformation (11.47) aus der Zeitdilatation $t = \gamma t'$ (12.12), indem Sie im System $O(x, t)$ die Ankunft eines Lichtsignals $x = ct - x_\circ$ in dem mit der Geschwindigkeit v gegen O bewegten System berechnen (► Abbildung 12.3).

Hinweis: Berechnen Sie ebenso ein rückwärts laufendes Lichtsignal und kombinieren Sie die beiden Ergebnisse.

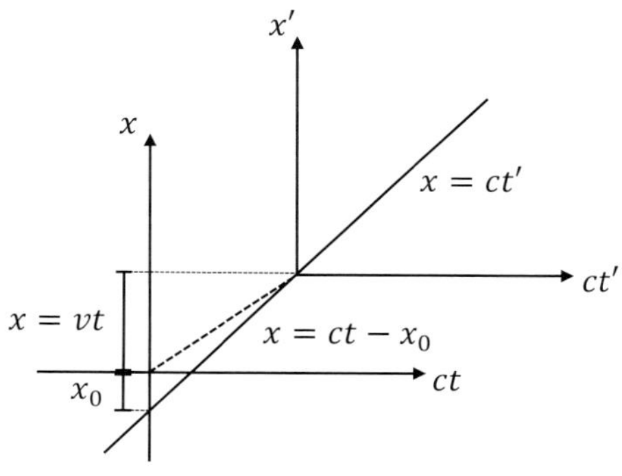

Abbildung 12.3 Ein Lichtsignal bewegt sich mit Lichtgeschwindigkeit in den Systemen (x, t) und (x', t'), deren Relativgeschwindigkeit v ist.

12.3 Längenkontraktion

Wegen der Verknüpfung von Raum und Zeit, die wir in der Lorentztransformation bereits erwähnt haben, müssen Längenmessungen zu einem festen Zeitpunkt durchgeführt werden.

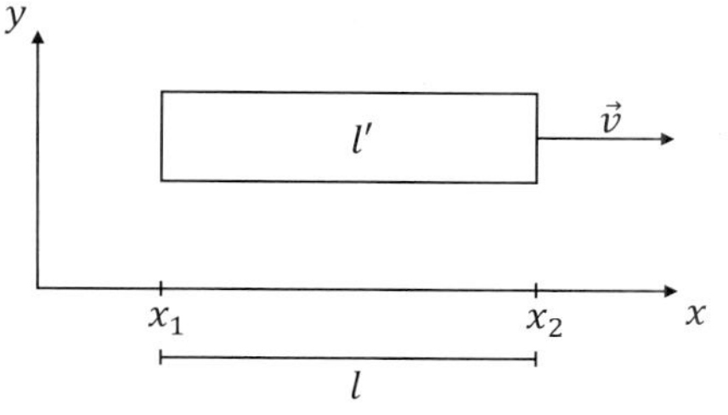

Abbildung 12.4 Längenmessung durch gleichzeitige Positionsbestimmung beider Endpunkte im Labor (x_1, x_2) und im mitbewegten System (x'_1, x'_2)

Um zum Beispiel die korrekte Länge eines Stabs zu bestimmen, müssen beide Positionen x_1 und x_2 gleichzeitig gemessen werden (▶ Abbildung 12.4). Wird der Stab mit der Geschwindigkeit v bewegt, so sind die Positionen nach der Lorentztransformation (11.46)

$$x'_1 = \gamma(x_1 - vt)\,, \tag{12.13}$$

$$x'_2 = \gamma(x_2 - vt)\,. \tag{12.14}$$

Die Differenz der beiden Gleichungen ergibt

$$x'_2 - x'_1 = \gamma(x_2 - x_1)\,. \tag{12.15}$$

Bezeichnen wir die Differenzen mit $l' \equiv x'_2 - x'_1$ und $l \equiv x_2 - x_1$, so erhalten wir die übliche Formulierung der **Längenkontraktion**:

$$l = \frac{1}{\gamma}\,l' \tag{12.16}$$

Bewegte Objekte wirken für den ruhenden Beobachter verkürzt.

12.4 Raumzeit-Diagramme

Aufgrund unserer täglichen Erfahrung verstehen wir Ort und Zeit als unabhängige Größen. Für hohe Geschwindigkeiten der Bezugssysteme sind aber beide Aspekte in der Raumzeit untrennbar miteinander verknüpft.

Ein sogenanntes **Raumzeit-Diagramm (Minkowski-Diagramm)** zeigt unsere Wissens- und Kommunikationsmöglichkeiten (▶ Abbildung 12.5).

Das Hier und Jetzt befindet sich im Koordinatenursprung der Raumzeitkoordinaten. Entlang der vertikalen Achse läuft die Zeit, hier formuliert als Produkt aus Lichtgeschwindigkeit

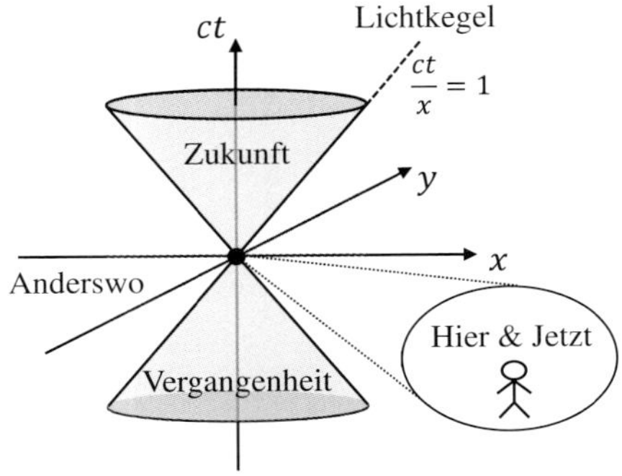

Abbildung 12.5 Raum-Zeit-Diagramm (hier nur zwei Raumdimensionen) mit dem Lichtkegel $x^2 + y^2 = c^2t^2$. Der Lichtkegel schneidet die (x, ct)-Ebene auf der Winkelhalbierenden $x = ct$.

und Zeit ct. Damit erhalten wir Längen als einheitliche Dimension für alle Raum- und Zeitkoordinaten. Die Vergangenheit ist nach unten und die Zukunft nach oben gerichtet.

Der sogenannte **Lichtkegel** in drei Raumdimensionen

$$x^2 + y^2 + z^2 = c^2t^2 \tag{12.17}$$

ist in ► Abbildung 12.5 für zwei Raumdimensionen x, y und die Zeit ct veranschaulicht. Die Winkelhalbierenden in den Ebenen (x, ct), (y, ct) und (z, ct) mit der Steigung 1 liegen auf dem Lichtkegel. Da die Lichtgeschwindigkeit c die maximale Signalübertragungsgeschwindigkeit ist, kann der Lichtkegel nach außen nicht überschritten werden. Innerhalb des Kegels können sich Objekte auf ihren sogenannten Weltlinien bewegen. Aber von innen aus dem Kegel heraus können wir weder Information in den äußeren Raumbereich (Anderswo) senden, noch Information vom Außenbereich erhalten.

Beispiel 36: Rakete

Eine Rakete fliegt auf einer Weltlinie mit der Geschwindigkeit v, die wir aus der Steigung $c(t_1 - t_\circ)/(x_1 - x_\circ) = c/v = 1/\beta$ ablesen können (► Abbildung 12.6).

12.4.1 Abstand in der Raumzeit

Im Rahmen der Längenkontraktion hatten wir bereits eine Längenmessung entlang der Ortskoordinaten durchgeführt. In diesem Abschnitt wollen wir allgemeine Längenmessungen in der Raumzeit untersuchen, die die Raum- und Zeitkoordinaten mit einschließen.

Abstandsmessungen in Räumen der reellen und komplexen Zahlen sind uns bereits bekannt.

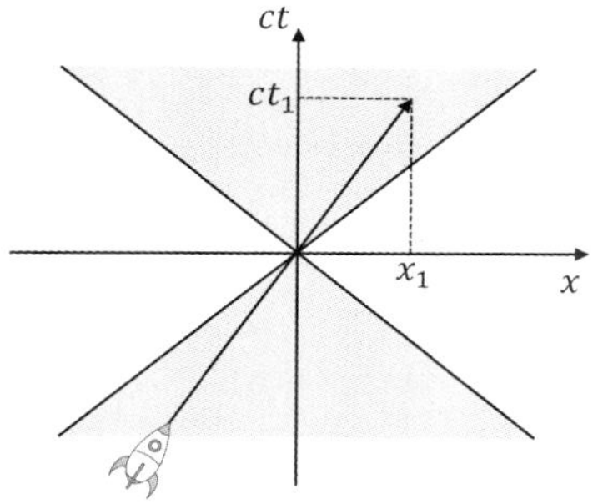

Abbildung 12.6 Weltlinie einer Rakete, die sich mit der Geschwindigkeit v entlang der x-Achse bewegt

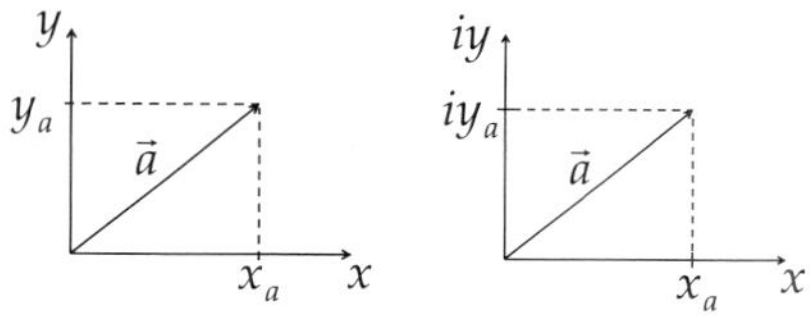

Abbildung 12.7 Länge $d = |\vec{a}|$ eines Vektors $\vec{a}$ im euklidischen Raum (links), in der komplexen Ebene (rechts)

Mathematischer Einschub: Abstandsmessungen

Im zweidimensionalen euklidischen Raum der reellen Zahlen (linke ► Abbildung 12.7) können wir Abstände d zwischen zwei Raumpunkten über das Skalarprodukt eines Vektors zwischen den beiden Punkten bestimmen:

$$d^2 = |\vec{a}|^2 = \vec{a}\,\vec{a} = \begin{pmatrix} x_a \\ y_a \end{pmatrix} \begin{pmatrix} x_a \\ y_a \end{pmatrix} \tag{12.18}$$

$$d^2 = x_a^2 + y_a^2 \tag{12.19}$$

$$d = \sqrt{x_a^2 + y_a^2} \tag{12.20}$$

In der komplexen Zahlenebene (rechte ► Abbildung 12.7) ist die Länge d durch das Skalarprodukt zwischen dem Vektor $\vec{a}$ mit dem konjugiert komplexen Vektor $\vec{a}^*$ definiert:

$$d^2 = |\vec{a}|^2 = \vec{a}\,\vec{a}^* = \begin{pmatrix} x_a \\ iy_a \end{pmatrix} \begin{pmatrix} x_a \\ -iy_a \end{pmatrix} \tag{12.21}$$

$$d^2 = x_a^2 + y_a^2 \tag{12.22}$$

$$d = \sqrt{x_a^2 + y_a^2} \tag{12.23}$$

In der vierdimensionalen **Minkowski-Raumzeit** verwendet man eine zunächst seltsam anmutende Abstandsdefinition, die wir im Folgenden genauer betrachten wollen.

Wir verwenden hier eine allgemein übliche Konvention für die Darstellung der **Minkowski-Raumzeit**. Darin definieren wir einen **Vierervektor** $\mathbf{a}$ der Raumzeit mit folgenden Koordinaten:

$$\mathbf{a} = \begin{pmatrix} ct_a \\ x_a \\ 0 \\ 0 \end{pmatrix} \tag{12.24}$$

Das Skalarprodukt zweier Vierervektoren $\mathbf{a}, \mathbf{b}$ wird mithilfe einer Matrix $\hat{g}$ berechnet, die man auch als **Metrik** bezeichnet:

$$\hat{g} = \begin{pmatrix} 1 & 0 & 0 & 0 \\ 0 & -1 & 0 & 0 \\ 0 & 0 & -1 & 0 \\ 0 & 0 & 0 & -1 \end{pmatrix} \tag{12.25}$$

Wir zeigen die Vektor-Matrixmultiplikation konkret anhand einer Längenmessung, bei der wir das Skalarprodukt des Vierervektors $\mathbf{a}$ mit sich selbst bilden (linke ► Abbildung 12.8). Damit erhalten wir das Quadrat der Länge von a in der Raumzeit:

$$\mathbf{a}^2 = (ct_a \; x_a \; 0 \; 0) \begin{pmatrix} 1 & 0 & 0 & 0 \\ 0 & -1 & 0 & 0 \\ 0 & 0 & -1 & 0 \\ 0 & 0 & 0 & -1 \end{pmatrix} \begin{pmatrix} ct_a \\ x_a \\ 0 \\ 0 \end{pmatrix} \tag{12.26}$$

$$= (ct_a \; x_a \; 0 \; 0) \begin{pmatrix} ct_a \\ -x_a \\ 0 \\ 0 \end{pmatrix} \tag{12.27}$$

$$\mathbf{a}^2 = c^2 t_a^2 - x_a^2 \tag{12.28}$$

Die mit der Matrix $\hat{g}$ berechnete Größe $d^2 = \mathbf{a}^2$ definiert den gesuchten **Abstand d in der Minkowski-Raumzeit**.

Die Besonderheit dieser zunächst seltsam anmutenden Abstandsdefinition liegt nun darin, dass die Größe $d^2 = \mathbf{a}^2$ invariant gegen Lorentztransformationen oder kurz **lorentzinvariant** ist. Das lässt sich mithilfe der Formeln (11.46) und (11.47) (oder später mit (12.31) und (12.32)) nachrechnen.

Raum-Zeit-Punkte mit gleichem Abstand zum Ursprung liegen auf folgender Kurve (rechte ► Abbildung 12.8):

$$d^2 = c^2 t^2 - x^2 \tag{12.29}$$

$$ct = \sqrt{d^2 + x^2} \tag{12.30}$$

Für die Vierervektoren in der Raumzeit gibt es eine nützliche Schreibweise der **Lorentztransformationen** (11.46) und (11.47) in **Matrixform**:

$$\begin{pmatrix} ct' \\ x' \\ y' \\ z' \end{pmatrix} = \begin{pmatrix} \gamma & -\gamma\beta & 0 & 0 \\ -\gamma\beta & \gamma & 0 & 0 \\ 0 & 0 & 1 & 0 \\ 0 & 0 & 0 & 1 \end{pmatrix} \begin{pmatrix} ct \\ x \\ y \\ z \end{pmatrix} \tag{12.31}$$

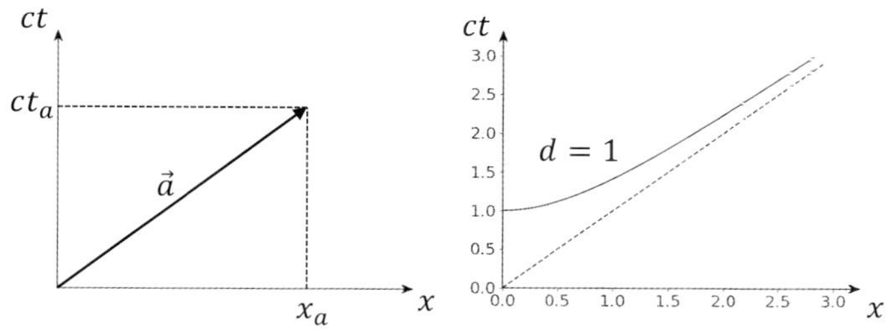

Abbildung 12.8 Links: Die Länge des Vektors $\vec{a}$ im Raum-Zeit-Diagramm wird mithilfe einer Metrik definiert. Rechts: Raum-Zeit-Punkte mit gleichem Abstand (z. B. obere Kurve für $d = 1$) nähern sich der Winkelhalbierenden.

$$\begin{pmatrix} ct \\ x \\ y \\ z \end{pmatrix} = \begin{pmatrix} \gamma & \gamma\beta & 0 & 0 \\ \gamma\beta & \gamma & 0 & 0 \\ 0 & 0 & 1 & 0 \\ 0 & 0 & 0 & 1 \end{pmatrix} \begin{pmatrix} ct' \\ x' \\ y' \\ z' \end{pmatrix} \tag{12.32}$$

Aufgabe 29: Lorentzinvarianz

Weisen Sie mithilfe der Lorentztransformationen nach, dass der Abstand d in (12.28) lorentzinvariant ist.

12.5 Energie-Impuls-Raum

In Analogie zur Raumzeit können wir einen Raum definieren, der die Gesamtenergie und den Impuls eines Objekts miteinander verbindet. Wir motivieren diesen Energie-Impuls-Raum über Einsteins Energie-Masse-Beziehung.

Albert Einsteins berühmte Formulierung

$$E = mc^2 \tag{12.33}$$

bringt zum Ausdruck, dass Masse eine Form von Energie ist. Diese Schreibweise ist für unsere weiteren Überlegungen unpraktisch, da die Masse m nicht eine Konstante bezeichnet, sondern mit dem γ Faktor (11.45) steigt (relativistische Masse):

$$m = \gamma m_\circ \tag{12.34}$$

Dabei bezeichnet $m_\circ$ die Ruhemasse des Teilchens, die in einem Bezugssystem gemessen wird, in dem das Teilchen unbewegt ist ($\gamma = 1$).

Beispiel 37: Ruhemasse von Teilchen

$$\text{Elektron} \quad m_e = 0{,}511\,\mathrm{MeV}/c^2$$
$$\text{Proton} \quad m_p = 938\,\mathrm{MeV}/c^2$$

$$\text{mit der Energieeinheit} \quad 1\,\mathrm{eV} = 1\,\text{Elektronenvolt} = 1{,}6 \cdot 10^{-19}\,\mathrm{J}$$

Zum besseren Verständnis von Einsteins Formel (12.33) schreiben wir sie für die Ruhemasse $m_\circ$ in folgender Weise um:

$$E^2 = m^2 c^4 \tag{12.35}$$
$$= \gamma^2 m_\circ^2 c^4 \tag{12.36}$$
$$= \gamma^2 m_\circ^2 c^2 (c^2 - v^2 + v^2) \tag{12.37}$$
$$= \gamma^2 m_\circ^2 c^2 (c^2 \underbrace{(1 - \frac{v^2}{c^2})}_{=\gamma^{-2}} + v^2) \tag{12.38}$$

$$(\underbrace{E}_{\text{Gesamtenergie}})^2 = (\underbrace{m_\circ c^2}_{\text{Ruheenergie}})^2 + \underbrace{(\underbrace{\gamma m_\circ v}_{\text{Impuls } p}\, c)^2}_{\text{Bewegungsenergie}} \tag{12.39}$$

$$E^2 = m_\circ^2 c^4 + p^2 c^2 \tag{12.40}$$

Die **Gesamtenergie** eines freien Teilchens, das keiner Wechselwirkung unterliegt, besteht nicht nur aus seiner **Bewegungsenergie** (seiner **kinetischen Energie**), sondern auch aus seiner **Ruheenergie**, die durch seine Masse zustande kommt.

Der Impuls des Teilchens ist dabei

$$\vec{p} = \gamma\, m_\circ\, \vec{v}\,. \tag{12.41}$$

Wir werden im Folgenden die Form (12.40) anstelle von (12.33) für Einsteins Gleichung verwenden.

Wir definieren den Energie-Impuls-Raum in analoger Weise wie zuvor die Raumzeit. Die Vierervektoren heißen hier Viererimpuls **p** und haben die Komponenten

$$\mathbf{p} = \begin{pmatrix} E/c \\ p_x \\ p_y \\ p_z \end{pmatrix} = \begin{pmatrix} E/c \\ \vec{p} \end{pmatrix}. \tag{12.42}$$

In Analogie zur Länge in der Raumzeit ergibt das Skalarprodukt von **p** mit sich selbst eine Länge, jetzt im Energie-Impuls-Raum. Zu ihrer Berechnung verwenden wir wieder die Metrik (12.25):

$$\mathbf{p}^2 = ((E/c)\; p_x\; p_y\; p_z) \begin{pmatrix} 1 & 0 & 0 & 0 \\ 0 & -1 & 0 & 0 \\ 0 & 0 & -1 & 0 \\ 0 & 0 & 0 & -1 \end{pmatrix} \begin{pmatrix} E/c \\ p_x \\ p_y \\ p_z \end{pmatrix} \tag{12.43}$$

$$= ((E/c)\; p_x\; p_y\; p_z) \begin{pmatrix} E/c \\ -p_x \\ -p_y \\ -p_z \end{pmatrix} \tag{12.44}$$

$$= (E/c)^2 - \vec{p}^{\,2} \tag{12.45}$$

Multiplizieren wir dieses Ergebnis mit dem Quadrat der konstanten Lichtgeschwindigkeit c^2 und vergleichen es mit Gleichung (12.40), erkennen wir die eigentliche Bedeutung des Skalarprodukts des Viererimpulses mit sich selbst:

$$\mathbf{p}^2\, c^2 = E^2 - \vec{p}^{\,2} c^2 = m_\circ^2\, c^4 \tag{12.46}$$

Die Länge im Energie-Impuls-Raum entspricht der Ruhemasse $m_\circ$ des Teilchens, multipliziert mit der Lichtgeschwindigkeit c:

$$|\mathbf{p}| = m_\circ\, c \tag{12.47}$$

Wenn wir die Gesamtenergie des Teilchens als Funktion der Bewegungsenergie auftragen, kann sich das Teilchen ausschließlich auf der in ► Abbildung 12.9 gezeigten Kurve bewegen, die durch die Ruhemasse $m_\circ$ des Teilchens festgelegt ist. Bei $\vec{p} = 0$ lesen wir die um c reduzierte Ruheenergie $E/c = m_\circ c$ des Teilchens ab.

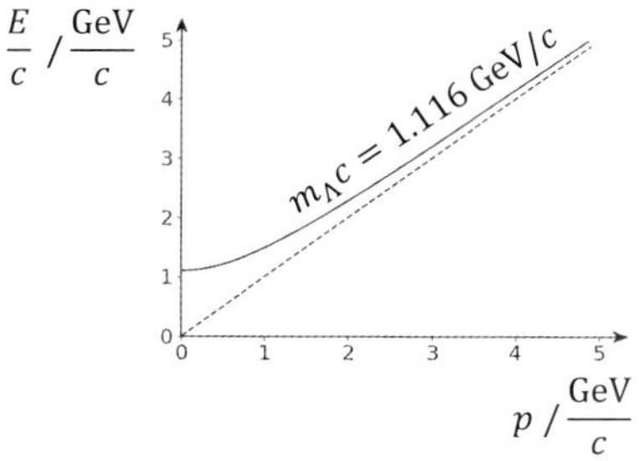

Abbildung 12.9 Energie-Impuls-Diagramm eines Λ-Teilchens der Masse $m_\Lambda = 1{,}116\,\mathrm{GeV}/c^2$

Die Viererimpulse unterliegen derselben Lorentztransformation (12.31) wie die Vierervektoren der Raumzeit. Ebenso ist die Ruheenergie invariant unter der Lorentztransformation.

Aufgabe 30: Lorentzinvarianz Energie-Impuls

Weisen Sie mithilfe der Lorentztransformationen nach, dass das Quadrat eines Energie-Impuls-Vektors $\mathbf{p}^2$ und damit die Ruheenergie $m_\circ c^2$ eines Teilchens lorentzinvariant ist.

12.6 Anwendung: Elementarteilchenphysik

In der Elementarteilchenphysik brauchen wir die spezielle Relativitätstheorie, um die Bewegungen der Teilchen zu beschreiben.

Die heute bekannten Materieteilchen sind 6 Quarks und 6 Leptonen. Diese 12 Materieteilchen haben 12 Antimateriepartner. Zusätzlich kennen wir für drei der vier bekannten fundamentalen Wechselwirkungen die Teilchen, die Wechselwirkungen zwischen den Materieteilchen vermitteln. Neu hinzugekommen ist 2012 das Higgs-Teilchen, das den sogenannten Higgs-Mechanismus zur Erzeugung von Teilchen-Ruhemassen anzeigt. Näheres dazu wird in späteren Vorlesungen zur Teilchenphysik gelehrt.

Wechselwirkung	Austauschteilchen
Starke	Gluon g
Elektromagnetische	Photon γ
Schwache	Boson Z, W
Gravitation	(Graviton?)
Erzeugung von Masse	Higgs

Quarks		Leptonen	
Up	u	Elektron	e
Down	d	Neutrino	ν_e
Charm	c	Myon	μ
Strange	s	Neutrino	ν_μ
Top	t	Tau	τ
Bottom	b	Neutrino	ν_τ

Tabelle 12.1 Elementarteilchen der Materie, Wechselwirkungen, Massenerzeugung

Während Leptonen als freie Teilchen in der Natur beobachtet werden, kommen Quarks nur in sogenannten Hadronen vor, das sind vorwiegend 2-er und 3-er Kombinationen: Baryonen wie z. B. das Proton uds, Neutron udd und das Lambda (Λ) uds haben drei Quarks, Mesonen (z. B. das Pion $u\bar{d}$) haben ein Quark und ein Antiquark. Antiquarks $\bar{u}$ werden von Quarks u durch einen Querstrich über dem Buchstaben unterschieden.

12.6.1 Produktion von Teilchen am Beschleuniger

Bei dem Large Electron-Positron Collider LEP am CERN in Genf wurden über viele Jahre Elektronen und Positronen auf eine hohe Geschwindigkeit $v \to c$ beschleunigt und dann zur Kollision gebracht (► Abbildung 12.10).

Abbildung 12.10 Kollision von Elektron-Positron-($e^- e^+$)-Paaren. Bei geeigneter Schwerpunktsenergie entstehen Z-Teilchen.

Wie bei den makroskopischen elastischen Stößen (Kapitel 8) müssen auch hier Energie- und Impulserhaltung gelten. Praktischerweise sind beide, Energie- und Impulserhaltung, in den Komponenten der Viererimpulse bereits enthalten. Wenn aus der Kollision z. B. ein Myon-Paar entstanden ist, gilt folgende Viererimpulserhaltung:

$$\mathbf{p}_{e^+} + \mathbf{p}_{e^-} = \mathbf{p}'_{\mu^+} + \mathbf{p}'_{\mu^-} \tag{12.48}$$

$$\begin{pmatrix} (E_{e^+} + E_{e^-})/c \\ \vec{p}_{e^+} + \vec{p}_{e^-} \end{pmatrix} = \begin{pmatrix} (E'_{\mu^+} + E'_{\mu^-})/c \\ \vec{p}'_{\mu^+} + \vec{p}'_{\mu^-} \end{pmatrix} \tag{12.49}$$

Elektronen und Positronen bei LEP hatten exakt dieselbe Energie, ihre jeweiligen Flugrichtungen waren entgegengesetzt, sodass ihre Impulse entgegengesetzt gleich groß waren $\vec{p}_{e^+} = -\vec{p}_{e^-}$. In diesem Fall ergänzen sich die Impulskomponenten jeweils auf beiden Seiten der Gleichung zu null, d. h. es ist auch $\vec{p}_{\mu^+} = -\vec{p}_{\mu^-}$.

Die Energiebilanz des Streuprozesses $e^+ + e^- \rightarrow \mu^{+\prime} + \mu^{-\prime}$ finden wir in der ersten Komponente von (12.49):

$$E_{e^+} + E_{e^-} = E'_{\mu^+} + E'_{\mu^-} \tag{12.50}$$

Nach Einsteins Gleichung (12.40) ist Masse eine Form der Energie. Das ermöglicht die Umwandlung von Bewegungsenergie in Masse und den umgekehrten Vorgang. Das Z-Teilchen ist ein Austauschteilchen der schwachen Wechselwirkung (▶ Tabelle 12.1). Es hat eine Ruhemasse von

$$m_Z = 91\,\mathrm{GeV}/c^2 \,. \tag{12.51}$$

Am LEP-Beschleuniger ist es gelungen, durch die Bewegungsenergie von kollidierenden Elektron-Positron-Paaren Z-Teilchen zu erzeugen und den Zerfall des Z-Teilchens z. B. in ein Myon-Paar zu beobachten. Der gesamte Prozess lautet dann $e^+ - e^- \rightarrow Z \rightarrow \mu^{+\prime} + \mu^{-\prime}$. Wird bei der Erzeugung $e^+ + e^- \rightarrow Z$ alle Bewegungsenergie des Elektrons und des Positrons in die Z-Masse umgewandelt, so ist das Z-Teilchen in Ruhe ($\vec{p}(Z) = 0$) und sein Viererimpuls beträgt

$$\mathbf{p}_Z = \begin{pmatrix} E/c \\ 0 \\ 0 \\ 0 \end{pmatrix} = \begin{pmatrix} m_Z c \\ 0 \\ 0 \\ 0 \end{pmatrix} . \tag{12.52}$$

Die Viererimpulserhaltung gilt auch einzeln für den Erzeugungsprozess des Z und genauso für seinen Zerfallsprozess. Für den Erzeugungsprozess lautet sie

$$\mathbf{p}_{e^+} + \mathbf{p}_{e^-} = \mathbf{p}_Z \,, \tag{12.53}$$

$$\begin{pmatrix} (E_{e^+} + E_{e^-})/c \\ \vec{p}_{e^+} + \vec{p}_{e^-} \end{pmatrix} = \begin{pmatrix} m_Z c \\ 0 \end{pmatrix} . \tag{12.54}$$

Wegen der symmetrischen Strahlenergien $E_{e^+} = E_{e^-}$ und wegen $\vec{p}_{e^+} = -\vec{p}_{e^-}$ folgt also, dass die Elektronen und Positronen jeweils die halbe Ruheenergie des Z-Teilchens in der Form von Bewegungsenergie mitbringen müssen:

$$E_e = \frac{m_Z c^2}{2} \tag{12.55}$$

Den Beweis dafür, dass der Zwischenzustand eines Z-Teilchens entstanden war, lieferten die Experimente am LEP-Beschleuniger durch Messungen der Zählrate der entstandenen Myon-Paare. Durch Variation der e^+e^--Strahlenergie fanden sie eine Resonanz in der Zählrate bei der Schwerpunktsenergie, die der Ruhemasse des Z-Teilchens entspricht (► Abbildung 12.11).

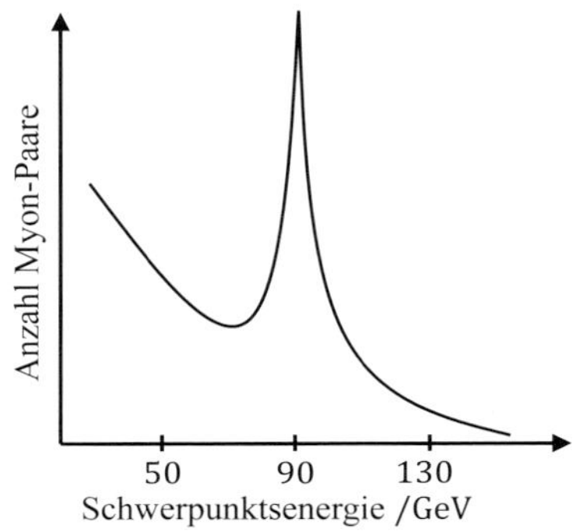

Abbildung 12.11 Anzahl der gemessenen Myon-Paare in Abhängigkeit von der Schwerpunktsenergie der kollidierenden e^+e^--Paare. An der Z-Resonanz steigt die Zahl stark an.

12.6.2 Zerfall eines Lambda-Teilchens

Auch beim Zerfall eines massiven Teilchens in Teilchen mit geringerer Masse müssen Energie und Impuls erhalten bleiben.

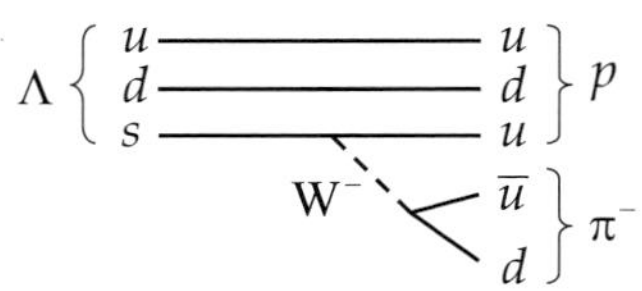

Abbildung 12.12 Zerfall des Lambda-Teilchens Λ in ein Proton p und ein Pion π, siehe auch Tabelle 12.1 für die Elementarteilchen u, d, s und W

Als Beispiel betrachten wir ein ungeladenes Lambda-Teilchen (Λ), das in ein positiv geladenes Proton p und ein negativ geladenes Pion π^- zerfällt (► Abbildung 12.12). Der Quark-Inhalt der Teilchen ist in Klammern gesetzt:

$$\Lambda\,(uds) \quad \rightarrow \quad p\,(uud) \quad + \quad \pi^-\,(\bar{u}d) \tag{12.56}$$

Wir untersuchen zunächst die Impulse der Zerfallsteilchen für ein ruhendes Λ-Teilchen (linke ► Abbildung 12.13). Anschließend analysieren wir den Λ-Zerfall im Flug und nutzen die Lorentztransformation, um die unterschiedlichen Zerfallswinkel des Protons und Pions zu begründen (rechte ► Abbildung 12.13).

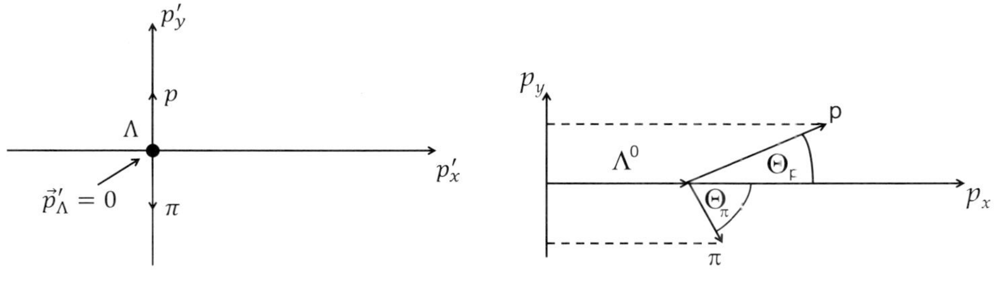

Abbildung 12.13 Links: Zerfall eines ruhenden Λ-Teilchens $\vec{p}\,'(\Lambda) = 0$ mit den entgegengesetzten Flugrichtungen seiner Zerfallsteilchen Proton und Pion entlang der p'_y-Achse. Rechts: Zerfall eines Λ-Teilchens, das sich mit dem Impuls $p_x > 0$ nach rechts bewegt. Die Zerfallswinkel Θ_p und Θ_π des Protons und des Pions ergeben sich aus der Lorentztransformation.

Impulse der Zerfallsteilchen Proton und Pion

Die Ruhemassen der einzelnen Teilchen sind

$$m_\Lambda = 1{,}116\,\mathrm{GeV}/c^2\,, \tag{12.57}$$

$$m_p = 0{,}938\,\mathrm{GeV}/c^2\,, \tag{12.58}$$

$$m_\pi = 0{,}140\,\mathrm{GeV}/c^2\,. \tag{12.59}$$

Die Masse des Λ-Teilchens ist größer als die Summe der Massen von Proton und Pion zusammen:

$$\Delta m \equiv m_\Lambda - (m_p + m_\pi) \approx 0{,}04\,\mathrm{GeV}/c^2 \tag{12.60}$$

Die Energieerhaltung erfordert, dass dieser Massenunterschied in Bewegungsenergie umgesetzt wird.

Zerfall des ruhenden Lambda-Teilchens

Wir betrachten zunächst den Zerfall im Ruhesystem des Λ-Teilchens (linke ► Abbildung 12.13). Viererimpulserhaltung impliziert, dass der Viererimpuls des ruhenden Λ-Teilchens und die Summe der Viererimpulse seiner Zerfallsteilchen gleich sein müssen. Wir verwenden hier Viererimpulse $\mathbf{p}'$ für das Ruhesystem zur späteren Unterscheidung von Impulsen **p** bei bewegtem Λ-Teilchen. Das Koordinatensystem wählen wir so, dass die p'_y-Achse in die Zerfallsrichtung des Protons zeigt (► Abbildung 12.13). Es gilt

$$\mathbf{p}'_\Lambda = \mathbf{p}'_p + \mathbf{p}'_\pi \tag{12.61}$$

$$\begin{pmatrix} m_\Lambda c \\ 0 \\ 0 \\ 0 \end{pmatrix} = \begin{pmatrix} E'_p/c \\ 0 \\ p'_{y,p} \\ 0 \end{pmatrix} + \begin{pmatrix} E'_\pi/c \\ 0 \\ p'_{y,\pi} \\ 0 \end{pmatrix} . \tag{12.62}$$

Die Impulserhaltung ist in den unteren drei Komponenten von (12.62) ablesbar. Mit dem absoluten Impulsbetrag $p'_y > 0$ erhalten wir

$$0 = \vec{p}'_p + \vec{p}'_\pi , \tag{12.63}$$

$$p'_{y,p} = p'_y , \tag{12.64}$$

$$p'_{y,\pi} = -p'_y . \tag{12.65}$$

Die Energieerhaltung steht in der oberen Komponente von (12.62):

$$m_\Lambda c^2 = E'_p + E'_\pi \tag{12.66}$$

Für die Energien der Zerfallsteilchen nutzen wir Gleichung (12.40) und setzen die Impulserhaltung (12.63) mit dem Impulsbetrag p'_y (12.64), (12.65) ein:

$$E'_p = \sqrt{\vec{p}'^2_p c^2 + m_p^2 c^4} = \sqrt{p'^2_y c^2 + m_p^2 c^4} \tag{12.67}$$

$$E'_\pi = \sqrt{\vec{p}'^2_\pi c^2 + m_\pi^2 c^4} = \sqrt{p'^2_y c^2 + m_\pi^2 c^4} \tag{12.68}$$

Die Bilanz der Ruheenergien des Zerfalls $\Delta m\, c^2 \approx 0{,}04\,\text{GeV}$ (12.60) wird in Bewegungsenergie $p'_y c$ der Zerfallsteilchen umgesetzt. Da $\Delta m\, c^2$ sehr viel kleiner als die Ruheenergie des Protons $m_p c^2 = 0{,}938\,\text{GeV}$ (12.58) ist, nähern wir den Wurzelterm in der Protonenergie (12.67) durch

$$p'_y c \ll m_p c^2 \tag{12.69}$$

$$E'_p = \sqrt{p'^2_y c^2 + m_p^2 c^4} \approx m_p c^2 . \tag{12.70}$$

Für die Pionenergie machen wir keine solche Näherung, da die Ruheenergie des Pions $m_\pi c^2 = 0{,}14\,\text{GeV}$ (12.59) nicht sehr viel größer als $\Delta m\, c^2$ ist.

Setzen wir in die Energieerhaltungsgleichung (12.66) die Protonenergie (12.70) und die Pionenergie (12.68) ein, so erhalten wir den Impulsbetrag p'_y der beiden Zerfallsteilchen:

$$m_\Lambda c^2 = E'_p + E'_\pi \tag{12.71}$$

$$m_\Lambda c^2 \approx m_p c^2 + \sqrt{p'^2_y c^2 + m_\pi^2 c^4} \tag{12.72}$$

$$(m_\Lambda c^2 - m_p c^2)^2 \approx p'^2_y c^2 + m_\pi^2 c^4 \tag{12.73}$$

$$p'^2_y \approx (m_\Lambda - m_p)^2\, c^2 - m_\pi^2 c^2 \tag{12.74}$$

$$p'_y \approx 0{,}11\, \frac{\text{GeV}}{c} \tag{12.75}$$

Zusammenfassend stellen wir fest: Aufgrund der **Impulserhaltung** $p'_{y,p} + p'_{y,\pi} = 0$ (12.63) sind Proton- und Pionimpuls im Betrag gleich groß und entgegengesetzt gerichtet (► Abbildung 12.13). Daher sind auch ihre Bewegungsenergien $E'_{\text{kin}} = p'_y c$ gleich groß. Aus der **Erhaltung der Gesamtenergie** $m_\Lambda c^2 = E'_p + E'_\pi$ (12.66) erhalten wir unter Berücksichtigung der Impulserhaltung den Betrag des Proton- und Pionimpulses zu $p'_y \approx 0{,}11\,\text{GeV}/c$.

Zerfall des bewegten Lambda-Teilchens

Bislang haben wir den Zerfall des ruhenden Λ-Teilchens betrachtet. Bei Laborexperimenten ist allerdings der typische Fall, dass sich das Λ-Teilchen bewegt und wie in der rechten ► Abbildung 12.13 im Flug zerfällt.

Der Zerfall des Λ-Teilchens in seinem Ruhesystem oder alternativ im Bezugssystem des Labors gemessen, ergibt verschiedene Bewegungsrichtungen der Zerfallsteilchen. Allerdings handelt es sich um denselben physikalischen Vorgang, der hier aus verschiedenen Bezugssystemen beobachtet wird. Im Folgenden zeigen wir den Transfer zwischen den beiden Bezugssystemen durch die Lorentztransformation und damit eine ihrer wichtigsten Anwendungen.

Aus unseren bisherigen Berechnungen sind uns die Energien und Impulse der Zerfallsteilchen Proton und Pion im Λ-Ruhesystem bekannt. Darüber hinaus sei die Energie des Λ-Teilchens im Laborsystem bekannt, das sich in p_x-Richtung bewegen soll (► Abbildung 12.13). Wir verwenden jetzt Viererimpulse $\mathbf{p}$ für das Laborsystem, um es von $\mathbf{p}'$ für das Ruhesystem des Λ-Teilchen zu unterscheiden.

Bei der Lorentztransformation vom Λ-Ruhesystem in das Laborsystem bezeichnen der γ-Faktor (11.45) und $\beta = v/c$ die Kinematik des Λ-Teilchens im Labor. Damit wird die Transformationsmatrix (12.32) gefüllt. Anschließend werden die Viererimpulse des Protons und des Pions einzeln wie folgt transformiert:

$$\begin{pmatrix} E/c \\ p_x \\ p_y \\ p_z \end{pmatrix} = \begin{pmatrix} \gamma & \gamma\beta & 0 & 0 \\ \gamma\beta & \gamma & 0 & 0 \\ 0 & 0 & 1 & 0 \\ 0 & 0 & 0 & 1 \end{pmatrix} \begin{pmatrix} E'/c \\ 0 \\ p'_y \\ 0 \end{pmatrix} \tag{12.76}$$

$$= \begin{pmatrix} \gamma E'/c \\ \beta\gamma E'/c \\ p'_y \\ 0 \end{pmatrix} \tag{12.77}$$

Damit sind die Energien und Impulse von Proton und Pion im Laborsystem bekannt.

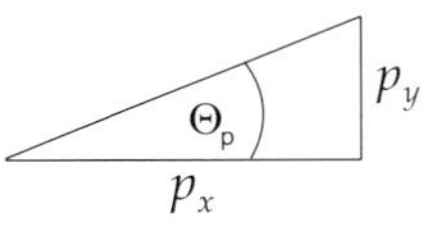

Abbildung 12.14 Zusammmenhang zwischen dem Zerfallswinkel Θ_p und den Impulskomponenten des Protons

Mit den Impulskomponenten p_x und p_y bestimmen wir die Winkel Θ beim Λ-Zerfall (► Abbildung 12.14). Ein größerer γ-Faktor des Λ-Teilchens entspricht wegen $E = \gamma\, m_\Lambda\, c$ (12.36) einer höheren Λ-Energie im Labor. Der Intuition entsprechend werden bei höherem γ-Faktor das Proton und das Pion unter kleinerem Winkel Θ zur ursprünglichen Bewegungsrichtung des Λ-Teilchens fliegen. Im Folgenden zeigen wir ein Anwendungsbeispiel.

Beispiel 38: Lorentztransformation beim Lambda-Zerfall

Ein Lambda-Teilchen Λ fliege im Labor in x-Richtung mit der Energie $E_\Lambda = 5{,}5\,\text{GeV}$ (► Abbildung 12.13). Der γ-Faktor und die relative Geschwindigkeit $\beta = v/c$ (11.45) lassen sich aus $E_\Lambda = \gamma\, m_\Lambda\, c^2$ (12.36) bestimmen:

$$\gamma = \frac{E_\Lambda}{m_\Lambda c^2} = \frac{5{,}5}{1{,}1} = 5 \tag{12.78}$$

$$\gamma = \frac{1}{\sqrt{1-\beta^2}} \Rightarrow \beta = \sqrt{1 - \frac{1}{\gamma^2}} = \sqrt{1 - \frac{1}{25}} \approx 1 \tag{12.79}$$

Wir setzen in die Lorentztransformation (12.77) ein und erhalten für das Proton im Laborsystem

$$\mathbf{p}_p = \begin{pmatrix} E \\ p_x c \\ p_y c \\ p_z c \end{pmatrix}_p = \begin{pmatrix} \gamma E'_p \\ \beta\gamma E'_p \\ p'_y c \\ 0 \end{pmatrix} . \tag{12.80}$$

Konkret erhalten wir für die Energie und den Impuls des Protons mit (12.75)

$$E_p = \gamma E'_p = \gamma\sqrt{m_p^2 c^4 + p'^2_y c^2} \approx 4{,}7\,\text{GeV} \tag{12.81}$$

$$\vec{p}_p = \begin{pmatrix} \beta\gamma\sqrt{m_p^2 c^4 + p'^2_y c^2} \\ p'_y c \\ 0 \end{pmatrix} \approx \begin{pmatrix} 4{,}7 \\ 0{,}1 \\ 0 \end{pmatrix} \frac{\text{GeV}}{c} . \tag{12.82}$$

Für das Pion ergibt sich der Viererimpuls im Laborsystem zu

$$\mathbf{p}_\pi = \begin{pmatrix} E \\ p_x c \\ p_y c \\ p_z c \end{pmatrix}_\pi = \begin{pmatrix} \gamma E'_\pi \\ \beta\gamma E'_\pi \\ -p'_y c \\ 0 \end{pmatrix} . \tag{12.83}$$

Pionenergie und -impuls sind dann

$$E_\pi = \gamma E'_\pi = \gamma\sqrt{m_\pi^2 c^4 + p'^2_y c^2} = 0{,}8\,\text{GeV} \tag{12.84}$$

$$\vec{p}_\pi = \begin{pmatrix} \beta\gamma\sqrt{m_\pi^2 c^4 + p'^2_y c^2} \\ -p'_y c \\ 0 \end{pmatrix} \approx \begin{pmatrix} 0{,}8 \\ -0{,}1 \\ 0 \end{pmatrix} \frac{\text{GeV}}{c} . \tag{12.85}$$

Der weitaus größere Anteil der Energie entfällt also auf das schwerere Proton. Die Energie ist mit $E_\Lambda = E_P + E_\pi = (4{,}7 + 0{,}8)\,\text{GeV} = 5{,}5\,\text{GeV}$ erhalten.

Beispiel 39: Winkel der Lambda-Zerfallsteilchen

Im Folgenden werden wir die Winkel im Laborsystem berechnen (linke ► Abbildung 12.15), unter denen das Proton und das Pion relativ zum Λ-Teilchen fliegen, das dieselbe Kinematik hat wie im vorigen Beispiel. Für das Proton erhalten wir den Zerfallswinkel durch Einsetzen von (12.80):

$$\tan\Theta_p = \frac{p_{y,p}}{p_{x,p}} = \frac{p'_y c}{\beta\gamma E'_p} = \frac{p'_y c}{\beta\gamma\sqrt{m_p^2 c^4 + p'^2_y c^2}} = \frac{1}{\beta\gamma\sqrt{\frac{m_p^2 c^4}{p'^2_y c^2} - 1}} \tag{12.86}$$

$$\approx \frac{1}{1\cdot 5{,}5\cdot\sqrt{\frac{1}{0{,}1^2}+1}} \approx \frac{1}{55} \Rightarrow \Theta_p \approx \frac{360^\circ}{2\pi}\cdot\frac{1}{55} \approx 1^\circ \tag{12.87}$$

Das Proton mit seiner großen Masse fliegt annähernd entlang der Originalflugrichtung des Λ-Teilchens. Für das Pion erhalten wir

$$\tan\Theta_\pi = \frac{p_{y,\pi}}{p_{x,\pi}} = \frac{-p'_y c}{\beta\gamma E'_\pi} = \frac{-1}{\beta\gamma\sqrt{\frac{m_\pi^2 c^4}{p'^2_y c^2} + 1}} \approx 0{,}11 \Rightarrow \Theta_\pi \approx -6^\circ\,. \tag{12.88}$$

Das sehr viel leichtere Pion fliegt im Vergleich zum Proton unter einem deutlich größeren Winkel relativ zur Flugrichtung des Λ-Teilchens:

$$|\Theta_\pi| \gg |\Theta_p| \tag{12.89}$$

In der rechten ► Abbildung 12.15 ist die Abhängigkeit der Zerfallswinkel vom Lorentzfaktor (Lorentz-Boost) dargestellt. Dabei wurde die Näherung $m_\pi = 0$ gewählt, um die besonders große Abhängigkeit für leichte Teilchen zu zeigen. Für genaue Zahlenwerte muss man natürlich die Ruhemasse des Pions berücksichtigen.

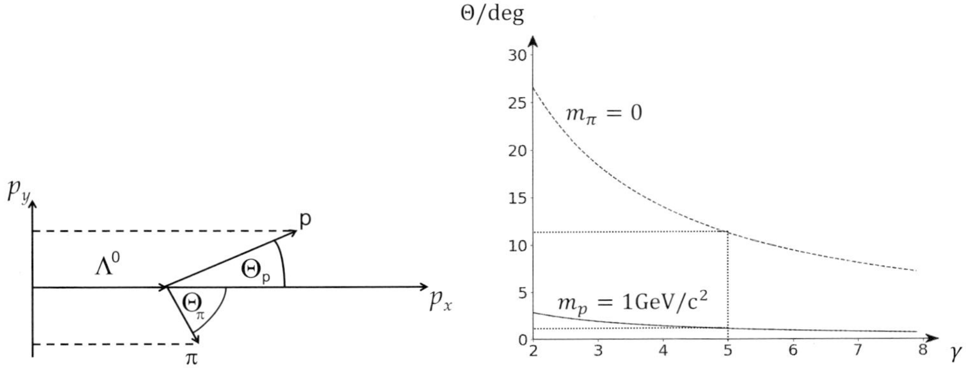

Abbildung 12.15 Links: Kinematik des Λ-Zerfalls in Pion und Proton, rechts: Die Zerfallswinkel des Pions und des Protons hängen vom Lorentzfaktor des Λ-Teilchens ab.

Das Wichtigste zum Mitnehmen

Zum Mitnehmen: Spezielle Relativitätstheorie 1

- Die Konstanz der Lichtgeschwindigkeit c im Vakuum

$$c = 299\,792\,458\,\frac{\mathrm{m}}{\mathrm{s}} \tag{12.6}$$

unabhängig vom Inertialsystem hat mit den Lorentztransformationen (11.46) für hohe Geschwindigkeiten weitgehende Auswirkungen in der **Speziellen Relativitätstheorie**.

- Zwischen den Zeitmessungen in gegeneinander bewegten Systemen ergibt sich eine **Zeitdilatation**

$$\Delta t = \gamma \Delta t' \tag{12.12}$$

die besagt, dass Uhren in schnell bewegten Systemen aus der Sicht eines ruhenden Beobachters langsamer laufen.

Dabei ist die **Eigenzeit** die Zeit $\Delta t'$ zwischen zwei Ereignissen in einem Bezugssystem am selben Ort. Ein eindringliches Beispiel für die Zeitdilatation sind Myonen, die trotz einer kurzen Lebensdauer von $\Delta t' \approx 5 \cdot 10^{-6}$ s in der kosmischen Höhenstrahlung sehr große Entfernungen zurücklegen können.

- Wegen der Raum-Zeit-Verknüpfung müssen die Endpunkte bei einer Längenmessung gleichzeitig bestimmt werden. Mit den Differenzen $l' \equiv x_2' - x_1'$ und $l \equiv x_2 - x_1$ erhalten wir die folgende Formulierung der **Längenkontraktion**:

$$l = \frac{1}{\gamma} l' \tag{12.16}$$

Bewegte Objekte wirken für den ruhenden Beobachter verkürzt.

- Für hohe Geschwindigkeiten sind Raum und Zeit in einem vierdimensionalen **Raum-Zeit-Diagramm** untrennbar miteinander verknüpft.

Das Hier und Jetzt befindet sich im Ursprung, entlang der Achsen laufen die Raumkoordinaten x, y, z und die Zeit als ct. Vergangenheit und Zukunft werden vom Vorzeichen der Zeit bestimmt.

Ereignisse werden durch **Weltlinien** dargestellt, die wegen der maximalen Geschwindigkeit c der Signalübertragung den **Lichtkegel**

$$x^2 + y^2 + z^2 = c^2 t^2 \tag{12.17}$$

nicht überschreiten können.

Zum Mitnehmen: Spezielle Relativitätstheorie 2

- Für die Längenmessung in der Raumzeit definiert man das Skalarprodukt mithilfe einer **Metrik**:

$$\hat{g} = \begin{pmatrix} 1 & 0 & 0 & 0 \\ 0 & -1 & 0 & 0 \\ 0 & 0 & -1 & 0 \\ 0 & 0 & 0 & -1 \end{pmatrix} \tag{12.25}$$

Für einen **Vierervektor** $\mathbf{a}$ der Raumzeit entlang der x-Achse ergibt sich durch Matrixmultiplikation mit der Metrik

$$\mathbf{a}^2 = (ct_a \; x_a \; 0 \; 0) \begin{pmatrix} 1 & 0 & 0 & 0 \\ 0 & -1 & 0 & 0 \\ 0 & 0 & -1 & 0 \\ 0 & 0 & 0 & -1 \end{pmatrix} \begin{pmatrix} ct_a \\ x_a \\ 0 \\ 0 \end{pmatrix}$$

Das Quadrat der Länge von a in der Raumzeit ist damit

$$\mathbf{a}^2 = c^2 t_a^2 - x_a^2 \,. \tag{12.28}$$

- Für die Vierervektoren in der Raumzeit gibt es eine nützliche Schreibweise der **Lorentztransformation** (11.46) **in Matrixform**:

$$\begin{pmatrix} ct' \\ x' \\ y' \\ z' \end{pmatrix} = \begin{pmatrix} \gamma & -\gamma\beta & 0 & 0 \\ -\gamma\beta & \gamma & 0 & 0 \\ 0 & 0 & 1 & 0 \\ 0 & 0 & 0 & 1 \end{pmatrix} \begin{pmatrix} ct \\ x \\ y \\ z \end{pmatrix} \tag{12.31}$$

$$\begin{pmatrix} ct \\ x \\ y \\ z \end{pmatrix} = \begin{pmatrix} \gamma & \gamma\beta & 0 & 0 \\ \gamma\beta & \gamma & 0 & 0 \\ 0 & 0 & 1 & 0 \\ 0 & 0 & 0 & 1 \end{pmatrix} \begin{pmatrix} ct' \\ x' \\ y' \\ z' \end{pmatrix} \tag{12.32}$$

Zum Mitnehmen: Spezielle Relativitätstheorie 3

- In Einsteins berühmter Formulierung

$$E = mc^2 \tag{12.33}$$

ist Masse eine Form von Energie, die mit dem Lorentzfaktor γ (11.45) zunimmt:

$$m = \gamma m_\circ \tag{12.34}$$

Der Impuls des Teilchens ist dann

$$\vec{p} = \gamma\, m_\circ\, \vec{v}\,. \tag{12.41}$$

Die **Ruhemasse** $m_\circ$ ist eine Teilcheneigenschaft, z. B.:

$$\begin{aligned} \text{Elektron} \quad & m_e = 0{,}511\,\mathrm{MeV}/c^2 \\ \text{Proton} \quad & m_p = 938\,\mathrm{MeV}/c^2 \\ \text{Elektronenvolt} \quad & 1\,\mathrm{eV} = 1{,}6 \cdot 10^{-19}\,\mathrm{J} \end{aligned}$$

Einsteins Formel (12.33) lässt sich mit (12.34) und (12.41) so schreiben:

$$E^2 = m_\circ^2 c^4 + p^2 c^2 \tag{12.40}$$

Die Gesamtenergie eines freien Teilchens besteht aus seiner Ruheenergie (Ruhemasse) und seiner Bewegungsenergie (Impuls).

- In Analogie zur Raumzeit kann man einen **Energie-Impuls-Raum** definieren. Die Vierervektoren heißen hier Viererimpulse **p**:

$$\mathbf{p} = \begin{pmatrix} E/c \\ p_x \\ p_y \\ p_z \end{pmatrix} = \begin{pmatrix} E/c \\ \vec{p} \end{pmatrix} \tag{12.42}$$

Durch die Metrik (12.25) ergibt das Skalarprodukt von **p** mit sich selbst

$$\mathbf{p}^2 = ((E/c)\ p_x\ p_y\ p_z) \begin{pmatrix} 1 & 0 & 0 & 0 \\ 0 & -1 & 0 & 0 \\ 0 & 0 & -1 & 0 \\ 0 & 0 & 0 & -1 \end{pmatrix} \begin{pmatrix} E/c \\ p_x \\ p_y \\ p_z \end{pmatrix}.$$

Vergleichen wir dieses Ergebnis mit Gleichung (12.40), sehen wir die eigentliche Bedeutung des Skalarprodukts des Viererimpulses mit sich selbst:

$$\mathbf{p}^2 c^2 = m_\circ^2 c^4 = E^2 - \vec{p}^{\,2} c^2 \tag{12.46}$$

Die invariante Länge $\mathbf{p}^2$ im Energie-Impuls-Raum entspricht hier der Ruhemasse $m_\circ$ des Teilchens, multipliziert mit dem Quadrat der Lichtgeschwindigkeit c.

Gemeinsam verstehen

13

Wie gut können Sie bereits über die Themen sprechen? Führen Sie zu zweit ein wissenschaftliches Gespräch miteinander und beachten Sie dabei die Hinweise in Kapitel 6.

Aufgabe 31: Partnerarbeit: Stoßprozesse

Welche physikalischen Größen sind bei Kollisionen einer Holzscheibe mit einer a) weiteren Holzscheibe, b) Eisenscheibe, c) Wand relevant?

Aufgabe 32: Partnerarbeit: Gravitationsgesetz

Erklären Sie die Form des Gravitationsgesetztes von Newton

$$\vec{F} = -G\,\frac{m\,M}{r^2}\,\vec{e}_r\,. \tag{13.1}$$

Welche Bedeutung hat die inverse quadratische Abstandsabhängigkeit? Wie stark ist die Wechselwirkung?

Aufgabe 33: Partnerarbeit: Gravitationskonstante

Erklären Sie ein Messverfahren zur Bestimmung der Gravitationskonstante.

Aufgabe 34: Partnerarbeit: Masse Himmelskörper

Wie kann man die Massen von Sonne und Erde bestimmen?

Aufgabe 35: Partnerarbeit: Effektives Potential

Skizzieren Sie das effektive Potential für die Bewegung der Planeten um die Sonne. Erklären Sie den Anteil kinetischer Energie an der Gesamtenergie für den Bindungszustand und für Streuprozesse. Welche Bahnkurvenformen sind möglich?

Aufgabe 36: Partnerarbeit: Foucault

Aus welchem Versuch konnte Foucault auf die Erdrotation schließen?

Aufgabe 37: Partnerarbeit: Ruhemasse

Welche Auswirkung hat die Ruhemasse eines Teilchens auf seine Gesamtenergie? Was bedeutet das für die Erzeugung des Teilchens an einem Beschleuniger?

Schwingungen – Wellen – Körperdrehung

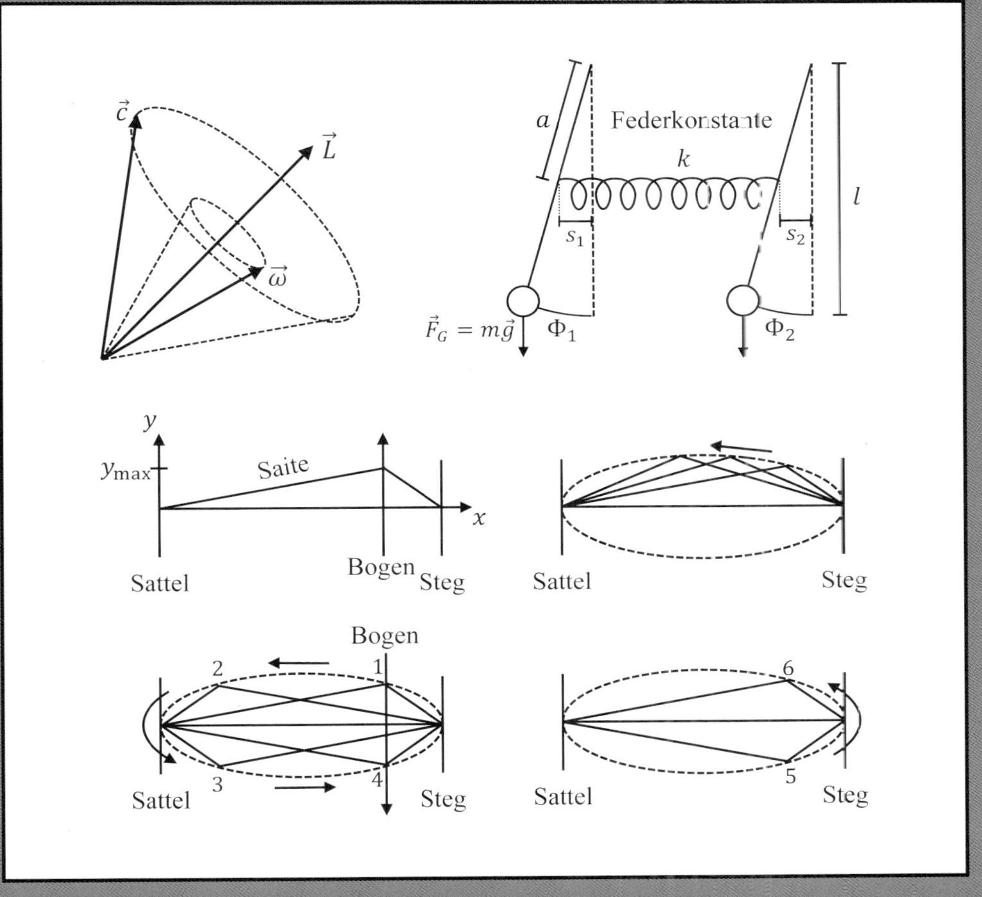

Teil III

Dynamik starrer Körper

14

14.1 Dichte . . . 259

14.2 Schwerpunkt . . . 260

14.3 Trägheitsmoment . . . 261

14.4 Drehschwingungen . . . 264

14.5 Satz von Steiner . . . 268

14.6 Rotationsenergie . . . 269
- 14.6.1 Trägheitsmomente Hohl- und Vollzylinder . . . 270
- 14.6.2 Potentielle Energie, Translations- und Rotationsenergie . . . 270
- 14.6.3 Beschleunigung des rollenden Zylinders . . . 272
- 14.6.4 Bahnkurve des Zylinders . . . 273

14.7 Trägheitstensor . . . 273

14.8 Kreisel . . . 278
- 14.8.1 Nutation . . . 279
- 14.8.2 Präzession . . . 283

ÜBERBLICK

» Die Beschreibung der Bewegung eines ausgedehnten Körpers können wir oft durch die Annahme vereinfachen, dass die gesamte Körpermasse im Schwerpunkt konzentriert ist (Massenpunkte). Für die Beschreibung von Drehbewegungen z. B. eines Karussells oder eines Kreisels funktioniert diese Näherung nicht.

In diesem Kapitel werden wir ausgedehnte, nicht deformierbare Körper untersuchen. Sie werden als starre Körper bezeichnet. Die Verteilung ihrer Masse hat entscheidende Konsequenzen für ihr Verhalten bei Drehbewegungen. «

14.1 Dichte

Wir zerlegen einen Körper in Volumenelemente dV mit entsprechender Masse dm (► Abbildung 14.1).

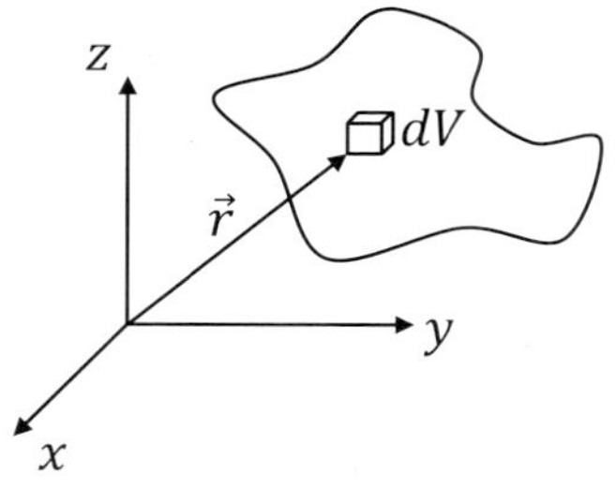

Abbildung 14.1 Zur Definition der Dichte ρ eines Körperelements dV

Das Volumen V eines kontinuierlichen Körpers mit Volumenelementen dV bestimmen wir durch Integration:

$$V = \int_V dV \tag{14.1}$$

Ebenso berechnen wir die Masse M des Körpers durch

$$M = \int_M dm \,. \tag{14.2}$$

Eine nützliche Definition ist die Massendichte

$$\rho = \frac{dm}{dV} \,. \tag{14.3}$$

Mit dieser Dichte können wir die Masse über ein Volumenintegral berechnen:

$$M = \int_V \rho \, dV \tag{14.4}$$

Wenn die Dichte des Materials im ganzen Körpervolumen gleich ist (»homogener Körper«), vereinfacht sich die Massenberechnung zu

$$M = \rho \int_V dV = \rho V \,. \tag{14.5}$$

Beispiel 40: Einige Dichtewerte

Material	Dichte in kg/m^3
Wasserstoff	0,08988
Luft (0 °C)	1,292
Balsaholz	ca. 200
Eichenholz	670

Material	Dichte in kg/m^3
Wasser (3,98 °C)	999,975
Meerwasser	1 025
Eisen	7 874
Platin	21 450

14.2 Schwerpunkt

Jeder Körper hat einen Schwerpunkt $\vec{r}_s$ bezüglich seiner Massenverteilung. Er lässt sich im diskreten Fall von Massenpunkten m_i über eine gewichtete Summe ermitteln:

$$\vec{r}_s = \frac{\sum_{i=1}^{n} m_i \vec{r}_i}{\sum_{i=1}^{n} m_i} \tag{14.6}$$

Im kontinuierlichen Fall ersetzen wir die Summe durch das Integral:

$$\vec{r}_s = \frac{\int \vec{r}\, dm}{\int dm} = \frac{\int \vec{r}\, \rho\, dV}{\int \rho\, dV} = \frac{1}{M} \int \vec{r}\, \rho\, dV \tag{14.7}$$

Für homogene Körper, d. h. für Körper mit $\rho(\vec{r}) = const. = \rho$, ist mit $\rho = M/V$ (14.5)

$$\vec{r}_s = \frac{\rho}{M} \int \vec{r}\, dV = \frac{1}{V} \int \vec{r}\, dV\,. \tag{14.8}$$

Beispiel 41: Schwerpunkt einer homogenen Halbkugel

In ► Abbildung 14.2 ist eine homogene Halbkugel mit Radius R gezeigt. Das Volumen einer Vollkugel ist $\frac{4}{3}\pi R^3$, dementsprechend ist das Halbkugelvolumen $V = \frac{2}{3}\pi R^3$. Aus Symmetriegründen liegt der Schwerpunkt bei $x_s = y_s = 0$, sodass wir nur die vertikale Komponente des Halbkugelschwerpunkts berechnen müssen:

$$z_s = \frac{1}{V} \int z\, dV \tag{14.9}$$

In Kugelkoordinaten (Abschnitt 9.3) ist $z = r \cos\vartheta$ (9.36). Das Volumenelement ist $dV = r^2 dr\, \sin\vartheta\, d\vartheta\, d\varphi$ (9.50). Damit erhalten wir nach Sortieren der Terme

$$z_s = \frac{1}{V} \iiint r \cos\vartheta\, r^2\, dr\, \sin\vartheta\, d\vartheta\, d\varphi \tag{14.10}$$

$$= \frac{1}{V} \cdot \int_0^R r^3 dr \cdot \int_0^{\frac{\pi}{2}} \cos\vartheta\, \sin\vartheta\, d\vartheta \cdot \int_0^{2\pi} d\varphi\,. \tag{14.11}$$

Das erste und das dritte Integral lassen sich sofort lösen. Das zweite Integral lässt sich mit $d\sin\vartheta/d\vartheta = \cos\vartheta$ berechnen:

$$\int_0^{\frac{\pi}{2}} \cos\vartheta\,\sin\vartheta\,d\vartheta = \int_{\sin(0)}^{\sin(\pi/2)} \sin\vartheta\,d\sin\vartheta = \frac{1}{2}\left[\sin^2(\vartheta)\right]\Big|_0^1 = \frac{1}{2} \tag{14.12}$$

Damit ergibt sich für den Schwerpunkt in z-Richtung

$$z_s = \frac{1}{\frac{2}{3}\pi R^3}\cdot\frac{1}{4}R^4\cdot\frac{1}{2}\cdot 2\pi = \frac{3}{8}R\,. \tag{14.13}$$

Der Schwerpunkt liegt also etwas unterhalb des halben Radius R auf der z-Achse.

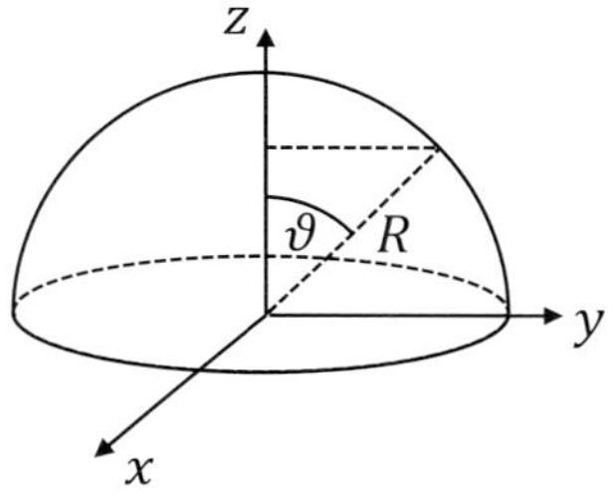

Abbildung 14.2 Zur Berechnung des Schwerpunkts einer homogenen Halbkugel mit Radius R

14.3 Trägheitsmoment

Als Ausgangspunkt für unsere Überlegungen zu Trägheitsmomenten verwenden wir den Drehimpuls eines Massenpunkts. Zur Erinnerung: Der Drehimpuls $\vec{L}$ (4.52) ist durch das Vektorprodukt aus dem Impuls $\vec{p}$ des Massenpunkts und seinem Abstand $\vec{r}$ von einem Bezugspunkt definiert:

$$\vec{L} = \vec{r}\times\vec{p} \tag{14.14}$$

Für Kreisbewegungen mit der Winkelgeschwindigkeit $\vec{\omega}$ können wir mit dem Impuls $\vec{p} = m\vec{v}$ des Massenpunkts und seiner Geschwindigkeit $\vec{v} = \vec{\omega}\times\vec{r}$ (4.48)

$$\vec{L} = m\vec{r}\times\vec{v} \tag{14.15}$$

$$= m\vec{r}\times(\vec{\omega}\times\vec{r}) \tag{14.16}$$

$$= \underbrace{mr^2}_{\equiv I}\vec{\omega} - m\vec{r}\underbrace{(\vec{r}\vec{\omega})}_{=0} \tag{14.17}$$

$$= I\vec{\omega} \tag{14.18}$$

schreiben. Bei Drehbewegungen ist das Trägheitsmoment $I = mr^2$ (4.60) eines einzelnen Massenpunkts in Analogie zu der trägen Masse m bei Translationsbewegungen definiert.

Für den starren Körper müssen wir das Trägheitsmoment in einer verallgemeinerten Form definieren. Wir legen das Koordinatensystem so, dass sich der Körper um die z-Achse dreht (► Abbildung 14.3) und damit der Vektor der Winkelgeschwindigkeit $\vec{\omega}$ parallel zum Einheitsvektor $\vec{e}_z$ steht: $\vec{\omega} \parallel \vec{e}_z$.

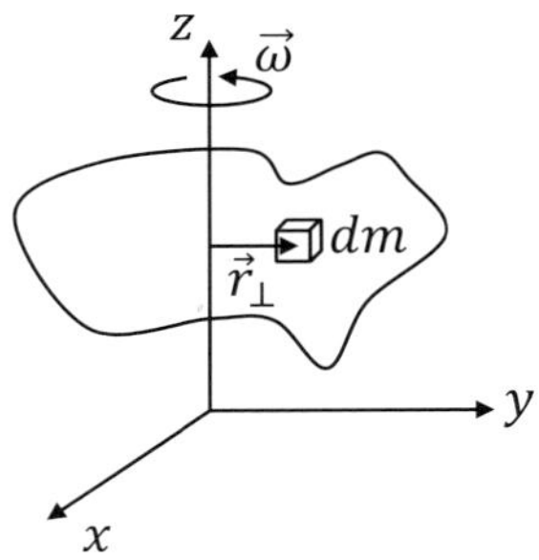

Abbildung 14.3 Trägheitsmoment eines starren Körpers, der sich mit der Winkelgeschwindigkeit $\vec{\omega}$ um die z-Achse dreht

Für einen Körper, der aus N diskreten Massenpunkten aufgebaut ist, lässt sich der Gesamtdrehimpuls $\vec{L}$ aus den N einzelnen Drehimpulsen $\vec{L}_i$ bezüglich der festen Drehachse ausdrücken:

$$\vec{L} = \sum_i^N \vec{L}_i \tag{14.19}$$

$$= \underbrace{\sum_i^N \left(m_i r_{\perp i}^2\right)}_{\equiv I_z} \vec{\omega} \tag{14.20}$$

Das Trägheitsmoment bezüglich der z-Achse ist hier die Summe über alle Massenelemente m_i des Körpers, gewichtet mit dem Quadrat ihres minimalen Abstands $r_\perp$ von der Drehachse.

Für eine kontinuierliche Massenverteilung ersetzen wir die Summe durch das Integral über alle Massenelemente dm des Körpers, gewichtet mit dem Quadrat ihres minimalen Abstands $r_\perp$ von der Drehachse:

$$I_z = \int r_\perp^2 \, dm \tag{14.21}$$

Häufig ist es für Berechnungen praktisch, dm durch die Dichte $\rho \, dV$ (14.3) auszudrücken:

$$I_z = \int r_\perp^2 \rho\left(\vec{r}\right) dV \tag{14.22}$$

Damit lässt sich der Drehimpuls weiterhin in der Form

$$\vec{L} = I_z \, \vec{\omega} \tag{14.23}$$

schreiben.

Beispiel 42: Trägheitsmoment einer homogenen Vollkugel

Die homogene Vollkugel in ► Abbildung 14.4 hat das Volumen $V = (4/3)\pi R^3$, wobei R den Kugelradius bezeichnet. Ihre Masse ist M und ihre Dichte ist $\rho = (M/V) =$ *const.* (14.5). Der senkrechte Abstand jedes Massenelements zur Drehachse ist in Kugelkoordinaten (9.36)

$$r_\perp = r \sin(\vartheta)\ . \tag{14.24}$$

Das Trägheitsmoment I_z (14.21) für die homogene Vollkugel bzgl. der z-Achse durch den Mittelpunkt berechnen wir mit den Massenelementen $dm = \rho\, dV$ (14.3) und dem Volumenelement in Kugelkoordinaten $dV = r^2 dr\ \sin\vartheta\, d\vartheta\, d\varphi$ (9.50):

$$I_z = \int r_\perp^2\, dm \tag{14.25}$$

$$= \int (r \cdot \sin(\vartheta))^2\ \rho\, dV \tag{14.26}$$

$$= \iiint r^2 \sin^2(\vartheta)\ \rho\, r^2\, dr\ \sin(\vartheta)\ d\vartheta\, d\varphi \tag{14.27}$$

Mit $\sin^2(\vartheta) = 1 - \cos^2(\vartheta)$, $d\cos(\vartheta)/d\vartheta = -\sin(\vartheta)$ und $\rho = M/V$ (14.5) erhalten wir

$$I_z = \rho \underbrace{\int_0^R r^4 dr}_{=\frac{1}{5}R^5} \quad \underbrace{\int_0^\pi \sin^3(\vartheta)\, d\vartheta}_{=\int_{\cos(0)}^{\cos(\pi)}(1-\cos^2(\vartheta))(-d\cos(\vartheta))} \quad \underbrace{\int_0^{2\pi} d\varphi}_{=2\pi} \tag{14.28}$$

$$= \frac{M}{V}\frac{1}{5}R^5 2\pi \cdot \left[-\cos(\vartheta) + \frac{1}{3}\cos^3(\vartheta)\right]_{\cos(0)=1}^{\cos(\pi)=-1} \tag{14.29}$$

$$= \frac{2\pi}{5}\frac{M}{V}R^5 \cdot \left[2 - \frac{2}{3}\right] = \frac{2\pi}{5}\frac{M}{\frac{4}{3}\pi R^3}R^5 \cdot \left[\frac{4}{3}\right] = \frac{2}{5}M\, R^2\ . \tag{14.30}$$

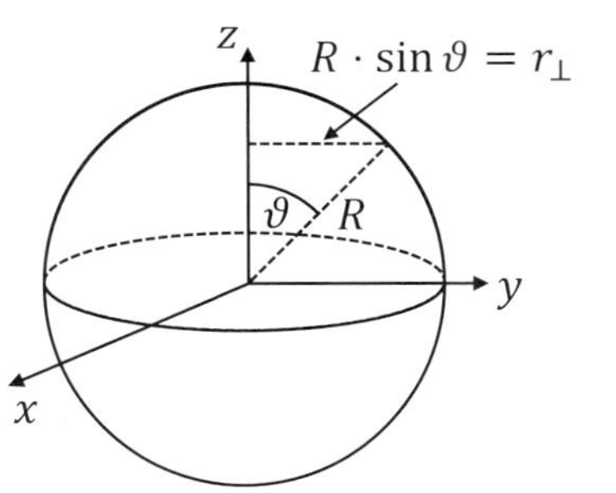

Abbildung 14.4 Zur Berechnung des Trägheitsmoments einer homogenen Vollkugel

Aufgabe 38: Trägheitsmoment einer dünnen Scheibe 1

Berechnen Sie das Trägheitsmoment einer dünnen, homogenen Scheibe bezüglich einer zur Scheibe senkrechten Achse durch den Scheibenmittelpunkt (z-Achse in ▶ Abbildung 14.5).

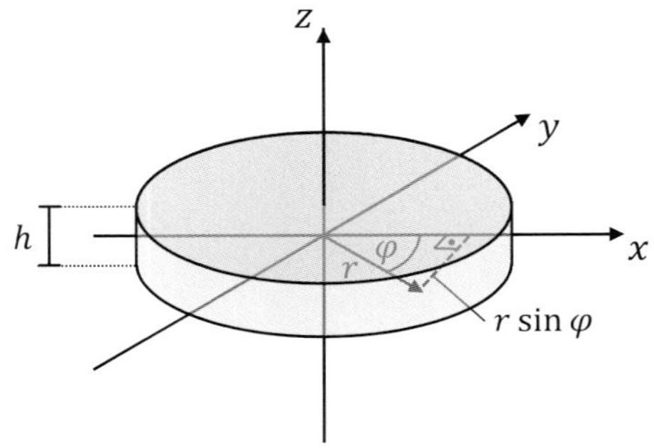

Abbildung 14.5 Zur Berechnung der Trägheitsmomente einer dünnen Scheibe

Aufgabe 39: Trägheitsmoment einer dünnen Scheibe 2

Berechnen Sie das Trägheitsmoment einer dünnen, homogenen Scheibe bezüglich einer zur Scheibe parallelen Achse durch den Scheibenmittelpunkt (y-Achse in ▶ Abbildung 14.5).

14.4 Drehschwingungen

Im folgenden Experiment wollen wir überprüfen, ob die Definition des Trägheitsmoments (14.21) eine sinnvolle Charakterisierung des trägen Verhaltens bei Drehungen eines starren Körpers darstellt.

An eine vertikal stehende Drehachse montieren wir eine Schneckenfeder so, dass Drehschwingungen durch Aufdrehen und Entspannen der Feder entstehen (▶ Abbildung 14.6).

Das rücktreibende Drehmoment durch die Feder ist für nicht zu extreme Winkelauslenkungen Φ linear in Φ mit dem Richtmoment D_r als Proportionalitätskonstante:

$$D = -D_r \, \Phi \tag{14.31}$$

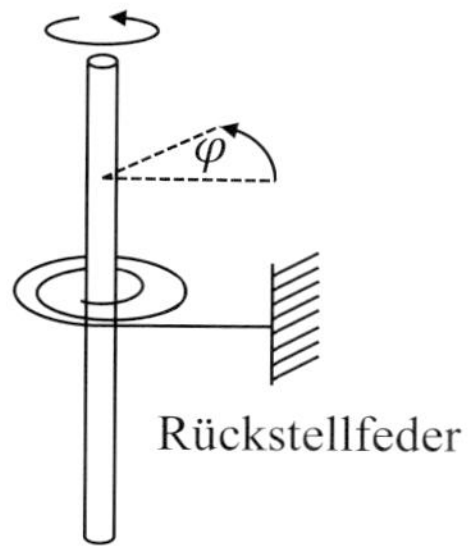

Abbildung 14.6 Drehschwingungen eines Stabes mit einer Rückstellfeder

Die Bewegungsgleichung für die Drehschwingungen können wir über das Bewegungsgesetz für Drehbewegungen ermitteln (4.70):

$$\frac{d\vec{L}}{dt} = \vec{D} \tag{14.32}$$

Dabei verwenden wir für den Drehimpuls $L = I\,\omega$ (14.23) mit der Winkelgeschwindigkeit $\omega = \dot{\Phi}$ und das rücktreibende Drehmoment D (14.31):

$$\frac{d}{dt}(I\,\dot{\Phi}) = -D_r \Phi \tag{14.33}$$

$$\ddot{\Phi} = -\underbrace{\frac{D_r}{I}}_{\equiv\omega^2}\Phi \tag{14.34}$$

Wir erkennen die Bewegungsgleichung des harmonischen Oszillators (4.25) wieder. Ihre Lösung hat die Form

$$\Phi(t) = A \cdot \cos(\omega t + \varphi_\circ) \tag{14.35}$$

mit der Winkelgeschwindigkeit

$$\omega = \sqrt{\frac{D_r}{I}}\,. \tag{14.36}$$

Die Schwingungsperiode $T = 2\pi/\omega$ (4.11) der Drehschwingung ist dementsprechend proportional zur Wurzel aus dem Trägheitsmoment:

$$T = 2\pi\sqrt{\frac{I}{D_r}} \tag{14.37}$$

Wir vergleichen dieses Ergebnis mit einem Massenpunkt der Masse m, der an einem Federpendel (Federkraft $F = -kx$) schwingt (4.30). Die Schwingungsperiode ist in diesem Fall $T = 2\pi\sqrt{m/k}$. Auch hier sehen wir, dass das Trägheitsmoment die Rolle der trägen Masse bei Translationsbewegungen übernimmt.

Experiment 22: Drehschwingungen einer Scheibe

Im Versuch verwenden wir eine dünne Scheibe (Dicke h, Radius R), die wir auf zwei verschiedene Arten auf der Drehachse montieren (► Abbildung 14.7). Die Trägheitsmomente $I = \int r_\perp^2\, dm$ bezüglich der Symmetrieachsen in der x, y- und z-Richtung lauten für $h \ll R$ (siehe Aufgaben zum Trägheitsmoment einer dünnen Scheibe im Abschnitt 14.3)

$$I_z = \frac{1}{2} M R^2 \quad \text{und} \tag{14.38}$$

$$I_x = I_y = \frac{1}{4} M R^2 \, . \tag{14.39}$$

Für die Schwingungsperiode T dieser Scheibe erwarten wir wegen Gleichung (14.37)

$$T_1 = 2\pi \sqrt{\frac{\frac{1}{2} M R^2}{D_r}} \, , \tag{14.40}$$

$$T_2 = 2\pi \sqrt{\frac{\frac{1}{4} M R^2}{D_r}} \, . \tag{14.41}$$

Das Verhältnis der erwarteten Schwingungsperioden

$$\frac{T_1}{T_2} = \sqrt{2} \approx 1{,}41 \tag{14.42}$$

wird im Versuch sehr gut bestätigt und zeigt, dass das Trägheitsmoment (14.21) des starren Körpers eine sinnvolle charakterisierende Größe für Drehbewegungen ist.

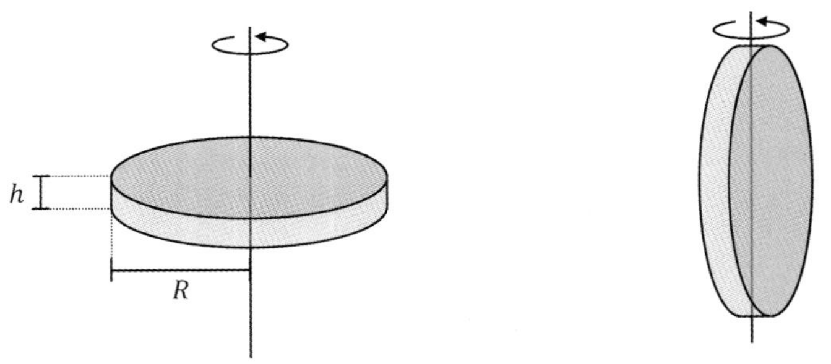

Abbildung 14.7 Drehschwingung einer Scheibe. Links: Schwingung um die z-Achse, rechts: Schwingung um eine Achse in der (x, y)-Ebene

Auch für kompliziertere Körper wie eine Hantel (► Abbildung 14.8) lassen sich die Trägheitsmomente sehr gut mit Drehschwingungen quantitativ überprüfen.

Das Trägheitsmoment eines Zylinders bezüglich der z-Achse durch das Zylinderzentrum beträgt [11]

$$I_{z,\text{Zylinder}} = \frac{1}{12} \left(m \left(h^2 + 3r^2 \right) \right) \, . \tag{14.43}$$

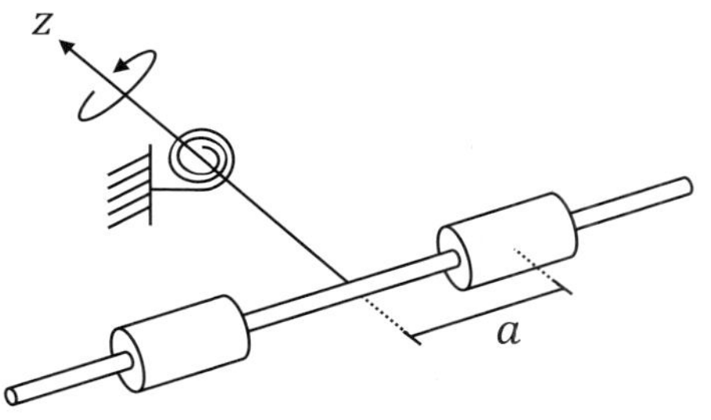

Abbildung 14.8 Drehschwingungen einer Hantel aus einem Stab mit zwei Zylindern

Das Trägheitsmoment der Stange, die die Zylinder hält, ist in der Näherung eines Zylinders mit $L = h \gg r$

$$I_{z,\text{Stange}} = \frac{1}{12} M L^2 . \tag{14.44}$$

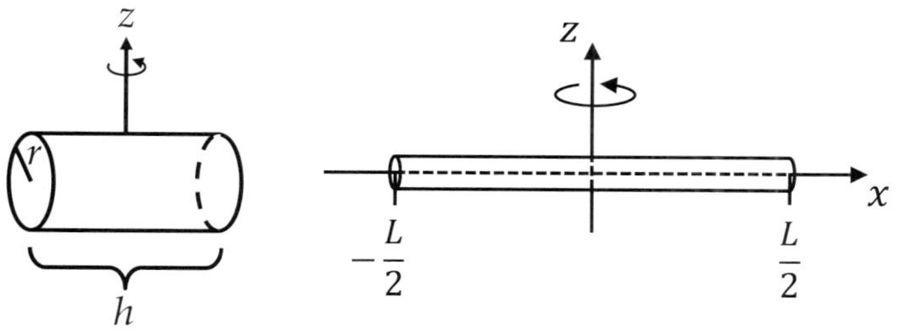

Abbildung 14.9 Trägheitsmoment eines Zylinders und einer Stange bezogen auf die Drehachse in ► Abbildung 14.8

Experiment 23: Drehschwingungen einer Hantel

Die in ► Abbildung 14.8 gezeigte Hantel regen wir zu Drehschwingungen an. Die Abstände a der beiden Zylinder auf der Stange von der Drehachse sind einstellbar. Wir vergleichen hier die Schwingungsdauern für zwei verschiedene (symmetrische) Zylinderabstände a_1 und a_2.

Die Stange mit der Masse $M = 0{,}131\,\text{kg}$ und der Länge $L = 0{,}61\,\text{m}$ hat ein Trägheitsmoment von (14.44)

$$I_{z,\text{Stange}} = \frac{1}{12} M L^2 = 4 \cdot 10^{-3}\,\text{kg}\,\text{m}^2 .$$

Die Zylinder mit Masse $m = 0{,}240\,\text{kg}$, Radius $r = 1{,}8\,\text{cm}$ und Höhe $h = 3{,}3\,\text{cm}$ haben als Trägheitsmoment bzgl. der z-Achse durch das Zylinderzentrum (14.43)

$$I_{z,\text{Zylinder}} = \frac{1}{12}\left(m\left(h^2 + 3r^2\right)\right) = 4 \cdot 10^{-5}\,\text{kg}\,\text{m}^2 . \tag{14.45}$$

Die beiden auf der Stange verschiebbaren Zylinder haben also ein vernachlässigbares Trägheitsmoment im Vergleich zum Trägheitsmoment der Stange. Wir betrachten die beiden Zylinder im Folgenden näherungsweise als Massenpunkte mit einem Trägheitsmoment, das nur vom Abstand a von der Drehachse abhängt. Für die Zylinderpositionen $a_1 = 0{,}1\,\text{m}$ und $a_2 = 0{,}2\,\text{m}$ sind ihre jeweiligen Trägheitsmomente in derselben Größenordnung wie das Trägheitsmoment der Stange:

$$I(a_1) = m\,a_1^2 = 0{,}24 \cdot 0{,}1^2\,\text{kg}\,\text{m}^2 = 2{,}4 \cdot 10^{-3}\,\text{kg}\,\text{m}^2$$
$$I(a_2) = m\,a_2^2 = 0{,}24 \cdot 0{,}2^2\,\text{kg}\,\text{m}^2 = 9{,}6 \cdot 10^{-3}\,\text{kg}\,\text{m}^2$$

Das Gesamtträgheitsmoment der Hantel beträgt

$$I_{z,\text{Hantel}}(a) = I_{z,\text{Stange}} + 2 \cdot m\,a^2 . \tag{14.46}$$

Wir setzen $I_{z,\text{Hantel}}$ in das Verhältnis der Schwingungsperioden ein und erhalten mit $T = 2\pi\sqrt{I/D_r}$ (14.37)

$$\frac{T_1}{T_2} = \sqrt{\frac{I_{z,\text{Hantel}}(a_1)}{I_{z,\text{Hantel}}(a_2)}} \approx \sqrt{\frac{I_{z,\text{Stange}} + 2 \cdot m a_1^2}{I_{z,\text{Stange}} + 2 \cdot m a_2^2}} \approx \sqrt{\frac{9}{23}} = 0{,}63 . \tag{14.47}$$

Dieses Ergebnis wird in unserem Experiment gut bestätigt.

14.5 Satz von Steiner

Im Prinzip müssen wir für jede neue Drehachse das Trägheitsmoment neu berechnen. Wenn wir die Ausrichtung des Körpers in Bezug auf eine Drehachse $\vec{B}$ beibehalten und nur den Körper parallel zu dieser Achse versetzen, können wir etwas Berechnungszeit sparen.

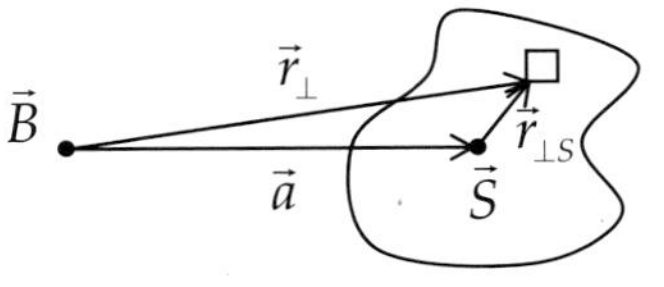

Abbildung 14.10 Satz von Steiner: Umrechnung des Trägheitsmoments von einer Drehachse $\vec{S}$ durch den Schwerpunkt auf eine andere Drehachse $\vec{B}$, beide senkrecht zur Zeichenebene

► Abbildung 14.10 zeigt die Projektion eines Körpers mit einer Drehachse $\vec{S}$ durch den Schwerpunkt und einer weiteren Drehachse $\vec{B}$ parallel zu $\vec{S}$. Der Körperschwerpunkt liegt in der Zeichenebene und die Vektoren $\vec{S}$ und $\vec{B}$ stehen senkrecht zur Zeichenebene.

Aus der Abbildung sehen wir, dass wir in der Definition des Trägheitsmoments (14.21) $\vec{r_\perp}$ durch $\vec{r_{\perp S}}$ bezüglich des Körperschwerpunkts und den Abstand $\vec{a}$ des Körperschwerpunkts von der Drehachse $\vec{B}$ ersetzen können:

$$I_B = \int \vec{r_\perp}^2 dm \tag{14.48}$$

$$= \int (\vec{r}_{\perp S} + \vec{a})^2 \; dm \tag{14.49}$$

$$= \underbrace{\int (\vec{r}_{\perp S})^2 \; dm}_{\equiv I_S} + \underbrace{\int \vec{a}^2 \; dm}_{= M\,\vec{a}^2} + \underbrace{2\int \vec{a}\,\vec{r}_{\perp S}\, dm}_{\text{siehe unten}} \tag{14.50}$$

Das erste Integral entspricht dem Trägheitsmoment I_S des Körpers bezüglich seines Schwerpunkts. Im zweiten Integral ist der Abstandsvektor $\vec{a}$ unabhängig von der Verteilung der Massenelemente dm. Wir können $\vec{a}$ vor das Integral ziehen und erhalten mit $\int dm = M$ die Masse des Körpers.

Das dritte Integral besteht durch das Skalarprodukt aus drei Integralen über die jeweilige Raumrichtung:

$$\int \vec{a}\,\vec{r}_{\perp S}\, dm = a_x \int r_{\perp S,x}\,\rho\, dx + a_y \int r_{\perp S,y}\,\rho\, dy + a_z \int r_{\perp S,z}\,\rho\, dz \tag{14.51}$$

Da wir den Schwerpunkt r_S so definiert haben, dass sich die Massenverteilungen bezüglich der drei Raumrichtungen balancieren, geben diese Integrale keinen Beitrag:

$$\int \vec{a}\,\vec{r}_{\perp S}\, dm = 0 \tag{14.52}$$

Damit bleibt der als »Satz von Steiner« bekannte Zusammenhang

$$I_B = I_S + M\,a^2\,. \tag{14.53}$$

14.6 Rotationsenergie

Ein sich drehender Körper besitzt Rotationsenergie, die mit seinem Trägheitsmoment zusammenhängt. Wir führen zunächst ein Experiment mit Zylindern durch, die eine schiefe Ebene hinabrollen (► Abbildung 14.11). Anschließend stellen wir die Rotationsenergie vor und berechnen die Bahnkurven der Zylinder.

Experiment 24: Zylinder-Rennen

Ein Hohlzylinder und ein Vollzylinder gleicher Masse rollen eine schräge Ebene hinab (► Abbildung 14.11). Welcher der beiden Zylinder gewinnt dieses Walzenrennen?

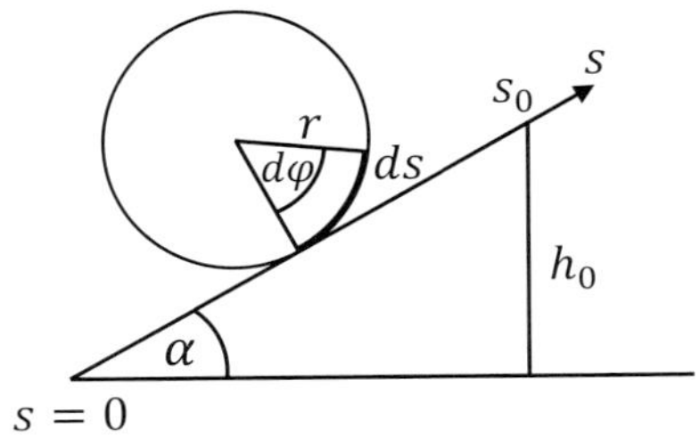

Abbildung 14.11 Die Zylinder rollen von der Höhe h_0 eine schiefe Ebene hinunter.

14.6.1 Trägheitsmomente Hohl- und Vollzylinder

Wir nehmen näherungsweise an, dass die Masse des Hohlzylinders vollständig im Zylindermantel konzentriert ist (► Abbildung 14.12) und erhalten mit (14.21) das Trägheitsmoment

$$I_H = m r^2 \,. \tag{14.54}$$

Für den Vollzylinder nehmen wir eine homogene Massenverteilung an und erhalten analog zur dünnen Scheibe (14.38) das Trägheitsmoment

$$I_V = \frac{1}{2} m r^2 \,. \tag{14.55}$$

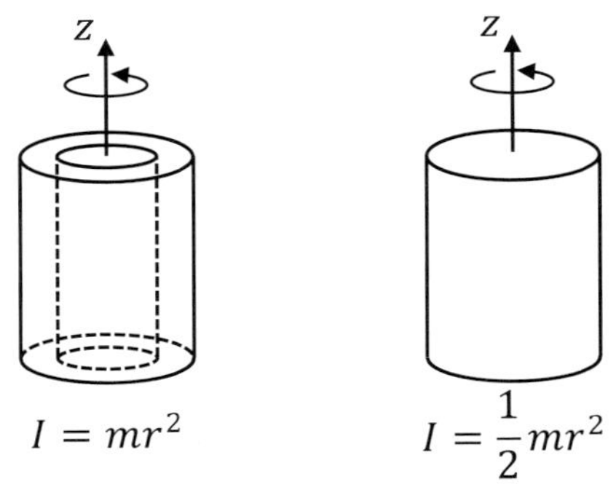

Abbildung 14.12 Trägheitsmomente eines Hohl- und eines Voll-Zylinders

14.6.2 Potentielle Energie, Translations- und Rotationsenergie

Wir beschreiben hier den Berührungspunkt des Zylinders mit der schiefen Ebene anhand der Koordinate s, die am tiefsten Punkt startet und entlang der schiefen Ebene ansteigt

(► Abbildung 14.11). Beide Zylinder starten von derselben Höhe $h_\circ = s_\circ \sin\alpha$ und besitzen damit dieselbe Gesamtenergie, die der potentiellen Energie (5.14) entspricht:

$$E = E_{\text{pot, max}} \tag{14.56}$$

$$= m\,g\,h_\circ \tag{14.57}$$

$$= m\,g\,s_\circ\,\sin\alpha \tag{14.58}$$

Wir vernachlässigen die geringe Reibung. Dann ist an jedem Ort auf der Rollstrecke die Gesamtenergie des Zylinders wegen der Energieerhaltung konstant (5.33) und teilt sich in die kinetische Energie der Translationsbewegung E_{kin}, die kinetische Energie der Rotationsbewegung E_{rot} und die verbleibende potentielle Energie E_{pot} auf:

$$E = E_{\text{kin}} + E_{\text{rot}} + E_{\text{pot}} = const. \tag{14.59}$$

Die Translationsbewegung mit der kinetischen Energie $E_{\text{kin}} = (1/2)\,m\,v^2$ (5.23) erfolgt parallel zur Koordinate s, sodass die Geschwindigkeit $v = ds/dt = \dot{s}$ beträgt:

$$E_{\text{kin}} = \frac{1}{2}\,m\,\dot{s} \tag{14.60}$$

Sie steigt, während der Zylinder hinunterrollt. Dagegegen nimmt die Höhe $h = s\,\sin\alpha$ ab, sodass die potentielle Energie sinkt:

$$E_{\text{pot}} = m\,g\,s\,\sin\alpha \tag{14.61}$$

Die Rotationsenergie erhalten wir mit folgender Überlegung: Für den Fall eines einzelnen Massenpunkts, der im Abstand r mit der Winkelgeschwindigkeit $\omega = \dot{\varphi}$ um eine feste Drehachse rotiert, beträgt die Rotationsenergie (5.25)

$$E_{\text{rot}} = \frac{1}{2}\,mr^2\,\omega^2\,. \tag{14.62}$$

Für einen ausgedehnten starren Körper müssen wir über alle Massenelemente $dm = \rho dV$ (14.3) integrieren und dabei ihren senkrechten Abstand $r_\perp$ von der gemeinsamen Drehachse berücksichtigen. Die Integration ergibt das Trägheitsmoment des starren Körpers (14.21):

$$E_{\text{rot}} = \frac{1}{2}\int r_\perp^2\,dm\,\omega^2 \tag{14.63}$$

$$= \frac{1}{2}\underbrace{\int r_\perp^2\,\rho\,dV}_{\equiv I}\,\omega^2 \tag{14.64}$$

Dementsprechend beträgt die **Rotationsenergie** eines starren Körpers mit seinem Trägheitsmoment I

$$E_{\text{rot}} = \frac{1}{2}\,I\,\omega^2\,. \tag{14.65}$$

Da wir alle Energieterme in der Streckenkoordinate s und ihrer zeitlichen Ableitung ausdrücken wollen, übertragen wir auch die Winkelgeschwindigkeit ω. In ► Abbildung 14.11 ist gezeigt, dass die Abrollstrecke auf der Zylinderoberfläche dem Bogenweg (4.4) entspricht:

$$ds = r\,d\varphi \tag{14.66}$$

Die Abrollgeschwindigkeit ist dann

$$\dot{s} = \frac{ds}{dt} = r\,\frac{d\varphi}{dt} = r\,\omega\,. \tag{14.67}$$

Damit beträgt die Rotationsenergie des Zylinders

$$E_{\text{rot}} = \frac{1}{2}\,I\,\frac{1}{r^2}\,\dot{s}^2\,. \tag{14.68}$$

Das Einsetzen der verschiedenen Beiträge (14.60), (14.61) und (14.68) in die Energieerhaltungsgleichung (14.59) für den hinunterrollenden Zylinder ergibt

$$\begin{aligned} E &= E_{\text{kin}} + E_{\text{rot}} + E_{\text{pot}} &(14.69)\\ &= \frac{1}{2}\,m\,\dot{s}^2 + \frac{1}{2}\,\frac{I}{r^2}\,\dot{s}^2 + m\,g\,s\,\sin\alpha &(14.70)\\ &= \frac{m}{2}\left(1+\frac{I}{mr^2}\right)\dot{s}^2 + (m\,g\,\sin\alpha)\,s\,. &(14.71) \end{aligned}$$

14.6.3 Beschleunigung des rollenden Zylinders

Unser nächstes Ziel ist, die Beschleunigung des hinunterrollenden Zylinders zu verstehen. Die Ableitung der Gesamtenergie E (14.71) des Zylinders nach der Zeit ergibt wegen der Energieerhaltung $E = const.$ keinen Beitrag. Allerdings können wir durch die zeitliche Ableitung der rechten Seite von (14.71) auf die Beschleunigung $\ddot{s}$ des Zylinders schließen:

$$\begin{aligned} \frac{dE}{dt} &= \frac{d}{dt}\left(\frac{m}{2}\left(1+\frac{I}{mr^2}\right)\dot{s}^2 + (m\,g\,\sin\alpha)\,s\right) &(14.72)\\ &= \frac{m}{2}\left(1+\frac{I}{mr^2}\right)2\,\dot{s}\,\ddot{s} + (m\,g\,\sin\alpha)\,\dot{s} &(14.73)\\ &= m\left(\left(1+\frac{I}{mr^2}\right)\ddot{s} + g\,\sin\alpha\right)\cdot\dot{s} &(14.74)\\ &= 0 &(14.75) \end{aligned}$$

Nach dem Start ist die Geschwindigkeit des Zylinders $\dot{s} > 0$, sodass der Term in den Klammern in Gleichung (14.74) verschwinden muss:

$$\left(1+\frac{I}{mr^2}\right)\ddot{s} + g\,\sin\alpha = 0 \tag{14.76}$$

Damit erhalten wir die effektive, konstante Beschleunigung des Zylinders:

$$\ddot{s} = -\frac{g\,\sin\alpha}{\left(1+\frac{I}{mr^2}\right)} \equiv -\tilde{g} = const. \tag{14.77}$$

14.6.4 Bahnkurve des Zylinders

Wegen der konstanten Beschleunigung $\ddot{s}$ (14.77) handelt es sich beim herunterrollenden Zylinder um eine gleichmäßig beschleunigte Bewegung (1.33). Seine Bahnkurve $s(t)$ beim Start aus der Anfangsposition $s(t=0) = s_\circ$ und mit der Anfangsgeschwindigkeit $\dot{s}(t=0) = 0$ lautet

$$s(t) = -\frac{1}{2}\tilde{g}t^2 + s_\circ\,. \tag{14.78}$$

Damit ist die Rollzeit t des Zylinders über die gesamte Strecke $s_\circ$ bis zum Ende der schiefen Ebene ($s = 0$) nach Einsetzen von $\tilde{g}$ (14.77)

$$t = \sqrt{\frac{2\,s_\circ}{\tilde{g}}} = \sqrt{\frac{2\,s_\circ}{g\,\sin\alpha}}\,\sqrt{1 + \frac{I}{mr^2}}\,. \tag{14.79}$$

Die Rollzeit t ist größer für den Körper mit dem größeren Trägheitsmoment.

Das Trägheitsmoment des Hohlzylinders $I_H = mr^2$ (14.54) ist größer als das Trägheitsmoment des Vollzylinders $I_V = (1/2)mr^2$ (14.55). Die Massenverteilung des Vollzylinders liegt im Mittel näher an der Drehachse als beim Hohlzylinder.

Also gewinnt der Vollzylinder das Rennen! Das Verhältnis der Laufzeiten beträgt mit (14.79)

$$\frac{t_V}{t_H} = \sqrt{1 + \frac{I_V}{mr^2}}\Bigg/\sqrt{1 + \frac{I_H}{mr^2}} = \sqrt{\frac{1 + 1/2}{1 + 1}} = \frac{\sqrt{3}}{2} \approx 0{,}87\,. \tag{14.80}$$

14.7 Trägheitstensor

Bisher haben wir nur Kreisbewegungen betrachtet, genauer gesagt Rotationsbewegungen um eine fest stehende Achse. Wir konnten dabei nutzen, dass der Radiusvektor und die Winkelgeschwindigkeit senkrecht aufeinander stehen: $\vec{r} \perp \vec{\omega}$. Weiterhin standen der Drehimpulsvektor und der Vektor der Winkelgeschwindigkeit parallel: $\vec{L} \parallel \vec{\omega}$ (siehe Gleichung (14.17)).

Bei allgemeinen Rotationsbewegungen stehen die Vektoren $\vec{r}$ und $\vec{\omega}$ nicht notwendigerweise senkrecht aufeinander, sodass auch der Drehimpulsvektor $\vec{L}$ und der Vektor der Winkelgeschwindigkeit $\vec{\omega}$ nicht parallel verlaufen. Für einen Massenpunkt berechnen wir den Drehimpuls aus (Abschnitt 14.3)

$$\vec{L} = \vec{r}\times\vec{p} = m\vec{r}\times\vec{v} = m\vec{r}\times(\vec{\omega}\times\vec{r}) \tag{14.81}$$

$$= m(\vec{\omega}\,\vec{r}^{\,2} - \vec{r}\,\underbrace{(\vec{r}\vec{\omega})}_{\text{im Allgemeinen}\neq 0}\,)\,. \tag{14.82}$$

Für starre Körper berechnen wir den Drehimpuls durch Integration von (14.82) über alle Massenelemente dm:

$$\vec{L} = \int (\vec{\omega}\,\vec{r}^{\,2} - \vec{r}\,(\vec{r}\vec{\omega}))\,dm \tag{14.83}$$

Dieses kompliziert aussehende Integral werden wir im Folgenden in einer eleganten Form aufschreiben. Dafür berechnen wir $\vec{L}$ komponentenweise in kartesischen Koordinaten. Für die Komponente L_x ergibt sich

$$L_x = \int \left[r^2 \omega_x - x\,(x\omega_x + y\omega_y + z\omega_z)\right] dm \tag{14.84}$$

$$= \int \left(r^2 - x^2\right) dm\,\omega_y - \int yx\,dm\,\omega_x - \int yz\,dm\,\omega_z \tag{14.85}$$

$$= \underbrace{\int \left(y^2 + z^2\right) dm}_{\equiv I_{xx}}\,\omega_x - \underbrace{\int xy\,dm}_{\equiv I_{xy}}\,\omega_y - \underbrace{\int xz\,dm}_{\equiv I_{xz}}\,\omega_z \tag{14.86}$$

$$= I_{xx}\omega_x + I_{xy}\omega_y + I_{xz}\omega_z\,. \tag{14.87}$$

Dabei bezeichnet I_{xx} das Trägheitsmoment des starren Körpers, der durch Drehung mit der Winkelgeschwindigkeit ω_x berücksichtigt werden muss. Der Anschauung entsprechend berechnen wir es durch die Quadrate der senkrechten Abstände y^2+z^2 der Massenelemente dm zur x-Achse. Die in den Koordinaten gemischten Terme, die wir mit I_{xy} und I_{xz} bezeichnen, erfassen durch ihre Dimension $\mathrm{m^2\,kg}$ ebenfalls Aspekte der Trägheit des Körpers. Sie heißen **Deviationsmomente** oder **Zentrifugalmomente**. Ihre Bedeutung werden wir weiter unten erläutern.

Analog berechnen wir die Drehimpulskomponenten L_y und L_z:

$$L_y = \int \left[r^2 \omega_y - y\,(x\omega_x + y\omega_y + z\omega_z)\right] dm \tag{14.88}$$

$$= \int \left(x^2 + z^2\right) dm\,\omega_y - \int yx\,dm\,\omega_x - \int yz\,dm\,\omega_z \tag{14.89}$$

$$= I_{yy}\omega_y + I_{yx}\omega_x + I_{yz}\omega_z \tag{14.90}$$

$$L_z = \int \left[r^2 \omega_z - z\,(x\omega_x + y\omega_y + z\omega_z)\right] dm \tag{14.91}$$

$$= \int \left(x^2 + y^2\right) dm\,\omega_z - \int zx\,dm\,\omega_x - \int zy\,dm\,\omega_y \tag{14.92}$$

$$= I_{zz}\omega_z + I_{zx}\omega_x + I_{zy}\omega_y \tag{14.93}$$

Die Berechnung der Drehimpulskomponenten $\vec{L}$ (14.87, 14.90, 14.93) können wir in Form einer Vektor-Matrix-Multiplikation zusammenfassen:

$$\begin{pmatrix} L_x \\ L_y \\ L_z \end{pmatrix} = \underbrace{\begin{pmatrix} I_{xx} & I_{xy} & I_{xz} \\ I_{yx} & I_{yy} & I_{yz} \\ I_{zx} & I_{zy} & I_{zz} \end{pmatrix}}_{\equiv \hat{I}} \cdot \begin{pmatrix} \omega_x \\ \omega_y \\ \omega_z \end{pmatrix} \tag{14.94}$$

$$\vec{L} = \hat{I} \cdot \vec{\omega} \tag{14.95}$$

Hier bezeichnet $\hat{I}$ eine Matrix, die wir **Trägheitstensor** nennen. Er erweitert das skalare Trägheitsmoment eines starren Körpers bzgl. einer festen Achse auf den Fall variabler Drehachsen, wie er z. B. beim Kreisel vorliegt. Wir finden die schon bewährte Schreibweise $\vec{L} = \hat{I} \cdot \vec{\omega}$ (14.18) wieder und können damit den Drehimpuls für Rotationsachsen in allen Raumrichtungen sofort berechnen. Der Trägheitstensor ist symmetrisch, d. h. für die Deviationsmomente neben der Hauptdiagonalen gilt

$$I_{xy} = \int xy\,dm = \int yx\,dm = I_{yx}\,, \quad I_{xz} = I_{zx}\,, \quad I_{yz} = I_{zy}\,. \tag{14.96}$$

Aufgabe 40: Drehimpulskomponenten

Verifizieren Sie die Zusammenfassung der Gleichungen (14.87), (14.90) und (14.93) in der Matrixschreibweise $\vec{L} = \hat{I} \cdot \vec{\omega}$ (14.94) durch Ausführen der Matrixmultiplikation.

Beispiel 43: Trägheitstensor eines homogenen Quaders

Wir legen den Ursprung des Koordinatensystems in den Schwerpunkt des Quaders (► Abbildung 14.13):

$$I_{xx} = \rho \int_{-\frac{a}{2}}^{\frac{a}{2}} \int_{-\frac{b}{2}}^{\frac{b}{2}} \int_{-\frac{c}{2}}^{\frac{c}{2}} \left(y^2 + z^2\right)\,dx\,dy\,dz \tag{14.97}$$

$$= \rho a \left(\frac{1}{3} y^3 \Big|_{-\frac{b}{2}}^{\frac{b}{2}} c + \frac{1}{3} z^3 \Big|_{-\frac{c}{2}}^{\frac{c}{2}} b \right) \tag{14.98}$$

$$= \frac{M}{V} a \left(\frac{1}{3} \left[\left(\frac{b}{2}\right)^3 - \left(-\frac{b}{2}\right)^3 \right] c + \frac{1}{3} \left[\left(\frac{c}{2}\right)^3 - \left(-\frac{c}{2}\right)^3 \right] b \right) \tag{14.99}$$

$$= \frac{M}{abc} a \left(\frac{1}{3} \frac{2}{2^3} b^3 c + b \frac{1}{3} \frac{2}{2^3} c^3 \right) = \frac{M}{12} \left(b^2 + c^2\right) \tag{14.100}$$

$$I_{xy} = -\rho \iiint x\,y\,dx\,dy\,dz = -\frac{M}{V} \frac{1}{2} x^2 \Big|_{-\frac{a}{2}}^{\frac{a}{2}} \frac{1}{2} y^2 \Big|_{-\frac{b}{2}}^{\frac{b}{2}} c \tag{14.101}$$

$$= -\frac{M}{abc} \frac{1}{2} \underbrace{\left(\left(\frac{a}{2}\right)^2 - \left(-\frac{a}{2}\right)^2 \right)}_{=0} \cdots = 0 \tag{14.102}$$

Aus analog durchgeführten Rechnungen folgen die anderen Elemente. Insgesamt lautet der Trägheitstensor des Quaders

$$\hat{I} = \frac{M}{12} \begin{pmatrix} b^2 + c^2 & 0 & 0 \\ 0 & a^2 + c^2 & 0 \\ 0 & 0 & a^2 + b^2 \end{pmatrix}\,. \tag{14.103}$$

Ist der Trägheitstensor, wie in diesem Beispiel, **diagonal**, so ist das Koordinatensystem entsprechend der Körpersymmetrie gewählt: Alle Achsen sind sogenannte **Hauptträgheitsachsen** des Körpers.

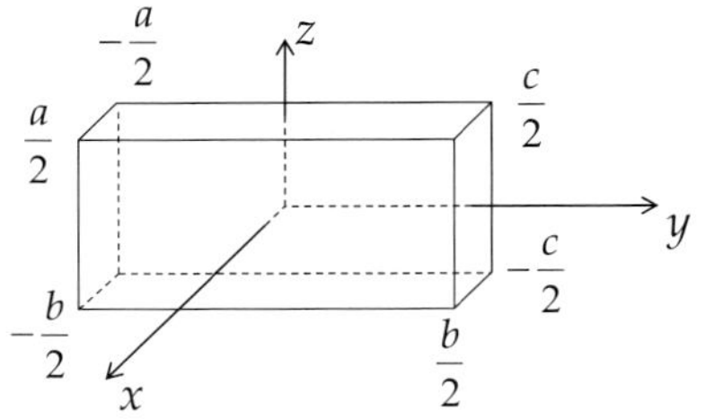

Abbildung 14.13 Trägheitsmoment eines homogenen Quaders bezüglich seines Schwerpunkts

Für den Quader mit den Kantenlängen $a > b > c$ lauten die Trägheitsmomente (14.103)

$$I_{xx} = b^2 + c^2 = I_{\min}\,, \tag{14.104}$$

$$I_{yy} = a^2 + c^2\,, \tag{14.105}$$

$$I_{zz} = a^2 + b^2 = I_{\max}\,. \tag{14.106}$$

Damit sind die Trägheitsmomente bezüglich der verschiedenen Hauptachsen geordnet:

$$I_{zz} > I_{yy} > I_{xx} \tag{14.107}$$

Beispiel 44: Drehimpulsvektor und Winkelgeschwindigkeitsvektor

Um zu sehen, dass der Drehimpuls und der Vektor der Winkelgeschwindigkeit nicht parallel stehen müssen, drehen wir den Quader um verschiedene Achsen. Sein Drehimpuls beträgt allgemein

$$\vec{L} = \begin{pmatrix} I_{xx} & 0 & 0 \\ 0 & I_{yy} & 0 \\ 0 & 0 & I_{zz} \end{pmatrix} \begin{pmatrix} \omega_x \\ \omega_y \\ \omega_z \end{pmatrix} = \begin{pmatrix} I_{xx}\,\omega_x \\ I_{yy}\,\omega_y \\ I_{zz}\,\omega_z \end{pmatrix}. \tag{14.108}$$

Für eine Drehung um eine Hauptträgheitsachse, z. B. die z-Achse, sind die Winkelgeschwindigkeit und der Drehimpuls parallel:

$$\vec{\omega} = \begin{pmatrix} 0 \\ 0 \\ \omega_z \end{pmatrix} \qquad \vec{L} = \begin{pmatrix} 0 \\ 0 \\ I_{zz}\,\omega_z \end{pmatrix} \tag{14.109}$$

Ist die Drehachse jedoch nicht entlang einer der Hauptträgheitsachsen, dann sind z. B. für $I_{xx} \neq I_{zz}$ die Vektoren $\vec{L}$ und $\vec{\omega}$ nicht parallel:

$$\vec{\omega} = \begin{pmatrix} \omega_x \\ 0 \\ \omega_z \end{pmatrix} \qquad \vec{L} = \begin{pmatrix} I_{xx}\,\omega_x \\ 0 \\ I_{zz}\,\omega_z \end{pmatrix} \tag{14.110}$$

Aufgabe 41: Trägheitstensor eines Quaders

Die exakte Form des Trägheitstensors des Quaders hängt von der Wahl des Koordinatensystems ab. Führen Sie die Berechnung für einen Quader durch dessen eine Ecke im Koordinatenursprung liegt.

Die Deviationsmomente auf den Nebendiagonalen eines Trägheitstensors (14.94) sind ein Maß für das Bestreben eines starren Körpers, seine Drehachse zu verändern. Die Folge ist, dass Drehachsen ausserhalb der Hauptachsen zu Instabilitäten führen, wie im folgenden Experiment gezeigt wird.

Experiment 25: Freie Rotation eines Quaders

Mit einem Quader mit verschiedenen Kantenlängen und farbigen Seitenflächen kann man durch Hochwerfen und gleichzeitiges Drehen beobachten, dass eine stabile freie Rotation nur um die Achsen des größten bzw. kleinsten Trägheitsmoments möglich ist.

Mathematischer Einschub: Hauptachsentransformation

Eine $(n \times n)$-Matrix wie den Trägheitstensor $\hat{I}$ (14.94) mit symmetrischen Nebendiagonalelementen $I_{ij} = I_{ji}$ (14.96) können wir in eine Diagonalmatrix $\hat{I}_{\text{diag}}$ umformen, d. h. alle Nichtdiagonalelemente sind null ($I_{ij} = 0$ mit $i \neq j$). Dafür nutzt man die sogenannte Hauptachsentransformation, die in der theoretischen Physik erklärt wird. Im dreidimensionalen Fall unserer Matrix $\hat{I}_{\text{diag}}$ besteht die Hauptdiagonale aus den sogenannten Eigenwerten der Matrix $I_{11} \equiv I_a$, $I_{22} \equiv I_b$ und $I_{33} \equiv I_c$. Durch eine solche Transformation erhalten wir ein neues Koordinatensystem mit neuen Einheitsvektoren $\vec{e}_a, \vec{e}_b$ und $\vec{e}_c$, die entlang der drei Hauptachsen zeigen.

In der Praxis ist es am einfachsten, die Körpersymmetrie für die Wahl des Koordinatensystems zu nutzen. Dadurch erreicht man meist, dass der Trägheitstensor $\hat{I}$ direkt diagonal wird ($I_{xy} = I_{xz} = I_{yz} = 0$) (siehe das Quaderbeispiel 43 (14.103)). Dessen drei Elemente (Eigenwerte) zeigen sogleich die Hauptträgheitsmomente des starren Körpers.

Ist eine solche Wahl nicht offensichtlich, verändern wir mit der Hauptachsentransformation das Koordinatensystem für den starren Körper in bestmöglicher Weise. Dabei verschwinden die eventuell vorhandenen Nichtdiagonalelemente (Deviationsmomente) I_{xy}, I_{xz}, I_{yz} und wir erreichen die gewünschte diagonale Matrix. Die neuen Einheitsvektoren $\vec{e}_a, \vec{e}_b, \vec{e}_c$ zeigen entlang der Symmetrieachsen des starren Körpers (z. B. eines Kreisels).

Den Drehimpulsvektor $\vec{L}$ und den Vektor der Winkelgeschwindigkeit $\vec{\omega}$ beschreiben wir ebenfalls in den neuen Koordinaten a, b, c und erhalten für $\vec{L} = \hat{I} \cdot \vec{\omega}$ (14.95)

$$\begin{pmatrix} L_a \\ L_b \\ L_c \end{pmatrix} = \begin{pmatrix} I_a & 0 & 0 \\ 0 & I_b & 0 \\ 0 & 0 & I_c \end{pmatrix} \cdot \begin{pmatrix} \omega_a \\ \omega_b \\ \omega_c \end{pmatrix} . \tag{14.111}$$

Die Hauptträgheitsmomente I_a, I_b, I_c reflektieren Symmetrieeigenschaften des Körpers in den Richtungen der Hauptachsen. Entlang jeder Hauptachse lässt sich die Drehimpulskomponente sofort hinschreiben:

$$L_i = I_i \, \omega_i \qquad i = a, b, c \tag{14.112}$$

14.8 Kreisel

Kreisel jeder Form sind wegen ihres scheinbaren Eigenlebens faszinierende Objekte.

Um die Ursache ihrer Bewegungen zu verstehen, konzentrieren wir uns zunächst auf den sogenannten kräftefreien, symmetrischen Kreisel. Ein Kreisel ist symmetrisch, wenn zwei seiner Hauptträgheitsmomente gleich groß sind. Als Trägheitstensor wählen wir hier mit $I_a = I_b$

$$\hat{I} = \begin{pmatrix} I_a & 0 & 0 \\ 0 & I_a & 0 \\ 0 & 0 & I_c \end{pmatrix} . \tag{14.113}$$

Die dritte Symmetrieachse mit dem Hauptträgheitsmoment I_c zeigt in Richtung der Figurenachse (► Abbildung 14.14), die wir mit dem Vektor $\vec{c}$ bezeichnen.

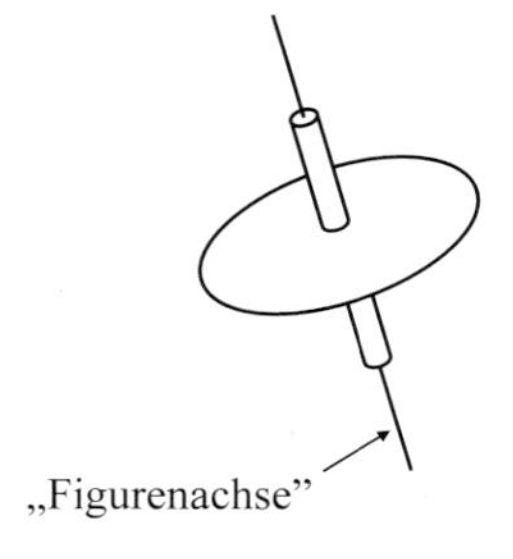

Abbildung 14.14 Symmetrischer Kreisel mit Figurenachse

Kräftefrei heißt ein Kreisel, wenn kein äußeres Drehmoment wirkt ($\vec{D} = 0$). Die Figurenachse kann eine kreisende Bewegung um eine feste Achse ausführen, die als Nutation bezeichnet wird und die wir im folgenden Abschnitt beschreiben.

Im zweiten Abschnitt danach lassen wir ein äußeres Drehmoment auf den symmetrischen Kreisel wirken, das zu einer weiteren Kreisbewegung führt, der sogenannten Präzession.

14.8.1 Nutation

Experiment 26: Kreisel

Wir lassen einen kräftefreien, symmetrischen Kreisel zunächst um seine Figurenachse drehen. Dabei steht die Figurenachse parallel zum Drehimpuls und zur Drehachse ($\vec{c} \parallel \vec{L} \parallel \vec{\omega}$).

Nach einem Schlag auf die Figurenachse beginnt die Figurenachse des Kreisels eine Rotationsbewegung um eine feste (virtuelle) Achse. Diese Bewegung wird als **Nutation** bezeichnet.

Im Anschluss an den Schlag wirkt kein äußeres Drehmoment auf den Kreisel, sodass die Gesamtenergie E und der Drehimpuls $\vec{L}$ des Kreisels erhalten bleiben

$$\text{Energieerhaltung} \quad E = E_{\text{rot}} = const. \tag{14.114}$$

$$\text{Drehimpulserhaltung} \quad \vec{L} = const. \tag{14.115}$$

Die Gesamtenergie steckt dabei in der Rotationsenergie $E = E_{\text{rot}}$ des Kreisels. In unserem Experiment zeigt die feste virtuelle Achse mit (14.115) in die Richtung des Drehimpulses $\vec{L}$.

Im Folgenden berechnen wir die Bewegung der Figurenachse $\vec{c}$ und die der Winkelgeschwindigkeit $\vec{\omega}$ mithilfe der beiden Erhaltungsgrößen $\vec{L}$ und E.

Vektor der Winkelgeschwindigkeit

Zunächst untersuchen wir den Vektor der Winkelgeschwindigkeit $\vec{\omega}$. Beim Experiment mit den rollenden Zylindern hatten wir die Rotationsenergie E_{rot} eines starren Körpers mit dem Trägheitsmoment I kennengelernt (14.65):

$$E_{\text{rot}} = \frac{1}{2} I \omega^2 = \frac{1}{2} \omega L \tag{14.116}$$

Im letzten Schritt haben wir den Drehimpuls und seinen Zusammenhang $L = I \omega$ (14.18) genutzt.

Um die Rotationsenergie bei beliebigen Raumrichtungen der Rotationsbewegung zu beschreiben, benötigen wir die Vektoren der Winkelgeschwindigkeit $\vec{\omega}$ und des Drehimpulses $\vec{L}$:

$$E_{\text{rot}} = \frac{1}{2} \vec{\omega} \vec{L} \tag{14.117}$$

Im Fall der rollenden Zylinder stehen $\vec{\omega}$ und $\vec{L}$ parallel, sodass sich wegen $\cos(0) = 1$ der maximale Wert $\vec{\omega} \vec{L} = \omega L$ ergibt. Für den Kreisel ergibt sich im Allgemeinen ein Öffnungswinkel γ zwischen $\vec{\omega}$ und $\vec{L}$.

Auf der linken und rechten Seite von (14.117) stehen die *Erhaltungsgrößen* Energie $E = const.$ (14.114) und Drehimpuls $L = const.$ (14.115). Damit ist auch der Betrag der Winkelgeschwindigkeit $|\vec{\omega}| = const.$. Die entscheidende Information steckt demnach im Skalarprodukt zwischen $\vec{\omega}$ und $\vec{L}$ in

$$E_{\text{rot}} = \frac{1}{2}\,\vec{\omega}\,\vec{L} = \frac{1}{2}\,|\vec{\omega}|\,|\vec{L}|\,\cos\gamma = const.\,, \tag{14.118}$$

das den Öffnungswinkel γ zwischen $\vec{\omega}$ und $\vec{L}$ anzeigt. Da der Drehimpulsvektor $\vec{L}$ im Laborsystem in eine feste Richtung zeigt, rotiert der Vektor der Winkelgeschwindigkeit $\vec{\omega}$ auf einem Kegel um die $\vec{L}$-Achse (▶ Abbildung 14.15).

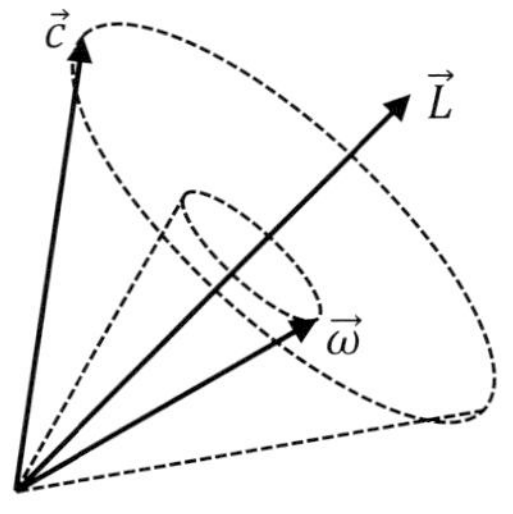

Abbildung 14.15 Drehimpuls $\vec{L}$, Winkelgeschwindigkeit $\vec{\omega}$ und Figurenachse $\vec{c}$

Im Folgenden werden wir die Rotationsenergie (14.117) wegen $\vec{L} = \hat{I}\vec{\omega}$ (14.95) in Komponenten des Trägheitstensors $\hat{I}$ (14.94) und der Winkelgeschwindigkeit $\vec{\omega}$ ausdrücken:

$$E_{\text{rot}} = \frac{1}{2}\begin{pmatrix}\omega_x & \omega_y & \omega_z\end{pmatrix}\begin{pmatrix}L_x\\ L_y\\ L_z\end{pmatrix} \tag{14.119}$$

$$= \frac{1}{2}\begin{pmatrix}\omega_x & \omega_y & \omega_z\end{pmatrix}\begin{pmatrix}I_{xx} & I_{xy} & I_{xz}\\ I_{yx} & I_{yy} & I_{yz}\\ I_{zx} & I_{zy} & I_{zz}\end{pmatrix}\begin{pmatrix}\omega_x\\ \omega_y\\ \omega_z\end{pmatrix} \tag{14.120}$$

Dabei benutzen wir die gleiche Matrixschreibweise wie in der Relativitätstheorie mit transponierten (Zeilen-)Vektoren. In Kurzschreibweise lautet diese Gleichung (der Index ^{T} steht für den transponierten Vektor)

$$E_{\text{rot}} = \frac{1}{2}\,\vec{\omega}^T\,\hat{I}\,\vec{\omega}\,. \tag{14.121}$$

Figurenachse

Um die Bewegung der Figurenachse $\vec{c}$ zu verstehen, wählen wir als Bezugssystem die Einheitsvektoren $\vec{e}_a$, $\vec{e}_b$ und $\vec{e}_c$, die in die Richtungen der Hauptträgheitsachsen des Kreisels zeigen.

Mit dem Trägheitstensor $\hat{I}$ (14.113) des symmetrischen Kreisels und dem Vektor der Winkelgeschwindigkeit $\vec{\omega}$ ausgedrückt in den Koordinaten a, b, c erhalten wir als Rotationsenergie des Kreisels (14.121)

$$E_{\text{rot}} = \frac{1}{2}\begin{pmatrix}\omega_a\,, & \omega_b\,, & \omega_c\end{pmatrix}\begin{pmatrix} I_a & 0 & 0 \\ 0 & I_a & 0 \\ 0 & 0 & I_c \end{pmatrix}\begin{pmatrix}\omega_a \\ \omega_b \\ \omega_c\end{pmatrix} \tag{14.122}$$

$$= \frac{1}{2}\left(I_a\omega_a^2 + I_a\omega_b^2 + I_c\omega_c^2\right) \tag{14.123}$$

$$= \frac{1}{2}\left(\frac{L_a^2}{I_a} + \frac{L_b^2}{I_a} + \frac{L_c^2}{I_c}\right) = const.\,. \tag{14.124}$$

Diese Gleichung hat die Form eines Ellipsoids $Ax^2 + By^2 + Cz^2 = const.$ (Rotationsellipsoid) mit $A = B$. Beachten Sie, dass dessen Koordinaten x, y, z hier den Drehimpulskoordinaten L_a, L_b, L_c entsprechen (linke ► Abbildung 14.16). Anschaulich impliziert die Erhaltung der Rotationsenergie demnach die Beschränkung der Kreiselbewegung auf eine Ellipsoidform im Drehimpulsraum.

Auch die Drehimpulserhaltung $\vec{L} = const.$ bzw. $|\vec{L}|^2 = const.$ können wir in diesen Drehimpulskoordinaten L_a, L_b, L_c beschreiben:

$$L_a^2 + L_b^2 + L_c^2 = L^2 = const. \tag{14.125}$$

Diese Gleichung hat die Form einer Kugel $x^2 + y^2 + z^2 = const.$. Drehimpulserhaltung beschränkt also die Kreiselbewegung auf eine Kugelform im Drehimpulsraum (rechte ► Abbildung 14.16).

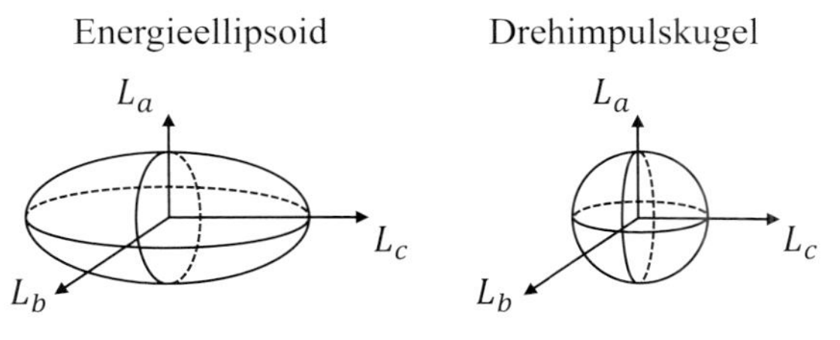

Abbildung 14.16 Links Energieerhaltung im Drehimpulsraum: L_a, L_b, L_c. Die erlaubten Rotationsenergien E_{rot} bilden ein Rotationsellipsoid im Drehimpulsraum. Rechts Drehimpulserhaltung: Die erlaubten Drehimpulsvektoren $\vec{L}$ beschreiben eine Kugel im Drehimpulsraum L_a, L_b, L_c.

Wir bringen die Gleichungen (14.124) und (14.125) auf eine Normalform, um sie direkt miteinander vergleichen zu können:

$$1 = \left(\frac{L_a^2}{2E_{\text{rot}}I_a} + \frac{L_b^2}{2E_{\text{rot}}I_a} + \frac{L_c^2}{2E_{\text{rot}}I_c}\right) \tag{14.126}$$

$$1 = \frac{L_a^2}{L^2} + \frac{L_b^2}{L^2} + \frac{L_c^2}{L^2} \tag{14.127}$$

Da die Energieerhaltung und die Drehimpulserhaltung gleichzeitig gelten müssen, liegen alle möglichen Werte auf der Schnittkurve des Ellipsoids und der Kugel (► Abbildung 14.17).

Gleichsetzen und umsortieren der Terme ergibt

$$\left(\frac{L_a^2}{2E_{\text{rot}}I_a}+\frac{L_b^2}{2E_{\text{rot}}I_a}+\frac{L_c^2}{2E_{\text{rot}}I_c}\right)=\frac{L_a^2}{L^2}+\frac{L_b^2}{L^2}+\frac{L_c^2}{L^2} \tag{14.128}$$

$$(L_a^2+L_b^2)\left(\frac{1}{L^2}-\frac{1}{2E_{\text{rot}}I_a}\right)=-L_c^2\left(\frac{1}{L^2}-\frac{1}{2E_{\text{rot}}I_c}\right) \tag{14.129}$$

$$(L_a^2+L_b^2)\left(E_{\text{rot}}-\frac{L^2}{2I_a}\right)=-L_c^2\left(E_{\text{rot}}-\frac{L^2}{2I_c}\right). \tag{14.130}$$

Die Größen E_{rot}, I_a, I_c, L^2 sind positiv, die Drehimpulskomponenten L_a^2, L_b^2, L_c^2 sind größer oder gleich null. Nach ► Abbildung 14.14 nehmen wir an, dass das Hauptträgheitsmoment entlang der Figurenachse am größten ist: $I_c > I_a$.

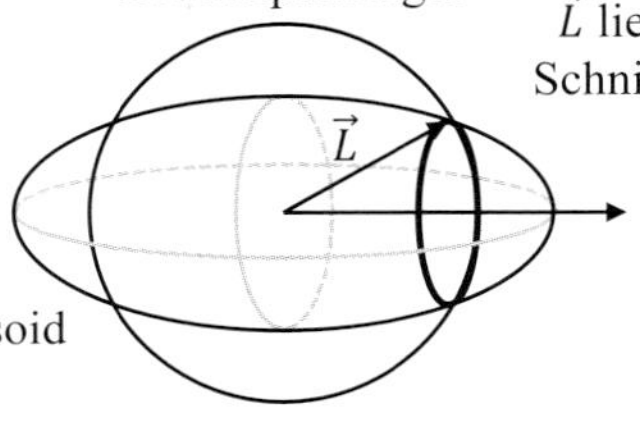

Abbildung 14.17 $\vec{L}$ liegt auf der Schnittkurve der Drehimpulskugel und des Energieellipsoids. Die Figurenachse $\vec{c}$ zeigt entlang des Einheitsvektors $\vec{e}_c$.

Die Schnittgleichung (14.130) ist z. B. erfüllt, wenn auf der rechten Seite $E_{\text{rot}} = L^2/(2I_c)$ ist. Dann müssen auf der linken Seite die Drehimpulskomponenten verschwinden $L_a = L_b = 0$, d. h. Drehimpuls $\vec{L}$ und Figurenachse $\vec{c}$ sind parallel. Die Drehimpulskugel und das Drehimpuls-Ellipsoid berühren sich in einem Punkt. Anschaulich beschreibt diese Rotationsenergie E_{rot} den ungestörten Kreisel mit dem Trägheitsmoment I_c der Figurenachse c.

Ist auf der linken Seite $E_{\text{rot}} = L^2/(2I_a)$, so muss rechts die Drehimpulskomponente wegfallen $L_c = 0$. In diesem Fall stehen Drehimpuls $\vec{L}$ und Figurenachse $\vec{c}$ senkrecht aufeinander.

Für alle weiteren Fälle ist (14.130) nur erfüllt, falls

$$\frac{L^2}{2I_c} < E_{\text{rot}} < \frac{L^2}{2I_a} \tag{14.131}$$

gilt. Den Nutationswinkel α zwischen der Figurenachse $\vec{c}$ und dem Drehimpuls $\vec{L}$ berechnen wir ebenfalls aus (14.130):

$$\tan\alpha=\sqrt{\frac{(L_a^2+L_b^2)}{L_c^2}}=\sqrt{-\left(E_{\text{rot}}-\frac{L^2}{2I_c}\right)\Big/\left(E_{\text{rot}}-\frac{L^2}{2I_a}\right)} \tag{14.132}$$

Der Winkel α zwischen der Figurenachse und dem Drehimpuls liegt also theoretisch zwischen $0°$ und $90°$. Mit zunehmender Energie, die dem Kreisel durch einen Stoß zugeführt wird, wird der Nutationskegel größer. Der Grenzfall $\alpha = 90°$, in dem der Kreisel senkrecht zur Figurenachse mit der Rotationsenergie E^a_{rot} und dem Trägheitsmoment I_a rotiert, ist nur mit einigem Aufwand (z. B. kardanischer Aufhängung) zu realisieren.

Gehen wir von den Drehimpulskoordinaten zurück ins Laborsystem: Hier bleibt die Richtung von $\vec{L}$ fest, sodass die Figurenachse $\vec{c}$ auf einem Kegel mit dem festem Öffnungswinkel α um den Drehimpulsvektor $\vec{L}$ rotiert (► Abbildung 14.15). Diese Bewegung der Figurenachse des Kreisels beobachten wir in dem oben vorgestellten Nutationsexperiment.

14.8.2 Präzession

Lassen wir ein äußeres Drehmoment $\vec{D} \neq 0$ auf den Kreisel wirken, so kommt es zu einer weiteren Drehbewegung, der sogenannten Präzession. Das Drehmoment kann z. B. durch die Gravitation verursacht werden.

Experiment 27: Kreiselpräzession

In ► Abbildung 14.18 ist ein rotierendes Rad gezeigt, das auf der einen Seite seiner Achse an einem Faden aufgehängt ist. Das Rad besitzt das Trägheitsmoment I_R bezüglich seiner Achse. Die Raddrehung hat die Winkelgeschwindigkeit $\vec{\omega}_R$, deren Vektor entlang der Radachse nach außen zeigt. Das Rad hat damit den Drehimpuls (14.18)

$$\vec{L} = I_R \vec{\omega}_R \,. \tag{14.133}$$

Senkrecht dazu wirkt das Drehmoment der Gravitation

$$\vec{D} = \vec{r} \times m_R \vec{g} \,. \tag{14.134}$$

Wir beobachten, dass durch das angreifende Drehmoment eine neue Drehung entsteht, die als **Präzession** bezeichnet wird: Während sich das Rad um seine Achse dreht, rotiert zusätzlich die Radachse um die Aufhängung mit einer Winkelgeschwindigkeit ω_p (»Präzessionsfrequenz«).

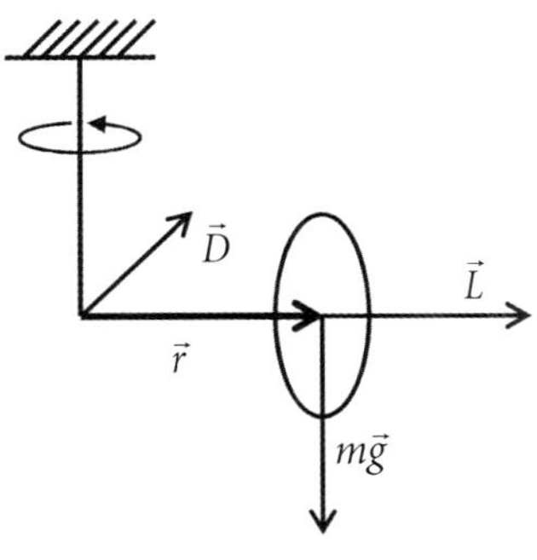

Abbildung 14.18 Kreiselpräzession: Ein Kreisel mit Drehimpuls $\vec{L}$ erfährt aufgrund der Schwerkraft $m\vec{g}$ mit dem Hebel $\vec{r}$ ein Drehmoment $\vec{D}$.

Nach dem Bewegungsgesetz für Drehbewegungen (4.70) bewirkt ein von außen angreifendes Drehmoment $\vec{D}$ (14.134) eine zeitliche Änderung des Drehimpulses $\vec{L}$ (14.133):

$$\frac{d\vec{L}}{dt} = \vec{D}\,. \tag{14.135}$$

In unserem Experiment stehen der Drehmomentvektor $\vec{D}$ und der Drehimpulsvektor $\vec{L}$ des Rads senkrecht aufeinander, wodurch der Betrag des Drehimpulses unverändert bleibt: $|\vec{L}| = const.$ Offenbar ändert sich aber durch das angreifende Drehmoment die Richtung des Drehimpulsvektors: Bei der Präzessionsbewegung rotiert $\vec{L}$ um die Richtung der Aufhängung.

Wir wählen hier Zylinderkoordinaten r, φ, z mit den Einheitsvektoren $\vec{e}_r$, $\vec{e}_\varphi$ und $\vec{e}_z$ (9.20). Dabei zeigt z nach oben zum Befestigungspunkt (▶ Abbildung 14.18). Der Drehimpulsvektor $\vec{L}$ wird anfangs ($\varphi = 0, z = 0$) durch

$$\vec{L} = L\,\vec{e}_r \tag{14.136}$$

beschrieben. Seine zeitliche Änderung führen wir auf die zeitliche Änderung des Einheitsvektors zurück, wobei wir die Kettenregel beachten:

$$\frac{d\vec{L}}{dt} = \frac{d}{dt}(L\,\vec{e}_r) \tag{14.137}$$

$$= L\,\frac{d\vec{e}_r}{d\varphi}\,\frac{d\varphi}{dt} \tag{14.138}$$

$$= L\,\frac{d\varphi}{dt}\,\frac{d}{d\varphi}\begin{pmatrix}\cos(\varphi)\\ \sin(\varphi)\\ 0\end{pmatrix} \tag{14.139}$$

$$= L\,\underbrace{\frac{d\varphi}{dt}}_{\equiv\omega_p}\underbrace{\begin{pmatrix}-\sin(\varphi)\\ \cos(\varphi)\\ 0\end{pmatrix}}_{=\vec{e}_\varphi} \tag{14.140}$$

$$= L\,\omega_p\,\vec{e}_\varphi \tag{14.141}$$

Die Richtung des Drehmomentvektors $\vec{D}$ (14.134) erhalten wir aus dem Kreuzprodukt:

$$\vec{D} = r\,\vec{e}_r \times m_R\,g\,\vec{e}_z \tag{14.142}$$

$$= r\,m_R\,g\,\vec{e}_\varphi \tag{14.143}$$

$$= D\,\vec{e}_\varphi \tag{14.144}$$

Der Drehmomentvektor $\vec{D}$ und die zeitliche Ableitung des Drehimpulsvektors $d\vec{L}/dt$ (14.141) zeigen demnach in dieselbe Richtung. Setzen wir (14.141) und (14.144) in die Bewegungsgleichung (14.135) ein, so erhalten wir die Präzessionsfrequenz ω_p direkt aus dem Verhältnis D zu L:

$$\frac{d\vec{L}}{dt} = \vec{D} \tag{14.145}$$

$$L\,\omega_p\,\vec{e}_\varphi = D\,\vec{e}_\varphi \tag{14.146}$$

$$\omega_p = \frac{D}{L} \tag{14.147}$$

Konkret ergibt sich die Präzessionsfrequenz mit $L = I_R\,\omega_R$ (14.133) und $D = r\,m_R\,g$ (14.134) zu

$$\omega_p = \frac{r\,m_R\,g}{I_R\,\omega_R}\,. \tag{14.148}$$

Diese Drehbewegung können wir ebenfalls als »Präzessionsperiode« mit der Periode des Rads $T_R = 2\pi/\omega_R$ (4.11) ausdrücken:

$$T_p = \frac{4\pi^2\,I_R}{m\,g\,r}\,\frac{1}{T_R} \tag{14.149}$$

Je schneller sich das Rad dreht, desto langsamer ist die Präzessionsbewegung.

Das Wichtigste zum Mitnehmen

Zum Mitnehmen: Dynamik starrer Körper 1

- Mit der **Massendichte**

$$\rho = \frac{dm}{dV} \tag{14.3}$$

  lässt sich die die Masse über ein Volumenintegral berechnen:

$$M = \int_V \rho \, dV \tag{14.4}$$

- Der **Schwerpunkt** eines Körpers $\vec{r}_s$ bezüglich seiner kontinuierlichen Massenverteilung ist gegeben durch

$$\vec{r}_s = \frac{\int \vec{r} \, dm}{\int dm} = \frac{\int \vec{r} \rho \, dV}{\int \rho \, dV} = \frac{1}{M} \int \vec{r} \rho \, dV \, . \tag{14.7}$$

- Für kontinuierlich verteilte Massenelemente dm eines Körpers mit minimalem Abstand $r_\perp$ von der Drehachse z lautet das **Trägheitsmoment I_z**

$$I_z = \int r_\perp^2 \, dm \, . \tag{14.21}$$

  Damit ist der Drehimpuls bei einer Winkelgeschwindigkeit $\vec{\omega}$ bezüglich der z-Achse

$$\vec{L} = I_z \, \vec{\omega} \, . \tag{14.23}$$

- Zur Umrechnung des Trägheitsmoments I_S eines Körpers der Masse M von einer Drehachse $\vec{S}$ durch den Schwerpunkt auf eine andere parallele Drehachse $\vec{B}$ im Abstand a gilt der als **»Satz von Steiner«** bekannte Zusammenhang

$$I_B = I_S + M \, a^2 \, . \tag{14.53}$$

Zum Mitnehmen: Dynamik starrer Körper 2

- An einer drehbaren Achse wird mit einer Feder mit Richtmoment D_r ein rücktreibendes Drehmoment D erzeugt:

$$D = -D_r \Phi \tag{14.31}$$

Mit dem Trägheitsmoment I des Drehkörpers entstehen **Drehschwingungen** mit der Bewegungsgleichung eines harmonischen Oszillators:

$$\ddot{\Phi} = -\underbrace{\frac{D_r}{I}}_{\equiv \omega^2} \Phi \tag{14.34}$$

Ihre Lösung lautet für den Drehwinkel Φ

$$\Phi(t) = A \cdot \cos(\omega t + \varphi_\circ) \tag{14.35}$$

mit einem Anfangswinkel $\varphi_\circ$, der Kreisfrequenz ω

$$\omega = \sqrt{\frac{D_r}{I}} \tag{14.36}$$

und der Schwingungsperiode $T = 2\pi/\omega$

$$T = 2\pi\sqrt{\frac{I}{D_r}} \,. \tag{14.37}$$

- Für einen starren Körper mit dem Trägheitsmoment I ist die **Rotationsenergie** dementsprechend

$$E_{\text{rot}} = \frac{1}{2} I \omega^2 \,. \tag{14.65}$$

Zum Mitnehmen: Dynamik starrer Körper 3

- Für einen starren Körper mit variabler Drehachse ersetzt man in (14.18) das Trägheitsmoment I durch einen **Trägheitstensor** $\hat{I}$:

$$\vec{L} = \hat{I} \cdot \vec{\omega} \tag{14.95}$$

Damit kann man den Drehimpuls für beliebige Rotationsachsen $\vec{\omega}$ berechnen. Der Trägheitstensor ist symmetrisch und lässt sich in die diagonale **Hauptachsenform** transformieren:

$$\begin{pmatrix} L_a \\ L_b \\ L_c \end{pmatrix} = \begin{pmatrix} I_a & 0 & 0 \\ 0 & I_b & 0 \\ 0 & 0 & I_c \end{pmatrix} \cdot \begin{pmatrix} \omega_a \\ \omega_b \\ \omega_c \end{pmatrix} \tag{14.111}$$

I_a, I_b, I_c sind die **Hauptträgheitsmomente** in Richtung der Hauptachsen mit den Einheitsvektoren $\vec{e}_a$, $\vec{e}_b$, $\vec{e}_c$.

- Die **Rotationsenergie** zu einem Trägheitstensor lautet in Kurzschreibweise (der Index ^{T} steht für den transponierten Vektor)

$$E_{\text{rot}} = \frac{1}{2} \vec{\omega}^T \, \hat{I} \, \vec{\omega} \, . \tag{14.121}$$

Mit dem Drehimpuls $\vec{L} = \hat{I}\vec{\omega}$ gilt dann (der Index ^{T} wird im Skalarprodukt meist weggelassen)

$$E_{\text{rot}} = \frac{1}{2} \vec{\omega} \, \vec{L} \, . \tag{14.117}$$

Die Rotationsenergie beträgt in seinem Hauptachsensystem

$$E_{\text{rot}} = \frac{1}{2} \begin{pmatrix} \omega_a & \omega_b & \omega_c \end{pmatrix} \begin{pmatrix} I_a & 0 & 0 \\ 0 & I_b & 0 \\ 0 & 0 & I_c \end{pmatrix} \begin{pmatrix} \omega_a \\ \omega_b \\ \omega_c \end{pmatrix} \tag{14.122}$$

$$= \frac{1}{2} \left(I_a \omega_a^2 + I_b \omega_b^2 + I_c \omega_c^2 \right) . \tag{14.123}$$

Zum Mitnehmen: Dynamik starrer Körper 4

- Energie- und Drehimpulserhaltung bewirken, dass für einen **symmetrischen Kreisel** mit $I_a = I_b$ und der Figurenachse entlang $\vec{e}_c$ der Drehimpulsvektor $\vec{L} = const.$ (14.111) auf der Schnittkurve der Drehimpulskugel $\vec{L} = const.$ mit dem Energieellipsoid $E = const.$ (14.123) liegt.

  Bei einem kräftefreien Kreisel gibt es genau eine Lösung auf der Figurenachse und die Kreiselbewegung ist stabil. Wenn dem System jedoch durch einen Stoß Energie zugeführt wird, gibt es eine Ausweichbewegung auf einem **Nutationskegel**.

- Wirkt ein äußeres Drehmoment $\vec{D} \neq 0$ z. B. durch die Gravitation auf den Kreisel ein, so kommt es zu der sogenannten **Präzession**. Einsetzen der Drehimpulsänderung $dL/dt = L\,d\varphi/dt$ in die Bewegungsgleichung (4.70) ergibt mit $d\varphi/dt = \omega_p$

$$\omega_p = \frac{D}{L}\,. \tag{14.147}$$

  Für ein Drehmoment aus der Schwerkraft $D = mgr$ lässt sich eine »Präzessionsperiode« T_p mit der Periode des Kreisels $T_R = 2\pi/\omega_R$ ausdrücken:

$$T_p = \frac{4\pi^2 I_R}{m\,g\,r}\frac{1}{T_R} \tag{14.149}$$

  Je schneller der Kreisel dreht, desto langsamer ist die Präzessionsbewegung.

Zum Mitnehmen: Dynamik starrer Körper 5

Gegenüberstellung Translation-Rotation

Translation	Symbol	Symbol	Rotation
Verschiebung	x	φ	Winkel
Geschwindigkeit	$dx/dt = v$	$\omega = d\varphi/dt$	Winkelgeschwindigkeit
Beschleunigung	$dv/dt = a$	$\dot{\omega} = d\omega/dt$	Winkelbeschleunigung
Masse	m	$I = \int r_\perp^2 \, dm$	Trägheitsmoment
Kraft	$m\vec{a} = \vec{F}$	$\vec{D} = \vec{r} \times \vec{F}$	Drehmoment
Impuls	$m\vec{v} = \vec{p}$	$\vec{L} = I\vec{\omega}$	Drehimpuls
Arbeit	$\int \vec{F} d\vec{r} = W$	$W = \int \vec{D}\vec{\omega} dt$	Arbeit
Kinetische Energie	$\frac{1}{2}mv^2 = E_{\text{kin}}$	$E_{\text{rot}} = \frac{1}{2}I\omega^2$	Rotationsenergie
Leistung	$\vec{F}\vec{v} = P$	$P = \vec{D}\vec{\omega}$	Leistung

Verbindende Gleichungen

Geschwindigkeit	$\vec{v} = \vec{\omega} \times \vec{r}$	Winkelgeschwindigkeit
Tangentialbeschleunigung	$\vec{a}_t = \dot{\vec{\omega}} \times \vec{r}$	Winkelbeschleunigung
Radialbeschleunigung	$\vec{a}_r = \vec{\omega} \times \vec{v}$	Winkelgeschwindigkeit
Zentrifugalbeschleunigung	$\vec{a}_{ZF} = \vec{\omega} \times (\vec{r} \times \vec{\omega})$	Winkelgeschwindigkeit
Coriolisbeschleunigung	$\vec{a}_C = 2\,(\vec{v}\,' \times \vec{\omega})$	(Winkel)-Geschwindigkeit
Drehmoment	$\vec{D} = \vec{r} \times \vec{F}$	Kraft
Drehimpuls	$\vec{L} = \vec{r} \times \vec{p}$	Impuls

Schwingungen

15

15.1 Harmonischer Oszillator **294**
15.1.1 Ortsfunktion durch Überlagerung komplexer Lösungen 294
15.1.2 Ortsfunktion durch Projektion einer komplexen Lösung 295

15.2 Gedämpfte Schwingungen **298**
15.2.1 Bewegungsgleichung für gedämpfte Schwingungen 298
15.2.2 Ortsfunktion für gedämpfte Schwingungen 298

15.3 Erzwungene Schwingungen **303**
15.3.1 Bewegungsgleichung für erzwungene Schwingungen 304
15.3.2 Ortsfunktion für erzwungene Schwingungen 304
15.3.3 Resonanz bei erzwungenen Schwingungen 305

15.4 Gekoppelte Schwingungen **307**
15.4.1 Bewegungsgleichungen für gekoppelte Pendel 308
15.4.2 Ortsfunktionen des Doppelpendels: Winkelkoordinaten 312

15.5 Fouriersynthese, -analyse, -transformation **316**
15.5.1 Fouriersynthese 316
15.5.2 Fourieranalyse 318
15.5.3 Fouriertransformation 319
15.5.4 Unschärferelation 321

ÜBERBLICK

» Mit Schwingungen bzw. Oszillationen bezeichnen wir periodische Bewegungen von Objekten bzw. ganzen Systemen. In diesem Kapitel erweitern wir unsere bisherigen Kenntnisse über den harmonischen Oszillator zu Schwingungssystemen mit Antrieb, Dämpfung und zu gekoppelten Schwingungssystemen. Schließlich untersuchen wir aus vielen Frequenzen zusammengesetzte Schwingungen.

Periodische Bewegungen sind häufig einfacher mit komplexen als mit reellen Zahlen zu beschreiben. Daher beginnen wir mit einem kurzen Einschub über die wichtigsten Aspekte komplexer Zahlen. «

Mathematischer Einschub: Komplexe Zahlen

Eine komplexe Zahl enthält einen Realteil $a = Re(c)$ und einen Imaginärteil $b = Im(c)$ (linke ► Abbildung 15.1):

$$c = a + ib \quad \text{mit } a, b \in \mathbb{R} \text{ und } c \in \mathbb{C} \tag{15.1}$$

Die dazugehörende komplex konjugierte Zahl lautet

$$c^* = a - ib\,. \tag{15.2}$$

Der Betrag einer komplexen Zahl ergibt sich unter Berücksichtigung von $i^2 = -1$:

$$|c| = \sqrt{c \cdot c^*} = \sqrt{(a + ib)\,(a - ib)} = \sqrt{a^2 - (ib)^2} = \sqrt{c^2 + b^2} \tag{15.3}$$

Die Darstellung einer komplexen Zahl in Polarkoordinaten lautet (rechte ► Abbildung 15.1)

$$c = |c|\,(cos\varphi + i\sin\varphi)\,, \tag{15.4}$$
$$c^* = |c|\,(cos\varphi - i\sin\varphi)\,. \tag{15.5}$$

Die häufig verwendete **Eulersche Formel**

$$\cos\varphi \pm i\sin\varphi = e^{\pm i\varphi} \tag{15.6}$$

erlaubt uns eine praktische Kurzschreibweise für (15.4) und (15.5):

$$\begin{aligned} c &= |c|\,e^{i\varphi} \\ c^* &= |c|\,e^{-i\varphi} \end{aligned} \tag{15.7}$$

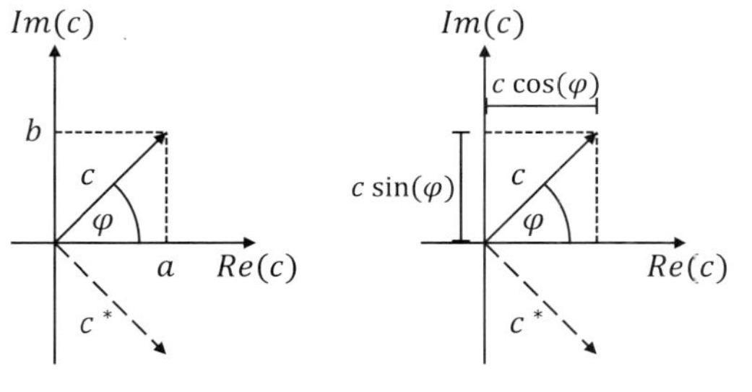

Abbildung 15.1 Komplexe Zahlenebene (links), komplexe Zahlen in Polarkoordinaten (rechts)

15.1 Harmonischer Oszillator

In diesem Abschnitt betrachten wir zunächst einen ungestört (insbesondere reibungslos) schwingenden harmonischen Oszillator. Ein Beispiel ist das Federpendel mit der bewegten Masse m und der rücktreibenden Federkraft (► Abbildung 15.2)

$$F = -kx\,. \tag{15.8}$$

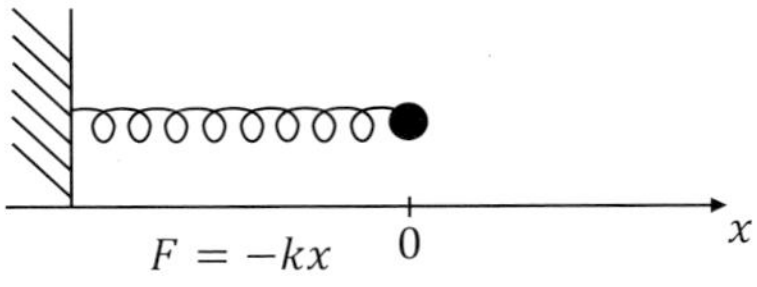

Abbildung 15.2 Federpendel: Körper der Masse m mit rücktreibender Kraft $F = -kx$

Mit dem Bewegungsgesetz von Newton (3.41) erhalten wir die Bewegungsgleichung

$$\frac{d\vec{p}}{dt} = \Sigma_i \vec{F}_i \tag{15.9}$$

$$ma = -kx \tag{15.10}$$

$$\ddot{x} = -\frac{k}{m}\,x\,. \tag{15.11}$$

Wir erhalten somit die Bewegungsgleichung des harmonischen Oszillators (4.30) und verwenden die Kreisfrequenz $\omega \equiv \sqrt{k/m}$ (4.32):

$$\ddot{x} = -\omega^2 x \tag{15.12}$$

Die Ortsfunktion

$$x(t) = x_\circ \cdot \cos\left(\omega t + \varphi_\circ\right) \tag{15.13}$$

löst die Bewegungsgleichung des Federpendels (15.12) mit der Kreisfrequenz ω und den beiden Anfangsbedingungen $x_\circ, \varphi_\circ \in \mathbb{R}$.

15.1.1 Ortsfunktion durch Überlagerung komplexer Lösungen

Mit den komplexen Zahlen versuchen wir nun einen alternativen Lösungsansatz zu verwenden:

$$x(t) = c\,e^{\lambda t} \qquad c, \lambda \in \mathbb{C} \tag{15.14}$$

Der Ansatz erscheint in diesem Fall zunächst unnötig aufwendig, hilft aber später bei anspruchsvollen Problemstellungen wie gedämpften oder gekoppelten Schwingungen.

Die zeitlichen Ableitungen für diesen Ansatz ergeben

$$x = c\,e^{\lambda t} \tag{15.15}$$

$$\dot{x} = \lambda\,c\,e^{\lambda t} \tag{15.16}$$

$$\ddot{x} = \lambda^2 \underbrace{c\,e^{\lambda t}}_{=x} \tag{15.17}$$

$$= \lambda^2\,x\,. \tag{15.18}$$

Unser Lösungsansatz erfüllt die Bewegungsgleichung (15.12) mit

$$\lambda^2 = -\omega^2\,. \tag{15.19}$$

Die beiden möglichen Lösungen für λ sind damit

$$\lambda_1 = i\omega \tag{15.20}$$

$$\lambda_2 = -i\omega\,. \tag{15.21}$$

Als allgemeine Lösung verwenden wir die Kombination der beiden Möglichkeiten:

$$x(t) = x_1(t) + x_2(t) \tag{15.22}$$

$$= c_1\,e^{i\omega t} + c_2\,e^{-i\omega t} \tag{15.23}$$

Wir zeigen im Folgenden, dass diese Lösung mit der Lösung (15.13) übereinstimmt. Zur Lösung der Differentialgleichung (15.12) mit der zweiten zeitlichen Ableitung benötigen wir zwei Anfangsbedingungen, die wir als Anfangsamplitude $|c|$ und Anfangswinkel $\varphi_\circ$ in $c_1 = (|c|/2)\,e^{i\varphi_\circ}$ und $c_2 = (|c|/2)\,e^{-i\varphi_\circ}$ in (15.23) einsetzen:

$$x(t) = \frac{|c|}{2}\,e^{i\varphi_\circ}e^{i\omega t} + \frac{|c|}{2}\,e^{-i\varphi_\circ}e^{-i\omega t} \tag{15.24}$$

$$= \frac{|c|}{2}\left(e^{i(\omega t+\varphi_\circ)} + e^{-i(\omega t+\varphi_\circ)}\right) \tag{15.25}$$

$$= \frac{|c|}{2}\left(\cos(\omega t+\varphi_\circ) + i\sin(\omega t+\varphi_\circ) + \cos(\omega t+\varphi_\circ) - i\sin(\omega t+\varphi_\circ)\right) \tag{15.26}$$

$$= |c|\,\cos(\omega t+\varphi_\circ) \tag{15.27}$$

Unser Ansatz (15.23) führt also zu derselben Lösung (15.13) wie der nicht-komplexe Ansatz, wobei sich die Amplituden entsprechen ($x_\circ = |c|$).

In ► Abbildung 15.3 ist das Vorgehen noch einmal visualisiert. Die komplexe Lösung $x_1(t)$ des harmonischen Oszillators wird durch einen gegen den Uhrzeigersinn rotierenden Zeiger dargestellt. Der Zeiger der komplexen Lösung $x_2(t)$ rotiert entsprechend im Uhrzeigersinn. Die Summe $x_1(t) + x_2(t)$ führt zur Kosinusfunktion als reelle Lösung (15.13).

15.1.2 Ortsfunktion durch Projektion einer komplexen Lösung

Physikalisch beobachtbare Lösungen sind immer reell. Erhalten wir die allgemeine Lösung in Form von komplexen Zahlen, so ergibt die Projektion auf die reelle Achse die physikalische Lösung. Daher genügt es häufig, entweder λ_1 (15.20) oder λ_2 (15.21) zu verwenden und anschließend eine Projektion auf die reelle Achse durchzuführen (► Abbildung 15.3).

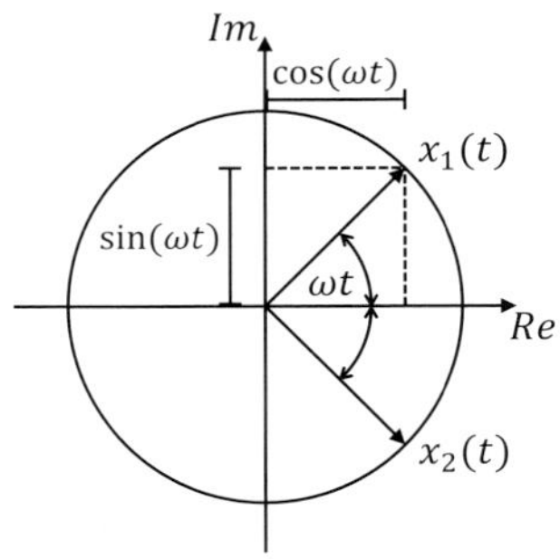

Abbildung 15.3 Lösungsansatz für den harmonischen Oszillator: Die komplexen Zeiger $x_{1,2}(t)$ rotieren gegensinnig mit der Winkelgeschwindigkeit ω in der komplexen Zahlenebene.

Wir wählen hier (15.20) als Lösung

$$x_1\,(t) = c_1 \cdot e^{i\omega t}\,. \tag{15.28}$$

Setzen wir für die Anfangsamplitude $x_\circ$ und den anfänglichen Winkel $\varphi_\circ$ in $c_1 = x_\circ\, e^{i\varphi_\circ}$ ein, so ergibt sich als komplexe Lösung

$$x_1(t) = x_\circ\, e^{i\varphi_\circ} e^{i\omega t} \tag{15.29}$$

$$= x_\circ\, e^{i(\omega t+\varphi_\circ)} \tag{15.30}$$

$$= x_\circ \left[\cos\left(\omega t + \varphi_\circ\right) + i \sin\left(\omega t + \varphi_\circ\right)\right]\,. \tag{15.31}$$

Der Realteil von $x_1(t)$ ist bereits physikalische Lösung von (15.12) mit den zwei Anfangsbedingungen $x_\circ$ und $\varphi_\circ$

$$Re\left(x_1\,(t)\right) = x_\circ \cos\left(\omega t + \varphi_\circ\right) \tag{15.32}$$

und entspricht der Lösung (15.13).

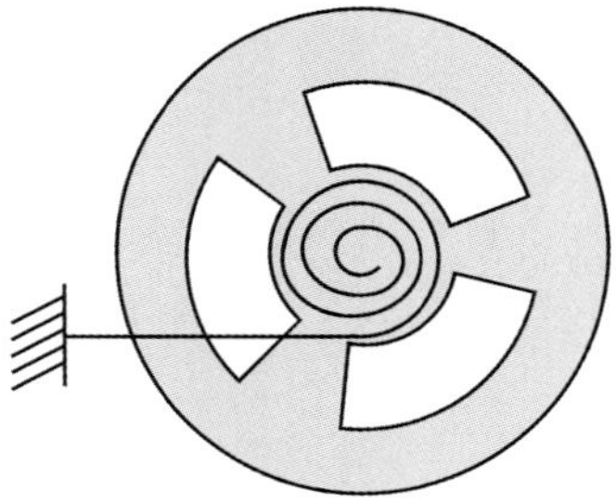

Abbildung 15.4 Drehschwingungen eines Pohlschen Rads durch eine Schneckenfeder

Experiment 28: Pohlsches Rad

Das **Pohlsche Rad** besteht aus einer vertikal positionierten Scheibe mit horizontaler Drehachse (► Abbildung 15.4). An ihr ist eine Schneckenfeder montiert, womit wir das Rad in Drehschwingungen versetzen können. Das Aufdrehen der Schneckenfeder bewirkt dabei ein rücktreibendes Drehmoment $D = -D_r\Phi$, das der Drehung der Scheibe entgegenwirkt. Das Richtmoment der Feder ist hier mit D_r bezeichnet.

Optional kann das Pohlsche Rad mit zusätzlichen Funktionen betrieben werden. Vorgesehen ist ein **periodischer Antrieb**, bei dem die links gezeigte Befestigung der Schneckenfeder hin und her bewegt und darüber das Rad zu Schwingungen angeregt wird.

Ebenso gibt es einen Mechanismus zur Dämpfung der Radschwingungen durch eine sogenannte **Wirbelstrombremse** (in der Abbildung nicht gezeigt). Dafür wird die elektrisch leitende Scheibe des Rades in das Magnetfeld einer Spule gebracht. Aufgrund der Bewegung werden in der Scheibe nach dem Induktionsgesetz Spannungen induziert, die zu Wirbelströmen im leitenden Material führen. Nach der Lenzschen Regel widersetzt sich dieser Vorgang seiner Ursache, also der Rad-Bewegung, und das Rad wird gebremst. Das Induktionsgesetz und die Lenzsche Regel werden im Band »Felder und Wellen« [4] näher ausgeführt.

Von Bedeutung für unser Experiment und seine Beschreibung ist, dass die Dämpfung der Radschwingung durch die Wirbelstrombremse geschwindigkeitsabhängig ist, weil die induzierte Spannung linear von der Winkelgeschwindigkeit abhängt. Eine linear von der Drehgeschwindigkeit $\dot{\Phi}$ des Rads abhängige Dämpfung führt zu einem bremsenden Drehmoment D_b der Form

$$D_b = -b\dot{\Phi}\,. \tag{15.33}$$

Die Stärke der Bremswirkung ist dabei in dem Faktor b erfasst und lässt sich durch Variieren des Magnetfeldes verändern.

Aufgabe 42: Pohlsches Rad

Die Scheibe des Pohlschen Rads (► Abbildung 15.4) hat das Trägheitsmoment I (14.21), die Winkelkoordinate seiner Drehung bezeichnen wir mit Φ. Der Drehimpuls des Rads beträgt $L = I\dot{\Phi}$ (14.23). Die Schneckenfeder hat das Richtmoment D_r und bewirkt ein rücktreibendes Drehmoment $D = -D_r\Phi$ (14.31).

1 Wie lautet die Bewegungsgleichung für das frei schwingende Pohlsche Rad?

2 Ist $\Phi(t) = c \cdot e^{i\omega t}$ mit $c \in \mathbb{C}$ eine Lösung?

3 Wie lautet die physikalische Lösung?

15.2 Gedämpfte Schwingungen

Als Beispiel für einen gedämpften Oszillator untersuchen wir das Pohlsche Rad mit eingebauter Wirbelstrombremse. Dabei steigt die Bremswirkung linear mit der Drehgeschwindigkeit $\dot{\Phi}$ der Scheibe (15.33), was sich mit einem konstanten Faktor b beschreiben lässt:

$$D_b = -b\dot{\Phi} \tag{15.34}$$

15.2.1 Bewegungsgleichung für gedämpfte Schwingungen

Wir verwenden das Bewegungsgesetz für Drehbewegungen (14.32) mit dem Drehimpuls des Pohlschen Rads $L = I\,\dot{\Phi}$ (14.23), dem rücktreibenden Drehmoment $D_s = -D_r\,\Phi$ der Schneckenfeder (14.31) und dem bremsenden Drehmoment $D_b = -b\dot{\Phi}$ (15.34):

$$\frac{d\vec{L}}{dt} = \sum_{i=1}^{n} \vec{D}_i \tag{15.35}$$

$$\frac{dL}{dt} = D_s + D_b \tag{15.36}$$

$$I\ddot{\Phi} = -D_r\,\Phi - b\dot{\Phi} \tag{15.37}$$

$$\ddot{\Phi} = -\underbrace{\frac{D_r}{I}}_{\equiv\omega_\circ^2}\Phi - \underbrace{\frac{b}{I}}_{\equiv 2\gamma}\dot{\Phi} \tag{15.38}$$

Dabei bezeichnen wir die Kreisfrequenz (14.36) des frei schwingenden Pohlschen Rads mit

$$\omega_\circ \equiv \sqrt{\frac{D_r}{I}}\,. \tag{15.39}$$

Die Stärke der Dämpfung beschreiben wir mit γ, wobei ein Faktor 2 in der Definition die später folgenden Berechnungen erleichtert:

$$\gamma \equiv \frac{b}{2\,I} \tag{15.40}$$

Die Bewegungsgleichung für die gedämpfte Schwingung lautet mit diesen Abkürzungen

$$\ddot{\Phi} + 2\gamma\dot{\Phi} + \omega_\circ^2\Phi = 0\,. \tag{15.41}$$

15.2.2 Ortsfunktion für gedämpfte Schwingungen

Als Lösungsansatz verwenden wir hier wie zuvor (15.14)

$$\Phi = c\,e^{\lambda t} \tag{15.42}$$

$$\dot{\Phi} = \lambda c\,e^{\lambda t} \tag{15.43}$$

$$\ddot{\Phi} = \lambda^2 c\,e^{\lambda t}\,. \tag{15.44}$$

Durch Einsetzen in die Bewegungsgleichung (15.41) erhalten wir

$$\lambda^2 + 2\gamma\lambda + \omega_\circ^2 = 0\,. \tag{15.45}$$

Die Lösung dieser Gleichung ergibt sich durch quadratische Ergänzung:

$$\lambda_{1,2} = -\gamma \pm \sqrt{\gamma^2 - \omega_\circ^2} \tag{15.46}$$

Als allgemeinen Lösungsansatz verwenden wir analog zu (15.23) die Summe beider Lösungen, die sich durch das Vorzeichen des Wurzelterms in λ_1 und λ_2 unterscheiden:

$$\Phi(t) = c_1\, e^{\lambda_1 t} + c_2\, e^{\lambda_2 t} \tag{15.47}$$

$$= c_1 \cdot e^{-\gamma t} \cdot e^{\sqrt{\gamma^2-\omega_\circ^2}\,t} + c_2 \cdot e^{-\gamma t} \cdot e^{-\sqrt{\gamma^2-\omega_\circ^2}\,t} \tag{15.48}$$

$$= e^{-\gamma t} \cdot \left[c_1 \cdot e^{\sqrt{\gamma^2-\omega_\circ^2}\,t} + c_2 \cdot e^{-\sqrt{\gamma^2-\omega_\circ^2}\,t}\right] \tag{15.49}$$

Da γ (15.40) eine reelle Zahl ist, zeigt der Term $e^{-\gamma t}$ einen exponentiellen Abfall in der Zeit. Dieser Term beschreibt demnach die Dämpfung der Anfangsamplituden $|c_1|$ und $|c_2|$.

Ob die Schwingung des Pohlschen Rads überhaupt zustande kommt entscheidet sich an den beiden Termen auf der rechten Seite:

$$e^{\pm\sqrt{\gamma^2-\omega_\circ^2}\,t} \tag{15.50}$$

Nur wenn sich aus dem Wurzelterm eine komplexe Zahl ergibt ($\gamma^2 - \omega_\circ^2 < 0$), erhalten wir über die Eulersche Formel (15.6) einen Sinus- oder Kosinusterm und damit Schwingungen. Ist der Wurzelterm eine reelle Zahl ($\gamma^2 - \omega_\circ^2 \geq 0$), so ergibt sich als zeitliches Verhalten lediglich ein exponentieller Anstieg oder Abfall oder eine Konstante (für $e^0 = 1$).

Wir unterscheiden im Folgenden drei mögliche Fälle je nach Art und Größe des Wurzelterms $\sqrt{\gamma^2 - \omega_\circ}$: schwache Dämpfung, starke Dämpfung und aperiodischen Grenzfall.

Schwache Dämpfung

Für die Bedingung einer schwachen Dämpfung $\gamma < \omega_\circ$ ist λ (15.46) eine komplexe Zahl, da der Term unter der Wurzel $\gamma^2 - \omega_\circ^2 < 0$ negativ ist. Es kommt also zu Schwingungen des Pohlschen Rads.

Wir definieren die neue Kreisfrequenz ω des gebremsten Pohlschen Rads über die Abweichung von der Kreisfrequenz $\omega_\circ$ (15.39) des frei schwingenden Rads durch

$$\omega^2 \equiv \omega_\circ^2 - \gamma^2 > 0\,. \tag{15.51}$$

Die Lösungen für λ_1 und λ_2 aus Gleichung (15.46) lauten damit

$$\lambda_{1,2} = -\gamma \pm \sqrt{\gamma^2 - \omega_\circ^2}\,, \qquad \lambda_1, \lambda_2 \in \mathbb{C} \tag{15.52}$$

$$= -\gamma \pm \sqrt{i^2\omega^2} \tag{15.53}$$

$$= -\gamma \pm i\omega\,. \tag{15.54}$$

Für die Differentialgleichung (15.41) benötigen wir wegen der zweiten zeitlichen Ableitung zwei Anfangsbedingungen. Wir setzen in die Lösungsfunktion (15.47) als Anfangsamplitude $\Phi_\circ$ und Anfangswinkel $\varphi_\circ$ in $c_1 = (\Phi_\circ/2)\, e^{i\varphi_\circ}$ und $c_2 = (\Phi_\circ/2)\, e^{-i\varphi_\circ}$ ein und erhalten ähnlich wie in (15.27) als physikalische Lösung

$$\Phi(t) = \frac{\Phi_\circ}{2} \left[e^{i\varphi_\circ}\, e^{\lambda_1 t} + e^{-i\varphi_\circ}\, e^{\lambda_2 t}\right] \tag{15.55}$$

$$= \frac{\Phi_\circ}{2} \left[e^{-\gamma t + i\omega\, t + i\varphi_\circ} + e^{-\gamma t - i\omega\, t - i\varphi_\circ}\right] \tag{15.56}$$

$$= \frac{\Phi_\circ}{2}\, e^{-\gamma t} \left[e^{i(\omega\, t + \varphi_\circ)} + e^{-i(\omega\, t + i\varphi_\circ)}\right] \tag{15.57}$$

$$= \Phi_\circ\, e^{-\gamma t} \cos\left(\omega t + \varphi_\circ\right). \tag{15.58}$$

Es handelt sich um eine Schwingung mit der Kreisfrequenz ω (15.51) und mit einer exponentiell abklingenden Amplitude (► Abbildung 15.5). Die Kreisfrequenz ω des gebremsten Pohlschen Rads ist dabei im Vergleich zu der Kreisfrequenz $\omega_\circ$ des ungedämpft schwingenden Rads geringer:

$$\Delta\omega = \omega - \omega_\circ = \sqrt{\omega_\circ^2 - \gamma^2} - \omega_\circ < 0 \tag{15.59}$$

Die Schwingungsperiode $T = 2\pi/\omega$ dauert also länger, wenn die Dämpfung eingeschaltet ist.

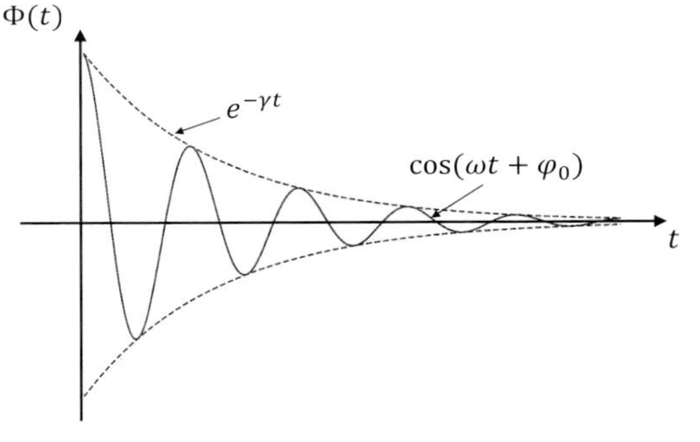

Abbildung 15.5 Gedämpfte Schwingung $\Phi(t)$ mit Kreisfrequenz ω, Dämpfung γ und Anfangsphase $\varphi_\circ = 0$

Experiment 29: *phyphox* Federpendel

Zur Durchführung dieses Experiments wird der Beschleunigungssensor des Smartphones benutzt. Befestigen Sie das Smartphone sicher an einem Gummiband.

1. Suchen Sie in der *phyphox*-App das Experiment »Federpendel«. Berühren des Drei-Punkte-Symbols rechts oben zeigt ein Menü mit Informationen zum Experiment (»Experiment Info«). Wählen Sie das Menü »Autokorrelation« zur Bestimmung der Schwingungsfrequenz.

2. Starten Sie die Aufzeichnung und lassen Sie das Smartphone an einem Gummiband frei schwingen. Stoppen Sie die Aufzeichnung. Das Gummiband dient gleichzeitig als rücktreibende Kraft und als Dämpfung.

3. Die Schwingungsfrequenz der gedämpften Schwingung wird angezeigt (► Abbildung 15.6).

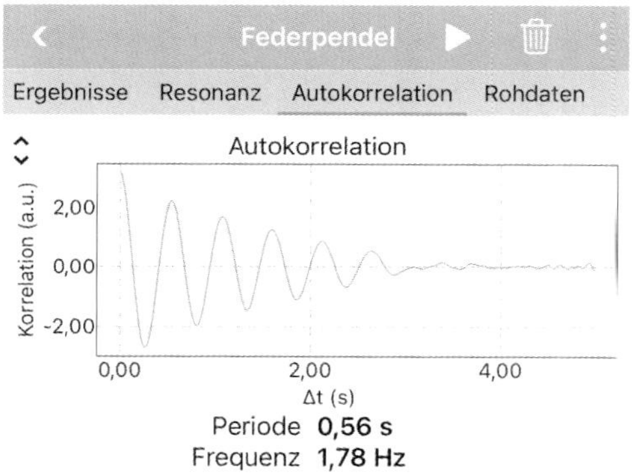

Abbildung 15.6 Gedämpfte Schwingung eines Smartphones, das an einem Gummiband hängt

Mathematischer Einschub: Hyperbelfunktionen

Die hyperbolischen Funktionen Cosinus hyperbolicus cosh, Sinus hyperbolicus sinh und Tangens hyperbolicus tanh sind über die Exponentialfunktion definiert:

$$\cosh x = \frac{e^x + e^{-x}}{2}, \qquad \cosh(0) = 1 \tag{15.60}$$

$$\sinh x = \frac{e^x - e^{-x}}{2}, \qquad \sinh(0) = 0 \tag{15.61}$$

$$\tanh x = \frac{\sinh x}{\cosh x} = \frac{e^x - e^{-x}}{e^x + e^{-x}}, \qquad \tanh(x \to \pm\infty) = \pm 1 \tag{15.62}$$

Starke Dämpfung

Für die Bedingung einer starken Dämpfung $\gamma > \omega_\circ$ sind die Lösungen λ_1 und λ_2 (15.46) reelle negative Zahlen:

$$\lambda_{1,2} = -\gamma \pm \sqrt{\gamma^2 - \omega_\circ^2} < 0 \tag{15.63}$$

Das System schwingt in diesem Fall *nicht*. Als Anfangsbedingungen setzen wir die Anfangsamplitude $\Phi_\circ$ ein und wählen hier zur Vereinfachung als Anfangsphase $\varphi_\circ = 0$. In der Lösungsfunktion (15.47) verbleiben dann mit $c_1 = c_2 = \Phi_\circ/2$ und unter Verwendung des Cosinus hyperbolicus cosh (15.60)

$$\Phi(t) = \frac{\Phi_\circ}{2}\, e^{-\gamma t} \left[e^{\sqrt{\gamma^2-\omega_\circ^2}\,t} + e^{-\sqrt{\gamma^2-\omega_\circ^2}\,t} \right] \tag{15.64}$$

$$= \Phi_\circ\, e^{-\gamma t} \cosh\left(\sqrt{\gamma^2 - \omega_\circ^2} \cdot t \right). \tag{15.65}$$

Somit erhalten wir den in der linken ► Abbildung 15.7 dargestellten **Kriechfall**, bei dem der exponentielle Abfall $e^{-\gamma t}$ durch den cosh-Term verzögert wird.

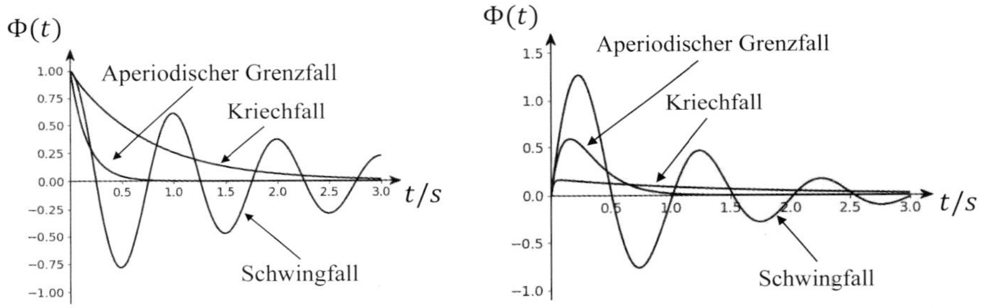

Abbildung 15.7 Gedämpfte Bewegungen in Einheiten von Sekunden s mit den Parametern: Schwingfall ($\omega_\circ = 2\pi/\mathrm{s}$, $\gamma = 1/\mathrm{s}$), aperiodischer Grenzfall ($\gamma = \omega_\circ$) und Kriechfall ($\gamma = 30/\mathrm{s}$). Links beim Start mit der Auslenkung $\Phi_\circ = 1$. Rechts beim Start durch Anstoßen aus der Ruhelage mit $\dot{\Phi}(t = 0) = v_\circ = 10/\mathrm{s}$.

Wir wollen auch den Fall untersuchen, dass das System aus der Ruhelage $\Phi(t = 0) = 0$ ausgelenkt wird (rechte ► Abbildung 15.7). Die interessante Frage dabei ist, wie schnell ein aus der Ruhe gebrachtes System gedämpft wird. Dazu wählen wir die Anfangsamplituden der beiden Lösungen (15.47) mit entgegengesetzten Vorzeichen $c_1 = \Phi_\circ/2 = -c_2$ und erhalten ähnlich zu (15.64) unter Verwendung des Sinus hyperbolicus sinh (15.61)

$$\Phi(t) = \frac{\Phi_\circ}{2}\, e^{-\gamma t} \left[e^{\sqrt{\gamma^2-\omega_\circ^2}\,t} - e^{-\sqrt{\gamma^2-\omega_\circ^2}\,t} \right] \tag{15.66}$$

$$= \Phi_\circ \cdot e^{-\gamma t} \cdot \sinh(\sqrt{\gamma^2 - \omega_\circ^2} \cdot t)\,. \tag{15.67}$$

Der $\sinh x$ beschreibt jetzt die Auslenkung und die Verzögerung des exponentiellen Abklingens $e^{-\gamma t}$ der Amplitude.

Aperiodischer Grenzfall

Wenn wir für die Lösungen $\lambda_{1,2} = -\gamma \pm \sqrt{\gamma^2 - \omega_\circ^2}$ (15.46) den Dämpfungsterm γ gegen die Kreisfrequenz $\omega_\circ$ des frei schwingenden Pohlschen Rads gehen lassen ($\gamma \to \omega_\circ$), nähern wir uns dem **aperiodischen Grenzfall** zwischen Kriechfall und gedämpfter Schwingung (► Abbildung 15.7). Mit dieser Annäherung hat der sinh-Term in (15.67) ein näherungsweise lineares Verhalten ($\sinh x \to x$ für $x \to 0$):

$$\Phi(t) = \Phi_\circ \cdot e^{-\gamma t} \cdot \sinh(\sqrt{\gamma^2 - \omega_\circ^2} \cdot t) \tag{15.68}$$

$$\underbrace{\approx}_{\gamma \to \omega_\circ} \underbrace{\Phi_\circ \sqrt{\gamma^2 - \omega_\circ^2}}_{\equiv \Phi_1} \cdot e^{-\gamma t} \cdot t \tag{15.69}$$

$$= \Phi_1 \cdot e^{-\gamma t} \cdot t \tag{15.70}$$

Dabei haben wir die Konstanten in Φ_1 mit der Einheit s^{-1} zusammengefasst. Die Lösung (15.70) lässt sich direkt durch Einsetzen in die Bewegungsgleichung (15.41) verifizieren (Aufgabe 43). Der Term t beschreibt den Anstieg mit der Anfangsgeschwindigkeit $\dot{\Phi}(t = 0) = \Phi_1$, der exponentielle Term beschreibt den Abfall mit der Dämpfung γ (► Abbildung 15.7).

Bei einer Auslenkung aus der Ruhelage nähert sich das System im aperiodischen Grenzfall am schnellsten ohne Überschwingen dem Ruhezustand. Diese optimale Dämpfung spielt in der Praxis eine große Rolle z. B. in der Messtechnik und im Fahrzeugbau.

Aufgabe 43: Aperiodischer Grenzfall

Zeigen Sie, dass neben $\Phi(t) = \Phi_\circ \cdot e^{-\gamma t}$ auch

$$\Phi(t) = \Phi_1 \cdot e^{-\gamma t} \cdot t \tag{15.70}$$

eine Lösung der Gleichung (15.41) für den aperiodischen Grenzfall $\lambda = -\gamma$ ist.

Experiment 30: Gedämpfte Schwingungen mit dem Pohlschen Rad

Mit dem Pohlschen Rad lassen sich unter Verwendung verschieden stark bremsender Drehmomente die gedämpften Schwingungen und der Kriechfall sehr schön demonstrieren.

15.3 Erzwungene Schwingungen

Wir statten das Pohlsche Rad mit einem Antrieb aus, indem wir das bislang fest montierte Ende der Schneckenfeder mithilfe eines Motors periodisch hin und her bewegen

(► Abbildung 15.4). Diese Bewegung hat die Kreisfrequenz ω_m und bewirkt ein Drehmoment

$$D_m = Re\left(D_\circ \cdot e^{i\omega_m t}\right) \tag{15.71}$$

$$= D_\circ \cdot \cos(\omega_m t)\ . \tag{15.72}$$

Die Antriebsfrequenz ω_m ist mit der Motorgeschwindigkeit variabel einstellbar und ist im Allgemeinen von der Kreisfrequenz $\omega_\circ$ der frei schwingenden Scheibe an der Schneckenfeder verschieden.

15.3.1 Bewegungsgleichung für erzwungene Schwingungen

Mit dem zusätzlichen Antriebsterm erweitern wir die Bewegungsgleichung (15.37) der gedämpften Schwingung um (15.72):

$$\frac{d\vec{L}}{dt} = \sum_{i=1}^{n} \vec{D}_i \tag{15.73}$$

$$I\ddot{\Phi} = -D_r\Phi - b\dot{\Phi} + D_\circ \cdot e^{i\omega_m t} \tag{15.74}$$

$$\ddot{\Phi} = -\underbrace{\frac{D_r}{I}}_{\equiv\omega_\circ^2}\Phi - \underbrace{\frac{b}{I}}_{\equiv 2\gamma}\dot{\Phi} + \underbrace{\frac{D_\circ}{I}}_{\equiv\kappa} e^{i\omega_m t} \tag{15.75}$$

Wir definieren die Antriebskonstante

$$\kappa \equiv \frac{D_\circ}{I} \tag{15.76}$$

und erhalten als Bewegungsgleichung für die erzwungene Schwingung

$$\ddot{\Phi} + 2\gamma\dot{\Phi} + \omega_\circ^2\Phi = \kappa\, e^{i\omega_m t}\ . \tag{15.77}$$

15.3.2 Ortsfunktion für erzwungene Schwingungen

Als Lösungsansatz verwenden wir hier ähnlich wie zuvor (15.14)

$$\Phi = c\, e^{i\omega_m t}\ , \tag{15.78}$$

$$\dot{\Phi} = i\,\omega_m\, c\, e^{i\omega_m t}\ , \tag{15.79}$$

$$\ddot{\Phi} = -\omega_m^2\, c\, e^{i\omega_m t}\ . \tag{15.80}$$

Einsetzen dieses Ansatzes in die Bewegungsgleichung (15.77) ergibt

$$\left(-\omega_m^2 + 2\gamma\, i\,\omega_m + \omega_\circ^2\right) c\, e^{i\omega_m t} = \kappa\, e^{i\omega_m t}\ . \tag{15.81}$$

Die komplexe Amplitude c berechnen wir somit aus

$$c = \frac{\kappa}{\omega_\circ^2 - \omega_m^2 + i\,2\,\gamma\,\omega_m} \tag{15.82}$$

$$= \frac{\kappa\left(\left(\omega_\circ^2 - \omega_m^2\right) - i\,2\,\gamma\,\omega_m\right)}{\left(\omega_\circ^2 - \omega_m^2\right)^2 - \left(i\,2\,\gamma\,\omega_m\right)^2} \tag{15.83}$$

$$= \underbrace{\frac{\kappa\left(\omega_\circ^2 - \omega_m^2\right)}{\left(\omega_\circ^2 - \omega_m^2\right)^2 + \left(2\,\gamma\,\omega_m\right)^2}}_{\equiv a} - i\underbrace{\frac{\kappa\left(2\,\gamma\,\omega_m\right)}{\left(\omega_\circ^2 - \omega_m^2\right)^2 + \left(2\,\gamma\,\omega_m\right)^2}}_{\equiv -d} \tag{15.84}$$

$$= a + id\,. \tag{15.85}$$

15.3.3 Resonanz bei erzwungenen Schwingungen

Der Betrag der physikalischen Amplitude $|c| \equiv \Phi_m(\omega_m)$ der erzwungenen Schwingung ist

$$\Phi_m(\omega_m) = |c| = \sqrt{a^2 + d^2} \tag{15.86}$$

$$= \frac{\kappa}{\left(\omega_\circ^2 - \omega_m^2\right)^2 + \left(2\,\gamma\,\omega_m\right)^2}\sqrt{\left(\omega_\circ^2 - \omega_m^2\right)^2 + \left(2\,\gamma\,\omega_m\right)^2} \tag{15.87}$$

$$= \kappa\left[\left(\omega_\circ^2 - \omega_m^2\right)^2 + \left(2\,\gamma\,\omega_m\right)^2\right]^{-1/2}\,. \tag{15.88}$$

Die Amplitude Φ_m wird dann maximal, wenn wir eine Erregerfrequenz ω_m einstellen, mit der die Ableitung $d\Phi_m/d\omega_m = 0$ wird. Die Ableitung lautet nach der Kettenregel

$$\frac{d\Phi_m}{d\omega_m} = -\kappa \cdot \frac{2\left(\omega_\circ^2 - \omega_m^2\right)\left(-2\,\omega_m\right) + \left(2\,\gamma\right)^2 2\,\omega_m}{2\sqrt{\left(\omega_\circ^2 - \omega_m^2\right)^2 + \left(2\,\gamma\,\omega_m\right)^2}^{\,3}}\,. \tag{15.89}$$

Die Ableitung ist dann null, wenn

$$\left(\omega_\circ^2 - \omega_m^2\right) - 2\,\gamma^2 = 0 \tag{15.90}$$

gilt. Die Erregerfrequenz $\omega_m = \omega_R$, bei der die Amplitude maximal wird, nennt man **Resonanzfrequenz**:

$$\omega_R = \sqrt{\omega_\circ^2 - 2\,\gamma^2} \tag{15.91}$$

Sie liegt etwas unterhalb der Kreisfrequenz $\omega_\circ$ des frei schwingenden Pohlschen Rads (► Abbildung 15.8). Als die natürliche Schwingungsfrequenz der freien Schwingung bezeichnet man $\omega_\circ$ auch als **Eigenfrequenz**.

Experiment 31: Resonanz

Variieren wir die Antriebsfrequenz des Motors und treiben das Pohlsche Rad mit einer Kreisfrequenz nahe der Resonanzfrequenz ω_R an, die etwas unterhalb der Kreisfrequenz $\omega_\circ$ der frei schwingenden Scheibe liegt, so kommt es zu einem resonanten Verhalten: Die Scheibenamplitude wird in der Nähe der Resonanzfrequenz auch bei kleinen Änderungen in der Erregerfrequenz ω_m drastisch größer, bis eine maximale Amplitude erreicht ist.

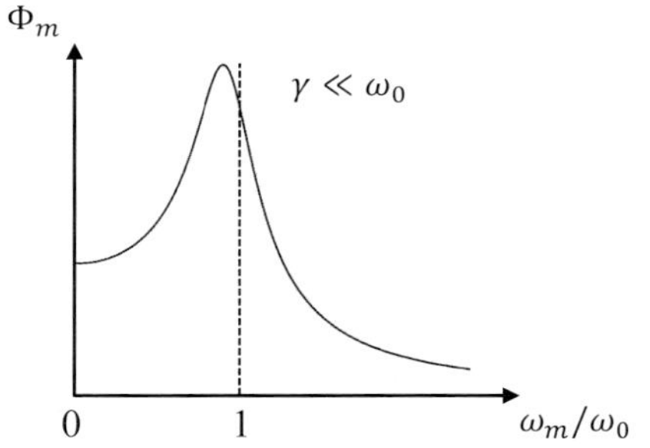

Abbildung 15.8 Die Resonanzfrequenz ω_R mit maximaler Amplitude Φ_m einer erzwungenen Schwingung ist gegenüber der Eigenfrequenz $\omega_\circ$ zu kleineren Werten verschoben.

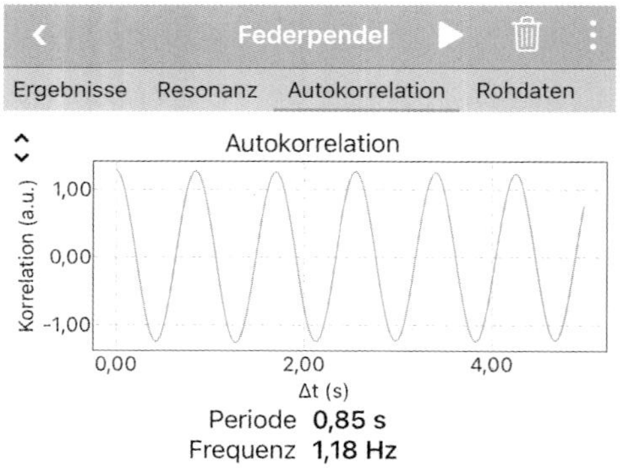

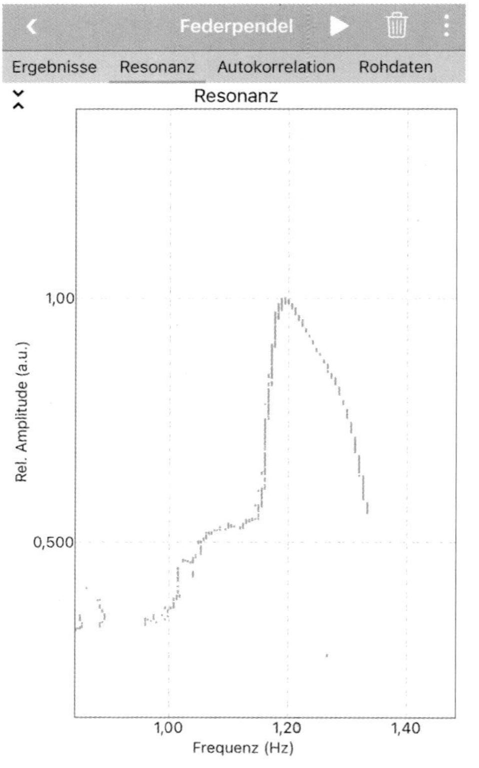

Abbildung 15.9 Eigenfrequenz des frei schwingenden Federpendels (links). Das Pendel wird per Hand zu erzwungenen Schwingungen steigender Frequenz angeregt (rechts). Die Resonanzfrequenz stimmt mit der Eigenfrequenz überein.

Experiment 32: *phyphox* Resonanz

Zur Durchführung dieses Experiments wird der Beschleunigungssensor des Smartphones benutzt. Als schwingendes System wurde hier der Sessel einer Kinderseilrutsche verwendet. In der Mitte des langen Seils hat das System sehr gute Elastizität mit geringer Dämpfung.

1. Suchen Sie in der *phyphox*-App das Experiment »Federpendel«. Berühren des Drei-Punkte-Symbols rechts oben zeigt ein Menü mit Informationen zum Experiment (»Experiment Info«). Wählen Sie das Menü »Autokorrelation« zur Bestimmung der Eigenfrequenz.
2. Starten Sie die Aufzeichnung und lassen Sie das Smartphone zusammen mit dem Pendel frei schwingen. Stoppen Sie die Aufzeichnung.
3. Der zeitliche Verlauf und die Eigenfrequenz der freien Schwingung werden angezeigt (► Abbildung 15.9, links). Man erkennt eine praktisch ungedämpfte Schwingung.
4. Wählen Sie das Menü »Resonanz« zur Bestimmung der Resonanzkurve und starten Sie die Aufzeichnung. Regen Sie das System mit unterschiedlichen Frequenzen an, am besten aufsteigend. Stoppen Sie die Messung.
5. Die rechte ► Abbildung 15.9 zeigt die entstandene Resonanzkurve. Die Resonanzfrequenz stimmt gut mit der Eigenfrequenz überein. Das bestätigt nach (15.91) die geringe Dämpfung des Systems.

15.4 Gekoppelte Schwingungen

Wir untersuchen nun zwei Pendel, die jeweils aus einer festen Stange mit einem Pendelgewicht an ihrem Ende bestehen und die über eine Feder miteinander gekoppelt sind (► Abbildung 15.10).

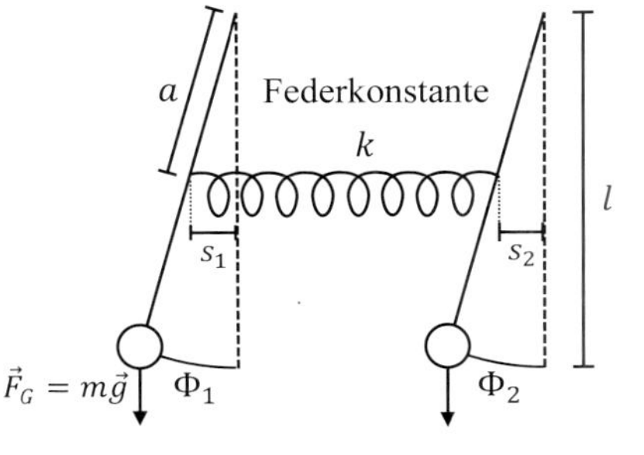

Abbildung 15.10 Zwei gleichartige Pendel (Länge l, Masse m) sind durch eine Feder mit Federkonstante k im Abstand a vom Aufhängepunkt gekoppelt.

Experiment 33: Gekoppelte Pendel

Lenkt man ein Pendel des gekoppelten Pendelsystems aus und lässt es schwingen, wird das zweite Pendel ebenfalls zu einer Pendelbewegung angeregt. Während das zweite Pendel an Amplitude gewinnt, verliert das erste Pendel Amplitude, bis es selbst zum Stillstand kommt. Dann kehrt sich der Vorgang um.

Im Folgenden werden wir zeigen, dass das beobachtete abwechselnde Schwingverhalten der Pendel der Überlagerung von zwei Schwingungen mit verschiedenen Frequenzen entspricht.

15.4.1 Bewegungsgleichungen für gekoppelte Pendel

Die Bewegungsgleichungen für die gekoppelten Pendel finden wir über das Bewegungsgesetz für Drehbewegungen (4.70):

$$\frac{d\vec{L}}{dt} = \sum_{i=1}^{n} \vec{D}_i \tag{15.92}$$

Jedes der beiden Pendel mit seinem Drehimpuls $L = I\dot{\Phi}$ (14.23) wird durch eine eigene Bewegungsgleichung beschrieben. Zunächst beschreiben wir die angreifenden Drehmomente und setzen sie anschließend in (15.92) ein.

Drehmoment durch Federkopplung

Die Dehnung s der Feder in ► Abbildung 15.10 verursacht die rücktreibende Kraft $F_F = -ks$. Da wir die Auslenkwinkel Φ_1, Φ_2 der beiden Pendel berechnen wollen, nähern wir die Dehnung s durch Φ_1 und Φ_2 für kleine Winkel:

$$s = s_1 - s_2 \tag{15.93}$$

$$= a\,\sin(\Phi_1) - a\,\sin(\Phi_2) \tag{15.94}$$

$$\approx a\,(\Phi_1 - \Phi_2) \tag{15.95}$$

Das von der Feder verursachte Drehmoment wirkt jeweils in der Entfernung a von der Drehachse, wobei wir den Winkel zwischen Aufhängung und Richtung der Federkraft berücksichtigen:

$$D_F = a\,F_F\,\sin\left(\frac{\pi}{2} - \Phi\right) \tag{15.96}$$

$$\approx -a\,k\,s \tag{15.97}$$

$$\approx -ka^2\,(\Phi_1 - \Phi_2) \tag{15.98}$$

Im zweiten Schritt haben wir den Term zweiter Ordnung in der Winkelnäherung vernachlässigt ($\sin(\pi/2 - \Phi) = \cos\Phi \approx 1 - \Phi^2/2 \approx 1$).

Drehmoment durch Gravitation

Das rücktreibende Drehmoment wird durch die Gravitation auf die Masse m des Pendels verursacht, dessen Schwerpunkt sich im Abstand l von der Drehachse befindet. Für kleine Auslenkungen Φ erhalten wir näherungsweise

$$\vec{D}_G = \vec{l} \times \vec{F}_G \tag{15.99}$$

$$D_G = -l \cdot m \cdot g \cdot \sin \Phi \tag{15.100}$$

$$\approx -l \cdot m \cdot g \cdot \Phi \,. \tag{15.101}$$

Gekoppelte Bewegungsgleichungen

In der ► Abbildung 15.10 zeigen die Drehmomente durch die Gravitation in dieselbe Richtung, während die Drehmomente durch die Feder in zueinander entgegengesetzte Richtungen zeigen. Die Bewegungsgleichungen für die Pendel lauten deshalb (15.92)

$$\dot{L}_1 = D_{G_1} + D_F \tag{15.102}$$

$$\dot{L}_2 = D_{G_2} - D_F \,. \tag{15.103}$$

Durch Einsetzen von (15.98) und (15.101) erhalten wir folgende Gleichungen für die beiden Pendel:

$$I\ddot{\Phi}_1 = -lgm\,\Phi_1 - ka^2\,(\Phi_1 - \Phi_2) \tag{15.104}$$

$$I\ddot{\Phi}_2 = -lgm\,\Phi_2 + ka^2\,(\Phi_1 - \Phi_2) \tag{15.105}$$

Die Winkelvariablen Φ_1 und Φ_2 der Pendel sind in beiden Gleichungen zu finden, es handelt sich um ein gekoppeltes Gleichungssystem. Die Bewegungsgleichungen müssten also in Verbindung gelöst werden, was wir im Folgenden durch eine passende Transformation der Variablen Φ_1 und Φ_2 umgehen werden.

Normalkoordinaten

Wir können die Kopplung der beiden Bewegungsgleichungen (15.104) und (15.105) auflösen, indem wir sie einmal addieren und einmal voneinander subtrahieren:

$$I(\ddot{\Phi}_1 + \ddot{\Phi}_2) = -lgm\,(\Phi_1 + \Phi_2) + ka^2\,[-(\Phi_1 - \Phi_2) + (\Phi_1 - \Phi_2)] \tag{15.106}$$

$$I(\ddot{\Phi}_1 - \ddot{\Phi}_2) = -lgm\,(\Phi_1 - \Phi_2) + ka^2\,[-(\Phi_1 - \Phi_2) - (\Phi_1 - \Phi_2)] \tag{15.107}$$

Anstelle der ursprünglich einzelnen Winkelvariablen Φ_1 und Φ_2 verwenden wir im Folgenden die beiden Kombinationen

$$\eta \equiv \frac{1}{2}\,(\Phi_1 + \Phi_2)\,, \tag{15.108}$$

$$\xi \equiv \frac{1}{2}\,(\Phi_1 - \Phi_2)\,. \tag{15.109}$$

Die Kombinationen und ihre zweiten zeitlichen Ableitungen

$$\Phi_1 + \Phi_2 = 2\eta\,, \quad \ddot{\Phi}_1 + \ddot{\Phi}_2 = 2\ddot{\eta} \tag{15.110}$$

$$\Phi_1 - \Phi_2 = 2\xi\,, \quad \ddot{\Phi}_1 - \ddot{\Phi}_2 = 2\ddot{\xi} \tag{15.111}$$

setzen wir in die Gleichungen (15.106) und (15.107) ein. Damit vereinfacht sich das ursprünglich gekoppelte Gleichungssystem ((15.104), (15.105)) zu zwei unabhängig voneinander lösbaren Differentialgleichungen:

$$I2\ddot{\eta} = -lgm\,2\eta \tag{15.112}$$

$$I2\ddot{\xi} = -lgm\,2\xi - 2ka^2\,2\xi \tag{15.113}$$

Isolieren wir jeweils die zweiten Ableitungen auf der linken Seite, ergeben sich zwei Bewegungsgleichungen vom Typ des harmonischen Oszillators (4.30):

$$\ddot{\eta} = -\underbrace{\frac{lgm}{I}}_{\equiv\omega_a^2}\eta \tag{15.114}$$

$$\ddot{\xi} = -\underbrace{\frac{lgm + 2ka^2}{I}}_{\equiv\omega_b^2}\xi \tag{15.115}$$

Die kombinierten Größen η und ξ, die zur Entkopplung der Bewegungsgleichungen führen, sind sogenannte **Normalkoordinaten**. Die konstanten Terme auf der rechten Seite dieser Gleichungen identifizieren wir als das Quadrat der Schwingungsfrequenzen.

Eigenschwingungen

Die Struktur der Frequenzterme können wir folgendermaßen verstehen: Für ein Pendel mit vernachlässigbarer Masse der Stange (mathematisches Pendel) ist das Trägheitsmoment $I = m\,l^2$. Damit entspricht die Kreisfrequenz ω_a in (15.114) näherungsweise der Frequenz $\omega_\circ$ eines frei schwingenden Pendels (4.86):

$$\omega_a = \sqrt{\frac{lgm}{I}} \approx \sqrt{\frac{lgm}{ml^2}} = \underbrace{\sqrt{\frac{g}{l}}}_{\equiv\omega_\circ} \tag{15.116}$$

Die Kreisfrequenz ω_b in (15.115) ist immer größer als ω_a, weil alle eingehenden Größen positiv sind. Wir interpretieren ω_b für eine schwache Federkopplung, d. h. $k \ll mgl/a^2$. Durch eine Taylor-Entwicklung ($\sqrt{1+x} \approx 1 + x/2$) sehen wir, dass die Frequenz ω_b dann ein wenig größer als ω_a ist:

$$\omega_b = \sqrt{\frac{lgm + 2ka^2}{I}} \tag{15.117}$$

$$= \sqrt{\frac{lgm}{ml^2} + \frac{2ka^2}{ml^2}} \tag{15.118}$$

$$\approx \sqrt{\frac{g}{l}}\sqrt{1 + \frac{2ka^2}{lgm}} \tag{15.119}$$

$$\approx \omega_\circ(1 + \underbrace{\frac{ka^2}{lgm}}_{\equiv 2\kappa}) \tag{15.120}$$

Als abkürzende Schreibweise verwenden wir im Folgenden κ und verlangen seinen kleinen numerischen Wert für die Näherung einer schwachen Federkopplung k:

$$\kappa \equiv \frac{ka^2}{2lgm} \ll 1 \tag{15.121}$$

Die Frequenzen ω_a und ω_b werden auch **Eigenfrequenzen** der Eigenschwingungen oder Normalschwingungen genannt (vergl. auch Abschnitt 16.2).

Anschaulich schwingen die Pendel bei den Normalkoordinaten η (15.108) exakt parallel und beeinflussen sich dabei nicht. Die Schwingung (15.114) gleicht also der des freien Pendels und die Frequenz ist $\omega_a = \omega_\circ$ (15.116).

In den Normalkoordinaten ξ (15.109) schwingen die Pendel exakt gegeneinander. Diese Schwingung (15.115) folgt der Schwingungsfrequenz ω_b (15.120), die durch die rücktreibende Federkraft im Vergleich zur freien Pendelschwingung erhöht ist: $\omega_b > \omega_\circ$

Allgemein kann man für ein schwingungsfähiges System alle möglichen Schwingungen aus seinen Eigenschwingungen aufbauen, im Fall der gekoppelten Pendel also aus einer Überlagerung (**Superposition**) von Eigenschwingungen der Frequenzen ω_a und ω_b. Eigenschwingungen und Eigenfrequenzen spielen in vielen Gebieten der Physik eine wichtige Rolle.

Experiment 34: Gedankenexperiment zur starken Kopplung

Für den Fall einer starken Federkopplung k führen wir ein Gedankenexperiment auf einem Planeten mit der Beschleunigung $g \ll kl/m$ durch. Wir montieren die Feder an der Stelle $a = l/\sqrt{2}$ und erhalten für

$$\omega_b = \sqrt{\frac{lgm + 2ka^2}{I}} \approx \sqrt{\frac{lgm + kl^2}{ml^2}} \approx \sqrt{\frac{k}{m}} \tag{15.122}$$

die Frequenz eines frei schwingenden Federpendels.

Ortsfunktionen in Normalkoordinaten

Mithilfe der Normalkoordinaten haben wir zwei entkoppelte harmonische Schwingungsgleichungen (15.114) und (15.115) erhalten und interpretiert:

$$\ddot{\eta} = -\omega_a^2 \eta \qquad \ddot{\xi} = -\omega_b^2 \xi \tag{15.123}$$

Um auch ihre Ortsfunktionen zu erhalten, setzen wir in bewährter Art (15.14) an:

$$\eta(t) = \eta_\circ \, e^{i\omega_a \, t} \qquad \text{mit} \quad \eta_\circ = \Phi_\circ \, e^{i\alpha} \tag{15.124}$$

$$\xi(t) = \xi_\circ \, e^{i\omega_b \, t} \qquad \text{mit} \quad \xi_\circ = \Phi_\circ \, e^{i\beta} \tag{15.125}$$

Dabei bezeichnet $\Phi_\circ$ die als gleich angenommene reelle Amplitude der beiden Pendel und α bzw. β ihre Anfangsphasen.

15.4.2 Ortsfunktionen des Doppelpendels: Winkelkoordinaten

Bei dem eingeschlagenen Lösungsweg wurden die gekoppelten Pendelgleichungen (15.104) und (15.105) durch Normalkoordinaten entkoppelt (15.123). Die Lösungen ihrer Bewegungsgleichungen liegen bislang in den Normalkoordinaten (15.124) und (15.125) vor. Bei der Beobachtung der gekoppelten Pendel im Experiment verfolgen wir ihre Winkelkoordinaten $\Phi(t)$, sodass wir eine Rücktransformation von den Normalkoordinaten in die Winkelkoordinaten benötigen.

Im Folgenden werden wir uns weiter auf den Fall einer schwachen Federkopplung k konzentrieren und dabei das wichtige Phänomen der Schwebung kennenlernen.

Mathematischer Einschub: Additionstheorem

Mithilfe von Additionstheoremen können wir Kosinus- und Sinusfunktionen zusammenfassen:

$$\cos(x) + \cos(y) = 2 \cos\left(\frac{x+y}{2}\right) \cdot \cos\left(\frac{x-y}{2}\right) \tag{15.126}$$

$$\cos(x) - \cos(y) = -2 \sin\left(\frac{x+y}{2}\right) \cdot \sin\left(\frac{x-y}{2}\right) \tag{15.127}$$

$$\sin(x) + \sin(y) = 2 \sin\left(\frac{x+y}{2}\right) \cdot \cos\left(\frac{x-y}{2}\right) \tag{15.128}$$

$$\sin(x) - \sin(y) = 2 \cos\left(\frac{x+y}{2}\right) \cdot \sin\left(\frac{x-y}{2}\right) \tag{15.129}$$

Ortsfunktion des 1. Pendels

Um die Ortsfunktion $\Phi_1(t)$ des ersten Pendels zu erhalten, verwenden wir die Rücktransformationen, die sich aus den Definitionsgleichungen für η (15.108) und ξ (15.109) ergeben.

Die physikalische Lösung für das 1. Pendel ergibt sich aus dem Realteil der transformierten Lösung:

$$\begin{aligned} \Phi_1 &= Re\,(\eta + \xi) && (15.130) \\ &= Re\,(\eta_\circ e^{i\omega_a t} + \xi_\circ e^{i\omega_b t}) && (15.131) \\ &= Re\left(\Phi_\circ \left[e^{i(\omega_a t+\alpha)} + e^{i(\omega_b t+\beta)}\right]\right) && (15.132) \\ &= \Phi_\circ \left[\cos(\omega_a t + \alpha) + \cos(\omega_b t + \beta)\right] && (15.133) \end{aligned}$$

Mit dem Additionstheorem (15.126) erhalten wir eine vergleichsweise gut interpretierbare Schreibweise der Lösung (15.133):

$$\Phi_1(t) = \underbrace{2\Phi_\circ \cos\left(\underbrace{\frac{\omega_b - \omega_a}{2}}_{\equiv\Omega_b} t + \underbrace{\frac{\beta-\alpha}{2}}_{\equiv\varphi_b}\right)}_{\text{Amplitudenmodulation}} \cdot \underbrace{\cos\left(\underbrace{\frac{\omega_b + \omega_a}{2}}_{\equiv\Omega_a} t + \underbrace{\frac{\beta+\alpha}{2}}_{\equiv\varphi_a}\right)}_{\text{Pendelschwingung}} \tag{15.134}$$

Die Bewegung des ersten Pendels können wir als eine Überlagerung von zwei Schwingungen verstehen.

Der zweite Term auf der rechten Seite der Gleichung zeigt den Mittelwert der beiden Frequenzen:

$$\Omega_a \equiv \frac{\omega_b + \omega_a}{2} \tag{15.135}$$

Er entspricht einer Pendelfrequenz, die mit den Näherungen (15.116), (15.120) und (15.121) im Bereich der Frequenz $\omega_\circ$ des frei schwingenden Pendels liegt:

$$\Omega_a \approx \frac{1}{2}\left(\omega_\circ(1 + 2\kappa) + \omega_\circ\right) = \omega_\circ(1 + \kappa) \approx \omega_\circ \tag{15.136}$$

Die Differenz der Frequenzen im ersten Term auf der rechten Seite

$$\Omega_b \equiv \frac{\omega_b - \omega_a}{2} \tag{15.137}$$

führt zu einer langsamen Amplitudenmodulation, die man als **Schwebung** bezeichnet. In den Näherungen einer vernachlässigbar kleinen Masse der Pendelstange (15.116) und einer schwachen Federkopplung k (15.120) ergibt sich eine Schwebungsfrequenz, die um einen Faktor κ (15.121) kleiner als die Frequenz $\omega_\circ$ des frei schwingenden Pendels ist:

$$\Omega_b \approx \frac{1}{2}\left(\omega_\circ(1 + 2\kappa) - \omega_\circ\right) = \kappa\,\omega_\circ \tag{15.138}$$

Die beiden Anfangsbedingungen φ_a und φ_b in Gleichung (15.134) setzen sich aus den anfänglichen Pendelauslenkungen zusammen.

Für eine schwache Federkopplung k und vernachlässigbarer Masse der Pendelstange finden wir also für das erste Pendel folgende Lösungsstruktur der Ortsfunktion:

$$\Phi_1(t) \approx 2\,\Phi_\circ\,\cos(\kappa\,\omega_\circ\,t + \varphi_b) \cdot \cos(\omega_\circ\,t + \varphi_a) \tag{15.139}$$

Die zugehörigen Schwingungsperioden T_a und T_b können wir sofort berechnen. Die kurze Schwingungsperiode

$$T_a = \frac{2\pi}{\Omega_a} \approx \underbrace{\frac{2\pi}{\omega_\circ}}_{\equiv T_\circ} \tag{15.140}$$

entspricht ungefähr dem mit der Periode $T_\circ = 2\pi/\omega_\circ$ frei schwingenden Pendel.

Die längere Periode T_b der Schwebung erhalten wir über

$$T_b = \frac{2\pi}{\Omega_b} = \frac{2\pi}{\kappa\omega_\circ} = \frac{T_\circ}{\kappa} \,. \tag{15.141}$$

Ortsfunktion des 2. Pendels

Die Lösung $\Phi_2\,(t)$ für das zweite Pendel erhalten wir durch Differenzbildung der Definitionsgleichungen (15.108) und (15.109) und Einsetzen der Lösungen $\eta(t)$ (15.124) und $\xi(t)$ (15.125). Die physikalische Lösung ergibt sich zu

$$\Phi_2 = Re\,(\eta - \xi) \tag{15.142}$$

$$= Re\left(\Phi_\circ\left[e^{i(\omega_a t+\alpha)} - e^{i(\omega_b t+\beta)}\right]\right) \tag{15.143}$$

$$= \Phi_\circ\left[\cos\left(\omega_a t + \alpha\right) - \cos\left(\omega_b t + \beta\right)\right] \,. \tag{15.144}$$

Wir verwenden hier das Additionstheorem (15.127) und erhalten die Ortsfunktion des zweiten Pendels

$$\Phi_2(t) \approx -2\,\Phi_\circ\,\sin(\kappa\,\omega_\circ\,t + \varphi_b)\cdot\sin(\omega_\circ\,t + \varphi_a)\,. \tag{15.145}$$

Seine Ortsfunktion enthält Sinusfunktionen im Vergleich zu den Kosinusfunktionen der Ortsfunktion des ersten Pendels (15.139). Damit sind die Argumente um $\pi/2$ gegeneinander verschoben ($\sin(x) = -\cos(x + \pi/2)$),

$$\Phi_2(t) \approx 2\,\Phi_\circ\cos\left(\kappa\,\omega_\circ\,t + \varphi_b + \frac{\pi}{2}\right)\cdot\cos\left(\omega_\circ\,t + \varphi_a + \frac{\pi}{2}\right)\,, \tag{15.146}$$

und bewirken das beobachtete abwechselnde Schwingen und zur Ruhe kommen der beiden Pendel. Ein konstanter Term wie z. B. $\pi/2$ im Argument einer Sinus- oder Kosinusfunktion wird auch als **Phase** bezeichnet. Dementsprechend spricht man bei dem abwechselnden Schwingverhalten von **Phasenverschiebung** oder Phasendifferenz zwischen den beiden Pendelschwingungen.

► Abbildung 15.11 zeigt die Bewegung des zweiten Pendels $\Phi = \Phi_2$ im Fall $\varphi_a = \varphi_b = 0$.

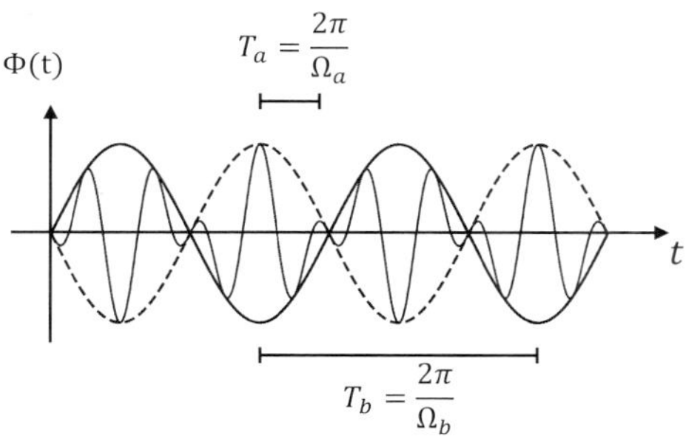

Abbildung 15.11 Überlagerung zweier Schwingungen mit ähnlichen Frequenzen $\omega_a < \omega_b$ zu einer Schwebung mit der Kreisfrequenz $\Omega_a = (\omega_b + \omega_a)/2$ und der Schwebungsfrequenz $\Omega_b = (\omega_b - \omega_a)/2$ und den zugehörigen Schwingungsperioden $T_{a,b}$.

Schwebung

Bei der Herleitung der Ortsfunktion (15.134) aus den Schwingunsgleichungen (15.124) und (15.125) wurden die Eigenschaften des Doppelpendels gar nicht benutzt. Das Phänomen der Schwebung tritt also ganz allgemein bei der Überlagerung von zwei Schwingungen mit ähnlicher Frequenz auf. Die Überlagerung der Schwingungen mit $\omega_b > \omega_a$

$$\Phi_a = \Phi_\circ \cos(\omega_a t + \alpha) \tag{15.147}$$

$$\Phi_b = \Phi_\circ \cos(\omega_b t + \beta) \tag{15.148}$$

führt zur Gesamtschwingung

$$\Phi = \Phi_a + \Phi_b \tag{15.149}$$

$$= \Phi_\circ \left[\cos(\omega_a t + \alpha) + \cos(\omega_b t + \beta)\right] . \tag{15.150}$$

(15.150) lässt sich wie (15.133) mithilfe der Additionstheoreme umformen:

$$\Phi(t) = \underbrace{2\Phi_\circ \cos(\Omega_b\, t + \varphi_b)}_{\text{Amplitudenmodulation}} \cdot \underbrace{\cos(\Omega_a\, t + \varphi_a)}_{\text{Schwingung}} \tag{15.151}$$

Wie beim Doppelpendel ergibt sich eine Schwingung mit der **mittleren Frequenz** Ω_a und der Phase φ_a aus den jeweiligen Mittelwerten

$$\Omega_a \equiv \frac{\omega_b + \omega_a}{2}, \quad \varphi_a \equiv \frac{\beta + \alpha}{2} \tag{15.152}$$

und einer **Amplitudenmodulation** mit der Frequenz Ω_b und der Phase φ_b aus den jeweiligen halben Differenzen

$$\Omega_b \equiv \frac{\omega_b - \omega_a}{2}, \quad \varphi_b \equiv \frac{\beta - \alpha}{2} . \tag{15.153}$$

Eine solche **Schwebung** ist in ► Abbildung 15.11 für $\varphi_a = +\pi/2$ und $\varphi_b = -\pi/2$ dargestellt. Die Schwebungsfrequenz Ω_a wird umso kleiner (und damit die Schwebungsdauer T_a umso länger) je näher die Frequenzen ω_a und ω_b beieinander liegen.

15.5 Fouriersynthese, -analyse, -transformation

Im vorherigen Abschnitt 15.4 haben wir bereits eine Überlagerung von zwei Schwingungen untersucht. ► Abbildung 15.12 zeigt ein weiteres Beispiel für die Überlagerung $\xi_1 + \xi_2$ zweier Schwingungen ξ_1 und ξ_2.

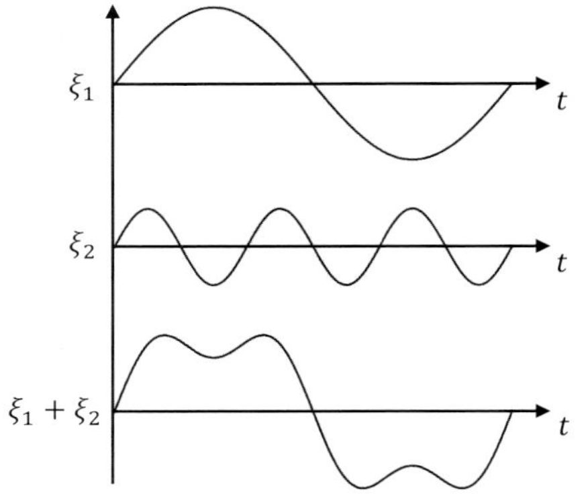

Abbildung 15.12 Überlagerung zweier Schwingungen $\xi_{1,2}$ zu einer Gesamtschwingung $\xi_1 + \xi_2$, z. B. als erster Schritt zur Konstruktion einer Rechteckschwingung (15.167)

15.5.1 Fouriersynthese

Hier verallgemeinern wir die Situation auf beliebig viele Schwingungen verschiedener Frequenzen. Dazu betrachten wir eine Überlagerung von N Schwingungen:

$$x(t) = \sum_{n=1}^{N} a_n \cos(\omega_n t + \varphi_n) \tag{15.154}$$

Hierbei bezeichnet a_n die Amplitude der n-ten Schwingung, ω_n ist ihre Kreisfrequenz und φ_n ist ihre Phasenverschiebung. Das Resultat ist wieder eine periodische Bewegung (musikalisch: Klang oder Ton).

Da wir dieses periodische Verhalten aus der Überlagerung vieler Einzelschwingungen erhalten haben, nennt man dieses Verfahren **Fouriersynthese**.

Mathematischer Einschub: Fourierreihen

Gleichung (15.154) lautet in allgemeiner Form mit konstantem Term und $N \to \infty$

$$x(t) = \frac{a_0}{2} + \sum_{n=1}^{\infty} a_n \cos(\omega_n t + \varphi_n) \ . \tag{15.155}$$

Anstelle des Phasenterms φ_n lassen sich solche **Fourierreihen** auch in der Form

$$x(t) = \frac{u_0}{2} + \sum_{n=1}^{\infty} (u_n \cos(\omega_n t) + v_n \sin(\omega_n t)) \tag{15.156}$$

darstellen. Mit $\omega_n = n\,\omega_\circ$ und $\omega_\circ = 2\pi/T$ ergeben sich die Fourierkoeffizienten zu

$$u_n = \frac{2}{T}\int_{-T/2}^{T/2} x(t)\cdot\cos(n\omega_\circ t)\,\mathrm{d}t \quad \text{für } n \geq 0 \tag{15.157}$$

$$v_n = \frac{2}{T}\int_{-T/2}^{T/2} x(t)\cdot\sin(n\omega_\circ t)\,\mathrm{d}t \quad \text{für } n \geq 1\,. \tag{15.158}$$

Die beiden Darstellungen (15.155) und (15.156) sind äquivalent und lassen sich ineinander umrechnen:

$$u_\circ = a_\circ \qquad u_n = a_n\cos\varphi_n \qquad v_n = a_n\sin\varphi_n \tag{15.159}$$

$$a_n = \sqrt{u^2+v^2} \qquad \tan\varphi_n = \frac{v_n}{u_n} \tag{15.160}$$

Die beiden Terme von (15.156) können wir auch in einer komplexen Schreibweise sehr elegant zusammenfassen, wobei wir die komplexen Koeffizienten c_n als diskrete Funktion $c_n(\omega_n)$ darstellen:

$$x(t) = \sum_{n=-\infty}^{\infty} c_n(\omega_n)e^{i\omega_n t}, \qquad c_n(\omega_n) = \frac{1}{T}\int_{-T/2}^{T/2} x(t)e^{-i\omega_n t}dt \tag{15.161}$$

$$\text{mit } c_\circ = \frac{u_\circ}{2}, \quad c_n = \frac{1}{2}(u_n - iv_n), \quad c_{-n} = \frac{1}{2}(u_n + iv_n) \tag{15.162}$$

$$\text{und } u_n = (c_n + c_{-n}), \quad v_n = i(c_n - c_{-n}) \tag{15.163}$$

Wir betrachten im Folgenden eine in der Zeit T periodische Rechteckschwingung der Form (► Abbildung 15.13)

$$x(t) = \begin{cases} -x_0 & \text{für } \pm T/2 < t \pm nT < \pm T/4\,, \\ +x_0 & \text{für } -T/4 < t \pm nT < T/4\,. \end{cases} \tag{15.164}$$

Die Berechnung der Koeffizienten gemäß den obigen Formeln zu Fourierreihen ergibt mit $\omega_\circ = 2\pi/T$

$$a_n = \frac{4x_\circ}{\pi}\frac{(-1)^{n-1}}{2n-1}, \quad \varphi_n = 0, \quad \omega_n = (2n-1)\,\omega_\circ\,. \tag{15.165}$$

Damit lautet die Fourierreihe

$$x(t) = \frac{4x_\circ}{\pi}\sum_{n=1}^{\infty}\frac{(-1)^{n-1}}{2n-1}\cos\left((2n-1)\,\omega_\circ t\right) \tag{15.166}$$

$$= \frac{4x_\circ}{\pi}\left(\cos(\omega_\circ t) - \frac{1}{3}\cos(3\omega_\circ t) + \frac{1}{5}\cos(5\omega_\circ t) - \ldots\right). \tag{15.167}$$

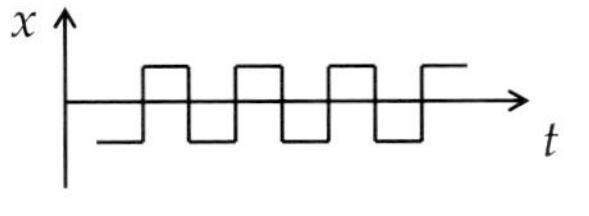

Abbildung 15.13 Eine Rechteckschwingung lässt sich aus sehr (∞) vielen Sinusschwingungen aufbauen.

Aufgabe 44: Fourierreihe

Berechnen Sie die Koeffizienten (15.165) der Fourierreihe (15.154) einer Rechteckfunktion (15.164).

In den folgenden Versuchen wird eine Rechteckschwingung (15.164) aus harmonischen Schwingungen aufgebaut (Fouriersynthese) und zerlegt (Fourieranalyse).

Experiment 35: Fouriersynthese

Mit einem Sinusgenerator, einem Lautsprecher und einem Oszilloskop können wir zunächst einen einzelnen Sinuston hörbar und sichtbar machen. Anschließend überlagern wir Sinustöne verschiedener Frequenzen. Der Ton verändert seine Klangqualität. Auf dem Oszilloskop werden verschiedene Schwingungsformen sichtbar.

Setzt man $a_n \propto (-1)^{n-1}/(2n-1)$ und $\omega_n = (2n-1)\,\omega_\circ$, so nähert man sich mit $n \to \infty$ für $x(t)$ immer mehr einer Rechteckschwingung (▶ Abbildung 15.12 und ▶ Abbildung 15.13).

Unter Verwendung von sinusförmigen Schwingungen ($\varphi_n = -\pi/2$) verschiebt sich die erzeugte Rechteckschwingung um $T/4$ gegenüber der Erwartung für Kosinusschwingungen mit $\varphi_n = 0$ (15.165).

15.5.2 Fourieranalyse

Betrachtet man den umgekehrten Fall, eine Zerlegung in Einzelschwingungen, so spricht man von einer **Fourieranalyse**. Sie hat große Bedeutung z. B. in der Akustik. Dort wird ein Klang in Sinusschwingungen verschiedener Amplituden und Frequenzen zerlegt.

Experiment 36: Fourieranalyse

Mit einem Mikrofon geben wir den Ton über eine Soundkarte in einen Computer ein, auf dem eine Fourieranalyse zur Bestimmung der Amplituden a_n zu den jeweiligen Frequenzen ω_n programmiert ist.

Die bildliche Darstellung des Klangs ist die jeweilige Amplitude als Funktion der zugehörigen Frequenz (▶ Abbildung 15.14). Und die Rechteckschwingung (▶ Abbildung 15.13) ergibt das Frequenzspektrum entsprechend $a_n \propto (-1)^{n-1}/(2n-1)$ und $\omega_n = (2n-1)\,\omega_\circ$.

Das Experiment kann auch mit der *phyphox*-App durchgeführt werden. Wir werden ein solches *phyphox*-Experiment zur Fourieranalyse im Abschnitt 16.2 im Zusammenhang von Klanganalysen mit Grundton und Oberschwingungen diskutieren.

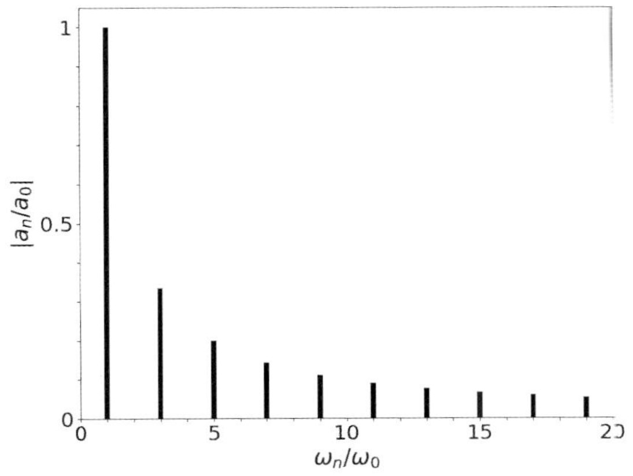

Abbildung 15.14 Umgekehrt zur Fouriersynthese liefert die Fourieranalyse einer Rechteckschwingung (► Abbildung 15.13) ein Spektrum mit ∞ vielen Frequenzen abnehmender Amplitude.

15.5.3 Fouriertransformation

Die wichtige Bedeutung der Gleichungen (15.155), (15.156) und (15.161) ist, dass wir sämtliche Information über die periodische Schwingung (Ton, Klang) gleichwertig durch folgende Alternativen erfassen können:

- Mit der linken Seite dieser Gleichungen beschreiben wir die Schwingung (Klang) durch die zeitliche Entwicklung der Ortsfunktion $x(t)$, die z. B. der Bewegung eines Lautsprechers entspricht.
- Mit der rechten Seite beschreiben wir die Frequenzen ω_n, die zum Klangspektrum beitragen. Jede Frequenz trägt mit einer individuellen Amplitude a_n und Phasenverschiebung φ_n bei.

Die Symmetrie der Betrachtungsweisen wird in der komplexen Darstellung (15.161) besonders deutlich. Wir können das Gleichungspaar

$$x(t) = \sum_{n=-\infty}^{\infty} c_n(\omega_n)e^{i\omega_n t}, \qquad c_n(\omega_n) = \frac{1}{T}\int_{-T/2}^{T/2} x(t)e^{-i\omega_n t} \tag{15.161}$$

als die beiden alternativen Beschreibungsformen desselben Phänomens durch eine Ortsfunktion $x(t)$ oder eine Frequenzverteilung (auch Spektralverteilung) mit Amplituden $c_n(\omega_n)$ verstehen.

Wenn wir nun in der Fourierreihe (15.161) $\omega_\circ \to 0$ gehen lassen, dehnen wir mit $T = 2\pi/\omega_\circ \to \infty$ die Periode T über den ganzen Zeitbereich aus. Die Gleichungen (15.161) gehen in das symmetrische Gleichungspaar (15.168) und (15.169) über. Um die Symmetrie zu unterstreichen, wird üblicherweise $F(\omega)$ statt $c(\omega)$ benutzt.

Mathematischer Einschub: Fouriertransformation

Wir definieren eine Funktion $F(\omega)$ mit der Frequenzvariablen ω durch folgendes Integral über eine Funktion $f(t)$ in der Zeit t:

$$F(\omega) = \frac{1}{\sqrt{2\pi}} \int_{-\infty}^{\infty} f(t)\, e^{-i\omega t}\, dt \qquad (15.168)$$

Die Rücktransformation lautet

$$f(t) = \frac{1}{\sqrt{2\pi}} \int_{-\infty}^{\infty} F(\omega)\, e^{+i\omega t}\, d\omega\,. \qquad (15.169)$$

Der Faktor $1/\sqrt{2\pi}$ sorgt als Konvention für die Symmetrie der Gleichungen. Die Gleichungen (15.161), (15.168) und (15.169) gelten auch für komplexe Funktionen, jedoch werden (wie hier dargestellt) reelle Funktionen $f(t)$ wieder in reelle Funktionen $F(\omega)$ transformiert.

Mit diesen sogenannten **Fouriertransformationen** können wir Funktionenpaare $f(t)$ und $F(\omega)$ ineinander überführen. Sie gilt außer für das Variablenpaar Zeit/Frequenz auch für Ort/Wellenlänge bei Wellen und für damit verwandte Paare von Zeit/Energie sowie Ort/Impuls in der Quantenmechanik.

Die kontinuierliche Funktion $f(t)$ kann periodisch oder nichtperiodisch sein, wie die folgenden Beispiele der Kosinusfunktion und Gaußverteilung zeigen. Die zugehörige Funktion $F(\omega)$ ist dann diskret (Kosinus) oder kontinuierlich (Gauß).

Beispiel 45: Fouriertransformation einer Kosinusfunktion

Als Funktion wählen wir

$$f(t) = \cos(\omega_\circ t)\,. \qquad (15.170)$$

Eine nicht ganz einfache Rechnung ergibt als Fouriertransformierte von $f(t)$ die Funktion

$$F(\omega) = \frac{\sqrt{2\pi}}{2} \left(\delta(\omega - \omega_\circ) + \delta(\omega + \omega_\circ)\right)\,. \qquad (15.171)$$

Die sogenannte Dirac-Funktion (bzw. Delta-Funktion) $\delta(\omega - \omega_\circ)$ gibt dabei nur bei der Frequenz $\omega = \pm\omega_\circ$ einen von null verschiedenen Beitrag. Natürlich hat nur der positive Wert eine physikalische Bedeutung. Es handelt sich also um eine Fourieranalyse entsprechend unseres obigen Experiments mit nur einer Schwingungsfrequenz.

Beispiel 46: Fouriertransformation einer Gaußverteilung

Als Funktion wählen wir die Gaußverteilung

$$f(t) = e^{-\frac{1}{2}\frac{t^2}{\sigma^2}} . \qquad (15.172)$$

Das Resultat der folgenden Aufgabe zeigt, dass die Fouriertransformierte von $f(t)$ auch im Frequenzraum eine Gaußverteilung ergibt:

$$F(\omega) = \sigma\, e^{-\frac{1}{2}\sigma^2\omega^2} \qquad (15.173)$$

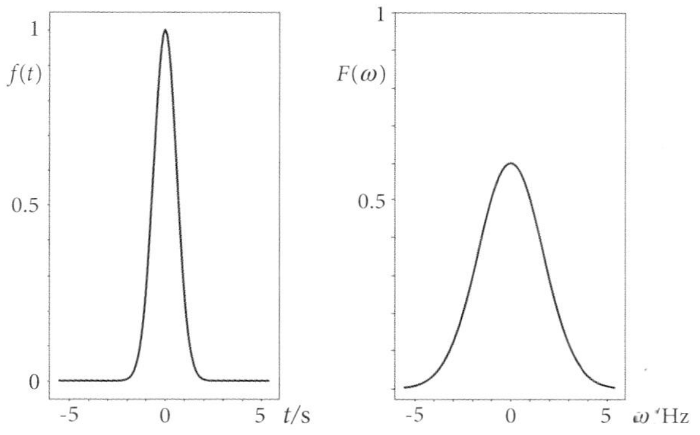

Abbildung 15.15 Die Fouriertransformierte einer Gaußverteilung $f(t)$ mit der Standardabweichung σ ist wieder eine Gaußverteilung $F(\omega)$ mit der Standardabweichung $1/\sigma$.

Aufgabe 45: Fouriertransformation Gauß

Die Rechnung für die Fouriertransformation der Gaußverteilung (15.172) in die Gaußverteilung (15.173) ist nicht sehr aufwendig, wenn man eine quadratische Ergänzung im Exponenten benutzt und $\int_{-\infty}^{+\infty} e^{x^2}\, dx = \sqrt{\pi}$ verwendet.

15.5.4 Unschärferelation

Aus der Fouriertransformation der Gaußverteilung ergibt sich eine grundsätzliche Grenze für die Genauigkeit mit der man zwei zusammenhängende Messgrößen gleichzeitig bestimmen kann.

Wir beobachten eine Schwingung innerhalb eines Zeitintervalls, dessen Zeitdauer wir mit der Genauigkeit Δt bestimmen. Dabei entspricht Δt der Standardabweichung σ im Nenner des Exponenten der Gaußverteilung (15.172):

$$\Delta t = \sigma \tag{15.174}$$

Die Genauigkeit $\Delta\omega$, mit der wir die Frequenz der Schwingung bestimmen können, lesen wir im Exponenten von (15.173) ab. Die Standardabweichung dieser Gaußverteilung entspricht

$$\Delta\omega = \frac{1}{\sigma}\,. \tag{15.175}$$

Setzen wir diese beiden Gleichungen ineinander ein, so erhalten wir als untere Grenze für die Messgenauigkeit des Zeitintervalls und der Frequenz

$$\Delta\omega \cdot \Delta t = 1\,. \tag{15.176}$$

Im Allgemeinen formuliert man diese beschränkte Messgenauigkeit als **Unschärferelation** in der Form

$$\Delta\omega \cdot \Delta t \geq 1\,. \tag{15.177}$$

In dem Experiment zur Fouriersynthese hatten wir gesehen, dass wir alternative Schwingungsformen zur Sinusform – und damit andere Klangformen – durch Überlagerung von Schwingungen verschiedener Frequenz erreichen können.

Auch die Erzeugung kurzer Klangpulse gelingt mit der Überlagerung von harmonischen Schwingungen. Die Unschärferelation (15.177) impliziert dabei folgende Einschränkung: Wenn wir einen sehr kurzen Klangpuls erzeugen, dessen Zeitdauer wir mit der Genauigkeit Δt kennen wollen, benötigen wir die Überlagerung von Schwingungen aus einem Frequenzspektrum von mindestens der Breite $\Delta\omega$.

Haben wir nur ein schmaleres Frequenzspektrum $\Delta\omega' < \Delta\omega$ für die Erzeugung des Pulses zur Verfügung, ist die Genauigkeit geringer, mit der wir die Pulslänge bestimmen können: $\Delta t' > \Delta t$

Einen theoretischen Extremfall der Unschärferelation sehen wir in dem Beispiel der Fouriertransformation der Kosinusfunktion (15.170, 15.171). Mit exakt einer einzigen Frequenz $\omega_\circ$ ($\Delta\omega \to 0$) hören wir nur einen dauerhaften Ton ($\Delta t \to \infty$).

Das Prinzip von Unschärfe spielt in der Quantenmechanik eine entscheidende Rolle. Wir werden die sogenannte Heisenbergsche Unschärferelation für Teilchen später kennenlernen.

Das Wichtigste zum Mitnehmen

Zum Mitnehmen: Schwingungen 1

- Ein Beispiel eines ungestört schwingenden **harmonischen Oszillators** ist das Federpendel mit der bewegten Masse m und der rücktreibenden Federkraft $F = -kx$. Die zugehörige Bewegungsgleichung

$$\ddot{x} = -\omega^2 x \tag{15.12}$$

hat mit der Frequenz $\omega = \sqrt{k/m}$ des Federpendels und mit den beiden Anfangsbedingungen $x_\circ, \varphi_\circ \in \mathbb{R}$ als Lösung

$$x(t) = x_\circ \cdot \cos(\omega t + \varphi_\circ)\ . \tag{15.13}$$

- Gleichung (15.12) lässt sich auch mit komplexen Zahlen lösen:

$$x(t) = c\,e^{\lambda t} \qquad c, \lambda \in \mathbb{C} \tag{15.14}$$

$$\text{mit} \quad \lambda_1 = i\omega \quad \text{und} \quad \lambda_2 = -i\omega$$

Physikalische Lösungen lassen sich mit $c_1 = x_\circ e^{i\varphi_\circ}$ als Realteil von (15.14) schreiben:

$$x(t) = Re\,(x_1(t)) = x_\circ \cos(\omega t + \varphi_\circ) \tag{15.32}$$

oder als Kombinationen von (15.14) mit geeigneten c_i:

$$x(t) = x_1(t) + x_2(t) = c_1\,e^{i\omega t} + c_2\,e^{-i\omega t} \tag{15.23}$$

(15.23) entspricht (15.13) mit $c_1 = (x_\circ/2)\,e^{i\varphi_\circ}$ und $c_2 \equiv c_1^* = (x_\circ/2)\,e^{-i\varphi_\circ}$.

Zum Mitnehmen: Schwingungen 2

- Bei Drehbewegungen sind die zur Reibung gehörenden Drehmomente der Bewegung entgegen gerichtet. Beim Pohlschen Rad führt eine linear von der Geschwindigkeit $\dot{\Phi}$ abhängige Wirbelstrombremse als **lineare Dämpfung** zu einem bremsenden Drehmoment D_b:

$$D_b = -b\dot{\Phi} \tag{15.34}$$

- Ein **gedämpfter Oszillator** wie das Pohlsche Rad hat dann zusätzlich zum Drehmoment $D = -D_r\Phi$ das Drehmoment D_b (15.34). Die Bewegungsgleichung für die gedämpfte Schwingung lautet dann mit Trägheitsmoment I, $\omega_\circ^2 = D_r/I$ und $2\gamma = b/I$

$$\ddot{\Phi} + 2\gamma\dot{\Phi} + \omega_\circ^2\Phi = 0\,. \tag{15.41}$$

Als Lösungsansatz verwenden wir analog zu (15.14)

$$\Phi = c\,e^{\lambda t}\,.$$

- Damit lautet die allgemeine Lösung

$$\lambda_{1,2} = -\gamma \pm \sqrt{\gamma^2 - \omega_\circ^2}\,. \tag{15.46}$$

Für $\gamma < \omega_\circ$ ergibt sich mit der Amplitude $\Phi_\circ$, der Phase $\varphi_\circ$ und der Kreisfrequenz ω mit

$$\omega^2 = \omega_\circ^2 - \gamma^2 \quad \text{die Lösung} \tag{15.51}$$

$$\Phi(t) = \Phi_\circ \cdot e^{-\gamma t} \cdot \cos(\omega t + \varphi_\circ)\,. \tag{15.58}$$

(15.58) beschreibt eine

ungedämpfte Schwingung für $\gamma = 0$ mit $\omega = \omega_\circ$ oder eine

gedämpfte Schwingung für $\omega_\circ > \gamma > 0$ mit $\omega = \sqrt{\omega_\circ^2 - \gamma^2} < \omega_\circ$.

Im **Kriechfall** für $\gamma \geq \omega_\circ$ mit $\omega^2 \leq 0$

kommt es bei einer Auslenkung ohne Schwingung zur Rückkehr in die Ruhelage

$$\Phi(t) = \Phi_\circ \cdot e^{-\gamma t} \cdot \sinh(\sqrt{\gamma^2 - \omega_\circ^2} \cdot t) \tag{15.67}$$

mit dem für viele Anwendungen wichtigen **aperiodischen Grenzfall** für $\omega_\circ = \gamma$ mit $\omega = 0$:

$$\Phi(t) = \Phi_1 \cdot e^{-\gamma t} \cdot t \tag{15.70}$$

Zum Mitnehmen: Schwingungen 3

- Wenn wir das Pohlsche Rad mit einem periodischen Antrieb der Frequenz ω_m ausstatten, bewirkt das zusätzliche Drehmoment

$$D_m = Re\left(D_\circ \cdot e^{i\omega_m t}\right)$$

eine **erzwungene Schwingung.** Mit dem zusätzlichen Antriebsterm mit der Antriebskonstanten $\kappa \equiv D_\circ/I$ erweitern wir die Bewegungsgleichung (15.41) der gedämpften Schwingung zu

$$\ddot{\Phi} + 2\,\gamma\,\dot{\Phi} + \omega_\circ^2\,\Phi = \kappa\,e^{i\omega_m t}\,. \qquad (15.77)$$

Als Lösungsansatz verwenden wir

$$\Phi = c\,e^{i\omega_m t}\,.$$

- Die physikalische Amplitude Φ_m der erzwungenen Schwingung beschreibt eine **Resonanzkurve**:

$$\Phi_m = \kappa\left(\left(\omega_\circ^2 - \omega_m^2\right)^2 + (2\,\gamma\,\omega_m)^2\right)^{-1/2} \qquad (15.88)$$

Die **Resonanzfrequenz** $\omega_m = \omega_R$ im Maximum der Resonanzkurve

$$\omega_R = \sqrt{\omega_\circ^2 - 2\,\gamma^2} \qquad (15.91)$$

liegt etwas unterhalb der Frequenz $\omega_\circ$ des frei schwingenden Pohlschen Rads.

- Zwei gleichartige Pendel (Länge l, Masse m) sind durch eine Feder mit Federkonstante k im Abstand a vom Aufhängepunkt gekoppelt.

Mit den Auslenkungen Φ_i der Pendel und ihrem Trägheitsmoment $I = m\,l^2$ erhalten wir folgendes gekoppelte Gleichungssystem für **gekoppelte Schwingungen**:

$$I\ddot{\Phi}_1 = -lgm\,\Phi_1 - ka^2(\Phi_1 - \Phi_2) \qquad (15.104)$$

$$I\ddot{\Phi}_2 = -lgm\,\Phi_2 + ka^2(\Phi_1 - \Phi_2) \qquad (15.105)$$

(15.104) und (15.105) werden mit $\eta \equiv \frac{1}{2}\,(\Phi_1 + \Phi_2)\,, \xi \equiv \frac{1}{2}\,(\Phi_1 - \Phi_2)$ in zwei harmonische Schwingungsgleichungen umgewandelt:

$$\ddot{\eta} = -\omega_a^2\eta \quad \text{und} \quad \ddot{\xi} = -\omega_b^2\xi \qquad (15.123)$$

Zum Mitnehmen: Schwingungen 4

- Die Frequenz ω_a liegt nahe der Frequenz $\omega_\circ$ eines frei schwingenden Pendels:

$$\omega_a \approx \sqrt{\frac{g}{l}} = \omega_\circ \tag{15.116}$$

Für eine schwache Federkopplung $k \ll mgl/a^2$ ist die Frequenz ω_b nur wenig größer als ω_a:

$$\omega_b \approx \omega_\circ \left(1 + \frac{ka^2}{lgm}\right) = \omega_\circ(1 + 2\kappa) \tag{15.120}$$

- Die Frequenzen ω_a und ω_b werden auch **Eigenfrequenzen** der **Eigenschwingungen** oder Normalschwingungen (15.123) genannt. Die Größen η und ξ sind die zugehörigen **Normalkoordinaten**.

Bei den Normalkoordinaten η schwingen die Pendel parallel mit der Frequenz $\omega_a = \omega_\circ$, bei ξ schwingen sie gegeneinander mit der Frequenz $\omega_b > \omega_\circ$.

Allgemein kann man für ein schwingungsfähiges System alle möglichen Schwingungen aus seinen Eigenschwingungen aufbauen, im vorliegenden Fall also aus einer Überlagerung (**Superposition**) von Eigenschwingungen der Frequenzen ω_a und ω_b.

Wir erhalten für die Auslenkung des ersten Pendels eine Überlagerung der Schwingungen mit ω_a und ω_b:

$$\Phi_1(t) = \underbrace{2\Phi_\circ \cos\left(\underbrace{\frac{\omega_b - \omega_a}{2}}_{\equiv\Omega_b} t + \underbrace{\frac{\beta - \alpha}{2}}_{\equiv\varphi_b}\right)}_{\text{Amplitudenmodulation}} \cdot \underbrace{\cos\left(\underbrace{\frac{\omega_b + \omega_a}{2}}_{\equiv\Omega_a} t + \underbrace{\frac{\beta + \alpha}{2}}_{\equiv\varphi_a}\right)}_{\text{Pendelschwingung}} \tag{15.134}$$

- Die halbe Differenz der Frequenzen

$$\Omega_b \equiv \frac{\omega_b - \omega_a}{2} \approx \kappa\,\omega_\circ \tag{15.137}$$

führt zu einer Amplitudenmodulation, der **Schwebung**. Für eine schwache Federkopplung k und vernachlässigbare Masse der Pendelstange ist die Schwebungsfrequenz um einen Faktor κ (15.121) kleiner als $\omega_\circ$

Zum Mitnehmen: Schwingungen 5

- Die Pendelfrequenz wird durch den Mittelwert

$$\Omega_a \equiv \frac{\omega_b + \omega_a}{2} \approx \omega_\circ \tag{15.135}$$

der beiden Frequenzen beschrieben. Sie liegt nahe der Frequenz $\omega_\circ$ des frei schwingenden Pendels.

Für das erste Pendel lautet die Ortsfunktion Φ_1

$$\Phi_1(t) \approx 2\,\Phi_\circ\,\cos(\kappa\,\omega_\circ\,t + \varphi_b) \cdot \cos(\omega_\circ\,t + \varphi_a)\,. \tag{15.139}$$

Φ_2 des zweiten Pendels ist um die Phase $\pi/2$ verschoben:

$$\Phi_2(t) \approx -2\,\Phi_\circ\,\sin(\kappa\,\omega_\circ\,t + \varphi_b) \cdot \sin(\omega_\circ\,t + \varphi_a) \tag{15.145}$$

- Eine **Schwebung** tritt ganz allgemein bei der Überlagerung von zwei Schwingungen mit ähnlicher Frequenz auf. Mit $\omega_b > \omega_a$ und

$$\Phi_a = \Phi_\circ\,\cos(\omega_a t + \alpha) \tag{15.147}$$
$$\Phi_b = \Phi_\circ\,\cos(\omega_b t + \beta) \tag{15.148}$$

ergibt sich für die Gesamtschwingung $\Phi = \Phi_a + \Phi_b$ mithilfe der Additionstheoreme

$$\Phi(t) = \underbrace{2\Phi_\circ\cos(\Omega_b\,t + \varphi_b)}_{\text{Amplitudenmodulation}} \cdot \underbrace{\cos(\Omega_a\,t + \varphi_a)}_{\text{Schwingung}}\,. \tag{15.178}$$

Wie beim Doppelpendel ist dies eine Schwingung mit der **mittleren Frequenz** Ω_a und der Phase φ_a aus den jeweiligen Mittelwerten

$$\Omega_a \equiv \frac{\omega_b + \omega_a}{2} \quad \text{und} \quad \varphi_a \equiv \frac{\beta + \alpha}{2} \tag{15.152}$$

sowie einer **Amplitudenmodulation** mit der Frequenz Ω_b und der Phase $\varphi_\circ$ aus den jeweiligen halben Differenzen

$$\Omega_b \equiv \frac{\omega_b - \omega_a}{2} \quad \text{und} \quad \varphi_b \equiv \frac{\beta - \alpha}{2}\,. \tag{15.153}$$

Die Schwebungsfrequenz Ω_a wird umso kleiner (und damit die Schwebungsdauer T_a umso länger), je näher die Frequenzen ω_a und ω_b beieinander liegen.

Zum Mitnehmen: Schwingungen 6

- Die **Fouriersynthese** ist eine Überlagerung von $N \to \infty$ Schwingungen zu einer periodischen Gesamtbewegung:

$$x(t) = \sum_{n=1}^{N} a_n \cos(\omega_n t + \varphi_n) \tag{15.154}$$

Hierbei bezeichnet a_n die Amplitude der n-ten Schwingung, ω_n ihre Kreisfrequenz und φ_n ihre Phasenverschiebung.

Umgekehrt lässt sich eine periodische Funktion $x(t)$ in einer **Fourieranalyse** in das Frequenzspektrum ihrer harmonischen Schwingungen zerlegen.

Die Koeffizienten können mit den Fourierintegralen (15.157) und (15.158) berechnet werden und die Koeffizienten verschiedener Darstellungen lassen sich ineinander umrechnen (15.159) und (15.160).

- Für eine beliebige auch nichtperiodische Funktion $f(t)$ geht die Fourierreihe in ein Integral über. Die Frequenzverteilung $F(\omega)$ und $f(t)$ sind durch **Fouriertransformationen** miteinander verbunden:

$$F(\omega) = \frac{1}{\sqrt{2\pi}} \int_{-\infty}^{\infty} f(t)\, e^{-i\omega t}\, dt \tag{15.168}$$

$$f(t) = \frac{1}{\sqrt{2\pi}} \int_{-\infty}^{\infty} F(\omega)\, e^{+i\omega t}\, d\omega \tag{15.169}$$

Die Gleichungen gelten auch für komplexe Funktionen, jedoch werden reelle Funktionen $f(t)$ wieder in reelle Funktionen $F(\omega)$ transformiert.

- Die Fouriertransformierte einer Gaußverteilung $f(t)$ mit der Standardabweichung σ ist wieder eine Gaußverteilung $F(\omega)$ mit der Standardabweichung $1/\sigma$.

Für die Genauigkeit der gleichzeitigen Messung der Zeit $\Delta t = \sigma$ und der Frequenz $\Delta\omega = 1/\sigma$ gibt es also die Grenze

$$\Delta\omega \cdot \Delta t = 1\,.$$

Im Allgemeinen formuliert man diese beschränkte Messgenauigkeit als **Unschärferelation** in der Form

$$\Delta\omega \cdot \Delta t \geq 1\,. \tag{15.177}$$

Wellen

16

16.1 Wellenausbreitung **331**
16.1.1 Ebene harmonische Welle 332
16.1.2 Energietransport 338
16.1.3 Kugelwelle 340

16.2 Stehende Wellen **340**
16.2.1 Eigenschwingungen 342
16.2.2 Schwingende Platten 344

16.3 Dispersion **345**
16.3.1 Phasengeschwindigkeit, Gruppengeschwindigkeit 345
16.3.2 Dispersion bei Wasserwellen 347

16.4 Interferenz **349**
16.4.1 Überlagerung von Wellen synchron schwingender Quellen . . 350
16.4.2 Interferenzeffekte 351
16.4.3 Interferenzmuster 352

16.5 Dopplereffekt **353**
16.5.1 Bewegter Beobachter 353
16.5.2 Bewegte Quelle 354
16.5.3 Nicht-relativistischer Dopplereffekt 356

16.6 Machscher Kegel **357**

16.7 Relativistischer Dopplereffekt **358**

16.8 Wellengleichung **360**

16.9 Transversale Wellen **361**
16.9.1 Saitenspannkraft in Richtung ihrer Auslenkung 361
16.9.2 Transversale Beschleunigung des Saitenstücks 363
16.9.3 Wellengleichung und Phasengeschwindigkeit der Saite 363
16.9.4 Streichinstrumente 364
16.9.5 Grundton und Obertöne 367

ÜBERBLICK

» Das besondere Interesse an Wellen beruht auf der Möglichkeit, Informationen mit hoher Geschwindigkeit über große Distanzen zu transportieren.

Mechanische Wellen basieren auf der Weitergabe eines physikalischen Zustands zwischen schwingenden Objekten. Die beteiligten Objekte werden dabei zwar bewegt, verbleiben aber an ihrem Ort.

In diesem Kapitel werden wir die Ausbreitung von Wellen untersuchen, ihre Bewegungsgleichung kennenlernen und Interferenzphänomene besprechen. «

16.1 Wellenausbreitung

Mechanische Wellen benötigen ein Medium, in dem sich eine Störung oder allgemein ausgedrückt ein physikalischer Zustand ausbreitet.

> *Wellenausbreitung bedeutet, dass ein physikalischer Zustand, der an einem Ort erzeugt wurde, sich auf andere Bereiche überträgt. Nicht die Materie wird transportiert, sondern nur der Bewegungszustand.*

Die Ausbreitung des Zustands kann als **longitudinale Welle** oder **transversale Welle** erfolgen. Bei Longitudinalwellen ist der physikalische Zustand entlang der Ausbreitungsrichtung orientiert, bei Transversalwellen senkrecht dazu.

Transversalwellen sind z. B. Seil- oder Saitenwellen (Abschnitt 16.9) Beispiele für Longitudinalwellen sind elastische Wellen in Festkörpern und Flüssigkeiten (Abschnitt 18.7) und in Gasen (Schallwellen, Abschnitt 18.7). Ein anschauliches Beispiel gibt auch das folgende Experiment.

Experiment 37: Longitudinale Wellenausbreitung in einer Feder

Wir spannen eine lange Feder zwischen zwei Halterungen ein (► Abbildung 16.1). Wir verformen die Feder durch Zusammendrücken von ca. fünf Windungen der Feder und lassen los. Das Resultat ist eine kurzzeitige Störung, die sich als mechanische Welle in der fest eingespannten Feder ausbreitet.

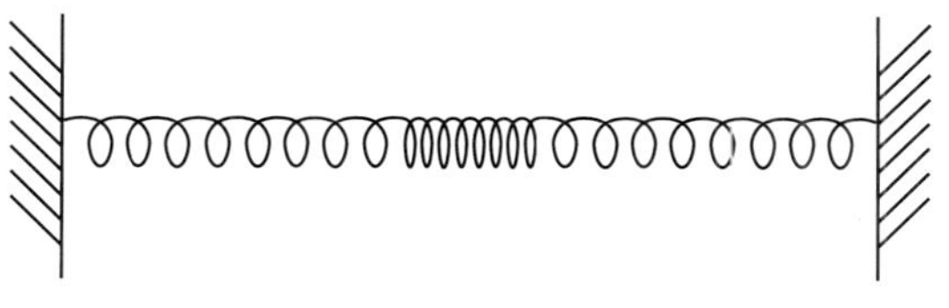

Abbildung 16.1 Eine kurzzeitige Störung an einer Feder breitet sich als Welle aus.

Im Folgenden überlegen wir, wie wir die Ausbreitung des Zustands mathematisch beschreiben. In Anlehnung an Koordinatentransformationen (11.1) können wir den Zustand entlang der x-Achse verschieben.

Eine lokale Verformung der Feder in ► Abbildung 16.1 beschreiben wir durch die Federelemente, die bei entspannter Feder an der Stelle x positioniert sind, deren aktuelle Auslenkung ξ aber von der Ruhestellung abweicht. Zu einem festen Zeitpunkt bilden die benachbarten Federelemente gemeinsam einen physikalischen Zustand $f(x)$, dessen Form beispielhaft in der Mitte der ► Abbildung 16.2 skizziert ist.

Für jedes Federelement an der Stelle x können wir dessen Auslenkung ξ aus der Verteilung $f(x)$ ablesen, die die Störung oder ganz allgemein den physikalischen Zustand beschreibt:

$$\xi(x) = f(x) \tag{16.1}$$

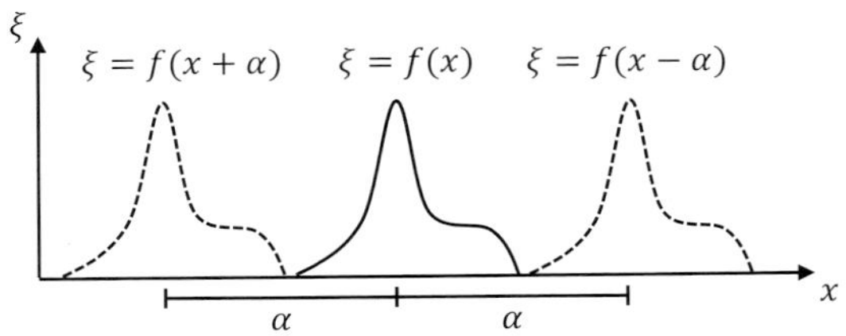

Abbildung 16.2 Ein physikalischer Zustand $f(x)$ mit den lokalen Auslenkungen $\xi(x)$ wird durch $\xi(x) = f(x+\alpha)$ nach links oder durch $\xi(x) = f(x-\alpha)$ nach rechts verschoben.

Wir verschieben nun den Zustand $f(x)$ entlang der x-Achse nach rechts oder links um die Strecke α und erhalten dementsprechend Auslenkungen von Federelementen an anderen Stellen der Feder (▶ Abbildung 16.2):

$$\xi(x) = f(x-\alpha) \quad \text{Verschiebung nach rechts} \tag{16.2}$$

$$\xi(x) = f(x+\alpha) \quad \text{Verschiebung nach links} \tag{16.3}$$

Anstelle von Verschiebungen um feste Strecken α wollen wir bei einer Wellenausbreitung den Zustand $f(x)$ kontinuierlich weiterbewegen, also den Abstand α mit der Zeit t vergrößern. Dafür nutzen wir die Ausbreitungsgeschwindigkeit v des Zustands, genauer gesagt die Verschiebung α pro verstrichene Zeit t, und ersetzen α in (16.2), (16.3) durch $\alpha = vt$.

Damit wandert der physikalische Zustand $f(x)$ in Abhängigkeit des Vorzeichens im Argument mit der Geschwindigkeit v nach rechts (−) oder links (+). Die Auslenkung ξ eines Federelements am Ort x wird damit zeitabhängig:

$$\xi(x,t) = f(x \mp vt) \tag{16.4}$$

Die Ausbreitungsgeschwindigkeit v nennen wir **Phasengeschwindigkeit**.

16.1.1 Ebene harmonische Welle

Wir stellen jetzt die wichtige sogenannte **ebene harmonische Welle** vor. Ihre Funktion f ist eine Sinus- oder Kosinusfunktion:

$$\underbrace{\xi(x,t)}_{\text{Auslenkung}} = \underbrace{\xi_0}_{\text{Amplitude}} \sin[\underbrace{k}_{\text{Wellenzahl}} \underbrace{(x-vt)}_{\text{Ausbreitung}}] \tag{16.5}$$

Darin beschreibt $\xi(x,t)$ die **momentane Auslenkung** des Mediums an der Stelle x zum Zeitpunkt t. Die **Amplitude**, also die maximale Auslenkung, bezeichnen wir mit ξ_0. Im Argument der Sinusfunktion erkennen wir den Term $(x - vt)$ (16.4) wieder, der eine Wellenausbreitung in positiver x-Richtung beschreibt.

Wie groß die räumliche Ausdehnung der Welle ist, wird interessanterweise nur innerhalb des Arguments der Sinusfunktion mit der sogenannten **Wellenzahl** k erfasst und funktioniert wie folgt.

In ► Abbildung 16.3 sind zwei ebene Wellen mit Wellenzahlen $k_1 > k_2$ gezeigt. Zum Zeitpunkt $t_\circ = 0$ ergibt sich für die erste Welle $\xi(x, 0) = \xi_\circ \sin(k_1 x)$ (16.5), sodass die erste Raumperiode bei $x = 0$ mit $\sin(0) = 0$ beginnt. Die nächste Periode startet bei $\sin(2\pi) = 0$, also an dem Ort x mit der Bedingung

$$k_1 x = 2\pi\,. \tag{16.6}$$

Im Vergleich der beiden Wellenzahlen führt die größere Wellenzahl $k_1 > k_2$ in der Sinusfunktion (16.5) zu einer kleineren räumlichen Ausdehnung der Welle (► Abbildung 16.3).

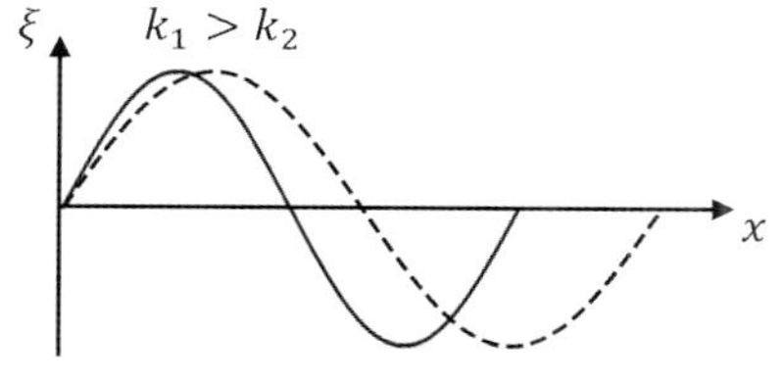

Abbildung 16.3 Räumliche Verteilung von harmonischen Wellen $\xi(x)$ zu einer festen Zeit t mit den Wellenzahlen k_1 (durchgezogen) $>$ k_2 (gestrichelt).

Im Folgenden werden wir das Argument der Sinusfunktion weiter motivieren und auf verschiedene Arten ausdrücken. Dabei werden wir die wesentlichen physikalischen Konzepte zur Beschreibung der Welle benennen und ihre Ausbreitung im Detail untersuchen.

Räumliche Periodizität: Die Welle zu einem festen Zeitpunkt

Die ebene harmonische Welle erstreckt sich mit ihren Wellenbergen und Wellentälern über den ganzen Raum. Wir zeigen zunächst, dass zu einem festen Zeitpunkt $t_\circ$ das Wellenmuster am Ort x in der linken ► Abbildung 16.4 auch am Ort

$$x + \frac{2\pi}{k} \tag{16.7}$$

zu finden ist. Dabei ist k die sogenannte Wellenzahl im Argument von (16.5), eine Bezeichnung, die wir weiter unten begründen werden.

Bei einer Orientierungsrichtung der Welle $\xi(x, t)$ nach rechts (16.2) machen wir die Ersetzung $x \to x - 2\pi/k$ im Argument der Wellenfunktion. Damit sehen wir, dass die Sinusfunktion der ebenen Welle (16.5) bereits die Periodizität ihrer räumlichen Ausdehnung enthält:

$$\xi\left(x - \frac{2\pi}{k}, t_\circ\right) = \xi_\circ \sin\left(k\left[\left(x - \frac{2\pi}{k}\right) - vt_\circ\right]\right) \tag{16.8}$$

$$= \xi_\circ \sin\left(k\left(x - vt_\circ\right) - 2\pi\right) \tag{16.9}$$

$$= \xi_\circ \sin\left(k\left(x - vt_\circ\right)\right) \tag{16.10}$$

$$= \xi\left(x, t_\circ\right) \tag{16.11}$$

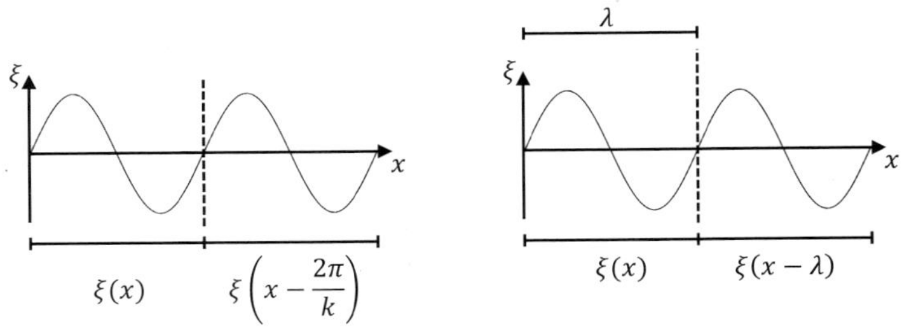

Abbildung 16.4 In einer harmonischen Welle wiederholt sich die räumliche Verteilung $\xi(x)$ mit einer Periode $2\pi/k$ mit der Wellenzahl k (links), alternativ mit der Wellenlänge λ (rechts).

Eine intuitive Größe für die räumliche Ausdehnung der Welle ist die **Wellenlänge** λ. Die Wellenlänge ist die Länge der Raumperiode in der Einheit Meter m, wie in der rechten ► Abbildung 16.4 gezeigt ist. Auch hier finden wir dasselbe Wellenmuster an der Stelle x und $x + \lambda$. Wir bringen die Muster überein, indem wir entsprechend (16.2) im Argument der ebenen Welle $\xi(x, t)$ (16.5) die Koordinate $x \to x - \lambda$ ersetzen:

$$\xi(x - \lambda, t_\circ) = \xi_\circ \sin(k\left[(x - \lambda) - vt_\circ\right]) \tag{16.12}$$

$$= \xi_\circ \sin\left(k\ (x - vt_\circ) - k\lambda\right) \tag{16.13}$$

Der konstante Term $k\lambda$ muss dabei $k\lambda = 2\pi$ ergeben, damit die nächste Periode der Sinusfunktion beginnt. Damit entsprechen sich $\xi(x - \lambda, t_\circ) = \xi(x, t_\circ)$.

Auch in den beiden ► Abbildungen 16.4 entsprechen sich die Längen $2\pi/k$ und λ, sodass wir den Zusammenhang zwischen der Wellenzahl k und der Wellenlänge λ ablesen können:

$$k = \frac{2\pi}{\lambda} \tag{16.14}$$

Die **Wellenzahl** k hat demnach die Einheit m^{-1}. Für eine Wellenlänge von z. B. $\lambda = 0{,}2\,\mathrm{m}$ erhalten wir $k/(2\pi) = 5$ Wellenlängen pro Meter. Wir interpretieren k daher als das 2π-fache der Anzahl Wellenlängen pro m.

Somit haben wir zunächst zwei äquivalente Darstellungen für harmonische Wellen:

$$\xi(x, t) = \xi_\circ \sin\left(k(x - vt)\right) \tag{16.15}$$

$$= \xi_\circ \sin\left(\frac{2\pi}{\lambda}(x - vt)\right) \tag{16.16}$$

Experiment 38: Wellenmaschine

Eine Wellenmaschine (► Abbildung 16.5) besteht aus vielen nebeneinanderliegenden Drehpendeln. Benachbarte Pendel sind jeweils miteinander gekoppelt. Jedes einzelne Pendel führt eine vertikale Schwingung aus. Durch die Kopplung wird eine am ersten Drehpendel eingebrachte Störung an die folgenden Pendel weitergegeben. Die räumliche Ausdehnung der Welle kann so mit dem Auge gut verfolgt werden.

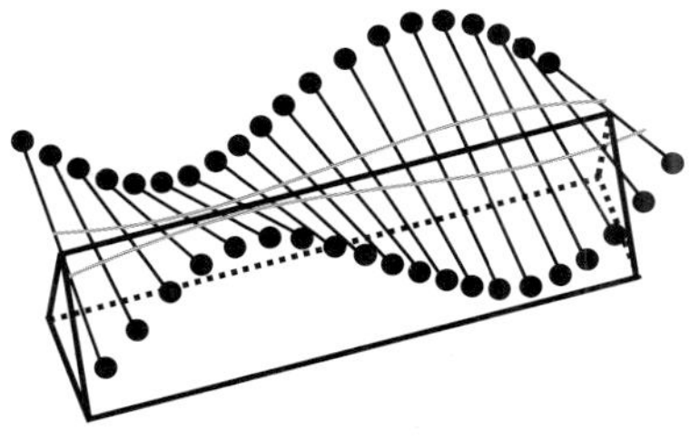

Abbildung 16.5 Prinzip einer Wellenmaschine mit gekoppelten Drehpendeln

Zeitliche Periodizität: Die Welle an einem festen Ort

Im Folgenden betrachten wir die Welle (16.15) an einem festen Ort $x_\circ$

$$\xi(x_\circ, t) = \xi_\circ \sin\left(k(x_\circ - vt)\right) \tag{16.17}$$

$$= \xi_\circ \sin(-\underbrace{kv}_{\equiv\omega}\, t + \underbrace{kx_\circ}_{\equiv\varphi_\circ}) \tag{16.18}$$

Wir erkennen die Ortsfunktion (15.13) des harmonischen Oszillators

$$\xi(x_\circ, t) = \xi_\circ \sin(-\omega t + \varphi_\circ) \tag{16.19}$$

wieder, wobei wir als Kreisfrequenz $\omega \equiv k\,v$ und als Phase $\varphi_\circ = k\,x_\circ$ identifizieren. Die Phase $\varphi_\circ$ hängt offenbar vom Ort $x_\circ$ ab.

Bei einer ebenen harmonischen mechanischen Welle wird das Medium an jedem Ort $x_\circ$ zu einer harmonischen Schwingung angeregt.

In ► Abbildung 16.6 können wir sehen, dass die Welle am Ort $x_\circ$ im Verlauf einer Schwingungsperiode (4.11)

$$T = \frac{2\pi}{\omega} \tag{16.20}$$

um eine Wellenlänge λ weiterrückt.

Für die Auslenkung $\xi(x_\circ, t)$ (16.18) an der Stelle $x_\circ$ mit $\varphi_\circ = 0$ ergibt sich zu den Zeitpunkten t und $t + T$ dasselbe Bild, da die Sinusfunktion sich periodisch mit 2π fortsetzt (► Abbildung 16.6):

$$\xi(x_\circ, t + T) = \xi_\circ \sin\left(-\omega\,(t + T)\right) \tag{16.21}$$

$$= \xi_\circ \sin\left(-\frac{2\pi}{T}\,(t + T)\right) \tag{16.22}$$

$$= \xi_\circ \sin\left(-\frac{2\pi}{T}\,t - 2\pi\right) \tag{16.23}$$

$$= \xi(x_\circ, t) \tag{16.24}$$

Die zeitliche Periodizität ist demnach in der ebenen Welle (16.15) bereits enthalten.

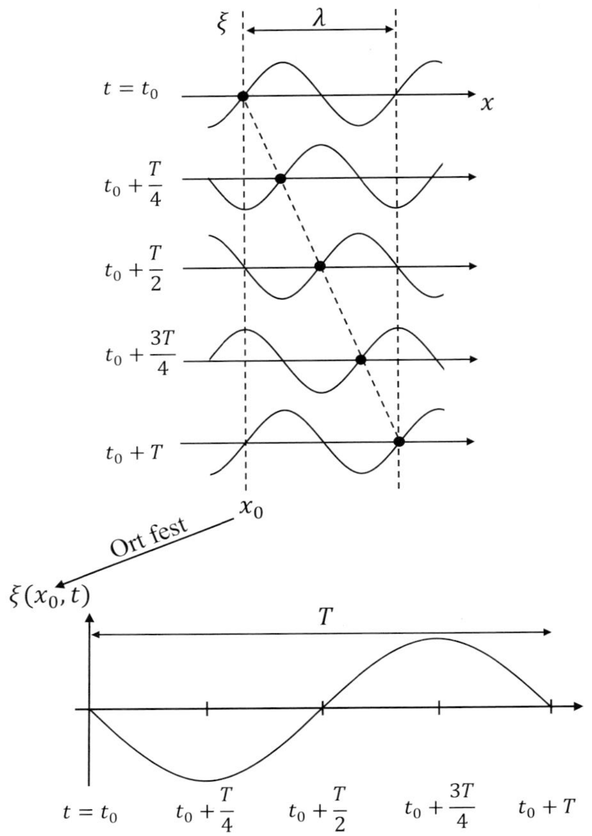

Abbildung 16.6 Eine in x-Richtung fortschreitende Welle zu verschiedenen Zeiten von $t = t_\circ$ bis $t = t_\circ + T$ (oben) und für einen festen Ort $x = x_\circ$ (unten)

Physikalische Größen zur Beschreibung ebener Wellen

Zur Beschreibung der räumlichen Ausdehnung einer ebenen Welle (16.15) hatten wir die Wellenzahl k und alternativ die Wellenlänge λ (16.14) eingeführt. Für den Zeitanteil ωt in der ebenen Welle $\xi\,(x, t)$ (16.18) können wir mit der **Frequenz** ν ebenfalls eine weitere Schreibweise verwenden. Die Frequenz beschreibt die Schwingungsanzahl n pro Sekunde $\Delta t = 1\,\mathrm{s}$ und hat die Einheit $1/\mathrm{s} = \mathrm{Hz}$. Auch Bruchteile einer Schwingung werden gezählt, sodass $n \in \mathbb{R}$ eine reelle Zahl ist. Da n Schwingungsperioden T in $\Delta t = 1\,\mathrm{s}$ passen, entspricht die Frequenz ν dem Kehrwert der Schwingungsperiode T:

$$\nu = \frac{n}{\Delta t} = \frac{1}{T} \tag{16.25}$$

Anstelle des Buchstabens ν wird häufig auch f benutzt. Mit dem Zusammenhang zwischen Kreisfrequenz und Schwingungsperiode $\omega = 2\pi/T$ (16.20) ergibt sich hier der

Zusammenhang zwischen Kreisfrequenz und Frequenz

$$\omega = 2\pi\nu \, . \tag{16.26}$$

Wir kommen noch einmal auf den Term $\omega = k\,v$ in Gleichung (16.18) zurück. Mit ihm verbinden wir die Phasengeschwindigkeit mit der Kreisfrequenz und der Wellenzahl:

$$v = \frac{\omega}{k} \tag{16.27}$$

Erweitern wir Zähler und Nenner mit 2π, finden wir mit $k = 2\pi/\lambda$ und $\omega = 2\pi\nu$ die Phasengeschwindigkeit in der ebenfalls häufigen Schreibweise

$$v = \lambda \cdot \nu \, . \tag{16.28}$$

Viele Schreibweisen für harmonische Wellen (16.15) sind demnach möglich. Eine Formulierung, die wir besonders oft gebrauchen werden, ist wegen $\omega = kv$ (16.27)

$$\xi(x,t) = \xi_\circ \, \sin\left(kx - \omega t\right) \, . \tag{16.29}$$

Ebenso werden wir die komplexe Schreibweise verwenden. Sie lautet wegen der Eulerschen Formel (15.6)

$$\xi(x,t) = \xi_\circ \, e^{i(kx-\omega t)} \, . \tag{16.30}$$

Ausbreitungsrichtung einer ebenen Welle

Die ebene harmonische Welle pflanzt sich in einer bestimmten Ortsrichtung fort, z. B. in x-Richtung. In der Richtung senkrecht zu x finden wir Flächen, in denen die Welle mit gleicher Phase schwingt. Wir sprechen deswegen von einer **ebenen Welle**. Die Flächen werden mit **Phasenflächen** bezeichnet.

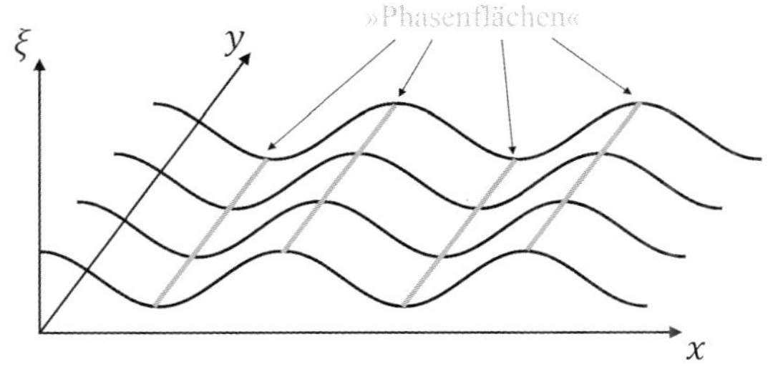

Abbildung 16.7 »Phasenflächen«: Orte gleicher Phase einer ebenen Welle $\xi(x, y, t)$ sind in zwei Dimensionen Linien, in drei Dimensionen Flächen.

Um beliebige Ausbreitungsrichtungen ebener Wellen zu beschreiben, definieren wir einen **Wellenvektor** $\vec{k}$ so, dass er in der Ausbreitungsrichtung der Welle steht und den Betrag entsprechend (16.14) hat:

$$|\vec{k}| = \frac{2\pi}{\lambda} \tag{16.31}$$

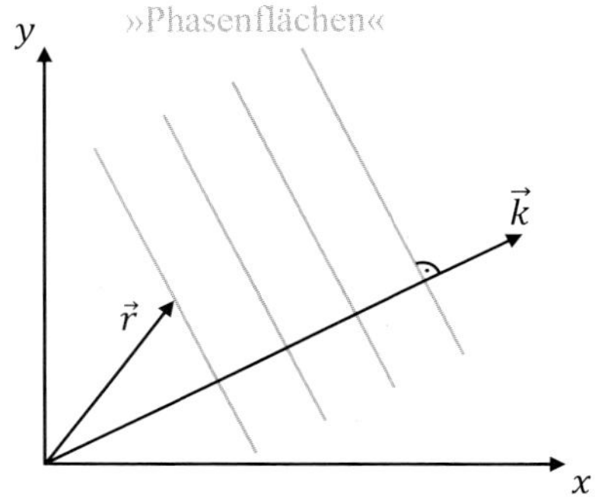

Abbildung 16.8 Phasenflächen einer ebenen Welle $\vec{\xi}(x,y,t) = \vec{\xi}_\circ \sin(\vec{k}\,\vec{r} - \omega t)$, hier in der x, y-Ebene mit $z = 0$, stehen senkrecht auf dem Wellenvektor $\vec{k}$.

Für jeden Vektor $\vec{r}$, der auf die Phasenfläche zeigt, ist das Skalarprodukt mit dem Wellenvektor k konstant:

$$\vec{k} \cdot \vec{r} = const. \tag{16.32}$$

Wir schreiben damit für die ebenen Wellen mit der Ausbreitungsrichtung $\vec{k}/|\vec{k}|$

$$\vec{\xi} = \vec{\xi}_\circ \sin\left(\vec{k}\,\vec{r} - \omega t\right) \tag{16.33}$$

oder auch in komplexer Schreibweise

$$\vec{\xi} = \vec{\xi}_\circ\, e^{i\left(\vec{k}\,\vec{r} - \omega t\right)} . \tag{16.34}$$

16.1.2 Energietransport

Die Ausbreitung einer mechanischen Welle verursacht Schwingungen der Materie, die auf ihrem Weg liegt. Mit einer Welle wird daher Energie transportiert.

Eine harmonische Welle (16.33) verursacht am festen Ort $\vec{r}_\circ$ eine harmonische Schwingung des Materials mit der Phase $\varphi_\circ = \vec{k}\,\vec{r}_\circ$:

$$\vec{\xi} = \vec{\xi}_\circ \sin(-\omega t + \varphi_\circ) \tag{16.35}$$

Wir stellen uns an diesem Ort eine Feder mit der Federkonstanten D vor (anstelle des zuvor im Buch verwendeten Buchstabens k verwenden wir hier D, um eine Verwechslung mit der Wellenzahl auszuschließen). Die Feder wird durch die ankommende Welle bis zur momentanen Auslenkung ξ gedehnt. Die potentielle Energie in der Feder ergibt sich aus der von der Welle gegen die Federkraft $F_F = -D\xi$ geleisteten Arbeit:

$$E_{\text{pot}} = \int_0^{\xi} -F_F \, d\xi \tag{16.36}$$

$$= \int_0^{\xi} -(-D\,\xi) \, d\xi \tag{16.37}$$

$$= \frac{1}{2} D\,\xi^2 \tag{16.38}$$

$$= \frac{1}{2} D\,\xi_\circ^2 \sin^2(-\omega t + \varphi_\circ) \tag{16.39}$$

Die kinetische Energie in der Feder ist

$$E_{\text{kin}} = \frac{1}{2} m\,\dot{\xi}^2 \tag{16.40}$$

$$= \frac{1}{2} m\,\omega^2\,\xi_\circ^2 \cos^2(-\omega t + \varphi_\circ)\,. \tag{16.41}$$

Die Gesamtenergie der angeregten Federschwingung ist mit $\omega^2 = D/m$ (Gleichung (15.11))

$$E = E_{\text{kin}} + E_{\text{pot}} \tag{16.42}$$

$$= \frac{1}{2} D\,\xi_\circ^2 \left(\cos^2(-\omega t + \varphi_\circ) + \sin^2(-\omega t + \varphi_\circ)\right) \tag{16.43}$$

$$= \frac{1}{2} D\,\xi_\circ^2 \tag{16.44}$$

$$= E_{\text{pot, max}} = E_{\text{kin, max}}\,. \tag{16.45}$$

Um eine Schwingung mit der Gesamtenergie E anzuregen, muss die Welle diese Energie transportieren. Anstelle der Feder können wir uns irgendein elastisches Material vorstellen. Entscheidend ist, dass die transportierte Energie proportional zum Quadrat der Schwingungsamplitude ist:

$$E \propto \xi_\circ^2 \tag{16.46}$$

Zur Beschreibung des Energietransports verwenden wir die **Intensität** I (5.65). Sie ist definiert als die Energie, die pro Zeiteinheit Δt durch eine zur Ausbreitungsrichtung senkrecht stehende Flächeneinheit ΔA transportiert wird:

$$I = \frac{E}{\Delta t \cdot \Delta A} \tag{16.47}$$

Damit ist die Intensität ebenfalls proportional zum Quadrat der Schwingungsamplitude. Da wir später auch komplexe Amplituden betrachten werden, schreiben wir den Zusammenhang zwischen Amplitude und Intensität in allgemeiner Form mit Beträgen:

$$I \propto |\xi_\circ|^2 \tag{16.48}$$

16.1.3 Kugelwelle

Bei einer Kugelwelle bilden die Phasenflächen Kugelflächen mit der Oberfläche $4\pi r^2$. Die Intensität einer punktförmigen Quelle, die Wellen in alle Richtungen gleichförmig aussendet, sinkt dann im Abstand r proportional zu

$$I \propto |\xi_\circ^2| \propto \frac{1}{r^2} \,. \tag{16.49}$$

Die momentane Auslenkung ist $\xi \propto 1/r$. Die Kugelwelle beschreiben wir damit durch

$$\xi(r,t) = \frac{\xi_\circ}{r} \sin(kr - \omega t) \tag{16.50}$$

bzw. in komplexer Schreibweise

$$\xi(r,t) = \frac{\xi_\circ}{r} e^{i(kr-\omega t)} \,. \tag{16.51}$$

16.2 Stehende Wellen

Stehende Wellen können durch Reflexion einer Welle oder allgemein durch Überlagerung der Wellen zweier synchron schwingenden Quellen entstehen. Wir untersuchen zunächst im Experiment 39 in ► Abbildung 16.9 die Erzeugung stehender Wellen durch Reflexion am festen Ende in einer Wellenmaschine.

Experiment 39: Stehende Welle

Bei der bereits im Experiment 38 beschriebenen Wellenmaschine koppeln wir die Störung auf dem ersten Drehpendel ein und fixieren das letzte Drehpendel. Durch Variation der Störfrequenz an der Eingabeseite können wir stehende Wellen mit verschiedener Anzahl von Bäuchen und Knoten erzeugen.

In der ► Abbildung 16.9 ist die Überlagerung der beiden Wellen zu verschiedenen Zeitpunkten gezeigt.

Eine von rechts einfallende Welle ξ_1 wird an der linken Wand bei $x = 0$ reflektiert. Die zurücklaufende Welle nennen wir ξ_2. Die Beschreibung der nach links und rechts laufenden Wellen lautet nach den Gleichungen (16.2), (16.3) und (16.29)

$$\xi_1 = \xi_\circ \sin(kx + \omega t) \tag{16.52}$$

$$\xi_2 = \xi_\circ \sin(kx - \omega t + \varphi) \,. \tag{16.53}$$

Der Term φ berücksichtigt einen eventuellen Phasensprung bei der Reflexion der Welle.

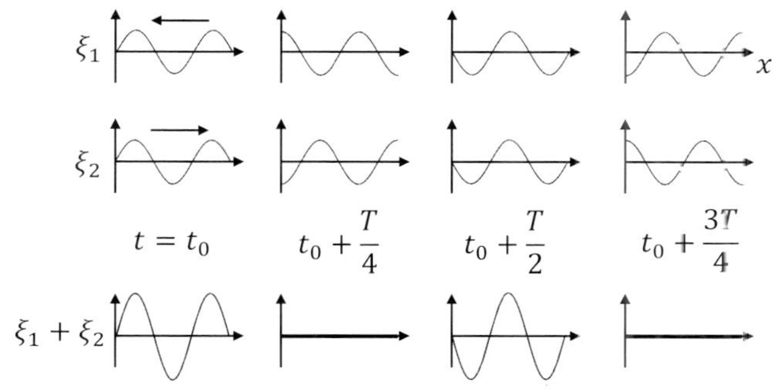

Abbildung 16.9 Erzeugung einer stehenden Welle durch Reflexion an einem festen Ende. Die einlaufende (ξ_1) und rücklaufende (ξ_2) Welle überlagern sich zur stehenden Welle $\xi_1 + \xi_2$.

Mit dem Additionstheorem (15.128) ergibt sich als Überlagerung beider Wellen

$$\begin{aligned} \xi(x,t) &= \xi_1 + \xi_2 && (16.54)\\ &= \xi_\circ \left[\sin(kx+\omega t) + \sin(kx - \omega t + \varphi)\right] && (16.55)\\ &= 2\xi_\circ \cdot \sin\left(\frac{kx+\omega t + kx - \omega t + \varphi}{2}\right) \cdot \cos\left(\frac{kx + \omega t - (kx - \omega t + \varphi)}{2}\right) && (16.56)\\ &= \underbrace{2\xi_\circ \sin\left(kx + \frac{\varphi}{2}\right)}_{\text{Amplitude}} \cdot \underbrace{\cos\left(\omega t - \frac{\varphi}{2}\right)}_{\text{Schwingung}} . && (16.57) \end{aligned}$$

Das Ergebnis $\xi(x,t)$ der Überlagerung beider Wellen ist eine sogenannte stehende Welle, die wir als Produkt aus einer harmonischen Schwingung und einer ortsabhängigen Amplitude verstehen können.

Interessant sind Stellen im Raum, an denen die Amplitude in (16.57) verschwindet:

$$\sin\left(kx + \frac{\varphi}{2}\right) = 0 \qquad (16.58)$$

Die Orte dieser sogenannten **Schwingungsknoten** erhalten wir aus dem Argument der Sinusfunktion. Wir bringen sie in einen intuitiven Zusammenhang mit der Wellenlänge λ (16.14):

$$kx + \frac{\varphi}{2} = n\pi\,, \quad n = 0, 1, 2, \ldots \qquad (16.59)$$

$$x = \frac{1}{k}\left[n\pi - \frac{\varphi}{2}\right] = \lambda\left[\frac{n}{2} - \frac{\varphi}{4\pi}\right] \underbrace{=}_{\varphi=0} n\,\frac{\lambda}{2} \qquad (16.60)$$

Ohne Phasensprung (d. h. $\varphi = 0$) gibt es Schwingungsknoten z. B. für $n = 0, 1, 2$ an Orten im Abstand der halben Wellenlänge: $x = 0, \lambda/2, \lambda$

Analog dazu erhalten wir **Schwingungsbäuche** bei Maxima im Betrag der Amplitude (16.57):

$$\left|\sin\left(kx + \frac{\varphi}{2}\right)\right| = 1 \tag{16.61}$$

$$kx + \frac{\varphi}{2} = \frac{2n+1}{2}\pi\,, \quad n = 0, 1, 2, \ldots \tag{16.62}$$

$$x = \frac{1}{k}\left[\frac{2n+1}{2}\pi - \frac{\varphi}{2}\right] = \lambda\left[\frac{2n+1}{4} - \frac{\varphi}{4\pi}\right] \underbrace{=}_{\varphi=0} \frac{2n+1}{2}\frac{\lambda}{2} \tag{16.63}$$

Hier gibt es für $\varphi = 0$ Schwingungsbäuche z. B. für $n = 0, 1, 2$ an den Orten $x = \lambda/4, 3\lambda/4, 5\lambda/4$. Sie sind gegenüber den Schwingungsknoten (16.60) um eine Viertelwellenlänge verschoben.

16.2.1 Eigenschwingungen

Bislang haben wir den Ausbreitungsbereich der Welle an nur einer Stelle begrenzt und dort die Welle reflektieren lassen (► Abbildung 16.9). Jetzt lassen wir als Schwingungsbereich einen zweiseitig begrenzten Ortsbereich zu (z. B. ► Abbildung 16.10). Durch die Randbedingungen auf beiden Seiten sind dann nicht mehr alle Wellenlängen λ möglich. Die Enden können im folgenden Sinn geschlossen oder offen sein:

- Bei festen Enden ergeben sich Schwingungsknoten wie z. B. bei fest eingespannten Saiten (► Abbildung 16.10) oder dem geschlossenen (»gedackten«) Ende einer Flöte. Dabei entsteht kein Phasensprung ($\varphi = 0$).
- An einem offenen oder losen Ende ergeben sich Schwingungsbäuche. Beispiele sind das freie Pendel an der Endposition der Wellenmaschine (Experiment 38) oder das offene Ende einer Flöte. Hier gibt es einen Phasensprung um $\varphi = \pi/2$.

Wir betrachten den Fall einer schwingenden Saite in ► Abbildung 16.10 mit den Randbedingungen, dass die Enden fest eingespannt sind. Daher verschwinden an den Enden der Saite mit Länge L die Auslenkungen für alle Zeiten t:

$$\xi(0, t) = 0 \quad \text{und} \quad \xi(L, t) = 0 \tag{16.64}$$

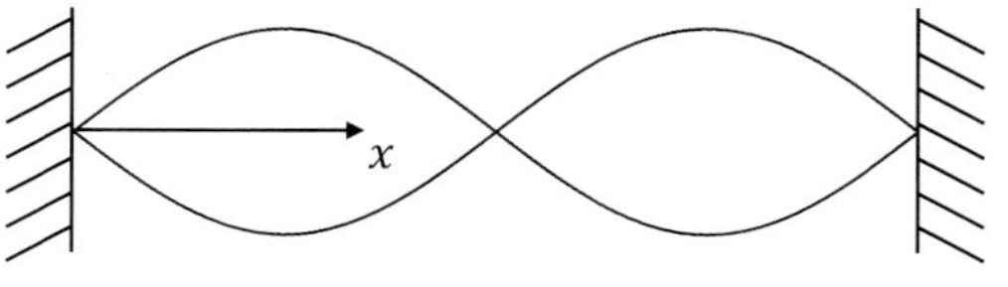

Abbildung 16.10 Erzeugung einer Eigenschwingung durch Reflexion an zwei festen Enden

Schwingungsknoten befinden sich im Abstand halber Wellenlängen λ voneinander (16.60). Daher muss auf die Saitenlänge L eine ganze Zahl $n \in \mathbb{N}$ von Schwingungsknoten passen:

$$n \cdot \frac{\lambda_n}{2} = L\,, \quad n = 1, 2, 3 \ldots \tag{16.65}$$

Die möglichen Wellenlängen unterscheiden wir hier durch den Index n. Mit steigender Anzahl n der Knoten werden passende Wellenlängen λ_n kürzer:

$$\lambda_n = \frac{2\,L}{n} \tag{16.66}$$

Die Schwingungen mit diesen Wellenlängen bezeichnen wir als **Eigenschwingungen** der Saite.

Mit welchen zugehörigen Frequenzen ν_n die Saite schwingt, erhalten wir aus der Phasengeschwindigkeit v der Welle entlang der Saite zusammen mit $v = \lambda\,\nu$ (16.28). Für die Saitenlänge L ergeben sich die Eigenschwingungen der Saite mit den **Eigenfrequenzen**

$$\nu_n = v \cdot \frac{n}{2\,L}\,, \quad n = 1, 2, 3 \ldots \tag{16.67}$$

Für $n = 1$ erhalten wir die Grundschwingung (Grundton), $n > 1$ sind die Oberschwingungen (Obertöne). In Abschnitt 16.8 werden wir auch die Phasengeschwindigkeit v der Welle in Abhängigkeit der Spannkraft F der Saite und ihrer Massenbelegung μ berechnen (16.165).

Experiment 40: *phyphox* Audio Spektrum

Zur Durchführung dieses Experiments wird das Mikrofon des Smartphones benutzt.

1 Suchen Sie in der *phyphox*-App das Experiment »Audio Spektrum«. Berühren des Drei-Punkte-Symbols rechts oben zeigt ein Menü mit Informationen zum Experiment (»Experiment Info«). Wählen Sie das Menü »Spektrum« zur Messung der Spektralverteilung.

2 Starten Sie die Aufzeichnung und erzeugen Sie sofort einen Klang oder ein Geräusch. Stoppen Sie die Aufzeichnung.

3 Das Frequenzspektrum sowie die Frequenzspitze mit der zugehörigen Note und die Abweichung vom Normwert dieser Note in Cent (100 Cent = 1 gleichstufiger Halbton) werden angezeigt. Häufig sind auch noch die Oberschwingungen zu erkennen.

4 Unter »Rohdaten« ist das Schwingungsbild und unter »Spektrum« das zugehörige Frequenzspektrum zu sehen (► Abbildung 16.11).

5 Die Genauigkeit einer Frequenzmessung $\Delta\nu$ steigt mit der Messzeit Δt (vergleiche mit (15.177)). Details in der Anwendung werden in der Nachrichtentechnik gelehrt.

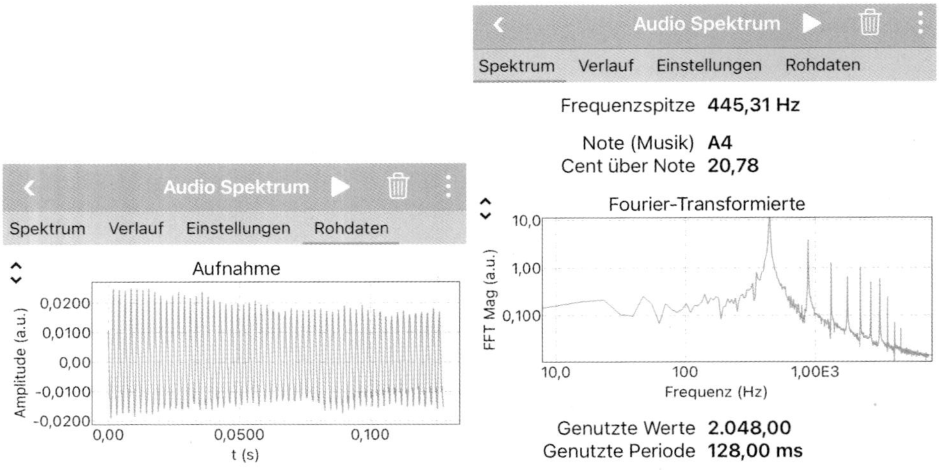

Abbildung 16.11 Der zeitliche Ablauf der Schwingungen (links) und das zugehörige Frequenzspektrum für den Kammerton (rechts) an einem Klavier. Die Oberschwingungen sind sehr gut zu erkennen.

16.2.2 Schwingende Platten

Auch in zwei Dimensionen gibt es das Phänomen stehender Wellen und Eigenschwingungen.

Experiment 41: Klangplatten

Wir streuen Sand auf eine Metallplatte und regen diese mit einem Violinbogen oder einem Lautsprecher zu Schwingungen an (► Abbildung 16.12). Die Platte bewegt sich nicht als Ganzes, sondern einzelne Flächenbereiche bewegen sich gegeneinander. Der Sand sammelt sich in den Knotenlinien, da sich die Platte dort nicht bewegt.

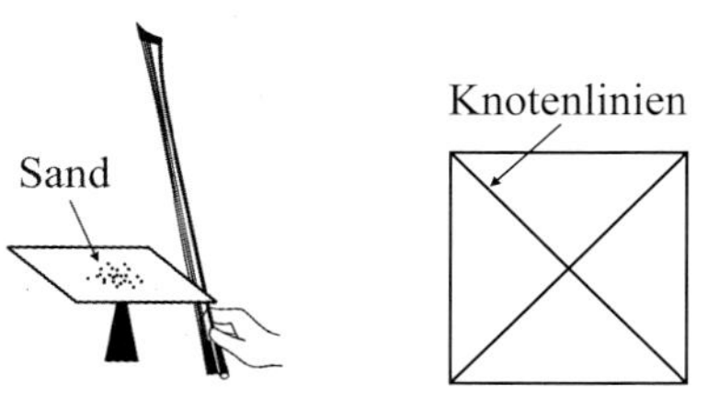

Abbildung 16.12 Eine mit Sand bestreute Klangplatte wird mit einem Violinbogen angeregt (links). Die Eigenschwingungen der Platte erzeugen Knotenlinien, an denen sich der Sand sammelt (rechts).

16.3 Dispersion

Wir untersuchen zwei ebene harmonische Wellen (16.5) gleicher Amplitude $\xi_\circ$, die wir hier mithilfe der Kosinusfunktion beschreiben. Ihre Kreisfrequenzen $\omega_1 \approx \omega_2$ sind ungefähr gleich groß und ihre Wellenzahlen $k_1 \approx k_2$ unterscheiden sich nur gering:

$$\xi_1 = \xi_\circ \cos(k_1 x - \omega_1 t) \tag{16.68}$$

$$\xi_2 = \xi_\circ \cos(k_2 x - \omega_2 t) \tag{16.69}$$

Ähnlich wie bei der Überlagerung von Schwingungen (Abschnitt 15 4.2) verwenden wir für die Überlagerung der beiden Wellen das Additionstheorem (15.126):

$$\xi(x,t) = \xi_1 + \xi_2 \tag{16.70}$$

$$= \xi_\circ \left[\cos(k_1 x - \omega_1 t) + \cos(k_2 x - \omega_2 t)\right] \tag{16.71}$$

$$= 2\xi_\circ \cdot \cos\left(\frac{(k_1 x - \omega_1 t) + (k_2 x - \omega_2 t)}{2}\right) \cdot \cos\left(\frac{(k_1 x - \omega_1 t) - (k_2 x - \omega_2 t)}{2}\right) \tag{16.72}$$

$$= 2\xi_\circ \cdot \cos\left(\frac{k_1 + k_2}{2}x - \frac{\omega_1 + \omega_2}{2}t\right) \cdot \cos\left(\frac{k_1 - k_2}{2}x - \frac{\omega_1 - \omega_2}{2}t\right) \tag{16.73}$$

Definieren wir Mittelwerte und halbe Differenzen für die beiden Kreisfrequenzen und Wellenzahlen

$$\omega \equiv \frac{\omega_1 + \omega_2}{2} \qquad \Delta\omega \equiv \frac{\omega_1 - \omega_2}{2} \tag{16.74}$$

$$k \equiv \frac{k_1 + k_2}{2} \qquad \Delta k \equiv \frac{k_1 - k_2}{2}\,, \tag{16.75}$$

so können wir die Überlagerung der Wellen folgendermaßen schreiben:

$$\xi = 2 \cdot \underbrace{\cos(\Delta k\, x - \Delta\omega\, t)}_{\text{Amplitudenmodulation}} \cdot \underbrace{\xi_\circ \cos(kx - \omega t)}_{\approx \xi_1 \approx \xi_2} \tag{16.76}$$

Die Struktur dieser Gleichung erinnert an die Schwebung bei Schwingungen (15.151), bei der die Argumente in der Kosinusfunktion zeitabhängig sind. Allerdings sind in (16.76) beide Argumente der Kosinusfunktionen orts- *und* zeitabhängig sodass die Schwingungszustände nicht lokal verbleiben, sondern sich als harmonische, überlagerte Wellen ausbreiten.

16.3.1 Phasengeschwindigkeit, Gruppengeschwindigkeit

Der rechte Term in (16.76) beschreibt eine harmonische Welle mit der Wellenzahl k und Kreisfrequenz ω, die ungefähr den ursprünglichen Wellen ξ_1 und ξ_2 entspricht. Die zugehörige Phasengeschwindigkeit (16.27) dieser Welle notieren wir hier mit dem Index p:

$$v_p = \frac{\omega}{k} \tag{16.77}$$

Der mittlere Term moduliert die Amplitude und zeigt ebenfalls die Struktur einer harmonischen Welle. Genauso wie wir die Phasengeschwindigkeit $v_p = \omega/k$ aus dem rechten Term in Gleichung (16.76) sofort ablesen können, können wir die Geschwindigkeit der Modulation im mittleren Term ablesen:

$$v_g = \frac{\Delta\omega}{\Delta k} = \frac{\omega_1 - \omega_2}{k_1 - k_2} \tag{16.78}$$

Die Modulation breitet sich demnach mit einer anderen Geschwindigkeit als v_p (16.77) aus. Im Allgemeinen schreiben wir diese sogenannte **Gruppengeschwindigkeit** als Ableitung:

$$v_g = \frac{d\omega}{dk} \tag{16.79}$$

In der ► Abbildung 16.13 sind die beiden unterschiedlichen Geschwindigkeiten v_p und v_g durch die zeitliche Entwicklung der beiden periodischen Strukturen visualisiert. Die Gruppengeschwindigkeit v_g ist hier geringer als die Phasengeschwindigkeit v_p.

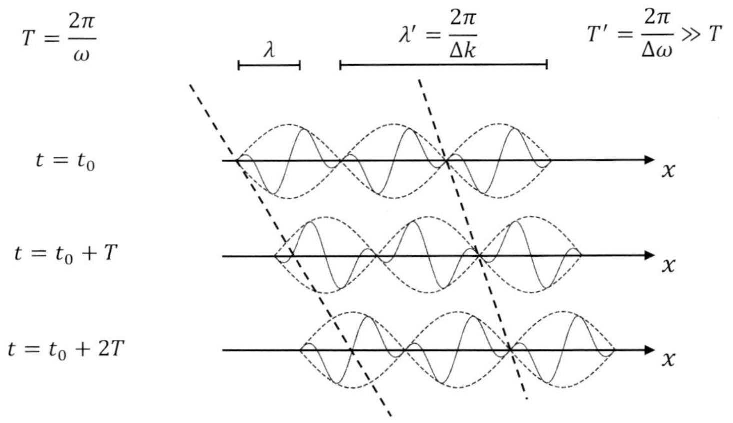

Abbildung 16.13 Die Gruppengeschwindigkeit v_g der Amplitudenmodulation mit der Wellenlänge λ' (rechte gestrichelte Linie) ist langsamer als die Phasengeschwindigkeit v_p der Schwingung mit der Wellenlänge λ (linke gestrichelte Linie). Die Abbildung zeigt den Fall $\lambda' \approx 4\,\lambda$, $T' \approx 8\,T \;\Rightarrow\; v_p \approx 2v_g$.

Die Gruppengeschwindigkeit spielt bei der Signalübertragung eine wichtige Rolle. In den meisten Fällen ist sie die Geschwindigkeit, mit der Signale übertragen werden.

Mit den Gleichungen (16.77) und (16.79) erhalten wir folgenden Zusammenhang zwischen der Phasen- und der Gruppengeschwindigkeit:

$$v_g = \frac{d}{dk}\,\omega \tag{16.80}$$

$$= \frac{d}{dk}\,(k\,v_p) \tag{16.81}$$

$$= v_p + k\frac{dv_p}{dk} \tag{16.82}$$

Alternativ wollen wir die Gruppengeschwindigkeit nicht nur in Abhängigkeit der Wellenzahl k ausdrücken, sondern auch direkt durch die Wellenlänge λ. Wir berechnen zunächst den Zusammenhang zwischen den Ableitungen:

$$\frac{dk}{d\lambda} = \frac{d}{d\lambda}\left(\frac{2\pi}{\lambda}\right) = -\frac{2\pi}{\lambda^2} \tag{16.83}$$

$$\frac{d}{dk} = -\frac{\lambda^2}{2\pi}\frac{d}{d\lambda} \tag{16.84}$$

Die Gruppengeschwindigkeit (16.82) ist damit

$$v_g = v_p + \frac{2\pi}{\lambda}\left(-\frac{\lambda^2}{2\pi}\right)\frac{dv_p}{d\lambda} \tag{16.85}$$

$$v_g = v_p - \lambda\frac{dv_p}{d\lambda}\,. \tag{16.86}$$

Falls die Phasengeschwindigkeit $v_p = v_p\,(\lambda)$ von der Wellenlänge λ abhängt, ist im zweiten Term die Ableitung nicht null:

$$\frac{dv_p}{d\lambda} \neq 0 \Rightarrow v_g \neq v_p\,. \tag{16.87}$$

Dann unterscheiden sich Phasen- und Gruppengeschwindigkeit. Bei einer Abhängigkeit der Phasengeschwindigkeit von der Wellenlänge sprechen wir von **Dispersion**. Tabelle 16.1 zeigt die drei Möglichkeiten für Dispersion und die zugehörigen Phasen- und Gruppengeschwindigkeiten.

Normale Dispersion	$dv_p/d\lambda > 0$ ist der häufigste Fall mit	$v_g < v_p$.
Keine Dispersion	$dv_p/d\lambda = 0$ erhält die Signalform bei	$v_g = v_p$.
Anomale Dispersion	$dv_p/d\lambda < 0$ ist eher selten mit	$v_g > v_p$.

Tabelle 16.1 Zusammenhang der Dispersion mit der Phasen- und Gruppengeschwindigkeit

16.3.2 Dispersion bei Wasserwellen

Wasserwellen sind hervorragend geeignet, die unterschiedlichen Arten von Dispersion zu untersuchen. Wasserwellen haben die Phasengeschwindigkeit

$$v_p = \sqrt{\left(\frac{g\lambda}{2\pi} + \frac{2\pi\sigma}{\rho\lambda}\right)\tanh\left(\frac{2\pi h}{\lambda}\right)}\,. \tag{16.88}$$

Hier sind die physikalischen Größen die Erdbeschleunigung g, die Dichte von Wasser $\rho \approx 1\,\mathrm{g/cm^3}$, die Flüssigkeitstiefe h und die Oberflächenspannung σ. Oberflächenspannung von Wasser entsteht an der Grenzfläche zu Luft durch die zur Luft hin fehlende Anziehung zwischen den Wassermolekülen. Sie beschreibt die erforderliche Energie pro Fläche, um die Wasseroberfläche zu vergrößern und beträgt $\sigma = 72\,\mathrm{mN/m}$ bei der Temperatur $20°\mathrm{C}$.

Die Funktion Tangens hyperbolicus tanh ist in (15.62) definiert. Die recht aufwendige Herleitung von (16.88) ist z. B. in [1] zu finden.

Durch die Abhängigkeit der Phasengeschwindigkeit von der Wellenlänge λ unterliegen die Wasserwellen Dispersionseffekten. Die drei Terme der Gleichung (16.88) stehen für verschiedene Ausprägungen von Wasserwellen, je nachdem welcher Term dominiert.

Rippelwellen

Wir wollen als Erstes die Näherung untersuchen, in der der zweite Term von (16.88) überwiegt. Sie werden **Rippelwellen** oder auch Kräuselwellen oder Kapillarwellen genannt. Hier dominiert der Term mit der Oberflächenspannung $\sigma \approx 72\,\mathrm{m\,N/m}$ gegenüber dem Term mit der Erdbeschleunigung g:

$$\frac{g\lambda}{2\pi} \ll \frac{2\pi\sigma}{\rho\lambda} \tag{16.89}$$

$$\lambda \ll \sqrt{\frac{4\pi^2\sigma}{\rho g}} \approx 5\,\mathrm{cm} \tag{16.90}$$

Bei dieser kleinen Wellenlänge λ ist eine Vergleichsbedingung $\lambda \ll h$ zur Wassertiefe h leicht zu erfüllen. Dann ist der dritte Term in (16.88) $\tanh(2\pi h/\lambda) \approx 1$. Die Phasengeschwindigkeit von Rippelwellen ergibt sich damit zu

$$v_p = \sqrt{\frac{2\pi\sigma}{\rho\lambda}} = \sqrt{\frac{2\pi\sigma}{\rho}}\,\lambda^{-\frac{1}{2}}\,. \tag{16.91}$$

Für die Dispersion leiten wir entsprechend (16.86)

$$\frac{dv_p}{d\lambda} = \sqrt{\frac{2\pi\sigma}{\rho}}\left(-\frac{1}{2}\lambda^{-\frac{3}{2}}\right) = -\sqrt{\frac{\pi\sigma}{2\rho\lambda^3}} \tag{16.92}$$

ab, sodass sich für die Gruppengeschwindigkeit

$$v_g = v_p + \sqrt{\frac{\pi\sigma}{2\rho\lambda}} \tag{16.93}$$

ergibt. Es liegt also **anomale Dispersion** mit $v_g > v_p$ vor.

Schwerewellen: Tiefwasserwellen

Umgekehrt wollen wir jetzt sogenannte **Schwerewellen** mit relativ großer Wellenlänge $\lambda \gg 5\,\mathrm{cm}$ (16.90) untersuchen, bei denen der zweite Term mit der Oberflächenspannung in (16.88) vernachlässigt werden kann. Zunächst verwenden wir als Näherung $h > \lambda/(2\pi)$ für die sogenannten **Tiefwasserwellen**. Bei ihnen nähert sich der dritte Term $\tanh(2\pi h/\lambda)$ erneut dem Wert 1, sodass in der Phasengeschwindigkeit (16.88) nur der erste Term relevant ist:

$$v_p = \sqrt{\frac{g\lambda}{2\pi}} = \sqrt{\frac{g}{2\pi}}\,\lambda^{\frac{1}{2}} \tag{16.94}$$

Die Phasengeschwindigkeit ist damit von der Wassertiefe h unabhängig. Aus der Wellenlängenabhängigkeit ergibt sich die Dispersion

$$\frac{dv_p}{d\lambda} = \sqrt{\frac{g}{2\pi}}\frac{1}{2}\lambda^{-\frac{1}{2}} = \frac{1}{2}\sqrt{\frac{g}{2\pi\lambda}} \tag{16.95}$$

und die Gruppengeschwindigkeit

$$v_g = v_p - \lambda\frac{dv_p}{d\lambda} = \sqrt{\frac{g\lambda}{2\pi}} - \frac{1}{2}\sqrt{\frac{g\lambda}{2\pi}} = \frac{v_p}{2}\,. \tag{16.96}$$

Es liegt also **Normale Dispersion** mit $v_g < v_p$ vor.

Schwerewellen: Flachwasserwellen

Erneut wollen wir Schwerewellen mit relativ großer Wellenlänge $\lambda \gg 5\,\text{cm}$ (16.90) betrachten, bei denen der Term mit der Oberflächenspannung fortfällt. Hier untersuchen wir im Gegensatz zu den Tiefwasserwellen die Näherung kleiner Wassertiefen $h \ll \lambda$.

Bei diesen sogenannten **Flachwasserwellen** ist der dritte Term in (16.88) entscheidend. Das Argument von $\tanh(2\pi h/\lambda)$ ist klein, sodass wir die Näherung $\tanh x \approx x$ anwenden können. Damit erhalten wir eine Phasengeschwindigkeit, die von der Wellenlänge λ unabhängig ist:

$$v_p \approx \sqrt{\frac{g\lambda}{2\pi}\tanh\left(\frac{2\pi h}{\lambda}\right)} \approx \sqrt{\frac{g\lambda}{2\pi}\frac{2\pi h}{\lambda}} = \sqrt{g\,h} \tag{16.97}$$

Es liegt also **keine Dispersion** mit $v_g = v_p$ vor.

Obwohl bei den Flachwasserwellen (der Name ist hier irreführend, es kommt nur auf $h \ll \lambda$ an) keine Dispersion bezüglich λ vorliegt, führt die h-Abhängigkeit zu einem dispersionsartigen Phänomen, das sich im **Tsunami** (Japanisch für »Hafenwelle«) in katastrophaler Weise auswirkt. Ursache ist ein Seebeben in großer Wassertiefe, welches eine sehr großflächige Störung (einige 100 km) auslöst, die sich als Wellenberg mit hoher Geschwindigkeit $v_g = v_p \approx 1000\,\text{km/h}$ zunächst ohne nennenswerte Dispersion ausbreitet. In Festlandnähe wirkt sich die abnehmende Wassertiefe h so aus, dass wegen (16.97) der vordere Teil des Wellenbergs schneller Geschwindigkeit verliert als der hintere und die Welle sich auftürmt. Der Effekt wird jetzt durch die Dispersion (16.95) noch verstärkt.

16.4 Interferenz

Zwei punktförmige Quellen S_1 und S_2 erzeugen kreisförmige Wellen auf einer Oberfläche, z. B. Wasserwellen. Im Folgenden wollen wir die Überlagerung der Wellen an einem Beobachtungspunkt P berechnen (► Abbildung 16.14).

Die Auslenkungen $\xi_i(r_i, t)$ der beiden Wellen ($i = 1, 2$) im jeweiligen Abstand $r_i = P$ zur Zeit t beschreiben wir mit der Projektion von Kugelwellen (16.50) auf die zweidimensionale Oberfläche:

$$\xi_i(r_i, t) = \frac{\xi_0}{r_i}e^{i(kr_i - \omega t)} \tag{16.98}$$

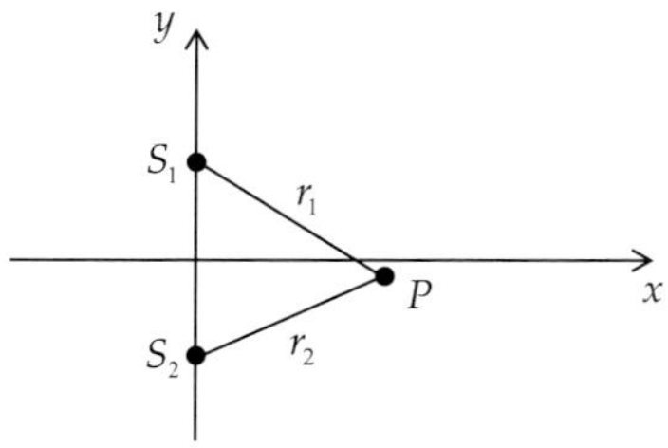

Abbildung 16.14 Zwei kreisförmige Wellen ξ_1 und ξ_2 werden von synchron schwingenden Quellen S_1 und S_2 gesendet und an einem Punkt P beobachtet.

Beide Quellen sollen synchron mit gleicher Amplitude $\xi_\circ$ und Kreisfrequenz ω schwingen. Synchron heißt, an beiden Quellorten gibt es zwischen den Auslenkungen keinen Phasenunterschied. Wir setzen $\varphi = 0$, sodass Phasenterme $e^{i\varphi} = e^0 = 1$ in den ursprünglich komplexen Amplituden $\xi_\circ$ verschwinden und die Amplituden reell sind. Die Verkleinerung der Amplituden mit dem Kehrwert $1/r_i$ des Abstands von der Quelle ist für die folgende Diskussion nicht entscheidend, wir erfassen sie in der Amplitudendefinition $A_i \equiv \xi_\circ / r_i$.

Die kreisförmigen Wellen der beiden synchron schwingenden Quellen S_1 und S_2 in ► Abbildung 16.14 lauten damit

$$\xi_1(r_1, t) = A_1\, e^{i(kr_1 - \omega t)}\,, \tag{16.99}$$

$$\xi_2(r_2, t) = A_2\, e^{i(kr_2 - \omega t)}\,. \tag{16.100}$$

16.4.1 Überlagerung von Wellen synchron schwingender Quellen

Den Effekt der Überlagerung beider Wellen können wir in einer Gesamtamplitude erfassen, während einer der beiden Terme $\exp\left(i(kr_1 - \omega t)\right)$ zur Beschreibung der Wellenbewegung genügt:

$$\xi(r_1, r_2, t) = \xi_1(r_1, t) + \xi_2(r_2, t) \tag{16.101}$$

$$= A_1\, e^{i(kr_1 - \omega t)} + A_2\, e^{i(kr_2 - \omega t)} \tag{16.102}$$

$$= \left(A_1 + A_2\, e^{ik(r_2 - r_1)}\right) e^{i(kr_1 - \omega t)} \tag{16.103}$$

$$= \left(A_1 + A_2\, e^{i\delta}\right) e^{i(kr_1 - \omega t)} \tag{16.104}$$

In der Amplitude bezeichnet der Term

$$\delta \equiv k\,(r_2 - r_1) \tag{16.105}$$

die Phasenverschiebung zwischen den beiden Wellen am Punkt P, der allein durch die verschiedenen Abstände der beiden Quellen zustande kommt.

In Abschnitt 15.1 hatten wir die komplexe Lösung des harmonischen Oszillators als Rotation eines Zeigers in der Ebene der komplexen Zahlen verstanden. Unsere beiden Wellen ξ_1 und ξ_2 erzeugen am Beobachtungspunkt P Schwingungen, die der Rotation von zwei Zeigern gegen den Uhrzeigersinn entsprechen. Die Zeiger sind um den Phasenwinkel δ relativ zueinander verschoben (► Abbildung 16.15).

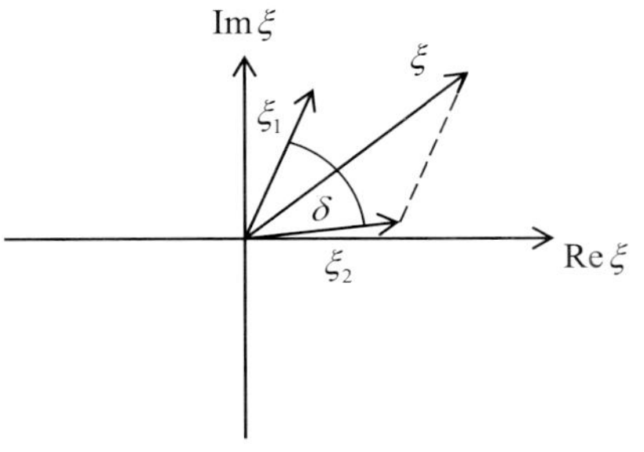

Abbildung 16.15 Komplexes Zeigerdiagramm der Wellen ξ_1 und ξ_2 und ihrer Summe ξ. Die drei Zeiger rotieren synchron.

16.4.2 Interferenzeffekte

Für die Überlagerung der beiden Wellen an einem Punkt P ergeben sich zwei Extremfälle für die resultierende Amplitude (16.104). Für $\delta = 0$ ist $e^0 = 1$ und die resultierende Amplitude $A_1 + A_2$ (▶ Abbildung 16.16). Für $\delta = \pi$ ist $e^{i\pi} = \cos\pi + i\sin\pi = -1$ (15.6), sodass die Gesamtamplitude minimal wird und für gleiche bzw. sehr große Abstände von den Quellen verschwindet.

Verstärkungs- und Auslöschungseffekte bei Überlagungerungen (Superposition) von Wellen bezeichnen wir als **Interferenz**.

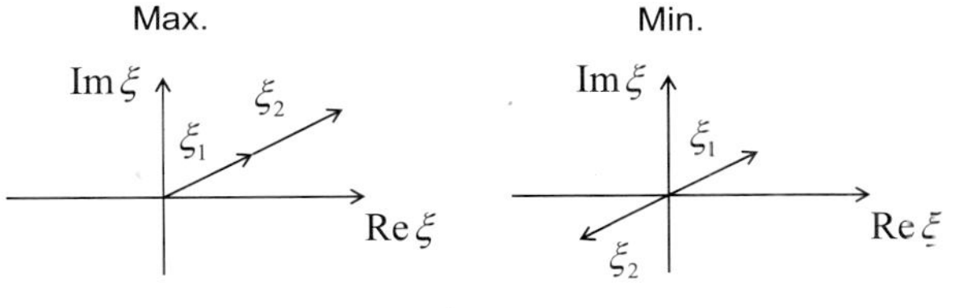

Abbildung 16.16 Die Extremfälle der Gesamtamplitude: $A_1 + A_2$ für gleichgerichtete ξ_1, ξ_2 (links) und 0 für $\xi_1 = -\xi_2$ (rechts).

Im Folgenden interessiert uns die Intensität I (16.47), die wir am Punkt P beobachten (▶ Abbildung 16.14). Wegen Gleichung $I \propto |\xi_\circ|^2$ (16.48) berechnen wir als relatives Maß für die Intensität das Quadrat der Amplitude der beiden überlagerten Wellen (16.104):

$$I \propto |\xi_\circ|^2 \tag{16.106}$$

$$= \left(A_1 + A_2\, e^{i\delta}\right)\left(A_1 + A_2\, e^{-i\delta}\right) \tag{16.107}$$

$$= (A_1 + A_2\,\cos\delta + i\,A_2\,\sin\delta)\,(A_1 + A_2\,\cos\delta - i\,A_2\,\sin\delta) \tag{16.108}$$

$$= (A_1 + A_2\,\cos\delta)^2 + (A_2\,\sin\delta)^2 \tag{16.109}$$

$$= A_1^2 + A_2^2\,\cos^2\delta + 2A_1A_2\,\cos\delta + A_2^2\,\sin^2\delta \tag{16.110}$$

$$= A_1^2 + A_2^2 + 2A_1A_2\cos\delta \tag{16.111}$$

$$\equiv A^2 \tag{16.112}$$

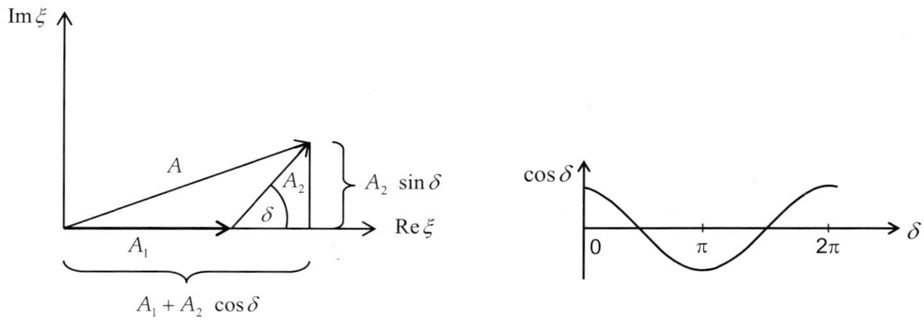

Abbildung 16.17 Links: Geometrische Konstruktion der Gesamtamplitude A aus den Teilamplituden A_1 und A_2 und der relativen Phase δ zwischen ξ_1 und ξ_2. Rechts: Die Funktion $\cos\delta$ für $0 < \delta < 2\pi$

In der linken ► Abbildung 16.17 ist diese Berechnung mit den Zeigern visualisiert.

Mit dem Verlauf der Kosinusfunktion (rechte ► Abbildung 16.17) treten damit Maxima und Minima für folgende Phasenverschiebungen δ (16.105) auf:

Maxima	Minima	
$\delta = 0, 2\pi, \cdots = n\,2\pi$	$\delta = \pi, 3\pi, \cdots = (2n+1)\,\pi$	(16.113)
$A = A_1 + A_2$	$A = A_1 - A_2$	(16.114)

Bei den Minima kann die Gesamtamplitude negativ werden. Im Experiment beobachten wir aber die Intensität I der Wellen. Sie wird nie negativ, da $I \propto A^2 \geq 0$ ist.

16.4.3 Interferenzmuster

Wir berechnen nun Intensitätsmuster der überlagerten Wellen auf der Ausbreitungsoberfläche. Die Muster ergeben Kurven, in denen die Welle (16.104) die gleiche Phase $\delta = const.$ hat. Mit Gleichung (16.105) erhalten wir Kurven, die wir zur besseren Vorstellung in Abhängigkeit der Wellenlänge $\lambda = 2\pi/k$ (16.14) ausdrücken:

$$\delta = k(r_1 - r_2) \tag{16.115}$$

$$= \frac{2\pi}{\lambda}(r_1 - r_2) \tag{16.116}$$

Kurven, auf denen die Intensität der überlagerten Wellen maximal ist, bezeichnen wir als Bauchlinien. Intensitätsmaxima treten wegen Gleichung (16.113) bei ganzzahligen Vielfachen der Wellenlänge auf:

$$\delta = n\,2\pi = \frac{2\pi}{\lambda}(r_1 - r_2) \tag{16.117}$$

$$r_1 - r_2 = n\lambda = const. \tag{16.118}$$

Minimale Intensität finden wir entlang von sogenannten Knotenlinien. Intensitätsminima liegen wegen Gleichung (16.113) genau zwischen den Maxima:

$$\delta = (2n+1)\pi = \frac{2\pi}{\lambda}(r_1 - r_2) \tag{16.119}$$

$$r_1 - r_2 = \frac{2n+1}{2}\lambda = const. \tag{16.120}$$

Beide Kurvenformen der Bauchlinien (16.118) und Knotenlinien (16.120) entsprechen Hyperbeln ((10.87), ► Abbildung 10.15).

Experiment 42: Interferenz mit Wasserwellen

Zwei kleine Tupfer, die in einer Glaswanne mit etwas Wasser montiert sind, produzieren synchron zwei Wasserwellen (► Abbildung 16.18). Das Interferenzmuster zeigt Hyperbeln der Knotenlinien oder Bauchlinien (Gleichungen (16.118) und (16.120)). Dasselbe Experiment gelingt ebenso beeindruckend beim Wurf mit zwei Steinen am Ententeich.

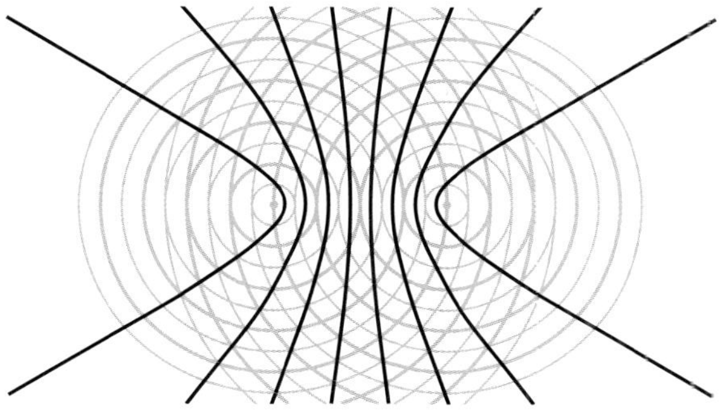

Abbildung 16.18 Das Interferenzmuster zweier synchroner Wasserwellen zeigt Hyperbeln als Bauch- oder Knotenlinien. Hier sind die meist besser sichtbaren Knotenlinien eingezeichnet.

16.5 Dopplereffekt

Wenn sich der Sender einer Welle und ihr Beobachter gegeneinander bewegen, so nimmt der Beobachter eine von der ursprünglichen Senderfrequenz ν verschiedene Frequenz ν_B wahr.

Wir betrachten eine punktförmige Quelle mit Kugelwellen der Wellenlänge λ, die sich mit der Phasengeschwindigkeit v ausbreiten. Wir unterscheiden im Folgenden zwei Fälle darin, ob sich der Beobachter oder die Quelle bewegt.

16.5.1 Bewegter Beobachter

Die Wellenfronten einer feststehenden Quelle breiten sich mit der Wellenlänge $\lambda = v/\nu$ (16.28) aus. Ein Beobachter bewege sich mit der Geschwindigkeit v_B direkt auf die Quelle zu und registriert die Wellenfronten mit der Relativgeschwindigkeit $v - v_B$. Folglich misst er die Frequenz

$$\nu_B = \frac{v - v_B}{\lambda} . \tag{16.121}$$

In ► Abbildung 16.19 ist die allgemeine Situation skizziert, in der die Quelle ruht und der Beobachter sich mit der Geschwindigkeit v_B unter dem Winkel ϑ_B zum Verbindungsvektor $\vec{r}$ von der Quelle zum Beobachter bewegt.

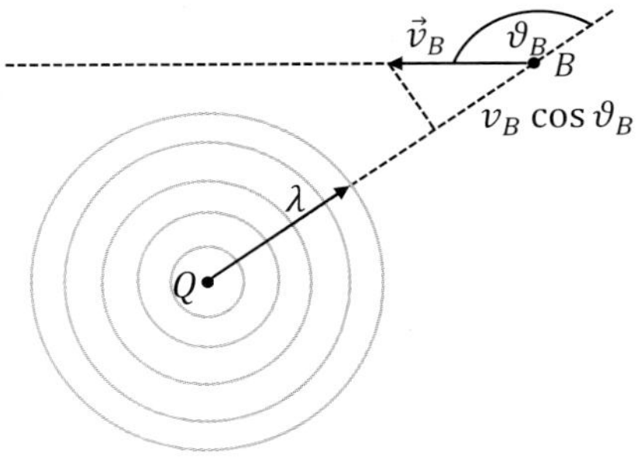

Abbildung 16.19 Ein Beobachter bewegt sich mit der Geschwindigkeit $\vec{v}_B$ relativ zu einer ruhenden Quelle. ϑ_B ist der (variable) Winkel zwischen $\vec{v}_B$ und dem Verbindungsvektor $\vec{r}$ Quelle-Beobachter. λ ist die Wellenlänge der emittierten Welle.

Relevant ist hier die Projektion der Geschwindigkeit v_B auf den Vektor $\vec{r}$. Mit der Definition des Winkels ϑ_B nach ► Abbildung 16.19 wird der Term $\cos\vartheta_B$ negativ. Die beobachtete Frequenz ν_B ist daher im Vergleich zur originalen Frequenz ν ($\lambda = v/\nu$ (16.28)) höher:

$$\nu_B = \frac{v - v_B \cos\vartheta_B}{\lambda} \tag{16.122}$$

$$= \nu \cdot \frac{v - v_B \cos\vartheta_B}{v} \tag{16.123}$$

Bewegt sich der Beobachter von der Quelle weg, ist $\cos\vartheta_B > 0$, was die im Beobachtersystem gemessene Frequenz erniedrigt.

16.5.2 Bewegte Quelle

Eine unbewegte Punktquelle sendet Kugelwellen aus, wie in ► Abbildung 16.19 visualisiert ist. Die Kreise zeigen Flächen gleicher Phase $\delta = const.$ der Welle und liegen im Abstand einer Wellenlänge λ voneinander. Jede Phasenfläche stammt von einem anderen Zeitpunkt $t = 0, T, 2T, \ldots$, wobei T die Schwingungsperiode ist.

Bewegt sich die Punktquelle, so sind die Phasenflächen aus den Zeiten $t = 0, T, 2T, \ldots$ nicht mehr zentriert. Abbildung 16.20 zeigt diese Situation für eine geringe Quellgeschwindigkeit v_Q im Vergleich zur Phasengeschwindigkeit $v > v_Q$ (den Fall $v \leq v_Q$ besprechen wir im folgenden Abschnitt 16.6).

Bewegt sich die Quelle auf den ruhenden Beobachter zu, so folgen die Phasenflächen schneller aufeinander. Die vom Beobachter gemessene Wellenlänge λ_B erscheint um $\Delta\lambda$ gegenüber der Wellenlänge λ der ruhenden Quelle verkürzt.

Zunächst betrachten wir den Fall, dass die Quelle direkt auf den Beobachter zukommt (Sonderfall $\vartheta_Q = 0$ in ► Abbildung 16.20). Die Abfolge zwischen zwei Phasenflächen verkürzt sich entsprechend der Strecke $\Delta\lambda = v_Q\, T$, die die Quelle in der Zeitdauer T einer Schwingungsperiode zurücklegt:

$$\lambda_B = \lambda - \Delta\lambda \tag{16.124}$$

$$= \lambda - v_Q\, T \tag{16.125}$$

Damit berechnen wir die beobachtete Frequenz unter Berücksichtigung von $v = \lambda\, \nu$ (16.28) und $\nu = 1/T$ (16.25):

$$\nu_B = \frac{v}{\lambda_B} \tag{16.126}$$

$$= \frac{v}{\lambda - v_Q\, T} \tag{16.127}$$

$$= \frac{v}{\frac{v}{\nu} - \frac{v_Q}{\nu}} \tag{16.128}$$

$$= \nu \cdot \frac{v}{v - v_Q} \tag{16.129}$$

Der allgemeine Fall für eine bewegte Quelle und einen Beobachter am festen Ort ist ebenfalls in ► Abbildung 16.20 dargestellt. Der Geschwindigkeitsvektor $\vec{v}_Q$ bildet den Winkel ϑ_Q mit der Verbindungslinie zum Beobachter. Für den Beobachter verkürzt sich die Wellenlänge λ um den Weg $\Delta\lambda = T v_Q \cos\vartheta_Q$, den die Quelle längs der Verbindungslinie Quelle-Beobachter in der Schwingungsperiode T zurücklegt. Beobachtet wird die Frequenz

$$\nu_B = \nu \cdot \frac{v}{v - v_Q \cos\vartheta_Q}\,. \tag{16.130}$$

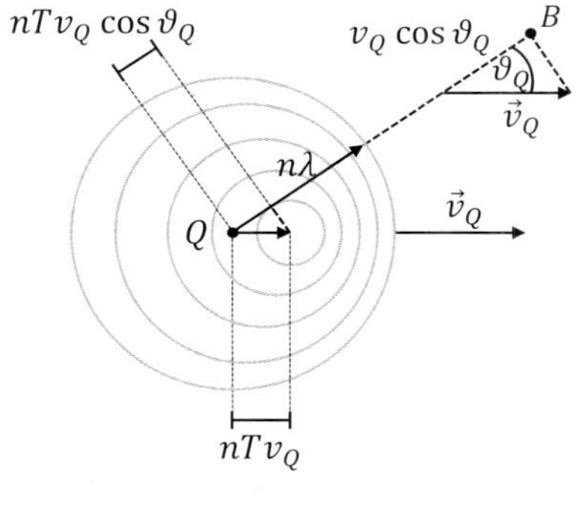

Abbildung 16.20 Die Quelle bewegt sich mit der Geschwindigkeit $\vec{v}_Q$ gegen den ruhenden Beobachter. Dabei ist ϑ_Q der Winkel zwischen $\vec{v}_Q$ und dem Verbindungsvektor $\vec{r}$ von der Quelle zum Beobachter.

16.5.3 Nicht-relativistischer Dopplereffekt

Die beiden Gleichungen (16.123) und (16.130) lassen sich zu einer Beschreibung für den allgemeinen Fall einer Bewegung von Quelle und Beobachter zum **nicht-relativistischen Dopplereffekt** zusammenfassen:

$$\nu_B = \nu \cdot \frac{v - v_Q \cos\vartheta_Q}{v - v_B \cos\vartheta_B} \qquad (16.131)$$

Wenn der Beobachter sich also aus großer Entfernung an der Quelle vorbei bewegt, wird sich die Frequenz mit dem Verlauf der Kosinus-Terme $\cos\vartheta_Q$ und $\cos\vartheta_B$ kontinuierlich verändern. Kommen Quelle und Beobachter zunächst einander entgegen und entfernen sich anschließend wieder voneinander, so ist die beobachtete Frequenz anfangs erhöht $\nu_B = \nu(v + v_Q)/(v - v_B) > \nu$, entspricht kurz der Originalfrequenz $\nu_B = \nu$ und sinkt danach ab $\nu_B = \nu(v - v_Q)(v + v_B) < \nu$. Die Situation ist geläufig, wenn man einem Krankenwagen oder Zug mit Signal begegnet.

Experiment 43: *phyphox* Dopplereffekt

Dieses Experiment benutzt das Mikrofon Ihres Smartphones. Außerdem braucht man eine Schallquelle konstanter Frequenz, z. B. ein zweites Smartphone.

1. Suchen Sie in der *phyphox*-App das Experiment »Dopplereffekt«. Das Drei-Punkte-Symbol rechts oben zeigt ein Menü mit Informationen zum Experiment (»Experiment Info«). Wählen Sie das Menü »Ergebnisse« zur Messung der Dopplerfrequenz und der resultierenden Geschwindigkeit (16.131).
2. Starten Sie auf dem zweiten Smartphone den »Tongenerator«. Stellen Sie unter »Einfach« die gleiche Frequenz wie auf der App »Dopplereffekt« ein (hier 1 000 Hz).
3. Befestigen Sie das Smartphone sicher an einem Band. Starten Sie den »Tongenerator«. Starten Sie die Messung »Dopplereffekt« und lassen Sie das Smartphone wie ein Pendel vor dem Tongenerator hin und her schwingen.

- Das Experiment ist nicht ganz einfach. Wiederholen Sie das Experiment gegebenenfalls für verschiedene »Einstellungen« der Frequenzspanne und des Zeitintervalls.
- Vermeiden Sie Nebengeräusche. Manchmal ist es einfacher, die Schallquelle statt des Empfängers schwingen zu lassen.
- Sie erkennen unter »Ergebnisse« eine Schwingung gemäß (16.131) mit einer Korrelation zwischen der Dopplerfrequenz und der sich daraus ergebenden Geschwindigkeit (► Abbildung 16.21).

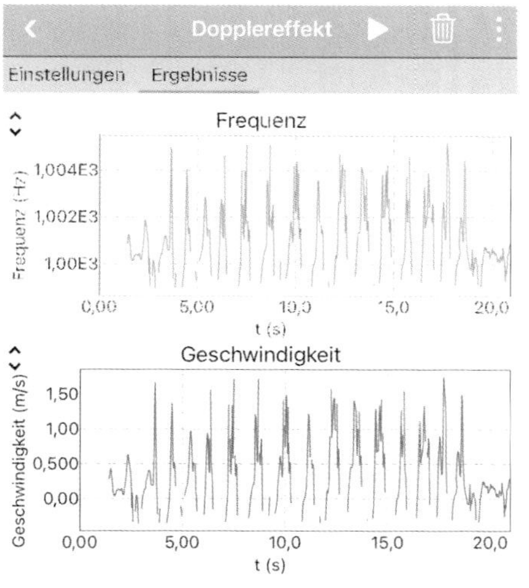

Abbildung 16.21 Experiment zum Dopplereffekt mit der *phyphox*-App: Messung der Dopplerfrequenzen für ein Smartphone in Ruhe, vor dem ein Tongenerator pendelt.

16.6 Machscher Kegel

Wir erweitern jetzt unsere Diskussion des Dopplereffekts bei bewegten Quellen auf Quellgeschwindigkeiten $v_Q \geq v$ oberhalb der Phasengeschwindigkeit v der Wellen.

Mit zunehmender Geschwindigkeit v_Q der Quelle rücken die in ► Abbildung 16.20 gezeigten Phasenflächen der Wellen immer dichter zusammen. Wenn die Quelle die Phasengeschwindigkeit $v_Q = v$ erreicht und alle Phasenflächen durch einen Punkt verlaufen, überlagern sich die Flächen und bilden einen Wellenberg.

Die Situation bei noch höheren Geschwindigkeiten $v_Q > v$ der Quelle ist in ► Abbildung 16.22 gezeigt. Mit den bisher ausgesandten Wellen entsteht eine neue gemeinsame Phasenfläche, die sich kegelförmig in Bewegungsrichtung der Quelle zuspitzt. Die kegelförmige Wellenfront wird als Machsche Welle oder **Machscher Kegel** bezeichnet.

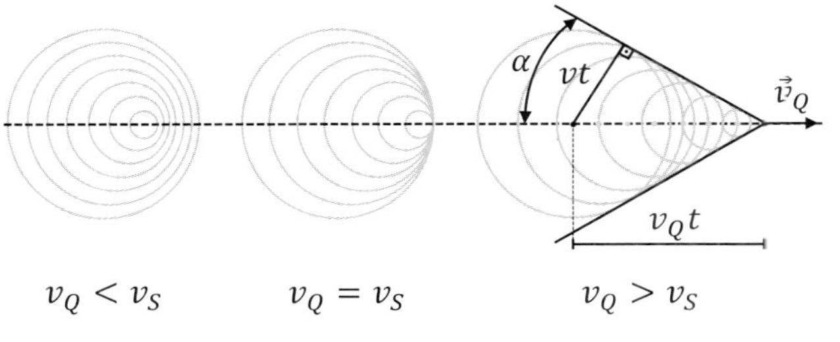

Abbildung 16.22 Eine mit der Geschwindigkeit $v_Q \geq v$ bewegte Quelle erzeugt einen Machschen Kegel mit halbem Öffnungswinkel α.

Die Öffnung des Machschen Kegels ist dabei durch den halben Öffnungswinkel α gegeben. Wir berechnen ihn aus dem Weg $v_Q\,t$ der Quelle und der Strecke $v\,t$, die die Welle in der Zeit t zurücklegt:

$$\sin\alpha = \frac{v\,t}{v_Q\,t} = \frac{v}{v_Q} \tag{16.132}$$

Als vergleichende Geschwindigkeitsangabe führt man die **Machsche Zahl Ma** ein:

$$\mathrm{Ma} = \frac{1}{\sin\alpha} = \frac{v_Q}{v}\,. \tag{16.133}$$

Im Fall der Schallgeschwindigkeit v durchstößt die Quelle bei $v_Q = v$ die »Schallmauer« und bewegt sich darüber mit Überschallgeschwindigkeit, z. B. Ma $= 2$ für zweifache Schallgeschwindigkeit. Da auf dem Kegelmantel der lokale Luftdruck der Luftmoleküle durch die Überlagerung der Phasenflächen sehr hoch ist, hört ein Beobachter, über den diese Druckfront hinwegzieht, einen explosionsartigen Knall. Dieser sogenannte Überschallknall tritt z. B. bei Geschossen und Überschallflugzeugen auf.

Bei Wasserwellen lässt sich ebenfalls eine Kopfwelle erzeugen, indem das Erregungszentrum mit $v_Q > v$ bewegt wird. Man spricht hier von einer **Bugwelle**. Diese bildet sich auch dann aus, wenn sich ein Hindernis an der Wasseroberfläche bewegt, z. B. bei einer Bugwelle an einem Schiff.

16.7 Relativistischer Dopplereffekt

Beim nicht-relativistischen Dopplereffekt (16.131) breitet sich die Welle in einem Medium aus, z. B. Schallwellen in Luft oder Wasserwellen in Wasser. Anders ist die Situation für Lichtwellen. Sie breiten sich im Vakuum mit konstanter Phasengeschwindigkeit c aus, ohne dass dafür ein Medium erforderlich ist. Allerdings gibt es auch hier Frequenzverschiebungen, wenn das Licht in verschiedenen Inertialsystemen beobachtet wird, die sich relativ zueinander mit der Geschwindigkeit v bewegen.

Wir konzentrieren uns jetzt auf Bewegungen der Inertialsysteme entlang der direkten Verbindungslinie zwischen Quelle und Beobachter. Wir messen Lichtwellen sowohl in einem Inertialsystem S als auch in einem zweiten System S', das sich mit der Geschwindigkeit v von S wegbewegt. Zum Zeitpunkt $t = t' = 0$ fallen die Ursprünge der beiden Systeme zusammen (► Abbildung 16.23).

Im System S wird am Ort $x = 0$ zu den Zeiten $t = 0, T, 2T, \ldots > 0$ jeweils ein Lichtsignal ausgesendet. Die Sendefrequenz beträgt demnach $\nu = 1/T$ (16.25). Der Zeitpunkt beim Aussenden des ersten Lichtsignals wird in beiden Systemen gleich bestimmt.

Das zweite Lichtsignal wird erst ausgesendet, wenn das System S' aus Sicht von S schon eine Entfernung $v\,T$ erreicht hat. Nach welcher Zeit t im System S erreicht das Signal S'? Wir nutzen dafür einerseits den Lichtkegel, um die Entfernung x von S' bei Ankunft des Lichtsignals zu berechnen (► Abbildung 16.23, [3]):

$$x = c\,(t - T) \tag{16.134}$$

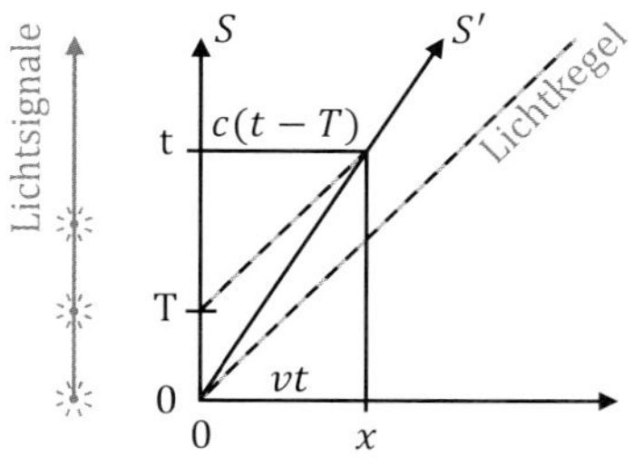

Abbildung 16.23 Raumzeit-Diagramm für das Inertialsystem S und Weltlinie des Inertialsystems S' zur Berechnung der Zeit t für die Ankunft eines bei $t = T$ gesendeten Lichtsignals in S'.

Andererseits können wir die Geschwindigkeit v von S' nutzen, um seine Entfernung x beim Eintreffen des Signals zu bestimmen:

$$x = v\,t \tag{16.135}$$

Durch Gleichsetzen der Entfernungen x erhalten wir die im System S ermittelte Zeit t, bei der das Lichtsignal vom Zeitpunkt T mit Verzögerung bei S' eintrifft. Wir verwenden hierbei $\beta = v/c$ (11.45), die Geschwindigkeit v des S'-Inertialsystems relativ zur Lichtgeschwindigkeit c:

$$c\,(t - T) = v\,t \tag{16.136}$$

$$t\,(c - v) = c\,T \tag{16.137}$$

$$t = \frac{T}{1 - \beta} \tag{16.138}$$

Die Zeit zwischen dem ersten und dem zweiten Lichtsignal ist demnach für das System S' um $1/(1 - \beta)$ länger als die Periode T der Lichtblitze.

Zusätzlich gehen die Uhren des bewegten Beobachters im System S' wegen der Zeitdilatation $t = \gamma\,t'$ (12.12) langsamer. Seine Zeitmessung für die Periode der Lichtblitze beträgt mit beiden Effekten zusammen

$$t' = \frac{t}{\gamma} \tag{16.139}$$

$$= t\,\sqrt{1 - \beta^2} \tag{16.140}$$

$$= \frac{T}{1 - \beta}\sqrt{(1 + \beta)(1 - \beta)} \tag{16.141}$$

$$= T\,\sqrt{\frac{1 + \beta}{1 - \beta}}\,. \tag{16.142}$$

Daher ist die Frequenzmessung im von der Lichtquelle wegbewegten System S' im Vergleich zur ausgesendeten Frequenz $\nu = 1/T$ zu kleineren Frequenzen hin verschoben:

$$\nu' = \nu\,\sqrt{\frac{1 - \beta}{1 + \beta}} \tag{16.143}$$

Diese Frequenzverschiebung hat in der Astronomie große Bedeutung und wird dort als »Rotverschiebung« bezeichnet. Bewegen sich die Systeme aufeinander zu, so gelten die umgekehrten Vorzeichen und die Frequenz erhöht sich:

$$\nu' = \nu \sqrt{\frac{1+\beta}{1-\beta}} \tag{16.144}$$

16.8 Wellengleichung

Bahnkurven von Körpern haben wir bereits über Newtons Bewegungsgleichungen (3.41) und deren Lösungen berechnet. Im Folgenden wollen wir eine Bewegungsgleichung auch für Wellen aufstellen. Wir betrachten dafür eine Welle ξ, die sich mit der Phasengeschwindigkeit v nach rechts bewegt (► Abbildung 16.24):

$$\xi\,(x,t) = f\,(x - vt) \tag{16.145}$$

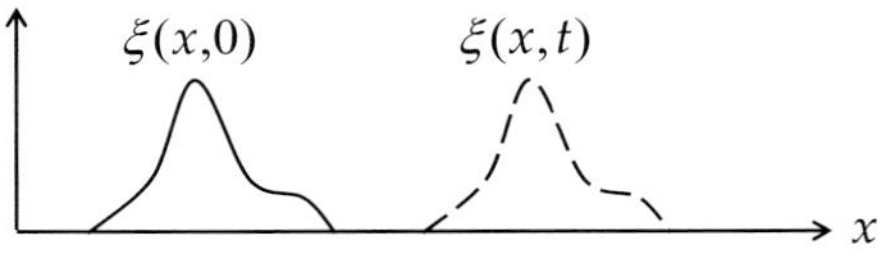

Abbildung 16.24 Eine Welle $\xi(x,t)$ bewegt sich entlang der x-Achse.

Um eine Bewegungsgleichung für Wellen zu erhalten, leiten wir ξ zwei Mal partiell nach der Zeit, sowie zwei Mal partiell nach dem Ort ab. Die ersten beiden äußeren Ableitungen der Funktion f bezeichnen wir hier mit f' und f''. Formal erhalten wir damit die folgenden Gleichungen:

$$\frac{\partial \xi}{\partial x} = f'\,(x-vt)\cdot\frac{\partial}{\partial x}\,(x-vt) \qquad \frac{\partial \xi}{\partial t} = f'\,(x-vt)\cdot\frac{\partial}{\partial t}\,(x-vt) \tag{16.146}$$

$$= f'\,(x-vt)\cdot 1 \qquad = f'\,(x-vt)\cdot(-v) \tag{16.147}$$

$$\frac{\partial^2 \xi}{\partial x^2} = f''\,(x-vt) \qquad \frac{\partial^2 \xi}{\partial t^2} = f''\,(x-vt)\cdot(-v)^2 \tag{16.148}$$

Durch Gleichsetzen von f'' ergibt sich die sogenannte **Wellengleichung**

$$\frac{\partial^2 \xi}{\partial x^2} = \frac{1}{v^2}\frac{\partial^2 \xi}{\partial t^2}\,. \tag{16.149}$$

Den Zusammenhang der beiden zweifachen partiellen Ableitungen in der Auslenkung $\xi(x,t)$ mit der Phasengeschwindigkeit v der Welle werden wir im folgenden Abschnitt 16.9 für die Bestimmung von v nutzen. Die Wellengleichung gilt allgemein z. B. für mechanische und elektromagnetische Wellen.

Aufgabe 46: Wellengleichung

Verifizieren Sie durch zweifache partielle Ableitung, dass harmonische Wellen (16.29)

$$\xi(x,t) = \xi_\circ \sin(kx - \omega t)$$

Lösungen der Wellengleichung (16.149) sind. Wie groß ist die Phasengeschwindigkeit v der harmonischen Welle?

16.9 Transversale Wellen

Wir legen eine Stahlsaite über ein Monochord und spannen sie mit der Kraft F, z. B. durch Anhängen eines Gewichts über eine Umlenkrolle (► Abbildung 16.25).

Abbildung 16.25 Monocord: Erzeugung von Wellen an einer gespannten Stahlsaite

Wie bei jedem Saiteninstrument vernehmen wir durch Zupfen an der Saite einen Ton mit der Frequenz ν. Vergrößern wir die Spannkraft F, erhöht sich die wahrgenommene Frequenz entsprechend. Gleichzeitig vergrößert sich die Phasengeschwindigkeit $v = \lambda\,\nu$ (16.28) entlang der Saite. Im Folgenden berechnen wir die Phasengeschwindigkeit v mithilfe der Wellengleichung (16.149) und schließen damit auf die Frequenz einer Saite.

16.9.1 Saitenspannkraft in Richtung ihrer Auslenkung

Wir lenken die Stahlsaite an einer Stelle transversal (senkrecht) zur ursprünglichen Saitenrichtung aus. An jedem Ort entlang der Saite ergibt sich dadurch eine transversale Komponente der Spannkraft (► Abbildung 16.26).

Zunächst berechnen wir die resultierende transversale Kraft auf das Saitenstück zwischen x und $x + dx$. Durch die Beugung der Saite ergeben sich an den Stellen x und $x + dx$ unterschiedliche transversale Kraftkomponenten. An der Stelle x ist der Betrag der Kraftkomponente

$$F_\xi(x) = F \cdot \sin(\alpha(x)) \,. \tag{16.150}$$

Bei der Stelle $x + dx$ beträgt die Kraftkomponente

$$F_\xi(x + dx) = F \cdot \sin(\alpha(x + dx)) \,. \tag{16.151}$$

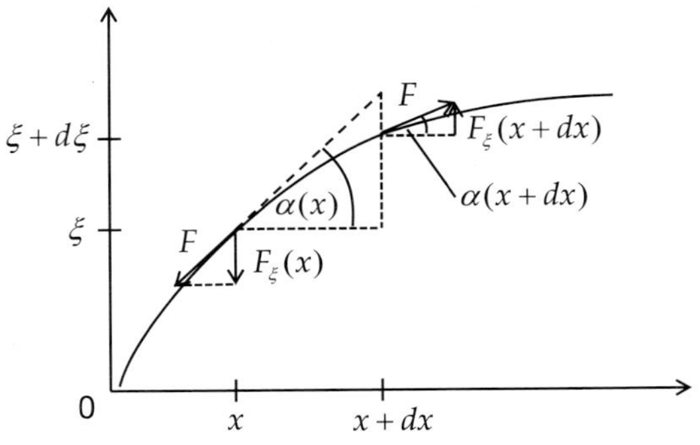

Abbildung 16.26 Zusammenhang zwischen der Auslenkung ξ und der transversalen Komponente der Spannkraft $F_\xi(x)$ bei einem Winkel α bezüglich der x-Achse

Die Differenz der transversalen Kräfte

$$\Delta F_\xi = F_\xi(x + dx) - F_\xi(x) \tag{16.152}$$

entspricht der Kraft, die eine Beschleunigung des kleinen Saitenstücks bewirkt.

Wir führen nun eine Taylor-Entwicklung (Abschnitt 3.4.2) der Kraftdifferenz (16.152) an der Stelle x durch und erhalten

$$\Delta F_\xi = F \cdot [\,\sin(\alpha(x + dx)) - \sin(\alpha(x))\,] \tag{16.153}$$

$$= F \cdot [\,\underbrace{\sin(\alpha(x)) + \frac{\partial \sin(\alpha)}{\partial x}\,dx}_{\text{Taylor-Entwicklung}} - \sin(\alpha(x))\,] \tag{16.154}$$

$$= F \cdot \frac{\partial \sin\alpha}{\partial x}\,dx\,. \tag{16.155}$$

Passend zur Wellengleichung (16.149), in der die Auslenkung der Saite am Ort x zur Zeit t erfasst wird, wollen wir anstelle des Winkels α die Auslenkung $\xi(x,t)$ schreiben. Ihren Zusammenhang lesen wir aus der Steigung in ► Abbildung 16.26 ab. Danach lautet die Ableitung der Auslenkung am Ort x in der Näherung kleiner Winkel $\tan\alpha \approx \alpha$

$$\frac{d\xi}{dx} = \tan\alpha \approx \alpha\,. \tag{16.156}$$

Wir führen die Näherung kleiner Winkel $\sin\alpha \approx \alpha$ ebenfalls in (16.155) durch und setzen für α die Ableitung der Auslenkung (16.156) ein. Damit erhalten wir die transversale Kraft auf das Saitenstück, die sich aus der Spannkraft F und der zweiten örtlichen Ableitung in der Auslenkung ξ ergibt:

$$\Delta F_\xi = F \cdot \frac{\partial \sin\alpha}{\partial x}\,dx \tag{16.157}$$

$$\approx F \cdot \frac{\partial \alpha}{\partial x}\,dx \tag{16.158}$$

$$= F\,\frac{\partial}{\partial x}\left(\frac{\partial \xi}{\partial x}\right)dx \tag{16.159}$$

$$= F\,\frac{\partial^2 \xi}{\partial x^2}\,dx \tag{16.160}$$

16.9.2 Transversale Beschleunigung des Saitenstücks

Ein Saitenstück, das um die transversale Auslenkung ξ aus der Ruheposition verschoben ist, erfährt durch die Kraft ΔF_ξ (16.160) eine Beschleunigung. Nach Newtons Bewegungsgleichung $ma = F$ (3.41) beträgt die Beschleunigung $\partial^2\xi/\partial t^2$ des Saitenstücks mit der Masse dm

$$dm\,\frac{\partial^2\xi}{\partial t^2} = \Delta F_\xi\,. \tag{16.161}$$

Wir definieren eine Massenbelegung μ pro Saitenlänge, also

$$\mu = \frac{dm}{dx}\,. \tag{16.162}$$

Die Massenbelegung μ können wir z. B. experimentell durch Messung der Länge L und der Masse M einer Saite bestimmen. Alternativ wird eine Berechnung von μ aus der Massendichte ρ des Saitenmaterials und der Zylindergeometrie der Saite im folgenden Experiment 44 vorgeführt.

16.9.3 Wellengleichung und Phasengeschwindigkeit der Saite

In die Bewegungsgleichung für die Auslenkung ξ (16.161) setzen wir die Massenbelegung (16.162) und die transversale Kraft ΔF_ξ (16.160) aufgrund der Spannkraft F der Saite ein und erhalten

$$\mu\,dx\,\frac{\partial^2\xi}{\partial t^2} = F\,\frac{\partial^2\xi}{\partial x^2}\,dx \tag{16.163}$$

$$\underbrace{\frac{\mu}{F}}_{\equiv v^{-2}}\,\frac{\partial^2\xi}{\partial t^2} = \frac{\partial^2\xi}{\partial x^2}\,. \tag{16.164}$$

Wir erkennen die Struktur der Wellengleichung (16.149) wieder. Im Vergleich der Terme sehen wir, dass der Vorfaktor dem inversen Quadrat der Phasengeschwindigkeit der Welle entspricht:

$$v = \sqrt{\frac{F}{\mu}} \tag{16.165}$$

Die Phasengeschwindigkeit v der Welle, die zwischen den beiden eingespannten Enden der Saite hin- und herläuft, ist also von der Spannkraft F und der Massenbelegung μ der Saite abhängig.

Experiment 44: Tonhöhe von Saiten

Zunächst zupfen wir an der Stahlsaite eines Monochords (oder Gitarre). Die Welle läuft mit der Phasengeschwindigkeit v zwischen den Enden der Monochord-Saite hin und her und erzeugt stehende Wellen. Über die Stege überträgt sich die Schwingung auf den Holzkörper des Monochords und wird dadurch verstärkt hörbar.

Für die Grundschwingung $n = 1$ bei einer Saitenlänge L ergibt sich nach (16.67) folgender Zusammenhang zwischen der Phasengeschwindigkeit und der Tonhöhe:

$$v = 2L\nu \tag{16.166}$$

Setzen wir diese Gleichung in Gleichung (16.165) ein, ergibt sich die Tonhöhe als Funktion der Saitenlänge, Spannkraft und Massenbelegung des Saitenmaterials:

$$2L\nu = \sqrt{\frac{F}{\mu}} \tag{16.167}$$

$$\nu = \frac{1}{2L}\frac{1}{\sqrt{\mu}}\sqrt{F} \tag{16.168}$$

Wir erhöhen das angehängte Gewicht F und spannen damit die Saite stärker. Der Ton klingt höher.

Ersetzen wir die Stahlsaite durch eine Angelschnur aus Plastik klingt der Ton bei gleichem angehängten Gewicht höher, was konsistent mit der Massenbelegung $\sqrt{\mu}$ im Nenner ist: Die Massendichte des Plastikmaterials ist viel geringer als die Dichte von Stahl.

Bauen wir ein größeres Instrument (Cello, Kontrabass), wird automatisch durch die größere Saitenlänge L der Ton tiefer.

Um ein Gefühl für die Größenordnungen zu bekommen, betrachten wir als Beispiel die A-Saite einer Geige. Die typische Dichte einer Darm-Saite ist $\rho = 1\,276\ \mathrm{kg/m^3}$. Bei einem Durchmesser von $0{,}85\,\mathrm{mm}$ und einer Länge von $L = 32\,\mathrm{cm}$ ergibt sich eine Massenbelegung von $\mu = 0{,}724\,\mathrm{g/m}$. Mit Gleichung (16.67) und $n = 1$ für die Grundschwingung der A-Saite von $\nu = 440\,\mathrm{Hz}$ benötigen wir eine Phasengeschwindigkeit von

$$v = 2\,L\,\nu = 2 \cdot 0{,}32 \cdot 440 = 282\,\frac{\mathrm{m}}{\mathrm{s}}\,, \tag{16.169}$$

die nach (16.165) mit einer Spannkraft von

$$F = v^2 \cdot \mu = 282^2 \cdot (0{,}724 \cdot 10^{-3}) = 57\,\mathrm{N} \tag{16.170}$$

erreicht wird.

16.9.4 Streichinstrumente

Während bei Zupfinstrumenten der Ton durch einmaliges Auslenken der Saite und Loslassen erreicht wird, erfährt die Saite eines Streichinstruments durch den Bogen eine dauerhafte Anregung zum Schwingen.

Der Instrumentenkörper in ► Abbildung 16.27 hält die Saiten und verstärkt die Schwingungen der Saite auf ein gut hörbares Niveau. Der Bogen erzeugt durch abwechselndes Haften und Gleiten an der Saite eine Schwingung der Saite. Die Haftreibung überwiegt, wenn der Violinbogen die Saite mitnimmt. Bei der Gleitreibung rutscht die Saite unter dem Bogen durch. Diese periodische Saitenbewegung verläuft parallel zur Instrumentendecke.

Die Saite versetzt dabei dem Steg periodisch kurze Schläge, sodass die Stegfüße auf der Holzdecke des Instruments eine kleine Bewegung durchführen, die den Klangkörper in Schwingungen versetzt.

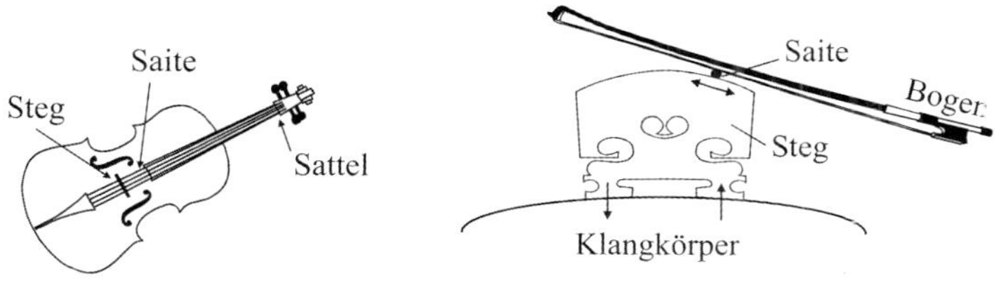

Abbildung 16.27 Geige bestehend aus Klangkörper mit Saiten, Steg, Sattel sowie einem Geigenbogen

Experiment 45: Pendel auf Scheibe

Den Wechsel von Haft- und Gleitreibung können wir uns an einem Pendel verdeutlichen, das im Mittelpunkt einer fast vertikal stehenden Scheibe aufgehängt ist. Ist die Scheibe in Ruhe, so bekommt man durch die Reibung des Pendels an der Scheibe eine gedämpfte Pendelschwingung.

Drehen wir die Scheibe, so kommt es zu einer erzwungenen Schwingung mit abwechselnden Haft- und Gleitphasen (► Abbildung 16.28).

Variieren wir die Drehzahl der Scheibe oder den Winkel der Scheibe und damit die Reibungskraft des Pendels auf der Scheibe, stellen wir folgendes fest: Die Amplitude steigt mit zunehmender Drehzahl und mit zunehmender Reibungskraft (Kapitel 3.2.4). Dafür erhöhen wir den Winkel, in dem die Scheibe steht, und damit die Auflagekraft des Pendels. Die Schwingungsperiode ist unabhängig von der Drehzahl der Scheibe. Diese Eigenschaft ist typisch für sogenannte **selbsterregte Schwingungen**.

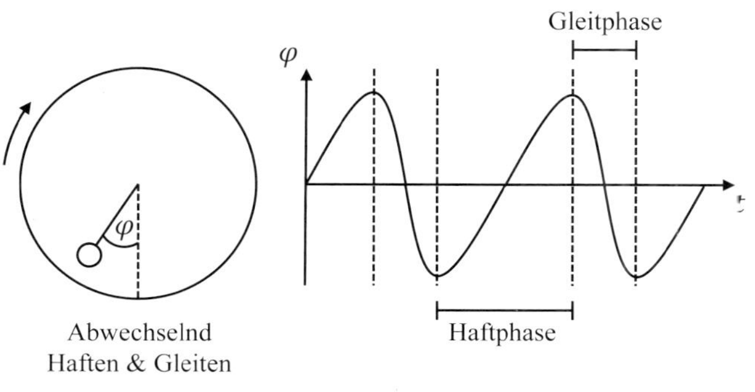

Abbildung 16.28 Das Pendel auf einer geneigten rotierenden Scheibe erzeugt eine Sägezahnschwingung.

Die gestrichene Saite bildet einen Knick, der entlang der Saite umläuft (► Abbildung 16.29). Dieser Knick läuft beim Kammerton A 440 Mal pro Sekunde vom Steg zum Sattel und zurück. Die Bewegung des umlaufenden Knicks wird auch nach seinem Entdecker als »Helmholtz-Bewegung« bezeichnet.

Der Bogen muss sich in die Helmholtz-Bewegung des umlaufenden Knicks einfügen. Er muss einerseits den umlaufenden Knick anregen, darf ihn aber andererseits in der Bewegung nicht stören.

In den Punkten 1 bis 4 haftet die Saite am Bogen (► Abbildung 16.29). An der Stelle $x = x_B$ haben also Bogen und Saite dieselbe Geschwindigkeit. Um von dem Punkt 5 zum Punkt 6 zu kommen, rutscht die Saite unter dem Bogen durch. Dies geschieht in ca. 2% der Schwingungsperiode. Bei diesem Durchrutschen der Saite erfährt der Steg einen seitwärts gerichteten Schlag, den er auf den Instrumentenkörper unter seinen Steg-Füßen weitergibt.

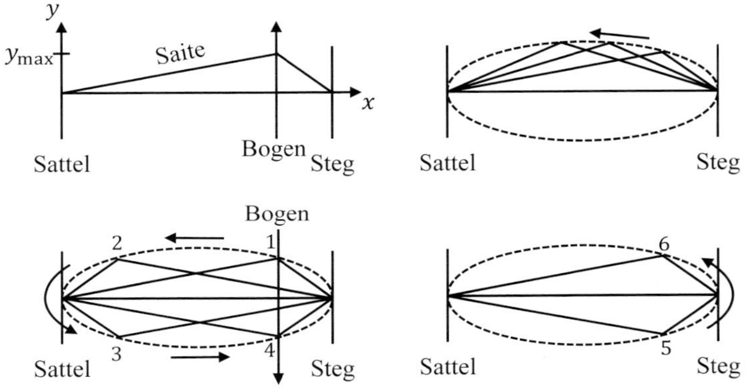

Abbildung 16.29 Links oben: Der Bogen übt eine Kraft auf die zwischen Sattel und Steg gespannte Saite aus. Rechts oben: Ein Knick in der Saite läuft mit der Frequenz f zwischen Steg und Sattel um. Links unten: Der Bogenstrich muss den Umlauf anregen ohne ihn zu behindern: Vom Punkt 1 bis 4 nimmt der Bogen die Saite mit. Rechts unten: An der Stelle 5 bis 6 gleitet die Saite unter dem Bogen weg.

Den umlaufenden Knick der gestrichenen Saite können wir mit einem Stroboskop sichtbar machen. Das funktioniert bereits mit einer von Hand gestrichenen Violine sehr gut.

Um die Helmholtz-Bewegung des umlaufenden Knicks entlang der Saite zu erzeugen und damit einen schönen Ton auf Streichinstrumenten hörbar zu machen, muss die Kombination der wichtigsten Spielerparameter Bogengeschwindigkeit, Bogen-Auflagekraft und Abstand des Bogens vom Steg zusammenpassen. Überraschend klein sind dabei die möglichen Parameterbereiche, in denen der umlaufende Knick losläuft und erhalten bleibt. Daher ist vom physikalischen Standpunkt aus gesehen das Musizieren mit Violine, Viola, Cello oder Kontrabass – fast – unmöglich.

16.9.5 Grundton und Obertöne

Die schlagartige Bewegung des Stegs in dem Moment, in dem der Knick der Saite am Steg reflektiert, ist für den Violinklang von entscheidender Bedeutung. Der Klang resultiert aus einer periodischen Sägezahnbewegung, die wir über eine Fouriersynthese (Kapitel 15.5) aus dem Grundton und den Obertönen mit den entsprechend verschiedenen Intensitäten im Prinzip auch künstlich zusammensetzen können.

Die Amplituden nehmen bei der Sägezahnbewegung mit der Tonhöhe des n-ten Tons wie $1/n$ ab, die Intensitäten entsprechend mit $1/n^2$:

$$\text{Amplitude} \propto \frac{1}{n} \tag{16.171}$$

$$\text{Intensität} \propto \frac{1}{n^2} \tag{16.172}$$

Für den relativen Intensitätsunterschied zwischen dem 9. Oberton und dem Grundton erwarten wir

$$\frac{I_{10}}{I_1} = \left(\frac{1}{10}\right)^2 = \frac{1}{100}\,. \tag{16.173}$$

Häufig werden in der Akustik Intensitätsunterschiede über die logarithmische Skala von Bel ausgedrückt, meistens in der Einheit Dezibel (1 dB = 0,1 Bel). Unser erwarteter Intensitätsunterschied Δ beträgt nach dieser Definition

$$\Delta = 10 \log_{10}\left(\frac{I_{10}}{I_1}\right) dB = -20\,\text{dB}\,. \tag{16.174}$$

In ► Abbildung 16.30 sind links Messungen der Obertonspektren verschiedener Streichinstrumente mithilfe einer Fourieranalyse gezeigt. Alle Streichinstrumente zeigen eine Intensitätsabnahme im Obertonspektrum ähnlich der erwarteten Sägezahnbewegung. Die genauen Details sind vom Instrumentenkörper und der Kopplung zwischen Steg und Holzkörper abhängig.

Zum Vergleich sind im rechten Bild Messungen mit Blasinstrumenten gezeigt, deren Obertonspektren sich deutlich vom Sägezahnspektrum unterscheiden. Entsprechend klingen diese Blasinstrumente anders.

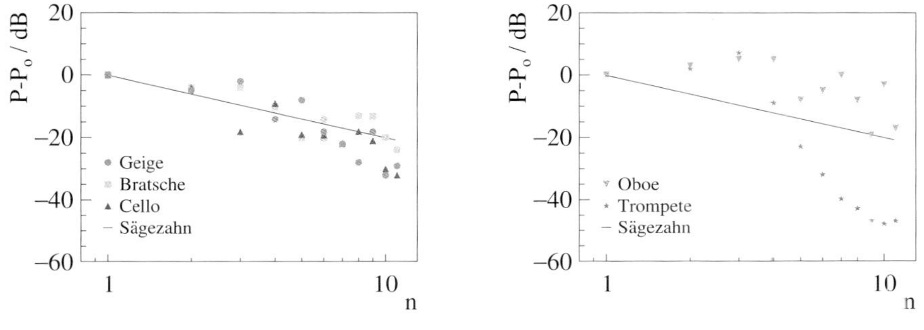

Abbildung 16.30 Grund und Obertöne links für verschiedene Streichinstrumente, rechts für Blasinstrumente im Vergleich zur Sägezahnschwingung

Das Wichtigste zum Mitnehmen

Zum Mitnehmen: Wellen 1

- Eine **ebene harmonische Welle** hat die Form

$$\underbrace{\xi(x,t)}_{\text{Auslenkung}} = \underbrace{\xi_0}_{\text{Amplitude}} \sin[\underbrace{k}_{\text{Wellenzahl}} \underbrace{(x - vt)}_{\text{Ausbreitung}}] . \qquad (16.5)$$

Dabei beschreibt $\xi(x,t)$ die Auslenkung zum Zeitpunkt t an der Stelle x mit der Amplitude $\xi_\circ$. Folgende Größen sind für die Beschreibung von Wellen wichtig:

– Die **Wellenzahl** k (Einheit m^{-1}) ist über die **Wellenlänge** λ (Einheit m) definiert als

$$k = \frac{2\pi}{\lambda} . \qquad (16.14)$$

– Die **Frequenz** ν (Einheit $\mathrm{s}^{-1} = \mathrm{Hz}$) hängt mit der **Schwingungsdauer oder -periode** T (Einheit s)

$$T = \frac{1}{\nu}$$

zusammen und ebenso mit der **Kreisfrequenz** ω (Einheit s^{-1})

$$\omega = 2\pi\nu .$$

– Die **Phasengeschwindigkeit** v (Einheit $\mathrm{m/s}^{-1}$) ergibt sich aus ω und k zu

$$v = \frac{\omega}{k} \qquad (16.27)$$

oder mit (16.27) und (16.14) zu

$$v = \lambda \cdot \nu .$$

Mit $k\,v = \omega$ aus (16.27) lautet die **harmonische Welle** (16.5) dann

$$\xi(x,t) = \xi_\circ \sin(kx - \omega t) \qquad (16.29)$$

oder in komplexer Schreibweise

$$\xi(x,t) = \xi_\circ\, e^{i(kx - \omega t)} . \qquad (16.30)$$

Zum Mitnehmen: Wellen 2

- Für beliebige Ausbreitungsrichtungen ebener Wellen definiert man einen **Wellenvektor** $\vec{k}$ senkrecht auf der Ausbreitungsrichtung der Welle und mit dem Betrag entsprechend (16.14):

$$|\vec{k}| = \frac{2\pi}{\lambda}$$

Da für jeden Vektor $\vec{r}$, der auf die Phasenfläche zeigt, das Skalarprodukt mit dem Wellenvektor k konstant ist,

$$\vec{k} \cdot \vec{r} = const.\,,$$

kann man für eine ebene Welle mit der Ausbreitungsrichtung $\vec{k}/|\vec{k}|$

$$\vec{\xi} = \vec{\xi}_\circ \sin\left(\vec{k}\,\vec{r} - \omega t\right) \tag{16.33}$$

schreiben oder in komplexer Schreibweise auch

$$\vec{\xi} = \vec{\xi}_\circ\, e^{i\left(\vec{k}\,\vec{r} - \omega t\right)}\,. \tag{16.34}$$

- Die **Intensität** I ist proportional zum Quadrat der Schwingungsamplitude. Um komplexe Werte einzuschließen, schreiben wir die Beträge:

$$I \propto |\xi_\circ|^2 \tag{16.48}$$

- Bei einer **Kugelwelle** bilden die Phasenflächen Kugelflächen und die Intensität sinkt proportional zum Quadrat des Abstands r, die Auslenkung ist damit $\xi \propto \frac{1}{r}$. Die Kugelwelle lautet damit

$$\xi = \frac{\xi_\circ}{r} \sin\left(kr - \omega t\right) \tag{16.175}$$

oder in komplexer Schreibweise

$$\xi = \frac{\xi_\circ}{r}\, e^{i(kr-\omega t)}\,. \tag{16.176}$$

Zum Mitnehmen: Wellen 3

- Eine einfallende Welle ξ_1 und die reflektierte Welle ξ_2 mit einem eventuellen Phasensprung φ, d. h.

$$\xi_1 = \xi_\circ \sin(kx + \omega t)\,,$$
$$\xi_2 = \xi_\circ \sin(kx - \omega t + \varphi)\,,$$

überlagern sich zu einer **stehenden Welle**

$$\xi(x,t) = \underbrace{2\xi_\circ \sin\left(kx + \frac{\varphi}{2}\right)}_{\text{Amplitude}} \cdot \underbrace{\cos\left(\omega t - \frac{\varphi}{2}\right)}_{\text{Schwingung}}\,, \tag{16.57}$$

die wir als Produkt aus einer harmonischen Schwingung und einer ortsabhängigen Amplitude verstehen können.

Die Stellen im Raum, an denen die Amplitude verschwindet, nennt man **Schwingungsknoten**:

$$kx + \frac{\varphi}{2} = n\pi\,, \qquad n = 0, 1, 2, \ldots \tag{16.59}$$

Ohne Phasensprung (d. h. $\varphi = 0$) gibt es z. B. für $n = 0, 1, 2$ an den Orten $x = 0, \frac{\lambda}{2}, \lambda$ Knoten.

Analog dazu erhält man **Schwingungsbäuche** bei

$$kx + \frac{\varphi}{2} = \frac{2n+1}{2}\pi\,, \qquad n = 0, 1, 2, \ldots\,. \tag{16.62}$$

Hier gibt es Schwingungsbäuche z. B. für $n = 0, 1, 2$ an den Orten $x = \frac{\lambda}{4}, \frac{3\lambda}{4}, \frac{5\lambda}{4}$.

- An einer schwingenden Saite der Länge L, deren Enden fest eingespannt sind, gilt für alle t

$$\xi(0,t) = 0 \quad \text{und} \quad \xi(L,t) = 0\,.$$

Mit der Phasengeschwindigkeit $v = \lambda\,\nu$ (16.28) führt das zu **Eigenschwingungen** mit den **Eigenfrequenzen** ν_n der Saite:

$$\nu_n = v \cdot \frac{n}{2\,L}\,, \qquad n = 1, 2, 3 \ldots \tag{16.67}$$

In Abschnitt 16.8 (16.165) wird die Phasengeschwindigkeit v der Welle in Abhängigkeit von der Spannkraft F und der Massenbelegung μ berechnet.

Zum Mitnehmen: Wellen 4

- Für zwei harmonische Wellen gleicher Amplitude $\xi_\circ$, deren Kreisfrequenzen $\omega_1 \approx \omega_2$ und Wellenzahlen $k_1 \approx k_2$ sich nur wenig unterscheiden, ergibt sich mit

$$\omega \equiv \frac{\omega_1 + \omega_2}{2} \qquad \Delta\omega \equiv \frac{\omega_1 - \omega_2}{2}$$
$$k \equiv \frac{k_1 + k_2}{2} \qquad \Delta k \equiv \frac{k_1 - k_2}{2}$$

die folgende Überlagerung der Wellen:

$$\xi = 2 \cdot \underbrace{\cos(\Delta k\, x - \Delta\omega\, t)}_{\text{Amplitudenmodulation}} \cdot \underbrace{\xi_\circ \cos(kx - \omega t)}_{\approx \xi_1 \approx \xi_2} \tag{16.76}$$

- Der zweite Term in (16.76) zeigt die **Phasengeschwindigkeit** $v \approx v_1 \approx v_2$ (16.27)

$$v_p = \frac{\omega}{k}\,. \tag{16.77}$$

Der erste Term in (16.76) zeigt die Geschwindigkeit der Modulation $v_g = \Delta\omega/\Delta k$. Im Allgemeinen schreiben wir diese sogenannte **Gruppengeschwindigkeit** als Ableitung:

$$v_g = \frac{d\omega}{dk} \tag{16.79}$$

- Die Gruppengeschwindigkeit v_g ist in den meisten Fällen die Geschwindigkeit der Signalübertragung. Zwischen v_p und v_g ergibt sich mit (16.77) und (16.79) der Zusammenhang

$$v_g = v_p + k\frac{dv_p}{dk} \tag{16.82}$$

oder in Abhängigkeit von der Wellenlänge λ

$$v_g = v_p - \lambda\frac{dv_p}{d\lambda}\,. \tag{16.86}$$

Falls die Phasengeschwindigkeit $v_p = v_p(k) = v_p(\lambda)$ von der Wellenlänge bzw. der Wellenzahl abhängt, unterscheiden sich Phasen- und Gruppengeschwindigkeit. Wir sprechen dann von **Dispersion**.

Zum Mitnehmen: Wellen 5

- Wir unterscheiden drei Fälle für die Gruppengeschwindigkeiten:
  - **Normale Dispersion**: $dv_p/d\lambda > 0$, $v_g < v_p$, häufigster Fall,
  - **Keine Dispersion**: $dv_p/d\lambda = 0$, $v_g = v_p$, Signalform bleibt erhalten,
  - **Anomale Dispersion**: $dv_p/d\lambda < 0$, $v_g > v_p$, eher selten.

  Wasserwellen (16.88) sind hervorragend geeignet, diese unterschiedlichen Arten von Dispersion zu untersuchen.

- Zwei synchron schwingende Quellen S_1 und S_2 erzeugen kreisförmige Wellen mit den reellen Amplituden A_i, die die Abstandsabhängigkeit r_i von den Quellen bereits enthalten:

  $$\xi_1 = A_1 e^{i(kr_1 - \omega t)} \qquad \xi_2 = A_2 e^{i(kr_2 - \omega t)}$$

  Die Überlagerung oder **Interferenz** der beiden Wellen ist

  $$\xi(r_1, r_2, t) = \xi_1 + \xi_2 = e^{i(kr_1 - \omega t)} \left(A_1 + A_2\, e^{i\delta}\right) \tag{16.104}$$

  mit der **Phasenverschiebung** δ zwischen den Wellen:

  $$\delta \equiv k(r_2 - r_1) \tag{16.105}$$

- Die **Intensität** I (16.47) am Beobachtungspunkt P ist wegen (16.48) das Quadrat der Amplitude der Überlagerung:

  $$I \propto A^2 = A_1^2 + A_2^2 + 2A_1A_2 \cos\delta \tag{16.111}$$

  Damit treten Maxima und Minima für folgende Phasenverschiebungen auf:

  $$\begin{array}{ll} \text{Maxima} & \text{Minima} \\ \delta = 0, 2\pi, \cdots = n\,2\pi & \delta = \pi, 3\pi, \cdots = (2n+1)\,\pi \\ A = A_1 + A_2 & A = A_1 - A_2 \end{array} \tag{16.113}$$

  Obwohl die Gesamtamplitude negativ werden kann, ist die Intensität $I \propto A^2 \geq 0$. Mit (16.113) ergeben sich aus (16.105) Knoten- und Bauchlinien als **Interferenzmuster**:

  $$\text{Bauchlinien} \qquad r_1 - r_2 = n\lambda = const. \tag{16.118}$$

  $$\text{Knotenlinien} \qquad r_1 - r_2 = \frac{2n+1}{2}\lambda = const. \tag{16.120}$$

  Die Kurven (16.118) und (16.120) sind **Hyperbeln** ($r_1 - r_2 =$ **const.**).

Zum Mitnehmen: Wellen 6

- Wir betrachten eine Quelle, die in einem Medium Wellen zu einem Beobachter aussendet. Bewegt sich die Quelle mit der Geschwindigkeit v_Q und der Beobachter seinerseits mit v_B relativ zum Medium, so verschiebt sich die abgestrahlte Frequenz ν im **nicht-relativistischen Dopplereffekt** zur beobachteten Frequenz ν_B:

$$\nu_B = \nu \frac{v - v_Q \cos \vartheta_Q}{v - v_B \cos \vartheta_B} \tag{16.131}$$

  Wenn sich ein Beobachter an der Quelle vorbei bewegt, wird sich die Frequenz von $\nu_B = \nu(v + v_Q)/(v - v_R) > \nu$ über $\nu_B = \nu$ nach $\nu_B = \nu(v - v_Q)/(v + v_R) < \nu$ kontinuierlich verändern. Die Situation ist geläufig, wenn man einem bewegten Signalton begegnet.

- Mit zunehmender Geschwindigkeit v_Q der Quelle rücken die Wellenflächen in Bewegungsrichtung der Quelle immer dichter zusammen, bis sie sich schließlich für $v_Q \geq v$ konstruktiv überlagern. Der dabei entstehende **Machsche Kegel** wird beschrieben durch den halben Öffnungswinkel α mit

$$\sin \alpha = \frac{v}{v_Q} . \tag{16.132}$$

  Die **Machsche Zahl Ma** wird zur Beschreibung großer Geschwindigkeiten z. B. von Strömungen benutzt:

$$\mathrm{Ma} = \frac{1}{\sin \alpha} = \frac{v_Q}{v} \tag{16.133}$$

  $\mathrm{Ma} = 2$ steht beispielsweise für zweifache Schallgeschwindigkeit.

- Beim **relativistischen Dopplereffekt** (z. B. für Licht) spielt nur die Relativgeschwindigkeit $\beta = v/c$ zwischen Quelle und Beobachter eine Rolle. Die beobachtete Frequenz ν' verschiebt sich gegenüber der abgestrahlten Frequenz ν:

$$\nu' = \nu \frac{\sqrt{1 \mp \beta}}{\sqrt{1 \pm \beta}} \tag{16.143}$$

  Dabei gelten die oberen (unteren) Vorzeichen für den Fall, dass sich Quelle und Beobachter von einander weg (aufeinander zu) bewegen.

Zum Mitnehmen: Wellen 7

- Für eine Welle ξ, die sich mit der Phasengeschwindigkeit v nach rechts bewegt, d. h.

$$\xi(x,t) = f(x - vt)\,, \tag{16.177}$$

ergibt sich die **Wellengleichung**

$$\frac{\partial^2 \xi}{\partial x^2} = \frac{1}{v^2}\frac{\partial^2 \xi}{\partial t^2}\,. \tag{16.149}$$

Diese Wellengleichung gilt für alle Arten von Wellen.

- Für ein Seil oder eine Saite, die mit der Kraft F gespannt sind und die Massenbelegung $\mu = dm/dx$ haben, folgen die Auslenkungen ξ der Wellengleichung (16.149)

$$\underbrace{\frac{\mu}{F}}_{\equiv v^{-2}} \frac{\partial^2 \xi}{\partial t^2} = \frac{\partial^2 \xi}{\partial x^2}\,.$$

Der Vorfaktor ist das inverse Quadrat der Phasengeschwindigkeit

$$v = \sqrt{\frac{F}{\mu}}\,. \tag{16.165}$$

Die Phasengeschwindigkeit v der Welle, die zwischen den beiden eingespannten Enden der Saite hin- und herläuft, hängt also von der Spannkraft F und der Massenbelegung μ der Saite ab.

- Bei Streichinstrumenten wird durch die Bogenbewegung ein Knick in der Saite angeregt, der der Frequenz ν entsprechend ν-mal pro Sekunde zwischen Steg und Sattel hin- und zurückläuft (Helmholtz-Bewegung). Passiert der Knick den Steg, so erfährt der Steg einen plötzlichen Schlag, den er auf den Instrumentenkörper überträgt. Dadurch entsteht das charakteristische Frequenzspektrum der Streichinstrumente, das einem Sägezahnspektrum entspricht.

17 Gemeinsam verstehen

Wie gut können Sie bereits über die Themen sprechen? Führen Sie zu zweit ein wissenschaftliches Gespräch miteinander und beachten Sie dabei die Hinweise in Kapitel 6.

Aufgabe 47: Partnerarbeit: Drehschwingungen

Welche Kreisfrequenz bei Drehschwingungen einer dünnen homogenen Scheibe ist größer: Drehachse durch den Mittelpunkt a) senkrecht zur Scheibe oder b) quer zur Scheibe? Warum?

Aufgabe 48: Partnerarbeit: Kreisel

Beschreiben Sie physikalisch ein Holzei, das auf dem Boden gerollt, wie ein Kreisel gedreht, weit geworfen wird.

Aufgabe 49: Partnerarbeit: Gedämpfte Schwingungen

Ein Pendel wird durch eine Reibungskraft gedämpft, die linear mit der Geschwindigkeit wächst: Unter welchen Bedingungen schwingt das System?

Aufgabe 50: Partnerarbeit: Erzwungene Schwingungen

Geben Sie Beispiele für erzwungene Schwingungen und diskutieren Sie die zugehörigen physikalischen Effekte.

Aufgabe 51: Partnerarbeit: Schwebung

Was ist eine Schwebung? Erläutern Sie anhand eines gekoppelten Pendelsystems.

Aufgabe 52: Partnerarbeit: Welle

Was ist eine harmonische Welle? Was unterscheidet eine Welle von einer Schwingung?

Hydromechanik – Wärme

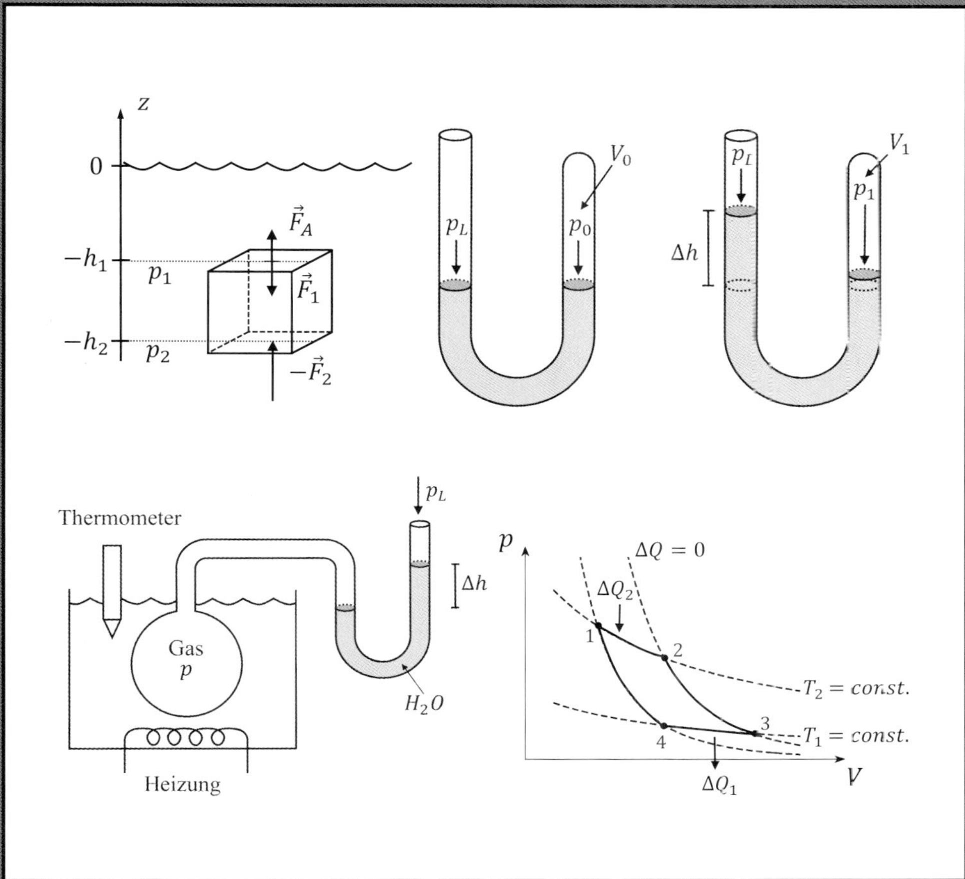

Teil IV

Hydromechanik

18

18.1 Druck **381**

18.2 Auftrieb **382**

18.3 Luftdruck **384**
18.3.1 Kompressibilität von Gas 386
18.3.2 Barometrische Höhenformel 387

18.4 Kontinuitätsgleichung **390**
18.4.1 Differentielle Form der Kontinuitätsgleichung 391

18.5 Bernoulli-Gleichung **393**
18.5.1 Dynamischer Auftrieb 395
18.5.2 Hydrodynamisches Paradoxon 396

18.6 Innere Reibung in Flüssigkeiten **397**
18.6.1 Reibung bei Flüssigkeitsströmung durch ein Rohr 399
18.6.2 Hagen-Poiseuille-Gesetz 400
18.6.3 Druckgefälle entlang des Rohrs 401
18.6.4 Stokessches Gesetz 402
18.6.5 Allgemeine Strömungen in Fluiden 403

18.7 Druckwellen **404**
18.7.1 Längenänderung 404
18.7.2 Volumenänderung 405
18.7.3 Wellen in festen Körpern und Flüssigkeiten 405
18.7.4 Herleitung der Wellengleichung 405
18.7.5 Druck- bzw. Dichtewellen in Gasen 407

ÜBERBLICK

» Zentrales Thema der Hydromechanik ist die Beschreibung von Flüssigkeiten und Gasen, die ruhen oder sich bewegen.

Wir werden einerseits statische Phänomene untersuchen und dabei den Druck in Flüssigkeiten kennenlernen, mit dem wir den Auftrieb erklären können. Weiterhin werden wir die Höhenabhängigkeit des Luftdrucks untersuchen und die Druckverhältnisse bei Flüssigkeiten und bei Gasen miteinander vergleichen.

Ebenso werden wir Bewegungsphänomene von Flüssigkeiten und Gasen thematisieren. Wir werden den dynamischen Auftrieb als Ursache für das Fliegen mit Tragflächen kennenlernen, Reibungsphänomene untersuchen und schließlich die Ausbreitung von Druckwellen in Festkörpern und Gasen beschreiben. «

18.1 Druck

Die physikalische Größe Druck p ist definiert durch die Kraft F, die auf eine Fläche A wirkt:

$$p = \frac{F}{A} \tag{18.1}$$

Druck wird in den Einheiten Pascal gemessen $1\,\text{Pa} = 1\,\text{N/m}^2$.

Da es in einem Gas- oder einem Flüssigkeits-Volumen keine Vorzugsrichtungen gibt, breitet sich der Druck innerhalb des Volumens in alle Richtungen gleichmäßig aus. An Grenzflächen, auch innerhalb des Volumens, erzeugt ein Druck p jedoch gerichtete Kräfte $\vec{F}$.

Beispiel 47: Wassersäule

Zur Beschreibung der Wassersäule in ► Abbildung 18.1 wählen wir ein Koordinatensystem, dessen Ursprung auf der Wasserbodenfläche liegt und dessen positive z-Achse nach oben zeigt. Die Gewichtskraft einer Wassersäule entspricht dem Integral über alle Massenelemente dm des Wassers:

$$F = \int g\, dm \tag{18.2}$$

Wir verwenden die Dichte $\rho = dm/dV$ (14.3) des Wassers und berechnen den Druck, den die Wassersäule auf die Bodenfläche ausübt:

$$p = \frac{F}{A} \tag{18.3}$$

$$= \frac{\int g\, \rho\, dV}{\int dA} \tag{18.4}$$

$$= \frac{\iint \rho\, g\, dA\, dz}{\int dA} \tag{18.5}$$

$$= \int_0^h \rho(z)\, g\, dz \tag{18.6}$$

Da die Kompressibilität von Wasser gering ist, ist seine Dichte innerhalb der Wassersäule näherungsweise konstant: $\rho\,(z) \approx$ *const.* In der Nähe der Erdoberfläche ist $g \approx$ *const.*. Mit diesen Näherungen vereinfacht sich die Integration und wir erhalten für den Druck der Wassersäule auf den Boden

$$p = \rho\, g \int_0^h dz \tag{18.7}$$

$$p = \rho\, g\, h\,. \tag{18.8}$$

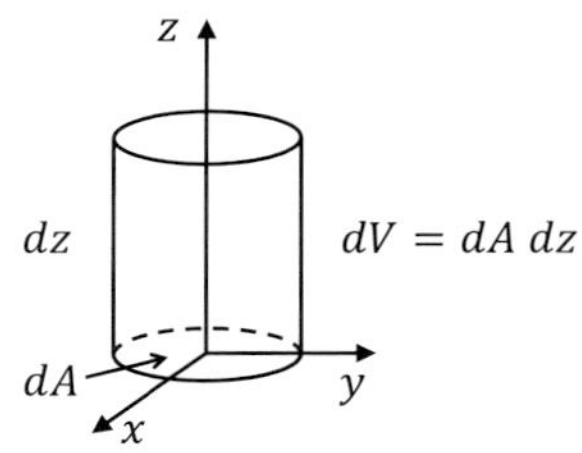

Abbildung 18.1 Die z-Koordinate zeigt nach oben und der Boden der Wassersäule liegt im Ursprung der z-Achse.

> **Aufgabe 53: Druck im Philippinengraben**
>
> Der Philippinengraben hat eine Meerestiefe von $h = 10\,000\,\text{m}$. Die Dichte von Wasser beträgt $\rho = 10^3\,\text{kg/m}^3$. Berechnen Sie den Druck auf den Meeresboden. Verwenden Sie für die Erdbeschleunigung näherungsweise $g \approx 10\,\text{m/s}^2$.

18.2 Auftrieb

Für einen Gegenstand unter Wasser sind der Druck in der Höhe der Oberseite des Gegenstands und der Druck auf der Höhe seiner Unterseite verschieden groß.

Wir denken uns einen Würfel, den wir in eine Flüssigkeit der Dichte ρ_{Fl} eintauchen (► Abbildung 18.2). In der Höhe der Oberseite des Würfels wirkt nach Gleichung (18.7) der Druck

$$p_1 = \rho_{Fl}\, g \int_{-h_1}^{0} dz \tag{18.9}$$

$$= \rho_{Fl}\, g\, (0 - (-h_1)) \tag{18.10}$$

$$= \rho_{Fl}\, g\, h_1 \,. \tag{18.11}$$

Der Druck in der Höhe der Unterseite ist

$$p_2 = \rho_{Fl}\, g\, h_2 \,. \tag{18.12}$$

Wegen der größeren Wassertiefe ist der Druck p_2 größer als p_1.

Die entsprechenden Kräfte $F_2 = p_2\, A$ und $F_1 = p_1\, A$ (18.1) auf die Würfeloberflächen der Größe A sind ebenfalls verschieden groß. Auf die Ober- und Unterseite des Würfels wirken die Kräfte $\vec{F}_1$ und $-\vec{F}_2$. Das bewirkt einen Auftrieb mit der Kraft

$$\vec{F}_A = \vec{F}_1 - \vec{F}_2 \tag{18.13}$$

$$= (p_2 - p_1)\, A\, \vec{e}_z \,. \tag{18.14}$$

Dabei ist $\vec{e}_z$ der Einheitsvektor in die nach oben zeigende z-Richtung. Die Kräfte auf den vier senkrechten Seiten des Würfels sind gleich groß und heben sich auf.

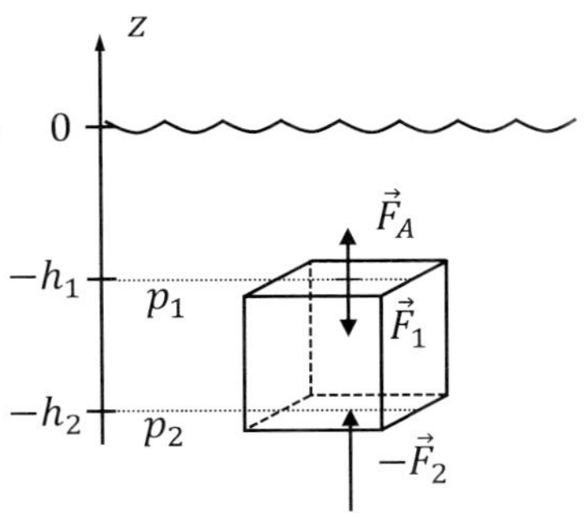

Abbildung 18.2 Die unterschiedlichen Kräfte F_i des Wasserdrucks in den Tiefen $-h_i$ resultieren in einer Auftriebskraft F_A.

Durch Einsetzen der Gleichungen (18.11) und (18.12) können wir den Auftrieb über die Gewichtskraft der verdrängten Flüssigkeit verstehen:

$$|\vec{F}_A| = (p_2 - p_1)\, A \tag{18.15}$$

$$= \underbrace{g \; \underbrace{\rho_{Fl} \; \underbrace{(h_2 - h_1)\, A}_{\text{Volumen des Körpers}}}_{\text{Masse der verdrängten Flüssigkeit}}}_{\equiv F_F \text{ Gewichtskraft der verdrängten Flüssigkeit}} \tag{18.16}$$

Die Auftriebskraft F_A entspricht der Gewichtskraft F_F der von dem Körper verdrängten Flüssigkeit. Dieses Prinzip wurde bereits von Archimedes formuliert.

Die Auftriebskraft F_A reduziert die Gewichtskraft F_G eines Körpers. Zum Beispiel ist die effektive Gewichtskraft F_G' eines vollständig eingetauchten homogenen Körpers mit dem Volumen $V_K = M_K \,/\, \rho_K$ in der Flüssigkeit mit der Dichte ρ_{Fl}

$$F_G' = F_G - F_A \tag{18.17}$$

$$= F_G - F_F \tag{18.18}$$

$$= \rho_K\, V_K\, g - \rho_{Fl}\, V_K\, g \tag{18.19}$$

$$= (\rho_K - \rho_{Fl})\, V_K\, g\,. \tag{18.20}$$

Falls die Massendichte ρ_K des Körpers kleiner als die Dichte ρ_{Fl} der Flüssigkeit ist ($\rho_K < \rho_{Fl}$), schwimmt der Körper auf, bis die Masse der von ihm verdrängten Flüssigkeit seiner Eigenmasse entspricht.

Experiment 46: Prinzip von Archimedes

Ein Eisenzylinder hängt an einer Federwaage. Die Waage zeigt eine Gewichtskraft von $F = 5\,\text{N}$. Der Zylinder hat einen Durchmesser von $d = 4\,\text{cm}$ und eine Höhe von $h = 5\,\text{cm}$. Sein Volumen beträgt demnach $V_Z = \pi\,(d/2)^2\,h \approx 63\,\text{cm}^3$.

Wir tauchen den Zylinder nun vollständig in ein mit Wasser gefülltes Glas ein und messen erneut die Gewichtskraft, die jetzt $F' = 4{,}3$ N beträgt.

Mit den Dichtewerten für Eisen $\rho_{\text{Fe}} = 7{,}9\,10^3\,\text{kg/m}^3$ und für Wasser $\rho_{\text{H}_2\text{O}} = 10^3\,\text{kg/m}^3$ erwarten wir nach Gleichung (18.20)

$$F'_G = (\rho_{\text{Fe}} - \rho_{\text{H}_2\text{O}})\,V_Z\,g \tag{18.21}$$

$$= (7{,}9 - 1)\,10^3 \cdot 63 \cdot 10^{-6} \cdot 10\,\text{kg m/s}^2 \tag{18.22}$$

$$= 4{,}3\,\text{N}\,. \tag{18.23}$$

Diese Erwartung stimmt mit unserer Messung gut überein.

Das Prinzip von Archimedes gilt auch für Gase. Eine Ballonfahrt ist dann möglich, wenn die Gesamtmasse aus Ballonhülle, Gas und Gondel

$$M = m_{\text{Ballon}} + m_{\text{Gondel}} + m_{\text{Gas}} \tag{18.24}$$

kleiner ist als die Masse der vom Ballon verdrängten Luft:

$$M < \rho_{\text{Luft}}\,V_{\text{Ballon}} \tag{18.25}$$

Aufgabe 54: Heißluftballon

Ein Heißluftballon hat vor dem Füllen seiner kugelförmigen Ballonhülle eine Gesamtmasse inklusive der Gondel, der Ballonhülle und den Reisenden von $m = 500\,\text{kg}$. Luft hat bei 20 °C eine Dichte von $\rho_\circ = 1{,}2\,\text{kg/m}^3$. Nach dem Aufblasen der Ballonhülle und dem Heizen der Luft in der Ballonhülle sinkt die durchschnittliche Dichte der Luft in der Hülle auf $\rho_h = 2/3 \cdot \rho_\circ$.

Wie groß ist der erforderliche Durchmesser der Ballonhülle für das Abheben des Ballons?

18.3 Luftdruck

Die Masse aller Moleküle, die sich in der Erdatmosphäre befinden, verursacht im Gravitationseinfluss der Erde einen Druck, den sogenannten Luftdruck. Wir verwenden den folgenden Versuchsaufbau, um diesen Luftdruck zu messen.

Experiment 47: Torricellische Röhre

Die Torricellische Röhre (► Abbildung 18.3) besteht aus einer U-förmigen Glasröhre, die auf der einen Seite offen, auf der anderen Seite geschlossen ist. Sie wird z. B. liegend mit etwas Quecksilber gefüllt und dann so aufgerichtet, dass die geschlossene Seite evakuiert bleibt. Auf der offenen rechten Seite des U-Rohrs wirkt nun der Luftdruck p_L. Auf der geschlossenen linken Seite wirkt der Druck der Quecksilbersäule p_{Hg}.

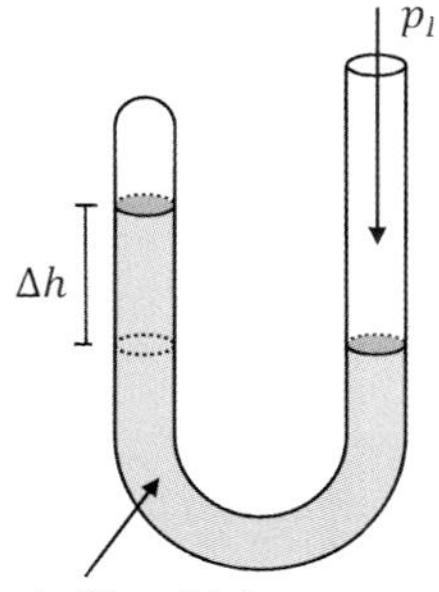

Abbildung 18.3 Die Torricellische Röhre ist mit Quecksilber Hg gefüllt. Der Luftdruck p_L erzeugt im evakuierten Teil einen Anstieg Δh der Hg-Säule.

Relevant für die Messung des Luftdrucks mit der Torricellischen Röhre (► Abbildung 18.3) ist die Quecksilbersäule Δh auf der geschlossenen linken Seite. Unterhalb von Δh sind die Pegelstände auf der linken und rechten Seite gleich groß, die Druckverhältnisse durch das Quecksilber sind hier ausgeglichen.

Nach Gleichung (18.8) beträgt der Druck der Quecksilbersäule Δh

$$p_{\text{Hg}} = \rho_{\text{Hg}}\, g\, \Delta h\,. \tag{18.26}$$

Im statischen Fall, d. h. das Quecksilber bewegt sich nicht, ist der Druck auf beiden Seiten gleich groß und der **Luftdruck** auf der rechten Seite der Torricellischen Röhre beträgt

$$p_{\text{Luft}} = p_{\text{Hg}} = \rho_{\text{Hg}}\, g\, \Delta h\,. \tag{18.27}$$

Physikalische Normalbedingungen herrschen definitionsgemäß bei einer Temperatur von 0 °C und einem Druck von

$$1 \text{ physikalischen Atmosphäre} = 1\,\text{atm} = 101\,325\,\text{Pa} \approx 10^5\,\text{Pa}\,. \tag{18.28}$$

Dabei ist Pascal Pa = N/m^2 die SI-Einheit des Drucks. 1 atm ist an den mittleren Luftdruck auf Meereshöhe angepasst. Alternativ kann auch die Einheit bar verwendet werden, wobei 1 bar = 10^5 Pa entspricht.

Im Vergleich mit dem Luftdruck auf Meereshöhe von 1 atm ist der Druck im Wasser in 10 km Tiefe des Philippinengrabens sehr groß: Er beträgt $\approx 1\,000$ atm (siehe vorherige Aufgabe 53 zum Philippinengraben).

Vor der SI-Einheit Pascal gab es andere Einheiten, die wir hier einordnen: Bei unserem Torricelli-Rohr mit Quecksilber der Dichte $\rho_{\mathrm{Hg}} = 13{,}546\,10^3\,\mathrm{kg/m^3}$ entspricht der Luftdruck einer Höhendifferenz von

$$\Delta h_{\mathrm{Hg}} = 760\,\mathrm{mm}. \tag{18.29}$$

Darauf gründet sich die veraltete (nicht-SI-) Einheit 1mm Hg = 1 Torr. Für Wasser anstelle von Quecksilber ist die Höhendifferenz

$$\Delta h_{\mathrm{H_2O}} = \Delta h_{\mathrm{Hg}} \frac{\rho_{\mathrm{Hg}}}{\rho_{\mathrm{H_2O}}} = 760\,\frac{13{,}595}{1}\,\mathrm{mm} = 10{,}33\,\mathrm{m}\,. \tag{18.30}$$

Dies hat Bezug zu der häufig noch verwendeten Einheit der »technischen Atmosphäre at«, sie entspricht einer Wassersäule von 10 m.

18.3.1 Kompressibilität von Gas

Anders als Festkörper und Flüssigkeiten lassen sich Gase gut komprimieren, wie wir z. B. vom Aufpumpen unserer Fahrradreifen wissen. Im folgenden Experiment messen wir die Kompressibilität von Luft.

Experiment 48: Gesetz von Boyle-Mariotte

Ein halbseitig geschlossenes U-Rohr wird mit etwas Wasser befüllt (linke ► Abbildung 18.4). Dabei entspricht der Druck auf beiden Seiten zunächst dem Luftdruck, was wir z. B. durch ein Ventil auf der geschlossenen Seite des U-Rohrs erreichen können:

$$p_\circ = p_L \tag{18.31}$$

Nun komprimieren wir das mit Luft gefüllte Volumen $V_\circ$ (rechte ► Abbildung 18.4). Dazu füllen wir weiteres Wasser in das U-Rohr, bis das Luftvolumen auf der rechten Seite um 1% des ursprünglichen Volumens verkleinert ist ($V_1 = 0{,}99\,V_\circ$). Bei dem Prozess lassen wir die Temperatur des Gases unverändert.

Wir messen die Druckänderung durch den Unterschied des Pegelstands auf der linken und rechten Seite im U-Rohr:

$$p_1 = p_L + \rho\, g\, \Delta h \tag{18.32}$$

Für das Komprimieren des Gasvolumens um 1% benötigen wir eine Wassersäule der Höhe $\Delta h \approx 10\,\mathrm{cm}$.

Der anfängliche Luftdruck im eingeschlossenen Volumen entsprach nach (18.30) einer ca. 10 m hohen Wassersäule. Die $\Delta h \approx 10\,\mathrm{cm}$ hohe Wassersäule nach dem Auffüllen entspricht

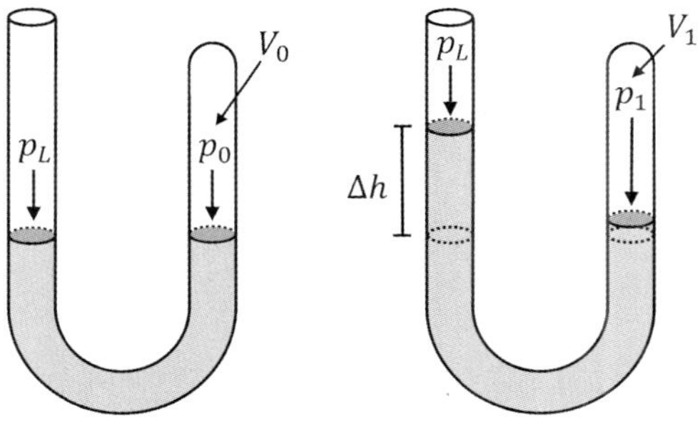

Abbildung 18.4 Das Luftvolumen V_0 wird unter dem zusätzlichen Druck einer Δh hohen Wassersäule auf V_1 zusammengedrückt.

daher einer Druckvergrößerung um $0{,}1\,\text{m}/10\,\text{m} = 1\%$. Sie bewirkt die Reduktion des Luftvolumens um 1%.

Für eine prozentuale Verkleinerung des Luftvolumens benötigen wir die entsprechende prozentuale Vergrößerung des Drucks.

Dieser inverse Zusammenhang zwischen dem Druck und dem Volumen eines Gases bei konstant gehaltener Temperatur ist als Gesetz von **Boyle-Mariotte** bekannt:

$$pV = const. \tag{18.33}$$

18.3.2 Barometrische Höhenformel

Wir stellen uns eine Luftsäule aus Gasmolekülen vor, die vom Erdboden hoch in die Atmosphäre reicht (senkrechte Koordinate $z > 0$, Erdboden $z = 0$). Der Luftdruck sinkt mit steigender Höhe z, wobei es hier zwei ursächliche Beiträge gibt:

1. Wie bei der Wassersäule verringert sich mit zunehmender Höhe die Gewichtskraft $F = mg$ der Luftsäule oberhalb einer vorgegebenen Höhe $z = h$.
2. Im Gegensatz zu Wasser ist Gas gut kompressibel, sodass die Dichte ρ der Gasmoleküle im unteren Bereich der Luftsäule viel größer ist als weiter oben.

Die Änderung der Gewichtskraft dF mit zunehmender Höhe dz beträgt

$$dF = -g\,dm \tag{18.34}$$
$$= -g\,\rho(z)\,dV \tag{18.35}$$
$$= -\rho(z)\,g\,A\,dz\,. \tag{18.36}$$

Das Minuszeichen berücksichtigt die Abnahme der Gewichtskraft mit steigender Höhe dz. Die Dichte $\rho(z)$ ist eine Funktion der Höhe. Das Volumenelement mit der Einheitsfläche A bezeichnen wir mit $dV = A\,dz$.

Der Luftdruck nimmt dementsprechend mit steigender Höhe ab (Druckdefinition (18.1)):

$$dp = \frac{dF}{A} \tag{18.37}$$

$$= -\rho(z)\, g\, dz \tag{18.38}$$

Um die Höhenabhängigkeit des Luftdrucks zu bestimmen, benötigen wir einen quantitativen Zusammenhang zwischen dem Druck p und der Dichte ρ. Ein solcher Zusammenhang ist im Gesetz von Boyle-Mariotte enthalten (18.33).

Der Luftdruck p in der Höhe h und der Luftdruck $p_\circ$ auf Meereshöhe stehen dabei in folgendem Zusammenhang mit den Volumina V und $V_\circ$, die von den Luftmolekülen der vorgegebenen Masse $M = const.$ eingenommen werden:

$$p\, V = p_\circ\, V_\circ \tag{18.39}$$

Für die Verwendung dieses Zusammenhangs sollen die Temperaturvariationen in der Luftsäule gering ausfallen. Dabei müssen wir beachten, dass die Temperatur auf der absoluten Temperaturskala in Kelvin gemessen werden muss. Die Kelvinskala ($0\,\mathrm{K} = -273{,}15\,^\circ\mathrm{C}$) stellen wir in Abschnitt 20.3 vor. Bis ca. $50\,\mathrm{km}$ Höhe variiert die Temperatur bis ca. $-60\,^\circ\mathrm{C} \approx 213\,\mathrm{K}$, sodass wir mit (18.39) eine auf ca. 20% genaue Näherung erhalten.

Für die vorgegebene Masse M der Gasmoleküle können wir die Gleichung (18.39) für die Dichte auf Meereshöhe ($\rho_\circ = M/V_\circ$) und die Dichte in der Höhe h ($\rho = M/V$) umschreiben:

$$p\,\frac{M}{\rho} = p_\circ\,\frac{M}{\rho_\circ} \tag{18.40}$$

Damit ist die Dichte ρ proportional zum Druck p:

$$\rho = p\,\frac{\rho_\circ}{p_\circ} \tag{18.41}$$

Eingesetzt in Gleichung (18.38) erhalten wir

$$dp = -\rho\, g\, dz \tag{18.42}$$

$$= -\,p\,\frac{\rho_\circ}{p_\circ}\, g\, dz\,. \tag{18.43}$$

Diese Differentialgleichung können wir durch Trennung der Variablen und anschließende Integration lösen:

$$\int_{p_\circ}^{p(h)} \frac{dp}{p} = -\frac{\rho_\circ}{p_\circ} g \int_0^h dz \tag{18.44}$$

$$\ln\left(p\left(h\right)\right) - \ln\left(p_\circ\right) = -\frac{\rho_\circ}{p_\circ}\, g\, h \tag{18.45}$$

$$\ln\left(\frac{p\left(h\right)}{p_\circ}\right) = -\frac{\rho_\circ}{p_\circ}\, g\, h \tag{18.46}$$

Durch Anwendung der Exponentialfunktion auf beide Seiten der Gleichung erhalten wir die sogenannte **barometrische Höhenformel**

$$p(h) = p_\circ \cdot e^{-\frac{\rho_\circ}{p_\circ}\, g\, h}\,. \tag{18.47}$$

Für die Konstante in der Exponentialfunktion ergibt sich experimentell der Wert $\rho_\circ\, g/p_\circ = 0{,}1256\,\mathrm{km}^{-1}$.

Vergleich des Drucks unter Wasser und in der Erdatmosphäre

Wir vergleichen hier noch einmal die Höhenabhängigkeit des Drucks unter Wasser und in der Erdatmosphäre.

Luft ist sehr gut kompressibel, Wasser dagegen kaum. Der Unterschied zeigt sich in der Abhängigkeit des Drucks von der Höhe beziehungsweise Tiefe (▶ Abbildung 18.5):

$$p_{\text{Luft}}(h) = p_\circ \cdot e^{-\frac{\rho_\circ}{p_\circ} g h} \qquad p_{H_2O}(h) = -\rho\, g\, h \tag{18.48}$$

Es gibt *kein* scharfes Ende der Lufthülle. Die Wasseroberfläche zeigt dagegen eine klare Grenzfläche.

Der auf der Erdoberfläche herrschende Luftdruck $p_\circ$ halbiert sich erst in großer Höhe h:

$$\frac{p_\circ}{2} = p_\circ \cdot e^{-\frac{\rho_\circ}{p_\circ} g\, h\left(\frac{1}{2}p_\circ\right)} \tag{18.49}$$

$$h\left(\frac{p_\circ}{2}\right) = \frac{\ln 2}{0{,}1256\,\text{km}^{-1}} = 5{,}5\,\text{km} \tag{18.50}$$

Im Wasser erhalten wir eine Verdopplung im Vergleich zum Luftdruck $p_\circ$ schon nach einer Wassertiefe h von

$$2p_\circ = -\rho\, g\, h(2p_\circ) + p_\circ \tag{18.51}$$

$$h(2p_\circ) = -\frac{p_\circ}{\rho\, g} = -\frac{10^5}{10^3\, 10}\,\text{m} = -10\,\text{m}\,. \tag{18.52}$$

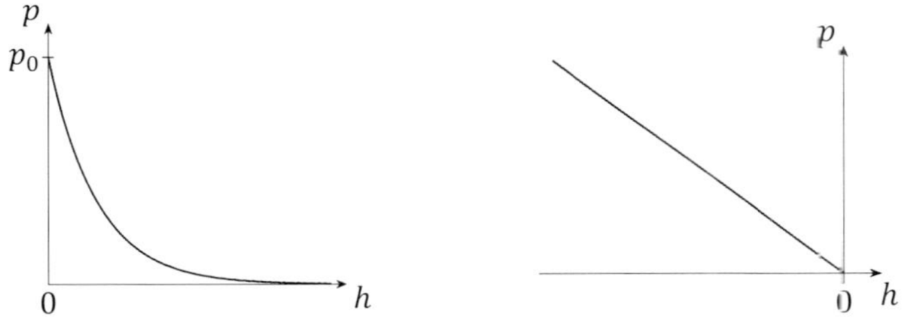

Abbildung 18.5 Links: Der barometrische Druck fällt exponentiell mit der Höhe. Rechts: Der Wasserdruck nimmt linear mit der Tiefe zu.

Aufgabe 55: Wetterballon

Wir füllen einen Wetterballon mit Helium, bis er einen Durchmesser von $d = 1\,\text{m}$ hat. Er steigt daraufhin bis zu einer Höhe von 25 km und zerplatzt (▶ Abbildung 18.6). Wie groß ist der Ballondurchmesser in 25 km Höhe?

Kugelvolumen:	$V = \frac{4}{3}\pi\left(\frac{d}{2}\right)^3$	Druck (h in km):	$p = p_\circ\, e^{-0{,}12\, h}$
Druck & Volumen:	$p\,V = p_\circ\, V_\circ$	Eulersche Zahl:	$e = 2{,}7$

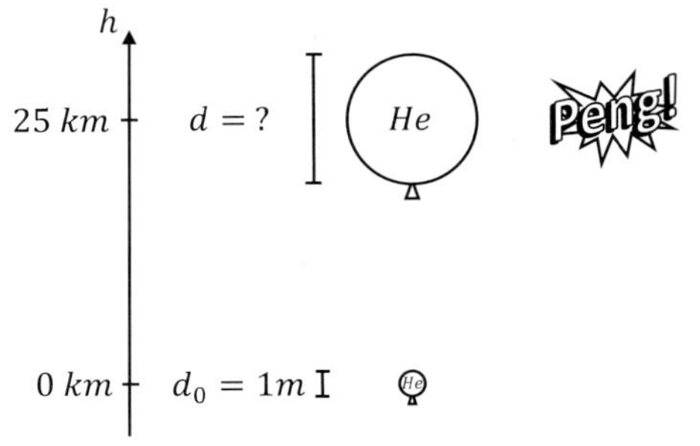

Abbildung 18.6 Heliumballon zerplatzt in großer Höhe.

18.4 Kontinuitätsgleichung

Masse ist eine Eigenschaft von Materie, z. B. eines Atoms, Moleküls oder Körpers. Die Masse dieser Objekte bleibt erhalten. Wir können diese Erhaltung dazu nutzen, um Aussagen über Fließgeschwindigkeiten von Flüssigkeiten in Rohren zu treffen.

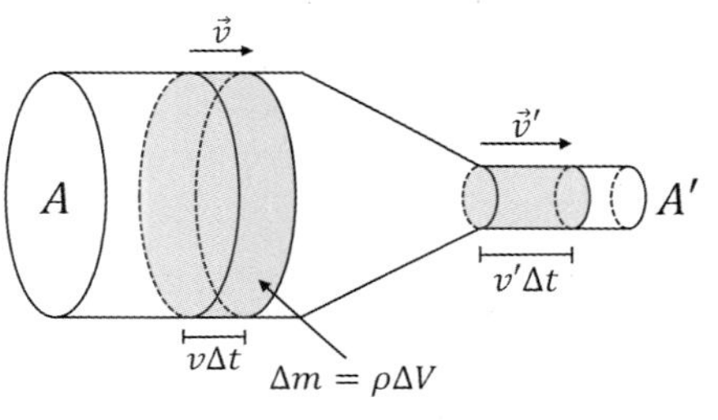

Abbildung 18.7 Kontinuitätsgleichung: Eine inkompressible Flüssigkeit erhöht bei einer Querschnittsverengung von A auf A' ihre Geschwindigkeit von v auf v'.

Die Masse Δm der Flüssigkeitsscheibe in dem Rohrteil mit der Querschnittsfläche A bewege sich mit der Geschwindigkeit v (► Abbildung 18.7). Das Scheibenvolumen beträgt $\Delta V = A\,\Delta x$. Die Koordinate Δx entlang der Bewegung können wir durch die Fließgeschwindigkeit v und das Zeitintervall Δt ausdrücken, in dem sich die Scheibe um die Strecke $\Delta x = v \cdot \Delta t$ nach rechts verschiebt. Zusammen mit der Dichte $\rho = \Delta m/\Delta V$ (14.3) der Flüssigkeit beschreiben wir die Masse Δm durch

$$\Delta m = const. \tag{18.53}$$

$$= \rho\, A\, \Delta x \tag{18.54}$$

$$= \rho\, A\, v\, \Delta t\,. \tag{18.55}$$

Bei dem kleineren Rohrdurchmesser A' ist die Geschwindigkeit für Δm modifiziert:

$$\Delta m = \rho'\, A'\, v'\, \Delta t \tag{18.56}$$

Für inkompressible Flüssigkeiten gilt $\rho = \rho'$. Bei kleinerem Rohrquerschnitt und gleichem Massendurchsatz hat die Flüssigkeit also eine höhere Geschwindigkeit:

$$A\,v = A'\,v' \tag{18.57}$$

Diese Gleichung wird im Sinn der Erhaltung der Masse als **Kontinuitätsgleichung** bezeichnet.

Um den Geschwindigkeitsverlauf von Strömungen zu visualisieren verwenden wir **Stromlinien** (► Abbildung 18.8).

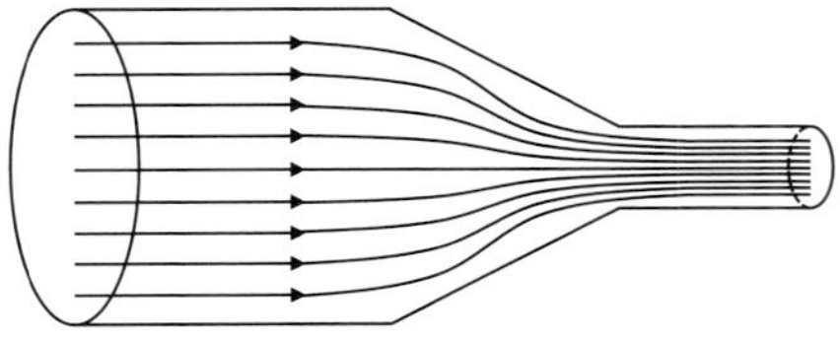

Abbildung 18.8 Die Stromlinien werden an einer Engstelle komprimiert.

Eine Stromlinie ist die Bahnkurve eines Flüssigkeitselements, die wir durch ein Korkstückchen sichtbar machen können. Dichtere Stromlinien implizieren eine größere Geschwindigkeit der Flüssigkeit.

Man unterscheidet **laminare** (lat. lamina »Blatt«) Strömungen, in denen sich die Stromlinien nicht überschneiden, und **turbulente** Strömungen mit Wirbeln. Wir werden, wenn nicht anders erwähnt, laminare Strömungen untersuchen.

18.4.1 Differentielle Form der Kontinuitätsgleichung

Im Folgenden wollen wir eine allgemeine Formulierung der Kontinuitätsgleichung (18.57) erhalten. Dafür führen wir die sogenannte **Massenstromdichte** $\vec{j}$ als Produkt aus der Dichte und der Geschwindigkeit der Flüssigkeit ein:

$$\vec{j} \equiv \rho\,\vec{v} \tag{18.58}$$

Mathematischer Einschub: Ringintegral, Oberflächenintegral

Integrale über eine geschlossene Kurve oder eine geschlossene Oberfläche kennzeichnen wir durch ein Kreissymbol auf dem Integralzeichen:

$$\oint \tag{18.59}$$

In der linken ► Abbildung 18.9 betrachten wir ein Volumen V mit der Masse $M = \int \rho\, dV$ zu einem festen Zeitpunkt t. Masse fließe nun kontinuierlich durch die Oberfläche S in das Volumen V hinein oder aus ihm heraus.

Für beliebig geformte Oberflächen verwenden wir das Oberflächenelement $d\vec{S}$, dessen Ausrichtung durch den Normalenvektor auf das Flächenelement beschrieben wird (rechte ► Abbildung 18.9). Die Geschwindigkeit $\vec{v}$ der Flüssigkeit durch diese Oberfläche kann im Allgemeinen einen Winkel mit der Oberflächennormalen bilden.

Wir interessieren uns für die effektive Oberflächengröße, die senkrecht zur Richtung der Geschwindigkeit durchflossen wird und verwenden deswegen das Skalarprodukt zwischen $\vec{v}$ und $d\vec{S}$.

Abbildung 18.9 Links: Ein Volumen V mit der Oberfläche S zu einem festen Zeitpunkt t. Rechts: Der Fluss ergibt sich als das Skalarprodukt aus der Geschwindigkeit $\vec{v}$ und dem Oberflächenelement $d\vec{S}$ senkrecht zur Oberfläche.

Für geschlossene Oberflächen soll $d\vec{S}$ nach außen zeigen. Mit den Gleichungen (18.55) und (18.58) ist dann die Massenänderung im Volumen pro Zeiteinheit durch den Fluss der Masse durch die gesamte Oberfläche gegeben, wobei $\vec{v}\, d\vec{S} < 0$ für den Zufluss gilt und $\vec{v}\, d\vec{S} > 0$ für das Ausfließen:

$$-\frac{\partial m}{\partial t} = \oint_s \rho\, \vec{v}\, d\vec{S} \tag{18.60}$$

$$= \oint_s \vec{j}\, d\vec{S} \tag{18.61}$$

Mathematischer Einschub: Integralsatz von Gauß

Die Divergenz oder Quellstärke eines Vektorfelds $\vec{j}$ ist definiert durch die Summe der partiellen Ableitungen

$$\nabla\vec{j} = \frac{\partial j_x}{\partial x} + \frac{\partial j_y}{\partial y} + \frac{\partial j_z}{\partial z}\,. \tag{18.62}$$

Für das Vektorfeld der Vektoren $\vec{j}$ bilden wir das Volumenintegral über die Divergenz $\nabla\vec{j}$. Nach dem Satz von Gauß entspricht es dem über die geschlossene Oberfläche des Volumens integrierten Skalarprodukt aus dem Oberflächennormalenvektor $d\vec{S}$ und $\vec{j}$:

$$\oint_S \vec{j}\, d\vec{S} = \int_V \nabla\vec{j}\, dV \tag{18.63}$$

Mit dem Integralsatz von Gauß können wir das Oberflächenintegral aus Gleichung (18.61) in ein Volumenintegral umwandeln und erhalten

$$-\frac{\partial m}{\partial t} = \int_V \nabla\vec{j}\,dV\,. \tag{18.64}$$

Die linke Seite von Gleichung (18.64) lässt sich mit den Größen Dichte und Volumen umschreiben ($m = \rho V$):

$$-\frac{\partial}{\partial t}\int \rho\,dV = \int \nabla\vec{j}\,dV \tag{18.65}$$

$$-\int \frac{\partial \rho}{\partial t}\,dV = \int \nabla\vec{j}\,dV \tag{18.66}$$

Durch Vergleich der Integranden folgt die **differentielle Form der Kontinuitätsgleichung**:

$$\frac{\partial \rho}{\partial t} + \nabla\vec{j} = 0 \tag{18.67}$$

Beispiel 48: Gasflasche

Eine Gasflasche enthalte Gas unter hohem Druck und mit dementsprechend hoher Dichte $\rho \propto p$ (18.41). Zum Zeitpunkt $t = 0$ wird das Ventil geöffnet und Gas strömt mit der Massenstromdichte $\vec{j} = \rho\,\vec{v}$ aus der Flasche. Mit der Zeit nimmt der Druck in der Flasche ab, dementsprechend sinkt die Massenstromdichte. Die Kontinuitätsgleichung (18.67) beschreibt diesen Zusammenhang zwischen Gasverlust und Fluss: Die sich ändernde Quellstärke (Divergenz) $\nabla\vec{j}$ führen wir auf die mit der Zeit abnehmende Dichte des Gases in der Flasche $\partial\rho/\partial t$ zurück.

18.5 Bernoulli-Gleichung

Die Bernoulli-Gleichung beschreibt die Druckverhältnisse bei fließenden Materialien. Sie basiert auf der Energieerhaltung.

In ► Abbildung 18.10 sind Flüssigkeitsscheiben gezeigt, die wir gegen einen Druck p verschieben wollen. Um die Flüssigkeitsscheibe mit dem Volumen $\Delta V_1 = A_1\,\Delta x_1$ im

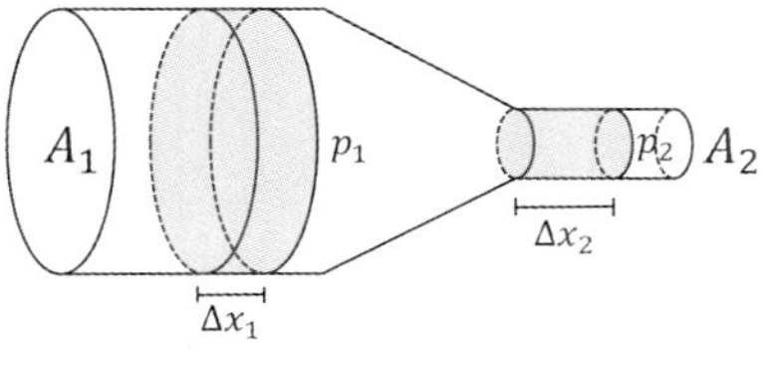

Abbildung 18.10 Bernoulli-Gleichung: Bei einer Verengung von A_1 nach $A_2 < A_1$ verringert sich der statische Druck von p_1 nach $p_2 < p_1$.

Rohr mit der Querschnittsfläche A_1 um Δx_1 zu verschieben, müssen wir Arbeit gegen den Druck p_1 verrichten. Die entsprechende Kraft ist $F_1 = p_1\, A$ (18.1) und die geleistete Arbeit ΔW_1 (5.1) beträgt

$$\Delta W_1 = F_1\, \Delta x_1 \tag{18.68}$$

$$= p_1\, A_1\, \Delta x_1 \tag{18.69}$$

$$= p_1\, \Delta V_1\,. \tag{18.70}$$

Die Flüssigkeitsscheibe gewinnt dabei die potentielle Energie

$$E_{\text{pot},1} = p_1\, \Delta V_1\,. \tag{18.71}$$

Im Rohrbereich mit der Querschnittsfläche A_2 sind die Arbeit und die gewonnene potentielle Energie entsprechend

$$\Delta W_2 = p_2\, \Delta V_2 \tag{18.72}$$

$$= E_{\text{pot},2}\,. \tag{18.73}$$

Die kinetischen Energien $E_{\text{kin}} = mv^2/2$ (5.23) der Flüssigkeitsscheiben sind in den beiden Rohrabschnitten

$$E_{\text{kin},1} = \frac{1}{2}\,\rho_1\, \Delta V_1\, v_1^2 \tag{18.74}$$

$$E_{\text{kin},2} = \frac{1}{2}\,\rho_2\, \Delta V_2\, v_2^2\,. \tag{18.75}$$

Das Gesamtsystem sei abgeschlossen, dementsprechend ist die Gesamtenergie erhalten:

$$E = E_{\text{pot}} + E_{\text{kin}} \tag{18.76}$$

An beiden Stellen des Rohrs gilt dann für die Gesamtenergien

$$E_1 = E_2 \tag{18.77}$$

$$E_{\text{kin},1} + E_{\text{pot},1} = E_{\text{kin},2} + E_{\text{pot},2}\,. \tag{18.78}$$

Durch Einsetzen der Terme (18.71), (18.73), (18.74) und (18.75) erhalten wir

$$\frac{1}{2}\,\rho_1\, \Delta V_1\, v_1^2 + p_1\, \Delta V_1 = \frac{1}{2}\,\rho_2\, \Delta V_2\, v_2^2 + p_2\, \Delta V_2\,. \tag{18.79}$$

Für inkompressible Flüssigkeiten ist die Dichte $\rho = m/V = const.$. Dann sind auch die Volumina V bezogen auf dieselbe Masse m der Flüssigkeit gleich groß $\Delta V_1 = \Delta V_2$. Damit vereinfacht sich die Gleichung zur sogenannten **Bernoulli-Gleichung**:

$$\underbrace{p_\circ}_{\text{Gesamtdruck}} = \underbrace{\frac{1}{2}\rho v^2}_{\text{Staudruck}} + \underbrace{p}_{\text{statischer Druck}} = const. \tag{18.80}$$

Der **Gesamtdruck** eines abgeschlossenen Systems bleibt konstant. Er setzt sich aus dem kinetischen **Staudruck** und dem **statischen Druck** zusammen.

Nach der Kontinuitätsgleichung (18.57) wird die Geschwindigkeit an engen Stellen im Rohr größer. Damit steigt die kinetische Energie bzw. der Staudruck. Wegen der Erhaltung des Gesamtdrucks nimmt der statische Druck entsprechend ab.

Experiment 49: Bernoulli-Gleichung

Durch ein Glasrohr mit variabler Querschnittsfläche lassen wir Wasser fließen (▶ Abbildung 18.11). Das Rohr hat an drei Stellen mit verschiedenen Querschnittsflächen Steigrohre, mit denen der statische Druck angezeigt wird. Zumindest qualitativ lässt sich mit diesem Experiment demonstrieren, dass der statische Druck in Rohrbereichen mit kleiner Querschnittsfläche im Vergleich zu den Bereichen mit großer Querschnittsfläche geringer ist.

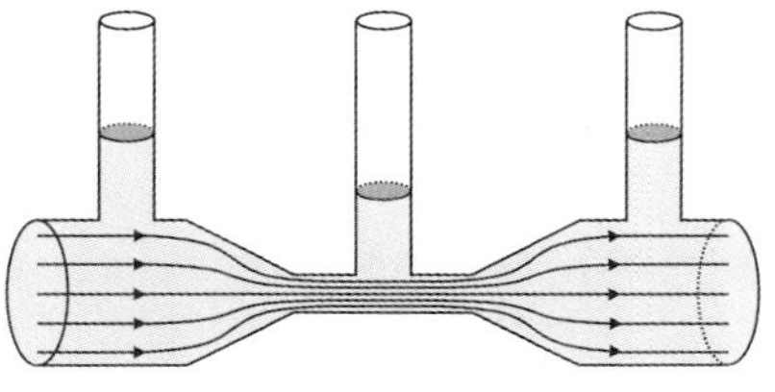

Abbildung 18.11 Bernoulli: An der Verengung sinkt der statische Druck bei erhöhter Strömungsgeschwindigkeit.

18.5.1 Dynamischer Auftrieb

An der Tragfläche eines Flugzeugs kommt es durch unterschiedliche Strömungsgeschwindigkeiten zum Auftrieb. In der linken ▶ Abbildung 18.12 sind die Stromlinien an der Oberseite des Profils dichter als an der Unterseite. Nach der Kontinuitätsgleichung (18.57) ist die Geschwindigkeit der Luftmoleküle auf der Oberseite höher:

$$v_1 > v_2 \tag{18.81}$$

Nach der Bernoulli-Gleichung (18.80) wird damit der Druck oberhalb der Tragfläche geringer:

$$p_1 < p_2 \tag{18.82}$$

Es kommt also aufgrund des Druckunterschieds zu der Auftriebskraft $\vec{F}$, die das Fliegen ermöglicht (rechte ▶ Abbildung 18.12).

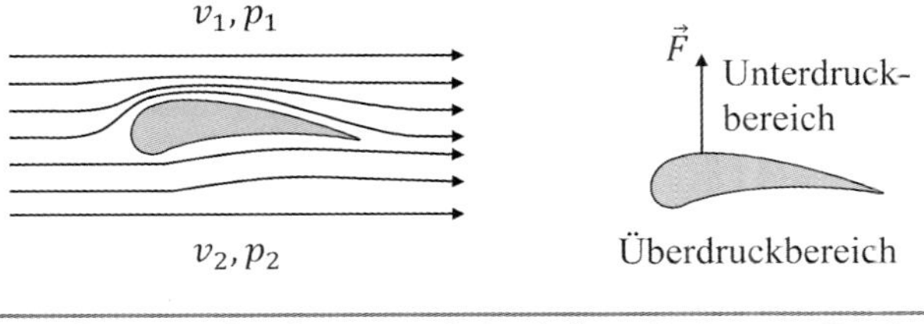

Abbildung 18.12 Links: Unterdruck $p_1 < p_2$ bei erhöhter Strömungs-Geschwindigkeit $v_1 > v_2$ an einer Tragfläche, rechts: Auftrieb $\vec{F}$ an einer Tragfläche durch Unter- und Überdruck

18.5.2 Hydrodynamisches Paradoxon

Im folgenden Experiment lassen wir Druckluft auf eine Platte strömen, die daraufhin *gegen* die Strömung angezogen wird.

Experiment 50: Hydrodynamisches Paradoxon

Wir lassen Druckluft mit geringem Überdruck p_D durch ein Rohr mit der Querschnittsfläche A_1 strömen (► Abbildung 18.13). Die Luft tritt durch die Öffnung der Größe A_2 zwischen dem Rohrende und einer Scheibe mit der Querschnittsfläche S und der Masse m aus.

Die Scheibe ist dabei durch Führungsstifte in kleinem Abstand zum Rohr gehalten. Lassen wir die Druckluft durch das Rohr strömen, hören wir ein periodisches Schlagen der Scheibe.

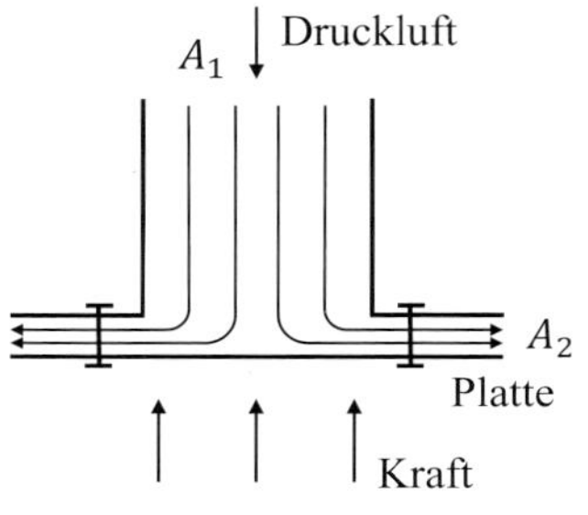

Abbildung 18.13 Bei kleiner Austrittsöffnung A_2 und damit hoher Austrittsgeschwindigkeit v_2 wird die Scheibe durch den Unterdruck angezogen und fällt wieder herunter, wenn sie das Rohr verschlossen hat.

Wir bezeichnen den Luftdruck in der Umgebung des Experiments mit p_L (► Abbildung 18.13). Nach der Bernoulli-Gleichung (18.80) werden die Druckverhältnisse innerhalb des senkrechten Rohrs mit dem Querschnitt A_1, Staudruck $(1/2)\,\rho v_1^2$, statischem Druck p_1 und dem Überdruck p_D beschrieben durch

$$p_L + p_D = \frac{1}{2}\,\rho\, v_1^2 + p_1\,. \tag{18.83}$$

Im Rohr baut sich der Überdruck p_D ab, sodass an der Öffnung A_2 der Luftdruck p_L herrscht (Genaueres dazu erklären wir in Abschnitt 18.6 und ► Abbildung 18.17). Zwischen der Öffnung A_2 und der unteren Scheibe können wir den ohnehin kleinen Druck p_D daher vernachlässigen und erhalten entsprechend

$$p_L = \frac{1}{2}\,\rho\, v_2^2 + p_2\,. \tag{18.84}$$

Die Scheibe mit der Querschnittsfläche S erfährt dann eine Auftriebskraft F_A, wenn der statische Druck p_2 im Austrittsbereich A_2 kleiner als der äußere Luftdruck $p_L > p_2$ ist:

$$F_A = (p_L - p_2)\, S \tag{18.85}$$

Ist die Auftriebskraft sogar größer als die Gewichtskraft der Scheibe so wird die Scheibe angehoben:

$$F_A > F_G = m \cdot g \tag{18.86}$$

Die Bedingung für das Anheben der Scheibe wollen wir auf die Fließgeschwindigkeit v_1 im senkrechten Rohr zurückführen (▶ Abbildung 18.13). Nach der Kontinuitätsgleichung (18.57) gilt

$$\frac{v_2}{v_1} = \frac{A_1}{A_2}\,. \tag{18.87}$$

Wenn die Öffnung A_2 kleiner als der Rohrquerschnitt A_1 ist, wird die Geschwindigkeit v_2 der ausströmenden Druckluft bei A_2 größer als die Geschwindigkeit v_1 im senkrechten Rohr ($v_2 > v_1$). Wir berechnen die Druckdifferenz $p_L - p_2$ aus (18.84) und erhalten mit (18.87)

$$p_L - p_2 = \frac{1}{2}\,\rho\, v_2^2 \tag{18.88}$$

$$= \frac{1}{2}\,\rho\, v_1^2 \left(\frac{A_1}{A_2}\right)^2\,. \tag{18.89}$$

Mit der Auftriebskraft (18.85) und der Schwerkraftsbedingung (18.86) erhalten wir die Bedingung für das Anheben der Scheibe

$$\frac{1}{2}\,\rho\, v_1^2 \left(\frac{A_1}{A_2}\right)^2 \cdot S > m\, g\,. \tag{18.90}$$

Die Scheibe muss also möglichst dünn sein, um für relativ großes S die Masse m klein zu halten, und sie muss in kleinem Abstand vom Rohrende gehalten werden ($A_2 \ll A_1$), um (18.90) erfüllen zu können.

Demnach wird bei genügend kleiner Öffnung A_2 und geeigneter Wahl des Überdrucks p_D die Scheibe durch den Unterdruck angehoben. Dadurch wird A_2 noch kleiner und der Auftrieb weiter verstärkt, bis die Scheibe die Austrittsöffnung verschließt. Druckluft und Schwerkraft stoßen die Scheibe jetzt zurück in ihre Ausgangsposition und der Prozess startet erneut, sodass sich ein periodisches, lautes Schlagen der Scheibe zwischen Verschluss- und Ausgangsposition ergibt.

Die Bernoulli-Gleichung und das hydrodynamische Paradoxon haben viele Anwendungen wie das Prandtl- oder Venturi-Rohr bei der Geschwindigkeitsmessung oder Wasserstrahlpumpen und Zerstäuber.

18.6 Innere Reibung in Flüssigkeiten

In einem Gedankenexperiment betrachten wir zwei parallele Platten, zwischen denen sich eine Flüssigkeit befindet (▶ Abbildung 18.14). Eine Platte wird mit einer konstanten Kraft

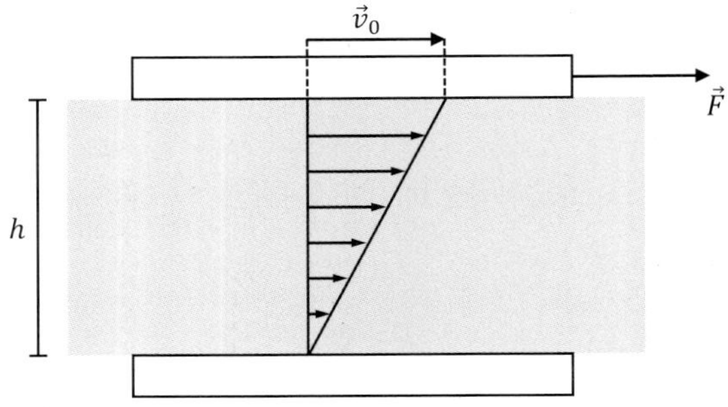

Abbildung 18.14 Newtonsche Reibung: Zwischen den gegeneinander bewegten Platten bildet sich in der Flüssigkeit ein Geschwindigkeitsgefälle aus.

$\vec{F}$ parallel an der anderen entlanggezogen. Nach einer kurzen Anfangsphase bewegt sich die Platte mit konstanter Geschwindigkeit $v = v_\circ$.

Zwischen den Platten bildet sich ein Geschwindigkeitsgefälle dv/dh. Dabei gleiten Flüssigkeitsschichten mit Reibung aneinander entlang, weil die Moleküle an den Platten und untereinander haften. Die Bewegung von Schichten bedeutet, dass es sich um eine laminare Strömung handelt.

Aufgrund der konstanten Geschwindigkeit $v_\circ$ ergibt sich zwischen der Zugkraft $\vec{F}$ und einer Reibungskraft $\vec{F_R}$ ein Kräftegleichgewicht $\vec{F_R} = -\vec{F}$. Empirisch findet man für die Reibungskraft F_R:

- F_R ist proportional zur Fläche A der Platten: $F_R \propto A$
- F_R ist proportional zum Geschwindigkeitsgefälle: $F_R \propto dv/dh$

Zusammengefasst führt dies zum **Newtonschen Reibungsgesetz**

$$F_R = \eta \, A \, \frac{dv}{dh} \, . \qquad (18.91)$$

Die Materialkonstante η wird **Koeffizient der inneren Reibung** und auch **Viskosität** oder dynamische Zähigkeit genannt.

Die SI-Einheit der Viskosität ist $\mathrm{N\,s/m^2} = \mathrm{Pa\,s}$. Der Kehrwert der Viskosität $1/\eta$ wird **Fluidität** genannt. Die Viskosität von Flüssigkeit ist stark temperaturabhängig. Als Beispiel sind in ► Tabelle 18.1 die Temperaturabhängigkeit der Viskosität von Wasser und Glyzerin angeführt.

Temperatur T	$\eta(T)$ [Pa s] von Wasser	$\eta(T)$ [Pa s] von Glycerin
0 °C	1,79	12 100
20 °C	1,002	1 480
100 °C	0,282	14,8

Tabelle 18.1 Temperaturabhängigkeit der Viskosität von Wasser und Glyzerin

18.6.1 Reibung bei Flüssigkeitsströmung durch ein Rohr

Das Newtonsche Reibungsgesetz kann genutzt werden, um die Geschwindigkeit einer **laminaren Strömung** durch ein Rohr zu berechnen. Dazu betrachten wir ein Rohr mit dem Radius R und der Länge l (► Abbildung 18.15). Zwischen den beiden Enden des Rohres liegt eine Druckdifferenz $p_1 - p_2$ vor, die eine Flüssigkeitsströmung durch das Rohr verursacht.

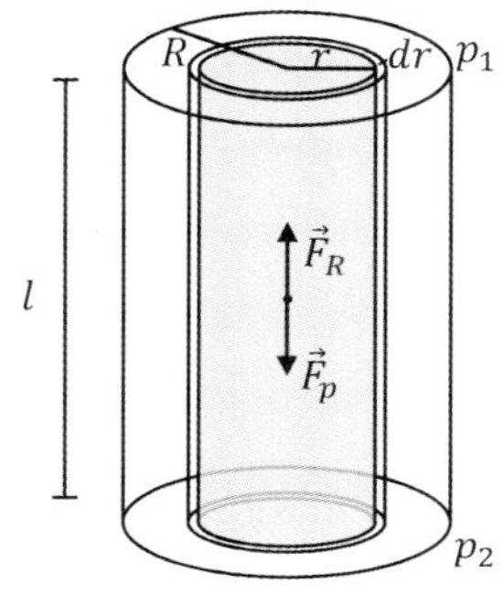

Abbildung 18.15 Auf ein zylindrisches Flüssigkeitselement mit Radius r und Dicke dr in einem Rohr mit Radius R wirkt ein Druck $p_1 > p_2$. Im stationären Gleichgewicht gilt für Reibungs- und Druckkräfte $\vec{F_p} = -\vec{F_R}$.

Konzentrisch im Rohr liegt ein Flüssigkeitszylinder mit dem Radius r und der Länge l. Aufgrund der Druckdifferenz wirkt die Kraft $F_p = (p_1 - p_2)\,A$ (18.1), wobei $A = \pi\,r^2$ die Querschnittsfläche des Flüssigkeitszylinders bezeichnet:

$$F_p = \pi\,r^2\,(p_1 - p_2) \tag{18.92}$$

Zunächst berechnen wir mit dem Newtonschen Reibungsgesetz (18.91) die Reibungskraft an der Mantelfläche $A_M = 2\pi\,r\,l$ des Flüssigkeitszylinders:

$$F_R = \eta\,A_M\,\frac{dv}{dr} \tag{18.93}$$

$$= \eta\,2\pi\,r\,l\,\frac{dv}{dr}\,. \tag{18.94}$$

Diese Reibungskraft F_R wirkt der Druckkraft F_p (18.92) entgegen, mit der die Druckdifferenz $p_1 - p_2$ auf den Flüssigkeitszylinder wirkt. Nach einer Anfangsphase stellt sich eine sogenannte stationäre Strömung mit konstanter Fließgeschwindigkeit ein, bei der die Reibungskraft F_R entgegengesetzt gleich der Druckkraft F_p ist:

$$F_p = -F_R \tag{18.95}$$

$$\pi\,r^2\,(p1 - p2) = -\eta\,2\pi\,r\,l\,\frac{dv}{dr} \tag{18.96}$$

Die Geschwindigkeitsverteilung dv/dr ist dabei noch unbekannt. Um zu berechnen, wie sich die Strömungsgeschwindigkeit mit der Entfernung vom Rohrmantel verändert, lösen wir diese Differentialgleichung durch Trennung der Variablen:

$$\frac{dv}{dr} = -\frac{p_1 - p_2}{2\,\eta\,l}\,r \tag{18.97}$$

$$\int_{v_r}^{v_R} dv = -\frac{p_1 - p_2}{2\,\eta\,l}\int_r^R r\,dr \tag{18.98}$$

Mit der Randbedingung $v_R = 0$, denn die äußersten Flüssigkeitsmoleküle haften am Rohr, ergibt sich daraus die Geschwindigkeitsverteilung in Abhängigkeit des Abstands r zum Rohr:

$$0 - v_r = -\frac{p_1 - p_2}{2\,\eta\,l}\frac{1}{2}(R^2 - r^2) \tag{18.99}$$

$$v(r) = \frac{p_1 - p_2}{4\,\eta\,l}(R^2 - r^2) \tag{18.100}$$

Es handelt sich um eine parabolische Geschwindigkeitsverteilung $v(r)$ (► Abbildung 18.16) mit den Extremwerten

$$v(R) = 0 \quad \text{und} \quad v(0) = v_\circ = \frac{p_1 - p_2}{4\eta l}R^2\,. \tag{18.101}$$

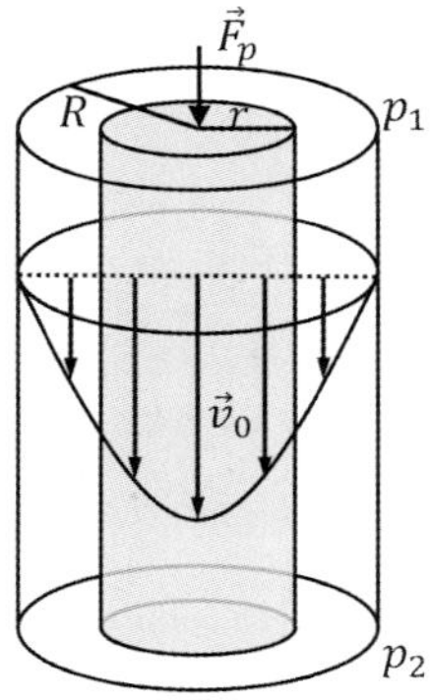

Abbildung 18.16 Das Druckgefälle $p_1 - p_2$ führt in einem Rohr mit viskoser Flüssigkeit zu einer parabolischen Geschwindigkeitsverteilung.

18.6.2 Hagen-Poiseuille-Gesetz

Welche Flüssigkeitsmenge können wir in einer vorgegebenen Zeit durch das Rohr in ► Abbildung 18.16 transportieren? Für den Gesamtstrom durch das Rohr berechnen wir zunächst die Flüssigkeitsmenge dV, die in einer vorgegebenen Zeit t durch den Rohrquerschnitt zwischen r und $r + dr$ fließt. Dieses Flüssigkeitsvolumen dV können wir uns als Strohhalm mit Wanddicke dr vorstellen, der die Querschnittsfläche $dA = 2\pi\,r\,dr$ und

Länge $l = v(r)\,t$ aufweist. Mit dieser Längenbeschreibung berücksichtigen wir, dass die Geschwindigkeit $v(r)$ (18.100) vom Radius r abhängt:

$$dV = dA\,v(r)\,t \tag{18.102}$$

$$dV = 2\pi\,r\,dr\,v(r)\,t\,. \tag{18.103}$$

Um die in der Zeit t transportierte Flüssigkeitsmenge V zu erhalten, integrieren wir über den Rohrradius:

$$V = \int_0^{V(R)} dV \tag{18.104}$$

$$= 2\pi\,t \int_0^R v(r)\,r\,dr \tag{18.105}$$

$$= 2\pi\,t \int_0^R \frac{p_1 - p_2}{4\,\eta\,l}\,(R^2 - r^2)\,r\,dr \tag{18.106}$$

$$= 2\pi\,t\,\frac{p_1 - p_2}{4\,\eta\,l}\left(R^2 \int_0^R r\,dr - \int_0^R r^3 dr\right) \tag{18.107}$$

$$= \pi\,t\,\frac{p_1 - p_2}{2\,\eta\,l}\left(R^2\,\frac{1}{2}\,R^2 - \frac{1}{4}\,R^4\right) \tag{18.108}$$

Dividieren wir durch die Zeit t, ergibt sich das von dem Arzt J. L. Poiseuille und dem Ingenieur G. H. L. Hagen (1838–1840) aufgestellte Gesetz von **Hagen-Poiseuille** für den Flüssigkeitsstrom $J = V/t$:

$$J = \frac{V}{t} = \frac{\pi\,(p_1 - p_2)}{8\,\eta\,l}\,R^4 \tag{18.109}$$

Es liefert eine starke R^4-Abhängigkeit der Stromstärke und eine lineare Abhängigkeit vom Druckgefälle pro Länge $(p_1 - p_2)/l$.

18.6.3 Druckgefälle entlang des Rohrs

Wenn die Details eines Geschwindigkeitsprofils $v(r)$ (18.100) nicht wichtig sind, können wir auch eine effektive oder mittlere Strömungsgeschwindigkeit $\bar{v}$ entlang des Rohrs der Länge l definieren (► Abbildung 18.15). Dafür nutzen wir die Stromstärke $J = V/t = A\,l/t$ dividiert durch die durchflossene Fläche $A = \pi R^2$:

$$\bar{v} = \frac{J}{\pi\,R^2} \tag{18.110}$$

Einsetzen von J aus (18.109) führt zu einem Kräftevergleich:

$$J = \pi\,R^2\,\bar{v} \tag{18.111}$$

$$\frac{\pi\,(p_1 - p_2)}{8\,\eta\,l}\,R^4 = \pi\,R^2\,\bar{v} \tag{18.112}$$

$$\underbrace{\pi\,R^2\,(p_1 - p_2)}_{=F_p} = \underbrace{8\,\pi\,\eta\,l\,\bar{v}}_{\equiv -F_R} \tag{18.113}$$

Auf der linken Seite sehen wir die Kraft F_p (18.92), die durch das Druckgefälle entsteht. Für eine stationäre Strömung mit konstanter Geschwindigkeit steht ihr eine Reibungskraft F_R entgegen. Diese Reibungskraft lautet in vektorieller Schreibweise

$$\vec{F}_R = -8\pi\, l\, \eta\, \vec{\bar{v}}\,. \tag{18.114}$$

Interessant ist dabei die Abhängigkeit von der Länge l des Rohrs. Sie führt zu einem Druckabfall, den wir aus (18.113) berechnen:

$$F_p = -F_R \tag{18.115}$$

$$\pi\, R^2\,(p_1 - p_2) = 8\,\pi\,\eta\, l\,\bar{v} \tag{18.116}$$

$$p_1 - p_2 = \frac{8\eta\bar{v}}{R^2}\cdot l \propto l \tag{18.117}$$

Für ein Rohr mit konstantem Radius ist nach der Kontinuitätsgleichung (18.57) die Fließgeschwindigkeit $\bar{v} = const.$. Der Reibungswiderstand ist also für das lineare Druckgefälle in einem Rohr verantwortlich (► Abbildung 18.17).

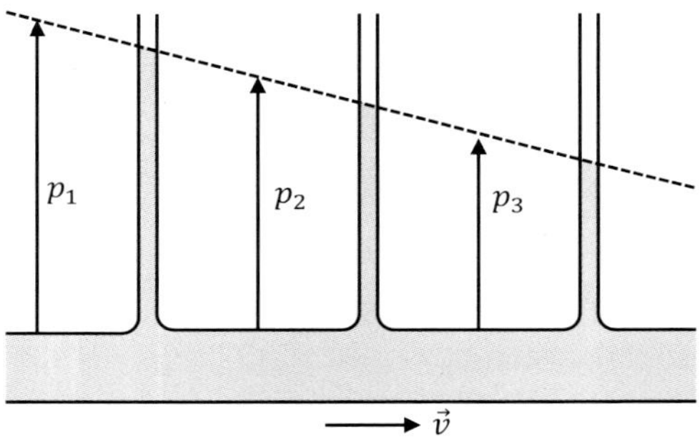

Abbildung 18.17 Aufgrund der inneren Reibung bildet sich in einer viskosen Flüssigkeit ein Druckgefälle aus.

18.6.4 Stokessches Gesetz

Eine ähnliche, aber kompliziertere Rechnung liefert für eine Kugel vom Radius r in einer laminaren Strömung das Reibungsgesetz

$$\vec{F}_R = -6\pi\, r\, \eta\, \vec{v}\,, \tag{18.118}$$

wobei v die Relativgeschwindigkeit der Kugel gegen die Flüssigkeit ist (► Abbildung 18.18). Die Reibungskraft der **Stokesschen Reibung** ist also proportional zur Geschwindigkeit. Im Folgenden betrachten wir zu diesem Reibungsgesetz einen Versuch.

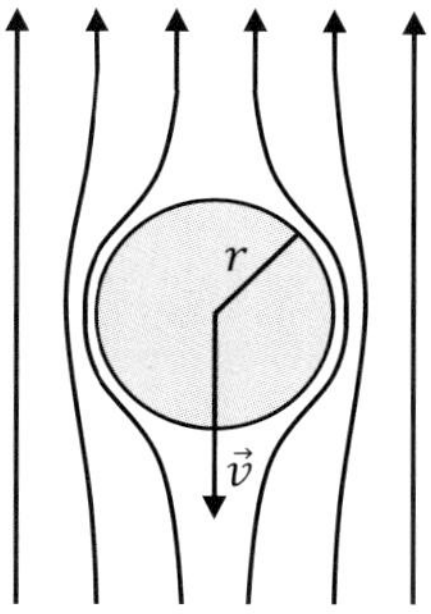

Abbildung 18.18 An einer Kugel vom Radius r, die sich in einer laminaren Strömung mit der Relativgeschwindigkeit $\vec{v}$ bewegt, greift die Reibungskraft $\vec{F}_R = -6\pi\, r\, \eta\, \vec{v}$ an.

Experiment 51: Stokessche Reibung

Bei diesem Versuch soll die Sinkgeschwindigkeit v verschieden großer Kugeln in einem Zylinder gemessen werden, der mit einer zähen Flüssigkeit gefüllt ist. Auf dem Zylinder sind zwei Messmarken angebracht. Während des Versuchs wird die Zeit gestoppt, die die Kugeln brauchen, um den Weg zwischen den beiden Markierungen zurückzulegen. Die erste Markierung ist unterhalb der Oberfläche der Flüssigkeit angebracht, damit sich erst ein Gleichgewicht einstellen kann. Die Kugeln werden von der Gewichtskraft F_G nach unten angezogen. Dem entgegen wirken die Reibungs- und Auftriebskraft F_R und F_A, entsprechend wählen wir die Vorzeichen. Für die Kräfte gilt

$$F_G = \rho_K\, g\, V_K \tag{18.119}$$

$$F_R = -6\pi\, \eta\, r\, v \tag{18.120}$$

$$F_A = -\rho_{Fl}\, g\, V_K\,, \tag{18.121}$$

wobei der Index K sich auf die Kugel und der Index Fl auf die Flüssigkeit bezieht. Für die Sinkgeschwindigkeit ergibt sich im Kräftegleichgewicht

$$v = \frac{2g(\rho_K - \rho_{Fl})}{9\eta} {r_K}^2\,. \tag{18.122}$$

Der Versuch mit verschiedenen Kugeln bestätigt, dass die Sinkgeschwindigkeit proportional zum Quadrat der Radien der Kugeln steigt.

18.6.5 Allgemeine Strömungen in Fluiden

Wir haben bisher nur **laminare Strömungen** in Flüssigkeiten betrachtet. Die meisten Überlegungen lassen sich auch auf Gase anwenden, man spricht deshalb auch häufig allgemein von Fluiden. Bei Gasströmungen muss man jedoch die Kompressibilität berücksichtigen, indem man z. B. den Massenstrom (18.58) statt des Volumenstroms verwendet.

Bei höheren Geschwindigkeiten treten Wirbel auf, die Strömungen werden **turbulent**. Der Strömungswiderstand wird dann durch das empirische Gesetz

$$F_W = c_W \cdot A \, \frac{1}{2} \, \rho \, v^2 \tag{18.123}$$

beschrieben. Dabei ist A die umströmte Querschnittsfläche und $\frac{1}{2}\,\rho\,v^2$ der Staudruck (18.80) mit der Dichte ρ des Gases oder der Flüssigkeit und der Relativgeschwindigkeit v zwischen Körper und Strömung. Die Luftwiderstandszahl c_W ist eine empirische Größe, die von den Eigenschaften der angeströmten Oberfläche abhängt.

18.7 Druckwellen

In diesem Abschnitt untersuchen wir die Wellenausbreitung in festen Körpern, Flüssigkeiten und Gasen. Wir fokussieren auf Longitudinalwellen, die auf periodischen Druckschwankungen entlang der Ausbreitungsrichtung beruhen. Offensichtliche Beispiele für solche Wellen sind Schallwellen.

Ein anschauliches Beispiel einer elastischen Longitudinalwelle hatten wir bereits mit der Federwelle in ► Abbildung 16.1 vorgestellt. Eine elastische Verformung breitet sich entlang der Feder aus. Die dabei wirkende Rückstellkraft der Feder (15.8) kennen wir von der Federwaage und vom Federpendel:

$$F = -k\,x \tag{18.124}$$

18.7.1 Längenänderung

Im Fall eines Festkörpers können wir uns die Atome oder Moleküle durch elastische Kräfte ähnlich der Federkraft (18.124) verbunden denken. Um die Wirkung einer Kraft auf ein Volumenelement zu beschreiben, nutzen wir die **Druck- oder Zugspannung**

$$\sigma = \frac{F}{A}\,, \tag{18.125}$$

die die Kraft pro Fläche senkrecht zur Kraftrichtung beschreibt (vergleiche (18.1) und ► Abbildung 18.19). Dabei setzen wir voraus, dass die Verformungen nicht über den linearen elastischen Bereich hinausgehen und erhalten das **Hookesche Gesetz** (R. Hooke 1676)

$$\sigma = E \cdot \varepsilon = E \cdot \frac{\Delta l}{l}\,. \tag{18.126}$$

Dabei bezeichnet E das **Elastizitätsmodul** und ε ist die relative Längenänderung $\varepsilon = \Delta l / l$. Das positive Vorzeichen bedeutet, dass im Unterschied zur rücktreibenden Kraft in (18.124), die durch die Verformung der Feder verursacht wird, in (18.126) die Spannung σ die Ursache für die Verformung Δl ist. Dabei sind die Vorzeichen so gewählt, dass eine positive Zugspannung $\Delta l > 0$ und eine negative Druckspannung $\Delta l < 0$ bewirken.

Typische Werte für E sind $10^9\,\mathrm{N/m^2}$ für Holz, $2 \cdot 10^{11}\,\mathrm{N/m^2}$ für Stahl und $10^{12}\,\mathrm{N/m^2}$ für Diamant.

18.7.2 Volumenänderung

In Flüssigkeiten, in geringerem Maße auch in Festkörpern, gibt es die elastische Volumenänderung unter dem Druck $p = F/A$ (18.1). Auch für den Druck auf ein Volumenelement eines Festkörpers oder einer Flüssigkeit gilt ein Hookesches Gesetz

$$p = K \cdot \frac{\Delta V}{V} \tag{18.127}$$

mit der relativen Volumenänderung $\Delta V/V$ und dem **Kompressionsmodul** K. Die Größenordnung von K variiert sehr stark von $10^5\ \mathrm{N/m^2}$ für Gase, $10^9\ \mathrm{N/m^2}$ für Flüssigkeiten und $10^{11-12}\ \mathrm{N/m^2}$ für Festkörper.

18.7.3 Wellen in festen Körpern und Flüssigkeiten

Eine elastische Longitudinalwelle lässt sich erzeugen, indem man z. B. mit einem Hammer auf die Stirnfläche eines langen, geraden, elastischen Stabes schlägt. Dabei entsteht nur *annähernd* eine Longitudinalwelle, da vor allem Punkte außerhalb der Stabachse zusätzlich eine transversale Bewegung erfahren. Im Folgenden wollen wir die transversalen Effekte vernachlässigen.

Wir stellen uns nun den Stab aus Molekülen bestehend vor, welche in Analogie zu Schraubenfedern über elastische Kräfte verbunden sind. Ein Schlag auf die Stirnseite des Stabes bewirkt eine Deformation des angrenzenden Volumenelementes, die sich infolge der Kopplung, hier ist es die Elastizität, über den gesamten Stab fortsetzt.

18.7.4 Herleitung der Wellengleichung

Gegeben sei ein Stab mit der Querschnittsfläche A, der Dichte ρ und dem Elastizitätsmodul E. Wir betrachten das Volumenelement $A\,dx$ (► Abbildung 18.19) zunächst (oben) im undeformierten Stab, dann (unten) zum Zeitpunkt t unter dem Einfluß einer elastischen Longitudinalwelle.

Die Ebenen, die das betrachtete Volumenelement seitlich begrenzen, werden durch die Spannungskräfte der Druckwelle verschoben, die erste Ebene um u, die zweite um $u' = u + du$. Dann ist die relative Längenänderung $\varepsilon = \partial u/\partial x$. In der ► Abbildung 18.19 wählen wir zur Verdeutlichung eine (positive) Zugspannung $\sigma > 0$ und eine Dehnung des Volumenelementes $du > 0$ und damit $\sigma' > \sigma$, um die Vorzeichen übersichtlicher zu halten.

Die auf die erste Ebene ausgeübte Kraft berechnen wir aus der Spannung (18.125) und dem Hookeschen Gesetz (18.126)

$$F = A \cdot \sigma = A \cdot E \cdot \varepsilon\,. \tag{18.128}$$

Die Kraft wird als

$$F' = F + dF \tag{18.129}$$

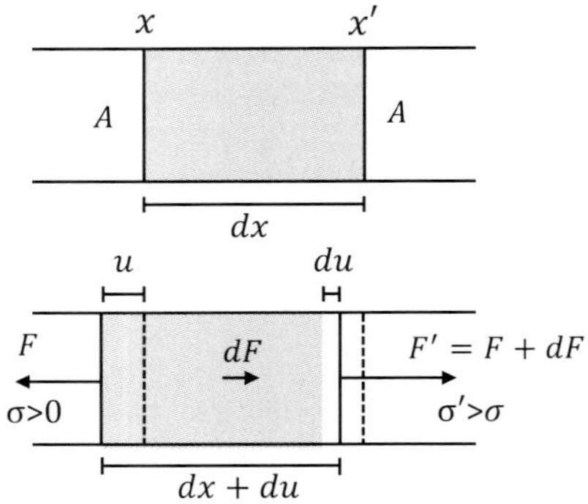

Abbildung 18.19 Ein Volumenelement $A\,dx$ wird durch die Zugspannung $\sigma > 0$ um u verschoben und um du verformt (hier gedehnt) und gibt die Spannung $\sigma' > \sigma$ an den Nachbarbereich weiter. Die Kräfte F und F' wirken auf das Volumenelement und beschleunigen es mit der Resultierenden dF.

an die zweite Ebene weitergegeben (▶ Abbildung 18.19). Die resultierende Kraft auf das Volumenelement ist dann mit der Ableitung dF/dx

$$dF = A \cdot E \cdot \frac{\partial \varepsilon}{\partial x}\, dx \;. \tag{18.130}$$

Mit $\varepsilon = \partial u/\partial x$ erhalten wir

$$dF = A \cdot E \cdot \frac{\partial^2 u}{\partial x^2}\, dx \,. \tag{18.131}$$

Das Massenelement $dm = \rho \cdot A\,dx$ erfährt die beschleunigende Kraft

$$dF_a = dm \cdot a = \rho \cdot A \cdot \frac{\partial^2 u}{\partial t^2}\, dx \,. \tag{18.132}$$

Durch Vergleich von (18.131) und (18.132) erhalten wir die Bewegungsgleichung

$$\rho \cdot A \cdot \frac{\partial^2 u}{\partial t^2}\, dx = A \cdot E \cdot \frac{\partial^2 u}{\partial x^2}\, dx \tag{18.133}$$

und daraus die **Wellengleichung für longitudinale elastische Wellen** (vergl. mit (16.149))

$$\frac{\partial^2 u}{\partial t^2} = \frac{E}{\rho} \cdot \frac{\partial^2 u}{\partial x^2} \,. \tag{18.134}$$

Daraus ergibt sich die Fortpflanzungsgeschwindigkeit elastischer Longitudinalwellen, die **»Schallgeschwindigkeit« im Festkörper**:

$$v = \sqrt{\frac{E}{\rho}} \tag{18.135}$$

Über eine analoge Herleitung mit (18.127) lässt sich auch die Ausbreitungsgeschwindigkeit von Longitudinalwellen in Flüssigkeiten, die **»Schallgeschwindigkeit« in Flüssigkeiten**, ermitteln:

$$v = \sqrt{\frac{K}{\rho}} \tag{18.136}$$

Dabei bezeichnet K das sogenannte Kompressionsmodul.

Schallwellen sind in Festkörpern schneller als in Flüssigkeiten, und dort wiederum schneller als in Gasen. Als Beispiel ergibt sich mit $K_{\mathrm{H_2O}} = 2{,}08 \cdot 10^9\,\mathrm{N/m^2}$ ein Wert von $v_{\mathrm{H_2O}} = 1\,440\,\mathrm{m/s}$, was der tatsächlichen (von Eigenschaften wie Druck, Temperatur und Salzgehalt stark abhängigen) Schallgeschwindigkeit in Wasser gut nahekommt.

18.7.5 Druck- bzw. Dichtewellen in Gasen

Für Wellen in Gasen ist die Wellengeschwindigkeit (18.136) oft keine gute Näherung, wie wir am Beispiel von Luft sehen: Mit $K_{\mathrm{Luft}} = 10^5\,\mathrm{N/m^2}$ und $\rho_{\mathrm{Luft}} = 1{,}225\,\mathrm{kg/m^3}$ ergibt sich ein viel zu kleiner Wert von $v_{\mathrm{Luft}} = 286\,\mathrm{m/s}$. Für Gase müssen wir deren Komprimierbarkeit berücksichtigen, der wir in einer ersten Form beim Boyle-Mariotte-Gesetz (18.33) begegnet sind.

Im Folgenden betrachten wir eine Druckwelle, die sich in einem gasgefüllten Raum in x-Richtung ausbreitet. In Analogie zur Diskussion der Stabwelle sehen wir uns den Zustand eines Massenelementes des Gases vor und während der Wirkung der Welle an und leiten daraus die Wellengleichung her.

Im Unterschied zur Betrachtung von Wellen in Festkörpern kann bei der Diskussion einer Druckwelle in einem Gas aufgrund der guten Komprimierbarkeit des Gases die Dichte nicht mehr als konstant angenommen werden.

Wir betrachten im Gleichgewichtszustand das Volumen $A\,dx$ (► Abbildung 18.20). Im ungestörten Gas herrsche der Druck p_0 und die Dichte sei ρ_0. Die Druckwelle bewirkt eine

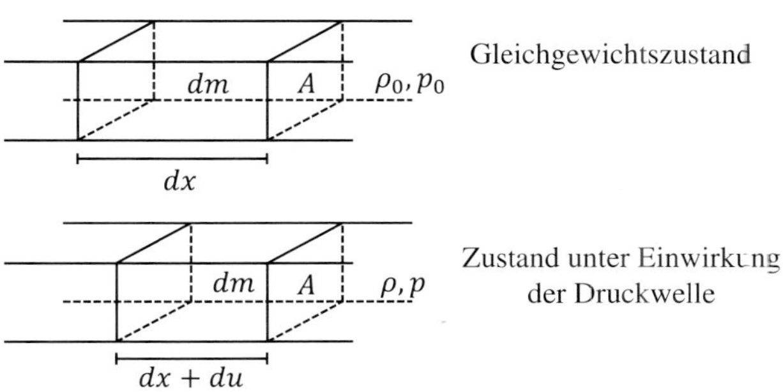

Abbildung 18.20 Ein Volumenelement $A\,dx$ der Masse dm wird um dx verschoben und um du verformt (gestaucht). ρ_0 (ρ) bezeichnen die Dichte, p_0 (p) den Druck vor (nach) der Verformung.

Verschiebung sowie Längenänderung des Elementes längs der x-Richtung. Dichte und Druck verändern sich zu ρ und p.

Beide Volumina $A\,dx$ und $A\,(dx + du)$ enthalten die gleiche Masse dm. Es gilt also

$$dm = \rho_0 \cdot A\,dx = \rho \cdot A\,(dx + du) \tag{18.137}$$

$$\frac{\rho}{\rho_0} = \frac{dx}{dx + du} \xrightarrow{dx \to 0} \frac{\rho}{\rho_0} = \left(1 + \frac{\partial u}{\partial x}\right)^{-1} . \tag{18.138}$$

An die Stelle der Spannung σ im Festkörper tritt bei Gasen der Druck $p(x,t) = F/A$. Um die Komprimierbarkeit des Gases korrekt zu berücksichtigen, greifen wir auf Energiebetrachtungen bei Gasen vor, die wir in Kapitel 22 erklären werden. Kompression bzw. Dilatation eines thermisch isolierten Gases führt zu Erwärmung bzw. Abkühlung. Bei der Druckwelle erfolgen die Druckänderungen $dp = (\partial p(x,t)/\partial x)\,dx$ so schnell, dass kein Wärmeaustausch mit der Umgebung stattfindet. Es handelt sich um einen sogenannten adiabatischen Prozeß. Für adiabatische Zustandsänderungen (vergl. Abschnitt 22.3.3) gilt mit dem Adiabatenkoeffizienten κ (22.30)

$$p \cdot V^{\kappa} = p_0 \cdot V_0^{\kappa} = const. \tag{18.139}$$

$$\frac{p}{p_0} = \left(\frac{V_0}{V}\right)^{\kappa} = \left(\frac{\rho}{\rho_0}\right)^{\kappa} \xrightarrow{dx \to 0} \frac{p}{p_0} = \left(1 + \frac{\partial u}{\partial x}\right)^{-\kappa} . \tag{18.140}$$

Für kleine $\partial u/\partial x$ können wir die Näherung

$$\left(1 + \frac{\partial u}{\partial x}\right)^{-\kappa} = 1 - \kappa \cdot \frac{\partial u}{\partial x} \tag{18.141}$$

verwenden. Damit erhalten wir

$$p = p_0 \cdot \left(1 - \kappa \cdot \frac{\partial u}{\partial x}\right) . \tag{18.142}$$

Die partielle Ableitung nach x ergibt

$$\frac{\partial p}{\partial x} = -\kappa \cdot p_0 \cdot \frac{\partial^2 u}{\partial x^2} . \tag{18.143}$$

Damit wird die Druckänderung beschrieben durch

$$dp = -\kappa \cdot p_0 \cdot \frac{\partial^2 u}{\partial x^2}\,dx . \tag{18.144}$$

Auf das Volumenelement $A\,dx$ wirkt die Kraft $d\vec{F} = \vec{F} - \vec{F'}$ (► Abbildung 18.21).

Für $\partial p/\partial x < 0$ ist $F' < F$ und der Betrag der resultierenden Druckkraft lautet

$$dF = -A\,dp = -A \cdot \frac{\partial p}{\partial x}\,dx = +A \cdot \kappa \cdot p_0 \cdot \frac{\partial^2 u}{\partial x^2}\,dx . \tag{18.145}$$

Diese Druckkraft erzeugt eine Beschleunigung $a = \partial^2 u/\partial t^2$ des zugehörigen Massenelementes:

$$dF_a = dm \cdot a = A \cdot \rho_0 \cdot \frac{\partial^2 u}{\partial t^2}\,dx \tag{18.146}$$

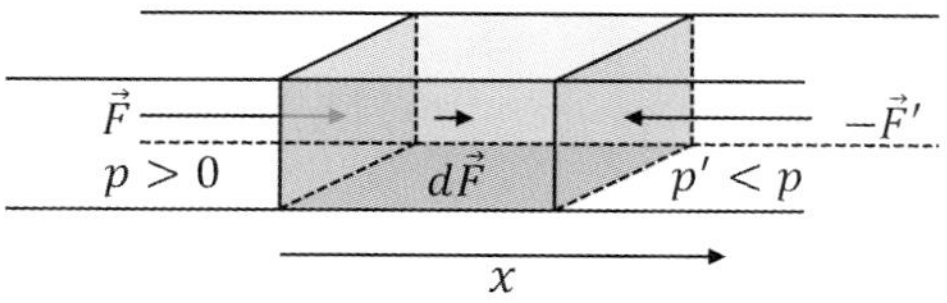

Abbildung 18.21 Am Volumenelement angreifende Druckkräfte $\vec{F} = \vec{F}' + d\vec{F}$ und $-\vec{F}'$. Das Element wird mit der resultierenden Kraft $d\vec{F}$ beschleunigt.

Durch Gleichsetzen von (18.145) und (18.146) erhalten wir die **Wellengleichung für Druckwellen** (vergl. mit (16.149))

$$\frac{\partial^2 u}{\partial t^2} = \frac{p_0 \cdot \kappa}{\rho_0} \cdot \frac{\partial^2 u}{\partial x^2} \,. \tag{18.147}$$

Daraus lesen wir $v^2 = p_0 \cdot \kappa / \rho_0$ ab. Allgemein gilt (mit p, ρ beliebig) für die **Schallgeschwindigkeit in Gasen**

$$v = \sqrt{\frac{p \cdot \kappa}{\rho}} \,. \tag{18.148}$$

Beachten sollten wir, dass die Schallgeschwindigkeit v mit der Dichte ρ auch von der *Temperatur* T abhängt.

Beispiel 49: Berechnung der Schallgeschwindigkeit in Luft

Mit Gleichung (18.148) können wir die Schallgeschwindigkeit in Luft bestimmen:

$$\left.\begin{aligned} \kappa &= 1{,}4 \\ \rho_0 &= 1{,}293\,\mathrm{kg/m^3} \\ p_0 &= 10^5\,\mathrm{Pa}\,(0\,^\circ\mathrm{C}) \end{aligned}\right\} \implies v_{\mathrm{Luft}} = 331\,\mathrm{m/s}\,(\,0\,^\circ\mathrm{C}\ \text{bzw.}\ T = 273{,}15\,\mathrm{K}\,)$$

Für die Temperaturabhängigkeit von v_{Luft} gilt $\Delta v_{\mathrm{Luft}}/\Delta T \approx 0{,}6\,(\mathrm{m/s})/\mathrm{K}$

$$\implies \quad v_{\mathrm{Luft}} = 331\,\frac{\mathrm{m}}{\mathrm{s}} + 0{,}6\,\frac{\mathrm{m}}{\mathrm{s}} \cdot (T/\mathrm{K} - 273{,}15) \,.$$

Experiment 52: Schallgeschwindigkeit

Mit einer Schreckschußpistole wird auf zwei im Abstand $\Delta s = 10\,\mathrm{m}$ befindliche Mikrofone ein Schuß abgegeben. Beide Mikrofone sind mit einer Stoppuhr verbunden. Die Zeitzählung wird aktiviert, sobald die Schallwelle das erste Mikrofon erreicht. Das zweite Mikrofon sendet entsprechend den Stoppimpuls aus. Die Messung liefert $\Delta t = 0{,}0117\,\mathrm{s}$ bei $18\,^\circ\mathrm{C}$.

Das Ergebnis $v_{\mathrm{Luft}} = \Delta s/\Delta t = 341\,\mathrm{m/s}$ stimmt sehr gut mit dem berechneten Wert von $(331 + 0{,}6 \cdot 18)\,\mathrm{m/s} \approx 342\,\mathrm{m/s}$ überein.

Das Wichtigste zum Mitnehmen

Zum Mitnehmen: Hydromechanik 1

- Der **Druck** p ist definiert durch die Kraft F, die auf eine Fläche A wirkt:

$$p = \frac{F}{A} \tag{18.1}$$

Druck wird in den Einheiten Pascal gemessen: $1\,\mathrm{Pa} = 1\,\mathrm{N/m^2}$.

- Für Wasser, dessen Kompressibilität gering ist, ist die Dichte innerhalb einer Wassersäule $\rho(h) \approx \mathit{const.}$. In der Nähe der Erdoberfläche mit $g \approx \mathit{const.}$ gilt für den **Druck einer Wassersäule** auf den Boden

$$p = \rho g h\,. \tag{18.8}$$

- Nach dem **Archimedischen Prinzip** entspricht die Auftriebskraft F_A der Gewichtskraft F_F der von dem Körper verdrängten Flüssigkeit:

$$|\vec{F}_A| = \underbrace{g \underbrace{\rho_{\mathrm{Fl}} \underbrace{(h_2 - h_1)\,A}_{\text{Volumen des Körpers}}}_{\text{Masse der verdrängten Flüssigkeit}}}_{\equiv F_F \text{ Gewichtskraft der verdrängten Flüssigkeit}}$$

Die Auftriebskraft F_A reduziert die Gewichtskraft F_G eines Körpers. Die effektive Gewichtskraft F'_G eines vollständig eingetauchten homogenen Körpers mit dem Volumen $V_K = M_K/\rho_K$ in der Flüssigkeit mit der Dichte ρ_{Fl} ist

$$F'_G = (\rho_K - \rho_{Fl})\,V_K\,g\,. \tag{18.20}$$

Falls die Gewichtskraft des Körpers kleiner als die Auftriebskraft ist, $F_G \leq F_A$, schwimmt der Körper in der Flüssigkeit.

Zum Mitnehmen: Hydromechanik 2

- **Physikalische Normalbedingungen** herrschen bei einer Temperatur von 0 °C und einem Druck von

$$1 \text{ physikalischen Atmosphäre} = 1\,\text{atm} = 101\,325\,\text{Pa}\,, \tag{18.28}$$

mit Pa = N/m^2(= 10^{-5} bar) als SI-Einheit des Drucks. 1 atm ist an den mittleren **Luftdruck** auf Meereshöhe angepasst.

Der Luftdruck lässt sich im Torricelli-Rohr mit der Länge der Quecksilbersäule Δh bestimmen:

$$p_{\text{Luft}} = p_{\text{Hg}} = \rho_{\text{Hg}}\, g\, \Delta h \tag{18.27}$$

1 atm entspricht einer Quecksilbersäule Δh_{Hg} von

$$\Delta h_{\text{Hg}} = 760\,\text{mm}\,. \tag{18.29}$$

Für Wasser anstelle von Quecksilber ist die Höhendifferenz

$$\Delta h_{\text{H}_2\text{O}} = \Delta h_{\text{Hg}} \frac{\rho_{\text{Hg}}}{\rho_{\text{H}_2\text{O}}} = 760\, \frac{13{,}595}{1}\,\text{mm} = 10{,}32\,\text{m}\,. \tag{18.30}$$

- Zwischen dem Druck p und dem Volumen V eines Gases gilt bei konstanter Temperatur das Gesetz von **Boyle-Mariotte**:

$$p\,V = \mathit{const.} \tag{18.33}$$

- Der Luftdruck $p(h)$ nimmt mit der Höhe h gemäß der **barometrischen Höhenformel** gegenüber dem Druck auf Meereshöhe $p_\circ$ exponentiell ab:

$$p(h) = p_\circ \cdot e^{-\frac{\rho_\circ}{p_\circ}\, g\, h} \tag{18.47}$$

Für die Konstante in der Exponentialfunktion gilt $\rho_\circ/(p_\circ\, g) = 0{,}1256\,\text{km}^{-1}$.

- Eine inkompressible Flüssigkeit erfährt an einer Querschnittsverengung $A' < A$ eine Geschwindigkeitserhöhung $v' > v$:

$$A\,v = A'\,v' \tag{18.57}$$

Diese Gleichung wird im Sinn der Erhaltung der Masse als **Kontinuitätsgleichung** bezeichnet.

Zum Mitnehmen: Hydromechanik 3

- Für eine allgemeine Formulierung von (18.57) führt man die **Massenstromdichte** $\vec{j}$ als Produkt aus der Dichte ρ und der Geschwindigkeit $\vec{v}$ ein:

$$\vec{j} \equiv \rho \vec{v} \tag{18.58}$$

Damit ergibt sich die **differentielle Form der Kontinuitätsgleichung**

$$\frac{\partial \rho}{\partial t} + \nabla \vec{j} = 0\,. \tag{18.67}$$

(18.67) besagt, dass die Dichteänderung in einem Volumenelement dem Massenzustrom oder Massenabstrom entspricht.

- Für inkompressible strömende Flüssigkeiten sind mit Geschwindigkeits- auch Druckänderungen verbunden. Es gilt die **Bernoulli-Gleichung**

$$\underbrace{p_\circ}_{\text{Gesamtdruck}} = \underbrace{\frac{1}{2}\rho v^2}_{\text{Staudruck}} + \underbrace{p}_{\text{statischer Druck}} = const.\,. \tag{18.80}$$

Der **Gesamtdruck** eines abgeschlossenen Systems bleibt konstant. Er setzt sich aus dem von der Geschwindigkeit herrührenden **Staudruck** und dem **statischen Druck** zusammen.

Nach der Kontinuitätsgleichung (18.57) wird die Geschwindigkeit an engen Stellen im Rohr größer. Damit steigt die kinetische Energie bzw. der Staudruck. Wegen der Erhaltung des Gesamtdrucks nimmt der statische Druck entsprechend ab.

Der Staudruck ist die Ursache für den **dynamischen Auftrieb** und das **hydrodynamische Paradoxon**. Bei beiden entsteht durch erhöhte Geschwindigkeiten nach (18.80) ein Unterdruck.

- Die **innere Reibung von Flüssigkeiten** wird durch das **Newtonsche Reibungsgesetz** für die Reibungskraft F_R beschrieben:

$$F_R = \eta\, A\, \frac{dv}{dh} \tag{18.91}$$

Dabei ist A die Reibungsfläche, dv/dh das Geschwindigkeitsgefälle in der Flüssigkeit und η als **Viskosität** oder dynamische Zähigkeit eine Materialkonstante.

Zum Mitnehmen: Hydromechanik 4

- In einem von einer Flüssigkeit mit Viskosität η **laminar** durchströmten Rohr mit dem Radius R ergibt sich bei einer Druckdifferenz $p_1 - p_2$ aufgrund der Reibung (18.91) eine paraboloide Geschwindigkeitsverteilung $v(r)$:

$$v(r) = \frac{p_1 - p_2}{4\eta l}(R^2 - r^2) \qquad (18.100)$$

Dabei ist r die Entfernung von der Rohrmitte und $v(r = R) = 0$. Daraus ergibt sich das Gesetz von **Hagen-Poiseuille** für den Flüssigkeitsstrom $J = V/t$

$$J = \frac{V}{t} = \frac{\pi(p_1 - p_2)}{8\eta l} R^4 . \qquad (18.109)$$

Es liefert eine starke R^4-Abhängigkeit des Stroms J und eine lineare Abhängigkeit vom Druckgefälle pro Länge $(p_1 - p_2)/l$. Mit der mittleren Strömungsgeschwindigkeit $\bar{v}$ gilt $J = \pi R^2 \bar{v}$. Im stationären Gleichgewicht $F_p = \pi R^2(p_1 - p_2) = -F_R$ ergibt sich dann für die **Reibungskraft** F_R in vektorieller Schreibweise

$$\vec{F}_R = -8\pi\, l\, \eta\, \vec{\bar{v}} . \qquad (18.114)$$

- Für eine Kugel vom Radius r in einer laminaren Strömung gilt das Reibungsgesetz

$$\vec{F}_R = -6\pi\, r\, \eta\, \vec{v} , \qquad (18.118)$$

wobei v die Relativgeschwindigkeit der Kugel gegen die Flüssigkeit ist. Die Reibungskraft der **Stokesschen Reibung** ist also proportional zur Geschwindigkeit.

- Bei höheren Geschwindigkeiten treten Wirbel auf, die Strömungen werden **turbulent**. Der Strömungswiderstand wird dann durch das empirische Gesetz

$$F_W = c_W \cdot A\, \frac{1}{2}\, \rho\, v^2 \qquad (18.123)$$

beschrieben. Dabei ist A die umströmte Querschnittsfläche und $\frac{1}{2}\rho v^2$ der Staudruck (18.80). Die Luftwiderstandszahl c_W ist eine empirische Größe, die von den Eigenschaften der angeströmten Oberfläche abhängt.

Zum Mitnehmen: Hydromechanik 5

- In einem Festkörper können wir uns die Atome oder Moleküle durch elastische Kräfte ähnlich der Federkraft (18.124) verbunden denken. Mit **Druck- oder Zugspannung** $\sigma = F/A$ ergeben sich Verformungen im linearen elastischen Bereich nach dem **Hookeschen Gesetz** (R. Hooke 1676)

$$\sigma = E \cdot \varepsilon = E \cdot \frac{\Delta l}{l} \tag{18.126}$$

  mit der relativen Längenänderung $\varepsilon = \Delta l/l$ und dem **Elastizitätsmodul** E. Typische Werte für E sind $10^9\,\mathrm{N/m^2}$ für Holz, $2 \cdot 10^{11}\,\mathrm{N/m^2}$ für Stahl und $10^{12}\,\mathrm{N/m^2}$ für Diamant.

- In Flüssigkeiten, in geringerem Maße auch in Festkörpern, gibt es die elastische Volumenänderung unter Druck $p = F/A$ (18.1)

$$p = K \cdot \frac{\Delta V}{V} \tag{18.127}$$

  mit der relativen Volumenänderung $\Delta V/V$ und dem **Kompressionsmodul** K. Die Größenordnung von K variiert sehr stark von $10^5\,\mathrm{N/m^2}$ für Gase, $10^9\,\mathrm{N/m^2}$ für Flüssigkeiten und $10^{11-12}\,\mathrm{N/m^2}$ für Festkörper.

- Mit (18.126) lässt sich die Fortpflanzungsgeschwindigkeit elastischer Longitudinalwellen (**»Schallgeschwindigkeit« im Festkörper**) zu

$$v = \sqrt{\frac{E}{\rho}} \tag{18.135}$$

  herleiten, mit dem Elastizitätsmodul E und der Dichte ρ. Analog kann man mit (18.127) die Ausbreitungsgeschwindigkeit von Longitudinalwellen in Flüssigkeiten (**»Schallgeschwindigkeit« in Flüssigkeiten**) als

$$v = \sqrt{\frac{K}{\rho}} \tag{18.136}$$

  ermitteln, mit dem Kompressionsmodul K und der Dichte ρ.

  Schallwellen sind in Festkörpern schneller als in Flüssigkeiten, und dort wiederum schneller als in Gasen. Als Beispiel ergibt sich mit $K_{\mathrm{H_2O}} = 2{,}08 \cdot 10^9\,\mathrm{N/m^2}$ ein Wert von $v_{\mathrm{H_2O}} = 1\,440\,\mathrm{m/s}$, was der tatsächlichen Schallgeschwindigkeit in Wasser gut nahekommt.

Zum Mitnehmen: Hydromechanik 6

- Im Gegensatz zu Festkörpern und Flüssigkeiten gibt (18.136) für Gase keine befriedigenden Ergebnisse. Bei der Druckwelle erfolgen die Druckänderungen $dp = (\partial p(x,t)/\partial x)\,dx$ so schnell, dass die Gleichung für adiabatische Zustandsänderungen (22.30) (vergl. Abschnitt 22.3.3) mit dem **Adiabatenkoeffizienten** κ verwendet werden muss:

  $$p \cdot V^{\kappa} = p_0 \cdot V_0^{\kappa} = const.$$

  Damit lässt sich mit dem Druck p und der Dichte ρ die **Schallgeschwindigkeit in Gasen** recht genau berechnen:

  $$v = \sqrt{\frac{p \cdot \kappa}{\rho}} \tag{18.148}$$

  Dabei ist zu beachten, dass die Schallgeschwindigkeit v mit der Dichte ρ auch von der *Temperatur* T abhängt.

- Für die **Schallgeschwindigkeit in Luft** liefert (18.148) mit $\kappa = 1{,}4$, $\rho_0 = 1{,}293\,\mathrm{kg/m^3}$, $p_0 = 10^5\,\mathrm{Pa}\,(0\,°\mathrm{C}) \Longrightarrow v_{\mathrm{Luft}} = 331\,\mathrm{m/s}\,(0\,°\mathrm{C})$. Mit einer Temperaturabhängigkeit von $\Delta v_{\mathrm{Luft}}/\Delta T \approx 0{,}6\,\mathrm{m/s/K}$ ergibt sich

  $$\Rightarrow v_{\mathrm{Luft}} = 331\,\mathrm{m/s} + 0{,}6\,\mathrm{m/s} \cdot (T/\mathrm{K} - 273{,}15)\,.$$

Thermische Ausdehnung

19

19.1 Längenausdehnung fester Körper 419

19.2 Volumenausdehnung fester und flüssiger Stoffe 422

19.3 Thermische Eigenschaften von Gasen 423

19.3.1 Temperaturbedingte Volumenausdehnung von Gasen 423

19.3.2 Temperaturabhängigkeit des Gasdrucks 423

19.3.3 Zusammenhang: Druck, Volumen, Temperatur bei Gasen . . . 425

19.3.4 Stoffmenge 426

19.4 Ideales Gasgesetz 427

19.4.1 Ausdehnungskoeffizient und absolute Temperatur 429

ÜBERBLICK

» Die Ausdehnung von Festkörpern, Flüssigkeiten und Gasen hängt von der Temperatur ab. Für Festkörper und Flüssigkeiten ist diese temperaturabhängige Ausdehnung zusätzlich stark vom Material abhängig.

Für Gase finden wir für die Temperaturabhängigkeit des Gasvolumens und des Gasdrucks einen gemeinsamen Koeffizienten, der nur geringfügig materialabhängig ist. Daraus erhalten wir einen Zusammenhang zwischen dem Druck, dem Volumen und der Temperatur eines Gases, der uns zum sogenannten idealen Gasgesetz führt. «

19.1 Längenausdehnung fester Körper

Die Längenausdehnung fester Körper (▶ Abbildung 19.1) ist bei nicht zu großen Temperaturänderungen proportional zur Temperatur:

$$l_2 = l_1(1 + \alpha(T_2 - T_1)) \tag{19.1}$$

Die Konstante α heißt **linearer Ausdehnungskoeffizient**:

$$\alpha = \frac{l_2 - l_1}{l_1\,(T_2 - T_1)} = \frac{\Delta l}{l_1\,\Delta T} \quad \text{oder differentiell} \quad \alpha = \frac{1}{l}\,\frac{\partial l}{\partial T} \tag{19.2}$$

Er beschreibt die relative Längenänderung $\Delta l/l$ pro Grad Temperaturänderung gemessen in Celsius (°C) oder Kelvin (K) (Abschnitt 20.3). Die Ausdehnung muss z. B. beim Bau von Brücken, Schienen und Freileitungen berücksichtigt werden.

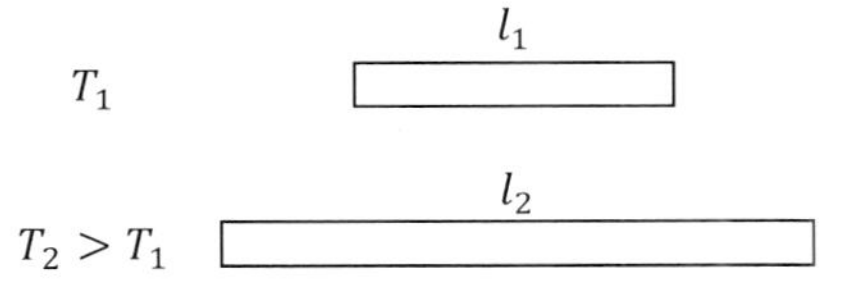

Abbildung 19.1 Feste Körper dehnen sich mit steigender Temperatur $T_2 > T_1$ aus: $l_2 > l_1$.

Beispiel 50: Längenausdehnungskoeffizient

Bei 20 °C ist der Längenausdehnungskoeffizient für

$$\alpha\,(\text{Eisen}) = 11{,}8 \cdot 10^{-6}\,\frac{1}{\text{K}}\,,$$
$$\alpha\,(\text{Aluminium}) = 23{,}1 \cdot 10^{-6}\,\frac{1}{\text{K}}\,.$$

Beispiel 51: Eisenbahnschiene

Eine Eisenbahnschiene habe bei 20 °C eine Länge von 20 m.

Bei 40 °C ist die Schiene $\Delta l = 20\,\text{m} \cdot 20\,\text{K} \cdot 11{,}8 \cdot 10^{-6}\,/\text{K} \approx 0{,}5\,\text{cm}$ länger.

Experiment 53: Längenausdehnung

Durch ein Messingrohr mit der anfänglichen Länge $l_1 = 500\,\text{mm}$ (bei $T_1 = 20\,°\text{C}$) wird kochendes Wasser der Temperatur $T_2 = 100\,°\text{C}$ geleitet (▶ Abbildung 19.2). Wir messen die Längenausdehnung des Rohrs, die $\Delta l = 0{,}9\,\text{mm}$ beträgt. Die Temperaturänderung ist $\Delta T = 100\,°\text{C} - 20\,°\text{C} = 80\,°\text{C} = 80\,\text{K}$. Damit ergibt sich als Längenausdehnungskoeffizient von Messing

$$\alpha = \frac{\Delta l}{l_1}\frac{1}{\Delta T} = \frac{0{,}9}{500}\frac{1}{80}\frac{1}{\text{K}} = 22{,}5 \cdot 10^{-6}\,\frac{1}{\text{K}}\,.$$

Diese Messung stimmt mit dem erwarteten Wert für Messing $\alpha = 22{,}5 \cdot 10^{-6}\,/\text{K}$ überein.

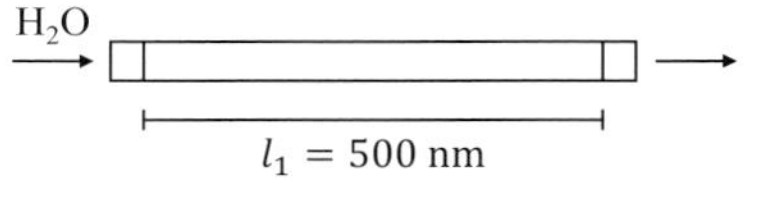

Abbildung 19.2 Längenausdehnung eines Messingrohrs mit kochendem Wasser

Experiment 54: Bimetallschalter

Ein Bimetall besteht aus zwei miteinander verbundenen Metallblechen, die aus verschiedenen Metallsorten mit unterschiedlichen Ausdehnungskoeffizienten hergestellt sind (▶ Abbildung 19.3). Die Erwärmung des Bimetalls führt zu seiner Verbiegung. Mit einem Bimetall können wir elektrische Schaltkreise wärmeabhängig steuern.

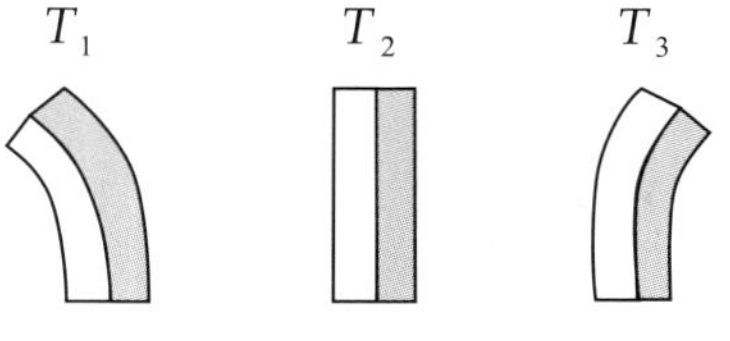

Abbildung 19.3 Ein Bimetallstab aus zwei Metallen mit verschiedenen Ausdehnungskoeffizienten verformt sich je nach Temperatur.

Experiment 55: Bolzensprenger

Eine große Metallschraube wird in einem Stahlgestell befestigt und unter Erwärmung nur der Schraube immer fester angezogen (► Abbildung 19.4).

Anschließend wird die Schraube mit kaltem Wasser übergossen. Das Schraubenmaterial zieht sich gegen seine Halterung so stark zusammen, dass die Metallschraube zerspringt.

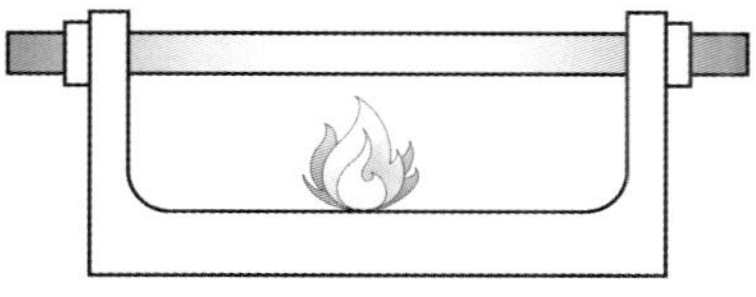

Abbildung 19.4 Bolzensprenger: Eine Schraube wird in einer stabilen Vorrichtung erhitzt und fest gespannt. Sie zerreißt beim Abkühlen.

Als mikroskopisches Bild der Längenausdehnung können wir uns ein Potentialbild der Atome im Festkörper vorstellen, wie es in ► Abbildung 19.5 gezeigt ist. Der mittlere Abstand der Atome hängt von der Temperatur ab. Bei erhöhter Temperatur befinden sich die Atome auf höheren Energieniveaus. Durch die Asymmetrie des Potentials wachsen auf der mikroskopischen Skala die mittleren Atomabstände. Damit vergrößert sich die makroskopische Ausdehnung des Körpers.

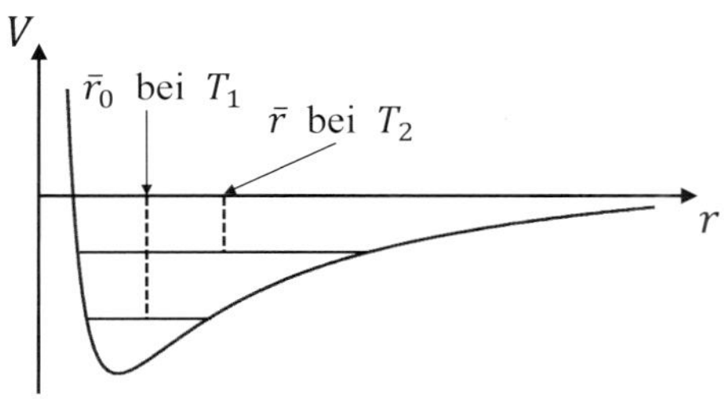

Abbildung 19.5 Bei höherer Temperatur $T_2 > T_1$ werden die Atome auf höhere Energieniveaus mit größeren mittleren Potentialradien $\bar{r} > \bar{r}_0$ gehoben.

19.2 Volumenausdehnung fester und flüssiger Stoffe

In ► Abbildung 19.6 betrachten wir ein Volumen mit den Kantenlängen l_1, l_2, l_3.

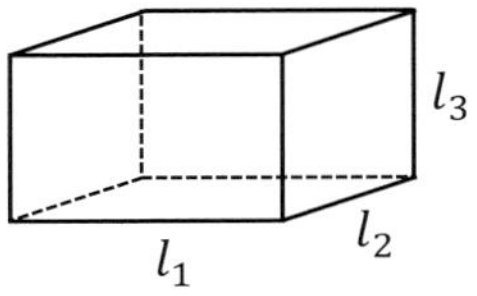

Abbildung 19.6 Volumenausdehnung eines Quaders mit den Kantenlängen l_i.

Bei einer Erhöhung der Temperatur werden alle Kantenlängen nach Gleichung (19.1) gedehnt und das ursprüngliche Volumen V_1 vergrößert sich entsprechend auf V_2:

$$V_1 = l_1 \cdot l_2 \cdot l_3 \tag{19.3}$$

$$V_2 = l_1\,(1 + \alpha\,\Delta T) \cdot l_2\,(1 + \alpha\,\Delta T) \cdot l_3\,(1 + \alpha\,\Delta T) \tag{19.4}$$

$$= V_1\,(1 + \alpha\,\Delta T)^3 \tag{19.5}$$

$$= V_1\,(1 + 3\,\alpha\,\Delta T + \underbrace{3\,(\alpha\,\Delta T)^2 + (\alpha\,\Delta T)^3}_{\text{klein für } \alpha\,\Delta T \ll 1}) \tag{19.6}$$

Wir nähern die Volumenänderung zur ersten Potenz im Ausdehnungskoeffizienten α und erhalten mit $\gamma \approx 3\alpha$

$$V_2 = V_1(1 + \gamma\Delta T)\,. \tag{19.7}$$

Der Faktor

$$\gamma = \frac{V_2 - V_1}{V_1\,\Delta T} \quad \text{oder differentiell} \quad \gamma = \frac{1}{V}\,\frac{\partial V}{\partial T} \tag{19.8}$$

heißt **Volumenausdehnungskoeffizient** für Festkörper und Flüssigkeiten.

Beispiel 52: Ausdehnungskoeffizient Flüssigkeit

Bei 20 °C sind die Ausdehnungskoeffizienten für Quecksilber und Ethanol

$$\gamma\,(\text{Quecksilber}) = 1{,}811 \cdot 10^{-4}\,\frac{1}{\text{K}}\,,$$

$$\gamma\,(\text{Ethanol}) = 1{,}1 \cdot 10^{-3}\,\frac{1}{\text{K}}\,.$$

19.3 Thermische Eigenschaften von Gasen

19.3.1 Temperaturbedingte Volumenausdehnung von Gasen

Wenn wir die Temperatur eines Gases ändern und gleichzeitig den Gasdruck p konstant halten, so beobachten wir auch für Gase eine lineare Volumenausdehnung mit der Temperaturänderung ΔT:

$$V_2 = V_1(1 + \gamma_p \, \Delta T) \tag{19.9}$$

$$\gamma_p = \frac{\Delta V}{V_1 \, \Delta T} \quad \text{oder differentiell} \quad \gamma_p = \frac{1}{V}\left(\frac{\partial V}{\partial T}\right)_p \tag{19.10}$$

Der Index p des Ausdehnungskoeffizienten γ_p zeigt an, dass es sich um eine Volumenausdehnung bei konstantem Druck handelt.

19.3.2 Temperaturabhängigkeit des Gasdrucks

Mit einer Temperaturänderung ändert sich im Allgemeinen auch der Gasdruck p. Im folgenden Experiment 56 werden wir den Einfluss der Temperatur des Gases auf den Gasdruck untersuchen.

Experiment 56: Gasthermometer

Wir heizen ein Wasserbad, in dem sich ein Glaskolben mit Luft befindet (► Abbildung 19.7). Das Gas (die Luft) im Kolben wird dabei ebenfalls erwärmt. Den steigenden Druck des Gases messen wir mit einem dünnen, wassergefüllten U-Rohr. Die Volumenänderung des Gases, die durch das Druckmessverfahren entsteht, sei dabei vernachlässigbar klein.

Die Druckänderung im Vergleich zum äußeren Luftdruck p_L ist direkt proportional zur Differenz der Steighöhe im U-Rohr (18.8):

$$\underbrace{p}_{\text{Gas}} = \rho_{\mathrm{H_2O}} \, g \, \Delta h + \underbrace{p_L}_{p_{\text{Luft}}} \tag{19.11}$$

In diesem Experiment sehen wir, dass der Gasdruck und die Temperatur proportional zueinander sind (rechte ► Abbildung 19.7):

$$p \propto T \tag{19.12}$$

Die gemessene Steigung betrug in unserem Experiment

$$\delta = 0{,}03 \, \frac{\mathrm{m}}{\mathrm{K}} \, . \tag{19.13}$$

Der lineare Zusammenhang zwischen Druckänderung und Temperaturänderung bei konstantem Volumen $V = const.$ ist im **Gay-Lussac-Gesetz** formuliert:

$$p_2 = p_1(1 + \gamma_V \, \Delta T) \tag{19.14}$$

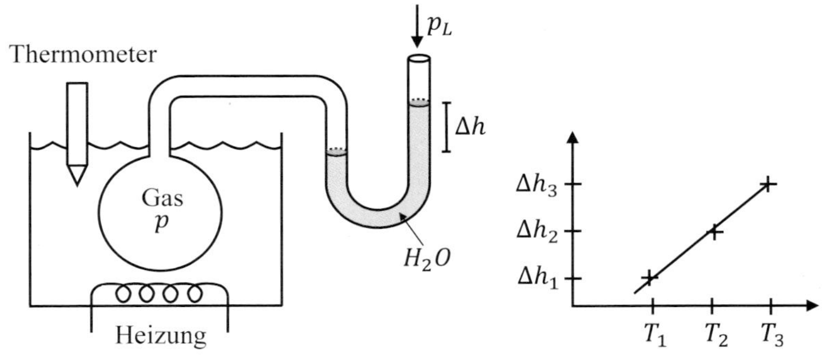

Abbildung 19.7 Links: Der steigende Druck p eines Gasvolumens beim Erwärmen wird über eine Wassersäule Δh gemessen. Rechts: Der Druck $p \propto \Delta h$ steigt proportional zur Temperatur T.

Der Index V des Koeffzienten γ_V indiziert hier die Druckänderung bei konstantem Volumen:

$$\gamma_V = \frac{\Delta p}{p_1\,\Delta T} \quad \text{oder differentiell} \quad \gamma_V = \frac{1}{p}\left(\frac{\partial p}{\partial T}\right)_V \tag{19.15}$$

Experimentell finden wir keinen Unterschied zwischen dem Ausdehnungskoeffizienten bei konstantem Druck (19.10) und dem Ausdehnungskoeffizienten bei konstantem Volumen (19.15):

$$\gamma_p = \gamma_V \equiv \gamma \tag{19.16}$$

Mithilfe unseres Experiments mit dem Gasthermometer können wir den Wert des Ausdehnungskoeffizienten γ bestimmen. Wir vernachlässigen dabei die kleine Volumenänderung des Gases durch das Messverfahren.

Ausgehend von Gleichung (19.11) setzen wir das Gay-Lussac-Gesetz (19.14) ein, um die Höhendifferenz Δh als Funktion der Temperaturänderung ΔT zu erhalten:

$$\Delta h = \frac{p - p_L}{\rho_{\mathrm{H_2O}}\, g} \tag{19.17}$$

$$= \frac{p_L(1 + \gamma\,\Delta T) - p_L}{\rho_{\mathrm{H_2O}}\, g} \tag{19.18}$$

$$= \underbrace{\frac{p_L\,\gamma}{\rho_{\mathrm{H_2O}}\, g}}_{\equiv\delta}\,\Delta T \tag{19.19}$$

Aus der gemessenen Steigung δ der Geraden (siehe 19.13) erhalten wir mit

$$\delta = \frac{p_L\,\gamma}{\rho_{\mathrm{H_2O}}\, g} = 0{,}03\,\frac{\mathrm{m}}{\mathrm{K}} \tag{19.20}$$

den Wert

$$\gamma = \frac{\rho_{\mathrm{H_2O}}\, g}{p_L}\,\delta \approx \frac{10^3\,10}{10^5}\,0{,}03\,\frac{1}{\mathrm{K}} = 3\cdot 10^{-3}\,\frac{1}{\mathrm{K}} \tag{19.21}$$

in der Nähe der typischen Ausdehnungskoeffizienten für Gase, die in dem folgenden Beispiel aufgeführt sind.

Beispiel 53: Ausdehnungskoeffizient für Gase

Bei $T = 0\,°\mathrm{C}$ ergeben sich folgende Ausdehnungskoeffizienten für Helium, Sauerstoff und ideales Gas (Definition in Abschnitt 20.1.1):

$$\gamma\,(\mathrm{He}) = 3{,}66 \cdot 10^{-3}\,\frac{1}{\mathrm{K}} \tag{19.22}$$

$$\gamma\,(\mathrm{O_2}) = 3{,}672 \cdot 10^{-3}\,\frac{1}{\mathrm{K}}$$

$$\gamma\,(\text{ideales Gas}) = 3{,}661 \cdot 10^{-3}\,\frac{1}{\mathrm{K}} \tag{19.23}$$

19.3.3 Zusammenhang: Druck, Volumen, Temperatur bei Gasen

Die Gleichheit der Koeffizienten γ (19.16) in der Ausdehnungsgleichung (19.9) und im Gay-Lussac-Gesetz (19.14) impliziert einen direkten Zusammenhang zwischen dem Druck p, dem Volumen V und der Temperatur T bei Gasen.

Nach dem Gesetz von Boyle-Mariotte $p\,V = const.$ (18.33) verhalten sich Druck und Volumen bei konstanter Temperatur invers zueinander. Bei der logarithmischen Auftragung im Druck-Volumen-Diagramm der ► Abbildung 19.8 haben diese Hyperbeln Geradenform:

$$\lg(p\,V) = \lg(const.) \tag{19.24}$$

$$\lg(p) = -\lg(V) + \lg(const.) \tag{19.25}$$

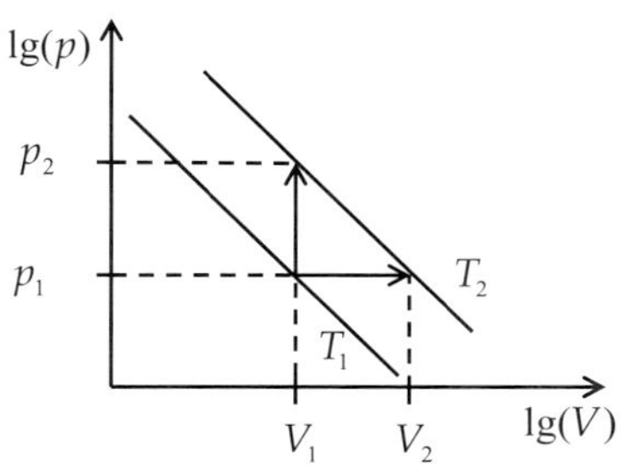

Abbildung 19.8 Im doppellogarithmischen $p - V$-Diagramm sind $p\,V = const.$ Geraden, die sich mit steigender Temperatur $T_2 > T_1$ nach rechts verschieben.

Um das Temperaturverhalten des Gases zu verstehen, nehmen wir in ► Abbildung 19.8 als Ausgangspunkt (p_1, V_1, T_1) und erhöhen im Sinn des Gay-Lussac-Gesetzes (19.14) die Temperatur bei konstantem Volumen V_1:

$$p_2 - p_1 = p_1\,\gamma\,(T_2 - T_1) \quad \text{bei} \quad V_1 = const. \tag{19.26}$$

Am Ende des Prozesses befindet sich das Gassystem am Punkt (p_2, V_1, T_2).

Genauso können wir durch Temperaturerhöhung das Volumen bei konstantem Druck vergrößern (19.9) und das System zum Punkt (p_1, V_2, T_2) bringen:

$$V_2 - V_1 = V_1\, \gamma\, (T_2 - T_1) \quad \text{bei} \quad p_1 = const. \tag{19.27}$$

Für eine bestimmte Temperaturänderung von $\Delta T = T_2 - T_1 > 0$ beobachten wir also eine *relative Vergrößerung des Drucks* bei konstantem Volumen $V_1 = const.$, die genauso groß ist wie eine relative *Vergrößerung des Volumens* bei konstantem Druck $p_1 = const.$:

$$\frac{p_2 - p_1}{p_1} = \frac{V_2 - V_1}{V_1} \quad \text{für} \quad \Delta T = T_2 - T_1 = const. \tag{19.28}$$

Daher stehen die Hyperbeln in ► Abbildung 19.8 in direktem Zusammenhang mit der Temperatur.

Bevor wir die relativen Änderungen von Druck und Volumen eines Gassystems in einen proportionalen Zusammenhang mit der Temperatur bringen, beziehen wir die Abhängigkeit von der Menge des Gases ein, das sich im betrachteten System befindet.

19.3.4 Stoffmenge

Die Anzahl N der Atome bzw. Moleküle können wir über die sogenannte Stoffmenge ausdrücken. Die Stoffmenge vereinfacht das Experimentieren und die Übertragung unserer Überlegungen zwischen verschiedenen Gasen.

Seit der Neudefinition des SI-Einheitensystems über physikalische Konstanten wird die **Stoffmenge Mol** über die **Avogadro-Konstante** N_A bestimmt (Abschnitt 2.1):

$$N_A = 6{,}02214076 \cdot 10^{23}\, \frac{1}{\text{mol}} \tag{19.29}$$

Die Avogadro-Konstante wird dabei auf die Zahl in (19.29) festgelegt. Ein Mol eines Stoffes enthält dann genau $6{,}022\,140\,76 \cdot 10^{23}$ Teilchen, z. B. Atome oder Moleküle.

Die Anzahl N der Teilchen im Volumen V erhält man somit aus der Anzahl ν der Mole und der Avogadro-Konstanten:

$$N = \nu\, N_A \tag{19.30}$$

Für die meisten praktischen Anwendungen ist eine **Näherung**, die auf der früheren Definition des Mols beruht, ausreichend genau und einfacher zu benutzen (siehe auch das folgende Beispiel): 1 mol ist die Stoffmenge eines Systems, das aus genau so vielen Teilchen besteht, wie Atome in $0{,}012$ kg des Kohlenstoffnukleids ^{12}C enthalten sind.

Beispiel 54: Molmasse eines Gases

- 1 mol Heliumgas He entspricht der Molmasse $M_{\text{mol}} = 4\,\text{g He}$.
- 1 mol Wasserstoffgas H_2 entspricht $M_{\text{mol}} = 2\,\text{g}\,H_2$.

Molvolumen von Gas

Die Stoffmenge 1 mol eines idealen Gases nimmt unter gleichen Bedingungen (Temperatur, Druck) das gleiche Volumen ein. Unter physikalischen **Normalbedingungen** beträgt das Molvolumen

$$V_{\text{mol}}(p = 1\,\text{atm},\ T = 0\,°\text{C}) = 22{,}413\,969\,54\ldots\ \text{dm}^3/\text{mol}\,. \tag{19.31}$$

Dabei beziehen wir uns auf *physikalische Normalbedingungen* mit $1\,\text{atm} = 101\,325\,\text{Pa}$. Bei *chemischen Standardbedingungen* herrscht ein Druck von $100\,000\,\text{Pa}$, dann ist $V_{\text{mol}} = 22{,}71\ldots\ \text{dm}^3/\text{mol}$.

Die Anzahl N der Teilchen im Gasvolumen V lässt sich so alternativ über das Volumenverhältnis ausdrücken:

$$N = \nu\, N_A \tag{19.32}$$

$$= \frac{V}{V_{\text{mol}}}\, N_A \tag{19.33}$$

Anzahl der Mole

Die Anzahl der Mole eines Gases können wir damit aus verschiedenen Größen berechnen. Bekannt sein müssen entweder die Anzahl der Teilchen N im Gas, sein Volumen V bei $1\,\text{atm}$ und $T = 0\,°\text{C}$ oder seine Masse M:

$$\nu = \frac{N}{N_A} = \frac{V}{V_{\text{mol}}} = \frac{M}{M_{\text{mol}}} \tag{19.34}$$

19.4 Ideales Gasgesetz

In ► Abbildung 19.8 haben wir alternativ Druck oder Volumen eines Gases durch Temperaturerhöhung vergrößert. Um einen Zusammenhang zwischen diesen drei physikalischen Größen zu erhalten, machen wir folgenden Ansatz mit einer noch zu bestimmenden Größe r:

$$p\,V = r\,T \tag{19.35}$$

> **Aufgabe 56: Gasgesetz**
>
> Zeigen Sie, dass der Ansatz $p\,V = r\,T$ (19.35) für eine gegebene Temperaturänderung ΔT die Gleichheit der relativen Druck- und Volumenänderung (19.28) richtig beschreibt.

Wir wollen im Folgenden untersuchen, wie unser Ansatz $p\,V = r\,T$ (19.35) von der Menge des Gases abhängt. Dazu führen wir folgendes Gedankenexperiment durch.

Experiment 57: Größe eines Gassystems

Wir füllen zwei gleich große Behälter mit Volumina V mit jeweils derselben Stoffmenge ν von Heliumatomen. Beide Gassysteme haben Zimmertemperatur $T_\circ$ und den gleichen Gasdruck $p_\circ$.

Anschließend vergrößern wir das Gassystem, indem wir die beiden Behälter zu einem einzigen Behälter mit dem Volumen $2V$ zusammenfügen und damit die Gasatome der gesamten Stoffmenge 2ν zusammenbringen.

Volumen und Stoffmenge sind bei diesem Vorgang proportional zueinander ($V \propto \nu$), während gleichzeitig die Temperatur $T_\circ$ und der Druck $p_\circ$ unverändert bleiben.

Die Größe r im Ansatz (19.35) ist also proportional zur Gasmenge ν, sodass wir mit einer neuen Konstanten R

$$p_\circ V = \nu R T_\circ \tag{19.36}$$

schreiben können. Anstelle der Stoffmenge ν wird häufig auch die Anzahl N der Gasatome verwendet. Mit dem Zusammenhang $\nu = N/N_A$ (19.34) können wir entsprechend

$$\nu R = N \underbrace{\frac{1}{N_A} R}_{\equiv k} \tag{19.37}$$

umrechnen. Die Größe k bezeichnet die sogenannte **Boltzmann-Konstante**, eine der physikalischen Konstanten des neuen SI-Systems (Abschnitt 2.1). Mit der Boltzmann-Konstante k werden wir in Abschnitt 20.2 die Basiseinheit der Temperatur Kelvin (K) definieren. Sie hat den Wert

$$k = 1{,}380\,649 \cdot 10^{-23}\,\frac{\mathrm{J}}{\mathrm{K}}\,. \tag{19.38}$$

Die Boltzmann-Konstante k (19.38) steht über die Avogadro-Konstante N_A (19.29) in direktem Zusammenhang mit der Größe R (19.37), die wir als **allgemeine Gaskonstante** bezeichnen:

$$R = k\, N_A \tag{19.39}$$

Sie hat den Wert

$$R = 8{,}314\,462\,618\,153\,24\,\frac{\mathrm{J}}{\mathrm{mol} \cdot \mathrm{K}}\,. \tag{19.40}$$

Mit den Gleichungen (19.36) und (19.37) erhalten wir zwei alternative, übliche Formulierungen für das **ideale Gasgesetz**:

$$pV = N\,k\,T = \nu\,R\,T \tag{19.41}$$

Die Eigenschaft »ideales« Gas bedeutet genügend große Abstände zwischen den Konstituenten (Gasatome bzw. -moleküle) und nur elastische Stoßprozesse. Genaueres erklären wir in Abschnitt 20.1.1.

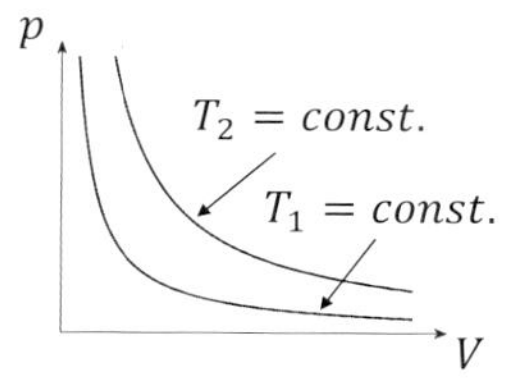

Abbildung 19.9 Im linearen p-V-Diagramm sind $p\,V = const.$ Hyperbeln, die sich mit steigender Temperatur $T_2 > T_1$ nach rechts oben verschieben.

Eine übliche Visualisierungsmöglichkeit des idealen Gasgesetzes ist die folgende Darstellung auf linearen Achsen (► Abbildung 19.9): Trägt man den Druck p als Funktion des Volumens V auf und hält die Temperatur $T = const.$ konstant, so erhält man Hyperbeln.

Der im idealen Gasgesetz (19.41) formulierte Zusammenhang zwischen dem Gasdruck p, dem Gasvolumen V, der Temperatur T und der Anzahl ν der Mole des Gases spielt im Rahmen der Thermodynamik eine wichtige Rolle, die wir in Kapitel 22 thematisieren.

Die Temperatur T in (19.41) ist die absolute Temperatur und wird in Kelvin (K) gemessen. Ihre physikalische Bedeutung werden wir in Abschnitt 20.2 untersuchen. Schon jetzt stellen wir fest: Während wir bislang nur relative Änderungen des Drucks, des Volumens, der Temperatur und der Gasmenge untersucht haben, sind mit dem idealen Gasgesetz auch absolute Vorhersagen möglich.

19.4.1 Ausdehnungskoeffizient und absolute Temperatur

Wir betrachten ein ideales Gas nach dem Gasgesetz (19.41)

$$p\,V = N\,k\,T\,. \tag{19.42}$$

Sowohl der Druck bei konstantem Volumen als auch das Volumen bei konstantem Druck ändern sich bei fester Anzahl N der Gasatome linear mit der Temperatur. Für den Ausdehnungskoeffizienten γ_V bei konstantem Volumen heißt das mit (19.15)

$$\gamma_V = \frac{1}{p}\left(\frac{\partial p}{\partial T}\right)_V = \frac{V}{N\,k\,T}\,\frac{N\,k}{V} = \frac{1}{T}\,. \tag{19.43}$$

Entsprechend gilt für den Ausdehnungskoeffizienten γ_p bei konstantem Druck mit (19.10)

$$\gamma_p = \frac{1}{V}\left(\frac{\partial V}{\partial T}\right)_p = \frac{p}{N\,k\,T}\,\frac{N\,k}{p} = \frac{1}{T}\,. \tag{19.44}$$

Konsistent mit (19.16) finden wir keinen Unterschied zwischen den beiden Ausdehnungskoeffizienten (19.43) und (19.44):

$$\gamma_V = \gamma_p \equiv \gamma = \frac{1}{T} \tag{19.45}$$

Der **Ausdehnungskoeffizient** γ für ideale Gase ist also gleich der inversen absoluten Temperatur T und damit stark temperaturabhängig. Für 1 mol ideales Gas erhalten wir γ unter Normalbedingungen ($p = 1\,\text{atm}$, $T_1 = 0\,°\text{C}$):

$$\gamma = \frac{1}{T} = \frac{N_A\, k}{p\, V_{\text{mol}}} \tag{19.46}$$

$$= \frac{6{,}0221\,10^{23} \cdot 1{,}3806\,10^{-23}\,\text{J/K}}{101\,325\,\text{Pa} \cdot 22{,}414\,10^{-3}\text{m}^3} = 3{,}661 \cdot 10^{-3}\,\frac{1}{\text{K}} \tag{19.47}$$

Wir hatten diesen Wert schon in (19.23) eingeführt und mit Messungen verglichen. Der theoretische Wert wird durch Messungen z. B. an Helium (19.22) sehr gut bestätigt.

Aus dem Ausdehnungskoeffizienten γ für ideale Gase ergibt sich die absolute Temperatur in Kelvin bei Normalbedingungen ($T_1 = 0\,°\text{C}$)

$$T_1 = \frac{1}{\gamma} = 273{,}15\,\text{K}\,. \tag{19.48}$$

Damit haben wir einen Wert der Kelvinskala benannt, in Abschnitt 20.2 erklären wir die Kelvin-Temperaturskala und bringen sie in Abschnitt 20.3 in den Zusammenhang mit der Celsius-Temperaturskala.

Das Wichtigste zum Mitnehmen

Zum Mitnehmen: Thermische Ausdehnung 1

- Die **Längenausdehnung** fester Körper ist näherungsweise proportional zur Temperaturänderung

$$l_2 = l_1(1 + \alpha(T_2 - T_1)) \tag{19.1}$$

mit dem Ausdehnungskoeffizienten α. Für kleine Volumenänderung bei Festkörpern und Flüssigkeiten gilt mit $\gamma \approx 3\alpha$

$$V_2 = V_1(1 + \gamma\Delta T)\,. \tag{19.7}$$

Der Faktor γ heißt **Volumenausdehnungskoeffizient**.

- Bei **Gasen** muss man zwischen Volumen- und Druckänderungen unterscheiden.
  - Bei konstantem Gasdruck p beobachtet man auch für Gase eine lineare Volumenausdehnung mit der Temperaturänderung ΔT:

$$V_2 = V_1(1 + \gamma_p\,\Delta T) \tag{19.9}$$

$$\gamma_p = \frac{1}{V}\left(\frac{\partial V}{\partial T}\right)_p \tag{19.10}$$

Der Index p steht für Volumenänderung bei konstantem Druck.

  - Bei konstantem Volumen $V = const.$ nennt man den linearen Zusammenhang zwischen Druckänderung und Temperaturänderung **Gay-Lussac-Gesetz**:

$$p_2 = p_1(1 + \gamma_V\,\Delta T) \tag{19.14}$$

Das V indiziert hier die Druckänderung bei konstantem Volumen

$$\gamma_V = \frac{1}{p}\left(\frac{\partial p}{\partial T}\right)_V \tag{19.15}$$

Experimentell ergeben sich gleiche Werte für die Ausdehnungskoeffizienten bei konstantem Druck (19.10) und bei konstantem Volumen (19.15):

$$\gamma_p = \gamma_V \equiv \gamma \tag{19.16}$$

Zum Mitnehmen: Thermische Ausdehnung 2

- **Die Stoffmenge mol** ist eine weitere **Basiseinheit des SI-Systems**: Bei der Neudefinition des SI-Einheitensystems über physikalische Konstanten wird das Mol über die **Avogadro-Konstante** N_A definiert:

$$N_A = 6{,}022\,140\,76 \cdot 10^{23}\,\frac{1}{\text{mol}} \tag{19.29}$$

  N_A ist als **Anzahl der Teilchen pro Mol** exakt festgelegt. Ein Mol eines Stoffes enthält dann genau $6{,}022\,140\,76 \cdot 10^{23}$ Teilchen, z. B. Atome oder Moleküle.

  Die Anzahl N der Teilchen im Volumen V erhält man somit aus der **Anzahl ν der Mole** und der Avogadro-Konstanten durch

$$N = \nu\, N_A\,. \tag{19.30}$$

- Das **Molvolumen** ist das Volumen, das die Stoffmenge 1 mol eines idealen Gases unter gleichen Bedingungen (Temperatur, Druck) einnimmt, insbesondere gilt unter **Normalbedingungen**

$$V_{\text{mol}}(p = 101\,325\,\text{Pa},\, T = 0\,°\text{C}) = 22{,}41\ldots\ \text{dm}^3/\text{mol}\,. \tag{19.31}$$

  Die Anzahl der Mole ν eines Gases kann man dann entweder aus der Anzahl der Teilchen N, seinem Volumen V bei Normalbedingungen oder seiner Masse M bestimmen:

$$\nu = \frac{N}{N_A} = \frac{V}{V_{\text{mol}}} = \frac{M}{M_{\text{mol}}} \tag{19.34}$$

- Der Zusammenhang von Druck p, Volumen V und Temperatur T eines idealen Gases hat zwei alternative übliche Formulierungen im **idealen Gasgesetz**:

$$pV = NkT = \nu RT \tag{19.41}$$

  Die Bezeichnung als »ideales« Gas bezieht sich auf die Wechselwirkungseigenschaften der Gasatome bzw. -moleküle, die wir im Abschnitt 20.1.1 benennen.

  Die Darstellung des Drucks p als Funktion des Volumens V bei konstanter Temperatur $T = const.$ ergibt Hyperbeln.

Zum Mitnehmen: Thermische Ausdehnung 3

- Die Größe k ist als **Boltzmann-Konstante** eine weitere physikalische Konstante des SI-Systems. Sie hat den exakten Wert

$$k = 1{,}380\,649 \cdot 10^{-23}\,\frac{\mathrm{J}}{\mathrm{K}}\,. \qquad (19.38)$$

  Aus der Einheit $\mathrm{J} = \mathrm{kg\,m^2/s^2}$ kann man erkennen, dass sich mit der Boltzmann-Konstanten unter Verwendung der Einheiten Sekunde, Meter und Kilogramm die **Temperatureinheit K** definieren lässt (siehe Abschnitt 20.2 und Abschnitt 20.3).

- Die **allgemeine Gaskonstante** R (19.40) und die Boltzmann-Konstante k (19.38) stehen über die Avogadro-Konstante N_A (19.29) in direktem Zusammenhang:

$$R = k\,N_A \qquad (19.39)$$

  Mit k und N_A hat auch R einen exakten Wert:

$$R = 8{,}314\,462\,618\,153\,24\,\frac{\mathrm{J}}{\mathrm{mol}\cdot\mathrm{K}} \qquad (19.40)$$

Kinetische Gastheorie

20

20.1 Energie der Gasatome 437
20.1.1 Ideales Gas 437
20.1.2 Wandstöße eines Atoms 437
20.1.3 Gasdruck 439
20.1.4 Kinetische Energie 440

20.2 Absolute Temperatur und Energie 441
20.2.1 Freiheitsgrade 441
20.2.2 Innere Energie 442

20.3 Celsius- und Kelvinskala 442

20.4 Geschwindigkeitsverteilung 443

ÜBERBLICK

» Die mikroskopische Bewegung von Atomen bzw. Molekülen eines Gases spielt eine entscheidende Rolle für die makroskopischen Größen, mit denen wir dieses Gas charakterisieren.

Wir werden sehen, dass die absolute Temperatur eines Gases bis auf eine Konstante der mittleren kinetischen Energie seiner Gasatome entspricht.

Zur Beschreibung der individuellen Geschwindigkeiten von Gasatomen werden wir die Maxwell-Boltzmann-Verteilung als Wahrscheinlichkeitsverteilung für die Geschwindigkeiten vorstellen. «

20.1 Energie der Gasatome

20.1.1 Ideales Gas

Wir betrachten zunächst in der linken ► Abbildung 20.1 ein Gas in einem würfelförmigen Volumen. Das Gas bestehe aus N Atomen oder Molekülen, die sich in dem Volumen mit den Geschwindigkeiten $\vec{v}_i$ bewegen.

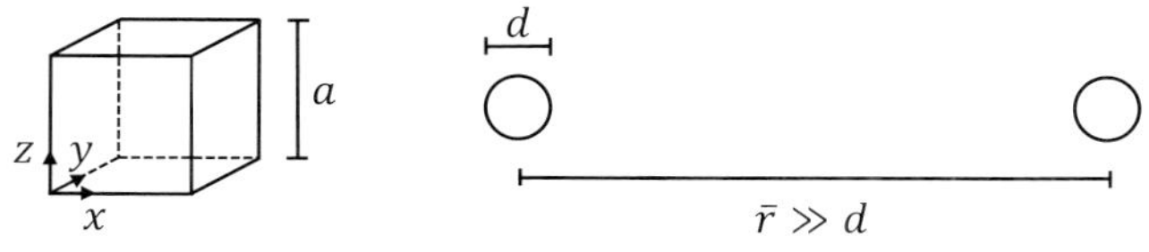

Abbildung 20.1 Links: Ein würfelförmiges Gasvolumen der Kantenlänge a. Rechts: Für ein ideales Gas ist der Atomdurchmesser d klein gegen den mittleren Abstand der Atome $\bar{r}$.

Der Durchmesser der Atome ist $d \approx 10^{-10}$ m. Der mittlere Abstand zwischen zwei Atomen sei $\bar{r}$. Man spricht von einem **idealen Gas**, falls $\bar{r} \gg d$ ist und eine Wechselwirkung zwischen Atomen untereinander und mit den Wänden nur durch elastische Stöße erfolgt (rechte ► Abbildung 20.1).

Wir werden im Folgenden die Stöße der Gasatome an den Wänden untersuchen und den Druck berechnen, den N Atome auf die Wände des Volumens ausüben.

20.1.2 Wandstöße eines Atoms

Durch ihre Bewegungen innerhalb des Volumens stoßen die Gasatome wiederholt gegen die Wände und üben dabei eine Kraft auf die Wände aus.

Wir betrachten hier die eindimensionale Bewegung des i-ten Atoms mit der Geschwindigkeit $\vec{v}_i$ und berechnen seine Kraft auf eine Wand (linke ► Abbildung 20.2).

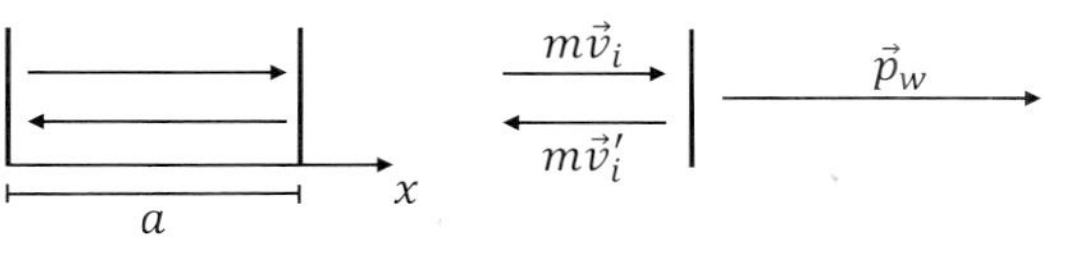

Abbildung 20.2 Links: Elastische Stöße an den Wänden im Abstand a. Rechts: Beim elastischen Stoß wird das Doppelte des Atom-Impulses auf die Wand übertragen.

Seine Flugzeit zwischen zwei Stößen mit derselben Wand beträgt

$$|\vec{v}_i| = \frac{2\,a}{\Delta t}\,, \tag{20.1}$$

$$\Delta t = \frac{2\,a}{|\vec{v}_i|}\,. \tag{20.2}$$

Bei jedem Stoß an der Wand wird der Impulsvektor $\vec{p}_i = m\,\vec{v}_i$ des Atoms um 180° gedreht ((8.33), rechte ► Abbildung 20.2). Dabei bezeichnet m die Masse des Gasatoms. Wir nehmen an, dass alle Stöße elastisch sind, d. h. $m = const.$ und $|\vec{v}_i| = |\vec{v}_i'|$. Der Impuls $\vec{p}_w$, der dabei auf die Wand übertragen wird, entspricht der Impulsänderung des Atoms:

$$\vec{p}_w = \vec{p}_i - \vec{p}_i' \tag{20.3}$$

$$= m\,(\vec{v}_i - \vec{v}_i') \tag{20.4}$$

$$= 2\,m\,\vec{v}_i \tag{20.5}$$

Den Betrag der Kraft $\vec{F}_i$, die das i-te Atom bei jedem Stoß auf die Wand ausübt, berechnen wir unter Berücksichtigung der Gleichungen (20.2) und (20.5):

$$F_i = \frac{|\vec{p}_w|}{\Delta t} \tag{20.6}$$

$$= \frac{2\,m\,v_i^2}{2\,a} \tag{20.7}$$

$$= \frac{m\,v_i^2}{a} \tag{20.8}$$

Wir visualisieren die Wandstöße von Atomen in einem makroskopischen Experiment mit kleinen Stahlkugeln und messen die Kraft, die die Kugeln auf eine Wand ausüben.

Experiment 58: Wandstöße von Atomen

Wir lassen Kugeln der Masse $m = 2{,}1\,\mathrm{g}$ in konstanten Zeitabständen $\Delta t \approx 0{,}5\,\mathrm{s}$ aus der Höhe $h = 0{,}22\,\mathrm{m}$ auf eine Balkenwaage fallen (► Abbildung 20.3). Damit erzeugen wir einen Impulsübertrag pro Zeiteinheit und messen die entsprechende Kraft auf der Waage.

Mithilfe der Waage bestimmen wir die durch den Kügelchenaufprall erzeugte effektive Kraft mittels der Masse $M_{\text{eff}} = 1{,}6\,\mathrm{g}$, die auf der anderen Seite der Waage diese Kraft kompensiert. Die experimentell gemessene Kraft $F = g\,M_{\text{eff}}$ ist also mit $g = 10\,\mathrm{m/s^2}$

$$F = 1{,}6 \cdot 10^{-2}\,\mathrm{N}\,. \tag{20.9}$$

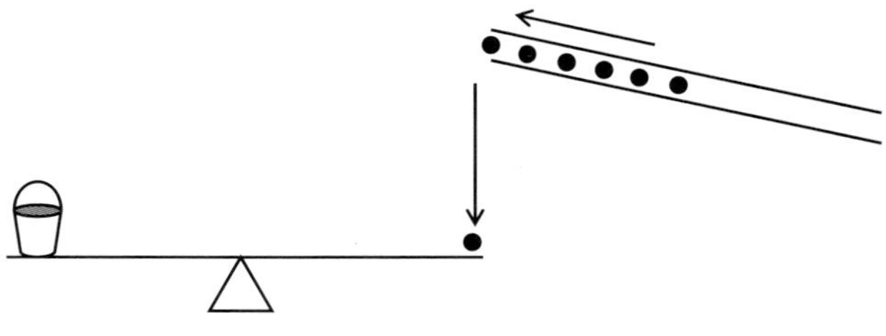

Abbildung 20.3 Modellversuch zur Wirkung von sehr vielen Wandstößen

Im Rahmen unserer Überlegungen zu Wandstößen würden wir folgende Kraft erwarten: Die Geschwindigkeit eines Kügelchens, das aus der Höhe h auf die Waage fällt, ist wegen der konstanten Beschleunigung g und wegen $h = g\,t^2/2$ (1.33) bzw. $t = \sqrt{2\,h/g}$ und $v = gt$: $v = \sqrt{2\,g\,h}$. Der Impulsübertrag eines Kügelchens auf die Waage beträgt nach Gleichung (20.5)

$$p_w = 2\,m\,v \tag{20.10}$$

$$= 2\,m\,\sqrt{2\,g\,h}\,. \tag{20.11}$$

Die erwartete Kraft ergibt sich aus der Impulsänderung pro Zeiteinheit:

$$F = \frac{\Delta p}{\Delta t} = \frac{2\,m\,\sqrt{2\,g\,h}}{\Delta t} = \frac{2 \cdot 0{,}0021\,\mathrm{kg} \cdot \sqrt{2 \cdot 10\,\mathrm{m/s^2} \cdot 0{,}22\,\mathrm{m}}}{0{,}5\,\mathrm{s}} = 1{,}8 \cdot 10^{-2}\,\mathrm{N} \tag{20.12}$$

Dieser Wert liegt nahe am Messwert (20.9) und ist im Rahmen der moderaten Messgenauigkeit der Apparatur mit unserer Erwartung konsistent.

20.1.3 Gasdruck

Wir berechnen jetzt den Druck p, den die Gasatome auf die Wände unseres würfelförmigen Volumens V mit der Kantenlänge a ausüben. Dazu nehmen wir jetzt eine regellose Bewegung der Gasatome an. Betrachten wir nun die rechte senkrechte Wand des Würfels und legen das Koordinatensystem so, dass die y-z-Ebene längs der angestoßenen Wand verläuft und die x-Achse senkrecht dazu steht (linke ► Abbildung 20.4).

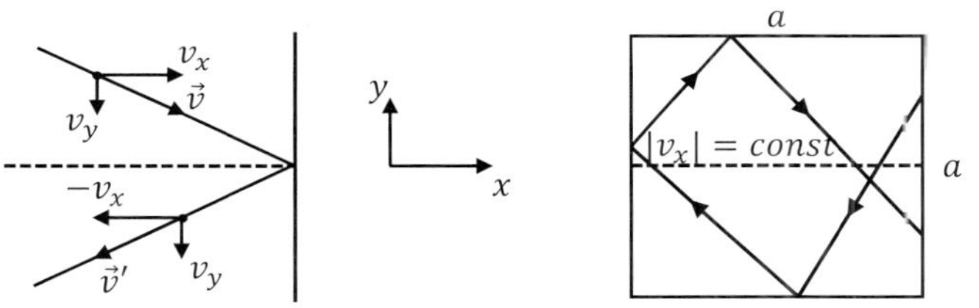

Abbildung 20.4 Links: x-y-Projektion eines Stoßes an der rechten Wand des Würfels. Rechts: Auch bei regelloser elastischer Reflexion durchquert das Teilchen den Würfel mit konstanter x-Komponente der Geschwindigkeit.

Wir zerlegen den Geschwindigkeitsvektor in y- und z-Komponenten, die beim Stoß erhalten bleiben, während die x-Komponente ihr Vorzeichen ändert (8.33). Diese Geschwindigkeitsänderung führt zu einer Impulsänderung wie in (20.5):

$$p_w = 2\,m\,v_x \tag{20.13}$$

Da alle Reflexionen elastisch sind, bleibt der Betrag der Geschwindigkeit des Atoms in x- Richtung konstant. Bis das Atom wieder an dieselbe Wand stößt, benötigt es die Zeit $\Delta t = 2a/v_{x,i}$ (rechte ► Abbildung 20.4).

Den Betrag der Kraft $\vec{F}_{x,i}$, die das i-te Atom bei jedem Stoß auf die Wand ausübt, berechnen wir unter Berücksichtigung der Gleichung (20.13):

$$F_{x,i} = \frac{|\vec{p}_w|}{\Delta t} = \frac{m\,v_{x,i}^2}{a} \tag{20.14}$$

Dabei gilt für den Mittelwert der quadratischen Geschwindigkeiten der Atome

$$\overline{v_x^2} = \frac{1}{N}(v_{x,1}^2 + v_{x,2}^2 + \ldots + v_{x,N}^2)\,. \tag{20.15}$$

Für die Geschwindigkeit des i-ten Atoms haben wir

$$v_i^2 = v_{x_i}^2 + v_{y_i}^2 + v_{z_i}^2\,. \tag{20.16}$$

Damit gilt auch für die Mittelwerte

$$\overline{v^2} = \overline{v_x^2} + \overline{v_y^2} + \overline{v_z^2}\,. \tag{20.17}$$

Wir nehmen jetzt an, dass in (20.17) die Komponenten der mittleren Geschwindigkeitsquadrate der Atome in den drei Raumrichtungen gleichverteilt sind:

$$\overline{v_x^2} = \overline{v_y^2} = \overline{v_z^2} = \frac{1}{3}\overline{v^2} \tag{20.18}$$

Für die mittlere Gesamtkraft auf eine Wand folgt mit (20.14)

$$\overline{F} = \frac{1}{3}\sum_{i=1}^{N} F_i \tag{20.19}$$

$$= \frac{1}{3}\,\frac{m}{a}\,(v_1^2 + v_2^2 + \ldots + v_N^2) \tag{20.20}$$

$$= \frac{1}{3}\,N\,\frac{m}{a}\,\overline{v^2}\,. \tag{20.21}$$

Der Druck p der Gasatome auf die Wand mit der Fläche a^2 beträgt dann nach der Definitionsgleichung (18.1) und mit dem würfelförmigen Volumen $V = a^3$

$$p = \frac{\overline{F}}{a^2} \tag{20.22}$$

$$= \frac{1}{3}\,\frac{N\,m}{a^3}\,\overline{v^2} \tag{20.23}$$

$$= \frac{1}{3}\,\frac{N\,m}{V}\,\overline{v^2}\,. \tag{20.24}$$

Fassen wir Druck und Volumen auf einer Seite der Gleichung zusammen, erhalten wir

$$p\,V = \frac{1}{3}\,N\,m\,\overline{v^2}\,. \tag{20.25}$$

20.1.4 Kinetische Energie

Alternativ können wir auf der rechten Seite von (20.25) auch die mittlere kinetische Energie (5.23) der Atome

$$\overline{E}_{\text{kin}} = \frac{1}{2}\,m\,\overline{v^2} \tag{20.26}$$

verwenden. Durch Einsetzen in die Gleichung (20.25) erhalten wir folgenden theoretischen Zusammenhang zwischen dem Druck p, dem Volumen V und der mittleren kinetischen Energie $\overline{E}_{\text{kin}}$ zusammen mit der Anzahl N der Atome:

$$p\,V = \frac{2}{3}\,N\,\overline{E}_{\text{kin}} \tag{20.27}$$

Bei fester mittlerer kinetischer Energie $\overline{E}_{\text{kin}}$ der N Gasatome erwarten wir, dass das Produkt aus Gasdruck und Gasvolumen konstant bleibt. Dieser Zusammenhang ist uns bereits als Gesetz von Boyle-Mariotte bekannt (18.33).

20.2 Absolute Temperatur und Energie

Die experimentellen Beobachtungen über die Druck-, Volumen- und Temperaturabhängigkeit von Gasen haben uns in Abschnitt 19.4 zum idealen Gasgesetz geführt. Zusammen mit unseren theoretischen Überlegungen zu idealen Gasen implizieren sie, dass sich der makroskopische Temperaturbegriff auf mikroskopische Bewegungsenergie zurückführen lässt.

Wir vergleichen das ideale Gasgesetz (19.41) und die Gleichung (20.27):

$$pV = \frac{2}{3}\,N\,\overline{E}_{\text{kin}} \tag{20.28}$$

$$pV = N\,k\,T \tag{20.29}$$

Der Vergleich ergibt

$$\overline{E}_{\text{kin}} = \frac{3}{2}\,k\,T\,. \tag{20.30}$$

> *Wir erhalten damit einen direkten Zusammenhang zwischen der mittleren kinetischen Energie der Atome und der Temperatur!*

Da die Bewegungsenergie $\overline{E}_{\text{kin}} > 0$ immer positiv ist, ist auch die **absolute Temperatur** T auf einer positiven Skala zu messen. Die Boltzmann-Konstante k (19.38) gibt dabei den grundlegenden Umrechnungsfaktor zwischen Temperatur in Kelvin und Energie in Joule an. Die Temperatureinheit ist **Kelvin** (K) und wurde nach dem Physiker Lord Kelvin (1824–1907) benannt. Wie die Kelvin-Temperaturskala und die übliche Temperaturangabe in Celsius zusammenhängen, erklären wir in Abschnitt 20.3.

20.2.1 Freiheitsgrade

Bislang haben wir den Spezialfall eines 1-atomigen Gases betrachtet, das sich in die drei Raumrichtungen x, y und z bewegen kann. In einer Verallgemeinerung des Zusammenhangs zwischen Energie und Temperatur für Gasmoleküle werden auch Energiegrößen innerhalb der Moleküle berücksichtigt. Diese werden als Freiheitsgrade bezeichnet.

Beispiel 55: Freiheitsgrade

Die Größe f bezeichnet die Anzahl der **Freiheitsgrade** der Atome bzw. Moleküle.

- Atome:

  $f = 3$ (Translationsfreiheitsgrade in x-, y- und z-Richtung)

- Hantelförmige Moleküle (z. B. N_2):

  Bei sehr niedrigen Temperaturen ist auch für Hantelmoleküle $f = 3$.

  Bei höheren Temperaturen werden Rotationen der Moleküle möglich, d. h. $f = 5$ mit 3 Translations- und 2 Rotationsfreiheitsgraden. Bei den Rotationen liefert eine Rotation um die Achse des Moleküls wegen des verschwindenden Trägheitsmoments keinen Beitrag.

Bei noch höheren Temperaturen gibt es Schwingungsbewegungen der beiden Atome gegeneinander, sodass $f = 7$ ist. Wir addieren dafür 3 Translations- plus 2 Rotations- plus 2 Vibrationsfreiheitsgrade, bei den letztgenannten wird ein Freiheitsgrad für die kinetische und die potentielle Energie gezählt.

Die Erhöhung der Zahl der Freiheitsgrade mit steigender Temperatur ist darauf zurückzuführen, dass Zustände höherer Energie erst bei größeren Temperaturen »angeregt« werden. Solche Energieanregungen werden später in der Quantenmechanik erklärt.

Damit lautet der allgemeine Zusammenhang zwischen der mittleren kinetischen Energie und der Temperatur unter Berücksichtigung der Freiheitsgrade des Atoms bzw. Moleküls (vergleiche mit (20.30))

$$\overline{E}_{\text{kin}} = \frac{f}{2}\, k\, T\,. \tag{20.31}$$

20.2.2 Innere Energie

Die innere Energie U eines Gases im Volumen V ist die gesamte kinetische Energie seiner N Atome bzw. Moleküle (vergleiche mit (20.31)):

$$U = N\,\frac{f}{2}\, k\, T \tag{20.32}$$

20.3 Celsius- und Kelvinskala

Für die Temperatur gibt es mehrere Messskalen. In den europäischen und vielen anderen Länder ist die Celsiusskala verbreitet. Die beiden Fixpunkte auf dieser Skala sind durch den Gefrierpunkt ($T_1 = 0\,°\text{C}$) und den Siedepunkt ($T_2 = 100\,°\text{C}$) von Wasser bei einem Luftdruck von 1 atm gegeben. Die Ausdehnung von Quecksilber definiert dann die Temperaturdifferenz von $1\,°\text{C}$ als den Hundertsten Skalenteil zwischen den beiden Fixpunkten.

Die SI-Einheit der Temperatur ist jedoch das Kelvin K (Abschnitt 2.1). Mit dem Zusammenhang $\overline{E}_{\text{kin}} = (3/2)\, k\, T$ (20.30) sind Energie und Temperatur eines idealen Gases über die **Boltzmann-Konstante** k miteinander verknüpft. Aus den Einheiten von k

$$k = 1{,}380\,649 \cdot 10^{-23}\,\frac{\text{J}}{\text{K}} \tag{19.38}$$

mit $\text{J} = \text{kg}\,\text{m}^2/\text{s}^2$ kann man erkennen, dass sich über die Boltzmann-Konstante unter Verwendung der Einheiten Sekunde, Meter und Kilogramm die **Temperatureinheit K** an das SI-System anschließen lässt. In den Jahren vor der Neudefinition des Kelvin 2019 ist es gelungen, die Genauigkeit in der Messung von k unter 1 ppm zu senken. Daraufhin wurde k als einer weiteren physikalischen Konstante des SI-Systems der exakte Wert (19.38) zugewiesen.

Nach Konvention sind die Temperaturdifferenzen ΔT gemessen in Kelvin (K) und in Celsius (°C) gleich groß.

Wir weisen hier für praktische Anwendungen auf die frühere und weiterhin hilfreiche Definition der Kelvinskala mit seinen Fixpunkten bei dem absoluten Nullpunkt $T = 0\,\mathrm{K}$ (Gleichung (20.31)) und beim Tripelpunkt von Wasser ($T = 273{,}16\,\mathrm{K} = 0{,}01\,°\mathrm{C}$) hin. Mit dem Tripelpunkt werden die Bedingungen definiert, bei denen Wasser in seinen festen, flüssigen und dampfförmigen Phasen gleichzeitig existieren kann (Abschnitt 23.3).

Für die Umrechnung zwischen der Celsius- und der Kelvinskala nutzen wir aus, dass der Tripelpunkt von Wasser in der Celsiusskala bei 0,01 °C und in der Kelvinskala bei 273,16 K liegt und dass der Skalenschritt von 1 °C definitionsgemäß in beiden Skalen gleich ist.

Damit ergibt sich für die Umrechnung zwischen den Temperaturskalen

$$T^C(°\mathrm{C}) = T(\mathrm{K}) - 273{,}15\,\mathrm{K}\,. \tag{20.33}$$

Die Konstante zeigt den absoluten Nullpunkt der Kelvinskala und ist konsistent mit den Überlegungen zum Kehrwert des Ausdehnungskoeffizienten γ (19.48).

20.4 Geschwindigkeitsverteilung

Die Geschwindigkeit von Atomen bzw. Molekülen in einem Gas folgt der **Maxwell-Boltzmann-Verteilung**. Mit ihr wird eine Wahrscheinlichkeitsdichte (2.10) für die Geschwindigkeiten der Atome formuliert:

$$f(v) = 4\,\pi\,v^2 \left(\frac{m}{2\,\pi\,k\,T}\right)^{\frac{3}{2}} e^{-\frac{m\,v^2}{2\,k\,T}} \tag{20.34}$$

Die exakte Form wird in der statistischen Physik begründet. Wir interessieren uns hier für das Verständnis der verschiedenen Terme.

Den Exponentialterm in (20.34) bezeichnet man als **Boltzmann-Faktor**:

$$e^{-\frac{m\,v^2}{2\,k\,T}} \tag{20.35}$$

Er enthält im Zähler des Arguments die kinetische Energie $E_{\mathrm{kin}} = mv^2/2$ des individuellen Gasatoms. Im Nenner steht das Produkt aus der Boltzmann-Konstanten k und der absoluten Temperatur T, das nach Gleichung (20.31) bis auf eine Konstante der mittleren kinetischen Energie der Gasatome entspricht.

Der Boltzmann-Faktor dominiert bei großen Geschwindigkeitswerten v (► Abbildung 20.5). Bei kleinen Geschwindigkeiten v ist der erste Term (v^2) in Gleichung (20.34) dominant.

Der Normierungsterm in der Verteilung (20.34)

$$\left(\frac{m}{2\,\pi\,k\,T}\right)^{\frac{3}{2}} \tag{20.36}$$

folgt aus der Bedingung, dass die Gesamtwahrscheinlichkeit P, alle Gasatome mit einer Geschwindigkeit anzutreffen, $P = 1$ sein muss (2.11).

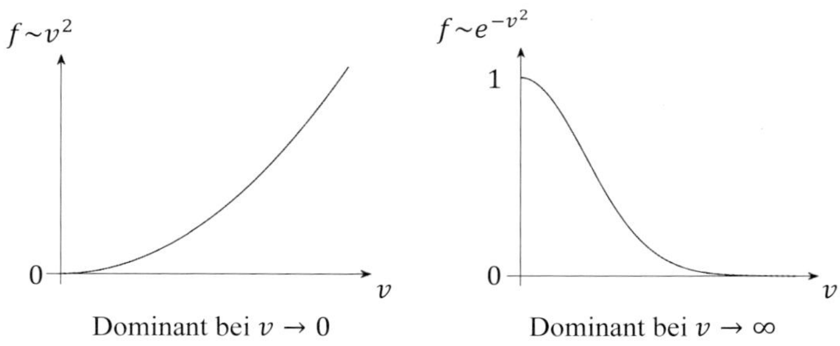

Abbildung 20.5 Maxwell-Boltzmann-Geschwindigkeitsverteilung (20.34): Bei kleinen Geschwindigkeiten überwiegt die kinetische Energie $f \propto v^2$, bei großen der Boltzmann-Faktor $f \propto e^{-v^2}$.

Das Maximum der Verteilung (20.34) befindet sich bei der Geschwindigkeit

$$v_{\max} = \sqrt{\frac{2kT}{m}}, \tag{20.37}$$

was man durch den Extremwert der Ableitung $df(v)/dv = 0$ nachrechnen kann. Wenn wir die Temperatur des Gases erhöhen, verschiebt sich die Verteilung zu höheren Geschwindigkeiten, insbesondere steigt $v_{\max} \propto 1/T^{1/2}$ (► Abbildung 20.6). Gleichzeitig flacht die Verteilung ab, da der Normierungsterm kleiner wird:

$$T_2^{-\frac{3}{2}} < T_1^{-\frac{3}{2}} \tag{20.38}$$

Somit verschiebt sich die gesamte Wahrscheinlichkeitsdichteverteilung mit höherer Temperatur zu höheren Geschwindigkeiten und verbreitert sich.

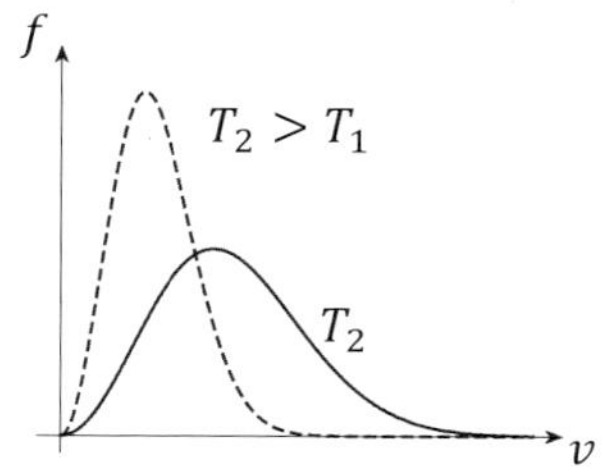

Abbildung 20.6 Maxwell-Boltzmann-Verteilungen $f(v)$ der Geschwindigkeiten v für zwei Temperaturen $T_2 > T_1$

Das Wichtigste zum Mitnehmen

Zum Mitnehmen: Kinetische Gastheorie 1

- Im Modell der **kinetischen Gastheorie** bewegen sich N Atome oder Moleküle im Volumen V mit den Geschwindigkeiten $\vec{v}_i$.

  In einem **idealen Gas** ist der Durchmesser der Atome d sehr klein gegen den mittleren Abstand $\overline{r}$ zwischen ihnen und es treten nur elastische Stöße auf.

  Der Druck p der Gasatome auf die Wand beträgt dann

$$p = \frac{1}{3}\,\frac{N\,m}{V}\,\overline{v^2}\,.$$

  Fassen wir Druck und Volumen zusammen, erhalten wir einen Zusammenhang wie im Gesetz von Boyle-Mariotte (18.33):

$$p\,V = \frac{1}{3}\,N\,m\,\overline{v^2} \qquad (20.25)$$

- Setzt man die mittlere kinetische Energie der Atome

$$\overline{E}_{\text{kin}} = \frac{1}{2}\,m\,\overline{v^2}$$

  in (20.25) ein, so erhält man einen theoretischen Zusammenhang zwischen dem Druck p, dem Volumen V, der mittleren kinetischen Energie $\overline{E}_{\text{kin}}$ und der Anzahl N der Atome:

$$p\,V = \frac{2}{3}\,N\,\overline{E}_{\text{kin}} \qquad (20.27)$$

- Der Vergleich des idealen Gasgesetzes $p\,V = N\,k\,T$ (19.41) mit Gleichung (20.27) ergibt

$$\overline{E}_{\text{kin}} = \frac{3}{2}\,k\,T\,. \qquad (20.39)$$

  Wir erhalten damit einen direkten Zusammenhang zwischen der mittleren kinetischen Energie der Atome und der Temperatur!

Zum Mitnehmen: Kinetische Gastheorie 2

- Neben der Translationsenergie treten vor allem bei höheren Temperaturen auch andere Energieformen auf. Die Energie verteilt sich dann gleichmäßig auf die verfügbaren voneinander unabhängigen Bewegungsmoden oder **Freiheitsgrade** f.

  Damit lautet der Zusammenhang zwischen der mittleren kinetischen Energie und der Temperatur unter Berücksichtigung der Freiheitsgrade des Atoms bzw. Moleküls (vergleiche mit (20.39)):

$$\overline{E}_{\text{kin}} = \frac{f}{2}\, k\, T \tag{20.31}$$

  Die **innere Energie** U ist die gesamte kinetische Energie seiner N Atome bzw. Moleküle (vergleiche mit (20.31)):

$$U = N\, \frac{f}{2}\, k\, T\,. \tag{20.32}$$

- Mit (20.31) und der Boltzmann-Konstante k (19.38) lässt sich eine **absolute Temperatur** T auf einer positiven Skala definieren. Ihre Einheit ist **Kelvin** (K) nach dem Physiker Lord Kelvin.

  Dementsprechend beruht die neue **SI-Definition der Temperatur** T auf der Festlegung der **Boltzmann-Konstante** k anstelle der alten (aber noch nützlichen) Definition über die Fixpunkte beim **absoluten Nullpunkt** $T = 0\,\text{K}$ und beim Tripelpunkt von Wasser $T = 273{,}16\,\text{K} = 0{,}01\,°\text{C}$ (Abschnitt 23.3).

  Gebräuchlich ist noch die Celsiusskala. Fixpunkte sind dabei der Gefrierpunkt ($T^C = 0\,°\text{C}$) und der Siedepunkt ($T^C = 100\,°\text{C}$) von Wasser bei einem Luftdruck von 1 atm.

  Nach Konvention sind die Temperaturdifferenzen ΔT gemessen in Kelvin (K) und in Celsius (°C) gleich groß. Für die Umrechnung zwischen den Temperaturskalen gilt dann

$$T^C(°\text{C}) = T(\text{K}) - 273{,}15\,\text{K}\,. \tag{20.33}$$

- Die Wahrscheinlichkeitsdichte für die Geschwindigkeit von Atomen bzw. Molekülen in einem Gas folgt der **Maxwell-Boltzmann-Verteilung**

$$f(v) = 4\,\pi\, v^2 \left(\frac{m}{2\,\pi\, k\, T}\right)^{\frac{3}{2}} e^{-\frac{m\, v^2}{2\, k\, T}} \tag{20.34}$$

  mit dem Maximum bei

$$v_{\max} = \sqrt{\frac{2kT}{m}}\,. \tag{20.37}$$

Wärme

21

21.1 Wärmeenergie **449**

21.2 Spezifische Molwärme **451**

21.2.1 Spezifische Molwärme von Gasen 451

21.2.2 Spezifische Molwärme von Festkörpern 454

21.3 Wärmetransport **454**

21.3.1 Konvektion 455

21.3.2 Wärmeleitung 456

21.3.3 Wärmestrahlung 457

21.3.4 Wärmedämmung 458

ÜBERBLICK

» In diesem Kapitel definieren wir den Begriff der Wärme als eine Form von Energie und entwickeln eine Vorstellung von den mikroskopischen Mechanismen, die eine Wärmeaufnahme durch ein Gas oder einen Festkörper ermöglichen.

Außerdem werden wir die Transportmechanismen für Wärme vorstellen: Konvektion, Wärmeleitung und Wärmestrahlung. «

21.1 Wärmeenergie

Im folgenden Experiment 59 erzeugen wir Wärme durch mechanische Reibung.

Experiment 59: Reibungswärme

Ein Gewicht mit der Masse m hängt an einem Kupferband, das über einem drehbaren Metallzylinder liegt. Das Band wird über eine Federwaage gehalten (► Abbildung 21.1). Beim Drehen der Kurbel stellt sich ein Gleichgewicht zwischen Reibungskraft (Abschnitt 3.2.4, Gleichung (3.29))

$$\vec{F}_R = -\mu \cdot F_N \cdot \frac{\vec{v}}{v}, \tag{21.1}$$

Gewichtskraft $F = mg$ und Federkraft F_F ein. Wir messen den Betrag der Reibungskraft $F_R = F - F_F$ mit der Federwaage. Mit einem Thermometer bestimmen wir die Temperatur des Zylinders, die sich durch die erzeugte Reibungswärme leicht erhöht.

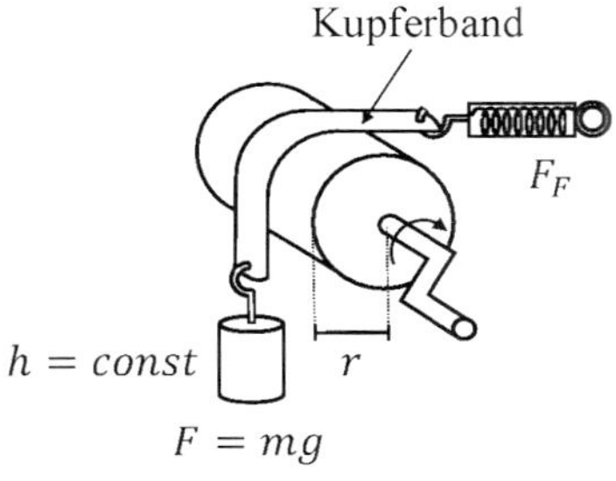

Abbildung 21.1 Erzeugung von Reibungswärme

Beim Drehen verrichten wir die Arbeit

$$\Delta W = s\, F_R\,. \tag{21.2}$$

Dabei bezeichnet $s = 2\,\pi\, r\, N$ die zurückgelegte Wegstrecke auf der Zylinderoberfläche bei insgesamt N Umdrehungen und F_R den Betrag der Reibungskraft (21.1).

Die mechanische Arbeit berechnen wir damit zu

$$\Delta W = 2\,\pi\, r\, N\, F_R\,. \tag{21.3}$$

Durch das Drehen wird Reibungswärme produziert, die wir durch den Temperaturanstieg ΔT des Zylinders messen.

Experimentell zeigt sich, dass die geleistete Arbeit linear mit der Temperaturerhöhung ΔT des Materials zusammenhängt und auch proportional zu der Masse M des erwärmten Materials ist. Außerdem ist die Arbeit, die man für eine Temperaturerhöhung leisten muss, stark materialabhängig. Insgesamt ergibt sich mit einer Materialkonstante c folgender Zusammenhang zwischen der mechanisch geleisteten Arbeit und der Temperaturänderung:

$$\Delta W = c\,M\,\Delta T \tag{21.4}$$

Im Experiment wird die mechanisch geleistete Arbeit ΔW vom Körper als Wärmemenge ΔQ aufgenommen:

$$\Delta Q = \Delta W \tag{21.5}$$

Wärme *ist eine Form der Energie, die in der Einheit Joule gemessen wird.*

Im Alltag ist noch die alte Wärmeeinheit Kalorien cal bzw. kcal gebräuchlich. Eine Wärmemenge von 1 cal ist erforderlich, um 1 g Wasser um 1 °C zu erwärmen. Die Umrechnung ergibt sich aus der spezifischen Wärme von Wasser (siehe Beispiel 56):

$$1\,\text{cal} = 4{,}187\,\text{J} \tag{21.6}$$

Wir können also eine Wärmeänderung bei Kenntnis der Materialkonstanten c und der Masse M eines Körpers makroskopisch durch die Temperaturänderung des Körpers bestimmen:

$$\Delta Q = c\,M\,\Delta T \tag{21.7}$$

Die Materialkonstante c heißt **spezifische Wärme**. Sie bezeichnet die benötigte Wärmemenge, um in 1 kg des Materials eine Temperaturerhöhung von 1 K zu erreichen.

Beispiel 56: Spezifische Wärme

In ► Tabelle 21.1 sind typische Werte für die spezifische Wärme beim Druck von 1 atm und Temperatur $T = 20\,°\text{C}$ aufgeführt.

	Wasser	Eisen	Aluminium
$c \,/\, \frac{\text{kJ}}{\text{kg K}}$	4,187	0,45	0,896

Tabelle 21.1 Werte für die spezifische Wärme verschiedener Materialien

Aufgabe 57: Reibungswärme

Ein drehbarer Eisenzylinder hat den Radius $r = 2\,\text{cm}$ und die Länge $l = 5\,\text{cm}$. Eisen hat die Dichte $\rho \approx 8 \cdot 10^3\,\text{kg/m}^3$. Seine spezifische Wärme beträgt $c = 0{,}45\,\text{kJ/(kg K)}$. Die Masse des am Band hängenden Gewichts sei $m = 1\,\text{kg}$. Das Band wird mit einer Federwaage gehalten, die beim Drehen 3 N anzeigt (► Abbildung 21.2).

Wie oft muss Roland kurbeln, um die Temperatur des Eisenzylinders um $\Delta T = 10\,\text{K}$ zu erhöhen?

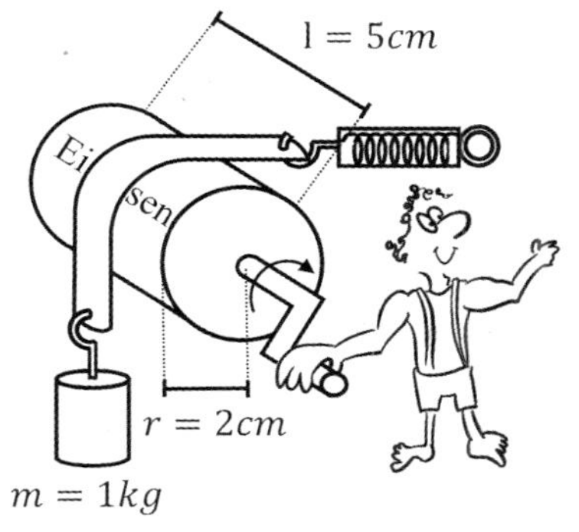

Abbildung 21.2 Aufgabe zur Reibungswärme

21.2 Spezifische Molwärme

Bislang haben wir die spezifische Wärme auf die Masse (Temperaturänderung pro Kilogramm Material) bezogen. Alternativ können wir uns auf die Molmasse beziehen. Dieser Weg führt zu einem mikroskopischen Verständnis, wie Stoffe Wärme aufnehmen. Wir betrachten zunächst Gase und dann Festkörper.

21.2.1 Spezifische Molwärme von Gasen

Bei Gasen unterscheiden wir die spezifische Wärme darin, ob wir das Volumen oder den Druck konstant halten.

Spezifische Wärme bei konstantem Volumen

Die innere Energie U eines Gases (20.32) können wir durch Energiezufuhr in der Form von Wärme ΔQ erhöhen. Wenn wir das Volumen konstant halten, entspricht die Änderung der inneren Energie ΔU der zugeführten Wärmemenge ΔQ:

$$\Delta U = \Delta Q \tag{21.8}$$

Beide Energieformen verhalten sich proportional zur Temperaturdifferenz (20.32), (21.7):

$$\Delta U = N \, \frac{f}{2} \, k \, \Delta T \tag{21.9}$$

$$\Delta Q = c \, M \, \Delta T \tag{21.10}$$

Die Umrechnung auf die Anzahl der Mole erhalten wir mit $\nu = M/M_{\text{mol}} = N/N_A$ (19.34):

$$\Delta U = \nu \, N_A \, \frac{f}{2} \, k \, \Delta T \tag{21.11}$$

$$\Delta Q = \nu \, \underbrace{c \, M_{\text{mol}}}_{\equiv c_V} \, \Delta T \tag{21.12}$$

Hierbei ist $\mathcal{C}_V$ die **spezifische Molwärme**. Der Index V dient zur Verdeutlichung, dass das Volumen V konstant ist. Aus beiden Gleichungen erhalten wir wegen $\Delta U = \Delta Q$ (21.8)

$$\mathcal{C}_V \equiv c\,M_{\text{mol}} = N_A\,k\,\frac{f}{2}\,. \tag{21.13}$$

Alternativ verwenden wir die Gaskonstante R (19.40) und erhalten

$$\mathcal{C}_V = \frac{f}{2}\,R\,. \tag{21.14}$$

Die Wärmeeigenschaften eines Gases werden also durch die Anzahl seiner Freiheitsgrade und damit durch die molekularen Eigenschaften des Gases bestimmt!

Ein 1-atomiges Gas (z. B. Helium) besitzt $f = 3$ Freiheitsgrade der Translation (Abschnitt 20.2.1, Beispiel 55). Ein Hantelmolekül hat abhängig von der Temperatur bis zu $f = 7$ Freiheitsgrade (z. B. Stickstoff N_2). Komplexere Moleküle haben zusätzliche Rotations- und Vibrationsfreiheitsgrade (z. B. $f(NO_2) = 12$).

Um die Freiheitsgrade z. B. der Rotation oder Vibration eines Gasmoleküls anzuregen, benötigen wir nach der Quantenmechanik jeweils bestimmte Mindestenergien. ► Abbildung 21.3 zeigt eine stufenförmige Temperaturabhängigkeit der spezifischen Molwärme verschiedener Gase, die Stufen entsprechen den Mindestenergien (Temperaturen) für die verschiedenen Freiheitsgrade.

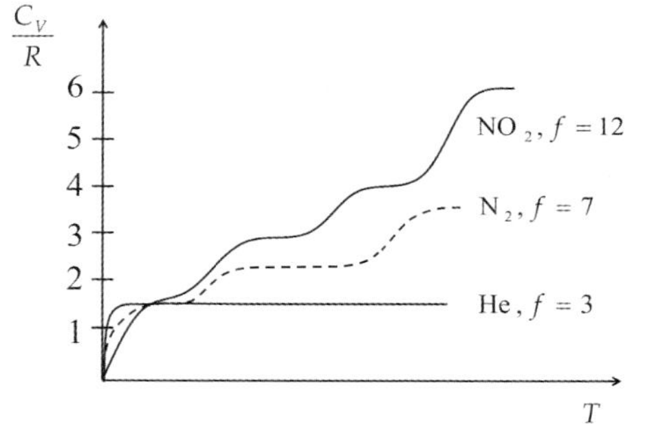

Abbildung 21.3 Die spezifische Molwärme $\mathcal{C}_V$ normiert auf die allgemeine Gaskonstante R als Funktion der Temperatur für verschiedene Freiheitsgrade f

Oberhalb der Mindestenergien stehen die jeweiligen Freiheitsgrade für die Wärmekapazität des Gases zur Verfügung. Da wir entsprechend der Maxwell-Boltzmann-Verteilung (20.34) ein Ensemble von Molekülen mit verschiedenen kinetischen Energien im Gas haben, sind die Stufen als Funktion der Temperatur ausgewaschen.

Spezifische Wärme bei konstantem Druck

Nach dem idealen Gasgesetz $p\,V = \nu\,R\,T$ (19.41) erhöht sich bei konstantem Volumen $V = const.$ der Druck p, wenn sich die Temperatur T erhöht: $\Delta p \propto \Delta T$.

Wenn wir umgekehrt den Druck p konstant halten und eine Temperaturerhöhung im Gas durch Wärmezufuhr verursachen, muss sich entsprechend (19.41) das Gasvolumen vergrößern $\Delta V \propto \Delta T$.

Im Folgenden werden wir die spezifische Molwärme für den Fall des konstanten Drucks $p = const.$ bestimmen. Das Volumen V eines bis auf eine Seite geschlossenen Gasbehälters kann durch einen Stempel mit der Querschnittsfläche A variiert werden. Der Druck des Gases auf den Stempel ist

$$p = \frac{F}{A} = \frac{\text{Kraft auf Stempel}}{\text{Stempelfläche}} \,. \tag{21.15}$$

Das Gas drückt mit konstantem Druck p gegen den Stempel und leistet dabei Arbeit (▶ Abbildung 21.4):

$$\Delta W = F\,\Delta x \tag{21.16}$$
$$= p\,A\,\Delta x \tag{21.17}$$
$$= p\,\Delta V \tag{21.18}$$

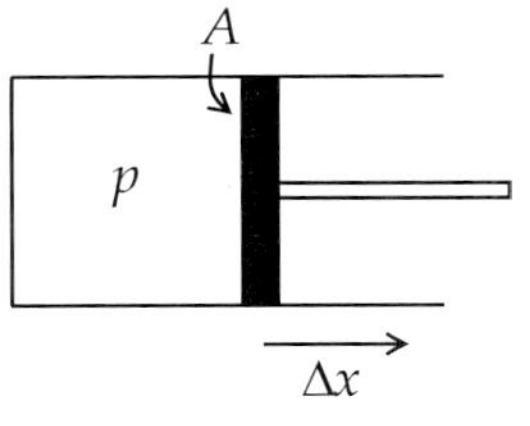

Abbildung 21.4 Ein Stempel mit Querschnitt A wird durch den Druck p um Δx bewegt.

Wegen des idealen Gasgesetzes $p\,V = \nu\,R\,T$ (19.41) können wir für $p = const.$ diese Volumenarbeit umschreiben:

$$\Delta W = \nu\,R\,\Delta T \tag{21.19}$$

Die Energie, die dem Gas durch die Wärmemenge ΔQ zugeführt wird, wird zum einen nach Gleichung (21.12) in Temperaturerhöhung ΔT umgewandelt. Zum anderen wird sie nach Gleichung (21.19) in Ausdehnungsarbeit ΔW umgesetzt, um den Druck $p = const.$ zu halten:

$$\Delta Q = \nu\,\mathcal{C}_V\,\Delta T + \nu\,R\,\Delta T \tag{21.20}$$
$$= \nu\,\underbrace{(\mathcal{C}_V + R)}_{\equiv \mathcal{C}_p}\,\Delta T \tag{21.21}$$

Die Struktur der Gleichung ist dieselbe wie (21.12), sodass wir als spezifische Molwärme $\mathcal{C}_p$ bei konstantem Druck p identifizieren:

$$\mathcal{C}_p = \mathcal{C}_V + R \tag{21.22}$$

Diese Gleichung schreiben wir alternativ mit Gleichung (21.14) in Abhängigkeit der Gaskonstanten R und der Zahl der Freiheitsgrade f:

$$\mathcal{C}_p = \frac{f+2}{2} R \tag{21.23}$$

Der Quotient der spezifischen Molwärmen heißt **Adiabatenkoeffizient**:

$$\kappa = \frac{\mathcal{C}_p}{\mathcal{C}_V} = \frac{f+2}{f} \tag{21.24}$$

Er hängt nur noch von der Anzahl der Freiheitsgrade des Gases ab. Eine Methode zur experimentellen Bestimmung des Adiabatenkoeffizienten κ ergibt sich über die Messung der Schallgeschwindigkeit $v = \sqrt{p\,\kappa/\rho}$ (18.148).

21.2.2 Spezifische Molwärme von Festkörpern

Die Gitterstruktur fester Körper ermöglicht Schwingungen in drei Raumrichtungen, aber keine Translation oder Rotation. Wir zählen für Vibrationsfreiheitsgrade doppelt, jeweils ein Freiheitsgrad für die kinetische und die potentielle Energie, sodass wir in diesem vereinfachten Modell für Festkörper insgesamt $f = 6$ Freiheitsgrade erwarten.

Das Volumen des Festkörpers ändert sich nur minimal, sodass wir für die spezifische Molwärme eines Stoffs bei konstantem Volumen (21.14) einen konstanten Wert erhalten:

$$\mathcal{C}_V = \frac{f}{2}R = 3R \tag{21.25}$$

Dieser Zusammenhang heißt das **Dulong-Petit-Gesetz**.

Experimentell wird das Gesetz bei genügend hohen Temperaturen für verschiedene Festkörpermaterialien bestätigt (► Abbildung 21.5).

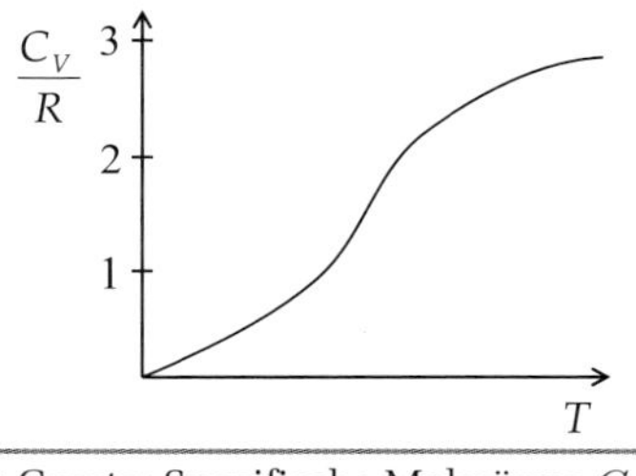

Abbildung 21.5 Dulong-Petit-Gesetz: Spezifische Molwärme C_V normiert auf die allgemeine Gaskonstante R als Funktion der Temperatur für Festkörper

21.3 Wärmetransport

Wir kennen drei Mechanismen, mit denen Wärmeenergie transportiert wird:

- Konvektion = Transport durch Bewegung von Materie,
- Wärmeleitung = Transport durch Materie hindurch und
- Wärmestrahlung = Transport durch elektromagnetische Wellen.

21.3.1 Konvektion

Wir visualisieren im Experiment 60 den Wärmetransport durch Konvektion.

Experiment 60: Konvektion

Wir füllen einen Wasserbehälter und färben die untere Wasserschicht mit einer geeigneten Tinte (► Abbildung 21.6). Dann heizen wir den Behälter von unten und beobachten das Aufsteigen der gefärbten Wasserschicht.

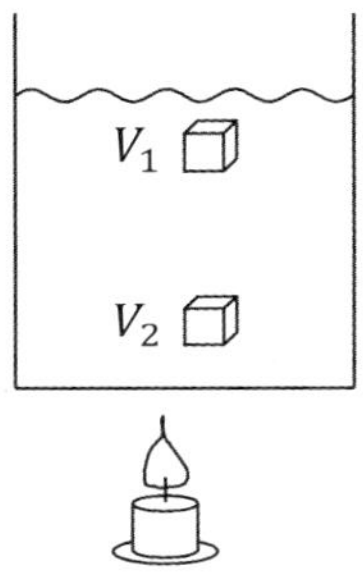

Abbildung 21.6 Konvektion: In den unteren Schichten gefärbtes Wasser wird von unten erhitzt und steigt auf.

Vor dem Heizen hat das Wasser die Temperatur T_1 und die Dichte $\rho_1 = m_\circ/V_1$. Durch das Erwärmen der unteren Wasserschichten auf die Temperatur T_2 dehnt sich das Wasser der Masse $m_\circ$ in einem unten liegenden Testvolumen V_1 nach Gleichung (19.7) aus:

$$V_2 = V_1(1 + \gamma\, \Delta T) \tag{21.26}$$

In der unteren Wasserschicht ist damit die Dichte $\rho_2 = m_\circ/V_2$ geringer als die Dichte ρ_1 in der oberen Wasserschicht.

Nach dem Prinzip von Archimedes reduziert sich die Gewichtskraft des Wassers im Volumen V_2 der unteren Wasserschicht nach Gleichung (18.20) auf

$$F = (\rho_2 - \rho_1)\, V_2\, g \tag{21.27}$$

$$= \left(\frac{m_\circ}{V_2} - \frac{m_\circ}{V_1}\right) V_2\, g \tag{21.28}$$

$$= \left(1 - \frac{V_2}{V_1}\right) m_\circ\, g\,. \tag{21.29}$$

Wir berechnen die Temperaturabhängigkeit dieser Gewichtskraft durch Einsetzen von Gleichung (21.26):

$$F = (1 - (1 + \gamma\, \Delta T))\, m_\circ\, g \tag{21.30}$$

$$= -\,\gamma\, \Delta T\, m_\circ\, g \tag{21.31}$$

Die erwärmte Flüssigkeit steigt also nach oben, gleichzeitig sinkt Flüssigkeit mit der Temperatur T_1 aus den kälteren Zonen nach unten. Der Effekt nimmt linear mit dem Ausdehnungskoeffizienten γ und der Temperaturdifferenz ΔT zu.

Beispiel 57: Konvektion

Der Ausdehnungskoeffizient für Wasser ist $\gamma(Wasser) = 2{,}1 \cdot 10^{-4}\, 1/\mathrm{K}$. Für eine Temperaturdifferenz von $\Delta T = 10\,\mathrm{K}$ ergibt sich nach Gleichung (21.31) eine nach oben gerichtete Kraft von

$$F = -\,2{,}1 \cdot 10^{-3}\, m\, g\,. \tag{21.32}$$

Sie entspricht zwei Promille der Gewichtskraft von Wasser.

Derselbe Konvektionsmechanismus tritt auch bei Gasen auf und spielt zum Beispiel beim Wetter eine Rolle. Der Erdboden wird durch die Sonne erwärmt, sodass die erwärmte Luft aufsteigt. Von der Seite fließt kalte Luft nach, es entstehen Windbewegungen.

21.3.2 Wärmeleitung

Im Experiment 61 messen wir den Wärmetransport durch ein Material.

Experiment 61: Wärmeleitung

Wir verbinden zwei Wasserbehälter durch einen Metallstab mit der Querschnittsfläche A (► Abbildung 21.7). Wir heizen den linken Behälter auf $T_2 = 100\,°\mathrm{C}$ und halten den rechten Behälter auf Zimmertemperatur $T_1 = 20\,°\mathrm{C}$. Entlang des Stabs entsteht ein Temperaturgefälle dT/dx, das wir mit einem Thermometer nachweisen können.

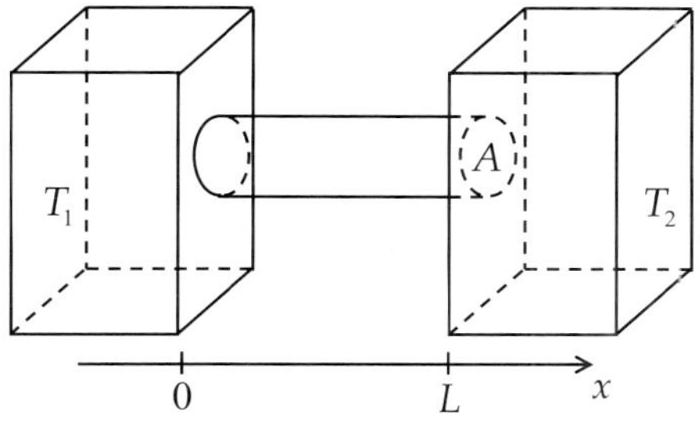

Abbildung 21.7 Zwei Wasserbehälter mit Temperaturen $T_2 > T_1$ werden durch einen Metallstab der Länge L und mit dem Querschnitt A verbunden.

Durch die Querschnittsfläche A des Stabs fließt pro Zeiteinheit Δt die Wärmeenergie ΔQ. Diese Energiestromdichte

$$J = \frac{\Delta Q}{A\,\Delta t} \tag{21.33}$$

verursacht ein näherungsweise lineares Temperaturgefälle $\Delta T/\Delta x$ entlang des Stabs.

In differentieller Schreibweise erhalten wir also folgende Gleichung für die Wärmeleitung:

$$\frac{dQ}{A\,dt} = -\lambda\,\frac{dT}{dx} \tag{21.34}$$

Die entsprechende Proportionalitätskonstante λ heißt **Wärmeleitvermögen** oder auch Wärmeleitzahl. Sie hängt vom Material des Stabs ab und kann aus dem obigen Experiment direkt bestimmt werden.

Beispiel 58: Wärmeleitvermögen

Typische Werte für das Wärmeleitvermögen verschiedener Materialien sind in ► Tabelle 21.2 gezeigt. Luft ist zur Wärmedämmung sehr gut geeignet.

	unlegierter Stahl	Beton	Glas	Luft
$\lambda / \frac{\mathrm{W}}{\mathrm{m \cdot K}}$	50	1,5	0,8	0,025

Tabelle 21.2 Wärmeleitvermögen verschiedener Materialien

21.3.3 Wärmestrahlung

Selbst im Vakuum gibt ein warmer Körper Wärme ab, zum Beispiel wärmt die Sonne die Erde. Hier gibt es keine Materie, mit deren Hilfe die Wärme transportiert wird, sondern die Wärme wird durch elektromagnetische Wellen übertragen.

Die Intensität J (Energiestromdichte) der elektromagnetischen Wellen wird durch das **Stefan-Boltzmannsche Strahlungsgesetz** beschrieben:

$$J = \frac{dE}{A\,dt} = \sigma T^4 \tag{21.35}$$

Dabei bezeichnet E die Energie der Strahlung, A ist die Oberflächengröße des abstrahlenden Körpers und T ist die Temperatur, die in dieses Gesetz zur *vierten Potenz* eingeht!

Das Gesetz wurde 1879 von J. Stefan experimentell und 1889 von L. Boltzmann theoretisch begründet. Erst die Quantentheorie liefert ein tiefes Verständnis insbesondere der **Stefan-Boltzmann-Konstante** σ, die mit den SI-Konstanten c, h und k zusammenhängt und daher einen festen Wert (mit vielen Nachkommastellen) hat:

$$\sigma = 5{,}67037\ldots \cdot 10^{-8}\,\frac{\mathrm{W}}{\mathrm{m}^2 \cdot \mathrm{K}^4}\,. \tag{21.36}$$

Gleichung (21.35) bezieht sich auf einen Körper, der in idealer Weise Strahlung absorbieren und emittieren kann, den sogenannten »schwarzen Körper«.

Befindet sich der Körper nicht im Vakuum, so konkurrieren die drei Mechanismen des Wärmetransports untereinander. Je höher die Temperatur ist, desto größer wird der relative Beitrag der Strahlung, da dieser Beitrag mit T^4 wächst.

21.3.4 Wärmedämmung

Um bei dem Bau eines Gebäudes minimale Wärmeverluste zu erreichen, sind kritische Elemente

- die Wände und die Fenster aufgrund der Wärmeleitung,
- undichte Fugen wegen der Konvektion und
- Fenster wegen der Wärmestrahlung.

Es ist in der Baubranche üblich, Material und Dicke zu einem sogenannten u-Wert zusammenzufassen:

$$\frac{dQ}{dt} = \underbrace{\frac{\lambda}{L}}_{\equiv u}\,\underbrace{A}_{=1\,\mathrm{m}^2}\,\underbrace{(T_2 - T_1)}_{=1\,\mathrm{K}} \tag{21.37}$$

Der u-Wert gibt die transportierte Wärmeleistung pro m^2 bei einer Temperaturdifferenz von 1 K an und sollte möglichst klein sein.

Beispiel 59: Wärmedämmung Wände

Die Wärmedämmung von Außenwänden ist stark materialabhängig (► Tabelle 21.3, ► Abbildung 21.8). Geeignete Kombinationen verschiedener Materialien werden durch die Summe

$$u = \left(\sum_{i=1}^{n} \frac{L_i}{\lambda_i}\right)^{-1}$$

berechnet und können zu guter Dämmung führen.

	Dicke	Wärmeleitvermögen	u-Wert
	$L/\,\mathrm{cm}$	$\lambda/\,\frac{\mathrm{W}}{\mathrm{m\cdot K}}$	$u/\,\frac{\mathrm{W}}{\mathrm{m^2\cdot K}}$
Porenbeton	30	0,1	0,3
Mineralwolle	10	0,04	0,4
Kombination			0,18

Tabelle 21.3 Wärmedämmung

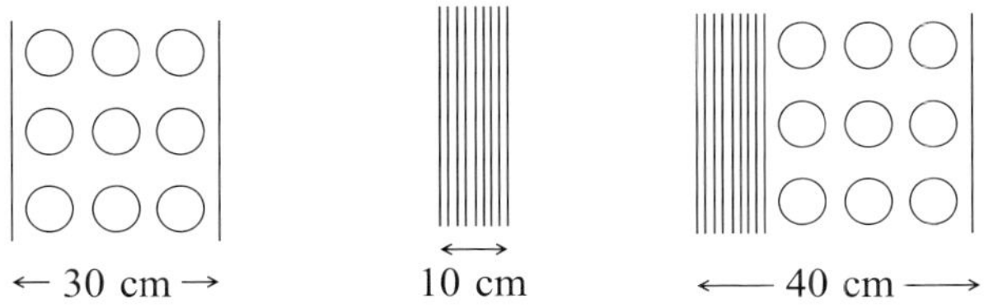

Abbildung 21.8 Wärmedämmung durch 30 cm Porenbeton, 10 cm Mineralwolle oder eine Kombination von beiden.

Das Wichtigste zum Mitnehmen

Zum Mitnehmen: Wärme 1

- Bei der Reibung wird mechanisch geleistete Arbeit ΔW vom Körper als Wärmemenge ΔQ aufgenommen:

$$\Delta Q = \Delta W \tag{21.5}$$

Wärme *ist eine Form der Energie, die in der Einheit Joule gemessen wird.*

Die aufgenommene Wärmemenge ΔQ ist proportional zur Masse M und zur Temperaturänderung des ΔT des Körpers:

$$\Delta Q = c\,M\,\Delta T \tag{21.7}$$

Die Materialkonstante c heißt **spezifische Wärme**. Sie bezeichnet die benötigte Wärmemenge, um in 1 kg des Materials eine Temperaturerhöhung von 1 K zu erreichen.

- Die innere Energie U eines Gases

$$\Delta U = \nu\,N_A\,\frac{f}{2}\,k\,\Delta T \tag{21.11}$$

lässt sich durch Energiezufuhr in Form von Wärme ΔQ

$$\Delta Q = \nu\,\underbrace{c\,M_{\text{mol}}}_{\equiv \mathcal{C}_V}\,\Delta T \tag{21.12}$$

erhöhen. Bei konstantem Volumen ist die Änderung der inneren Energie ΔU gleich der zugeführten Wärmemenge ΔQ:

$$\Delta U = \Delta Q \tag{21.8}$$

Mit der Anzahl der Mole $\nu = M/M_{\text{mol}} = N/N_A$ (19.34) erhalten wir durch Vergleich von (21.11) und (21.12) mit der Gaskonstanten R (19.40) die **spezifische Molwärme bei konstantem Volumen**

$$\mathcal{C}_V = \frac{f}{2}\,R\,. \tag{21.14}$$

Die Wärmeeigenschaften eines Gases werden also durch die Anzahl seiner Freiheitsgrade und damit durch seine molekularen Eigenschaften bestimmt.

Zum Mitnehmen: Wärme 2

- Die Anzahl der Freiheitsgrade reicht von z. B. $f_{\mathrm{He}} = 3$ für die Translation bei 1-atomigen Gasen bis zu z. B. $f_{\mathrm{NO_2}} = 12$ für komplexere Moleküle mit zusätzlichen Rotations- und Vibrationsfreiheitsgraden (Abschnitt 20.2.1).

  Um die Freiheitsgrade z. B. der Rotation oder Vibration eines Gasmoleküls anzuregen, benötigt man nach der Quantenmechanik jeweils bestimmte Mindestenergien. Dadurch ergibt sich eine stufenförmige Temperaturabhängigkeit der spezifischen Molwärme.

- Wenn man bei konstantem Druck p durch Wärmezufuhr die Temperatur erhöht, muss sich nach dem idealen Gasgesetz $pV = \nu RT$ (19.41) das Gasvolumen vergrößern $\Delta V \propto \Delta T$.

  Für die Arbeit $\Delta W = p\,\Delta V$ zur Volumenausdehnung gilt mit dem idealen Gasgesetz $p\,V = \nu\,R\,T$ (19.41)

  $$\Delta W = \nu\,R\,\Delta T\,. \tag{21.19}$$

  Die Energie, die dem Gas durch die Wärmemenge ΔQ zugeführt wird, muss gleichzeitig nach Gleichung (21.12) die Temperaturerhöhung ΔT bewirken und nach Gleichung (21.19) die Ausdehnungsarbeit ΔW leisten, um den Druck $p =$ *const.* zu halten:

  $$\Delta Q = \nu\,\underbrace{(\mathcal{C}_V + R)}_{\equiv \mathcal{C}_p}\,\Delta T$$

  Die Struktur der Gleichung ist dieselbe wie (21.12), sodass wir als **spezifische Molwärme $\mathcal{C}_p$ bei konstantem Druck** p erhalten

  $$\mathcal{C}_p = \mathcal{C}_V + R \tag{21.22}$$

  oder in Abhängigkeit der Gaskonstanten R und der Zahl der Freiheitsgrade f

  $$\mathcal{C}_p = \frac{f+2}{2}R\,. \tag{21.23}$$

  Der Quotient der spezifischen Molwärmen heißt **Adiabatenkoeffizient**:

  $$\kappa = \frac{\mathcal{C}_p}{\mathcal{C}_V} = \frac{f+2}{f} \tag{21.24}$$

  Er hängt nur noch von der Anzahl der Freiheitsgrade des Gases ab. Eine Methode zur experimentellen Bestimmung des Adiabatenkoeffizienten κ ergibt sich über die Messung der Schallgeschwindigkeit (18.148).

Zum Mitnehmen: Wärme 3

- Die Gitterstruktur fester Körper ermöglicht Schwingungen in drei Raumrichtungen, aber keine Translation oder Rotation. Vibrationsfreiheitsgrade zählen doppelt (für die kinetische und die potentielle Energie) sodass man insgesamt $f = 6$ Freiheitsgrade erhält.

  Das Volumen des Festkörpers ändert sich nur minimal, sodass sich für die **Spezifische Molwärme von Festkörpern**

  $$C_V = \frac{f}{2}R = 3R \qquad (21.25)$$

  ergibt. Dieser Zusammenhang ist als das **Dulong-Petit-Gesetz** bekannt und wird experimentell bei genügend hohen Temperaturen für verschiedene Festkörpermaterialien bestätigt.

- Es gibt verschiedene Arten des **Wärmeetransports**:
  - **Konvektion**: Transport durch Bewegung von Masse, z. B. durch die Schwerkraft (Auftrieb).

    Erwärmt man den unteren Bereich einer Flüssigkeitssäule, steigt die erwärmte Flüssigkeit auf, gleichzeitig sinkt die kältere Flüssigkeit ab.
  - **Wärmeleitung**: Transport durch Materie hindurch.

    In differentieller Schreibweise gilt für den Wärmestrom dQ/dt durch einen Querschnitt A mit einem Temperaturgefälle dT/dx

    $$\frac{dQ}{A\,dt} = -\lambda \frac{dT}{dx}\,. \qquad (21.34)$$

    Die Proportionalitätskonstante λ heißt **Wärmeleitvermögen** oder auch Wärmeleitzahl.
  - **Wärmestrahlung**: Abstrahlung elektromagnetischer Wellen.

    Die Intensität J (Energiestromdichte) der elektromagnetischen Wellen wird durch das **Stefan-Boltzmannsche Strahlungsgesetz** beschrieben:

    $$J = \frac{dE}{A\,dt} = \sigma T^4 \qquad (21.35)$$

    Dabei bezeichnet E die Energie der Strahlung, A ist die Oberflächengröße des abstrahlenden Körpers und T ist die Temperatur, die in dieses Gesetz zur *vierten Potenz* eingeht!

Thermodynamik

22

22.1 Thermodynamische Zustände **465**
22.1.1 Zustandsgrößen 465
22.1.2 Zustandsänderungen 465

22.2 Erster Hauptsatz der Thermodynamik **466**

22.3 Spezielle Zustandsänderungen **467**
22.3.1 Isochore Prozesse 467
22.3.2 Isotherme Prozesse 468
22.3.3 Adiabatische Prozesse 469

22.4 Kreisprozesse **471**
22.4.1 Carnot-Kreisprozess 472
22.4.2 Stirlingmotor 475
22.4.3 Wärmepumpe und Kühlschrank 477

22.5 Entropie **479**
22.5.1 Reversible Prozesse 479
22.5.2 Irreversible Prozesse 481
22.5.3 Wahrscheinlichkeit und Entropie 483

22.6 Zweiter Hauptsatz der Thermodynamik **486**

22.7 Dritter Hauptsatz der Thermodynamik **487**

ÜBERBLICK

» Zentrales Thema der Thermodynamik sind Änderungen des Zustands eines Gassystems, z. B. eine Volumenänderung.

Wir werden die drei Hauptsätze der Thermodynamik vorstellen, in denen als Basis für den Ablauf von Prozessen die Energieerhaltung, die natürliche Prozessrichtung durch die Entropie und das Verhalten am Temperaturnullpunkt formuliert sind.

Weiterhin werden wir Zustandsänderungen klassifizieren und periodisch ablaufende Zustandsänderungen für den Bau thermodynamischer Maschinen kennenlernen. «

22.1 Thermodynamische Zustände

Um den Zustand eines Gassystems zu charakterisieren, fügen wir den bereits bekannten physikalischen Größen weitere hinzu. Zustandsänderungen werden danach benannt, welche dieser Größen bei einem Prozess verändert wird.

22.1.1 Zustandsgrößen

Zustandsgrößen sind Messgrößen, mit denen wir den Zustand eines Gassystems beschreiben. Beispiele für Zustandsgrößen sind:

- Druck p
- Volumen V
- Temperatur T
- Innere Energie U
- Entropie S
- Chemische Zusammensetzung

Ist ein System im Gleichgewicht, so ändern sich diese Größen nicht mit der Zeit. Die **Zustandsgrößen** beschreiben dann den **stationären** Zustand, in dem sich das System befindet. Sie sind unabhängig vom Weg, auf dem es zu diesem Zustand gekommen ist. Im Unterschied dazu beschreiben **Prozessgrößen** wie Arbeit und Wärme den Verlauf einer Zustandsänderung.

22.1.2 Zustandsänderungen

Namenskonventionen für Spezialfälle von Änderungen am System sind in ► Tabelle 22.1 aufgeführt.

Messgröße	Bedingung	Zustandsänderung
Temperatur	$T = const.$	isotherm
Druck	$p = const.$	isobar
Volumen	$V = const.$	isochor
Entropie	$S = const.$	isentrop
Wärmeaustausch System & Umgebung	$\Delta Q = 0$	adiabatisch

Tabelle 22.1 Zustandsänderungen eines thermodynamischen Systems

22.2 Erster Hauptsatz der Thermodynamik

In der Mechanik hatten wir die Erhaltung der Gesamtenergie eines Körpers kennengelernt, die sich aus seiner kinetischen und potentiellen Energie (5.33) zusammensetzt. Das Grundprinzip der Energieerhaltung ist aber viel weitergehend, denn neben mechanischer Energie sind Energieformen auch Wärme (21.5), elektromagnetische Energie, chemische Energie und Kernenergie. Eine grundlegende Erweiterung fand weiterhin durch A. Einstein statt, der mit $E = mc^2$ (12.33) auch die Masse als eine weitere Form von Energie einschloss.

Dass Wärme eine Form von Energie ist und mechanische und thermische Energien ineinander umwandelbar sind, erkannten in den Jahren 1840 bis 1849 J. R. Mayer, J. P. Joule und H. v. Helmholtz. Die Gesamtenergie eines abgeschlossenen Systems bleibt bei Umwandlungsprozessen erhalten. In der heutigen Formulierung sind diese Erkenntnisse im wichtigen **1. Hauptsatz der Thermodynamik** zusammengefasst:

$$\Delta U = \Delta Q + \Delta W \tag{22.1}$$

Die Gesamtenergie aller Formen von Energien eines abgeschlossenen Systems ist erhalten.

Die innere Energie U (20.32) eines Systems nimmt um ΔU zu, wenn eine Wärmemenge ΔQ von außen zugeführt wird oder mechanische Arbeit ΔW am System geleistet wird.

In der Thermodynamik wird üblicherweise unsere **Vorzeichenkonvention** verwendet (vergl. Abschnitt 5.2): Mechanische Arbeit, die dem System zugeführt wird, wird positiv gerechnet. Arbeit, die das System leistet, wird negativ gerechnet. Genauso werden dem System zugeführte Wärmemengen positiv gerechnet und vom System abgegebene Wärmemengen negativ.

Beispiel 60: Komprimieren eines Gases

Wir verrichten mechanische Arbeit an einem wärmeisolierten System, indem wir das im Volumen eingeschlossene Gas mit einem Stempel mit der Querschnittsfläche A komprimieren (► Abbildung 22.1). Wir verrichten mit dem Stempel Arbeit gegen den Druck p des Gases bzw. gegen die Kraft (18.1)

$$\vec{F} = p\,A\,\vec{e}_x\,. \tag{22.2}$$

Unsere von außen am System verrichtete Arbeit mit der Kraft $-\vec{F}$ entlang des Weges $\vec{e}_x\,dx$ zählt im Einklang mit unserer Vorzeichenkonvention positiv, denn dx verkleinert das Volumen:

$$dW = -\vec{F}\,\vec{e}_x\,dx \tag{22.3}$$

$$= -p\,A\,dx \tag{22.4}$$

$$= -p\,dV \tag{22.5}$$

Für den 1. Hauptsatz der Thermodynamik (22.1) ergibt sich wegen des wärmeisolierten Systems ohne Wärmeaustausch $dQ = 0$

$$dU = dW \tag{22.6}$$

$$= -p\,dV\,. \tag{22.7}$$

Die z. B. durch Kompression $dV < 0$ des Gassystems von außen geleistete mechanische Arbeit dW führt zu einer Erhöhung der inneren Energie dU

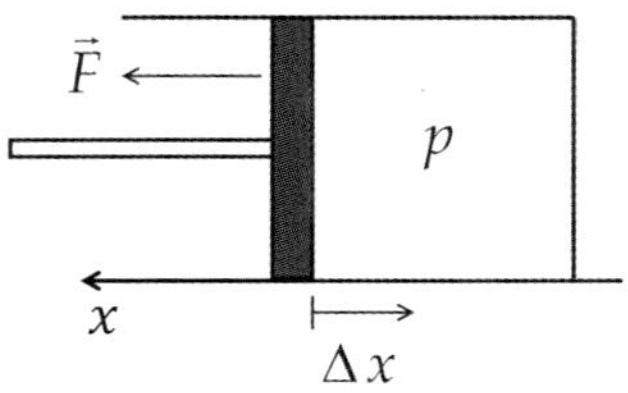

Abbildung 22.1 Ein Gas wird gegen seinen Druck p und die daraus resultierende Kraft $\vec{F}$ um Δx komprimiert.

22.3 Spezielle Zustandsänderungen

Wir untersuchen hier die Konsequenzen des 1. Hauptsatzes der Thermodynamik für mehrere Spezialfälle von Zustandsänderungen eines Gases. In den folgenden Abschnitten betrachten wir einfachheitshalber jeweils ein Mol eines idealen Gases Natürlich können alle Ergebnisse für ein größeres System sofort mit der Anzahl der Mole ν umgerechnet werden.

22.3.1 Isochore Prozesse

Isochore Prozesse sind Prozesse, bei denen sich das Volumen des Gassystems nicht ändert ($V = const.$). Diese Bedingung impliziert nach Gleichung (22.5), dass keine mechanische Arbeit geleistet wird:

$$dW = 0 \tag{22.8}$$

Damit lautet der 1. Hauptsatz der Thermodynamik (22.1) für isochore Prozesse

$$dU = dQ + \underbrace{dW}_{=0}\,. \tag{22.9}$$

Wird zum Beispiel dem System von außen eine Wärmemenge dQ zugeführt, wird sie vollständig in zusätzliche innere Energie dU umgesetzt.

Nach Gleichung (21.12) folgt der zugeführten Wärme eine Temperaturerhöhung des Systems entsprechend seiner spezifischen Molwärme C_V:

$$dQ = C_V\,dT \tag{22.10}$$

Damit erhalten wir einen Zuwachs an innerer Energie im System mit der Steigung $\mathcal{C}_V$ (► Abbildung 22.2):

$$dU = \mathcal{C}_V\,dT \tag{22.11}$$

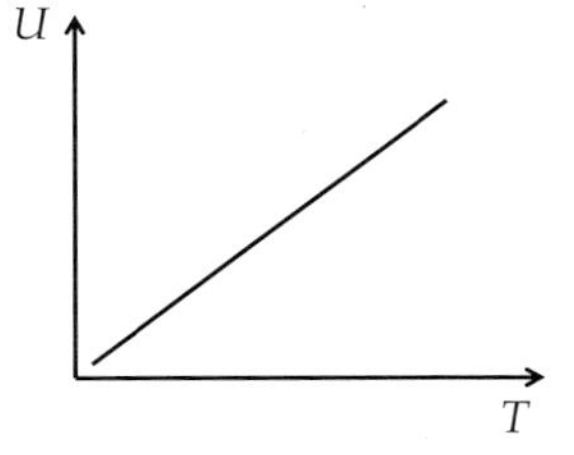

Abbildung 22.2 Die innere Energie eines idealen Gases steigt linear mit der Temperatur T.

Diesen Zusammenhang erhalten wir auch nach Gleichung (20.32) für die innere Energie von einem Mol ($\nu = 1$) Gas:

$$U = N\,\frac{f}{2}\,k\,T = \nu\,\frac{f}{2}\,N_A\,k\,T = \underbrace{\frac{f}{2}}_{=\mathcal{C}_V}\,R\,T \tag{22.12}$$

Mit $\mathcal{C}_V = (f/2)\,R$ nach Gleichung (21.14) erhalten wir konsistent

$$dU = \mathcal{C}_V\,dT\,. \tag{22.13}$$

22.3.2 Isotherme Prozesse

Bei einem isothermen Prozess bleibt die Temperatur des Gassystems unverändert ($T =$ *const.*). Wegen der Gleichung (22.11) gibt es im ersten Hauptsatz (22.1) damit keine Änderung der inneren Energie des Systems:

$$dU = 0 \tag{22.14}$$

Eine von außen am System verrichtete Kompressionsarbeit dW wird vollständig als Wärmeenergie dQ vom System wieder abgegeben:

$$\underbrace{dU}_{=0} = dQ + dW \tag{22.15}$$

Wir berechnen diese mechanische Arbeit nach Gleichung (22.5) und erhalten für die abgegebene Wärmemenge

$$-dQ = dW \tag{22.16}$$

$$= -p\,dV\,. \tag{22.17}$$

Die gesamte am Gassystem geleistete Arbeit berechnen wir über das entsprechende Volumenintegral und berücksichtigen dabei die Druckänderung durch Einsetzen der idealen Gasgleichung für ein Mol Gas, $pV = RT$ (19.41):

$$\Delta W = -\int_{V_1}^{V_2} p\,dV \tag{22.18}$$

$$= -\int_{V_1}^{V_2} \frac{RT}{V}\,dV \tag{22.19}$$

$$= RT \ln\left(\frac{V_1}{V_2}\right) \tag{22.20}$$

Da bei einer Kompression $V_2 < V_1$ ist, ist die am System geleistete mechanische Arbeit positiv: $\Delta W > 0$. Die gesamte Wärmemenge ΔQ, die das System daraufhin wieder abgibt, ist gemäß Vorzeichenkonvention nach Gleichung (22.16) negativ: $\Delta Q = -\Delta W < 0$.

Wenn umgekehrt das Gassystem die Wärmemenge $\Delta Q > 0$ aufnimmt und damit mechanische Arbeit ΔW durch Expansion des Gasvolumens leistet, ist $V_2 > V_1$. Die mechanische Arbeit des Systems ist dann nach Gleichung (22.20) negativ ($\Delta W < 0$) und ist mit der Vorzeichenkonvention der Thermodynamik ebenfalls konsistent.

In einem Druck-Volumen-Diagramm (▶ Abbildung 22.3) lässt sich der Betrag der mechanischen Arbeit (22.18) als Fläche unter der Kurve direkt ablesen. Diese geleistete Arbeit ist demnach vom Prozessweg abhängig, der vom System durchlaufen wird.

Im Einklang mit unserer Vorzeichenkonvention zählt die vom System geleistete Arbeit negativ, wenn der Prozess entlang der V-Achse durchlaufen wird. Umgekehrt zählt die am System geleistete Arbeit positiv, wenn der Prozess entgegen der V-Achse läuft.

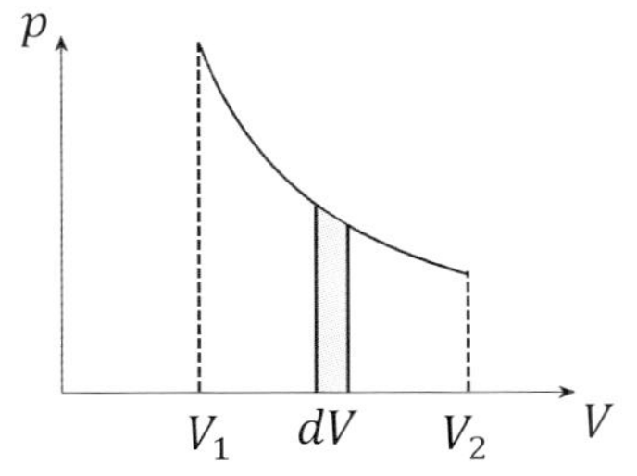

Abbildung 22.3 Die geleistete Arbeit entspricht der Fläche unter der Kurve im Druck-Volumen-Diagramm (p-V-Diagramm).

22.3.3 Adiabatische Prozesse

Bei adiabatischen Prozessen findet kein Wärmeaustausch mit der Umgebung statt:

$$dQ = 0 \tag{22.21}$$

Damit ändert sich die innere Energie U des Systems nach dem ersten Hauptsatz der Thermodynamik (22.1) nur aufgrund von mechanischer Arbeit W:

$$dU = \underbrace{dQ}_{=0} + dW \tag{22.22}$$

Für eine von außen dem System zugeführte Kompressionsarbeit erhalten wir mit Gleichung (22.5)

$$dU = -p\,dV\,. \tag{22.23}$$

Wir können diese drei Variablen auf zwei Variable reduzieren, indem wir auf der linken Seite dieser Gleichung den Zusammenhang $dU = \mathcal{C}_V\,dT$ (22.11) verwenden. Auf der rechten Seite können wir den Druck p über die ideale Gasgleichung $p\,V = R\,T$ (19.41) ersetzen und dann die Gaskonstante über die spezifische Molwärme ausdrücken, d. h. $R = \mathcal{C}_p - \mathcal{C}_V$ (21.22):

$$\mathcal{C}_V\,dT = -\frac{R\,T}{V}\,dV \tag{22.24}$$

$$\mathcal{C}_V\,dT = -\left(\mathcal{C}_p - \mathcal{C}_V\right)\,T\,\frac{dV}{V} \tag{22.25}$$

Diese Differentialgleichung können wir durch Trennung der Variablen und anschließende Integration lösen. Wir verwenden als Abkürzung den Adiabatenkoeffizienten $\kappa = \mathcal{C}_p/\mathcal{C}_V$ aus Gleichung (21.24):

$$\frac{dT}{T} = -\left(\frac{\mathcal{C}_p}{\mathcal{C}_V} - 1\right)\,\frac{dV}{V} \tag{22.26}$$

$$\int \frac{1}{T}\,dT = -(\kappa - 1)\int \frac{1}{V}\,dV \tag{22.27}$$

$$\ln(T) = -(\kappa - 1)\,\ln(V) + const. \tag{22.28}$$

$$\ln\left(T\,V^{\kappa-1}\right) = const. \tag{22.29}$$

Durch Anwenden der Exponentialfunktion erhalten wir eine Formulierung der sogenannten **Adiabatengleichungen**:

$$T\,V^{\kappa-1} = const. \tag{22.30}$$

Alternativ können wir diese Gleichung auch für den Druck p und das Volumen V formulieren. Ersetzen wir T durch die ideale Gasgleichung $p\,V = R\,T$, so erhalten wir

$$\frac{p\,V}{R}\,V^{\kappa-1} = const.\,, \tag{22.31}$$

beziehungsweise in der bekannten Schreibweise

$$p\,V^{\kappa} = const.\,. \tag{22.32}$$

Im Unterschied zu den Isothermen der idealen Gasgleichung (19.41) verknüpfen die Adiabatengleichungen (22.30) und (22.32) nur zwei Zustandsgrößen. Daher ist die Konstante *const.* in (22.28) und den folgenden Gleichungen zunächst frei und wird erst durch die Wahl eines Ausgangszustands festgelegt.

Im Druck-Volumen-Diagramm (► Abbildung 22.4) sind uns die isothermen Zustandsänderungen bereits wegen des idealen Gasgesetzes (19.41) als Hyperbeln der Form $p \sim 1/V$ bekannt. Zum Vergleich sind in ► Abbildung 22.4 auch adiabatische Prozesse eingezeichnet, deren Hyperbeln $p \sim 1/V^{\kappa}$ (22.32) etwas steiler verlaufen, da $\kappa = (f + 2)/f > 1$ (21.24) ist.

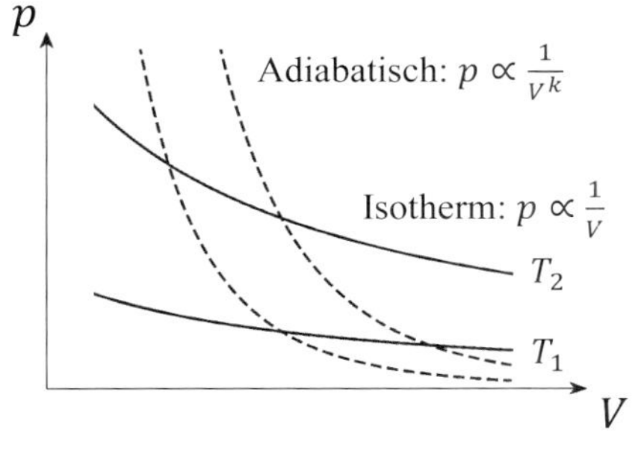

Abbildung 22.4 Isothermen und (wegen $\kappa > 1$ steilere) Adiabaten im p-V-Diagramm für Temperaturen $T_2 > T_1$

22.4 Kreisprozesse

Als Kreisprozess bezeichnet man einen Vorgang, in dem ein System verschiedene Zustände durchläuft und zum Anfangspunkt zurückkehrt (► Abbildung 22.5).

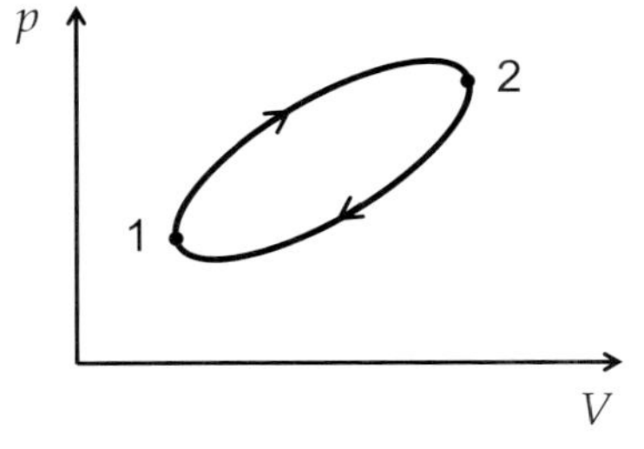

Abbildung 22.5 Ein Kreisprozess im p-V-Diagramm

Die **gesamte verrichtete Arbeit** entspricht der vom Weg eingeschlossenen Fläche. Die Fläche unterhalb der unteren Kurve hebt sich nämlich bei der Differenzbildung der Wege $1 \to 2$ und $2 \to 1$ auf.

Gemäß unserer Vorzeichenkonvention wird beim Umlauf im Uhrzeigersinn (► Abbildung 22.5) Arbeit *vom System* geleistet, beim Umlauf gegen den Uhrzeigersinn wird Arbeit aufgenommen (*am System* geleistet).

In einem sogenannten **reversiblen Prozess** kann der Prozess in beiden Richtungen verlustfrei durchlaufen werden.

Kreisprozesse unserer realen Welt sind **irreversible Prozesse**. Zum Beispiel verlieren ein Automotor oder eine Dampfmaschine beim Durchlaufen des Prozesses durch Reibung Energie. Um den Anfangszustand wieder zu erreichen, muss zusätzliche Energie investiert werden.

22.4.1 Carnot-Kreisprozess

In diesem Abschnitt wollen wir untersuchen, welcher Anteil der einem System zugeführten Wärmeenergie maximal in mechanische Arbeit umgewandelt werden kann. Wir betrachten dazu als Gedankenexperiment die sogenannte **Carnot-Maschine**, deren Zustandsänderungen in der ► Abbildung 22.6 gezeigt sind.

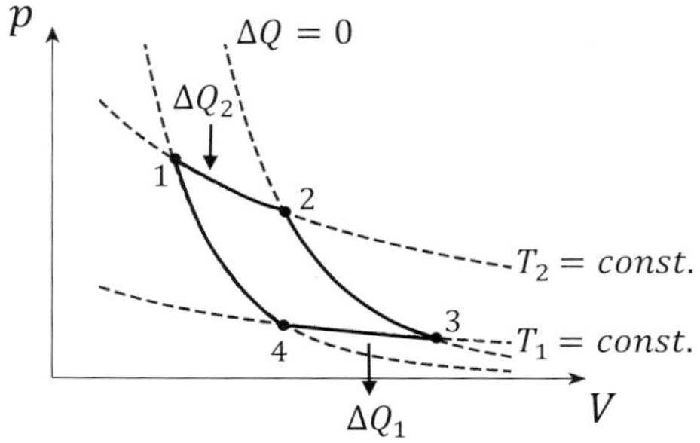

Abbildung 22.6 Carnot-Prozess: Kreisprozess mit Isothermen $T = const.$ und Adiabaten $\Delta Q = 0$ im p-V-Diagramm, $T_2 > T_1$

Die einzelnen Schritte dieses Kreisprozesses sind:

(1) Anfangs befindet sich das Gassystem im Zustand (V_1, p_1, T_2).

(1) → (2) Durch die Zufuhr der Wärmemenge ΔQ_2 von außen leistet das System mechanische Arbeit in Form einer **isothermen Expansion**, bei der die Temperatur $T_2 = const.$ bleibt.

(2) Das System ist im Zustand (V_2, p_2, T_2).

(2) → (3) In diesem Prozess vergrößert das jetzt wärmeisolierte System sein Volumen durch eine **adiabatische Expansion**. Dabei sinkt seine innere Energie und damit seine Temperatur auf T_1 ab.

(3) Das Gassystem befindet sich nun im Zustand (V_3, p_3, T_1).

(3) → (4) Durch von außen geleistete mechanische Arbeit erfährt das System eine **isotherme Kompression**. Das System gibt dabei die Wärmemenge ΔQ_1 bei der konstanten Temperatur T_1 ab.

(4) Das System ist im Zustand (V_4, p_4, T_1).

(4) → (1) In diesem Prozess wird das System wärmeisoliert und erfährt durch von außen geleistete Arbeit in Form einer **adiabatischen Kompression** eine Vergrößerung seiner inneren Energie und damit seiner Temperatur auf T_2.

Um die Gesamtbilanz der geleisteten mechanischen Arbeit zu erhalten, betrachten wir zunächst die beiden isothermen Prozesse und dann die adiabatischen Prozesse.

Isotherme Prozesse

Die mechanische Arbeit bei isothermen Prozessen hatten wir bereits aus dem ersten Hauptsatz (22.15) in Gleichung (22.20) berechnet.

In dem Prozessschritt $1 \to 2$ wird dem System von außen Wärme zugeführt, die das System unter Vergrößerung seines Volumens ($V_2 > V_1$) in mechanische Arbeit umsetzt:

$$\Delta Q_2 = -\Delta W_{12} \tag{22.33}$$

$$\Delta W_{12} = RT_2 \ln\left(\frac{V_1}{V_2}\right) < 0 \tag{22.34}$$

Im Prozessschritt $3 \to 4$ wird das Volumen durch mechanische Arbeit von außen komprimiert ($V_4 < V_3$). Dabei gibt das System Wärme ab:

$$-\Delta Q_1 = \Delta W_{34} \tag{22.35}$$

$$\Delta W_{34} = RT_1 \ln\left(\frac{V_3}{V_4}\right) > 0 \tag{22.36}$$

Adiabatische Prozesse

Für die adiabatischen Prozesse wird nach dem ersten Hauptsatz (22.22) mechanische Arbeit in innere Energie umgesetzt und umgekehrt.

In dem Prozessschritt $2 \to 3$ reduziert die Ausdehnungsarbeit des Systems seine innere Energie:

$$U(T_1) - U(T_2) = \Delta W_{23} \tag{22.37}$$

In dem Schritt $4 \to 1$ wird die innere Energie durch von außen geleistete Arbeit erhöht:

$$U(T_2) - U(T_1) = \Delta W_{41} \tag{22.38}$$

Daher sind die beiden in den adiabatischen Prozessen geleisteten Arbeiten entgegengesetzt gleich groß:

$$\Delta W_{23} = -\Delta W_{41} \tag{22.39}$$

Geleistete Arbeit

Damit können wir alle Beiträge zu der vom System geleisteten Arbeit ΔW aus den Gleichungen (22.34), (22.36), (22.37), (22.38) zusammentragen und mit Gleichung (22.39) vereinfachen:

$$\Delta W = \Delta W_{12} + \Delta W_{23} + \Delta W_{34} + \Delta W_{41} \tag{22.40}$$

$$= \Delta W_{12} + \Delta W_{34} \tag{22.41}$$

$$= RT_2 \ln\left(\frac{V_1}{V_2}\right) + RT_1 \ln\left(\frac{V_3}{V_4}\right) \tag{22.42}$$

Die isothermen Prozesse tragen zur insgesamt geleisteten Arbeit bei, nicht aber die adiabatischen Prozesse.

Die insgesamt sechs Variablen werden wir nun mithilfe der Adiabatengleichung (22.30) auf vier Variable reduzieren. Für die Prozesse $2 \to 3$ und $4 \to 1$ gilt

$$T_1\,V_3^{\kappa-1} = T_2\,V_2^{\kappa-1}\,, \tag{22.43}$$

$$T_1\,V_4^{\kappa-1} = T_2\,V_1^{\kappa-1}\,. \tag{22.44}$$

Die Volumina stehen damit in folgendem Zusammenhang:

$$\frac{V_3}{V_4} = \frac{V_2}{V_1} \tag{22.45}$$

$$\ln\left(\frac{V_3}{V_4}\right) = -\ln\left(\frac{V_1}{V_2}\right) \tag{22.46}$$

Durch Einsetzen in die Gleichung (22.42) erhalten wir für die vom System geleistete Gesamtarbeit

$$\Delta W = \underbrace{R\,\underbrace{\ln\left(\frac{V_1}{V_2}\right)}_{<1}}_{<0}\,\underbrace{(T_2 - T_1)}_{>0} < 0\,. \tag{22.47}$$

Andererseits lässt sich die geleistete Arbeit ΔW wegen (22.33), (22.35) und (22.45) auch durch die Wärmemengen ausdrücken (wird in der linken ► Abbildung 22.10 noch einmal verdeutlicht):

$$\Delta W = \Delta Q_2 - \Delta Q_1 \tag{22.48}$$

Die vom System geleistete Arbeit wird konsistent nach Vorzeichenkonvention negativ gezählt. Der Arbeitsbetrag $|\Delta W|$ entspricht der Fläche, die von den vier Kurven im obigen Druck-Volumen-Diagramm eingeschlossen wird.

Die Maschine hat die Wärme ΔQ_2 aus einem Reservoir entnommen und damit die mechanische Arbeit ΔW geleistet. Sie wird daher als **Wärme-Kraft-Maschine** bezeichnet. Die Wärme ΔQ_1 geht im Allgemeinen verloren.

Wirkungsgrad

Der **Wirkungsgrad** einer Wärme-Kraft-Maschine ist als das Energieverhältnis aus dem Betrag der verrichteten mechanischen Arbeit und der aufgenommenen Wärme definiert:

$$\eta = \left|\frac{\Delta W}{\Delta Q_2}\right| \tag{22.49}$$

Wir unterscheiden im Folgenden die unter realen Bedingungen erzielten Wirkungsgrade η vom idealen Wirkungsgrad η_C des Carnot-Prozesses.

Für den Carnot-Prozess ergibt sich der bestmögliche Wert von η mit (22.48), (22.47) und (22.33):

$$\eta_C = \left|\frac{\Delta W}{\Delta Q_2}\right| = \left|\frac{\Delta Q_2 - \Delta Q_1}{\Delta Q_2}\right| \tag{22.50}$$

$$= \left|\frac{R \ln \frac{V_1}{V_2} (T_2 - T_1)}{-RT_2 \ln \frac{V_1}{V_2}}\right| \tag{22.51}$$

Der Wirkungsgrad der Carnot-Maschine hängt also nur von der relativen Temperaturdifferenz ab:

$$\eta_C = \frac{T_2 - T_1}{T_2} \tag{22.52}$$

Da beide Temperaturen T_2 und T_1 größer als null sind (3. Hauptsatz der Thermodynamik, Abschnitt 22.7) und $T_2 > T_1$ ist, ist der Wirkungsgrad eine dimensionslose Zahl zwischen $0 < \eta < 1$. Der Wirkungsgrad wird größer bei höherer Temperaturdifferenz, z. B. beim Diesel-Motor größer als beim Otto-Motor.

In unserem Gedankenexperiment der Carnot-Maschine haben wir keine Reibung und sonstige Energieverluste berücksichtigt, sodass dieser berechnete Wirkungsgrad den bestmöglichen Wert darstellt. Während z. B. beim Diesel-Motor $\eta_C \approx 65\%$ ist, erreicht η bestenfalls 50%.

22.4.2 Stirlingmotor

Der Stirlingmotor ist eine Wärme-Kraft-Maschine, in deren Kreisprozess Luft als Arbeitsmedium zwischen zwei Wärmereservoiren hin- und herbewegt wird (▶ Abbildung 22.7).

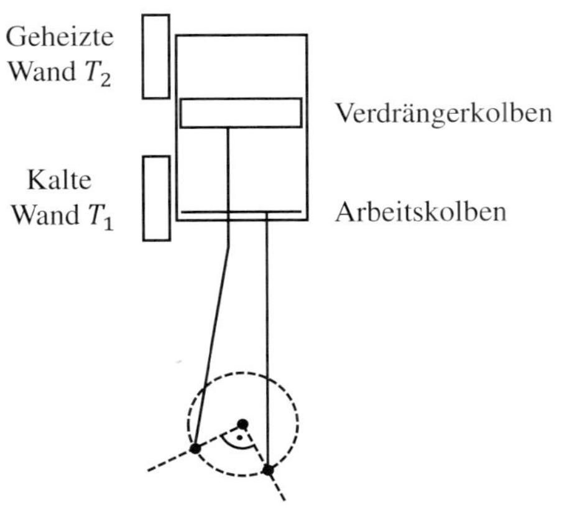

Abbildung 22.7 Sterlingmotor: Der Verdrängerkolben verschiebt das Gas zwischen den Bereichen mit $T_2 > T_1$ und treibt damit den phasenversetzten Arbeitskolben an.

Der Motor hat zwei Kolben:

1 Der Arbeitskolben begrenzt das Gasvolumen. Mit ihm werden isotherme Prozesse ($T = const.$) durchgeführt.

2 Der Verdrängerkolben verschiebt je nach seiner Stellung entweder geheizte Luft in den kalten Raumbereich oder kalte Luft in den Heizbereich. Mit diesem Kolben werden isochore Prozesse ($V = const.$) durchgeführt.

Beide Kolbenstangen sind an einer Scheibe befestigt und dort in einem Winkel von 90° montiert. Dadurch ist bei der Drehung der Scheibe jeweils einer der beiden Kolben fast unbewegt, während der andere Kolben gerade seine größte Bewegung durchführt.

Die Abfolge der Kolbenstellungen in den vier Zustandsänderungen des Stirlingmotors ist in ► Abbildung 22.8 skizziert.

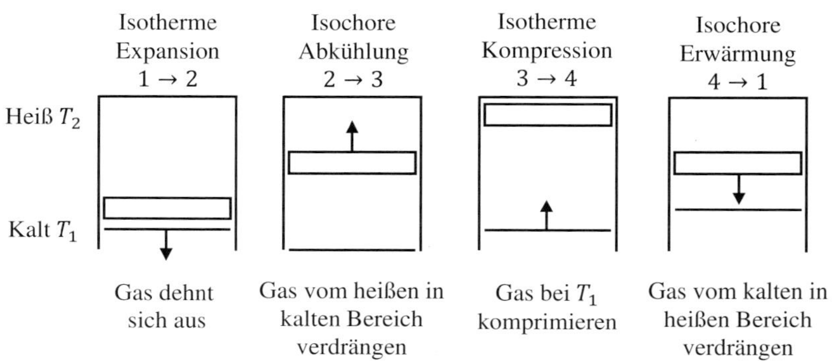

Abbildung 22.8 Arbeitstakte des Stirlingmotors: Der breite Verdrängerkolben eilt dem schmalen Arbeitskolben um 90° (1/4 Takt) voraus.

In ► Abbildung 22.9 sind die zugehörigen Prozessschritte dieses Kreisprozesses in einem Druck-Volumen-Diagramm dargestellt.

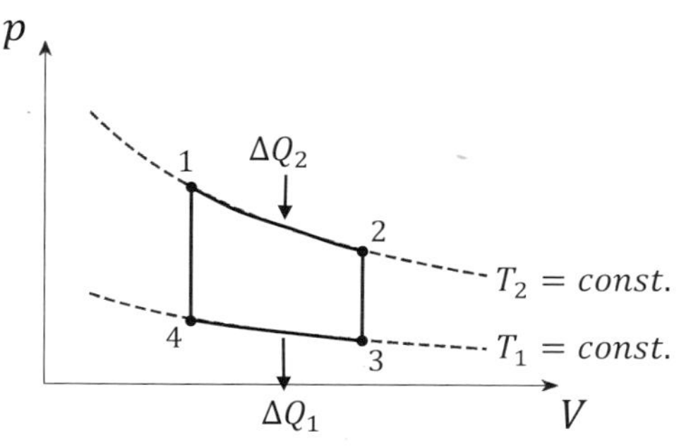

Abbildung 22.9 Arbeitstakte des Stirlingmotors im p-V-Diagramm: Die Isothermen sind jetzt durch Isochoren verbunden.

Die Maschine wird von der Wärmemenge ΔQ_2 angetrieben. Die isothermen Prozesse entsprechen Hyperbeln mit $p \propto V^{-1}$. Die isochoren Prozesse sind wegen des konstanten Volumens Prozesse, die entlang der senkrechten Linien verlaufen.

Die Nettoarbeit und der Wirkungsgrad ergeben sich wie bei der Carnot-Maschine (22.47, 22.52) aus den beiden isothermen Prozessen zu

$$\Delta W = R\,(T_2 - T_1)\,\ln\frac{V_1}{V_2}\,, \tag{22.53}$$

$$\eta_C = \frac{T_2 - T_1}{T_2}\,. \tag{22.54}$$

Experiment 62: Stirlingmotor

Stirlingmotoren für Demonstrationsexperimente bestehen aus einem auf der einen Seite geschlossenen Glaszylinder, in den die beiden Kolben eingebaut sind. Dadurch kann man die Abfolge der Kolbenbewegungen gut sehen.

Im Internet finden sich zahlreiche Bilder von Modellen begeisterter Anhänger dieses relativ einfach zu bauenden Motors.

22.4.3 Wärmepumpe und Kühlschrank

Bisher verliefen alle unsere Kreisprozesse im Uhrzeigersinn. Dadurch wurde einem System Wärme zugeführt, um mechanische Arbeit zu leisten (**Wärme-Kraft-Maschinen**, ► Abbildung 22.10 links). Bei einer **Kraft-Wärme-Maschine**, ► Abbildung 22.10 rechts und ► Abbildung 22.11, verläuft der Kreisprozess gegen den Uhrzeigersinn. Hierbei wird Arbeit aufgewendet, um einem System Wärme zuzuführen (**Wärmepumpe**) oder zu entziehen (**Kältemaschine oder Kühlschrank**).

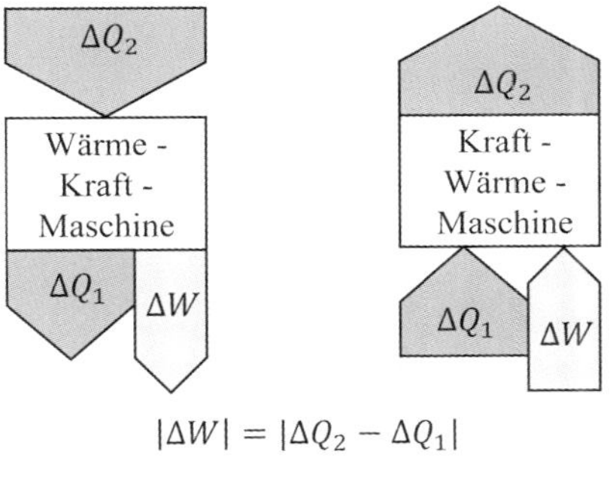

Abbildung 22.10 Energieverhältnisse beim reversiblen Carnot-Prozess. Links: In der Wärme-Kraft-Maschine wird $|\Delta W| = |\Delta Q_2 - \Delta Q_1|$ in Arbeit umgewandelt. Rechts: Bei einer Wärmepumpe wird die Energie $|\Delta W| = |\Delta Q_2 - \Delta Q_1|$ aufgewendet, um Wärme von 1 nach 2 zu pumpen.

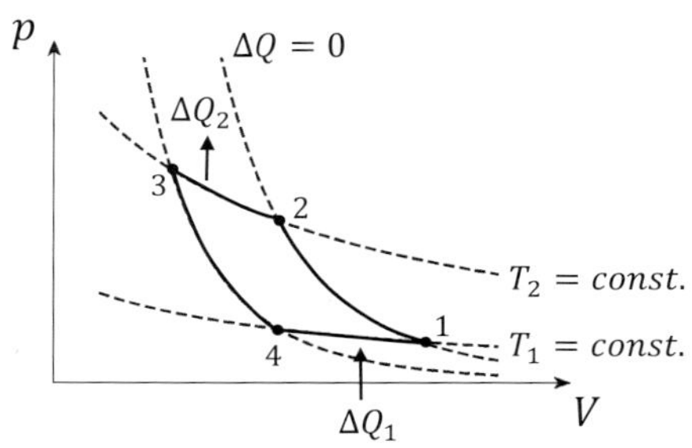

Abbildung 22.11 Bei einer Wärmepumpe wird der Carnot-Prozess von ► Abbildung 22.6 in umgekehrter Richtung durchlaufen.

Der Kreisprozess läuft durch folgende Zustandsänderungen, wobei das Wärmebad 2 mit Temperatur T_2 geheizt und das Wärmebad 1 mit Temperatur T_1 gekühlt wird:

(1)	Anfangs befindet sich das Gassystem im Zustand (V_1, p_1, T_1). Dann folgen
$(1) \to (2)$	eine adiabatische Kompression,
$(2) \to (3)$	eine isotherme Kompression, d. h. die Wärmemenge ΔQ_2 wird an das Wärmebad 2 bei der Temperatur T_2 abgegeben,
$(3) \to (4)$	eine adiabatische Expansion,
$(4) \to (1)$	eine isotherme Expansion, d. h. die Wärmemenge ΔQ_1 wird dem Wärmebad 1 bei der Temperatur T_1 entnommen. Das System kehrt in den Anfangszustand (V_1, p_1, T_1) zurück.

Beispiel 61: Kühlschrank

Beim Kühlschrank befindet sich das Wärmebad 1 innerhalb des Kühlschranksystems. Im isothermen Prozess $4 \to 1$ wird diesem Wärmebad die Wärmemenge ΔQ_1 entzogen, sodass sich das Wärmebad und damit der Innenraum des Kühlschranks abkühlt.

Das Wärmebad 2 befindet sich außerhalb des Kühlschranks. Beim isothermen Prozess $2 \to 3$ wird die Wärmemenge ΔQ_2 an die Umwelt abgegeben.

Die **Leistungszahl** ϵ wird als Verhältnis der zu- oder abgeführten Wärme zur aufgewendeten (elektrischen oder mechanischen) Energie definiert. Der bestmögliche Wert ϵ_C entspricht dem Kehrwert des Wirkungsgrades η_C des zugehörigen Carnot-Prozesses :

$$\epsilon_C = \frac{1}{\eta_C} = \frac{\Delta Q_i}{\Delta W} = \frac{T_i}{T_2 - T_1} \tag{22.55}$$

Dabei ist T_2 die wärmere und T_1 die kältere Temperatur und $\Delta Q_i = \Delta Q_2, T_i = T_2$ für die Wämepumpe und $\Delta Q_i = \Delta Q_1, T_i = T_1$ für die Kältemaschine. η_C wird umso größer, je geringer die Temperaturdifferenz ist. Technische Kältemaschinen erreichen für ϵ etwa die Hälfte von ϵ_C.

22.5 Entropie

Die Entropie S ist eine weitere wichtige Zustandsgröße zur Beschreibung eines thermodynamischen Systems. Die Änderung der Entropie entspricht der sogenannten reduzierten Wärmemenge

$$dS = \frac{dQ_{\text{rev}}}{T} . \tag{22.56}$$

Dabei ist es wichtig, die Wärmemenge dQ_{rev} über einen **reversiblen Prozess** und damit wegunabhängig zu definieren, damit S zu einer **Zustandsgröße** wird.

Im Folgenden werden wir zunächst die Entropie in den Zusammenhang der bereits bekannten Zustandsgrößen und Zustandsänderungen stellen. Wir werden dann die Entropieänderung für einen irreversiblen Prozess untersuchen. Anschließend werden wir Entropie als Wahrscheinlichkeitsinterpretation für die Realisierungsmöglichkeiten eines Zustands kennenlernen.

22.5.1 Reversible Prozesse

Die Definition der Entropie (22.56) werden wir zunächst für einen reversiblen Kreisprozess am Beispiel der Carnot-Maschine verdeutlichen (► Abbildung 22.12). Zur Vereinfachung lassen wir in diesem Abschnitt den Index »rev« weg.

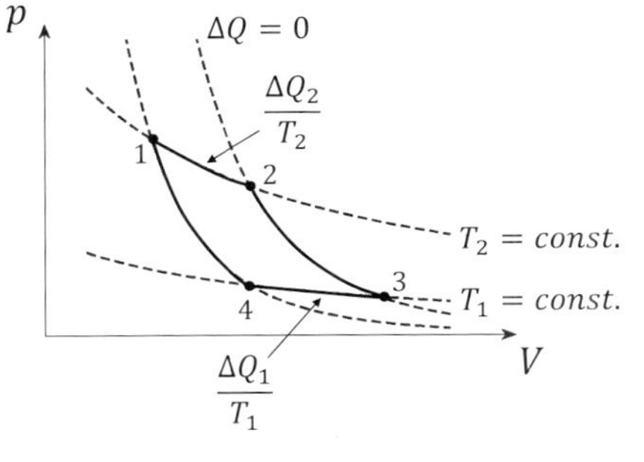

Abbildung 22.12 Reversibler Carnot-Prozess zur Verdeutlichung der Entropie

Für den isothermen Prozess 1 → 2 hatten wir bereits die Energiebilanz (22.33) und die mechanische Arbeit (22.34) bestimmt:

$$\Delta Q_2 = -\Delta W_{12} \tag{22.57}$$

$$\Delta W_{12} = R T_2 \ln \frac{V_1}{V_2} \tag{22.58}$$

Die Entropieänderung bzw. reduzierte Wärmemenge ist damit

$$\Delta S_2 = \frac{\Delta Q_2}{T_2} \tag{22.59}$$

$$= -R \ln \frac{V_1}{V_2} . \tag{22.60}$$

Ebenso hatten wir für den Prozess $3 \to 4$ die Energiebilanz (22.35) und die mechanische Arbeit (22.36) berechnet:

$$-\Delta Q_1 = \Delta W_{34} \tag{22.61}$$

$$\Delta W_{34} = RT_1 \ln \frac{V_3}{V_4} \tag{22.62}$$

Die entsprechende Entropieänderung bzw. reduzierte Wärmemenge beträgt

$$\Delta S_1 = \frac{\Delta Q_1}{T_1} \tag{22.63}$$

$$= -R \ln \frac{V_3}{V_4}\,. \tag{22.64}$$

Aus den Adiabatengleichungen folgte bereits die Gleichheit der Volumenverhältnisse (22.45)

$$\frac{V_3}{V_4} = \frac{V_2}{V_1}\,. \tag{22.65}$$

Damit ergibt sich für diese Entropieänderung

$$\Delta S_1 = \frac{\Delta Q_1}{T_1} \tag{22.66}$$

$$= R \ln \frac{V_1}{V_2}\,. \tag{22.67}$$

Die reduzierten Wärmemengen $\Delta Q_2/T_2$ (22.60) und $\Delta Q_1/T_1$ sind demnach betragsmäßig gleich groß, haben aber entgegengesetzte Vorzeichen:

$$\frac{\Delta Q_1}{T_1} = -\frac{\Delta Q_2}{T_2} \tag{22.68}$$

Die Beiträge zur Änderung der Entropie im Kreisprozess ($1 \to 2 \to 3 \to 4 \to 1$) entstehen nur aus diesen isothermen Anteilen, da für die adiabatischen Prozesse $dQ = 0$ gilt. Wegen Gleichung (22.68) ist

$$\Delta S = \underbrace{\frac{\Delta Q_2}{T_2}}_{\text{Wärmeaufnahme}} + \underbrace{\frac{\Delta Q_1}{T_1}}_{\text{Wärmeabgabe}} = 0\,. \tag{22.69}$$

Die Entropie bei reversiblen Kreisprozessen bleibt also erhalten.

Für den halben Weg $1 \to 2 \to 3$ mit der Aufnahme der Wärmemenge ΔQ_1 ist die Entropieänderung nach Gleichung (22.60)

$$\Delta S = \frac{\Delta Q_2}{T_2} + \underbrace{0}_{\text{Adiabate } 2 \to 3} \tag{22.70}$$

$$= \frac{\Delta Q_2}{T_2}\,. \tag{22.71}$$

Für den Weg $1 \to 4 \to 3$, der in der Gegenrichtung verläuft und damit eine Wärmeaufnahme erfordert, ist die Entropieänderung wegen Gleichung (22.68)

$$\Delta S' = -\frac{\Delta Q_1}{T_1} + \underbrace{0}_{\text{Adiabate } 1 \to 4} \tag{22.72}$$

$$= -\frac{\Delta Q_1}{T_1} \tag{22.73}$$

$$= \frac{\Delta Q_2}{T_2}\,. \tag{22.74}$$

Daher ist die Entropieänderung ΔS für reversible Prozesse vom Weg unabhängig und damit eine **Zustandsgröße**.

In einem Temperatur-Entropie-Diagramm entsprechen isotherme Prozesse den horizontalen Linien und isentrope Prozesse den vertikalen Linien (linke ► Abbildung 22.13).

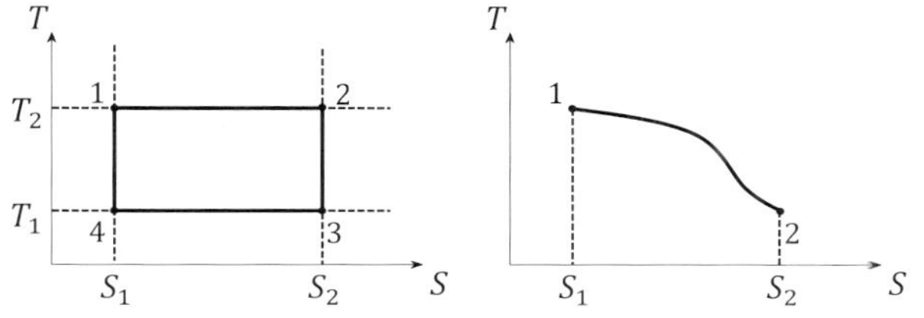

Abbildung 22.13 Links: In einem Temperatur-Entropie-Diagramm (T-S-Diagramm) sind die Isothermen horzontale und die Isentropen vertikale Geraden. Rechts: In einem T-S-Diagramm entspricht die Fläche unter der Kurve der ausgetauschten Wärme.

Durch Integration von (22.56)

$$\Delta Q = \int_{S_1}^{S_2} T \, dS \tag{22.75}$$

können wir die ausgetauschte Wärmemenge im Temperatur-Entropie-Diagramm als Fläche unter der Kurve ablesen (rechte ► Abbildung 22.13).

22.5.2 Irreversible Prozesse

Um die Entropie bei irreversiblen Prozessen zu verstehen, machen wir folgendes Gedankenexperiment:

Experiment 63: Wärmeübertrag

Zwei identische Würfel mit den Temperaturen T_1 und T_2 sind zunächst isoliert von ihrer Umgebung (Situation A in ► Abbildung 22.14). Der zweite Würfel habe die größere Temperatur $T_2 > T_1$. Mit der spezifischen Wärme c und der Masse M sind die Wärmemengen (21.7) der Würfel

$$Q_1 = c\, M\, T_1 \,, \tag{22.76}$$

$$Q_2 = c\, M\, T_2 \,. \tag{22.77}$$

In der Situation B in ► Abbildung 22.14 werden die beiden Würfel in Berührung gebracht, aber isoliert von der weiteren Umgebung gehalten.

In der Folge werden sich die beiden Temperaturen angleichen und eine mittlere Temperatur T_3 erreichen:

$$T_3 = \frac{T_1 + T_2}{2} \tag{22.78}$$

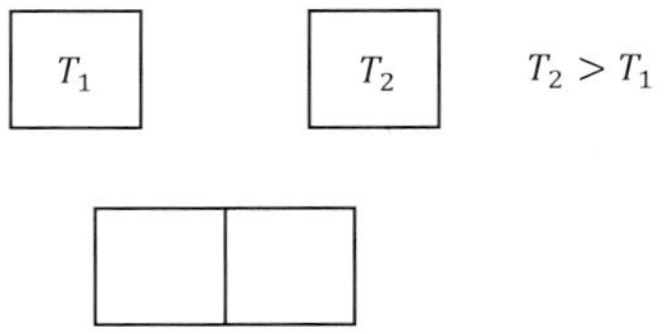

Abbildung 22.14 Zwei getrennte identische Würfel mit den Temperaturen $T_2 > T_1$ (A, oben) werden verlustfrei thermisch verbunden (B, unten).

Wir berechnen nun die entsprechenden Entropieänderungen bei dem Gedankenexperiment. Der Temperaturausgleich ist ja eindeutig irreversibel und im Gesamtsystem der beiden Würfel ist

$$\Delta Q = 0\,. \tag{22.79}$$

Man muss die Entropieänderung also über einen reversiblen »Ersatzprozess« definieren und berechnen.

Als reversiblen Vorgang kann man sich eine umkehrbare Erwärmung oder Abkühlung

$$\Delta Q_{\mathrm{rev},1} \quad \text{und} \quad \Delta Q_{\mathrm{rev},2} \tag{22.80}$$

der jeweiligen voneinander getrennten Würfel auf die mittlere Temperatur T_3 vorstellen. Erst wenn T_3 erreicht ist, werden die Würfel zusammengeführt.

Die Entropieänderung für den ersten Würfel ist dann

$$\Delta S_1 = \int_{T_1}^{T_3} \frac{1}{T}\, dQ_{\mathrm{rev},1} \tag{22.81}$$

$$= c\,M \int_{T_1}^{T_3} \frac{1}{T}\, dT \tag{22.82}$$

$$= c\,M\, \ln \frac{T_3}{T_1}\,. \tag{22.83}$$

Die entsprechende Entropieänderung für den zweiten Würfel beträgt

$$\Delta S_2 = c\,M\, \ln \frac{T_3}{T_2}\,. \tag{22.84}$$

Die gesamte Änderung der Entropie ist dann

$$\Delta S = \Delta S_1 + \Delta S_2 \tag{22.85}$$

$$= c\,M \left(\ln \frac{T_3}{T_1} + \ln \frac{T_3}{T_2} \right) \tag{22.86}$$

$$= c\,M\, \ln \frac{T_3^2}{T_1\, T_2}\,. \tag{22.87}$$

Mit dem Temperaturunterschied $T_1 \neq T_2$ können wir unter Verwendung von (22.78) das Vorzeichen der Entropieänderung untersuchen:

$$(T_1 - T_2)^2 > 0 \tag{22.88}$$

$$T_1^2 + T_2^2 - 2\,T_1\,T_2 > 0 \tag{22.89}$$

$$T_1^2 + T_2^2 + 2\,T_1\,T_2 > 4\,T_1\,T_2 \tag{22.90}$$

$$\left(\frac{T_1 + T_2}{2}\right)^2 > T_1\,T_2 \tag{22.91}$$

$$\frac{T_3^2}{T_1\,T_2} > 1 \tag{22.92}$$

Da $\ln(\tau)$ für $\tau \equiv T_3^2/(T_1\,T_2) > 1$ positiv ist, ist also bei einer Angleichung der Temperaturen zweier Systeme die Änderung der Entropie (22.87) des Gesamtsystems immer positiv:

$$\Delta S = c\,M\,\ln\frac{T_3^2}{T_1\,T_2} > 0 \tag{22.93}$$

In anderen Worten, die Entropie des Gesamtsystems nimmt bei einem solchen Prozess zu. Die Erfahrung vieler Experimente zeigt, dass die Temperatur der Körper ausgeglichen bleibt, es findet keine Umkehrung des Prozesses statt.

22.5.3 Wahrscheinlichkeit und Entropie

Ein ideales Gas befinde sich zum Zeitpunkt t_1 in einem Volumen V_1 (► Abbildung 22.15). Dieses Volumen grenzt an ein zweites leeres Volumen V_2, ist aber zunächst durch eine Tür davon getrennt. Beide Volumina befinden sich in einem Wärmebad, so dass die Temperatur konstant bleibt: $T = const.$

Zum Zeitpunkt t_2 öffnen wir die Verbindungstür der beiden Volumina und lassen das Gas aus dem kleineren in das größere Volumen diffundieren.

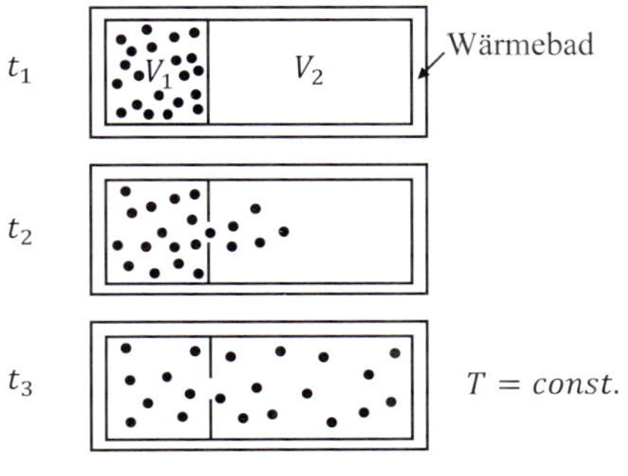

Abbildung 22.15 Isotherme Ausdehnung eines Gases vom Volumen V_1 zur Zeit t_1 ins Volumen $V_1 + V_2$ zur Zeit t_3.

Dies ist wieder ein irreversibler Prozess und wir können die Änderung der Entropie wie im vorigen Beispiel nur über einen reversiblen »Ersatzprozess« berechnen. Dazu wählen wir die isotherme Expansion des Gases von V_1 nach $V_1 + V_2$. Wir können die Rechnung direkt aus dem reversiblen Kreisprozess (22.60) übernehmen. Die Entropieänderung bzw. reduzierte Wärmemenge ist damit

$$\Delta S = \frac{\Delta Q_{\text{rev}}}{T} \tag{22.94}$$

$$= -R \ln \frac{V_1}{V_1 + V_2} \,. \tag{22.95}$$

Wir wollen uns jetzt einer statistischen Interpretation der Entropie zuwenden. Dazu bestimmen wir die Wahrscheinlichkeit, dass sich das System im Zeitpunkt t_3 nach Öffnen der Verbindung entweder vollständig im Volumen V_1 oder verteilt im Volumen $V_1 + V_2$ befindet.

Zum anfänglichen Zeitpunkt t_1 ist die Wahrscheinlichkeit P_1, das Molekül A_1 im Volumen V_1 zu finden,

$$P_1 = 1 \,. \tag{22.96}$$

Die entsprechende Wahrscheinlichkeit, A_1 zum Zeitpunkt t_3 in V_1 zu finden, beträgt

$$P_1' = \frac{V_1}{V_1 + V_2} \,. \tag{22.97}$$

Genauso ist für das Molekül A_2 die Wahrscheinlichkeit, sich zum Zeitpunkt t_3 im Volumen V_1 zu befinden,

$$P_2' = \frac{V_1}{V_1 + V_2} \,. \tag{22.98}$$

Die Wahrscheinlichkeit, beide Moleküle zum Zeitpunkt t_3 im Volumen V_1 zu finden, ist das Produkt beider Wahrscheinlichkeiten:

$$P = P_1' \cdot P_2' \tag{22.99}$$

Alle N Moleküle zur Zeit t_3 in V_1 zu finden, besitzt eine entsprechend geringe Wahrscheinlichkeit

$$P = \left(\frac{V_1}{V_1 + V_2}\right)^N \,. \tag{22.100}$$

Beispiel 62: Zustandswahrscheinlichkeit

Das ursprüngliche Volumen V_1 und das zusätzliche Volumen V_2 für die Ausdehnung seien gleich groß: $V_1 = V_2$. Die Wahrscheinlichkeit, dass sich 1 mol eines idealen Gases nach Öffnung des Zusatzvolumens komplett im ursprünglichen Volumen V_1 befindet, ist

$$P = \left(\frac{V_1}{V_1 + V_2}\right)^{N_A} = \left(\frac{1}{2}\right)^{6 \cdot 10^{23}} \tag{22.101}$$

$$= \left(10^{\lg 0{,}5}\right)^{6 \cdot 10^{23}} \approx 10^{-1{,}8 \cdot 10^{23}} \,. \tag{22.102}$$

Da der Exponent sehr klein wird, ist es übersichtlicher, den Logarithmus zu verwenden. Für 1 mol des idealen Gases ergibt sich dann unter Verwendung der Gaskonstanten $R = k\,N_A$ (19.39) als Maß für die Wahrscheinlichkeit P, alle Moleküle im ursprünglichen Volumen V_1 zu finden,

$$\ln P = N_A \ln\left(\frac{V_1}{V_1 + V_2}\right) \tag{22.103}$$

$$= \frac{1}{k} R \ln\left(\frac{V_1}{V_1 + V_2}\right) . \tag{22.104}$$

Über diese Wahrscheinlichkeit werden wir einen alternativen Zugang zum Entropiebegriff erhalten. Dafür untersuchen wir die Anzahl der Realisierungsmöglichkeiten Ω für einen Zustand. Wenn jede Realisierungsmöglichkeit dieselbe Wahrscheinlichkeit P besitzt, ist $\Omega = 1/P$ (siehe auch Aufgabe 58: Entropie). Damit ergibt sich

$$\ln \Omega = -\ln P \tag{22.105}$$

$$= \frac{1}{k} \underbrace{\left[-R \ln\left(\frac{V_1}{V_1 + V_2}\right)\right]}_{=\Delta S} . \tag{22.106}$$

Den rechten Term dieser Gleichung identifizieren wir wegen Gleichung (22.95) als Änderung ΔS der Entropie. Damit erhalten wir folgende **Wahrscheinlichkeitsinterpretation der Entropie**:

$$S = k \ln \Omega = -k \ln P \tag{22.107}$$

> *Die Entropie gibt ein Maß für die Wahrscheinlichkeit, dass ein Zustand von einem System eingenommen wird.*

Im Vergleich zur thermodynamischen Definition der Entropie über die reduzierten Wärmemengen reversibler Prozesse

$$dS = \frac{dQ_{\text{rev}}}{T} \tag{22.56}$$

legt (22.107) die Skala fest ($S = 0$ für $P = 1$). Bei einem thermodynamischen Prozess vom Zustand 1 in den Zustand 2 gibt die Änderung der Entropie

$$\Delta S = S_1 - S_2 \tag{22.108}$$

ein Maß für die Wahrscheinlichkeit, dass die Zustandsänderung stattfindet.

Aufgabe 58: Entropie

Vier Gasmoleküle werden auf zwei Raumbereiche verteilt (▸ Abbildung 22.16).

1. Wie viele Anordnungsmöglichkeiten Ω gibt es für die vier Moleküle in den Situationen A und B?
2. Welche Situation hat die größere Entropie?
3. Welcher Prozess ist wahrscheinlicher, A→B oder B→A?

Abbildung 22.16 Die Anordnungen A (3:1) und B (2:2).

22.6 Zweiter Hauptsatz der Thermodynamik

Der 2. Hauptsatz der Thermodynamik beschäftigt sich mit der alltäglichen Erfahrung, dass viele Prozesse wie Reibungsverlust oder Wärmetransport nur in einer Richtung ablaufen. Es gibt viele **empirische Formulierungen** für dieses wichtige Gesetz der Thermodynamik, darunter

- Wärme kann nicht von selbst von einem Körper niedriger Temperatur auf einen Körper höherer Temperatur übergehen.
- Wärme kann durch eine periodisch arbeitende Maschine nicht vollständig in Arbeit umgewandelt werden.
- Der Wirkungsgrad des Carnot-Prozesses kann nicht übertroffen werden.

Man fasst dies häufig als die Unmöglichkeit eines sog. »Perpetuum Mobile zweiter Art« zusammen (die »erste Art« bezieht sich auf den 1. Hauptsatz):

> *Es gibt keine periodisch arbeitende Maschine, die nichts weiter bewirkt als Arbeitsleistung und Abkühlung eines Wärmereservoirs.*

Mathematisch lassen sich all diese Formulierungen und Erfahrungen mithilfe der Entropie zusammenfassen. Der **2. Hauptsatz der Thermodynamik** besagt, dass die wahrscheinlichsten Vorgänge solche sind, in denen die Entropie zunimmt oder konstant bleibt

$$\Delta S \geq 0 \tag{22.109}$$

Oder schärfer formuliert:

> *In einem abgeschlossenen System nimmt die Entropie nie ab. Nur bei reversiblen Prozessen bleibt sie konstant.*

22.7 Dritter Hauptsatz der Thermodynamik

Die Aussage des dritten Hauptsatzes der Thermodynamik ist

$$T > 0\,\mathrm{K}\,. \tag{22.110}$$

> *Der absolute Temperaturnullpunkt $T = 0\,\mathrm{K}$ kann nicht erreicht werden.*

Die Ursache wird in der Quantenmechanik formuliert: Alle mikroskopischen Systeme haben auch im Grundzustand eine von null verschiedene Energie.

Beispiel 63: 3. Hauptsatz Thermodynamik

Für die Carnot-Maschine impliziert der 3. Hauptsatz, dass der Wirkungsgrad (22.52) der Maschine, d. h. die Umsetzung der zugeführten Wärme in mechanische Arbeit, wegen $T_2 > T_1$ und $T_1 > 0$ nicht 100% sein kann:

$$\eta = \frac{T_2 - T_1}{T_2} < 1 \tag{22.111}$$

Das Wichtigste zum Mitnehmen

Zum Mitnehmen: Thermodynamik 1

- Die **Thermodynamik** behandelt den Zustand eines Gassystems und seine Änderungen. Der Zustand wird mit **Zustandsgrößen** wie Druck p, Volumen V, Temperatur T und Entropie S beschrieben, bei Zustandsänderungen treten zusätzlich **Prozessgrößen** wie Arbeit W und Wärme Q auf.

  Ist ein System im Gleichgewicht, so ändern sich diese Größen nicht mit der Zeit, sie sind **stationär**.

  Für **Zustandsänderungen** gibt es folgende Namenskonventionen:

Messgröße	Bedingung	Zustandsänderung
Temperatur	$T = const.$	isotherm
Druck	$p = const.$	isobar
Volumen	$V = const.$	isochor
Entropie	$S = const.$	isentrop
Wärme	$\Delta Q = 0$	adiabatisch

- Der **1. Hauptsatz der Thermodynamik** beschreibt die **Energieerhaltung**:

$$\Delta U = \Delta Q + \Delta W \tag{22.1}$$

  Die innere Energie U (20.32) eines Systems nimmt um ΔU zu, wenn eine Wärmemenge ΔQ von außen zugeführt wird oder mechanische Arbeit ΔW am System geleistet wird.

  Wir verwenden die übliche **Vorzeichenkonvention**: Dem System zugeführte Arbeit oder Wärme wird positiv gerechnet, vom System geleistete Arbeit oder abgegebene Wärme negativ.

  Für die am System verrichtete Arbeit erhalten wir dann

$$dW = -p\,dV\,. \tag{22.5}$$

- Bei **Isochoren Prozessen** ändert sich das Volumen des Gassystems nicht (d. h. $V = const.$), wegen (22.5) wird dann keine mechanische Arbeit geleistet: $dW = 0$.

  Damit erhalten wir einen Zuwachs an innerer Energie im System mit der Steigung $\mathcal{C}_V$ (21.14):

$$dU = \mathcal{C}_V\,dT \tag{22.11}$$

Zum Mitnehmen: Thermodynamik 2

- Bei einem **isothermen Prozess** ist mit $T = const.$ wegen (22.11) $dU = 0$. Damit wird nach dem ersten Hauptsatz (22.1) eine von außen am System verrichtete Kompressionsarbeit dW vollständig als Wärmeenergie dQ vom System wieder abgegeben:

$$-dQ = dW = -p\,dV \tag{22.16}$$

Die gesamte am Gassystem geleistete Arbeit ist mit der idealen Gasgleichung für ein Mol Gas, $p\,V = R\,T$ (19.41),

$$-\Delta Q = \Delta W = R\,T\,\ln\left(\frac{V_1}{V_2}\right) . \tag{22.20}$$

Bei einer Kompression $V_2 < V_1$ ist $\Delta Q = -\Delta W < 0$, bei einer Expansion ist $\Delta Q = -\Delta W > 0$.

- Bei **Adiabatischen Prozessen** findet kein Wärmeaustausch mit der Umgebung statt, d. h. $dQ = 0$. Damit ändert sich die innere Energie U des Systems nach dem ersten Hauptsatz der Thermodynamik (22.1) nur aufgrund von mechanischer Arbeit W (22.5):

$$dU = -p\,dV \tag{22.23}$$

Mit $dU = C_V\,dT$ (22.11) und der idealen Gasgleichung $p = R\,T/V$ (19.41) erhält man die **Adiabatengleichung**

$$T\,V^{\kappa-1} = const.. \tag{22.30}$$

Man kann (22.30) mit $p\,V = R\,T$ auch für den Druck p und das Volumen V formulieren:

$$p\,V^{\kappa} = const. \tag{22.32}$$

In einem p-V-Diagramm verlaufen die Adiabaten etwas steiler als die isothermen Hyperbeln, da $\kappa = (f+2)/f > 1$ (21.24) ist.

- In einem **Kreisprozess** durchläuft ein System verschiedene Zustände und kehrt zum Anfangspunkt zurück. Die **gesamte verrichtete Arbeit** entspricht der vom Weg eingeschlossenen Fläche.

In einem **reversiblen Prozess** kann der Prozess in beide Richtungen verlustfrei durchlaufen werden.

Kreisprozesse unserer realen Welt sind **irreversible Prozesse**, in denen Energie z. B. durch Reibung als Wärme verloren geht.

Zum Mitnehmen: Thermodynamik 3

- In einem Gedankenexperiment wird an der sogenannten **Carnot-Maschine** untersucht, welcher Anteil einer zugeführten Wärme maximal in mechanische Arbeit umgewandelt werden kann.

  Dazu wird im **Carnot-Kreisprozess** ein Gas mit der Wärmemenge ΔQ_2 **isotherm expandiert** und verrichtet dabei mechanische Arbeit. Darauf wird es durch eine **adiabatische Expansion** auf die Temperatur $T_1 < T_2$ abgekühlt.

  Dann wird das Gas durch von außen geleistete mechanische Arbeit **isotherm komprimiert**. Das System gibt dabei die Wärmemenge ΔQ_1 ab. Anschließend kehrt das System durch **adiabatische Kompression** in den Ausgangszustand zurück.

  Die Maschine hat die Wärme ΔQ_2 aus einem Reservoir entnommen und damit die mechanische Arbeit ΔW (22.42) geleistet. Sie wird daher als **Wärme-Kraft-Maschine** bezeichnet. Die Wärme ΔQ_1 geht im Allgemeinen verloren.

  Der **Wirkungsgrad** einer Wärme-Kraft-Maschine ist definiert als das Verhältnis aus den Beträgen der verrichteten mechanischen Arbeit ΔW und der aufgenommenen Wärme ΔQ_2:

$$\eta = \left| \frac{\Delta W}{\Delta Q_2} \right| \tag{22.50}$$

  Der günstigste Wert von η wird in einem (reversiblen) Carnot-Prozess erreicht. Der **Carnot-Wirkungsgrad** η_C ist dann

$$\eta_C = \frac{T_2 - T_1}{T_2} . \tag{22.52}$$

  η_C hängt nur von den absoluten Temperaturen ab.

  Da beide Temperaturen T_2 und T_1 größer als null sind (3. Hauptsatz der Thermodynamik, Abschnitt 22.7) und $T_2 > T_1$ ist, ist der Wirkungsgrad eine dimensionslose Zahl zwischen $0 < \eta < 1$. Der Wirkungsgrad wird größer bei höherer Temperaturdifferenz!

  In unserem Gedankenexperiment der Carnot-Maschine haben wir keine Reibung und sonstige Energieverluste berücksichtigt, sodass dieser berechnete Wirkungsgrad den bestmöglichen Wert darstellt.

  Im **Stirlingmotor** ist ein Carnot-Prozess mit Isochoren statt Adiabaten mit dem Wirkungsgrad (22.52) näherungsweise realisiert.

Zum Mitnehmen: Thermodynamik 4

- Bei einer **Kraft-Wärme-Maschine** verläuft der Kreisprozess gegen den Uhrzeigersinn. Hierbei wird Arbeit aufgewendet, um einem System Wärme zuzuführen (**Wärmepumpe**) oder zu entziehen (**Kältemaschine oder Kühlschrank**).

  Die **Leistungszahl** ϵ wird als Verhältnis der zu- oder abgeführten Wärme zur aufgewendeten Arbeit definiert. Der bestmögliche Wert entspricht dem Kehrwert des Wirkungsgrades des zugehörigen Carnot-Prozesses:

$$\epsilon_C = \frac{1}{\eta_C} = \frac{T_i}{T_2 - T_1} \qquad (22.112)$$

  Dabei ist T_2 die wärmere und T_1 die kältere Temperatur und $T_i = T_2$ für die Wämepumpe und $T_i = T_1$ für die Kältemaschine.

- Die **Entropie** S ist als weitere wichtige Zustandsgröße zur Beschreibung eines thermodynamischen Systems über die **reduzierte Wärmemenge** dQ/T eines **reversiblen Prozesses** definiert:

$$dS = \frac{dQ_{\text{rev}}}{T} \qquad (22.56)$$

  Wir betrachten als Beispiel die Änderung der Entropie im Carnot-Kreisprozess. Da für die Adiabaten $dQ = 0$ gilt, tragen nur die Isothermen zu ΔS bei. Die zugehörigen reduzierten Wärmemengen

$$-\Delta S_1 = \Delta S_2 = \frac{\Delta Q_2}{T_2} = -R \ln \frac{V_1}{V_2}\,, \qquad (22.60)$$

  $\Delta Q_2/T_2$ und $\Delta Q_1/T_1$ sind betragsmäßig gleich groß, haben aber entgegengesetzte Vorzeichen und damit gilt

$$\Delta S = \underbrace{\frac{\Delta Q_2}{T_2}}_{\text{Wärmeaufnahme}} + \underbrace{\frac{\Delta Q_1}{T_1}}_{\text{Wärmeabgabe}} = 0\,. \qquad (22.69)$$

  Die Entropie bei reversiblen Kreisprozessen bleibt also erhalten und ist damit eine ***Zustandsgröße****.*

Zum Mitnehmen: Thermodynamik 5

- Um die Entropie bei **Irreversiblen Prozessen** zu verstehen, muss man die Entropieänderung nach (22.56) über einen reversiblen »Ersatzprozess« definieren.

  Dabei ergibt sich, dass für irreversible Prozesse die Entropie zunimmt:

  $$\Delta S > 0 \tag{22.93}$$

  Die Entropie des Gesamtsystems nimmt bei irreversiblen Prozessen zu.

- Die Entropie lässt sich auch statistisch über Wahrscheinlichkeiten formulieren. Ein ideales Gas diffundiert aus einem Volumen V_1 bei $T = const.$ in ein größeres Volumen V_2. Für die Anzahl der Realisierungsmöglichkeiten Ω für einen Zustand als Kehrwert der Wahrscheinlichkeit $\Omega = 1/P$ ergibt sich

  $$\ln \Omega = -\ln P = \frac{1}{k} \underbrace{\left[-R \ln \left(\frac{V_1}{V_1 + V_2}\right)\right]}_{=\Delta S} . \tag{22.106}$$

  Den rechten Term dieser Gleichung identifizieren wir wegen Gleichung (22.60) als Änderung ΔS der Entropie und erhalten als **Wahrscheinlichkeitsinterpretation der Entropie**

  $$S = k \ln \Omega = -k \ln P . \tag{22.107}$$

  Die Entropie gibt ein Maß für die Wahrscheinlichkeit, dass ein Zustand von einem System eingenommen wird.

  Bei einem thermodynamischen Prozess vom Zustand 1 in den Zustand 2 gibt die Änderung der Entropie

  $$\Delta S = S_1 - S_2$$

  ein Maß für die Wahrscheinlichkeit, dass die Zustandsänderung stattfindet.

Zum Mitnehmen: Thermodynamik 6

- **Zweiter Hauptsatz der Thermodynamik**

  Die wahrscheinlichsten Vorgänge sind solche, in denen die Entropie zunimmt oder konstant bleibt:

  $$\Delta S \geq 0 \tag{22.109}$$

  Oder schärfer formuliert:

  > *In einem abgeschlossenen System nimmt die Entropie nie ab. Nur bei reversiblen Prozessen bleibt sie konstant.*

- Nach dem **dritten Hauptsatz der Thermodynamik** kann der absolute Temperaturnullpunkt $T = 0\,\mathrm{K}$ nicht erreicht werden, d. h.

  $$T > 0\,\mathrm{K}, \tag{22.110}$$

  weil in der Quantenmechanik alle mikroskopischen Systeme auch im Grundzustand eine von null verschiedene Energie haben. Für die Carnot-Maschine bedeutet das, dass der Wirkungsgrad (22.52) nicht 100% sein kann:

  $$\eta = \frac{T_2 - T_1}{T_2} < 1 \tag{22.111}$$

Thermodynamik realer Gase und Flüssigkeiten

ÜBERBLICK

23

23.1 Van-der-Waals-Gleichung **497**

23.2 Verdampfungswärme **498**

23.2.1 Phasenübergang 501

23.2.2 Siedepunkt 502

23.3 Schmelzwärme **503**

23.3.1 Phasenübergänge, Tripelpunkt 504

» In der Natur finden wir Stoffe in festen, flüssigen oder gasförmigen Zuständen vor. Zur Beschreibung von realen Gasen werden wir in diesem Kapitel eine geeignete Modifikation der idealen Gasgleichung, die Van-der-Waals-Gleichung vorstellen. Außerdem werden wir Übergänge zwischen festen, flüssigen und gasförmigen Zuständen berechnen und Bedingungen kennenlernen, in denen nebeneinander verschiedene Zustände eines Stoffs koexistieren. «

23.1 Van-der-Waals-Gleichung

Bisher haben wir ausschließlich ideale Gase betrachtet, die für unser Grundverständnis thermodynamischer Vorgänge eine gute Näherung darstellen.

Für reale Stoffe kennen wir die drei verschiedenen Aggregatzustände: gasförmig, flüssig, fest. Diese Zustände bezeichnen wir auch als **Phasen**.

Reale Gase haben folgende Eigenschaften, die sich auf die Beschreibung von thermodynamischen Zuständen und Zustandsänderungen auswirken:

- Der Aggregatzustand eines Stoffs ist temperaturabhängig.
- Das freie Volumen ist durch das Eigenvolumen der Teilchen eingeschränkt.
- Die einzelnen Teilchen wechselwirken miteinander mit einer Stärke, die vom Abstand zwischen den Teilchen abhängt.

Die ideale Gasgleichung für 1 mol, $p\,V_{\text{mol}} = R\,T$ (19.41), muss deswegen modifiziert werden. Die entsprechende Gleichung heißt **Van-der-Waals-Gleichung**:

$$\left(p + \frac{a}{V_{\text{mol}}^2}\right)(V_{\text{mol}} - b) = R\,T \tag{23.1}$$

Die Zusatzterme haben folgende Bedeutung: Das **Kovolumen** $b = 4\,N_A\,V_{\text{Molekül}}$ berücksichtigt das Eigenvolumen der Moleküle und vermindert so das Volumen, das für die freie Bewegung der Moleküle verfügbar ist.

Der Term a/V_{mol}^2, der zum äußeren Druck p addiert wird, ist ein Maß für den zusätzlichen **Binnendruck**, der an den Grenzflächen des Volumens entsteht. Wenn die Umgebung eines Moleküls isotrop ist, addieren sich die ausgetauschten Kräfte mit den Umgebungsmolekülen wegen der Symmetrie zu null. An den Grenzflächen hat die Summe der anziehenden Wechselwirkungskräfte jedoch eine resultierende Kraft in Richtung des Mediums (► Abbildung 23.1).

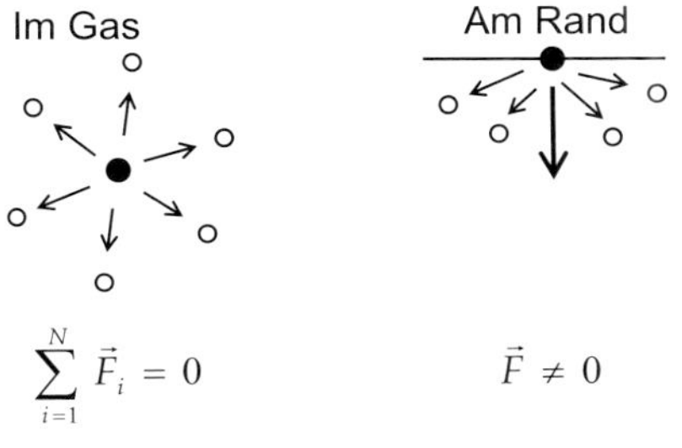

Abbildung 23.1 Die Kräfte $\vec{F}$ auf ein Molekül heben sich im Gas auf (links), am Rand nicht (rechts).

Die Van-der-Waals-Gleichung beschreibt die flüssige und gasförmige Phase eines realen Gases und ist als Druck-Volumen-Diagramm in ► Abbildung 23.2 skizziert.

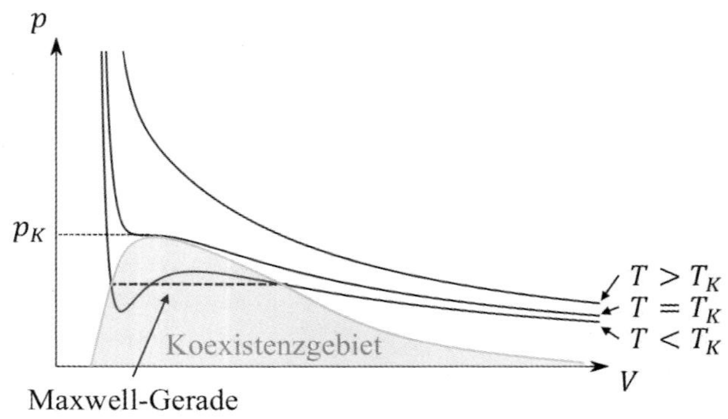

Abbildung 23.2 Die Van-der-Waals-Gleichung im p-V-Diagramm mit flüssiger und gasförmiger Phase und Maxwell-Gerade im Bereich der Verdampfung oder Kondensation (Koexistenzgebiet). Zu höheren Temperaturen T nähert sich die Kurve über die kritische Temperatur T_K einer Hyperbel für ideale Gase an.

Zu höheren Temperaturen T nähert sich die Gleichung einer Hyperbel für ideale Gase an. Hier ist der Druck p viel größer als das Verhältnis aus dem Parameter a und dem Quadrat des Molvolumens ($p \gg a/V_{\text{mol}}^2$), und der Parameter b ist viel kleiner als das Molvolumen des Stoffs ($b \ll V_{\text{mol}}$).

Oberhalb der kritischen Temperatur T_K gibt es nur die gasförmige Phase, die kinetische Energie der Moleküle ist größer als die Bindungsenergie zwischen den Molekülen.

Unterhalb von T_K ist die Phase abhängig vom Druck p und der Temperatur T. Es können gasförmige, flüssige Zustände, oder Mischzustände aus beiden Aggregatzuständen gleichzeitig auftreten. Der geschwungene Verlauf der Van-der-Waals-Gleichung im Bereich des Phasenübergangs ist nicht realistisch und wird für die Verdampfung oder Kondensation (vergl. Abschnitt 23.2) bei konstantem Druck durch die sog. **Maxwell-Gerade** ersetzt. Sie ist dadurch definiert, dass die Flächen zwischen Gerade und Kurve oberhalb und unterhalb der Geraden gleich groß sind.

23.2 Verdampfungswärme

Um einen Verdampfungs- bzw. einen Kondensationsprozess (Übergangsprozess zwischen flüssigem und gasförmigem Aggregatzustand) zu verstehen, betrachten wir folgendes Experiment.

Experiment 64: Sättigungsdampfdruck

Wir füllen in ein geschlossenes Gefäß eine Flüssigkeit, wobei wir einen Teil des Volumens frei lassen (► Abbildung 23.3). Wir messen die Temperatur T der Flüssigkeit mit einem Thermometer und den Druck p des gasförmigen Zustands mit einem Drucksensor. Bei fester Temperatur T steigt der Druck zunächst an und stabilisiert sich nach einer Weile bei einem festen Wert p_s.

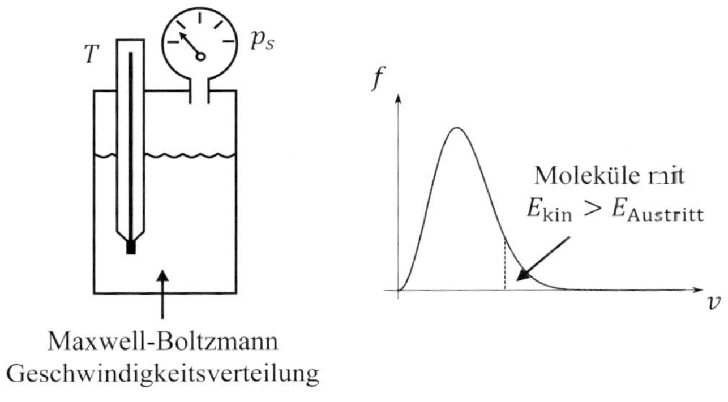

Abbildung 23.3 Experiment zur Bestimmung des Sättigungsdampfdrucks p_s (links) und Maxwellverteilung mit der Austrittsenergie E_{Austritt} (rechts)

Die Flüssigkeitsmoleküle bewegen sich mit einer Geschwindigkeitsverteilung, die wir durch die Maxwell-Boltzmann-Verteilung (20.34) annähern. Falls ein Molekül eine kinetische Energie E_{kin} hat, die größer als die wegen der Oberflächenspannung erforderliche Austrittsenergie E_{Austritt} ist, wird das Molekül ins Gasvolumen gelangen.

Das Gesamtsystem ist dann im Gleichgewicht, wenn genauso viele Moleküle aus der Flüssigkeit austreten und in den gasförmigen Zustand übergehen, wie umgekehrt. Dieser Gleichgewichtszustand stellt sich bei dem **Sättigungsdampfdruck** p_s ein. Der genaue Wert $p_s(T)$ ist temperaturabhängig, da die Maxwell-Boltzmann-Geschwindigkeitsverteilung stark von der Temperatur abhängt.

Unser Ziel ist zunächst die Berechung der notwendigen Wärmezufuhr für einen Verdampfungsprozess. Dafür führen wir einen Kreisprozess gerade oberhalb des Sättigungsdampfdrucks p_s durch (► Abbildung 23.4).

Wir erhöhen die Temperatur T infinitesimal um dT und damit den Druck um dp. Dann führen wir die Wärmemenge $dQ_2 = \Lambda$ zu, um die Flüssigkeit zu verdampfen.

Anschließend senken wir die Temperatur zurück auf T und den Druck auf p_s und führen schließlich die Wärmemenge dQ_1 aus dem System ab, um in den Ausgangszustand zu gelangen.

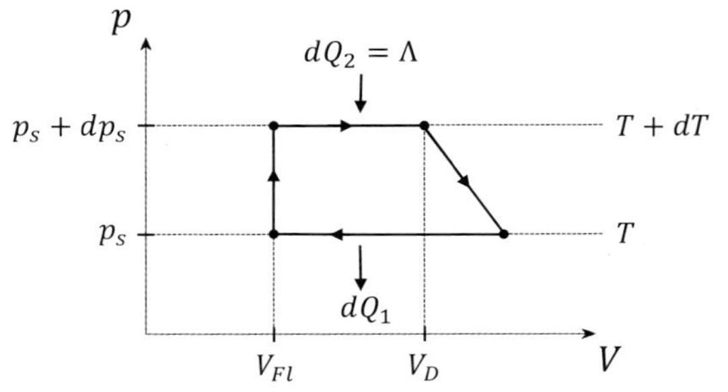

Abbildung 23.4 Kreisprozess zur Verflüssigung eines Dampfvolumens V_D in ein Flüssigkeitsvolumen V_{Fl} beim Sättigungsdampfdruck p_S mit der Verdampfungswärme Λ. $dV_D \ll V_{\mathrm{Fl}} \ll V_D$, nicht maßstabsgerecht dargestellt.

Die vom System geleistete Arbeit entspricht der von dem Kreisprozess eingeschlossenen Fläche im Druck-Volumen-Diagramm ($dV_D \ll V_{\mathrm{Fl}} \ll V_D$):

$$dW \approx dp_s \left(V_D - V_{\mathrm{Fl}}\right) \tag{23.2}$$

Dabei bezeichnet V_D das Volumen des Dampfs mit der Masse M und V_{Fl} das Volumen der Flüssigkeit mit derselben Masse. Der Wirkungsgrad des Prozesses beträgt laut Definition (22.50)

$$\eta = \left|\frac{dW}{dQ_2}\right| \tag{23.3}$$

$$= \frac{dp_s \left(V_D - V_{\mathrm{Fl}}\right)}{\Lambda} . \tag{23.4}$$

Näherungsweise können wir den Wirkungsgrad auch über die Temperaturänderung ausdrücken (22.52):

$$\eta = \frac{(T + dT) - T}{(T + dT)} \approx \frac{dT}{T} \tag{23.5}$$

Mit diesen beiden Gleichungen erhalten wir folgenden Zusammenhang zwischen dem Dampfdruck, der Volumenänderung und der Verdampfungswärme einerseits und der relativen Temperaturänderung andererseits:

$$\frac{dp_s \left(V_D - V_{\mathrm{Fl}}\right)}{\Lambda} = \frac{dT}{T} \tag{23.6}$$

Die erforderliche Verdampfungswärme Λ beträgt damit

$$\Lambda = \frac{dp_s}{dT} T \left(V_D - V_{\mathrm{Fl}}\right) . \tag{23.7}$$

Diese Gleichung heißt **Clausius-Clapeyron-Gleichung**.

Für den Verdampfungsprozess muss also Energie in Form der Verdampfungswärme Λ zugeführt werden. Sie wird benötigt, um die Vergrößerung der Molvolumina und die Vergrößerung der Molekülabstände gegen die Anziehungskräfte der Moleküle zu bewirken.

Zur Verallgemeinerung definieren wir die **spezifische Verdampfungswärme** λ als das Verhältnis der benötigten Wärmemenge pro Masse der verdampften Flüssigkeit:

$$\lambda = \frac{\Lambda}{M} \tag{23.8}$$

Beispiel 64: Spezifische Verdampfungswärme

Die spezifische Verdampfungswärme von Wasser beträgt

$$\lambda = 2\,257\,\frac{\text{kJ}}{\text{kg}} = 40{,}8\,\frac{\text{kJ}}{\text{mol}}\,.$$

23.2.1 Phasenübergang

Unter bestimmten Bedingungen können Aggregatzustände (Phasen) eines Stoffs simultan existieren. Wir berechnen im Folgenden den Sättigungsdampfdruck p_s als Funktion der Temperatur. Dieser Druckwert zeigt *den Übergang zwischen der flüssigen und der gasförmigen Phase* an.

Wir beginnen die Näherung mit der Clausius-Clapeyron-Gleichung (23.7) und nehmen an, dass das Volumen des gasförmigen Zustands sehr viel größer als das entsprechende Volumen des flüssigen Zustands ist $V_D \gg V_{\mathrm{Fl}}$. Wir beziehen die betrachtete Stoffmenge auf 1 mol und verwenden hier entsprechend V_D als das Molvolumen:

$$\Lambda \approx \frac{dp_s}{dT}\,T\,V_D \tag{23.9}$$

Bei ausreichend hoher Temperatur können wir anstelle der Van-der-Waals-Gleichung die ideale Gasgleichung (19.41) für 1 mol des Dampfs verwenden:

$$p_s\,V_D \approx R\,T \tag{23.10}$$

Um den funktionalen Zusammenhang $p_s(T)$ zu erhalten, setzen wir in Gleichung (23.9) für V_D die ideale Gasgleichung ein. Wir lösen dann die folgende Differentialgleichung durch Trennung der Variablen:

$$\Lambda = \frac{dp_s}{dT}\,T\,\frac{R\,T}{p_s} \tag{23.11}$$

$$\frac{dp_s}{p_s} = \frac{\Lambda}{R}\,\frac{dT}{T^2} \tag{23.12}$$

$$\int_{p_\circ}^{p_s} \frac{dp_s}{p_s} = \frac{\Lambda}{R}\int_{T_\circ}^{T} \frac{dT}{T^2} \tag{23.13}$$

$$\ln\frac{p_s}{p_\circ} = \frac{\Lambda}{R}\left(-\frac{1}{T} + \frac{1}{T_\circ}\right) \tag{23.14}$$

$$p_s = p_\circ\, e^{\frac{\Lambda}{R\,T_\circ}}\, e^{-\frac{\Lambda}{R\,T}} \tag{23.15}$$

Damit hat also der Sättigungsdampfdruck folgende exponentielle Temperaturabhängigkeit:

$$p_s \propto e^{-\frac{\Lambda}{R}\frac{1}{T}} \tag{23.16}$$

Entlang der exponentiellen Kurve im Druck-Temperatur-Diagramm (▶ Abbildung 23.5 links) finden wir sowohl den flüssigen als auch den gasförmigen Zustand. Oberhalb der kritischen Temperatur T_K sind die Moleküle so weit voneinander entfernt, dass nur noch der gasförmige Zustand existiert.

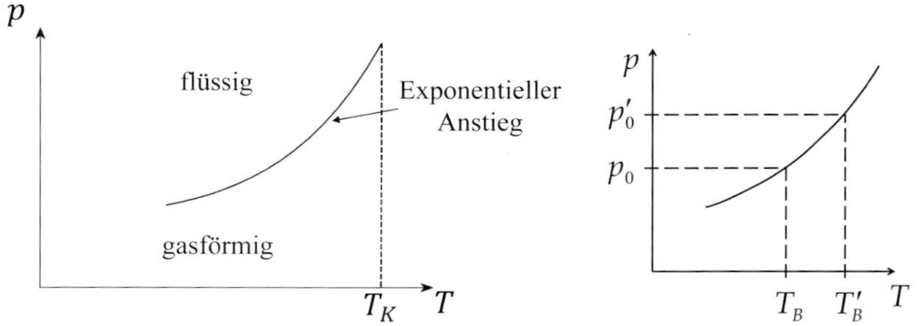

Abbildung 23.5 Links: Der Sättigungsdampfdruck p steigt exponentiell mit der Temperatur T bis zum kritischen Punkt T_K. Rechts: Die Siedetemperatur T_B in Abhängigkeit vom äußeren Druck $p_\circ$.

23.2.2 Siedepunkt

Wenn der äußere Druck p kleiner als der Sättigungsdampfdruck p_s ist, können im Inneren der Flüssigkeit Dampfblasen entstehen, d. h. die Flüssigkeit siedet (»kocht«).

Die Druckabhängigkeit der Siedetemperatur $T_B(p)$ können wir aus dem physikalischen Zusammenhang (23.14) herleiten. Ist die Siedetemperatur T_B bei dem Druck $p_\circ$ gegeben, so erhalten wir bei dem höheren Druck $p'_\circ$ die höhere Siedetemperatur T'_B:

$$\ln\frac{p'_\circ}{p_\circ} = \frac{\Lambda}{R}\left(-\frac{1}{T'_B} + \frac{1}{T_B}\right) \tag{23.17}$$

$$\frac{1}{T'_B} = \frac{1}{T_B} - \frac{R}{\Lambda}\ln\frac{p'_\circ}{p_\circ} \tag{23.18}$$

$$T'_B = \frac{1}{\dfrac{1}{T_B} - \dfrac{R}{\Lambda}\ln\dfrac{p'_\circ}{p_\circ}} \tag{23.19}$$

Die funktionale Abhängigkeit der Siedetemperatur vom äußeren Druck ist in der rechten ▶ Abbildung 23.5 dargestellt.

Beispiel 65: Dampfdrucktopf

Die Erhöhung der Siedetemperatur in Abhängigkeit des Drucks wird im Dampfdrucktopf genutzt. Während die Siedetemperatur bei

$$p_\circ = 1\,\text{atm} \quad T_B = 100\,^\circ\text{C}$$

beträgt, erhöht sich die Siedetemperatur bei

$$p_\circ' = 2\,\text{atm} \quad \text{auf} \quad T_B' = 120\,^\circ\text{C}\,.$$

23.3 Schmelzwärme

Wir erweitern unsere Überlegungen auf Übergänge zwischen festen und flüssigen Zuständen.

Experiment 65: Schmelzwärme und Verdampfungswärme

Wir füllen ein Kochgefäß mit Eis und geben anschließend etwas Wasser dazu. Wir messen die Temperatur des Wasser-Eis-Gemischs. Mit einem Tauchsieder heizen wir das Gemisch und beobachten dabei die Temperatur als Funktion der Zeit (► Abbildung 23.6).

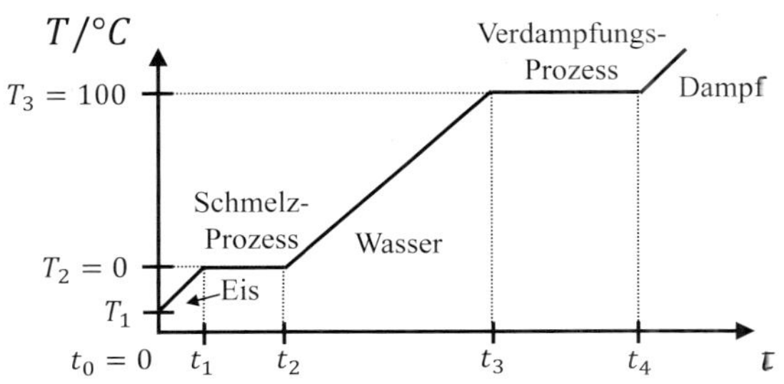

Abbildung 23.6 Zeitlicher Verlauf der Temperatur T von Wasser bei gleichmäßiger Wärmezufuhr mit Stufen für die Schmelzwärme bei $0\,^\circ C$ und die Verdampfungswärme bei $100\,^\circ C$

Zunächst steigt die Temperatur T an, bis der Schmelzprozess des Eises eintritt. Hier bleibt die Temperatur konstant, bis das Eis geschmolzen ist: Die Wärmezufuhr durch den Tauchsieder wird für das Schmelzen des Eises benötigt.

Nachdem das Eis geschmolzen ist, steigt die Temperatur wieder, bis der Verdampfungsprozess beginnt. Auch hier bleibt die Temperatur konstant, denn die Wärmezufuhr durch den Tauchsieder wird für den Verdampfungsprozess gebraucht.

Den Prozess des Schmelzens können wir analog zum Verdampfungsprozess verstehen. Wir gehen wieder von einem infinitesimalen Kreisprozess aus, der uns zur Clausius-Clapeyron-Gleichung (23.7) geführt hat:

$$\Lambda = \frac{dp}{dT} T \, (V_{\text{Fl}} - V_{\text{Fest}}) \tag{23.20}$$

Dabei bezeichnet V_{Fl} das Volumen des flüssigen Zustands und V_{Fest} das Volumen des festen Zustands bezogen auf dieselbe Masse M des Stoffs. Die Volumenänderung ist hier geringer als beim Verdampfungsprozess. Die erforderliche Schmelzwärme eines Stoffs ist im Allgemeinen geringer als seine Verdampfungswärme.

Beispiel 66: Spezifische Schmelzwärme

Die spezifische Schmelzwärme ($\lambda = \Lambda/M$) von Eis beträgt

$$\lambda = 332{,}5 \, \frac{\text{kJ}}{\text{kg}} \, .$$

23.3.1 Phasenübergänge, Tripelpunkt

Für die überwiegende Mehrheit der Stoffe gilt, dass die Volumina ihrer festen Phasen kleiner sind als die ihrer flüssigen Phasen:

$$\Delta V = V_{\text{Fl}} - V_{\text{Fest}} > 0 \tag{23.21}$$

Damit ist in Gleichung (23.20) die Steigung

$$\frac{dp}{dT} > 0 \, . \tag{23.22}$$

Wie bei der Abhängigkeit des Dampfdrucks von der Temperatur (23.16) ergibt sich auch für den Schmelzprozess eine ansteigende Funktion $p(T)$, auf der in diesem Fall die flüssige und die feste Phase simultan existieren können.

Eine Ausnahme bildet Wasser. Wasser hat seine größte Dichte bei $T = 4\,^\circ\text{C}$. Das Volumen verringert sich also beim Schmelzen von Eis, sodass die Volumenänderung in (23.20) negativ ist:

$$\Delta V = V_{\text{Fl}} - V_{\text{Fest}} < 0 \tag{23.23}$$

Damit ist auch die Ableitung des Drucks als Funktion der Temperatur negativ:

$$\frac{dp}{dT} < 0 \tag{23.24}$$

Anstelle einer steigenden Funktion ergibt sich demnach hier eine fallende Funktion $p(T)$, auf der die flüssige und die feste Phase gleichzeitig existieren können (► Abbildung 23.7).

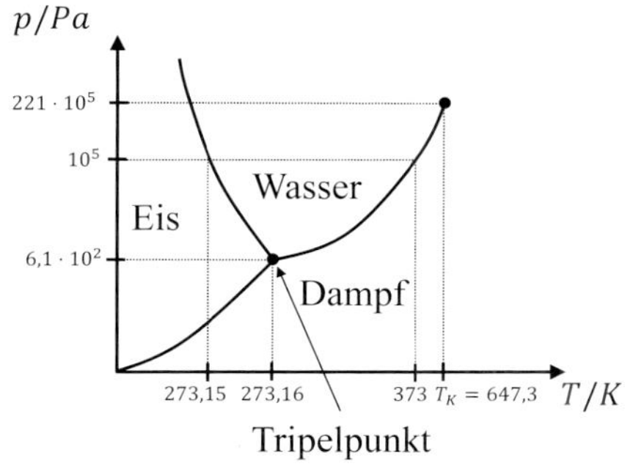

Abbildung 23.7 Am Tripelpunkt (Fixpunkt der Kelvinskala) existieren die 3 Phasen des Wassers gleichzeitig: fest, flüssig und gasförmig.

Diese fallende Kurve ermöglicht einen einzigen Phasenraumpunkt, in dem alle Aggregatzustände des Wassers gleichzeitig existieren können. Diesen Phasenraumpunkt bezeichnet man als **Tripelpunkt**. Er ist der zweite Fixpunkt der vorherigen, bis 2019 gültigen Kelvinskala mit

$$T = 273{,}16\,\mathrm{K}\,. \tag{23.25}$$

Diese Temperatur entspricht $0{,}01\,^\circ\mathrm{C}$ auf der Celsiusskala und ist zusammen mit dem Temperaturnullpunkt für den praktischen Gebrauch weiterhin sehr nützlich (siehe auch Abschnitt 20.3).

Das Wichtigste zum Mitnehmen

Zum Mitnehmen: Thermodynamik realer Gase und Flüssigkeiten 1

- Im Gegensatz zu idealen Gasen kommen reale Stoffe in verschiedenen **Aggregatzuständen (Phasen)** vor: gasförmig, flüssig, fest.

  Reale Gase haben folgende Eigenschaften:

  – Der Aggregatzustand ist temperaturabhängig.

  – Das freie Volumen ist durch das Eigenvolumen der Teilchen eingeschränkt.

  – Die Teilchen haben eine abstandsabhängige Wechselwirkung.

  Die ideale Gasgleichung für 1 mol, $p\,V_{\text{mol}} = RT$ (19.41), muss daher in die **Van-der-Waals-Gleichung** modifiziert werden:

  $$\left(p + \frac{a}{V_{\text{mol}}^2}\right)(V_{\text{mol}} - b) = RT \qquad (23.1)$$

  Das **Kovolumen** $b = 4\,N_A\,V_{\text{Molekül}}$ berücksichtigt das Eigenvolumen der Moleküle.

  Der **Binnendruck** a/V_{mol}^2 resultiert aus der anziehenden Wechselwirkung der Moleküle an den Grenzflächen des Mediums.

  Die Van-der-Waals-Gleichung beschreibt die flüssige und gasförmige Phase eines realen Gases. Der Bereich des Phasenübergangs wird durch die **Maxwell-Gerade** ersetzt. Zu höheren Temperaturen T nähert sich die Gleichung einer Hyperbel für ideale Gase an.

- Die Flüssigkeitsmoleküle bewegen sich mit Geschwindigkeiten der Maxwell-Boltzmann-Verteilung (20.34). Falls ihre kinetische Energie E_{kin} größer als die Austrittsenergie E_{Austritt} ist, gelangen sie ins Gasvolumen. Dort stellt sich ein **Sättigungsdampfdruck** p_s ein, wenn ebenso viele Moleküle aus der Flüssigkeit austreten wie zurückkehren. Die erforderliche **Verdampfungswärme** Λ ergibt sich aus der **Clausius-Clapeyron-Gleichung**:

  $$\Lambda = \frac{dp_s}{dT}\,T\,(V_D - V_{\text{Fl}}) \qquad (23.7)$$

  Λ wird für die Vergrößerung der Volumina V_{Fl} nach V_D gegen die Anziehungskräfte der Moleküle zugeführt. Die **spezifische Verdampfungswärme** λ ist mit der Masse der verdampften Flüssigkeit

  $$\lambda = \frac{\Lambda}{M}\,. \qquad (23.8)$$

Zum Mitnehmen: Thermodynamik realer Gase und Flüssigkeiten 2

- Der Sättigungsdampfdruck p_s beim **Phasenübergang** zwischen der flüssigen und der gasförmigen Phase ist stark temperaturabhängig. Mit $V_{\text{Fl}} \gg V_D$ und V_D als Molvolumen vereinfacht sich (23.7) zu

$$\Lambda \approx \frac{dp_s}{dT}\, T\, V_D\,. \tag{23.9}$$

Mit der idealen Gasgleichung (19.41) $p_s\, V_D \approx R\,T$ ergibt sich eine Differentialgleichung mit folgender Lösung für den Sättigungsdampfdruck:

$$p_s \propto e^{-\frac{\Lambda}{R}\frac{1}{T}} \tag{23.16}$$

p_S steigt also exponentiell mit der Temperatur. Die Proportionalitätskonstante in (23.16) lautet $p_\circ\, e^{\frac{\Lambda}{R\,T_\circ}}$ (23.15).

- Wenn der äußere Druck p kleiner/größer als der Sättigungsdampfdruck p_s ist, verschiebt sich der **Siedepunkt** zu kleineren/höheren **Siedetemperaturen** T_B. Ist der Druck $p_\circ$ gegeben, so erhalten wir bei dem höheren Druck $p'_\circ$ die Siedetemperatur T'_B:

$$T'_B = \frac{1}{\frac{1}{T_B} - \frac{R}{\Lambda}\ln\frac{p'_\circ}{p_\circ}} \tag{23.19}$$

- Das Schmelzen können wir ähnlich zum Verdampfen mit einer Gleichung analog zur Clausius-Clapeyron-Gleichung (23.7) beschreiben:

$$\Lambda = \frac{dp}{dT}\, T\, (V_{\text{Fl}} - V_{\text{Fest}}) \tag{23.20}$$

Dabei bezeichnen V_{Fl} und V_{Fest} das Volumen des flüssigen bzw. des festen Zustands bezogen auf dieselbe Masse M des Stoffs. Die Volumenänderung ist hier geringer als beim Verdampfungsprozess. Die erforderliche **Schmelzwärme** eines Stoffs ist im Allgemeinen geringer als seine Verdampfungswärme.

- Es gibt einen einzigen Phasenraumpunkt, den **Tripelpunkt**, in dem alle Aggregatzustände des Wassers gleichzeitig existieren können:

$$T = 273{,}16\,\text{K} \tag{23.24}$$

Mit dem **Tripelpunkt** und dem Temperaturnullpunkt wurde vor 2019 die Kelvinskala definiert (siehe Abschnitt 20.3).

Gemeinsam verstehen

24

Wie gut können Sie bereits über die Themen sprechen? Führen Sie zu zweit ein wissenschaftliches Gespräch miteinander und beachten Sie dabei die Hinweise in Kapitel 6.

Aufgabe 59: Partnerarbeit: Luftdruck

Wodurch entsteht der Luftdruck und wie misst man ihn? Was unterscheidet ihn vom Wasserdruck?

Aufgabe 60: Partnerarbeit: Fliegen

Erklären Sie, warum ein Flugzeug fliegen kann.

Aufgabe 61: Partnerarbeit: Gase

Die Temperatur schwankt im Sommer/Winter um ca. 20 °C. Welche Bedeutung hat das z. B. für den Reifendruck eines Autos?

Aufgabe 62: Partnerarbeit: Temperatur

Was ist Temperatur? Erklären Sie ihre Bedeutung mithilfe des idealen Gasgesetzes.

Aufgabe 63: Partnerarbeit: Wärme

Was beinhaltet der Wärmebegriff? Welche Stoffeigenschaften bestimmen die spezifische Molwärme?

Aufgabe 64: Partnerarbeit: Wärmetransport

Erklären Sie die verschiedenen Wärmetransport-Mechanismen.

Aufgabe 65: Partnerarbeit: Isotherme

Welche Arbeit leistet ein mit Gas gefülltes System bei isothermer Expansion?

Aufgabe 66: Partnerarbeit: Adiabaten

Wodurch sind adiabatische Prozesse stärker von Volumenänderungen abhängig, als isotherme Prozesse?

Aufgabe 67: Partnerarbeit: Kreisprozess

Erklären Sie Ablauf und Anwendungsmöglichkeiten eines Kreisprozesses in der Thermodynamik.

Aufgabe 68: Partnerarbeit: Stirlingmotor

Erklären Sie den Kreisprozess eines Stirlingmotors mit zwei isothermen und zwei isochoren Zustandsänderungen.

Aufgabe 69: Partnerarbeit: Entropie

Was bedeutet Entropie? Erklären Sie anhand eines Beispiels.

Aufgabe 70: Partnerarbeit: Verdampfung

Welche Arbeit leistet ein System beim Verdampfen einer Flüssigkeit?

Index

A

- Ableitung, 18
  - partielle, 58
- Absolute Temperatur, 429, 441, 445, 487
- Abstandsmessung, 234
  - Euklidische Ebene, 235
  - Invarianz Raumzeit, 236
  - Komplexe Ebene, 235
  - Raumzeit, 235 f., 249
- Actio = Reactio, 70, 89
- Additionstheoreme, 312
- Adiabatenkoeffizient, 454, 461
- Adiabatischer Prozess, 465, 469, 472, 478, 488 f.
- Aktionsprinzip, 69, 89
- Amplitude, 100, 315, 332
- Amplitudenmodulation, 313, 315, 327
- Aperiodischer Grenzfall, 303
- Äquipotentialflächen, 189
- Arbeit, 121
  - Kraft, 121
  - Vorzeichenkonvention, 122, 466
  - Weg, 121
- Archimedisches Prinzip, 383, 410, 456
- Atom
  - Freiheitsgrade, 441, 452, 461
  - Geschwindigkeitsverteilung, 443
- Aufgabe
  - Astronaut, 177
  - Brunnen, 27
  - Drehimpulskomponenten, 275
  - Druck Philippinengraben, 382
  - Endgeschwindigkeit, 29
  - Entropie, 486
  - Erdmasse, 177
  - Fall zweier Kugeln, 33
  - Freier Fall, 77
  - Gasgesetz, 427
  - Heißluftballon, 384
  - Kugelbahn, 112
  - Kugeloberfläche, 170
  - Mondbewegung, 177
  - Ortsfunktion, 20
  - Partnerarbeit
    - Adiabaten, 511
    - Aufstiegsarbeit, 138
    - Bergsteiger, 138
    - Bewegungsgesetz, 137
    - Drehschwingungen, 376
    - Effektives Potential, 252
    - Entropie, 511
    - Erzwungene Schwingungen, 376
    - Federpendel, 138
    - Fliegen, 510
    - Foucault, 253
    - Garnrolle, 137
    - Gase, 510
    - Gedämpfte Schwingungen, 376
    - Gravitationsgesetz, 252
    - Gravitationskonstante, 252
    - Isotherme, 511
    - Kreisel, 376
    - Kreisprozess, 511
    - Luftdruck, 510
    - Masse Himmelskörper, 252
    - Massenmessung, 136
    - Messung, 136
    - Reibung, 137
    - Ruhemasse, 253
    - Schwebung, 376
    - Schwingung, 137
    - Schwingungsdauer, 138
    - Stirlingmotor, 511
    - Stoßprozesse, 252
    - Temperatur, 510
    - Verdampfung, 511
    - Wärme, 510
    - Wärmetransport, 510
    - Weitwurf, 136
    - Welle, 376
  - Pohlsches Rad, 297
  - Rakete, 83
  - Reibungswärme, 450
  - Sonnenschaukel, 177
  - Strohhalm, 167
  - Tennisballmaschine, 57
  - Trägheitsgesetz, 70
  - Trägheitsmoment, 264
  - Trägheitstensor Quader, 277
  - Wellengleichung, 361
  - Wer arbeitet mehr?, 123
  - Wetterballon, 389
  - Wippe, 106
  - Zentraler Grenzwertsatz, 52
- Auftrieb, 382
  - Archimedisches Prinzip, 383, 410, 456
  - Dynamischer, 395, 397
  - Kraft, 382
- Ausdehnungskoeffizient, 419, 422
  - Festkörper, 422, 431
  - Flüssigkeit, 422, 431
  - Gas, 423, 429, 431
- Auswertung Experiment, 53
  - Angabe Ergebnis, 44
  - Fehler, 55
  - Fehler des Mittelwerts, 56, 64
  - Fehlerfortpflanzung, 58
  - Getrimmter Mittelwert, 53
  - Mittelwert, 53
  - Schätzwert, 53
  - Standardabweichung, 54 f.
  - Statistische Fehler, 54 ff., 58, 60, 64
  - Stichprobe, 50
  - Streuung, 54
  - Systematische Fehler, 60
  - Varianz, 55
- Avogadro-Konstante, 43, 426, 432

B

- Bahnkurve, 18
  - Allgemeine Lösung, 25
  - Freier Fall, 41
  - Gleichförmige Kreisbewegung, 94
  - Gleichmäßige Beschleunigung, 27, 37
  - Harmonischer Oszillator, 100, 115
  - Lösung Bewegungsgleichung, 77
  - Luftkissenbahn, 80
  - Ortsfunktion, 18
  - Pendel, 110, 117, 312
  - Rakete, 85
  - Wurf, 33 f., 38
- Barometrische Höhenformel, 387
- Beispiel
  - Arbeit, 122
  - Ausdehnung Flüssigkeiten, 422
  - Ausdehnung Gase, 424
  - Balkenwaage, 106
  - Bremsweg, 29
  - Dampfdrucktopf, 503
  - Dritter Hauptsatz Thermodynamik, 487, 493
  - Eisenbahnschiene, 419
  - Ellipse, 196
  - Energieerhaltung, 130
  - Fallexperiment, 41
  - Federdehnung, 126
  - Federentspannung, 127
  - Fehler des Mittelwerts, 56
  - Fehler Erdbeschleunigung, 59
  - Fehler Längenmessung, 59
  - Fehlerursachen, 45
  - Fouriertransformation (Gauß), 321
  - Fouriertransformation (Kosinus), 320
  - Freiheitsgrade, 441
  - Galileitransformation, 212
  - Gasflasche, 393
  - Gleichförmige Kreisbewegung, 166, 170
  - Harmonischer Oszillator, 99
  - Kühlschrank, 478
  - Komprimieren eines Gases, 466
  - Konvektion, 456, 462
  - Korrigierter Mittelwert, 55
  - Kräfte, 67

Längenausdehnung, 419
Lorentztransformation
Λ-Zerfall, 246
Luftkissenbahn, 80
Massendichte, 260
Molmasse eines Gases, 426
Planetenbahnen, 202
Rakete in Raumzeit, 234
Raketenstart, 87
Raumstation, 186
Ruhemasse, 238, 250
Scheinkraft, 213
Schwerpunkt Halbkugel, 260
Spezifische Schmelzwärme, 504
Spezifische Verdampfungswärme, 501
Spezifische Wärme, 450
Statistische Fehler, 60
Streuung Messwerte, 56
Trägheitsmoment Vollkugel, 263
Trägheitstensor Quader, 275
Wärmedämmung Wände, 459
Wärmeleitvermögen, 457
Wassersäule, 381
Weg zur Vorlesung, 19
Winkelgeschwindigkeitsvektor, 276
Wurf, 36
Würfel, 46
Zeitdilatation, 232, 248
Zentraler Grenzwertsatz, 51
Zerfallswinkel, 246
Zustandswahrscheinlichkeit, 484
Bel, 367
Bernoulli-Gleichung, 393
Gesamtdruck, 394, 412
Statischer, 394, 412
Staudruck, 394, 412
Beschleunigung, 16, 19 f., 25
Durchschnittsbeschleunigung, 19
Erdbeschleunigung, 41
Gleichmäßige, 26, 41
Kartesische Koordinaten, 162, 171
Kugelkoordinaten, 169, 172
Momentanbeschleunigung, 19
Vektor, 31
Vektor bei Kreisbewegung, 95
Zentripetalbeschleunigung, 96, 114
Zylinderkoordinaten, 165, 171
Beta β, 223
Lorentztransformation, 223, 226, 245
Bewegung
Allgemeine, 24
Beschleunigte, 25
Gleichförmige, 23
Gleichmäßig beschleunigte, 26
Gleichmäßig gebremste, 28
Überlagerung, 31
Bewegungsgesetz, 66–90
Beschleunigung, 69, 89
Kraft, 69, 89
Masse, 69, 89
Rotationsbewegung, 107
Translationsbewegung, 79, 90
Bewegungsgleichung, 77
Harmonischer Oszillator, 99, 115, 265, 287, 294, 310, 323
Luftkissenbahn, 80
Pendel, 110, 117
Präzession, 284
Rakete, 83, 90
Schwingungen
Drehschwingung, 265, 287
Erzwungene, 304, 325
Gedämpfte, 298, 324
Gekoppelte, 309, 325
Wellengleichung, 360, 363, 374
Bezugssystem
Beschleunigtes, 213
Inertialsystem, 211
Rotierendes, 214
Bindungszustand
Ellipse, 201, 207
Kreis, 202, 208
Bogenweg, 93, 113
Boltzmann-Konstante, 43, 428, 433, 441 f., 445
Boyle-Mariotte-Gesetz, 386 ff., 440, 445
Brennpunkt
Ellipse, 195
Hyperbel, 197
Kreis, 196

C

Carnot-Kreisprozess, 472, 490
Wirkungsgrad, 475, 490
Celsiusskala, 442, 446
Chemische Zusammensetzung
Zustandsgröße, 465
Clausius-Clapeyron-Gleichung, 500, 506
Corioliskraft, 217 f., 226

D

Dämpfung, 297
Lineare Dämpfung, 76, 297
Pohlsches Rad, 297
Delta-Funktion, 320
Deviationsmoment, 274
Dezibel, 367
Diagramm
Druck-Volumen, 425, 469
Ort-Geschwindigkeit, 129
Temperatur-Entropie, 481
Dichte, 259
Differentialgleichung, 77
Differentialrechnung, 18
Dirac-Funktion, 320
Dispersion, 345, 347, 371
Anomale, 347
Normale, 347
Wasserwellen, 347
Divergenz, 392
Dopplereffekt, 353
Relativistischer, 358
Drehbewegung, 92–117
Drehimpuls, 104, 116, 261
Trägheitsmoment, 262, 286
Trägheitstensor, 274, 288
Drehimpulserhaltung, 108, 144, 146, 191
Drehmoment, 105, 116
Torsion, 179
Drehschwingung, 264
Bewegungsgleichung, 265, 287
Ortsfunktion, 265, 287
Schwingungsperiode, 265, 287
Druck, 381
Barometrische Höhenformel, 387
Druck-Volumen-Diagramm, 425, 469
Druckabfall Hagen-Poiseulle, 401, 413
Druckwelle in Flüssigkeiten, 406, 414
Druckwelle in Gasen, 407
Gas, 439 f., 445
Gesamtdruck, 394, 412
Luftdruck, 384
Philippinengraben, 382, 386
Quecksilbersäule, 385
Sättigungsdampfdruck, 499, 506
Statischer Druck, 394, 412
Staudruck, 394, 412
Wassersäule, 381, 410
Zustandsgröße, 465
Dulong-Petit-Gesetz, 454, 462
Dynamik, 66–90
Dynamischer Auftrieb, 395

E

Ebene Welle, 332, 337
Effektives Potential, 192
Eigenfrequenz, 305, 310, 342
Eigenschwingung, 305, 310, 342
Eigenwert, 277
Eigenzeit, 231, 248
Einheitsvektoren
Kartesische Koordinaten, 161, 171
Kugelkoordinaten, 168, 172
Zylinderkoordinaten, 164, 171

Einstein
  $E = mc^2$, 237, 241, 250
  Postulat Lichtgeschwindigkeit, 221
  Träge = Schwere Masse, 75
Elastischer Stoßprozess, 149, 151
Ellipse, 195, 200
  Beispiel, 196
  Brennpunkt, 195
  Exzentrizität, 195
  Halbparameter, 195
  Normalform, 195
  Polarkoordinaten, 195
Elliptisches Integral, 199
Energie, 121
  Energie-Impuls-Raum, 238, 250
  Erhaltung, 126, 143, 145, 149, 192, 206, 241, 466
  Erster Hauptsatz, 466, 488
  Gesamtenergie, 126, 133, 238, 271
  Innere, 442
  Kinetische, 121, 124
  Masse, 237
  Potentielle, 121, 124, 126, 132, 270 f.
  Rotation, 125 f., 270 f., 280, 287 f.
  Ruheenergie, 238
  Temperatur, 441
  Translation, 124, 126, 270
  Transport, 338
  Wärme, 121, 449
Entropie, 479, 491
  Irreversibler Prozess, 481, 492
  Reversibler Prozess, 479
  Temperatur-Entropie-Diagramm, 481
  Zustandsgröße, 465
  Zweiter Hauptsatz, 486, 493
Eötvös, 74
Erdbeschleunigung, 41
Erde
  Beschleunigung g, 176, 205
  Masse, 177
  Rotation, 218
Erhaltungssätze, 142–146
  Drehimpuls, 108, 144, 146, 191
  Energie, 143, 145, 192, 206, 241
  Impuls, 143, 145, 241
Erzwungene Schwingungen, 303
  Bewegungsgleichung, 304, 325
  Ortsfunktion, 304
  Resonanz, 305
Eulersche Formel, 293
Experiment, 40–64
  phyphox, 21
    (In)elastischer Stoß, 155
    Audio Spektrum, 343
    Aufzug, 22
    Beschleunigungssensor, 44
    Dopplereffekt, 356
    Fadenpendel, 110
    Fallschnüre, 30
    Federpendel, 301
    Resonanz, 307
    Zentripetalbeschleunigung, 97
  Auftriebskraft, 384
  Auswertung, 53
  Bernoulli-Gleichung, 395
  Bimetallschalter, 420
  Bolzensprenger, 421
  Boyle-Mariotte-Gesetz, 386
  Drehimpulserhaltung, 108
  Drehmoment, 107
  Drehschwingungen Scheibe, 266
  Erzwungene Schwingungen, 305
  Fallexperiment, 45
  Fallschnüre, 29
  Federpendel, 100
  Fliehkraftregler, 217
  Foucaultsches Pendel, 220
  Fourieranalyse, 318
  Freie Rotation Quader, 277
  Galileitransformation, 213
  Gasthermometer, 423
  Gedämpfte Schwingungen, 303
  Gedankenexperiment Fahrstuhl, 75
  Gekoppelte Pendel, 307
  Gravitationskonstante, 178
  Größe eines Gassystems, 428
  Hydrodynamisches Paradoxon, 396
  Interferenz mit Wasserwellen, 353
  Konvektion, 455
  Kreiselnutation, 279
  Kreiselpräzession, 283
  Längenausdehnung, 420
  Lichtgeschwindigkeit, 229
  Longitudinale Welle, 331
  Luftkissenbahn, 72
  Newtons 2. Gesetz, 72
  Pendel auf Scheibe, 365
  Pohlsches Rad, 297
  Reibungswärme, 449
  Resonanz, 305
  Sättigungsdampfdruck, 499
  Schallgeschwindigkeit, 409
  Schmelzwärme, 503
  Schwingende Platten, 344
  Spielzeugrakete, 82
  Stehende Welle, 340
  Stirlingmotor, 477
  Stoß zweier Kugeln, 151
  Tonhöhe von Saiten, 363
  Torricellische Röhre, 385
  Verdampfungswärme, 503
  Wandstöße von Atomen, 438
  Wärmeübertrag, 481, 492
  Wärmeleitung, 456
  Wellenmaschine, 334
  Zeitdilatation, 230
  Zylinder-Rennen, 269
Exzentrizität
  Ellipse, 195
  Hyperbel, 197
  Kreis, 196
  Parabel, 197

F

Federpendel, 99
  Bewegungsgleichung, 99, 115
  Kreisfrequenz, 100, 115
  Ortsfunktion, 100, 115
Fehler, 48, 55, 62
  Absoluter, 59
  Fehlerfortpflanzung, 58
  Mittelwert, 56, 64
  Relativer, 59
  Statistischer, 44
  Systematischer, 44, 60
  Zentraler Grenzwertsatz, 51
Festkörper
  Dulong-Petit, 454, 462
  Freiheitsgrade, 454, 462
  Längenausdehnung, 419, 431
  Spezifische Molwärme, 454, 462
Flächenelement
  Kartesische Koordinaten, 162
  Kugelkoordinaten, 169, 172
  Zylinderkoordinaten, 165, 171
Flüssigkeit
  Flüssigkeitsstrom Hagen-Poiseulle, 401, 413
  Reale, 496–507
Foucaultsches Pendel, 218
Fourier
  Analyse, 316
  Reihe, 316
  Synthese, 316
  Transformation, 320, 328
    Gaußverteilung, 321
    Kosinusfunktion, 320
Freiheitsgrade, 441
  Rotation, 441, 452, 461
  Translation, 441, 452, 461
  Vibration, 441, 452, 454, 461 f.
Frequenz, 336, 368
  Kreisfrequenz, 337

G

Galilei, 74 f.
Galileitransformation, 211
  Beispiel, 212
Gamma γ, 223
  Lorentztransformation, 223, 226, 245
Ganze Zahl, 46

Gas
Ausdehnungskoeffizient, 423 f., 429, 431
Gasdruck, 439
Gaskonstante, 428, 433
Geschwindigkeitsverteilung, 443
Ideales Gas, 437
Ideales Gasgesetz, 427
Innere Energie, 442
Reale Gase, 496–507
Spezifische Molwärme, 451
Van-der-Waals-Gleichung, 497, 506
Volumenausdehnung, 423, 431
Gaußscher Integralsatz, 392
Gaußverteilung, 49, 63
Mittelwert, 49, 51, 63
Standardabweichung, 49, 63
Varianz, 49, 51, 63
Gay-Lussac-Gesetz, 423, 425, 431
Gedämpfte Schwingungen, 298, 324
Aperiodischer Grenzfall, 303
Bewegungsgleichung, 298, 324
Kriechfall, 302
Ortsfunktion, 299
Schwingfall, 300
Gekoppelte Pendel, 307
Gekoppelte Schwingungen
Amplitudenmodulation, 313, 327
Bewegungsgleichung, 309, 325
Ortsfunktion, 313, 326
Schwebung, 313, 327
Gesamtdruck, 394, 412
Gesamtenergie, 126, 133, 238
Gesamtenergieerhaltung, 126, 143
Geschwindigkeit, 16, 20, 93, 113
Durchschnittsgeschwindigkeit, 17
Geschwindigkeitsverteilung, 443
Gruppengeschwindigkeit, 346, 371
Kartesische Koordinaten, 162, 171
Kugelkoordinaten, 169, 172
Momentangeschwindigkeit, 18, 37
Phasengeschwindigkeit, 332, 337, 347, 368, 370
Vektor, 31
Vektor bei Kreisbewegung, 95
Winkelgeschwindigkeit, 93, 113
Wurf, 33
Zylinderkoordinaten, 165, 171
Gesetz
Boyle-Mariotte, 386 ff., 440, 445
Dulong-Petit, 454, 462
Gay-Lussac, 423, 425, 431
Ideales Gasgesetz, 427
Newton, 68 ff., 398
Stefan-Boltzmann, 458, 462
Gewichtskraft, 74, 89
Gleichförmige
Bewegung, 23
Kreisbewegung, 95
Gleichung
Bernoulli-Gleichung, 393
Clausius-Clapeyron-Gleichung, 500, 506
Kontinuitätsgleichung, 390, 393, 412
Van-der-Waals-Gleichung, 497, 506
Gradient, 188
Kartesische Koordinaten, 163, 172, 187, 205
Kraft und Potential, 189, 205
Kugelkoordinaten, 169, 172, 188
Zylinderkoordinaten, 166, 172, 188
Gravitation, 174–208
Gravitationsgesetz, 175, 205
Gravitationskonstante, 176, 178, 205
Gravitationswaage, 178
Konservative Kraft, 176, 205
Kraft, 175, 205
Kraftgesetz, 175, 205
Potential, 185, 187, 205
Zentralkraft, 189
Zentralpotential, 187, 205
Grundgrößen, 41
Länge, 42, 61
Masse, 43, 61
Stoffmenge, 43
Temperatur, 43
Zeit, 42, 61
Grundton, 367
Gruppengeschwindigkeit, 346, 371

H

Hagen-Poiseulle
Druckabfall, 401, 413
Flüssigkeitsstrom, 401, 413
Halbparameter
Ellipse, 195
Hyperbel, 197
Kreis, 196
Parabel, 197
Harmonischer Oszillator, 98
Bewegungsgleichung, 99, 115, 265, 287, 294, 310, 323
Federpendel, 99
Kreisfrequenz, 100, 115
Ortsfunktion, 100, 115, 265, 287, 294, 323, 335
Pendel, 110, 117
Harmonische Welle, 332, 337, 368
Hauptachsentransformation, 277
Hauptsatz Thermodynamik
1. Erster, 466, 488
2. Zweiter, 486, 493
3. Dritter, 487, 493
Hauptträgheitsachsen, 275
Homogener Körper, 259
Hookesches Gesetz, 71, 404
Hydrodynamisches Paradoxon, 396
Hyperbel, 197, 200
Brennpunkt, 197
Exzentrizität, 197
Halbparameter, 197
Normalform, 197
Polarkoordinaten, 197
Hyperbelfunktionen, 301

I

Ideales Gas, 437
Ideales Gasgesetz, 427
Impuls, 78
Erhaltung, 82, 90, 143, 145, 149, 241
Rakete, 80
Inelastischer Stoßprozess, 154
Inertialsystem, 211, 225
Lichtgeschwindigkeit, 221
Innere Energie, 442
Innere Reibung, 397, 412
Newtonsches Reibungsgesetz, 397, 412
Reibungsindex, 397, 412
Viskosität, 397, 412
Integral
Geschlossene Oberfläche, 391
Geschlossener Weg, 391
Integralrechnung, 23
Integralsatz von Gauß, 392
Intensität, 131, 133, 339
Interferenz, 349
Invarianz Abstand Raumzeit, 236
Irreversibler Prozess, 471, 481
Isentroper Prozess, 465, 481, 488
Isobarer Prozess, 465, 488
Isochorer Prozess, 465, 467, 476, 488
Isothermer Prozess, 465, 468, 472, 476, 478 f., 488 f.

K

Kartesische Koordinaten, 161
Beschleunigung, 162, 171
Einheitsvektoren, 161, 171
Flächenelement, 162
Geschwindigkeit, 162, 171
Gradient, 163, 172, 187, 205
Linienelement, 162
Ortsvektor, 161, 171
Projektion, 161
Skalarprodukt, 161
Vektorbetrag, 161
Volumenelement, 162

Kegelschnitt, 194
- Ellipse, 195, 200
- Hyperbel, 197, 200
- Kreis, 196, 200
- Parabel, 197, 200

Kelvin, 43
Kelvinskala, 442, 446
- Nullpunkt, 442, 446, 487, 493
- Tripelpunkt, 442, 446, 505, 507

Kepler-Gesetze, 203, 208
Kilogramm, 43, 61, 71
Kinematik, 16–38
Kinetische Energie, 121, 124
Kinetische Gastheorie, 436–446
Knotenlinie, 344
Komplexe Zahlen, 46, 293
Kompressibilität
- Gase, 386 f., 389
- Wasser, 381, 389, 410

Konservative Kraft, 124, 132, 176, 205
Kontinuitätsgleichung, 390–393, 412
Konvektion, 455
Koordinaten
- Kartesische, 161
- Kugelkoordinaten, 167, 172
- Normalkoordinaten, 309
- Polarkoordinaten, 93, 163
- Zylinderkoordinaten, 163

Koordinatensystem, 160–172
Körper im Gravitationspotential, 189
- Bahnkurve, 198
- Bindungszustand, 200
- Effektives Potential, 192
- Ellipse, 201, 207
- Historische Entwicklung, 203
- Hyperbel, 201, 207
- Kepler-Gesetze, 203, 208
- Komet, 200
- Kreis, 202, 208
- Parabel, 201, 207
- Planet, 193, 200
- Radiale Bewegung, 193
- Streuprozess, 200

Kraft, 66
- Auftrieb, 382, 455
- Corioliskraft, 217
- Definition, 69, 89
- Dynamischer Auftrieb, 395, 397
- Feder, 71, 404
- Gewichtskraft, 74, 89
- Hookesches Gesetz, 71, 404
- Konservative, 124, 132
- Schubkraft, 85
- Schwerkraft, 71
- Wandstöße, 437
- Zentrifugalkraft, 217
- Zentripetalkraft, 96, 114

Kraft-Wärme-Maschine, 477
Kraft und Potential, 189, 205
Kreis, 196, 200
- Exzentrizität, 196
- Halbparameter, 196
- Normalform, 196
- Polarkoordinaten, 196

Kreisbewegung, 92–117
- Beschleunigungsvektor, 95
- Bogenweg, 93, 113
- Geschwindigkeit, 93, 113
- Geschwindigkeitsvektor, 95
- Gleichförmige, 95
- Kreisfrequenz, 93, 100, 113, 115
- Ortsvektor, 93, 95, 113
- Polarkoordinaten, 93, 113
- Projektion, 98
- Schwingungsperiode, 94, 113
- Winkelgeschwindigkeit, 93, 113
- Zentripetalbeschleunigung, 96, 114
- Zentripetalkraft, 96, 114

Kreisel, 278
- Figurenachse, 278
- Kräftefrei, 278
- Nutation, 279
- Präzession, 283
- Symmetrischer, 278

Kreisfrequenz, 93, 113
- Federpendel, 100, 115
- Frequenz, 337
- Harmonischer Oszillator, 100, 115
- Pendel, 110, 117

Kreisprozess, 471
- Carnot, 472, 490
- Irreversibel, 471
- Reversibel, 471 f., 479, 490
- Stirlingmotor, 475
- Wärme-Kraft-Maschine, 472, 475, 490
- Wärmepumpe, 477, 491
- Wirkungsgrad, 475, 490

Kreuzprodukt, 102, 115
- Fläche, 103
- Parallele Vektoren, 103
- Regel bac-cab, 103
- Senkrechte Vektoren, 102

Kriechfall, 302
Kugelkoordinaten, 167, 172
- Beschleunigung, 169, 172
- Einheitsvektoren, 168, 172
- Flächenelement, 169, 172
- Geschwindigkeit, 169, 172
- Gradient, 169, 172, 188
- Linienelement, 168, 172
- Ortsvektor, 167, 172
- Umrechnung in kartesische, 168, 172
- Volumenelement, 169, 172

Kugelwelle, 340, 369

L

Laminare Strömung
- Strömungswiderstand, 399, 402, 413
- Stromlinien, 391, 403, 413

Länge
- Längenausdehnung, 419, 431
- Längenkontraktion, 233, 248
- Meter 42, 61

Leistung, 131, 133
Lichtgeschwindigkeit
- Experiment, 229
- Inertialsystem, 221

Linienelement
- Kartesische Koordinaten, 162
- Kugelkoordinaten, 168, 172
- Zylinderkoordinaten, 164, 171

Longitudinale Welle, 331, 404
Lorentztransformation, 220, 222, 224, 226, 232, 245
- Beta β, 223, 226, 246
- Gamma γ, 223, 226, 246
- Invarianz Abstand, 236
- Lambda-Zerfall, 246
- Matrixschreibweise, 236, 249

Luftdruck, 384
- Barometrische Höhenformel, 387

M

Machscher Kegel, 357
Machsche Zahl, 358
Maschine
- Carnot, 472
- Kraft-Wärme, 477
- Stirling, 475
- Wärme-Kraft, 477

Masse, 70, 237, 250
- Energie, 237
- Erde, 177
- Kilogramm, 43, 61, 71
- Reduzierte, 152, 157
- Relativistische, 237
- Ruheenergie, 238, 250
- Ruhemasse, 237
- Schwere, 71, 74, 89
- Teilchen-Ruhemasse, 233, 241, 243, 250
- Träge, 74, 89

Massendichte, 259
Massenpunkt, 17, 31
Massenstromdichte, 391, 412
Mathematisches Pendel, 108
Maxwell-Boltzmann-Verteilung, 443, 499
Maxwell-Gerade, 498, 506

Messgenauigkeit, 40–64
  Statistische Fehler, 44
  Systematische Fehler, 44
Messwert, 40–64
  Interpretation, 50
  Stichprobe, 50
Meter, 42, 61
Metrik, 236, 249
  Skalarprodukt, 239, 250
Minkowski-Raumzeit, 235
Mittelwert, 48, 53, 61 ff.
  Diskrete Verteilung, 47
  Fehler, 56, 64
  Gaußverteilung, 49
  Getrimmter, 53
  Poissonverteilung, 49, 62
  Stichprobe, 51
Mol
  Anzahl, 427, 432
  Avogadro-Konstante, 43, 426, 432
  Masse, 43, 426, 432
  Spezifische Molwärme, 451, 454, 462
  Stoffmenge, 43, 426, 432
  Teilchenzahl, 43, 426, 432
  Volumen, 427
Molekül
  Freiheitsgrade, 441, 452, 461
  Geschwindigkeitsverteilung, 443

N
Natürliche Zahl, 46
Naturkonstanten, 41
Newton, 66–90
  Actio = Reactio, 70, 89
  Aktionsprinzip, 69, 89
  Axiome, 68
  Bewegungsgesetz, 69, 76, 79, 89 f.
  Trägheitsgesetz, 68, 89
Normalbedingungen
  Physikalische, 385
Normalform
  Ellipse, 195
  Hyperbel, 197
  Kreis, 196
  Parabel, 197
Normalkoordinaten, 309 f.
Nullpunkt, Temperatur, 487
Nutation, 279

O
Oberflächenintegral, 391
Oberflächenspannung, 347, 499
Oberton, 367
Ort, 16
  Vektor, 31, 93, 113
  Vektor Kreisbewegung, 95
Orts-Geschwindigkeits-Diagramm, 129
Ortsfunktion, 18, 37, 312
  Bahnkurve, 18
Ortsvektor
  Kartesische Koordinaten, 161, 171
  Kugelkoordinaten, 167, 172
  Polarkoordinaten, 93
  Zylinderkoordinaten, 163
Oszillator, 294, 298, 323 f.
  Frei schwingender, 294, 298, 323 f.
  Gedämpfter, 298, 324
  Ungedämpfter, 294, 323

P
Parabel, 197, 200
  Exzentrizität, 197
  Halbparameter, 197
  Normalform, 197
  Polarkoordinaten, 197
Partielle Ableitung, 58
Partnerarbeit
  Adiabaten, 511
  Aufstiegsarbeit, 138
  Bergsteiger, 138
  Bewegungsgesetz, 137
  Drehschwingungen, 376
  Effektives Potential, 252
  Entropie, 511
  Erzwungene Schwingungen, 376
  Federpendel, 138
  Fliegen, 510
  Foucault, 253
  Garnrolle, 137
  Gase, 510
  Gedämpfte Schwingungen, 376
  Gravitationsgesetz, 252
  Gravitationskonstante, 252
  Isotherme, 511
  Kreisel, 376
  Kreisprozess, 511
  Luftdruck, 510
  Masse Himmelskörper, 252
  Massenmessung, 136
  Messung, 136
  Reibung, 137
  Ruhemasse, 253
  Schwebung, 376
  Schwingung, 137
  Schwingungsdauer, 138
  Stirlingmotor, 511
  Stoßprozesse, 252
  Temperatur, 510
  Verdampfung, 511
  Wärme, 510
  Wärmetransport, 510
  Weitwurf, 136
  Welle, 376
Pascal, 381
Pendel, 108, 294
  Bewegungsgleichung, 110, 117
  Federpendel, 294
  Gekoppelte, 307
  Kreisfrequenz, 110, 117
  Ortsfunktion, 110, 117
  Schwingungsperiode, 110, 117
Phase, 100, 497, 506
  Phasendifferenz, 314
  Phasenfläche, 337
  Phasengeschwindigkeit, 332, 337, 360, 368, 370
    Wasserwellen, 347
  Phasenverschiebung, 314
Phasenübergang, 501, 504
Physikalische Normalbedingungen, 385
Plancksches Wirkungsquantum, 43
Planet, 193
Pohlsches Rad, 297
  Dämpfung, 297
Poissonverteilung, 48, 62
  Mittelwert, 49, 62
  Standardabweichung, 49, 62
  Varianz, 49, 62
Polarkoordinaten, 93, 163
  Ellipse, 195
  Hyperbel, 197
  Kreis, 196
  Parabel, 197
Potential, 187, 205
  Effektives, 192
Potential und Kraft, 189, 205
Potentielle Energie, 121, 124, 126, 132, 185, 270
  Referenzwert, 186
Präzession, 283
  Frequenz, 284, 289
Projektion, Kartesische Koordinaten, 161
Prozess
  Adiabatisch, 465, 469, 472, 478, 488 f.
  Irreversibel, 471, 481
  Isentrop, 465, 481, 488
  Isobar, 465, 488
  Isochor, 465, 467, 476, 488
  Isotherm, 465, 468, 472, 476, 478 f., 488 f.
  Kreisprozess, 471
  Reversibel, 471, 479

Q
Quadratische Ergänzung, 35

R

Rakete, 80
  Bahnkurve, 85
  Senkrechter Start, 83
Raumzeit, 233, 248
  Anderswo, 234, 248
  Hier und Jetzt, 234, 248
  Vergangenheit, 234, 248
  Zukunft, 234, 248
Reale
  Flüssigkeit, 496–507
  Gase, 496–507
Rechte-Hand-Regel, 106
Reduzierte Masse, 152, 157
Reelle Zahl, 46
Reibung, 76
  Geschwindigkeitsabhängigkeit, 76
  Gleitreibung, 76, 365
  Haftreibung, 76, 365
  Innere, 397
  Newtonsches Reibungsgesetz, 397
  Reibungskoeffizient, 76
  Rollreibung, 76
Relativistische Masse, 237
Resonanz, 305
  Frequenz, 305, 325
Reversibler Prozess, 471, 479
Richtmoment, 264, 287
Ringintegral, 391
Rotationsenergie, 125 f., 269 ff., 287
  Trägheitsmoment, 271, 287
  Trägheitstensor, 280, 288
Rotierendes Bezugssystem, 214
Ruheenergie
  Masse, 238, 250
Ruhemasse, 237
  Elektron, 238, 250
  Lambda, 243
  Pion, 243
  Proton, 238, 243, 250
Ruhesystem, 243

S

Sättigungsdampfdruck, 499, 506
Satz von Steiner, 268
Schätzwert, 53
Schallgeschwindigkeit
  im Festkörper, 406
  in Flüssigkeiten, 407, 414
  in Gasen, 409, 415
  in Luft, 409, 415
Scheinkraft
  Corioliskraft, 217, 226
  Trägheitskraft, 213
  Zentrifugalkraft, 217, 226
Schmelzwärme, 503
  Spezifische, 504
Schubkraft, 85
Schwebung, 313, 315, 327
Schwerkraft, 71
Schwerpunkt, 260, 286
Schwerpunktsenergie, 240
Schwingende Platte, 344
Schwingende Saite, 342, 361
Schwingungen, 292–328
  Drehschwingung, 264
  Erzwungene, 303 f., 325
  Gedämpfte, 298, 324
  Gekoppelte, 307, 309, 325
  Harmonische, 294, 323
Schwingungsbäuche, 342, 370
Schwingungsknoten, 341, 370
Schwingungsperiode, 94, 113
  Pendel, 110, 117
Seilwelle, 361
Sekunde, 42, 61
Siedepunkt, 502, 507
Skalarprodukt, 101 f., 115
  Euklidischer Raum, 235
  Kartesische Koordinaten, 161
  Komplexe Ebene, 235
  Metrik, 236, 239, 249 f.
  Minkowski-Raumzeit, 235
  Parallele Vektoren, 101
  Projektion, 102
  Senkrechte Vektoren, 101
  Vektorkomponente, 102
  Viererimpuls, 239, 250
  Vierervektor, 236, 249
Spezielle Relativitätstheorie, 228–250
  Eigenzeit, 231, 248
  Energie-Impuls-Raum, 238, 250
  Gesamtenergie, 237 f., 250
  Impuls, 238, 250
  Kinetische Energie, 238, 250
  Längenkontraktion, 233, 248
  Masse, 237, 250
  Metrik, 236, 249
  Raumzeit, 233, 248
  Ruheenergie, 238, 250
  Ruhemasse, 237, 250
  Ruhesystem, 243
  Skalarprodukt, 236, 239, 249 f.
  Teilchenphysik, 240
  Viererimpuls, 238, 250
  Vierervektor, 236, 249
  Zeitdilatation, 230
Spezifische Molwärme
  Festkörper, 454, 462
  Gas, 451
  Konstanter Druck, 453, 461
  Konstantes Volumen, 452, 460
Spezifische Wärme, 450, 460
  Schmelzwärme, 504
  Verdampfungswärme, 501, 506
Standardabweichung, 48, 62 f.
  Gaußverteilung, 49
  Poissonverteilung, 49, 62
Starrer Körper, 253–290
Stationärer Zustand, 465
Statischer Druck, 394, 412
Statistische Fehler, 44
Staudruck, 394, 412
Stefan-Boltzmann-Gesetz, 458, 462
Stehende Wellen, 340
Stichprobe, 50
  Mittelwert, 51
  Varianz, 55
Stirlingmotor, 475
Stoffmenge, 426
  Mol, 43, 426, 432
Stokessches Gesetz
  Laminare Strömung, 402, 413
  Strömungswiderstand, 402, 413
Stoßprozess, 148–158
  Elastisch, 149, 151
  Gleiche Massen, 149
  Inelastisch, 154
  Ungleiche Massen, 151
  Wand, 154, 437
  Zentraler Stoß, 150
Strömung
  Laminare, 391, 403, 413
  Turbulente, 391, 403, 413
Strömungswiderstand
  Laminare Strömung, 399, 402, 413
  Stokessches Gesetz, 402, 413
  Turbulente Strömung, 403, 413
Streichinstrument, 364
  Bogen, 364
  Gestrichene Saite, 364
  Grundton, 367
  Instrumentenkörper, 364
  Oberton, 367
  Saite, 364
Streuprozess, 193
  Hyperbel, 201, 207
  Parabel, 201, 207
Streuung, 55
  Messwerte, 55
Stromlinien, 391
  Laminare Strömung, 391, 403, 413
  Turbulente Strömung, 391, 403, 413
Superposition, 310
Systematische Fehler, 44, 60

T

Taylor-Entwicklung, 85
Teilchenphysik, 240
  Antimaterie, 240
  Collider-Experiment, 240
  Elektron, 238, 240, 250

Hadron, 240
Lambda-Teilchen, 240, 242
Lepton, 240
Materie, 240
Myon, 232, 248
Pion, 240, 242
Positron, 240
Proton, 238, 240, 242, 250
Quark, 240
Ruhemasse, 237, 241, 243, 250
Ruhesystem, 243
Schwerpunktsenergie, 240
Wechselwirkungen, 240
Z-Boson, 240 f.
Zerfall, 242
Zerfallswinkel, 245
Teilchenzahl
Avogadro-Konstante, 43, 426, 432
Mol, 43, 426, 432
Temperatur
Absolute, 429, 441, 445, 487
Celsiusskala, 442, 446
Dritter Hauptsatz, 487, 493
Energie, 441
Kelvin, 43
Kelvinskala, 442, 446
Nullpunkt, 442, 446, 487, 493, 505, 507
Temperatur-Entropie-Diagramm, 481
Zustandsgröße, 465
Thaleskreis, 151
Thermische Ausdehnung, 418–433
Thermodynamik, 464–493
Dritter Hauptsatz, 487, 493
Erster Hauptsatz, 466, 488
Vorzeichenkonvention, 466
Zweiter Hauptsatz, 486, 493
Ton
Grundton, 367
Oberton, 367
Torsion, 179
Trägheit, 213
Trägheitsgesetz, 68, 89
Trägheitsmoment, 105, 125, 133, 261
Dünne Scheibe, 264
Hohlzylinder, 269
Satz von Steiner, 268
Scheibe, 266
Stab, 267
Vollkugel, 263
Vollzylinder, 266, 268 f.
Trägheitstensor, 273 f., 288
Kreisel, 278
Quader, 275, 277
Transformation Bezugssysteme, 210–226
Beschleunigte, 213
Galileitransformation, 211
Gradlinig beschleunigte, 213
Lorentztransformation, 236 f.
Rotierende, 214
Translationsenergie, 124, 126, 270
Tripelpunkt, 504 f., 507
Kelvinskala, 442
Turbulente Strömung
Strömungswiderstand, 403, 413
Stromlinien, 391, 403, 413

U

Unschärferelation, 322, 328

V

Van-der-Waals-Gleichung, 497, 506
Binnendruck, 497, 506
Kovolumen, 497, 506
Varianz, 48, 55, 62 f.
Gaußverteilung, 49
Poissonverteilung, 49, 62
Vektor, 67
Vektoraddition, 67
Vektorprodukt, 102, 115
Fläche, 103
Parallele Vektoren, 103
Regel bac-cab, 103
Senkrechte Vektoren, 102
Verdampfungswärme, 498
Spezifische, 501, 506
Verdrehung, 179
Verteilung, 46
Diskrete, 46
Gauß, 49, 63
Kontinuierliche, 47
Maxwell-Boltzmann, 443, 499
Mittelwert, 47, 61
Poisson, 48, 62
Viererimpuls, 238, 241, 250
Vierervektor, 236, 249
Violine, 364
Volumen
Mol, 427
Zustandsgröße, 465
Volumenausdehnung, 422 f., 431
Volumenausdehnungskoeffizient, 422 f., 431
Volumenelement
Kartesische Koordinaten, 162
Kugelkoordinaten, 169, 172
Zylinderkoordinaten, 165, 171
Vorzeichenkonvention
Arbeit, 122, 466
Kraft, 175
Thermodynamik, 466, 488

W

Wahrscheinlichkeit, 46
Dichte, 47
Wärme, 448–462
Dämmung, 458
Energie, 121, 449
Konvektion, 455
Schmelzwärme, 503
Spezifische Molwärme, 451, 454, 462
Spezifische Schmelzwärme, 504
Spezifische Verdampfungswärme, 501, 506
Spezifische Wärme, 450, 460
Transport, 454
Konvektion, 455
Wärmeleitung, 456
Wärmestrahlung, 457
u-Wert Dämmung, 458
Verdampfungswärme, 498
Wärmeleitung, 456
Wärmeleitvermögen, 457, 462
Wärmetransport, 454
Wärmepumpe, 477, 491
Wärmestrahlung, 457
Wärme-Kraft-Maschine, 472, 477, 490
Carnot, 472, 490
Stirlingmotor, 475
Wasserwellen, 347
Dispersion, 347
Phasengeschwindigkeit, 347
Wechselwirkung, 66
Welle, 330–374
Ausbreitung, 331
Ausbreitungsrichtung, 337
Dispersion, 347, 371
Druckwelle, 407
Ebene Welle, 332, 337
Elastische Welle, 405
Energie, 338
Frequenz, 336, 368
Gruppengeschwindigkeit, 346, 371
Harmonische, 332, 337, 368
Intensität, 339
Interferenz, 349
Knotenlinie, 344
Kugelwelle, 340, 369
Longitudinale Welle, 331, 404
Phasenfläche, 337
Phasengeschwindigkeit, 332, 337, 360, 368, 370
Schwingungsbäuche, 342, 370
Schwingungsknoten, 341, 344, 370
Seilwelle, 361
Stehende Welle, 340
Transversale Welle, 361

Wasserwelle, 347
Wellengleichung, 360
Wellenvektor, 337, 369
Wellenzahl, 332
Winkelgeschwindigkeit, 93, 113
Vektor, 104, 116
Wirbelstrombremse, 297
Wirkungsgrad, 475, 490
Wurf
Abwurfwinkel, 34
Anfangsgeschwindigkeit, 34
Bahnkurve, 33
Parabel, 34, 38
Scheitelpunkt, 34
Schiefer Wurf, 32
Wurfweite, 34 f.

Z

Zeit, 16
Sekunde, 42, 61
Zeitdilatation, 230, 232
Zentraler Grenzwertsatz, 51
Zentraler Stoß, 150
Zentralkraft, 176, 189
Zentralpotential, 187, 205
Zentrifugalkraft, 217, 226
Zentrifugalmoment, 274
Zentripetalbeschleunigung, 96, 114
Zentripetalkraft, 96, 114
Zustand, stationärer, 465
Zustandsänderung, 465, 467, 488
Adiabatisch, 465, 469, 472, 478, 488 f.
Irreversibel, 471
Isentrop, 465, 481, 488
Isobar, 465, 488
Isochor, 465, 467, 476, 488
Isotherm, 465, 468, 472, 476, 478 f., 488 f.
Kreisprozess, 471
Reversibel, 471
Zustandsgröße, 465
Chemische Zusammensetzung, 465
Druck, 465
Entropie, 465, 479, 491
Temperatur, 465
Volumen, 465
Zustandswahrscheinlichkeit, 483, 492
Zylinderkoordinaten, 163
Beschleunigung, 165, 171
Einheitsvektoren, 164, 171
Flächenelement, 165, 171
Geschwindigkeit, 165, 171
Gradient, 166, 172, 188
Linienelement, 164, 171
Ortsvektor, 163
Umrechnung in kartesische, 163, 171
Volumenelement, 165, 171

Literaturverzeichnis

[1] BIPM, Le Système international d'unités / The International System of Units (The SI Brochure). Bureau international des poids et mesures, ninth edition (2019), ISBN 978-92-822-2272-0, www.bipm.org.

[2] Bronstein, I., Semendjajev, K., Musiol, G., Mühlig, H., Taschenbuch der Mathematik, 11. Auflage. Verlag Europa-Lehrmittel, Haan-Gruiten (2020)

[3] Demtröder, W., Experimentalphysik 1, Mechanik und Wärme, 7. Auflage Springer Verlag, Berlin (2015)

[4] Embacher, F., https://homepage.univie.ac.at/franz.embacher/SRT/Dopplereffekt.html (29-Dez-2021)

[5] Erdmann, M., Flügge, G., Risse, M., Experimentalphysik 2, Felder und Wellen, Pearson Studium, ISBN 978-3-8689-4418-1 (2021)

[6] Erdmann, M., Hebbeker, T., Schmidt, A., Statistische Methoden in der Experimentalphysik, Pearson Studium, ISBN: 978-3-8689-4391-7 (2019)

[7] Fendt, W., Physik-Applets zum Ausprobieren verschiedener Fälle z.B. bei https://www.walter-fendt.de

[8] Feynman, R., Leighton, R., and Sands, M., The Feynman Lectures on Physics. 3 volumes 1964, 1966. Library of Congress Catalog Card No. 63-20717

[9] Mohr, P.J., Newell, D.B., Taylor, B.N., CODATA Recommended Values of the Fundamental Physical Constants: 2014, arxiv 1507.07956 [physics.atom-ph] (2015); Newell, D. B. et al., The CODATA 2017 values of h, e, k, and N_A for the revision of the SI, Metrologia 55, L13-L16 (2018); Mohr, P.J., Newell, D.B., Taylor, B.N., Tiesinga, E., Data and analysis for the CODATA 2017 special fundamental constants adjustment, Metrologia 55, 125-146 (2018); *NIST:* physics.nist.gov/constants (7-Sep-2020)

[10] NASA, Masse der Erde, (28-Dez-2021)

[11] Plag, R., Internetbasierter Rechner für u-Werte, www.u-wert.net (9-Jan-2022)

[12] Stöcker, H., Taschenbuch der Physik, 9. Auflage. Verlag Europa-Lehrmittel, Haan-Gruiten (2021)